# Management of Industrial Cleaning Technology and Processes

# Management of Industrial Cleaning Technology and Processes

John Durkee, Ph.D., P.E.

2006

ELSEVIER

AMSTERDAM • BOSTON • HEIDELBERG • LONDON • NEW YORK
OXFORD • PARIS • SAN DIEGO • SINGAPORE • SYDNEY • TOKYO

Elsevier
The Boulevard, Langford Lane, Kidlington, Oxford, OX5 1GB, UK
Radarweg 29, PO Box 211, 1000 AE Amsterdam, The Netherlands

First edition 2006



**British Library Cataloguing in Publication Data**
Durkee, John B.
Management of industrial cleaning technology and processes
1. Industrial equipment – Cleaning 2. Manufactures – Cleaning
I. Title
670.4'2

**Library of Congress Cataloging-in-Publication Data**
A catalog record for this book is available from the Library of Congress

ISBN 13: 978-0-08-044888-6
ISBN 10: 0-08-044888-7

Typeset by Charon Tec Ltd, Chennai, India
www.charontec.com
Printed and bound in United Kingdom

06 07 08 09 10 10 9 8 7 6 5 4 3 2 1

# Contents

# Preface

The challenge in preparing this book was deciding what material to omit. That the organization, research, and presentation required more than three years to complete speaks to how that challenge was addressed. No material necessary for managers responsible for cleaning work was left out.

Managing industrial cleaning processes and technologies requires knowledge of engineering and chemistry, environmental science and regulations, industrial equipment, statistical process control, and analytical testing. No less important is knowledge of health hazards and workplace safety, human relations and motivation, choosing cleaning equipment and chemistries, and dealing with suppliers. All are covered in this volume.

# About the Author

**John B. Durkee**, Ph.D. studied at Lehigh University (Chemical Engineering, 1962, 1964, 1969). Throughout a 25-year career with DuPont and Conoco, he managed industrial technologies and processes, including the development and implementation of environmentally friendly, commercially successful alternatives to CFCs. A professional consultant, his monthly columns appear in *Controlled Environments* (critical cleaning), *Galvanotechnik* (precision cleaning), and *Metal Finishing* (metal cleaning). Dr. Durkee is a member of AICHE, ACS, IEST, and ASTM.

# Dedication

I owe the managers who guided me and allowed me the freedom to learn and grow professionally over a 25-year career at Du Pont/Conoco: Ed Brugel, Tom Schrenk, Fred Radloff, Al Lundeen, Barry Coon, and Gene Harlacher, among others. Many of their lessons are communicated here.

I owe Gifford Pinchot, who motivated me to be an entrepreneur, and Janice Baker, who partnered with me as an independent consultant.

I owe Tom Robison and Ron Joseph, who encouraged me in development as an author.

Many acted as mentors as I began learning about cleaning technology and how to use it. I owe Kenny Dishart, Art Gillman, Joe McChesney, Rajiv Kohli, Mike Goodson, and many others.

I owe my parents for encouraging me to learn how things really work.

And I owe my wife, Dorothy Rosa Durkee, for her personal support and role as an editor. Without her help, my writing would be less clear – and completed sooner.

To all, my thanks for your needed and generous support.

JBD

# 1

# Modern cleaning technologies

**Chapter contents**



This chapter covers how cleaning technologies do that which is valued – manage soil. Also covered are the reasons why managers choose to implement these technologies.

## 1.1 WHAT CLEANING IS NOT

Cleaning work receives mixed reviews. There is a dichotomy of opinion. By many industrial managers, it isn't well thought of. By a minority of others, it's recognized as crucial to commercial success.

Why? Because there is a mixed understanding about what cleaning work is, and is not. One minor aim of this book is to clarify the information on which these conflicting opinions are based upon.

Cleaning is not:

- *Rocket science*: But aerospace technology depends upon successful cleaning operations. The same engineering and scientific fundamentals upon which cleaning is based also support manufacture and use of the parts upon which cleaning work is done.
- *Simple minded*: Granted, some solvents were and are capable of making some situations involving parts cleaning appear no more complex than dunking a doughnut into a cup of coffee. Those are exceptions.
- *Valueless*: Cleaning work allows parts to effectively perform in the next expected step of processing, or use. Few customers would want to purchase uncleaned parts. Few inspectors would accept the surfaces of parts as defect-free if they couldn't see all of the surfaces. Few operators would machine, form, or assemble parts which were contaminated with debris from previous operation.
- *Difficult to implement*: Yes, cleaning work can be poorly done so as to produce performance damaging to an enterprise. But it's easy to do it well.

A major aim of this book is to describe how to complete successful cleaning work, how to recognize when that outcome isn't achieved, and how to manage cleaning work to produce that outcome.

Generally, cleaning is not being outsourced. While there is a modest contract cleaning business, in the US cleaning work is done in-house. If your enterprise makes or repairs or tests, you must manage cleaning work.

### 1.1.1 The Nature of Cleaning Work

It's simple. Cleaning work is soil management.

Managers manage soil by causing it to be moved from where it is found (perhaps on the parts) to where it is wanted (perhaps in some container staged for disposal or treatment).

Cleaning work includes at least the five management tasks given in Table 1.1.

**Table 1.1 Tasks in Soil Management**

| Cleaning step | Description |
|---|---|
| 1. Cleaning | Managing *to remove* soil(s) from metal, glass, or plastic surfaces by emulsification with a surfactant or by solution with a solvent.[1] |
| 2. Rinsing[2] | Managing *to dilute* or displace the emulsion or solution of soil(s) wetting the part surfaces with more pure material.<br>This task is done until the amount of soil(s) remaining on the part surfaces are acceptably small when the emulsion or solution of soil(s) is removed (usually by evaporation). |
| 3. Relocation of soil within cleaning machine | Managing *to convey* the bulk soil-bearing emulsion or solution of oil(s) from where the parts are located to where the soil(s) can be separated from the emulsion or solution.<br>This task is done so that oil-laden emulsion or solvent is not stored within the cleaning machine and reapplied to cleaned parts. |
| 4. Drying[3] | Managing *to relocate* dilute emulsion or solution of soil(s) from the part surfaces so that they are available for the next step of use. |
| 5. Disposal of waste soil | Managing *to remove and concentrate* soil(s) from the emulsion or solution so that the detergent or solvent can be reused, *and* that the soil(s) can be recovered in a pure enough form for either efficient disposal, or possible reuse. |

In each of these five tasks, soil is managed to produce a set of acceptable ends: part quality, productivity, disposal impact, and operating cost.

Yet, cleaning work involves other management tasks, so that:

- No one gets injured or has their health impaired.
- No environmental regulations are violated.
- No better choice for cleaning work is ignored, which might be paying another firm to do this work (contract cleaning).

That's the nature of cleaning work – soil management.

### 1.1.2 The Nature of Soils

Soils are something managers don't want where they don't want them. The same chemical may or may not be a soil depending upon where it is and whether or not managers want it there:

- Managers desire soil(s) contaminated with a small amount of cleaning agent (solvent or surfactant) to be located in some container. These soils can be efficiently disposed as waste, reused, or perhaps sold for further reprocessing.
- Managers don't desire soil(s), diluted with a large amount of cleaning agent, to surround valuable parts. Additional cleaning agent will have to be used to further dilute or displace the soil(s) *and* convey the dilute stream away from these parts.

The nature of soils is that they must be *relocated.*

### 1.1.3 The Nature of Cleaning Processes

Cleaning work is about moving chemical materials from where they are not wanted to where they are so. The tools by which this is done are the components of or stages within a robust cleaning process.

Some cleaning agents almost function as their own process. Halogenated cleaning solvents (e.g. CFC-113 or 1,1,1-Trichloroethane) effectively and efficiently dissolve many other chemicals. Parts treated with these solvents dry quickly as the solvents evaporate rapidly without outside action.

---

[1] If this was the only step in cleaning: the cleaning machine would be full of oil-bearing fluid, the parts would still have diluted soil around them and still be wet with cleaning agent, the bill for waste disposal would have probably have cost someone for their job, and the surface quality of the parts would be out of control.

[2] Please note that the "soil" in this case is not the oil(s), but rather the relatively concentrated mixture of oil(s) in cleaning agents.

[3] Please note that the "soil" in this case is not the raw soil(s), but rather the dilute mixture of soil(s) in cleaning agents.

Other cleaning agents, such as aqueous cleaning agents, implement process equipment, space, and time to provide effective cleaning, rinsing, and drying. Aqueous cleaning agents[4] require mechanical force, controlled temperature, as well as considerable space and time when used to clean parts.

Still other cleaning agents, such as blast media, also implement process equipment, space, and time but there is no need for rinsing and drying *per se*. Blast media are worthless as cleaning agents until process equipment propels and aims a stream of them at contaminated parts.

The nature of cleaning processes is that they enable cleaning agents to *perform as desired*.

### 1.1.4 The Nature of Individual Process Steps

A "written picture" may help here:

- After cleaning, part surfaces are surrounded by cleaning agent saturated, or nearly so, with soil. Nothing is attached to these surfaces, but they are fully wetted with dirty liquid. In other words, in the cleaning step it is valued to separate parts from soils.
- After rinsing, the valued condition is the part surfaces being surrounded by pure cleaning agent (no soil). In other words, in the rinsing step it is wanted to flush the parts to remove all soluble, emulsified, entrained, or insoluble soil. All will become unwanted residue if not removed.
- After drying, the parts are surrounded by nothing. In other words, in this step it is valued to separate pristine cleaning agent from the parts via evaporative or non-evaporative drying.

The nature of cleaning process steps is that they are *all necessary*. All must be managed together or cleaning quality will suffer.

### 1.1.5 The Nature of Cleaning Agents

Cleaning agents are chemicals, as are soils. As soils are usually chosen for their properties in some upstream operation (e.g. lubrication, heat transfer, cutting, etc.), so are cleaning agents chosen for their performance in process cleaning equipment.

Solvents or detergent solutions which provide good rinsing have the following:

- Low surface tension (so they can penetrate into crevices or flush through sections with small clearances between components).
- Low viscosity (so frictional pressure drop does not limit flow volume).
- High specific gravity (so lighter materials are easily displaced).
- Either complete miscibility or complete immiscibility with the cleaning agent (so they can dilute or displace the cleaning agent, respectively).

Solvents or detergent solutions which provide poor cleaning can be described as follows:

- Having a strong affinity for a soil but having a low holding capacity for it (solubility).
- Only gradually penetrating and swelling the soil and so it can be removed by rinse fluids.
- Efficiently dissolving a soil only at a temperature above its boiling point. This is nearly useless, as pressurized contacting equipment is expensive.
- Having a low evaporation rate, without regard to its solubility for the soil. After all, any undried cleaning or rinsing solvent is just another soil on the parts.

The nature of cleaning agents is that they are *chosen for their properties relative to those of soils,* to the character of parts, and to the specification of the cleaning process machinery.

### 1.1.6 Food Fights

There is an analogy to the human body. Food plays multiple roles:

- It satisfies our need for good taste and texture, provides energy to support activity, and supplies nutrition for long-term stability. So-called junk food only satisfies one need – our taste buds.

[4] Early ones were called "soaps."

A cleaning agent also plays multiple roles:

- A cleaning agent with good affinity for the soil but with a high surface tension and a low evaporation rate is a poor choice for a process to clean complex parts. It won't penetrate the parts, or easily and uniformly leave them! That's a major reason why *n*-methyl pyrrolidone solvent has only found narrow acceptance in industrial cleaning applications. It satisfies only one need – solvency.

It is the process which provides good cleaning (washing, rinsing, and drying). The cleaning agent does play vital roles in that process. The process wouldn't function without it.

The attention of managers must be on *the overall cleaning process.*

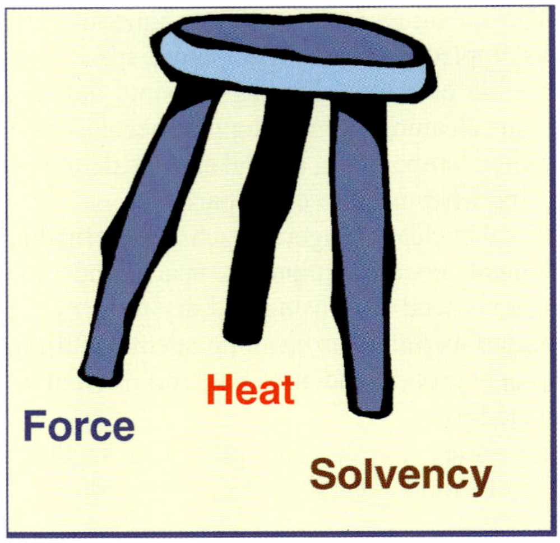


**Figure 1.1**

## 1.2 HOW IT'S DONE

Consultants are often asked to make sense of the varied options and outcomes associated with cleaning systems. Clients ask if there is some "structure" or methodology which can simplify options and outcomes.

The answer to that question is YES.

All cleaning systems depend on *one* or a *combination* of three basic actions:[5]

- A *mechanical* action, such as abrasive surface cleaning or spray agitation.
- A *thermal* action, such as where the environment is heated.
- A *chemical* action, such as:
  - a *dissolving* action (absorption and dilution effect such as an organic solvent dissolving an oil) or
  - a *surface active* action whereby soils are de-sorbed (the reverse of adsorption) from the part surfaces with the aid of surface active agents.

It doesn't matter if the cleaning process is: "dip-and-dunk" cold solvent cleaning, vaporization of debris by lasers, popular detergent-based aqueous cleaning, dislocation of particles by "energy storms" created by laser energy, ozone oxidation, or blast cleaning with hulls from vegetable products. The three actions are involved in all.

### 1.2.1 Said Another Way

The design of any cleaning system is supported by those three functions. This structure, shown in Figure 1.1 is called the "three legged stool." The legs are as follows:

- Mechanical force
- Heat or temperature
- Chemistry (detergency/solvency)

Solvency means choice of solvent (for solvent cleaning) or detergent (for aqueous cleaning).

Implicit in selection of temperature are reaction or solution rates, change in viscosity or fluidity (thinning), or formation/breakage of an emulsion.

Mechanical force means choice of spray system, use of ultrasonic transducers, or hand cleaning with a brush.

### 1.2.2 Examples of How It's Done

Aqueous, semi-aqueous, solvent cleaning, or other cleaning processes are all based on these three functions as shown in Table 1.2.

[5] Remember this covers cleaning. Rinsing, soil management, and drying are other issues which will be discussed below.

**Table 1.2 How Cleaning Work is Done**

| Cleaning Process | Role of Mechanical Force | Role of Heat or Temperature | Role of Chemistry |
|---|---|---|---|
| Aqueous | Dislocation of soil | Thinning of soil | Formation of emulsions |
| Semi-aqueous | | Breaking/making of emulsions | |
| Boiling solvent | Removal of particles | Solution of soil | Solvency of soil |
| Cold solvent | Dislocation of soil | Control of emissions | |
| "Energy storms" generated by laser | | Oxidation of soil | None |
| Vaporization by laser | Removal of debris | Vaporization | |
| Ozone oxidation | Removal of oxidation products | Optimizing rate versus ozone loss | Oxidant |
| Blast | Dislocation of soil | | None |
| "No-chemistry" | Emulsification of soil | Optimizing cleaning rate versus rusting | |

Two general cleaning processes (solvent and aqueous technology) and one specific situation involving both will be discussed in more detail below. They were chosen because of their frequency of use.

## 1.3 SOLVENT CLEANING

Not as simple as "dip-and-dunk" with your favorite chlorinated solvent. Not as vulnerable to environmental regulation as expressed by those with politically correct opinions.

### 1.3.1 The Past Decade

Three developments make solvent cleaning processes a more credible option than they were during the chlorofluoro carbon (CFC) phaseout of the 1990s.

The most important development supporting solvent cleaning processes is the various environmental regulations whose aim was to restrict solvent emissions from solvent cleaning processes. These regulations produced at least the following:

- Validation of an engineering approach(es) to control of emissions from open-top solvent cleaning machines via the US EPA's NESHAP[6] for halogenated solvents.
- Development of vacuum vapor degreasers which require significantly less than €100,000 of investment for purchase.

Said another way, environmental regulations produced the effect desired – solvent cleaning processes (and machines) which can comply with all but the most restrictive emission control regulations,[7] are affordable, and can produce clean parts.

The second most important development is the chemical identification and commercial production of "designer" cleaning[8] solvents.

If some halogenated solvents are considered to hold the extreme position of having excellent solvency but provoking concern about health and environmental issues, "designer" solvents are considered to hold the opposite extreme position of minor concern about health and environmental issues while having limited solvency.

These new products have survived expensive and lengthy health and environmental testing. Some are exempt from US EPA Volatile Organic Compound (VOC) regulations. It is the cost and uncertainty of developments and testing which make it unlikely

[6] The US EPA's National Emission Standard for Hazardous Air Pollutants.

[7] The most restrictive environmental regulations are those which either directly ban solvent cleaning processes or which indirectly do so.

[8] Granted some of these solvents also play commercial roles as heat transfer agents (HFE 7500 or PFPE ZT-85), flushing agents (the OS series), and high-voltage testing and dielectric fluids (HFC-43 10mee).

that chemical firms will produce significant new "designer" solvents for cleaning work.

The third most important development was the belated recognition that azeotropes[9] of existing solvents can fulfill technical demands of cleaning solvents while providing most of the safety and environmental qualities of the "designer" solvents.

The value of azeotropes is their number. More than 400 have been identified. Many include the "designer" solvents. Consequently, a great variety of cleaning problems can be solved because of the available variety of solvencies, boiling points, and other solvent properties.

Excellent management of solvent cleaning processes requires understanding and possible implementation of these three developments rather than the "wisdom" inherent in political correctness. Political correctness is a point of view[10] – not a method of solving cleaning problems. These three recent developments may do so if they are properly applied.

### 1.3.2 The Solvent Cleaning Process

A solvent cleaning process has three steps: wash, rinse, and dry.

1. The washing step brings parts and a chosen solvent together. Usually, the togetherness means immersion[11] of the parts in solvent. The choice of solvent is chiefly based on compatibility of the solvent with the soil to be removed.[12] Soil is removed only[13] when it dissolves in the solvent. The solvent is usually boiling, as within a vapor degreaser.[14]
2. The rinsing step brings fresh (or more soil-free) solvent together with the parts, using the same contact method used in the washing step. The aim is to dilute the soil-rich solvent. *A fundamental limitation on cleanliness is the cleanliness of the solvent material which last contacts the parts.* Soiled solvent can't ever produce perfectly cleaned parts.

   Washing and rinsing steps are usually separated in time and space because good cleaning can't be obtained if parts are being contacted with soil-rich solvent.
3. The drying step means separation of nearly clean solvent from parts. Almost always this is done by evaporation of the solvent.

Solvent cleaning is preferred by some because of the simplicity inherent in the above three steps.

### 1.3.3 Hidden Functions of a Solvent Cleaning Process

If any cleaning process was as simple as one described above, consultants would have to seek other employment. The situation is like that of a movie or a play. Activity outside the view of the camera or behind the curtains is vital to the performance, but is seldom seen. This means management of solvent cleaning is more complex than implied above.

---

[9] Azeotropes are mixtures (usually binary) of solvents. When heated, it is the multi-component azeotrope which is vaporized and not its individual components. Further, the mixture boiling point remains fixed as long as there is enough of both components present to complete the azeotropic composition.

[10] No criticism is intended here of the politically correct approaches which apply only certain solutions to problems. These approaches are responsible, often credible, and common. They are based on the point of view that the politically preferred approach should be tried first, and that it usually can be made to work. Approaches which are not politically preferred generally don't receive equal consideration despite their being based on positive experience, engineering and chemical fundamentals, and useful economics.

Judgements which are politically correct are common outside of cleaning work. The principle, currently politically correct, of continuous improvement (see Chapter 4) is based on taking action not justified in the short term in order to profit from improved quality in the long term.

[11] In some maintenance cleaning work parts are sprayed with solvent. This is done either to pre-soak the soil so that immersion time can be reduced or occasionally to dislodge the soil.

[12] Frequently, liquid physical properties, such as surface tension, viscosity, or density, are significant in the choice of solvent. In these cases the chosen solvent may not have maximum compatibility with the soil, but is more able to flow through restricted passages to reach all part surfaces.

[13] In critical cleaning applications, where soil load is light and probably includes particulate, mechanical force provided by ultrasonic transducers is used to dislodge tiny particles from surfaces. The particles are suspended in the flowing solvent.

[14] Within the US, there are thousands of solvent cleaning machines (called "sink-on-a-drum") in which the solvent is not heated. Worldwide, "sink-on-a-drum" machines are very common because of their cost, size, and simplicity. Cleaning is done at ambient temperature to minimize solvent emission and loss.

**Table 1.3 Hidden Functions of a Solvent Cleaning Process**

| Hidden Function | Purpose | Management Issue |
|---|---|---|
| Solvent storage | Reservoir for virgin and/or distilled solvent | 1. Contamination<br>2. Available volume for changeout |
| Solvent distillation | 1. Remove soil from machine<br>2. Prepare clean solvent for use | 1. Maintain required overhead and bottom temperatures<br>2. Maintain reservoir level |
| Heating coil | Boil solvent for cleaning | 1. Balance heat load versus cooling load<br>2. Balance heat load versus thermal load of parts |
| Water separator | Remove water from system | Stable water–solvent interface within separator vessel |
| Cover | Reduce emissions | 1. Closed when work not entering or leaving<br>2. No odor |
| Cooling coil | | Temperature above coil no more than 30% of solvent boiling point |
| Refrigeration coil | | 1. No permanent frost to inhibit heat transfer<br>2. No odor |
| Parts hoist | 1. Reduce emissions<br>2. Load and unload machine | 1. Speed less than 11 ft/min<br>2. Timing between steps sufficient for solvent drainage |
| Parts basket | Align solvent application with orientation of parts | Avoid retention of solvent within parts |
| Parts transport | Maintain productivity | Not overload capacity of degreaser |
| Stabilizer addition | Avoid solvent degradation ("acid degreaser") | 1. Water content<br>2. Stabilizer content |

Additional functions to be managed within a solvent cleaning process are described in Table 1.3.

Managing events occurring within the cleaning chamber is not enough. Cleaning is about soil management. That happens *throughout* the cleaning machine. Events *throughout* the entire machine must be managed as all are interconnected. One can't clean parts with soil-laden or degraded cleaning agents.

Solvent cleaning technology is described in complete detail in a companion book by this author.[15]

## 1.4 AQUEOUS CLEANING

Aqueous cleaning is not as user and environmentally friendly as "soap[16] and water." Yet this technology is the dominant approach to industrial cleaning used by the majority of global users.

### 1.4.1 Why Aqueous Cleaning?

Water is the ideal solvent for water-soluble soils – road salt, some food and beverage products, plating salts, organic compounds rich in hydroxyl[17] groups such as glycerin, and stable water emulsions such as water-based or latex paints and heat-transfer agents.

But that extensive and significant list of soils are minuscule compared to the depth and variety of soils found in global applications of industrial cleaning. For nearly all oils and greases, water is not the ideal solvent. In fact, it is usually the worst choice of solvents because the common hydrocarbon is not soluble in water.

The basic guidance is that if the oil or grease was derived from crude oil (hydrocarbons), it is not water soluble. If the oil or grease was produced synthetically

[15] Durkee, J.B., *On Solvent Cleaning*, to be published in 2007 by Elsevier, ISBN 185617 4328.

[16] The invention of soap relates to a desire for personal cleanliness. Generic soap dates several millenia before the formulation of Ivory Soap. Animal fats were boiled with ashes to produce soap. The chemical identity of soaps is that they are usually esters. An excellent reference is http://www.ccspa.org/conseducation/SDAC_soaps.html.

[17] The species composed of two atoms, Oxygen and Hydrogen, and a negative charge: $OH^-$ species.

or is derived from vegetable material, it may be water soluble.

Aqueous cleaning is the technology used to clean oils and greases which are not soluble in water. That's why it was developed.

### 1.4.2 How Aqueous Cleaning Works

Table 1.4 gives some simple principles for use of an aqueous cleaning system, and commentary about them. If your aqueous cleaning system isn't performing to your satisfaction, the odds are high that you are violating at least one of these principles.

Granted, all of these 14 principles are not equally important. Principles 5, 10, 11, and possibly 14 are probably of lesser importance.

But the point is that the quality, consistency, and production rate of cleaning with aqueous technology can be improved by applying and managing the above principles to the cleaning system for which a manager is responsible. This type of situation is found in many sites in industry, where conversions from solvents were completed.

Aqueous cleaning technology is described in complete detail in a companion book by this author.[18]

### 1.4.3 What's a Mixed Metaphor?

Examination of Table 1.4 reveals concern about a mistake too commonly made. It is to assume that an aqueous cleaning is the same as a solvent cleaning process, except that a detergent is used instead of a solvent.

The two processes have little in common outside of a hoped-for outcome (clean parts) and reliance upon the same three actions (mechanical, thermal, and chemical).

Said more simply:

- One probably can't do effective solvent cleaning work in a tank designed for aqueous cleaning.
- One probably can't do effective aqueous cleaning work in a tank designed for solvent cleaning.

That's right. A cleaning tank is not a cleaning tank. Consultants have made good income from helping those who have converted a solvent cleaning system to an aqueous cleaning system only to learn that the new system didn't perform as desired.

Some differences are described in Table 1.5 (also see Chapter 7, Section 7.4).

Using aqueous cleaning technology in equipment designed for solvent cleaning technology is like trying to fry a juicy steak in a cocktail blender. Facilities and methods, specific for one cleaning process, don't translate to the other.

## 1.5 MANAGEMENT OF CHOICES AMONG CLEANING PROCESS

Yet, some find it curious that *either* aqueous *or* solvent cleaning technology can successfully fulfill many parts cleaning challenges when the needed facilities are so different. That was shown to be true during the phaseout of CFC solvents in the 1990s. Many jobs done with solvent cleaning technology were ported to aqueous technology. Both solvent and aqueous cleaning effectively met the cleaning needs of much more than half of all cleaning problems.[19]

Said another way, the choice among aqueous and solvent cleaning technologies doesn't matter if one measures the outcome by the cleanliness of the produced parts.

### 1.5.1 Hot Air: Not Used for Parts Drying

For the past decade, or more, the aqueous versus solvent choice has dominated industrial cleaning. The associated spirited dialog has been characterized as political correctness (aqueous technology) versus practicality (solvent technology). Seemingly, the noun *solvent* is hyphenated with the adjective *toxic* and the adjective *simple* is hyphenated with the term *aqueous technology*.

The arena in which this dialog has (and is) taken place is environmental regulations. Speakers are regulators sincerely interested in reducing emissions and associated atmospheric damage and suppliers properly interested in retaining or increasing market

[18] Durkee, J.B., *On Aqueous Cleaning*, to be published in 2007 by Elsevier, ISBN.

[19] This means that a significant fraction of users should have a preference for aqueous or solvent cleaning technology based on the nature of their application.

**Table 1.4 Principles for use of Aqueous Technology**

| General Principles | Comments |
|---|---|
| 1. Aqueous cleaning is performed by a combination of the following:<br>• Detergents to lift the soils from the parts.<br>• Heat to make detergents more compatible with the soils and to soften the soils.<br>• Fluid force to dislodge the soils from the parts and collect the insoluble soils in some removal system. | Users obtain poor cleaning if they are operating an aqueous cleaning bath as if it were a solvent cleaning bath: chemistry, heat, but little or no fluid force.<br>Generally this produces inconsistent cleaning: OK with levels of some soils, but not with others.<br>This can be a very expensive problem to fix because the tank is designed to support the wrong cleaning method. |
| 2. Soils should be periodically or continuously removed from the cleaning bath, via some skimmer or purge system. | A cleaning bath isn't a waste pit where soils are discarded, it is the vessel in which soils are permanently separated from parts.<br>Soils aren't generally dissolved in an aqueous system, they are suspended or emulsified. Perhaps only ½% soil (or less) can be tolerated without reinfecting the parts.<br>Soils are dissolved in a solvent cleaning bath, so as much as 10% or 20% soil can be tolerated without reinfecting the parts. |
| 3. Time and motion should be provided to allow cleaning solution to be dislodged from the rack of parts and not to contaminate the rinse tank. | Operators are frequently trained to let the parts basket drain before moving it to the rinse tank. But they frequently forget to do it. Automation can be a cheap fix. This problem of poor drainage is called "dragout" when dirty cleaning agent adheres to supposedly cleaned parts (see Section 1.13.7). |
| 4. Parts should be removed from cleaning and rinsing tanks without being recontaminated with soil floating on or suspended in any of the tanks. | In the last decade, some clever technology has been commercialized keeping the top surface of an aqueous cleaning bath clean (see Section 7.4.3). |
| 5. Rinsing should be normally done at the temperature of the cleaning bath so that the detergents remain soluble, and don't come out of solution. | Rinsing at reduced temperatures normally isn't a good idea because solubility of detergents is reduced, surface tension of water is increased, and chances of foaming are increased. |
| 6. Monitoring should be done of concentration of detergents in the cleaning bath, pH, and temperature, and they must be adjusted to a consistent and correct level. | Many firms do little or no monitoring, and make adjustments by discarding the bath.<br>This is the largest possible adjustment, which makes the largest possible change in the quality of cleanliness. |
| 7. Typically, aqueous cleaning lines are not used for pre-clean, in-process, and finish cleaning in the same system. Typically, soils are not mixed. The reason is that aqueous detergents are not as forgiving of differences in soil chemistry as are solvents. | The fastener industry is a good example. Parts are made in three to five stages of work. In some plants a central washer for all parts is preferred for reasons of work flow. Cleaning quality may not get the priority that it deserves in this decision.<br>Often the result is that parts are wet with some soil from all previous stages of work! |
| 8. The rinse water must be clean, or the parts won't be clean – especially true for the final rinse bath. | Self-evident, but not universally followed. |
| 9. If spot-free drying is needed, the final rinse of cleaned parts is with pristine (mineral-free) water so that evaporation does not leave mineral deposits. | The consequences of violating principles eight and nine are often the same: stains on parts. Poor rinsing produces detergent-based stains; use of mineral-laden water produces mineral-based stains. |
| 10. Water replacement should be of mineral-free water into the last rinse tank, and cascaded forward. | As with principles eight and nine, the parts should be last contacted with the cleanest fluid. |

(Continued)

**Table 1.4 Principles for use of Aqueous Technology (Continued)**

| General Principles | Comments |
|---|---|
| 11. Drying, with forced hot air, should focus on those part sections which are most difficult to dry (interior volume, high-thermal mass, or narrow passages). | This principle is easy to apply by operating the final rinse bath at 160°F or above, and the parts tend to "dry themselves." Recently, some firms have begun to design their dryers to "pull" the hot air around the parts rather than "push" it at the parts. The reason is that high-velocity hot air striking a cold surface will rebound or reflect without heating the surface. This tendency is reduced if the air is "pulled" along the parts rather than aimed to impact them. |
| 12. Typically, a purge should be taken from aqueous cleaning and rinsing baths so that spent cleaners, soils, solids, and water-minerals do not build up (see Chapter 4, Section 4.12.3). | Sadly, often the only material consumption is of water by evaporation. |
| 13. Some cleaning test is routinely performed to assure cleaning quality is satisfactory (see Chapter 5, Section 5.2). | Sadly, as well, negligible time and resources are normally used to test either the cleaned parts or the cleaning process. |
| 14. Some validation process is occasionally performed to verify that testing is meaningful (see Chapter 5, Section 5.4). | |

**Table 1.5 Comparison of Cleaning Tanks**

| Feature of Tank | Aqueous Cleaning | Solvent Cleaning |
|---|---|---|
| Geometry | *Broad* to allow for high-velocity liquid circulation<br>*Significant surface area* to allow for skimming of soil | *Tall* to minimize emissions<br>*Low surface area* to minimize emissions |
| Volume | Many dozens to hundreds of gallons because of low soil loading in water | A few dozen because of high soil loading in solvent |
| Material of construction | Any material, but steel is common (and low cost) | Nearly always stainless steel |
| Needed floorspace | ~100 sq ft | ~10 sq ft |
| Pipe fittings | As provided | Often flanged to minimize leakage |
| Heating coil | Probably outside of the tank so as to not restrict interior liquid circulation | Probably inside of tank |
| Cooling coil | Probably none | Internal |
| Cover | Probably none | Essential to minimize emissions |

share. Listeners are users confused by unfounded or partially-true claims about efficacy. Often they are driven to make a choice so as to obtain an environmental permit.

The outcome has been dissatisfaction by all.[20] Regulators spend scarce resources eliminating few "units" of pollution relative to that emanating from other industrial operations (automobiles, dry cleaners, bakeries, power plants, etc.). Suppliers gain or lose share based on events outside their control. Users have choices made for them by regulators – some of which are poor from a performance standpoint.[21]

[20] An unpublished survey of several hundred users by this author in the late 1990s revealed that at least half of all were dissatisfied with the choice they had made of a recently purchased cleaning system.

[21] And some of those choices are absolutely excellent!

### 1.5.2 Parts and Soil, Soil and Parts

This author recommends avoidance of that spirited dialog.

Consider a unbiased method for selecting the cleaning process most likely to meet the needs of the application. This chapter examines that method for making a selection.

The basic idea is that the fundamentals of the application should be the basis for decision – if local environmental regulators allow users to choose between aqueous and solvent cleaning technology.[22]

The fundamentals are the nature of the following:

- *Parts*
- *Soil*

This means one should first evaluate the proposed solvent and aqueous cleaning processes based on the stated cleaning needs, standards, or practices.

But if there is no major flaw in either process, then downstream (secondary) consequences should be considered. They include floor space requirements, operating or capital cost, local water or air pollution regulations, soil management, stated preferences of current operating staff, cycle time, training needs or capability, or perhaps occasionally a guideline from a manager.

Said another way, the choice among aqueous and solvent cleaning processes should be considered based on which process:

- Best meets quality requirements to clean actual soil from actual parts.
- Achieves consequences after cleaning which are most compatible with the enterprise's goals, resources, and style (downstream issues).

### 1.5.3 Unbiased Process Selection

This selection method is based on experience of the author, his clients, and their industries.

Common cleaning problems or concerns are described in Tables 1.6 and 1.7. In each table characteristics of cleaning problems, or concerns about

**Table 1.6 Organization of Cleaning Choices: Based on Parts**

| Expected Cleaning Problem | Process Deals Best with that Problem | Reason(s) | Enterprise weighting factor |
|---|---|---|---|
| High initial soil level | Either aqueous or solvent | Solvent can carry large amounts of soil through solution. Pressurized water sprays can dislodge large amounts of soil – a two-stage process is likely. | |
| Low final soil level required | | Solvent rinsing is with pristine solvent. Remember, residual solvent can be a soil! | |
| Blind holes | | Best approach is to immerse and gently flush and reflush holes. Avoid high-velocity jets. If holes are narrow, a low-surface tension solvent is preferred. | |
| Packed and stacked parts | | For aqueous technology, use open rotating baskets and pressurized sprays. For solvent technology, use pump turbulation. | |
| Parts must not be exposed to high temperatures | | Either a low-boiling solvent, which likely will be expensive to buy and contain, or an aqueous-emulsion not stable at high temperatures. | |
| High production | | Either batch or continuous lines are common. | |
| Plastic parts | Aqueous | Especially if your solvent will attack the plastic. | |
| Parts can't take high force level | Solvent | Aqueous requires application of mechanical force. | |

[22] If the use of solvents is forbidden *per se*, with any level of emissions, users must comply with that dictum. This book doesn't advocate environmental anarchy.

However, technology exists (usually vacuum vapor degreasers) to efficiently conduct vapor degreasing operations with most of any solvent – under now current (2005) environmental regulations.

**Table 1.7 Organization of Cleaning Choices: Based on Soils**

| Expected Cleaning Problem | Process Deals Best with that Problem | Reason(s) | Weighting Factor |
|---|---|---|---|
| Soil is a known common chemical | Either aqueous or solvent | Likely, a solvent (see Footnote 15) or an aqueous emulsifying detergent (see Footnote 18) can be found. | |
| Soil is an unknown and/or uncommon chemical | Either aqueous or solvent, *but probably not both* | Use of solubility parameters can guide selection of a solvent (see Footnote 15). Suppliers will probably have to prepare a custom formulated detergent-based cleaner. | |
| Soils are polar acidic salts | Aqueous | Ease of solubility in water and difficulty in solvent. | |
| Dominant soils are particulate | | Ultrasonic transducers, which most effectively apply force to liberate particles, work best with water. | |
| Multiple soils of different character | Solvent | Unlikely to find detergent package capable of emulsifying both types of soil. | |
| Soils are low-molecular-weight oligomers, un-cured polymers, or monomers | | Ease of solubility in solvent, and difficulty in water. | |

cleaning quality, are the point of entry. For each characteristic or concern the process (aqueous or solvent) best suited to address this concern is given. Also given is a reason(s) for that choice and occasionally suggestions for implementation.

The two tables are organized around the major issues described in the previous section:

1. Table 1.6, where the nature of the *parts* is considered.
2. Table 1.7 where the nature of the *soils* is considered.

There is a third table, Table 1.8, in which downstream (secondary) considerations can be included in the selection method.

### 1.5.3.1 *Weighting Factors*

If a situation could be encapsulated by a single characteristic or consideration, choice of cleaning technology would be easy. But real-world situations have multiple, and often many, factors which must be recognized.

How should various factors be judged? This author suggests a weighting scale to measure the quantity that's most important and the one that's less so. One can use this simple scale[23] to describe the problem's significance: 1 = Deal Breaker, 2 = Concern, or 3 = Wish or Want.

The first table (Table 1.6) focuses on the *parts*.

Table 1.6 shows the basic dilemma. Both aqueous and solvent cleaning can do the majority of cleaning jobs for most part configurations. Other factors must be used to differentiate and make a choice.

The second table (Table 1.7) focuses on the *soils*.

The basic dilemma is repeated in Table 1.7. Both aqueous and solvent cleaning can do the majority of cleaning jobs for most soils, although in some cases, one will be preferred.

Tables 1.6 and 1.7 show why the choice between aqueous and solvent cleaning technologies does not have to be made based on characteristics of the parts or the soil.

Cleaning concerns alone may not be sufficient grounds for selection of a cleaning technology. The impact of *downstream issues* are shown in Table 1.8.

Again, downstream issues may not be a difference-maker in enabling a choice between aqueous and solvent cleaning technology.

### 1.5.3.2 *The Difference-Maker*

One lesson of this book is that *for most situations, there is no difference-maker in choice of cleaning technologies.*

[23] Or one which better suits the needs and values of your enterprise.

**Table 1.8 Organization of Cleaning Choices: Based on Downstream Issues**

| Expected Cleaning Problem | Process Deals Best with that Problem | Reason(s) | Weighting Factor |
|---|---|---|---|
| Reliability | Either aqueous or solvent | Solvent cleaning does have fewer control points or setpoints. | |
| Operating cost | | Usually this is a tradeoff between capital cost (favoring aqueous) and cost of ingredients (favoring solvent, because solvents are normally recovered). | |
| Capital cost | | | |
| Soil rejection, and recovery of cleaning agent | | Either can be effective, or a failure. **This is a soil-specific issue.** • Coalescence of multi-component soils (nearly all are) from water can be problematic. • Distillation of low-boiling soil components from a low-boiling solvent can also be problematic. | |
| Local water or air pollution regulations | | Air in Los Angeles and water in New England (in the US). **This is a local issue.** | |
| Cycle time (production rate) | | Solvent cleaning is usually slower because the parts have to be heated. | |
| Training needs | | Aqueous has more detailed steps, but solvent may have more hazards. | |
| Staff capability | | Some staff are not prepared to deal with issues associated with flammable solvents. | |
| Stated preferences of current operating staff | | A mandate from operating staff or management usually must be heeded. | |
| A management guideline | | | |
| Odor preference | Aqueous | Some persons are sensitive to solvent odors, or to skin irritation. | |
| Floorspace requirements | Solvent (usually) | Multiple stages can take place in the same footprint. Some new aqueous machines are vertically mounted and have small footprints. | |
| Soil management | Solvent | Distillation of soil from solvent is usually easier than breaking a chemically made emulsion. | |
| Final dryness quality | Solvent | Low-boiling solvents evaporate rapidly and completely. | |

Without prejudice, either technology can be made to work in the majority of applications, and probably made to work well.

The difference-maker is **YOU**, and **what compromises you are willing to make**.

Examine the weighting factors you and your staff entered in the right-hand columns of the above three tables. It is those columns where the difference-making characteristics are found. Note where the number 1 (Deal Breaker) has been entered. That's where focus belongs.[24]

Table 1.9 shows some examples from the previous tables where both processes could be made to perform acceptably, but one would be preferred.

1.5.3.2.1 *Examples of Non-Compromise* Some cleaning is nearly always done with aqueous technology. This is true even when a suitable solvent can be identified, flushing may be limited by high surface tension or viscosity of a solvent, environmental regulations do allow solvents, and higher operating cost/floorspace/control requirements aren't dominant.

[24] Since there are multiple opportunities for choice, some managers favor a mathematical selection sequence where the sum of the weighting factors is minimized – since the value 1 represents Deal Breaker.

**Table 1.9 Specific Preferences for Cleaning Processes**

| Deal Breaker (Significance = 1) | Preferred Process | Reason(s) for Preference | Compromises Necessary for Preference |
|---|---|---|---|
| High initial soil level | Aqueous | Pre-cleaning, or first stage cleaning, is usually well, cheaply, and safely done with high-velocity impact by water jets. | 1. Increased floorspace requirements<br>2. Jets must be aimed to strike soiled areas<br>3. Possible part damage<br>4. Potential need to dispose of large volumes of water |
| Low final soil level required | Solvent | 1. Final contact with parts can be pristine distilled solvent<br>2. Reduced floorspace<br>3. Reduced number of stages to control<br>4. No concern about impact damage | 1. Distillation system needed<br>2. Excellent control of distilled solvent quality<br>3. Minuscule solvent residues can be tolerated<br>4. Potential emission, health, and flammability problems<br>5. An acceptable solvent must be allowable |
| Soil rejection, and recovery of cleaning agent | Solvent | Distillation is usually more forgiving than emulsion breaking/decanting. | 1. Distillation system needed<br>2. Potential emission, health, and flammability problems<br>3. An acceptable solvent must be allowable |
| Floorspace requirements | Solvent | In nearly all cases, aqueous cleaning machines required more floorspace than do equivalent solvent cleaning machines. | 1. An acceptable solvent must be allowable<br>2. Potential emission, health, and flammability problems |

Such areas are critical cleaning (semiconductors, Oxygen tubing, MEMS[25], superconductive tape, disk drive components, etc.) and cleaning related to human activities (medical, pharmaceutical, food and beverage, etc.).

The reason is that these users recognize and the following:

- *No cleaning process is perfect*: There is always some residual from the cleaning process. These users would prefer that residue to be water rather than anything else.
- Use of water may be required, and use of solvent disallowed, by a commercial or regulatory specification.

This cleaning is probably done with pure water and not surfactants. The reason is to avoid residues that are anything except water.[26] Basically, these cleaning processes are ones where mechanical forces (jets or nozzles, ultrasonic or megasonic transducers) play a dominant role. A second dominant role is played by the action of displacement flushing.

There is no compromise here to achieve a simpler, less costly, and more compact cleaning process. Water is used because no solvent can be left behind – in Oxygen tubing, on food processing equipment, or where the nature of contamination at the molecular level is "critical."

[25] Microelectromechanical systems – so-called "works on a chip."

[26] The opposite of this point is also true. In many medical and biological applications, isopropanol (isopropyl alcohol or IPA) is always used because users and regulators know that IPA residues are not harmful or act as a disinfectant to the next contact by these parts.

### 1.5.4 Management Energy

A core principle of cleaning management is illustrated in Table 1.9, the three tables which preceded it, and the example is given in Section 1.5.3.2.1.

The principle covering selection of cleaning technology, is simple:

- *Spend management energy in the main to make a choice work.* If the choice doesn't work, it doesn't matter by what process it was made.

There is a corollary principle:

- *Spend little management energy on making the choice.* Either solvent or aqueous cleaning technology can be made to work.

Currently in the US, this principle is honored more in the avoidance. Consequences of this situation will be covered in Chapter 6, Sections 6.11 and 6.12.

## 1.6 REMOVAL OF PARTICLES

Whether trying to remove medical residue from glass, nuclear contamination from scrap metal, or CMP[27] byproducts from semiconductor stock, one has to be concerned about trends in managing removal of particles from surfaces.

This chapter is focused on trends and issues around particle removal, and the reasons for them. (see Chapter 6, Section 6.6 for details about specific processes.)

### 1.6.1 Deep Background

For many years, including the 1980s, cleaning involved two basic concepts:

1. A tank, in which ultrasonic transducers had been inserted, of warm water and a detergent at an elevated or neutral pH.
2. A tall tank of boiling solvent without ultrasonics.

Both technologies performed, and still do, in a very satisfactory manner. They can and are being used for pre-treatment.

One virtue of both ultrasonic-based and solvent-based cleaning process is that their successful use does not depend upon knowledge of the specific location of soil material on the substrate. Both processes are omni-directional. Cleaning action takes place in all directions. Managers once took this capability for granted. But they don't now.

Both technologies were heavily dependent upon chemical activity – detergents or solvency. Yet both technologies were augmented by physical action, with pressure waves producing cavitation or pressurized jets producing drag forces.

### 1.6.2 Factors Driving the Change

It isn't the capabilities of these two technologies which have changed. Two general factors have reduced their value in use:

1. Performance requirements are more severe. Aperture sizes in the structures being cleaned are becoming smaller, especially in production of semiconductors. Consequently, sizes of the residual contaminants *not* removed are now smaller. Further, there is increased concern about the amount of residual contaminant not removed. In summary, substrates must be cleaner as the debris being removed is smaller.
2. Environmental requirements are more stringent. Concern about emission of solvents as VOCs and replacement of consumed water has led to a search for replacement processes.

### 1.6.3 Smaller is Not Better

Most particles whose diameter is larger than 0.5 μm (500 nm) will settle down readily, and are more easily removed via filtration.

Debris particles of smaller size bring more significant problems. Particles whose diameter is from 0.01 μm (10 nm) to 2 μm (2,000 nm) are not easily removed, located, controlled, or managed.[28]

In an ordinary room there may be as much as $10^{+6}$ particles per cubic feet whose size is more than 20 nm (0.02 μm) in diameter.

[27] Chemical mechanical planarization/polishing which is how Silicon surfaces are prepared for next use.

[28] Note that there is no single particle size "barrier" below which removal is significantly more difficult.

There are at least four reasons for this:

1. Adhesion forces between debris (particles) and substrates change as an inverse power function of size. Removal of smaller particles requires significantly larger pressure forces.
2. Smaller debris are more easily concealed (hidden) in fluid boundary layers[29] adjacent to substrate surfaces.[30] In fact, debris smaller than about 1 μm (1,000 nm) are smaller than the boundary layer that is thick for nearly any hydrodynamic (flow) situation. Particles smaller than about 0.5 μm (500 nm) in diameter probably can't be removed by hydrodynamic methods.
3. Smaller debris are more difficult to locate so that any cleaning process can focus on them. They may also be more numerous.
4. Surface characterization is much more difficult via analytical efforts (particle measurements). Improvements can be nearly impossible to quantify without in-use evaluation which can be costly.

### 1.6.4 The Effect of Change

These reasons caused development, chiefly in the 1990s, of three different types of processes. Each process had the following characteristics:

1. It offered improved compliance with the two crucial factors (performance and environmental). That's why each was developed.
2. It was based on a different principle than the aqueous ultrasonic and solvent cleaning processes.
3. It involved a different balance between chemical and physical action than was seen with the ultrasonic-based or solvent-based processes. The new balance favors physical action over chemical action.

### 1.6.5 Processes for Removal of Particles: Today and Tomorrow

The three processes being developed and commercialized involve:

1. *Megasonic transducers*: Here pressure waves of a much higher frequency[31] produce short-range lower-intensity hydrodynamic forces which can liberate debris. Cavitation is not involved – no vapor bubbles are produced. Nearly all work is done in water.
2. *High-velocity impingement*, using a solid or liquid material. The materials are chemically inert: water droplets,[32] fragments of condensed $CO_2$,[33] or Argon aerosols.[34] Action is exactly that of a cue-ball on a nine-ball.[35] The material strikes the debris and the debris is dislodged (hopefully) from the substrate.
3. *Local release of energy*:[36] Here the key word is *laser*.[37] That supplies the energy to a specific site on the substrate. The energy release can produce vaporization of some debris, shock waves which dislodge debris, thermal expansion via pulsed beams,[38] and other effects catastrophic to debris.

All these technologies bring value. Some do so more than others. None is as accomplished an art as ultrasonic cavitation technology.[39]

---

[29] See Chapter 6, Section 6.6.2.1 for specific details.

[30] Durkee, J.B. and Baker, J., "C4: Hiding Particles in the Boundary Layer: Part 1," *$A^2C^2$ Magazine,* September 2001.

[31] The designation of frequency type here is artificial. But ultrasonic frequencies are typically those below 250,000 cycles/seconds (250 kHz). Megasonic frequencies are typically those somewhat above that level and less than 1,000,000 cycles/seconds (1,000 kHz). See Chapter 6, Section 6.6.2.1.

[32] US Patent 5,730,806, to NASA.

[33] Banerjee, S., Via, A., Chung, H.F. and Small, R.J., "Combining Aqueous and Cryogenic Post-CMP Cleaning," *Semiconductor International,* February, 2003. Also see Chapter 6, Section 6.1.5.

[34] Butterbaugh, J.W., "Using a Cryogenic Aerosol Process to Clean Copper, Low-K Materials Without Damage," *Micro Magazine*, February 2002.

[35] Except that no pocket is involved …

[36] Some centers of academic research in the US are University of Nebraska-Lincoln (Prof. Y.F. Lu), Clarkson University (Prof. C. Centinkaya), and Arkansas State University (Prof. S. Shukla). Some international centers of research are the Federal Institute for Materials Research and Testing, Berlin, Germany (W. Kautek) and POSTECH, Pohang, Korea (D. Kim).

[37] Durkee, J.B., "C4: Technology In Transition – Removal of Particles Part II," *$A^2C^2$* Magazine, February 2004.

[38] Cetinkaya, C., Vanderwood, R. and Rowell, M., "Nanoparticle Removal From Substrates With Pulsed-Laser Generated Plasma and Shock Waves," *Journal of Adhesion Science and Technology*, 2002, Vol. 16, No. 9, pp. 1201–1214.

[39] See Chapter 6, Section 6.6 for additional details about specific processes used to remove particles.

### 1.6.6 Knowledge of Location

None of these three techniques is omni-directional. Megasonic transducers produce fluid motion in a single dimension. Impingement techniques require open access without barrier. Laser techniques usually require some knowledge of which area of the substrate is contaminated.

If these were the only three cleaning methods that were ever available, users would accept these limitations. There would be no other methods for comparison. But the simplicity and forgiveness associated with ultrasonic-based and solvent-based cleaning makes one long for the past.[40]

### 1.6.7 The Change from Chemical to Physical

All cleaning processes are based on three factors: chemical action, physical or mechanical force, and heat (temperature). Through the 1980s the emphasis was on chemical action. The effect of these changes over the last decade or so is to replace the emphasis on chemical action with an emphasis on physical or mechanical force.

Chemical action brings cost, safety,[41] disposal, and environmental concerns. But there is usually not a concern about damage to substrates or access to debris.

In general, physical action reverses that situation.[42]

### 1.6.8 Technology Perspective

An author who pronounced in 1990 that exchange of technical information would be only by postal mailing of printed papers and pre-prints probably didn't become wealthy via investments in America Online, Inc. (AOL) stock.

This author didn't and doesn't believe that removal of nano-sized (sub-micron) debris will be limited to only use of pressure waves (generated by megasonic transducers), which can't penetrate boundary layers; impingement by high-speed particles, which must contact the debris and which can damage the substrate; or local energy release (produced by lasers), which also suffers from the latter two defects.

Someone will invent another useful technology (see Chapter 6, Section 6.6.4.4).

#### 1.6.8.1 *Future Issues Around Particle Removal*

Particle removal:

- This is going to get more expensive. Remember that cleaning to a higher standard always costs more.[43] The increase is more exponential than linear with decrease in the size (or amount) of residue.

  So as managers seek to eliminate nano-sized particles, they will be paying significantly more to remove each milligram of residue. Hence, pre-cleaning will become more important. Managers will use cheaper technology to remove the micron-sized, reserving the "dry (or new wet)" technologies for the nano-sized particles.
- This is or will soon be done with the same tools and techniques used in the cleanroom for processing the parts. In other words, critical cleaning will become processing. The technology used for cleaning will morph into the technology for processing (manufacturing).
- It may become a rate-limiting step.

## 1.7 MANAGEMENT OF CLEANING PROCESSES

There is a hierarchy within any organization including those who operate and those who manage cleaning processes. There are at least four roles within that hierarchy, relative to cleaning work:

- An *operator* observes automatic control or adjusts manual control of temperature, cleaning agent quality, reservoir level, part flow, time, or other

[40] See Chapter 6, Section 6.6.4.4.

[41] Wet cleaning produces significant quantities of waste and uses lots of water to do that. And many of the chemicals employed (e.g. HF, $H_2O_2$, $H_2SO_4$, $NH_4OH$) are hazardous – especially in semiconductor applications.

[42] See Chapter 6, Section 6.6.4.2.

[43] Durkee, J., "Now Cost is Becoming Critical. Part 1: The Cost/Quality Tradeoff," *$A^2C^2$ Magazine*, March 2003. See Chapter 6, Section 6.7.

parameters in cleaning machines to produce the required part cleanliness. Other roles are to:

- Recognize unusual performance, whether within the list of monitored behaviors or not.
- Recognize needs within or around the cleaning system.
- Take necessary action in the event of a threatened or realized emergency situation.

- A *supervisor* coordinates the work of an operating system which includes a cleaning process. Training/disciplining employees and helping them to solve problems so as to achieve those parameters are a major role of the supervisor (see Chapter 4, Sections 4.18 and 4.19). Other roles are to:

    - Monitor history of, need for, and capability to do maintenance of/replacement on the cleaning system.
    - Recognize and act in response to effects on the cleaning system from upstream operations.
    - Recognize and act in response to effects on downstream operations of the cleaning system.
    - Recognize and act when the cleaning process becomes a rate-limiting step in the overall operation.
    - Order necessary ingredients and other supplies.
    - Keep records of past operation and reviews current operation versus past operation.

- A *manager* coordinates the work of supervisors and other employees. A manager:

    - Provides direction and support to supervisors.
    - Sets goals and objectives such as part cleanliness (see Chapter 5, Section 5.1) and usage rate for cleaning ingredients.
    - Sets overall goals for the working organization.
    - Determines methods for process control (see Chapter 4, Section 4.12).
    - Establishes internal controls such as part production rate versus business demand and process equipment capability.
    - Decides about whether parts are to be cleaned after/during processing, or not (see Section 1.8).
    - Decides about selection of cleaning technology, including specific equipment and cleaning agents (see Chapter 6, Section 6.8).
    - Decides about selection of suppliers (see Chapter 6, Section 6.9).
    - Recognizes need for and selection of consultant for external support (see Chapter 6, Section 6.4).
    - With the enterprise leader and their marketing counterpart, chooses parts to be processed and the rate at which this is to be done.
    - Facilitates communication both upward and downward in the enterprise, especially about whether next use of cleaned parts is consistent with the current cleaning standards (see Chapter 5, Section 5.2 and Chapter 6, Section 6.7).

- A *leader* sets strategic direction and goals for managers to implement:

    - Approves (or rejects) expenditure of enterprise funds for new or replacement cleaning equipment, and annual budget.
    - Makes decision about whether cleaning is to be done in-house or via external contract.
    - Provides guidance to other staff in obtaining and complying with environmental permits.

This book is written for those participating in an hierarchy as a manager. But it should be apparent that a manager does not act without support of and for others.

### 1.7.1 Misorganization

The above is a written organizational chart and a scope-of-work[44] to be completed around a need for clean parts.

It is also a list of ingredients which if "stirred well" without instructions can produce a disaster.

The problem is that the elements of this or any other scope-of-work can enable well-meaning staff

[44] The phrase scope-of-work is relatively common in many industries and government agencies. At a minimum, it is a list of tasks to be done. It is generally understood that if a task is not explicitly written into the scope-of-work, it is not to be done. Obviously, this constraint is in conflict with the well-known dictum that "It is always better to ask for forgiveness than permission."
At a maximum, a scope-of-work includes what is to be and not to be done in addition to a schedule with timing and individual assignments.

to take action not in their sphere-of-responsibility.[45] Here are some examples[46] using the above roles:

- *The operator*, noticing a backlog of parts to be cleaned, should not arbitrarily shorten the cleaning cycle time to consume the backlog. The operator doesn't know the effect on cleaning quality, and that is ultimately for what the manager is responsible.
- The *supervisor*, being responsible for operation of the cleaning system, should not be making independent adjustments to control setpoints. This frustrates attempts by the operator to achieve on-aim control (see Chapter 4, Section 4.12.1).
- The *manager*, being responsible for budgets, should not be ordering supplies. It's a waste of their time, and a budget is not an order form.
- The *leader* should not be instructing the manager as to which cleaning technology should be adopted. The leader may have legitimate concerns relative to relations with local environmental regulators, but the leader is not likely to have the necessary technical experience to make this decision.

This misorganization (misuse of an organization) is found nearly everywhere. It exists because of the understandable desire of persons in the organization for success – that of the organization, and perhaps their own, and many other reasons.

### 1.7.2 Roles, Goals, and "Who's Got the 'D'?"

The cure for the disease of misorganization is to adopt the title of this sub-chapter as the workplace philosophy.

Every position in an organization has a **role**.[47] See Section 1.7 for the roles of four positions in an organization with a cleaning process.

The person holding each position has one or more **goals**. Usually, these are or should be metrics.

**Table 1.10 Examples of Common Goals for Positions in an Organization**

| Position | Example Goal |
|---|---|
| Operator | Cleaning quality at least "X" percent during every periodic (daily, weekly, monthly, or yearly) reporting period. |
| Supervisor | Zero exceedance of budget for cleaning operations. |
| Manager | Zero lost customers (downstream users) who are dissatisfied with the quality of cleaned parts. |
| Leader | Zero safety, health, and environmental incidents. |

Table 1.10 shows single example goals for the positions above.

Acknowledgment of a position **goal** is not in itself sufficient to avoid misorganization. Some other policy is necessary to keep the supervisor, manager, or leader from "tweaking" setpoints on the cleaning machine because they think it's helpful to the operator achieving their position goal.

That policy is to determine "**Who's got the 'D'?**" with the explicit understanding that everyone else does not! The "**D**" is the decision-making power necessary to achieve an organizational goal. "**D**" stands for **decision**.

An enterprise has a better chance to succeed when those responsible for meeting goals have the decision-making power necessary to achieve them.

Table 1.11[48] shows how the cure for misorganization should be used with the above goals.

Granted, modern organizations are shrinking. There are fewer positions[49] in a hierarchy. A manager may also be the operator, though probably not the leader. The title of this sub-chapter should be the policy adopted by every organization without regard to the span of control of each position. Why? Because it's proven to work!

---

[45] This is another phrase in somewhat common use. Sphere-of-responsibility is a more quantitative specification of a person's role in an organization.

[46] Obviously, each enterprise will have its own views as to how its operations will be organized.

[47] Obviously, a person (position) without a defined role is unneeded.

[48] The assignments of span of decision in Table 1.11 are reasonable based on the author's experience, but are purely arbitrary. Other assignments may be more suitable for specific organizations.

[49] The four positions above were chosen for illustration, and represent organizations which may now be considered as overstaffed.

**Table 1.11 Examples of "Roles, Goals, and Who's Got the 'D'?" for Positions in an Organization With Cleaning Operations**

| Position | Example Goal | Decision Power | No Power to Decide |
|---|---|---|---|
| Operator | Cleaning quality at least "X" percent during every periodic (daily, weekly, monthly, or yearly) reporting period. | • Work with "R," X-bar, and CUSUM charts to achieve on aim control.<br>• Recommend when operation is abnormal and common or special causes are in play. | • If level of cleaning quality is acceptable to downstream users (manager's power of decision).<br>• If a published safety rule is necessary (manager's power of decision). |
| Supervisor | Zero exceedance of budget for cleaning operations. | • Setpoints for on-aim control.<br>• Selection of items and quantities to be ordered with chosen timing. | • Choice of suppliers or negotiation of prices (manager's powers of decision).<br>• If an article should be presented at a technical conference about local operation (leader's power of decision). |
| Manager | Zero lost customers (downstream users) who are dissatisfied with the quality of cleaned parts. | • Cleaning tests to be used for both qualification and validation (see Chapter 5, Section 5.3).<br>• When, if, and how an end-use customer should tour the cleaning facility. | • If supplier should be consulted about equipment situation (supervisor's power of decision).<br>• If upstream operation should be shutdown to allow backlog of cleaning work to be dissipated (supervisor's power of decision). |
| Leader | Zero safety, health, and environmental incidents. | • When enterprise must cease operation and be retrained in safety practices because of incidents.<br>• Specify business schedule for product. | • Schedules maintenance (supervisor's power of decision).<br>• Propose budget (manager's power of decision). |

## 1.8 TWO NO-CLEAN CHOICES

As there are two meanings to the word flammable (see Chapter 3, Section 3.5), there are two distinct meanings to the phrase no-clean (NC).

To one extent, commercial pressures are responsible for blurring the distinction between these meanings. To another extent, one meaning of the phrase no-clean is incorrect. Here are the two common applications to which the NC phrase is applied:

- Not cleaned.
- "No-clean" in the electronics industries.

### 1.8.1 The Choice Not to Clean

A crucial aspect of the management of any endeavor, including cleaning work, is deciding when and whether not to do it.

Examples of this choice and a negative outcome when it is made are in Table 1.12 (see Table 1.13).

Finally, a manager can consider one of two opposites:

- Choose to eliminate the cleaning step for individual components, and then clean an assembly of components before packaging for customer use.

**Table 1.12 A Choice Not to Clean**

| When a Part Is: | A Negative Outcome Can Be That: |
|---|---|
| Inspected in its native condition. | Detritus from previous processing masks defects. |
| Abrasively polished without removal of the forming fluid used when its shape was produced. | The boundary layer of forming fluid acts as a barrier, keeping abrasive particles from effectively reaching the surface. |
| Treated with a water-soluble brightening agent after being milled with a water-soluble heat transfer agent. | The acidic corrosive attack of the brightening agent is diminished by dilution of the brightener with the heat transfer agent. |
| Cut with a saw while a protective coating previously applied for storage is still present. | The viscous protective coating fills the gaps between the teeth of the saw as it moves through the groove or plow. Consequently, there is no volume for a chip to be removed from the surface – the saw "binds" and doesn't cut. |
| Repeatedly drawn through successive dies without removal of debris produced in the preceding draw step. | The part surface is scored, as debris is forced into it through friction as the part passes through successive dies. |
| Tested for circuit integrity after soldering but without removal of excess flux and solder materials. | Excess materials form a conductive bridge which "fools" or voids the integrity test. |

- Choose to omit a cleaning step when a "finished" product is assembled from cleaned components, but the assembly is not cleaned prior to packaging.[50]

In this author's experience, all of the examples shown in Table 1.12 have successfully been completed by some users, and not successfully completed by others. As expected, the difference between successful and not successful lies in the details of the application.

This is not a trivial choice. Adoption of it can save cost, floorspace, labor, and time – as well as add unexpected risk of deterioration of quality. Give this tradeoff consideration.

But there may also be an intermediate choice – *elimination of only part of the cleaning step.*

## 1.8.2 When to Choose Not to Clean

Guidance about when and how to eliminate a cleaning step from operations must be general because success and not success are determined by the specific details of the application:

- Compatibility of fluids is crucial. If the cleaning step removes one fluid before another is applied in a successive step, both fluids must be compatible when the cleaning step is eliminated.
- Elimination of the cleaning step will probably require at least one or more modifications to existing operations. *Successful* elimination of the cleaning step won't happen by *ceasing to do it.*
- The cost of cleaning can be difficult to quantify (see Chapter 6, Section 6.7). Rather than attempting to fully understand the cost, accept an estimate.
- Thoroughly evaluate all impacts if the downstream operation is not successfully completed.
- The downside to this tradeoff is more significant. Quality is nearly always more significant than its cost. This is because the downstream user of the cleaned part will pay nothing for unacceptable quality. It's usually better to retain business at a higher price than to lose it.

## 1.8.3 Examples of Elimination of the Cleaning Step

The negative outcomes cited in Table 1.12 may (or may not) be successfully avoided by the following actions shown in Table 1.13.

[50] Welker, R.W., "Clean Then Assemble Versus Assemble Then Clean: Several Comparisons," a paper presented at *the Ninth International Symposium on Particles On Surfaces: Detection, Adhesion and Removal*, Philadelphia, PA, June 16–18, 2004.

**Table 1.13 Elimination of the Cleaning Steps**

| When a Part Is: | Changes Which May Allow Elimination of Cleaning: |
|---|---|
| Inspected in its native condition. | Flush the detritus, rather than fully wash, rinse, and dry the parts. |
| Abrasively polished without removal of the forming fluid used when its shape was produced. | • Suspend the abrasive particles in the forming fluid, possibly after dilution to reduce viscosity.<br>• Or, absorb the forming fluid in the abrasive material. |
| Treated with a water-soluble brightening agent after being milled with a water-soluble heat transfer agent. | Combine the brightening agent with the heat transfer agent. After milling, allow contact time for acidic corrosion (brightening), and then clean both materials with one cleaning step. |
| Cut with a saw while a protective coating previously applied for storage is still present. | • Replace the viscous protective coating with one more fluid.<br>• Or, operate the saw at a higher speed to reduce the viscosity of the coating.<br>• Or, change design of the sawing operation so that the viscous coating is mechanically displaced rather than chemically cleaned. |
| Repeatedly drawn through successive dies without removal of mechanical debris produced in the preceding draw step. | Use a fabric wiper, rather than a cleaning process, to remove mechanical debris. Discard the wiper. |
| Tested for circuit integrity after soldering but without removal of excess flux and solder materials. | • Improve the solder operation so that the integrity test isn't needed or can be done later.<br>• Or, delay the integrity test until after the next operational step. |

Notice that in each case in Table 1.13, it wasn't that the cleaning step was arbitrarily shut down, with the equipment being sold. One or more changes were necessary to allow that outcome, while preserving acceptable next use of the parts. And there was additional testing and qualification to establish that the next use wasn't compromised.

So no-clean isn't a choice without consequence. When the no-clean choice is considered it is the net outcome which must be accepted.

It must be noted, since this book is about management of cleaning, that it is relatively rare that the outcome of eliminating a cleaning operation is net positive.[51] It is unlikely that the choice to not clean will be made in applications to which the adjective critical[52] would be applied.

### 1.8.4 "No-Clean"[53] in Electronics Industries

It started in the 1980s.[54] The basic idea was to avoid the use of ozone-depleting chemicals[55] to clean electronic structures.[56] It continued through the 1990s when the value of cleaning was seen as providing differential reliability.[57] And it continues into this

[51] This means that the long-term effect on the downstream user must be included in the evaluation.

[52] These are applications where success or failure of the application depends upon the quality of the cleaning operation. Cleaning quality may be more or as important as dimensional tolerance or chemical composition. Examples are applications: with human contact, involving flammable materials such as pure Oxygen, or where surface character is significant at the molecular or atomic level.

[53] Please note the presence of quotation marks. In this book, "no-clean" refers to the technology where soil materials are carefully chosen to be removed by vaporization and not by cleaning; and no-clean refers to the technology where cleaning isn't done.

[54] Guth, L.A., "To Clean or Not To Clean?," *Circuits Manufacturing*, February 1989, pp. 59–63.

[55] Chiefly CFC-113, and 1,1,1-Trichloroethane (TCA).

[56] Some users migrated to semi-aqueous cleaning technology, which has become a standard approach for cleaning printed circuit boards (PCBs). Others migrated to aqueous cleaning and changed the flux and solder to materials which were mostly water soluble. Still others migrated to "no-clean."

[57] Bixenman, M., "The 'End' of Cleaning?," *Surface Mount Technology (SMT) Magazine*, September 1999.

decade when the dominant concern has been to avoid soldering/joining materials which do contain Lead.[58]

The defining issue is environmental management.

The choices in Table 1.13 have no direct[59] environmental impact. One wouldn't expect to need an environmental permit to eliminate a cleaning step in operations.

But "no-clean" technology as practiced in manufacture and use of electronic components involves replacement of a cleaning step with vaporization of a new chemical. That emission must be internally contained and or externally permitted.

### 1.8.4.1 *The "No-Clean" Concept*

It's really clever. A processed chemical which must be removed from surfaces by cleaning is replaced with another chemical which provides the same function but which is removed from surfaces by vaporization.

Surface treatment is replaced with area heating.

The process chemical is flux – used in soldering operations. Fluxes are used in electronics manufacturing to promote the wetability required to make a good solder joint. Flux improves distribution of heat, so hot spots are avoided, dissolves or reacts with surface oxides and metal salts, and reduces the interfacial tension between the solder material and the component surface.

Without "no-clean" materials, excess water-soluble solder paste and the flux are cleaned using solvents, aqueous detergents, or a semi-aqueous process. The surface oxides and metal salts are removed with the flux.

With "no-clean" technology, excess solder paste and the oxides/metal salts are retained on the surface while the flux is vaporized – often in the soldering step.[60]

Material recipes used in "no-clean" fluxes are changing and proprietary. As with elimination of cleaning steps from Table 1.13, compensating process and product changes are necessary:

- Oxide formation is retarded by conducting the solder operation in an inert atmosphere. "No-clean" fluxes are more dilute – contain much less solids (often by a factor of five) – and consequently must be applied at significantly higher volumes. And the acceptable "window" of process operating conditions is considerably less forgiving.
- Residue levels at component junctions and throughout the product are increased. Product reliability is changed, and is less under control. This can be a "deal-breaking" issue when the circuit board is to be used in a military weapon, a commercial airliner, or a product *you* own.

### 1.8.4.2 *Tradeoffs with "No-Clean" Technology*

Replacement of traditional solder chemistries with "no-clean" trades one set of environmental issues for another. That trade is:

- Easy to enable when the cleaning agent is subject to a global ban on manufacture, such as CFC-113.
- Understandable when the cleaning agent is a solvent with a low exposure limit such as *n*-propyl bromide.
- Driven by local environmental regulations when a water waste from an aqueous cleaning system is replaced with a VOC emission as "no-clean" flux.

There are other tradeoffs as well:

- Since cleaning of assembled circuit boards won't be done, increased cleaning work may have to be

[58] The favorable Lead-free alloys primarily comprise of Sn with Ag, Bi, Cu, Sb, In or Zn. It's no surprise that there is no absolute drop-in replacement for Tin–Lead with identical melting temperature, cost, wetting, and strength properties. An excellent reference is *The Lead-Free Soldering Cookbook Interactive CD-ROM*, by Robert Willis and the National Physical Laboratory. It is available via several Internet-based sellers.

[59] An indirect effect is possible. The wiper with mechanical debris and drawing agent might be viewed as a new waste stream for which a permit is required though the debris and the drawing agent were components of waste from the existing cleaning system. Or the protective coating with lower viscosity might be considered a new waste component even though another similar component was eliminated as a waste.

[60] In oven reflow soldering, the solder paste is "printed" via a stencil onto the circuit board at points where connections are desired. The board is heated in an oven and the solder melts (reflows) in position.

done by suppliers of components. In that case, the cleaning machine isn't eliminated – just relocated to the jurisdiction of another manager.

- There are/may be cost savings when ingredients, facilities, and procedures are all considered.[61] One estimate[62] is that the savings approximate 10% of the normal full cost of the joining process.
- There may or may not be simplification of the assembly process if the cleaning step is eliminated, but the standard of process control must be higher.
- Product reliability, always difficult to specify and measure, is not enhanced with "no-clean" materials.

Where does the balance lie? Probably in favor of "no-clean" for assembly of electronic components into circuit boards because it is commonly done.[63] Those considering such a choice would be advised to consult more current literature[64] than this book, allow at least one year to adopt and digest the change, and consider professional assistance.

## 1.9 DESIGN FOR CLEANING

There was a time when the paradigm for improved cleaning might have included stronger (whatever that meant) solvents, solvents with lower surface tension, aqueous spray cleaning systems with more nozzles discharging at a higher pressure, detergents capable of surviving at a higher temperature in hot water, or ultrasonic transducers which allowed more control of the pressure waveform.

That time is still now. But there is another paradigm – don't change the cleaning system, change the part to make the cleaning job more easy to do.

To some extent this may seem to be an irrational choice. After all, why move the target when the gun is much easier to move?

Others may see it as an inspired choice. After all, if you can legally change the rules in a game of poker so that only you are dealt an extra card, why not do so?

Inadequate cleaning quality is the usual reason for such consideration. To use the words above, one changes the rules (character of the cleaning job) in order to win the game (gain acceptable cleaning quality). Only occasionally there is another reason – cost.

### 1.9.1 A Change of Design

Two approaches are common:

- Change the character of structures within a part.
- Change the position of structures within a part.

This distinction is made because often a change in the character of a structure brings less pain than does a change in its location.

Some examples for consideration are shown in Table 1.4.[65] (See also Footnote 51). They can require changes in thinking, design, or performance. Obviously the latter, which is why the enterprise is in business, is not where compromise is desired.

Table 1.14 also shows a third approach, that is to change the cleaning process (machine). This approach usually mandates a custom machine with a higher price, and enables a discussion about the value produced by and the cost of good cleaning performance.

One seldom (or never) thinks about the ideas within Table 1.14 until after a problem (poor cleaning quality) has produced harm (unacceptable downstream performance). At that point, one has to contact a consultant, and then it is too late....

## 1.10 OUTCOMES OF CLEANING WORK

A successful election produces a clear winner. A successful concert produces enjoyment for the audience and profit for its producers. What does a successful

[61] Pacific Northwest Pollution Prevention Resource Center (PPRC), "Aqueous Cleaning Technology Review: Technical Issues and Aqueous Cleaning Systems." Available at http:www.pprc.org/pprc/p2tech/aqueous/aqtech.html

[62] http://www.protonique.com/plcom/files/whycl.htm

[63] Figure 4.1 of "The US Solvent Cleaning Industry and the Transition to Non-Ozone Depleting Substances," September 2004 claims that about 60% of those using ozone-depleting solvents transitioned to it. The reference is available at http://www.epa.gov/Ozone/snap/solvents/EPASolventMarketReport.pdf

[64] Keynon, W.G., "Regulations – Innovation Drivers or Hindrances?," *Surface Mount Technology Magazine*, April 2005, p. 16. Also see Footnote 28 of Chapter 2, Section 2.1.4.

[65] Considerations in Table 1.14 do conflict with one another and aren't meant to be considered as a package of options.

**Table 1.14 Considerations About Design for Cleaning**

| Change in Character | Reason for Change |
|---|---|
| Eliminate or minimize all crevices.<br>Mill corners round rather than square. | Corners can trap soil and usually are difficult to reach with mechanical force or solvents. |
| Eliminate threaded sections. Join with pins/holes, slots/keyways, or possibly adhesives. | Threads are simply multiple confined corners in this context. |
| Eliminate internal threaded sections. Where threaded sections are necessary, use external ones such as nuts which can be cleaned separately. | |
| Eliminate blind holes or replace them with a connection which fully penetrates the structure. | It is difficult to thoroughly flush debris from a blind hole. |
| Increase the diameter of holes. | The very high surface tension and high viscosity of water, relative to solvents, can limit volume penetration when flushing or spraying. |
| Increase the standoff distance between components. | |
| **Change in position** | **Reason for change** |
| Move obstructions which block spraying under structures. | Avoid structures with "dirty feet". |
| Eliminate "shadowing" where taller structures are placed in front of shorter structures. | Allow access of flush and spray to all structures. |
| Allow open space between structures for drainage. | Improved flushing. |
| **Change in Cleaning process** | **Reason for change** |
| Delay installation of some structures until they have been individually cleaned (clean, then assemble). | Assemblies are more difficult to clean, then are components. |
| Use two different types of spray nozzles and supply pumps:<br>1. To produce large droplets at high velocity (high energy).<br>2. To produce tiny droplets which can penetrate spaces between part structures (low energy). | • To apply mechanical force where needed.<br>• To rinse where force has been applied. |
| Rearrange parts in batch baskets or in continuous conveying machines for maximum "evacuation" of liquid from the part:<br>• Place long axis of parts perpendicular to direction of spray.<br>• Place small-outline components lengthwise along line of spray.<br>• Place leading edges of regular components perpendicular to direction of spray. | • Distinct open rows can allow efficient drainage and flushing.<br>• To improve access of fluid and force.<br>• See Section 1.12.2. |

cleaning operation produce? Said another way, how to know when you did it right?

There are (at least) five parameters which define a successful cleaning operation. They are:

- **Clean parts** as defined[66] prior to the start of cleaning work. Basically, this means that sequential operations can be conducted by the user, owner, or purchaser of the parts without regard to contamination related to previous operations.
- **No limitation** on production rate.
- **No damage** to parts.

[66] Cleaning tests, and validation of cleaning tests, are covered in Chapter 5.

- No current or expected future incidents involving **environment, safety, or worker health** consequences.[67]
- **Costs of cleaning** operations less than the costs of coping with contaminated parts.[68]

Selection of commercial alternatives should be to produce the most value within these five parameters defining a successful outcome. Weighting among these five parameters will be different among each situation.

### 1.10.1 Cycle Time

This is the metric by which cleaning processes or cleaning systems are often graded. Basically, this is the time cost to do the work. It is normally measured in minutes (and seconds).

Cycle time should be an important consideration – a process taking too long will be non-competitive. But the only significant metric on which cleaning processes should be graded is absence of soil, as recognized by the next user.

#### 1.10.1.1 *Components of Cycle Time*

Cycle time[69] is usually stated (or misstated) as the total time to complete the cleaning, rinsing, and drying stages – *including any delays involved between stages*. More important are the components of which cycle time is composed. They are:

- Cleaning time.
- Drainage time after cleaning to minimize dragout to the rinsing stage.
- Rinsing time.
- Drainage time after rinsing to minimize drying time.[70]
- Drying time.

Additional components of cycle time can be time necessary to fill and empty fixtures (baskets) holding the parts, transport parts between locations where the individual stages are implemented (batch process), and inspect between stages (batch process).[71]

Cycle time can sometimes be used as a fourth (after solvency, mechanical force, and heat) factor (point of support) to improve the cleaning outcome.

Yet the point here is simple: doubling the dwell time in the cleaning sump of a vapor degreaser won't affect the quality of rinsing or drying.

### 1.10.2 Rates of Performance Change

Applying the principle that "more is always better" induces human beings to lengthen cleaning cycles when the cleaning outcome is nearly satisfactory (but not fully so).

More often than not, there is a better idea – change one of the three factors (solvency, mechanical force, or heat). This is because the relationship between cleaning performance (quality) and cycle time is usually asymptotic.

Longer cycle times can yield improved value, but at a diminished rate (often greatly). A general relationship between cycle time and many types of performance is shown in Figure 1.2.[72]

Granted, increase of cycle time for immersion or spray cleaning, rinsing, drying, or draining will improve performance. But the gain will be small or

---

[67] See Chapter 2.

[68] Not less than that predicted within an assumed budget.

[69] Cycle time has the meaning stated here for both batch and continuous processes.

[70] See Chapter 6, Section 6.5.6.

[71] See Table 4.13.

[72] The process model assumed in Figure 1.2 is called one of "first order." The defining relationship for a first-order process is that the rate at which the process performs is proportional to the amount of performance remaining to be achieved.

This relationship approximates performance in many systems – including cleaning, rinsing, drying, and draining:

- For cleaning, the rate of solution of soil is proportional to the concentration of soil already in solution compared to the concentration of soil remaining on the parts. When the cleaning bath is full of soil, the rate of soil removal by solutioning is small. When the part is loaded with soil and the cleaning solvent is pure, the rate of soil removal by solution (cleaning) is at its highest.
- For rinsing, the rate of dilution of soil in rinse fluid is proportional to the concentration of soil already in solution. In other words, it is difficult to make progress in rinsing when the rinse fluid is already dirty.
- For drying, the rate of evaporation of solvent into hot air or vacuum is proportional to the concentration of solvent at the part surface compared to the concentration already in the environment. In other words, the last molecule of soil can only be rinsed by dilution into pristine solvent and the last molecule of solvent can only be dried into a solvent-free atmosphere.

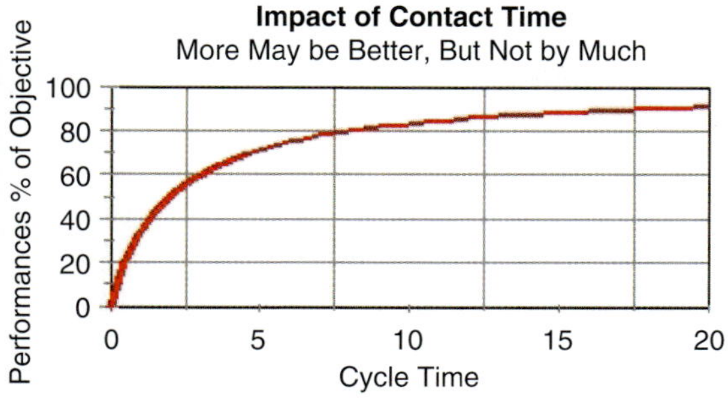


**Figure 1.2**

minuscule – especially if the level of performance is nearly complete.

A more powerful process change is likely to be one which directly affects one of the three factors upon which the process was designed (solvency, mechanical force, and heat). Yet it must be remembered that cycle times necessary to achieve nearly complete performance will always be multiples of cycle times necessary to achieve limited performance.

## 1.11 OTHER OPERATIONS ASSOCIATED WITH CLEANING

Though this book is about management of cleaning operations, those operations don't exist alone. They are integrated into a chain of operations for which the manager also has responsibility.

The manager's aim for their operations, including cleaning, should be that of the baseball umpire or football referee – not to be noticed. That means:

- Upstream operation, whether they be production or maintenance, produce dirty parts.
- Downstream operations perform as expected.

It isn't that the umpire or the referee or the cleaning process isn't valued, it is that the cleaning operations between them should be invisible. Cleaning might be called "The Cloak of Invisibility."[73]

In a sense, cleaning operations function as a filter to keep upstream mistakes from propagating to downstream operations.[74] This is especially true where the downstream user is a customer of the enterprise.

### 1.11.1 Compatibility of Operations

Cleaning operations don't exist in a vacuum – even though some portions of them may be completed in that environment. The cleaning process, no matter what the technical demands for it are, must be compatible with other enterprise operations. Here are some examples:

- All industrial plating work is done in tanks (baths) of water, acids, and other chemicals. A manager would need an unexpected reason to choose plasma cleaning, solvent cleaning, or any other process than aqueous cleaning technology. After all, why dry parts which are going to be next immersed in water?
- It will be difficult for a manager to justify a cleaning process which uses a flammable cleaning solvent (acetone, methyl ethyl ketone, hexane, etc.) in a shop in which welding or metal cutting is openly done.
- While blast cleaning may provide freedom from concern about many safety and environmental hazards, it will be difficult for a manager to select this cleaning approach for operations in a cleanroom or medical facility.
- A manager whose staff is composed of persons whose level of industrial experience is low will likely make a poor choice when they choose supercritical $CO_2$ cleaning – which involves high pressures and sophisticated facilities.

Said another way, whatever the technical demands are for a cleaning process, the enterprise makes additional demands which include financial limits, compatibility (or not) with "political correctness," staff capability, common sense, and safety. All those demands must be met by the manager's choice of cleaning process.

---

- For draining, the rate at which films of solvent flow by gravity off parts is proportional to the mass of solvent film already present (undrained).

The curvature displayed in Figure 1.2 is artificial, based on an arbitrary but realistic choice of proportionality constant, and not meant to represent any specific situation.

[73] Apologies to Las Vegas comic magician Mac King.

[74] For an example, see Footnote 83, Chapter 4, Section 4.13.4.

## 1.12 HOW RINSING IS DONE[75]

Parts poorly rinsed are and will never be clean. Rinsing takes more time, space, and consumables than does cleaning.

### 1.12.1 A Belief, and an Equation

Critical, precision, and general cleaning involve a belief in *equilibrium dilution under immersion*. That's rinsing.

While that's not always true, it is expected to be so; and there isn't a better approximation of reality. Equipment and processes are designed around this belief (see Chapter 6, Section 6.5.1).

Please consider an immersion rinsing situation:

- A certain part has a certain amount of soil on it.
- "Six Sigma"[76] minimum cleanliness is required.
- The soil is readily soluble or suspendable in a certain solvent or water.

As a manager, you want to know:

- How long will it take to achieve this cleanliness via rinsing?
- What's the minimum volume of totally clean solvent or water needed?
- How much faster can this work be done if a pump is purchased with twice the proposed rinse flow rate?

This is a typical problem in rinsing, or cleaning – faced by designers and managers of systems, aqueous or solvent.

The belief in equilibrium dilution can provide answers. Equilibrium dilution means that:

- All the soil will be diluted into all the water/solvent.
- The dilution rate will be proportional to the concentration of soil on the part.
- The concentration of soil in the water/solvent will be the average or equilibrium concentration.
- The parts are reinfected with soil to the extent that they contact that dirty water/solvent.

An engineering material balance based on the assumption of equilibrium rinsing and a soil removal rate proportional to the concentration of soil on the parts yields Equation (1.1):

$$\text{Fraction rinsed} = 1 - e^{[-k \times t]} \qquad (1.1)$$

where:

*k* A "rate constant" with the units of reciprocal time, minutes$^{-1}$ for example.
*k* is calculated as the system throughput ($T$) divided by gross system volume ($V$). If the rinse flow was ½ gallon per minute (gpm) and the rinse tank volume was 1 gal, *k* would (½)/1, or $k = 0.5$/minute. Note that, for simplicity, the volume of the parts and tank's internal piping are ignored.
Some refer to the reciprocal of *k* as the turnover time, holdup time, or space time. In other words, the reciprocal of *k* is the time to fill the tank. This would be calculated as ($V/T$) or 1 gal/(½ gpm) or 2 minutes. The nomenclature for holdup time is the Greek symbol $\theta$. With this convention, the exponential term is $[-t/\theta]$.

*t* Elapsed time in the rinsing cycle from start at zero time, minutes for example. The product of *k* and *t*, or *t* divided by $\theta$, should be dimensionless.

Fraction rinsed is the equilibrium concentration of soil on the parts divided by the initial concentration of soil on the parts. In other words, fraction rinsed is the ratio by which the dragout on the parts has been diluted – assuming the rinse fluid is pristine fluid.

Equation (1.1) describes the behavior of a first-order (concentration dependence to the first power) release of soil into a isothermal vessel which is mixed perfectly.

The behavior is that soil concentration declines with time – that is with more rinse tank volume ($V$) or rinsing at a higher rate of flow ($T$). But the decline is at a decreasing rate with additional rinsing.

#### 1.12.1.1 *About Disbelief*

There are good reasons to believe that equilibrium immersion rinsing does not occur in practical situations. The two necessary assumptions aren't quite true:

1. The first assumption is that perfect (complete) mixing exists within the rinsing chamber. This

[75] Please see Chapter 6, Section 5 for a discussion of the reasons why rinsing is necessary.
[76] The "number of sigma" refer to the number of standard deviations from the average which bound all valid observations of soil content.

exists only in vessels specially designed to produce this outcome.[77]

2. The second assumption is that all surfaces of the parts are completely rinsed with the well-mixed liquid in the rinse tank. This is unlikely. The parts normally act as obstructions to complete fluid turnover within the rinse vessel. In other words, the first assumption conflicts with the second assumption!

Nevertheless, equilibrium immersion rinsing is a reasonable and common assumption in designing a rinsing system or in projecting how it will perform.

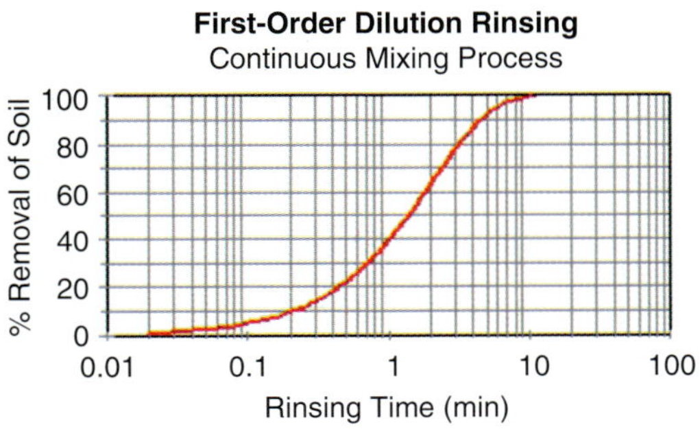


**Figure 1.3**

### 1.12.2 Requirements of Equilibrium Rinsing

Equation (1.1)[78] is plotted in Figure 1.3.

Note the asymptotic behavior; that is how the same amount of rinsing (time or volume of rinse fluid) that dilutes soil from 40% to 85% removal only dilutes soil from 85% to ~99% removal. Also note that in an equilibrium situation, one never gets 100% of the soil off the parts – it just can't be done (see Section 1.12.6).

In the nomenclature of this book, "Six Sigma" rinsing of soil is taken to mean that the initial level of dragout has been diluted, so that the diluted concentration of soils has been reduced by 99.8%. This is the same percentage used in conventional process control technology to reflect the percentage of data which must be within six standard deviations of the mean.

The challenge of reaching "Six Sigma" dilution of soil is shown in Table 1.15. The same information is plotted as Figure 1.4(a).[79]

The times given in Table 1.1 (for $\theta = 2$) are lowest estimates of time required to complete the chosen level of rinsing quality, since Equation (1.1) is an imperfect but acceptable representation of reality.

**Table 1.15 Rinse Holdup Calculations for a Single Tank**

| Number of Sigma | Number of Vessel Turnovers (holdup volumes) | Percent of All Data Inside of | Fraction Rinsed | Hold up Time, minutes (for Volume of 1 gallon and rinse flow of 0.5 gpm, or $\theta = 2$) |
|---|---|---|---|---|
| 1 | 1 | 63.2% | 63.2% | 2.0 |
| 2 | 2 | 86.5% | 86.5% | 4.0 |
| 3 | 3 | 95.0% | 95.0% | 6.0 |
| 4 | 4 | 98.2% | 98.2% | 8.0 |
| 5 | 5 | 99.3% | 99.3% | 10.0 |
| 6 | 6 | 99.8% | 99.8% | 12.0 |

[77] See, among many other references, Oldshue, J.Y., *Fluid Mixing Technology*, McGraw-Hill, New York, 1983, pp. 339–341. Perfect mixing implies that the incoming rinse fluid is completely and instantaneously dispersed among the contents of the rinse vessel. Thus, the soil content of the effluent is the content of all the volume of the rinse vessel. This is never completely true – some volume always is not fully diluted.

Vessel dimensions, length and diameter and their ratio, play a significant role. There should be properly designed agitation (mixing) facilities. The rinse fluid should be added at the proper point in the rinse tank. There should be no unmixed zones ("dead spots") in the rinse tank. And, there should be no structures obstructing complete fluid turnover – such as the parts being rinsed! Designers of cleaning systems almost never make full allowance for perfect mixing in the design of their facilities. Vessel parameters can be computed from rules given in the Oldshue reference.

[78] This equation means that removal rates of soil are related only to differences in soil concentration between the parts and the rinse agent. Details can be found in Perry's *Handbook of Chemical Engineering* (5th ed.), pp. 4–23, Table 4.11.

[79] Note that these values are arbitrary and should not be used for design purposes. They are based on arbitrary assumptions of tank volume, rinse flow rate, and parameter "k."

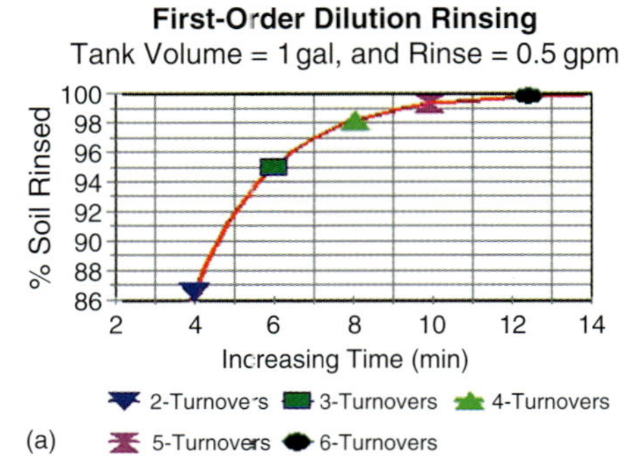


Figure 1.4(a)

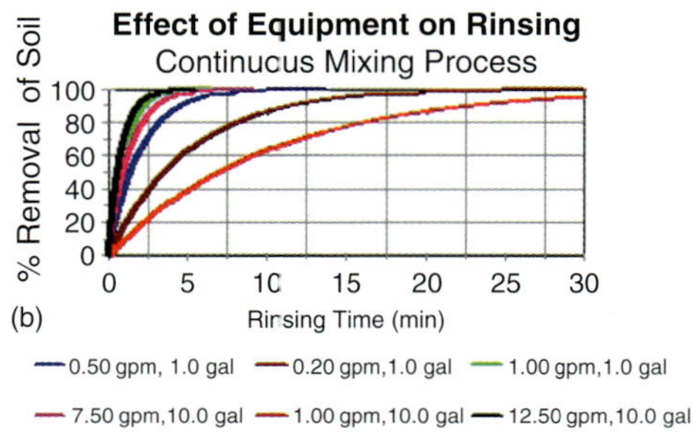


Figure 1.4(b)

Yet, commercial cleaning machines seldom provide even the minimum level of rinsing contact with fluid.

The minimum volume of solvent needed is the number of vessel turnovers times the vessel volume for the selected level of rinsing quality. In the example above, at least 12 gal of rinse fluid will be needed to dilute dragout by 99.8%.

Rinse time can be reduced if a larger pump is purchased, or extended if a larger rinse tank is used. This is shown in Figures 1.4(b) and (c).

At the same level of rinse quality (dilution), rinse time and pump capacity are inversely related as are rinse time and rinse tank volume. If the rinse pump delivers twice as much volume, the parts will be rinsed in half the time. If a manager wishes to achieve "Six Sigma" quality rinsing (dilution) in the same time required for four sigma rinse quality, the capacity of the rinse pump will have to be made 50% (12/8) larger[80] – despite the fact that only a tiny amount of dragout will be diluted (see Section 1.12.7.3).

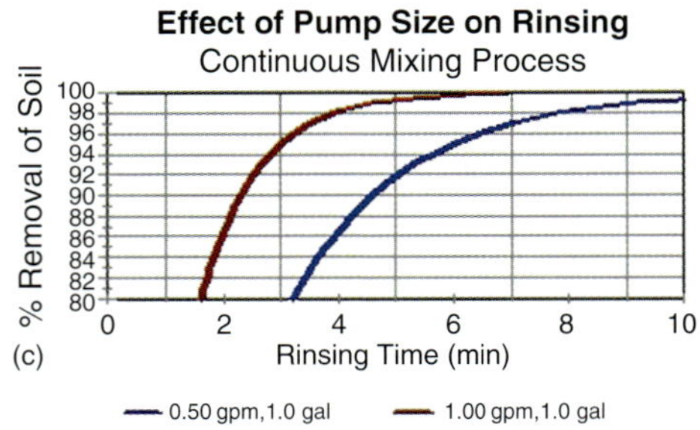


Figure 1.4(c)

## 1.12.3 Rinsing Mechanisms[81]

Rinsing is either dilution of water and soil with soil-free fluid, or displacement of the former with the latter. They mean different things, *dilution* and *displacement*. They represent different mechanisms of rinsing.

### 1.12.3.1 *Dilution Rinsing*

Dilution is the normal means of rinsing. It is the basis for Equation (1.1).

Dilution means that the concentration of soil is reduced by mixing the dirty material with soil-free liquid.

Perfect dilution rinsing means that all the soil-laden fluid is combined with all the rinse fluid so that there are no zones of soil-laden fluid whose concentration is different than the average concentration (0.01% for this example).

### 1.12.3.2 *Displacement Rinsing*

Displacement is not a normally used method of rinsing. But it adds to efficiency where it can be used.[82]

Here, one uses as a rinse fluid a different liquid than that used for cleaning.

The two fluids must be immiscible. If they are miscible (mix with one another), that's dilution rinsing.

In displacement rinsing, one uses a high-density liquid to displace an immiscible liquid of lower density from the volume whose concentration of soil must be reduced. The difference in density makes it easier to penetrate thin sections with the displacement fluid.

Some examples of displacement rinsing would be to flush oil or hydrocarbons with water, flush water with a halogenated solvent in which water isn't very

[80] Or for the same rinse flow rate, 0.5 gpm, the rinse tank will have to be made 9.2 times larger.

[81] See Section 1.13.

[82] See Chapter 7, Section 7.12.8, about displacement drying.

soluble, or flush halogenated solvents with pressurized $CO_2$.

### 1.12.4 More is Not Better

The rinsing job becomes more difficult when more soil remains. If 5 mil of dirty film are left on the parts (versus 1 mil of film), then the rinse volume for dilution must be increased by a factor of 5 to achieve the same level of diluted residue!

From another perspective, if more cleaning agent is used in your cleaning bath than is needed for cleaning, that's wasted money *and* time on the rinsing bill (as well as having to purchase the cleaning agent in the first place).

Every bit of soil must be removed from parts if they are to be truly clean. Soil can be dirt or cleaning agent. Consultants get hired to tell managers to reduce the detergent concentration to reduce the level of spots on metal parts.

### 1.12.5 Patience, Anyone?

Time spent allowing parts to drain *before* rinsing is nearly always time well spent:

- *If the time for liquid drainage is shortened, the time for rinsing will be lengthened.* The bill for rinsing materials will be increased. And the bill for disposal/recycle of rinsing materials will also be increased – perhaps the major cost element.

Transporting parts from the cleaning bath to the rinsing bath without pause is a recipe for failure.

Here is an example of how this failure can be avoided. Suppose production rate is limited to a cycle time of 4 minutes for washing and rinsing operations in a single-stage cleaning machine suppose it is production rate and not quality that is of paramount importance (Figure 1.5).

The question is about how should those 4 minutes be spent to achieve the best quality? If any other information is absent (such as about the character of the parts), this author's recommendation is to leave them in the cleaning bath for ~1 minute, allow them to drain for ~2 minutes, and rinse them for ~1 minute.

Obviously, all the standard qualifications apply: simple shapes will drain faster, higher soil levels will take more cleaning time, and there is a point of diminishing return with everything, including dragout removal by drainage.

**Figure 1.5**

### 1.12.6 The Central Rinsing Theorem

This is all a manager needs to know:

> *The cleanliness of your parts can be no better than that of the liquid which last contacts them.*[83]

Said another way, if a manager wants perfectly clean parts, they must rinse them with perfectly clean cleaning agent. Said another way, "Garbage in, garbage out".

### 1.12.7 Six Rules for Better Rinsing

The following guidance is derived from service at many plants conducting rinsing operations.

#### 1.12.7.1 *Good Rinsing Takes Time and Space*

Dilution of 1 liter (or 1 quart) of dragout from a collection of parts by a factor of 10,000 to 1 will require a single large tank or multiple smaller ones (stages). This is shown in Figure 1.6.

For example, for a rinse tank volume of 10 units (gal) and a dragout volume of 0.01 gal (~38 $cm^3$), Figure 1.6 shows that three consecutive rinse (tanks) stages will be required to dilute that dragout by a

[83] See Chapter 7, Section 7.12.10 for a discussion about how the quality of rinse water affects the quality of dried parts.

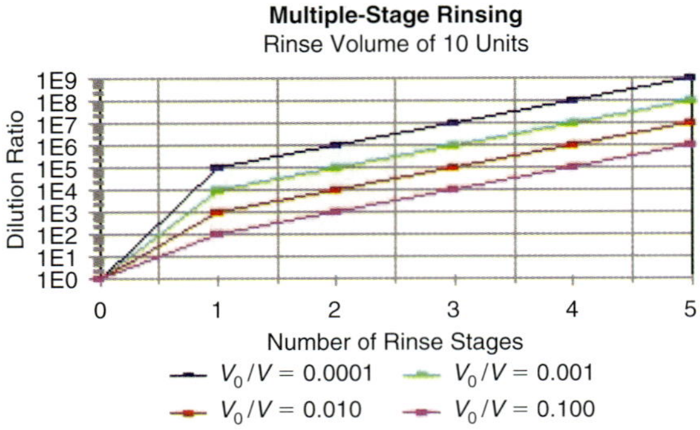


Figure 1.6

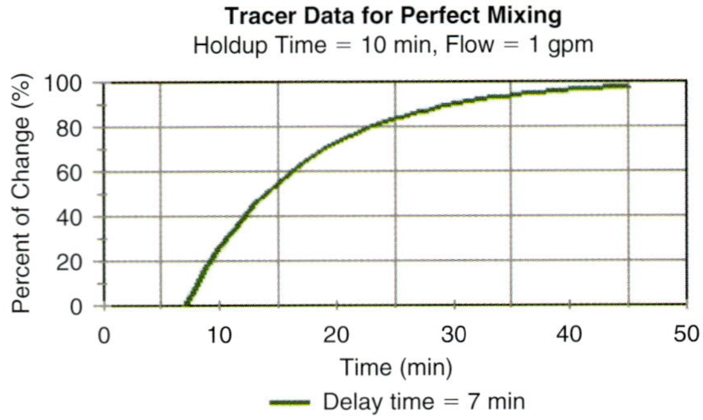


Figure 1.7

factor of 10,000 (more than six vessel turnovers as shown in Table 1.15) ($10^{+5}$ or 1E5). Here, $V_0/V = 0.01/10 = 0.001$.

This need for cleanliness changes floorspace requirements from one tank (for cleaning) to four tanks (for one cleaning, three for rinsing). Processing time will also be extended by a significant amount as well.

#### 1.12.7.2 *All Rinse Fluid Must Contact All Parts*

To no surprise, and similar to cleaning operations, surfaces not well contacted with rinse fluid will not be well rinsed:

- The parts basket must be positioned within the fluid volume so that all parts are thoroughly immersed and exposed to fluid.[84]
- Parts on hangers, hooks, or overhead conveyors must be sprayed from all three directions (dimensions) so that all surfaces are covered.

#### 1.12.7.3 *Good Rinsing = Good Mixing*

It is not how many stages of rinse contact that matter. Rinsing quality is controlled by mixing (dilution) of dragout with rinse fluid within each stage – good engineering.

There are no secrets to vessel designs that will produce good mixing and good rinsing. Rinsing outcomes are generally predictable. Equation (1.1) or Figure 1.6 can be used.

A manager seeking to be well-informed should request mixing data[85] from a supplier. Also refer to Figure 1.7.

#### 1.12.7.4 *Bad Sample = Wrong Conclusion*

Output from a poorly mixed rinse tank, depleted of or enriched with soil, may be the material sampled. Obviously, the wrong conclusion will be drawn from the analysis of that sample. The worst case is that a manager concludes that rinsing quality (mixing quality) is satisfactory when it is not.

---

[84] Poorly trained operators will occasionally seek to improve production by overfilling the cleaning or rinsing bath with parts or parts baskets so that some parts are not fully immersed (or rinsed).

[85] This data is not difficult to obtain. It is basically a tracer study. One compares the predicted concentration from mixing equations with measurements of concentration of chosen tracer compound. Parts should be within the tank during the study. Suitable tracer compounds, easily detectible in water, are food dyes which can be detected colorimetrically.

Note in Figure 1.7 that for the first 7 minutes (an arbitrary value) after injection of the tracer material into the incoming rinse flow, there is no measured concentration of tracer material in the output rinse material. This delay represents imperfect mixing, and is 0 for perfect mixing in a continuous-flow tank is Equation (1.2), and is similar to (Equation 1.1).

$$\frac{C}{C_0} = e\left\{\frac{-t}{\theta}\right\} \tag{1.2}$$

where:

$C$ Concentration of tracer at elapsed time in any units,
$C_0$ Initial concentration of tracer material, in same unit system,
$t$ Elapsed time, in any units,
$\theta$ Holdup time = Vessel volume/rinse flow rate, with vessel volume and rinse flow in the same units, and rinse flow rate in the same units used for elapsed time.

Comparison of sample data to that predicted from Equation (1.2) should identify if this is a concern. For the rinsing (mixing) quality to be satisfactory, no matter the sample point, it must be consistent with this equation.

#### 1.12.7.5 *Rinse Vessel Design Does Matter*

Most tanks are purchased as a component of a packaged cleaning machine. Most suppliers will provide tanks in their cleaning packages based only on cost to them.

The tanks in these cleaning machines usually have round cross-sections, but square corners.[86] This allows fluid to be trapped and not well mixed within the bulk volume. For more details on recognition of superior tanks provided in cleaning machines, see Chapter 7, Section 7.4, and Footnote 77 of this chapter.

#### 1.12.7.6 *Light Does Not Displace Heavy*

If all soil and cleaning agent components are soluble, rinsing is done by dilution. If some components are insoluble, rinsing is done by displacement.

A past client found that two minor soil components were heavy oils insoluble in water. Displacement rinsing of heavy insolubles (oils) with light solubles (water) is equivalent to pushing a chain uphill. Here, a better design of cleaning process was needed to account for this condition.

### 1.12.8 Cleaning Up from Rinsing

The unit operation of rinsing is as or more significant than the unit operation of cleaning. Rinsing is the process of cleaning up the mess made by the process of parts cleaning. Poorly rinsed parts are still dirty – with cleaning agent and soil. A shoddy job of managing dragout and rinsing will ruin an excellent cleaning job.

One completes good rinsing by dilution, but only after nearly all fluid left on from the cleaning bath is allowed to drain. That's crucial. Expect to consume ~100 to ~10,000 times the volume of rinse fluid as the volume of dirty film not drained from the parts. Expect to allow drainage for ~½ of the cleaning–rinsing cycle.

## 1.13 HOW DRYING IS DONE

For most managers, drying is synonymous with evaporation. Drying of water *by evaporation* can be the most costly and time-consuming stage in the cleaning of parts. It frequently causes more problems than does soil removal.

Drying can also be *non-evaporative*. There are several useful non-evaporative methods for getting all or most water off parts – but they aren't commonly used.

### 1.13.1 The Good Old Days

The phaseout of CFCs as cleaning agents caused a global revolution in the way products are manufactured and repaired. After January 1, 1996, it became illegal[87] to manufacture and sell CFCs identified as capable of depleting ozone from the Earth's stratosphere.

CFCs provided both cleaning and drying functions. The cleaning agent was also the drying agent. The drying function was fulfilled by evaporation of the cleaning agent. Dry parts could be obtained in just a few minutes without surface defects (residue) or directed action on the part of the user.

That capability essentially is gone – forever.[88] There are other quick-drying cleaning solvents,[89] but their use adds problems not faced then by users of the banned materials.

---

[86] Here the corner is the intersection between the tank sidewall and its bottom. Some welded tanks may have flanged and dished corners which are not square but somewhat rounded.

[87] In the US because of the Clean Air Act, and in industrialized countries per adherence to the Montreal Protocol.

[88] CFC-11, CFC-12, CFC-113, methyl chloroform (1,1,1-Trichloroethane, TCA, 111TRI, or MCF), halons, and carbon tetrachloride haven't been manufactured (for sale) in the US and other countries after 1995. Some have been manufactured as intermediates where they are consumed in the production of other products.

[89] *n*-propyl bromide (*n*-PB) dries as does TCA, but its use replaces concerns about depletion of the Earth's ozone layer with concerns about human toxicity. Manufacture of TCA is banned. Use of *n*-PB is limited in the US to where parts are dried in a piece of equipment, and not in the open air. Current exposure limit recommended by the ACGIH is 10 ppm.

HCFC-225 ca/cb, HFC-43 10mee, both types of HFEs (which are ethers), and the OS silicon-based solvents all force evaluation of a tradeoff between operating cost of use and investment in drying facilities because of their selling price.

Acetone-and methyl-acetate-free users in the US from concern about VOC regulation, but force learning of the electrical safety codes because they are flammable.

These choices, and many others, are described in more detail in the forthcoming book by this author: *On Solvent Cleaning*, to be published in 2007 by Elsevier, ISBN 185617 4328.

### 1.13.2 Today's Drying Problems

There are four types of problems with drying of cleaning agents:

- Aqueous and semi-aqueous cleaning agents dry (evaporate) and *leave surface residues* (often called "watermarks").
- Non-aqueous cleaning agents dry (evaporate) and leave an *environmental problem* (VOCs), *a safety problem* (flammability), a *human problem* (health effects), or *a personal problem* (odor) with the emissions.
- Aqueous, semi-aqueous, and non-aqueous cleaning agents don't dry (evaporate) well from internal part sections. Drying quality is often poor.
- Aqueous cleaning agents *evaporate slowly*, take *great quantities of energy* to do so, and can *damage parts* by heating them.

These problems occur when a cleaning agent, such as a CFC, is replaced, as the chemical structure of aqueous and acceptable solvent cleaning agents isn't the same as that of the replaced materials.

### 1.13.3 Drying of Water is Difficult

As above, drying generally means evaporation of water. It takes a lot of energy,[90] and a lot of time, to evaporate a little water.[91]

The rate of drying parts is limited by the rate at which heat can be transferred from hot air to the water, causing it to evaporate:

- Slow heat transfer from heated air to wet parts is normally the rate-limiting process step.
- Even worse, air doesn't have a high capacity to carry heat or water. Consequently, huge volumes of hot air can be required.
- Evaporative drying of water is psychologically slow. Operators believe that clean parts have been produced and may be anxious to use them.

### 1.13.4 A Demonstration of Evaporation

Please consider this example of evaporative drying of water, which should be an easy task. Please assume:

- A 1 qt. stainless steel saucepan, half-full of water, on an electric stove.
- It is desired to evaporate all the water in *5 minutes*.

The energy demand to do this evaporation of water is equivalent to 1 ton of refrigeration (12,000 BTU/h).[92] But since it is necessary to heat the stainless steel saucepan as well, to evaporate the contained water, the energy requirement is equivalent to the refrigeration requirement for cooling of a large home.

*Note*: Consider that this task should be done without heating the saucepan!

That's what's done when parts are dried. Managers don't normally want to heat the part to the temperature necessary to cause evaporation at a sufficient rate. That would be likely to damage most parts, and the parts would have to be cooled before use.

A conventional approach would be to use a hair dryer to blow hot air across the top of the saucepan to evaporate the top surface skin or film of water. That's how drying of parts is usually done. Hopefully, this fictitious example will demonstrate why drying of water from metal parts is consuming of energy and time.[93]

In summary, drying of water from parts, as it is normally and commonly done, is a very inefficient scheme.

#### 1.13.4.1 *The Chemical Engineering of Evaporative Drying*

This discipline of chemical engineering, of which this author is both a student and registered practitioner, involves what are known as transport phenomena. That is what occurs when cold parts are dried of water by exposure to heated air.

---

[90] The heat of vaporization of water is about five times higher than that of solvents – 1,000 BTU/lb versus 200 BTU/lb.

[91] For example, 100 SI of surface wetted with 10 mil of water film (a typical number for a wet part), contains about 15 g of water. To evaporate (dry) this small amount of water in 5 minutes from those 100 SI might require *7,500 CFM* (cubic feet per minute) of air heated to 212°F. This example is not a substantial drying task.

[92] Since the volume of water is one pint, and a pint weighs about one pound, and the heat of vaporization of water is about 1,000 BTU/lb, the heat transfer rate is 1,000 BTU/5 minutes or 12,000 BTU/h. Obviously, if the evaporation could take place in 50 minutes versus 5 minutes, the required rate of energy supply would be 1,200 BTU/h (nearly inconsequential). But in an operating plant, the trade of productivity for cost (energy) always favors productivity.

[93] Estimates by this author are that around 1,500 CFM of air, heated to at least the boiling point of water, in addition to at least 15–30 minutes, would be necessary.

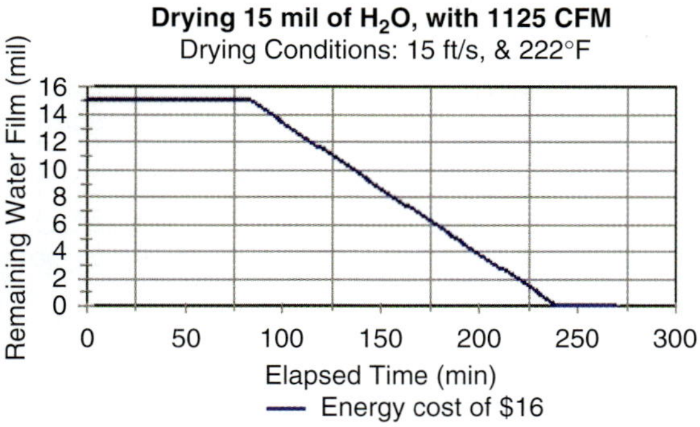


Figure 1.8

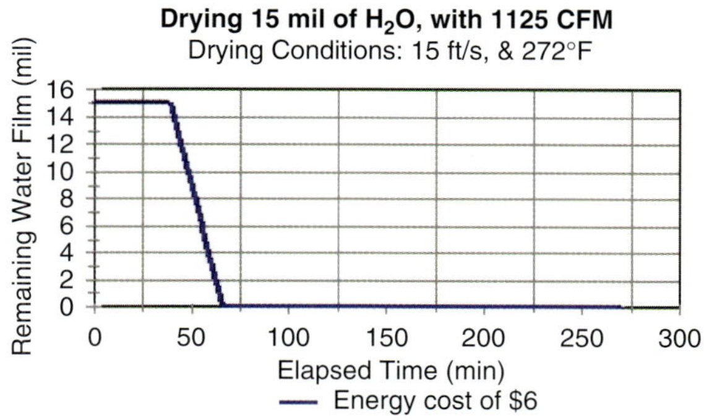


Figure 1.9

The following transport operations occur in sequence in this situation:

- Large volumes of hot air are transported from a source (a heater) to be flowing alongside the part surfaces.
- Heat is transferred from the rapidly moving air stream to heat the films of water which wet the parts.
- The water films are heated, and ultimately evaporate when they are heated to a high enough temperature for a long enough time.
- Heat is also transferred from the rapidly moving air stream to heat the parts.

The operation (step) which limits the rate of removal of water from parts by evaporation is the transfer of heat from the hot air to the water films. Two factors affect that rate of removal. They are:

1. The *temperature* of the hot air. Higher air temperatures produce higher rates of heat transfer because the rate of heat transfer is proportional to the difference between the temperature of the hot air and the temperature of the cold (less hot) water films covering the parts.
2. The *velocity* of the hot air as it moves across the part surface. Higher air velocities produce higher rates of heat transfer – chiefly by increasing the proportionality constant[94] between temperature difference and heat transfer rate.

Both factors require a cost for energy. Obviously, when the air is hotter more energy must be supplied to raise its temperature above the boiling point of water. Less obviously, there is a cost for power to drive a blower producing a higher velocity of air across part surfaces.

If there are two factors, which is more significant in producing the fastest drying rate at the cheapest cost? Some calculated examples are shown in Figures 1.8–1.10.

Figures 1.8 and 1.9 show the dominant effect of air temperature. Hotter supply air, at the same linear air velocity, evaporates the water film much more

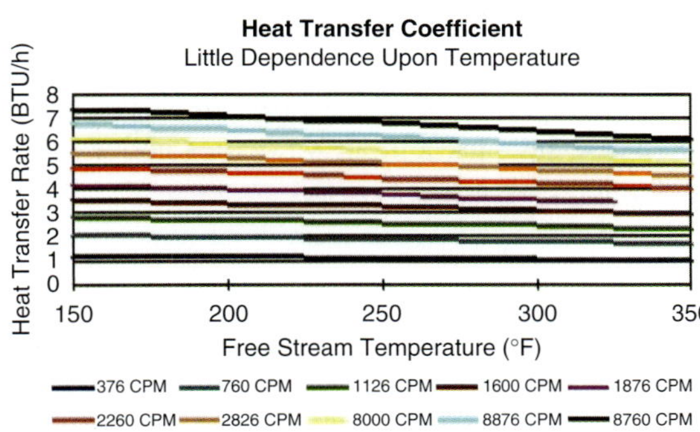


[94] Chemical engineers refer to this proportionality constant as a heat transfer coefficient. Its symbol is either U or h. Its units are heat flow per area per temperature difference, or in English units: BTU/h – SF – °F. Many empirical and theoretical equations exist for predicting values of heat transfer coefficient. An excellent general and available resource is *Perry's Handbook of Chemical Engineering*, Chapter 5.

In general, coefficients for heat transfer between moving hot air and cold surfaces increase as does the air velocity increase by some fractional power. Values for this exponent are commonly around 0.2. This means that the relationship between air velocity and heat transfer rate is not near being linear (exponent of 1).

Calculated values of the coefficient of heat transfer from hot air to large metal plates are shown in the figure. Note that the variation of physical properties with air temperature (horizontal axis) has little effect upon heat transfer coefficient (vertical axis). However, there is a substantial effect upon heat transfer coefficient as the parameter of free stream air flow is varied from around 375 cubic feet per minute to 10 times that volumetric flow rate.

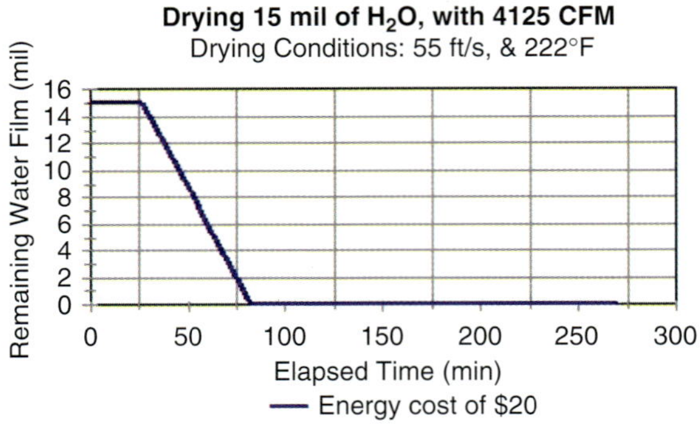


Figure 1.10

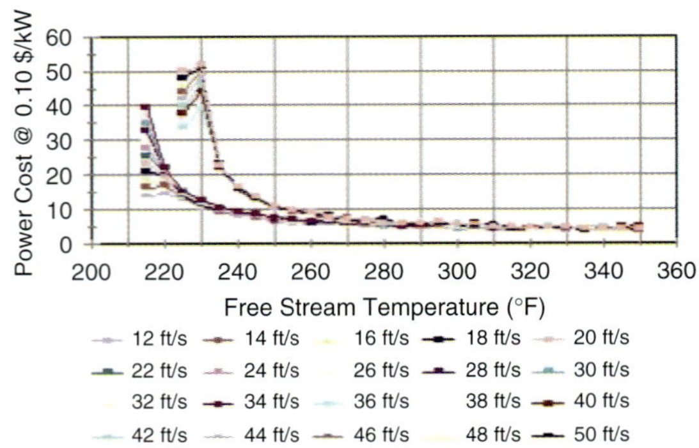


Figure 1.11

quickly, and therefore the cost of drying is lowered. This is a "double win," as drying cycle time is shortened simultaneously with the power cost being decreased.

The effect of linear air velocity at the same air temperature is seen by comparing Figures 1.8 and 1.10. Yes, the increase of air velocity does substantially shorten drying cycle times. But because all that extra air has to be heated, power costs increase somewhat.

The conclusion should be clear:

- Dry parts with heated air at the highest temperature which does not cause part damage or affect material handling after drying.
- Dry parts with air flow at the value which produces the required quality of dryness at the required time.

These calculated[95] effects are shown for a broad range of operating conditions in Figure 1.11. Note that evaporation of water with heated air temperatures above 250°F produces little *calculated* benefit as energy savings, and air velocities above around 40 ft/s increase energy costs.

There is a second conclusion which should also be clear:

- Drying of water using forced hot air is a slow process. The drying step will almost always take substantially longer than the cleaning step, and somewhat longer than the rinsing step. Figure 1.9 shows the calculated drying time to be more than 1 hour if the water film is as thick as 15 mil.[96]

### 1.13.5 Drying of Water without Evaporation

Methods of drying other than open evaporation should be considered when selection of a cleaning/rinsing/drying process is made. There are at least[97] six different methods, called non-evaporative:

1. Centrifugal force.
2. Displacement by insoluble material.[98]
3. Drainage (gravity force) enhanced by vibration.
4. Entrainment into moving stream of air (vacuum).
5. Dislodgement by high-velocity air.
6. Evaporation under vacuum where liquid is recovered.

[95] In Figures 1.8 through 1.10, the elapsed time during which energy is being added but all water film is remaining represents heating of the water to its normal boiling point.

[96] This value is somewhat larger than that expected in normal operation. It was chosen for illustration to make calculated outcomes of power cost more different. More typical values are in the range of 1–5 mil of water film.

[97] See Section 1.13.5.1.

[98] The solvents used for displacement drying can be PFs (perfluorinates), HFEs (hydrofluoroethers of which there are two flavors), HCFCs (hydrochlorofluorocarbons), OSs (methyl siloxanes), or HFCs (hydrofluorocarbons).

**Table 1.16 Non-evaporative Drying Methods for Water**

| Non-evaporative Drying Method | Application Positives and Negatives |
|---|---|
| Centrifugal force | Very low cost.<br>Practical only for parts that will fit into dryer, that is "smaller than a breadbox."<br>Liquid recovered for reuse.<br>Can dry some complicated internal sections.<br>Commercially available. |
| Displacement by insoluble materials | Uses PFs, HFEs, HCFCs, OSs, or HFCs.<br>Useful for "spot-free" drying of complicated sections.<br>Liquid recovered for reuse.<br>Commercially available, with some development work. |
| Drainage (gravity force) enhanced by vibration | Cheap and easy, but only removes and recovers ~2/3 of liquid.<br>May be useful for goods about to be stored.<br>Custom applications. |
| Entrainment into moving stream of air (vacuum) | Practical only for flat chapters, like sheets.<br>"Dry to the touch".<br>Liquid recovered for reuse.<br>Custom applications. |
| Blowoff by high-velocity air | Useful for small parts without internal sections.<br>"Dry to the touch".<br>Liquid recovered for reuse.<br>Custom applications. |
| Evaporation under vacuum where liquid is recovered | Very expensive.<br>Used only for drying last ~ hundred ppm of "moisture."<br>Commercially available. |

There is efficient commercial equipment to implement three of these methods. The others (#3, #4, and #5) can be implemented by your fabrication staff.[99] The most commonly used are centrifugal force, evaporation under vacuum, and displacement by insoluble material.

These six methods for drying of water without evaporation are described in Table 1.16.

Seldom will a manager find non-evaporative drying implemented in a commercial cleaning machine. The great majority of successful applications are designed and implemented by users.[100] See Sections 7.12.1–7.12.5 for specific guidance about how to implement the most common non-evaporative technology – air knives.

### 1.13.5.1 *Marangoni Drying*

His full name was Carlo Giuseppe Matteo Marangoni (1840–1925). An Italian physicist working in Paris, Marangoni studied the conditions for the spreading of one liquid onto another. He published about the phenomenon that liquid will flow along a gas–liquid or a liquid–liquid interface from areas having low surface tension to areas having higher surface tension.[101]

That's the *Marangoni effect* – that liquids flow from regions of low surface tension to regions of higher surface tension.

James Thomson (1822–1892), older brother of Lord Kelvin, had earlier discovered that gradients

[99] It is quite common for a site to construct its own drying equipment.
[100] Forgive the modification of a commercial message – you can do this, this book can help.
[101] C.G.M. Marangoni, "Sull' expansione dell goccie di un liquido galleggianti sulla superficie di altro liquido," Tipografia dei fratelli Fusi, Pavia, 1865.

in surface tension arise due to concentration differences[102] in solution. So, the Marangoni dryer might be correctly named as the Thomson dryer.

#### 1.13.5.2 *Stimulating Tension*

In critical cleaning for medical, dental, electronic, and pharmaceutical applications, the priority is movement of the water mass without leaving non-aqueous residues (water spots). That's why evaporation is often a poor approach toward removing water: non-volatile minerals get left behind as water spots.

Anything which reduces surface tension can be used to stimulate the Marangoni effect.[103] Small amounts of dissolved solute, increased temperature (thermocapillarity), and electric or magnetic fields can also influence flow at an interface by their influence on the surface tension.

Isopropanol (IPA) is conventionally used as the solute to create the gradient in surface tension. Adequate volume flow results from solution of a small amount of IPA. Acetone can also be used.

#### 1.13.5.3 *A Flat Plate of Marangoni*

A Marangoni dryer works well only with parts which are essentially flat surfaces (at the macro-level). The reason is that the fluid force produced by differential surface tension is diminished by competition with other forces such as gravity or convection. That's why most applications are with multiple flat wafers.

The flat surface is slowly (many seconds to minutes) withdrawn vertically (so as not to compete with gravity) from a DI water bath. The headspace is IPA in Nitrogen. IPA dissolves in the water, creating a zone of lower surface tension. Pure water flows (and diffuses) away from the flat surface, leaving it dry of water.[104] The pace of upward part removal from the bath must be synchronized with the pace of water removal from the flat surface to where the solute is dissolved.

Drying rates of a single piece can be ~1 to ~10 SI/min. That slow rate has limited application to simultaneous processing of multiple pieces rather than single pieces. Marangoni drying has

**Figure 1.12**

been scaled up to larger flat surfaces, such as flat panel displays.

As unwanted water is removed as a liquid, soluble materials, such as minerals, are removed as well, and not left behind as water spots. That's the great news.

The bad news is that productivity for single parts is low (as above), only flat surfaces need apply, emissions of VOC (IPA) can sometimes limit application, and undiluted IPA is a fire hazard.

Marangoni technology has achieved a dominant position with simultaneous drying of multiple pieces such as wafers (see Figure 1.12) of MEMS because it avoids the need for evaporative drying schemes which leave residues.

### 1.13.6 Drying of Solvents

Drying of solvents avoids the problems above, but adds another:

- Since the heat of vaporization of solvents is around 200 BTU/lb (one-fifth that of water), energy consumption is much less.
- Since heat transfer rate is often the limiting factor in drying operations, drying of

[102] Thomson, J., "On Certain Curious Motions Observable at the Surfaces of Wine and Other Alcoholic Liquors," *Philos. Mag*. Ser. 4, 10, 330, 1855.

[103] Molenkamp, T., PhD thesis: "Marangoni Convection, Mass Transfer and Microgravity," Rijksuniversiteit Groningen, 6 November 1998.

[104] Marra, J. and Huethorst, J.A.M., "Physical Principles of Marangoni Drying," 1991, *Langmuir*, 7, pp. 2748–2755.

solvents is rapid because less heat has to be transferred.

The problem is that:

- Emission of vaporized solvent cleaning agents usually requires an environmental permit, and compliance with same. This is because most cleaning solvents are VOCs.

The result is that solvent cleaning machines are usually vapor degreasers which by their design provide for drying internal to the machine. VOC emissions are limited by the constraints associated with that design.

#### 1.13.6.1 *Cold Cleaning*

Some cleaning operations with solvents are conducted in the ambient environment. These are called cold cleaning (or dip tank) operations – because the cleaning tank is not usually heated.[105]

Here, there is minimal drying technology. Parts are supported in air and allowed to dry by evaporation of retained solvent.

Obviously, selection of the solvent must include cleaning, safety (flammability), health (exposure), and environmental (VOC) properties. Globally significant today, this cleaning technology will be less frequently implemented in the future.

### 1.13.7 Selection of the Proper Drying Method

What is the right drying method for the situation you manage? The answer depends on two factors:

- The degree of dryness required (see Section 1.13.8).
- The nature of the parts (see Section 1.13.9).

Table 1.17[106] gives general recommendations for drying methods used in a variety of situations. Other

**Table 1.17 Recommendations for Non-evaporative and Evaporative Drying Processes**

| | Parts Can Have Some Remaining "Moisture" | | Dryness to "the Touch" is Satisfactory | | Very High Level of Dryness is Needed | |
|---|---|---|---|---|---|---|
| Nature of Parts | Aqueous or Semi-aqueous Cleaning Agents | Solvent Cleaning Agents | Aqueous or Semi-aqueous Cleaning Agents | Solvent Cleaning Agents | Aqueous or Semi-aqueous Cleaning Agents | Solvent Cleaning Agents |
| Parts "smaller than a breadbox" have simple or no internal sections | Drainage or centrifugal | Drainage or centrifugal | Centrifugal, hot air, or high-velocity air blowoff | Centrifugal, or high-velocity air blowoff | Hot air, followed by vacuum evaporation | Centrifugal, possibly followed by vacuum evaporation |
| Parts "larger than a breadbox" have simple or no internal sections | Drainage | Drainage | Hot air, or high-velocity air blowoff | High-velocity air blowoff, or hot air | Extended time with hot air | High-velocity air blowoff, followed by extended hot air |
| Parts "smaller than a desk" have very complicated internal sections | Displacement by insoluble materials | Displacement by insoluble materials | Displacement by insoluble materials | Displacement by insoluble materials | Displacement by insoluble materials | Displacement by insoluble materials |

[105] See Footnote 23 of Chapter 3, Section 3.7.4. NFPA 34 defines work in open tanks as that in which liquids are not heated above their boiling point. Consequently, cold cleaning can be done in heated tanks where the solvent is not boiled. To avoid safety and environmental problems, this is seldom done.

[106] Also see Table 7.16 in which recommendations are made for specific part configurations.

recommendations are possible based on additional information. This table reflects the belief that the cleaning agent should be chosen based on the nature of the soil, and the rest of the process be chosen based on the nature of the parts.

### 1.13.8 Dryness Specification: How Dry is Dry?

This first factor is easy to evaluate:

- A manager shouldn't dry parts any more than necessary, based on what will be done next with their parts.

*The reason for this is that drying investment and costs are almost exponentially dependent on the degree of dryness needed.*

If there are no externally required dryness specifications, assume "dry to the touch" is adequate:

- "To the touch" means remaining moisture is in the range of 1–5%. Said another way, it means what it says: a manager can't feel moisture on parts.

For example:

- If painting is the next step after cleaning, match the carrier in the paint to the carrier in the cleaning agent, i.e. water or solvent.
- If plating is the next step after cleaning, use an aqueous cleaning agent, rinse well, and don't be concerned about adding water to the water in the plating bath.
- If the parts are to be stored after drying, consider letting them air dry in storage.

If a very high dryness is needed ($\approx<25$ ppm moisture), the drying should be done in two steps: wet down to $\sim1\%$ "moisture" and $\sim1\%$ down to $\approx25$ ppm. The reason is that the costs of the "polishing" drying step are very dependent on the amount of "moisture" being removed.

### 1.13.9 Drying of Large Parts

This factor can be difficult to evaluate.

Since the aqueous cleaning agent (or water rinse) doesn't easily evaporate, and has high surface tension, the replacement drying processes must be able to remove liquid from ALL sections of parts. Both interior and exterior chapters can hold fluid droplets in corners, blind holes, threads, depressions, cavities, etc. Inside chapters of tubing can be very difficult to dry.

If hot air can directly contact the liquid, it can evaporate the liquid and dry that area. But if hot air can't access corners, blind holes, etc., then hot air must heat the part to a temperature where evaporation occurs. Heating the part takes time and adds cost, as well as raising concern about part damage.

If there is a continuous downward path where centrifugal force can pull liquid from interior chapters, the centrifugal dryer will likely be an excellent choice. For example, interior threads which are horizontally presented can be usually dried while interior threads which are vertically presented cannot.

Compressed air blowoff can only dry parts if ALL surfaces can be impacted by the high velocity air stream.

#### 1.13.9.1 *Very Large Parts*

If a manager's parts are larger than their desk, they have a difficult problem – especially if they cannot tolerate surface imperfections, such as "watermarks." A useful solution can be to use aqueous cleaning agents in a spray cabinet with the last spray rinse being with DI water.

For these large parts which can tolerate surface imperfections, hot air is probably the best recommendation. If the nature of the soil requires solvent cleaning of large parts, hot air drying can be used. However, there will be a VOC emission unless the solvent is VOC exempt.

### 1.13.10 Costs of Drying Systems

Predicting generalized costs of drying is an inexact science. The main cost element is energy that is electricity to drive a centrifugal dryer, natural gas or electricity to heat air, and electricity to power an air compressor.

So it makes sense to compare drying costs on the basis of energy equivalents. The values in Table 1.18 are ballpark comparative projections of the energy supply necessary to operate a modest-sized unit. See Chapter 7, Section 7.12.7.1.

**Table 1.18 Comparison of Projected Drying Costs**

| Method | Energy Requirements (kW) |
|---|---|
| Centrifugal | 2–4 |
| Air blowoff | 8–22 |
| Hot forced air | 10–30 |

### 1.13.11 Summary

Drying of parts is a critical part of industrial production and maintenance. If that processing step is not done properly, successive processing steps won't be finished as managers expect.

But there are many choices, and yet just one.[107] The visceral reaction of most managers is to choose as their only drying method to evaporate aqueous cleaning agents with forced hot air. This is often an unfortunate choice – condemning the enterprise to accept very high energy costs, slow processing cycle times, and large requirements for floorspace.

One aim of this book is to allow managers to explore the value to their enterprise of other choices for drying of parts.

[107] Durkee, J.B., "Why Is Drying So Hard with Aqueous Cleaning Technology?," *Products Finishing Magazine*, September 1995.

# 2

# US and global environmental regulations

**Chapter contents**



Through about 1990, users in the US, Europe, and Japan chose cleaning processes, and cleaning chemicals, based on criteria related to performance or business situations. Some users might have included these approaches of matching the following:

- Cleaning process to the parts.
- Overall process to the part transport.
- Cleaning agent to the soil.
- Rinsing step to the final cleaning specification.
- Drying step to the overall product specification.

What changed was the rules. Global and national, environmental regulations were legislated, defined, or promulgated.

And that caused attitudes to change because almost everyone supports, at least in principle, action to preserve the environment. The common expectation[1] changed from the use of chlorofluorocarbon (CFC) or similar solvents for nearly all applications of washing and drying to the use of aqueous cleaning technology for nearly all of the same applications.

The change of attitudes drove still other changes. Cleaning performance, and user's satisfaction with it were two of them:

- Performance changed because the choice[2] of cleaning technology was being made based on reasons other than the five items listed above – which might have been expected to produce the best performance.
- User satisfaction[3] changed because cleaning performance didn't meet that provided by the very forgiving solvent cleaning technology. While there are many reasons for this dissatisfaction, the phrase "ineffective communication" summarizes many of those reasons:
  - Aqueous and solvent cleaning technologies are very dissimilar implementations of common principles (see Chapter 1, Section 1.2).
  - That difference was not understood, so user expectations were often not fulfilled.
  - Change driven by fiat (regulation or requirement) is often less-well accepted than change driven by need or want.
  - The change seldom produced cost savings or benefits outside of environmental ones, so dissatisfaction with performance was exacerbated when it happened.

---

[1] Though there were, as always, a few "knuckleheads," the great majority of users changed over more than one-half decade from use of CFC and similar solvents to some implementation of aqueous technology.

[2] Prior to changes driven by environmental regulations, the author's estimate is that considerably more than one-half used some variant of solvent cleaning. Response to environmental regulations caused that distribution to reverse. Those in the US using some variant of aqueous cleaning technology now exceed three-quarters of the population of doing cleaning work.

[3] The author's estimate is that probably around one-half of population of those doing cleaning work became satisfied with their replacement cleaning system. The half not adequately satisfied can possibly benefit from this book.

This chapter covers the regulations which produced, are producing, and will produce change in management of cleaning processes. These regulations are global in scope though their authorization and implementation is local.[4]

There are six types of regulations about chemicals which need to be understood about their management. They are:

1. Ozone-depleting chemicals (ODCs) – Section 2.1
2. Volatile organic compounds (VOCs) – Section 2.2
3. Global warming – Section 2.3
4. Biologically active – Section 2.4
5. Criteria pollutants – Section 2.6.

## 2.1 CLEANING CHEMICALS AS OZONE-DEPLETING AGENTS

Various zones are identified in the Earth's atmosphere by altitude as described in Figure 2.1.[5]

Some compounds are so inert that they survive and populate the Earth's upper atmosphere.

A generation ago, scientific data showed that Chlorine atoms in these compounds could be liberated by reaction with high-intensity ultraviolet (UV) light from the sun. In the stratosphere, these chlorine atoms react with ozone and consume it. These chemicals, containing Chlorine (or Bromine) atoms, are called ODCs.

It is the Chlorine (or Bromine) that makes a substance ozone-depleting; CFCs and hydrochlorofluorocarbons (HCFCs) are a threat to the ozone layer but hydrofluorocarbons (HFCs) and hydrofluoroethers (HFEs) are not. The latter is because HFCs and HFEs don't contain Chlorine (or Bromine) atoms.

CFC-113 is a strong ODC not because it contains three Fluorine atoms, but because it contains three Chlorine atoms. Carbon tetrachloride is a very strong ODC (Ozone Depletion Potential [ODP] of 1.1), because it contains four Chlorine atoms.

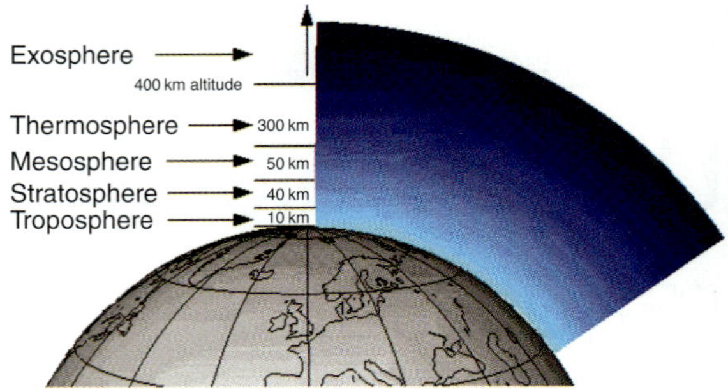


**Figure 2.1** Segregation within Earth's atmosphere

Fortunately, most molecules with Chlorine atoms are fairly reactive. They degrade within 6–8 days (trichloroethylene, known as TCE) and 5–6 months (perchloroethylene, known as PCE or perc, and methylene chloride, known as "meth").[6] They are regarded as low tropospheric ozone creators as well as insignificant (<0.5%) contributors to acid rain formation. Their ODP is negligible and they are not regulated as ozone-depleters.

The ban on manufacture of CFC-113 was necessary, consequential, and highly ironic. After all, CFC-113 was developed to be inert. Its inertness was valued, relative to TCE, because it had negligible effect on human health. But its inertness allowed emitted CFC-113 to reach the stratosphere and do its evil work!

In summary, two factors are necessary for a chemical to be of concern relative to its reactivity with the Earth's ozone layer:

- A chemical must be quite inert so it reaches the stratosphere, where
- Chlorine (or Bromine) atoms in the solvent molecule react with ozone.

### 2.1.1 Regulation of ODCs

The discovery of the annual depletion of ozone above the Antarctic was first announced in a paper by Joe Farman, Brian Gardiner, and Jonathan Shanklin which appeared in *Nature* in May 1975.

[4] The Montreal Protocol is local in neither scope or implementation. It is an international agreement banning, in phased change, production of substances which deplete the Earth's ozone layer.
[5] Created by Cambridge University. See http://www.theozonehole.com/atmosphere.htm
[6] http://www.eurochlor.org/chlorsolvents/generalinfo/info.htm

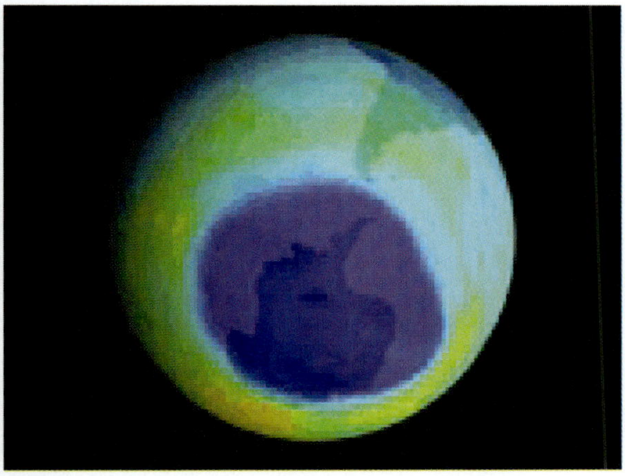

**Figure 2.2** The "ozone hole"

The British Antarctic Survey (BAS) is an excellent reference about the science and events which followed.[7]

Later, NASA scientists re-analyzed their satellite data and found that the whole of the Antarctic was affected. NASA produced a color-retouched photograph (of the Earth, Figure 2.2[8]) which clearly identified the "ozone hole."

In 1977, the Co-ordinating Committee on the Ozone Layer was established by the United Nations Environment Programme (UNEP), and UNEP's Governing Council adopted the World Plan of Action on the Ozone Layer. In 1981, UNEP acted on a proposal submitted by a meeting of legal experts, chaired by Canada, and decided to develop a global convention. Adding to the difficulty of achieving an agreement by the participants were questions about the validity of the atmospheric science and data. Many felt there wasn't the right technological capacity to respond to the challenge.[9]

In 1985, the Vienna Convention on the Protection of the Ozone Layer was signed. By September 1987, the disagreements and lack of understanding had given way to limited trust. In turn, the trust expanded and offered the prospect of consensus on control measures.

### 2.1.2 The Montreal Protocol

Thus it was on September 16, 1987 that the Montreal Protocol[10] on Substances that Deplete the Ozone Layer was signed by 24 countries. As of this writing, over 185 countries have signed it.[11] The Montreal Protocol, administered by UNEP,[12] is the international agreement to preserve the stratospheric ozone layer that protects the Earth from harmful radiation.

At least four unique characteristics of this agreement[13] were as follows:

- It is accepted that the environmentally undesirable CFC chemicals had real value in the industrial economies of all countries as refrigerants and solvents.
- It is recognized that there are historical and economic differences between developed and developing countries (see Section 2.3.5.2).
- It is recognized that a ban on local use would be unfair, difficult to enforce, and ultimately unproductive, and thus chose the more productive step of banning manufacture.
- Chose to emphasize some CFC chemicals versus others by both reactivity with sunlight and emitted volume.

The ban on manufacture was an inspired regulatory choice. With one phrase the number of firms which had to be "policed" was reduced from tens of thousands to a dozen or so. The use (versus manufacture) of only one cleaning solvent, HCFC 141b, was banned by the US Environmental Protection Agency (EPA).

[7] http://www.antarctica.ac.uk/index.php
[8] Image produced by NASA and is found at http://earthobservatory.nasa.gov/Newsroom/NewImages/images.php3?img_id=15776
[9] http://www.ec.gc.ca/press/ozone2_b_e.htm
[10] http://www.unep.org/ozone/montreal.htm
[11] For a complete list, see http://www.unep.org/ozone/ratif.shtml or http://www.iifiir.org/2enreglementation_ozone_Montreal.htm As of November 4, 2004, there are 139 Parties categorized as operating under Article 5 of the Montreal Protocol (developing countries). This list of developing countries include: Brazil, China, India, Iran, Republic of Korea, Saudi Arabia, and Viet Nam.
[12] See http://www.teap.org/ for information about the UNEP Technology and Economic Assessment Panel. They provide technical guidance about compliance to countries.
[13] "Perhaps the single most successful international agreement to date has been the Montreal Protocol," Kofi Annan, Secretary General of the United Nations. See http://www.theozonehole.com/montreal.htm

### 2.1.3 Specific Classification

The Montreal Protocol required acceptance of specific classifications by the signing parties. Classifications and actions regarding types of ODC are shown in Table 2.1.[14] Not all ODC types involve normally-used cleaning solvents.

Emphasis among ODCs by types was done via creation of classes and groups.

Distinction among classes was made based on stratospheric reactivity. Distinction among groups, within classes, was made based on structure:

- Class I substance: one of several groups of chemicals with an ODP[20] of 0.2 or higher. Class I substances listed in Chapter 602 of the 1990 Clean Air Act (CAA) include CFCs, halons, carbon tetrachloride, and methyl chloroform. The US EPA later added HBFCs and methyl bromide to the list by regulation.
- Class II substance: a chemical with an ODP of less than 0.2. Currently, all of the HCFCs are Class II substances.

#### 2.1.3.1 *Not Quite the Same, but Different*

Structure is all.

Even small differences in molecular structure of chemicals can have dominant effects upon their

**Table 2.1 Summary of Montreal Protocol Control Measures**

| Ozone Depleting Substances | Class | Developed Countries | Developing Countries |
|---|---|---|---|
| Chlorofluorocarbons | I | Phased out end of 1995[15] | Total phase out by 2010 |
| Halons | I | Phased out end of 1993[5] | Total phase out by 2010 |
| Carbon tetrachloride | I | Phased out end of 1995[5] | Total phase out by 2010 |
| Methyl chloroform | I | Phased out end of 1995[5] | Total phase out by 2015 |
| Hydrobromofluorocarbons (HBFCs) | I | Phased out end of 1995 | Phased out end of 1995 |
| Methyl bromide | I | Freeze in 1995 at 1991 base level[16] | Freeze in 2002 at average |
| | | 25% reduction by 1999 | 1995–1998 base level |
| | | 50% reduction by 2001 | 20% reduction by 2005[17] |
| | | 70% reduction by 2000 | Total phase out by 2015 |
| | | | Total phase out by 2005 |
| Hydrochlorofluorocarbons | II | Freeze from beginning of 1996[18] | |
| | | 35% reduction by 2004 | |
| | | 65% reduction by 2010 | Freeze in 2016 |
| | | 90% reduction by 2015 | At 2015 base level |
| | | Total phase out by 2020[19] | Total phase out by 2040 |

[14] http://www.theozonehole.com/montreal.htm

[15] With the exception of a very small number of internationally agreed essential uses that are considered critical to human health and/or laboratory and analytical procedures.

[16] All reductions include an exemption for pre-shipment and quarantine uses.

[17] Reviewed in 2003 to decide on interim further reductions beyond 2005.

[18] Based on 1989 HCFC consumption with an extra allowance (ODP weighted) equal to 2.8% of 1989 CFC consumption.

[19] Up to 0.5% of base level consumption can be used until 2030 for servicing existing equipment.

[20] The source of ODP and GWP data used in this book (Tables 2.2 and 2.3) is also the source used by the US EPA. See "The Scientific Assessment of Ozone Depletion", World Meteorological Association's Global Ozone Research and Monitoring Project, 2002, Table 1-5. Data in Table 1-5 of the 2002 report have not been updated since 1998 and are from "The Scientific Assessment of Ozone Depletion," 1998.

**Table 2.2 Class I Ozone-Depleting Substances**

| Product Name | Chemical Name | Lifetime Years | ODP | GWP | CAS Number |
|---|---|---|---|---|---|
| **Group I** | | | | | |
| CFC-11 ($CCl_3F$) | Trichlorofluoromethane | 45 | 1.00 | 4680 | 75-69-4 |
| CFC-12 ($CCl_2F_2$) | Dichlorodifluoromethane | 100 | 1.00 | 10 720 | 75-71-8 |
| CFC-113 ($C_2F_3Cl_3$) | 1,1,2-Trichlorotrifluoroethane | 85 | 1.00 | 6030 | 76-13-1 |
| CFC-114 ($C_2F_4Cl_2$) | Dichlorotetrafluoroethane | 300 | 0.94 | 9880 | 76-14-2 |
| CFC-115 ($C_2F_5Cl$) | Monochloropentafluoroethane | 1700 | 0.44 | 7250 | 76-15-3 |
| **Group II** | | | | | |
| Halon 1211 ($CF_2ClBr$) | Bromochlorodifluoromethane | 16 | 6.00 | 1860 | 353-59-3 |
| Halon 1301 ($CF_3Br$) | Bromotrifluoromethane | 65 | 2.00 | 7030 | 75-63-8 |
| Halon 2402 ($C_2F_4Br_2$) | Dibromotetrafluoroethane | 20 | <8.6 | 1620 | 124-73-2 |
| **Group III** | | | | | |
| CFC-13 ($CF_3Cl$) | Chlorotrifluoromethane | 640 | 1.00 | 14 190 | 75-72-9 |
| CFC-111 ($C_2FCl_5$) | Pentachlorofluoroethane | | 1.00 | | 354-56-3 |
| CFC-112 ($C_2F_2Cl_4$) | Tetrachlorodifluoroethane | | 1.00 | | 76-12-0 |
| CFC-211 ($C_3FCl_7$) | Heptachlorofluoropropane | | 1.00 | | 422-78-6 |
| CFC-212 ($C_3F_2Cl_6$) | Hexachlorodifluoropropane | | 1.00 | | 3182-26-1 |
| CFC-213 ($C_3F_3Cl_5$) | Pentachlorotrifluoropropane | | 1.00 | | 2354-6-5 |
| CFC-214 ($C_3F_4Cl_4$) | Tetrachlorotetrafluoropropane | | 1.00 | | 29255-31-0 |
| CFC-215 ($C_3F_5Cl_3$) | Trichloropentafluoropropane | | 1.00 | | 4259-43-2 |
| CFC-216 ($C_3F_6Cl_2$) | Dichlorohexafluoropropane | | 1.00 | | 661-97-2 |
| CFC-217 ($C_3F_7Cl$) | Chloroheptafluoropropane | | 1.00 | | 422-86-6 |
| **Group IV** | | | | | |
| $CCl_4$ | Carbon tetrachloride | 26 | 0.73 | 1380 | 56-23-5 |
| **Group V** | | | | | |
| Methyl chloroform ($C_2H_3Cl_3$) | 1,1,1-Trichloroethane | 5 | 0.12 | 144 | 71-55-6 |
| **Group VI** | | | | | |
| Methyl bromide ($CH_3Br$) | | 0.7 | 0.38 | 5 | 74-83-9 |
| **Group VII** | | | | | |
| $CHFBr_2$ | | | 1.00 | | |
| HBFC-12B1 ($CHF_2Br$) | | | 0.74 | | |
| $CH_2FBr$ | | | 0.73 | | |
| $C_2HFBr_4$ | | | 0.3–0.8 | | |
| $C_2HF_2Br_3$ | | | 0.5–1.8 | | |

(Continued)

**Table 2.2 Class I Ozone-Depleting Substances (Continued)**

| Product Name | Chemical Name | Lifetime Years | ODP | GWP | CAS Number |
|---|---|---|---|---|---|
| $C_2HF_3Br_2$ | | | 0.4–1.6 | | |
| $C_2HF_4Br$ | | | 0.7–1.2 | | |
| $C_2H_2FBr_3$ | | | 0.1–1.1 | | |
| $C_2H_2F_2Br_2$ | | | 0.2–1.5 | | |
| $C_2H_2F_3Br$ | | | 0.7–1.6 | | |
| $C_2H_3FBr_2$ | | | 0.1–1.7 | | |
| $C_2H_3F_2Br$ | | | 0.2–1.1 | | |
| $C_2H_4FBr$ | | | 0.1–0.1 | | |
| $C_3HFBr_6$ | | | 0.3–1.5 | | |
| $C_3HF_2Br_5$ | | | 0.2–1.9 | | |
| $C_3HF_3Br_4$ | | | 0.3–1.8 | | |
| $C_3HF_4Br_3$ | | | 0.5–2.2 | | |
| $C_3HF_5Br_2$ | | | 0.9–2.0 | | |
| $C_3HF_6Br$ | | | 0.7–3.3 | | |
| $C_3H_2FBr_5$ | | | 0.1–1.9 | | |
| $C_3H_2F_2Br_4$ | | | 0.2–2.1 | | |
| $C_3H_2F_3Br_3$ | | | 0.2–5.6 | | |
| $C_3H_2F_4Br_2$ | | | 0.3–7.5 | | |
| $C_3H_2F_5Br$ | | | 0.9–1.4 | | |
| $C_3H_3FBr_4$ | | | 0.1–1.9 | | |
| $C_3H_3F_2Br_3$ | | | 0.1–3.1 | | |
| $C_3H_3F_3Br_2$ | | | 0.1–2.5 | | |
| $C_3H_3F_4Br$ | | | 0.3–4.4 | | |
| $C_3H_4FBr_3$ | | | 0.0–0.3 | | |
| $C_3H_4F_2Br_2$ | | | 0.1–1.0 | | |
| $C_3H_4F_3Br$ | | | 0.1–0.8 | | |
| $C_3H_5FBr_2$ | | | 0.0–0.4 | | |
| $C_3H_5F_2Br$ | | | 0.1–0.8 | | |
| $C_3H_6FBr$ | | | 0.0–0.7 | | |

chemical, physical, and environmental properties. Here are three pertinent examples of chemicals useful in cleaning operations.

HCFC-141b, 1,1,1-Trichloroethane (TCA), and CFC-113 are all ODCs. The structures of these similar molecules[21] (HCFC 141b,[22] 1,1,1-TCA, and CFC-113) are shown in Table 2.4.

From the structures shown in Figure 2.3, one might speculate that they have consequential reactivity with ozone. And the EPA rates them as being

[21] All of whose manufacture is or will be banned, worldwide, because of concern about the Earth's ozone layer.

[22] Note that HCFC 141b is only different from 1,1,1-TCA in that one Chlorine atom has been replaced by one Fluorine atom. See Table 2.1 to see how their solvent properties are very different and Table 2.4 to see how their ODP values are quite different.

**Table 2.3 Class II Ozone-Depleting Substances**

| Product Name | Chemical Name | Lifetime Years | ODP | GWP | CAS Number |
|---|---|---|---|---|---|
| HCFC-21 ($CHFCl_2$) | Dichlorofluoromethane | 1.7 | 0.04 | 143 | 75-43-4 |
| HCFC-22 ($CHF_2Cl$) | Monochlorodifluoromethane | 12 | 0.05 | 1780 | 75-45-6 |
| HCFC-31 ($CH_2FCl$) | Monochlorofluoromethane | | 0.02 | | 593-70-4 |
| HCFC-121 ($C_2HFCl_4$) | Tetrachlorofluoroethane | | 0.01–0.04 | | 354-14-3 |
| HCFC-122 ($C_2HF_2Cl_3$) | Trichlorodifluoroethane | | 0.02–0.08 | | 354-21-2 |
| HCFC-123 ($C_2HF_3Cl_2$) | Dichlorotrifluoroethane | 1.3 | 0.02 | 76 | 306-83-2 |
| HCFC-124 ($C_2HF_4Cl$) | Monochlorotetrafluoroethane | 5.8 | 0.02 | 599 | 2837-89-0 |
| HCFC-131 ($C_2H_2FCl_3$) | Trichlorofluoroethane | | 0.007–0.05 | | 359-28-4 |
| HCFC-132b ($C_2H_2F_2Cl_2$) | Dichlorodifluoroethane | | 0.008–0.05 | | 1649-8-7 |
| HCFC-133a ($C_2H_2F_3Cl$) | Monochlorotrifluoroethane | | 0.02–0.06 | | 75-88-7 |
| HCFC-141b ($C_2H_3FCl_2$) | Dichlorofluoroethane | 9.3 | 0.12 | 713 | 1717-0-6 |
| HCFC-142b ($C_2H_3F_2Cl$) | Monochlorodifluoroethane | 17.9 | 0.07 | 2270 | 75-68-3 |
| HCFC-221 ($C_3HFCl_6$) | Hexachlorofluoropropane | | 0.015–0.07 | | 422-26-4 |
| HCFC-222 ($C_3HF_2Cl_5$) | Pentachlorodifluoropropane | | 0.01–0.09 | | 422-49-1 |
| HCFC-223 ($C_3HF_3Cl_4$) | Tetrachlorotrifluoropropane | | 0.01–0.08 | | 422-52-6 |
| HCFC-224 ($C_3HF_4Cl_3$) | Trichlorotetrafluoropropane | | 0.01–0.09 | | 422-54-8 |
| HCFC-225ca ($C_3HF_5Cl_2$) | Dichloropentafluoropropane | 1.9 | 0.02 | 120 | 422-56-0 |
| HCFC-225cb ($C_3HF_5Cl_2$) | Dichloropentafluoropropane | 5.8 | 0.03 | 586 | 507-55-1 |
| HCFC-226 ($C_3HF_6Cl$) | Monochlorohexafluoropropane | | 0.02–0.1 | | 431-87-8 |
| HCFC-231 ($C_3H_2FCl_5$) | Pentachlorofluoropropane | | 0.05–0.09 | | 421-94-3 |
| HCFC-232 ($C_3H_2F_2Cl_4$) | Tetrachlorodifluoropropane | | 0.008–0.1 | | 460-89-9 |
| HCFC-233 ($C_3H_2F_3Cl_3$) | Trichlorotrifluoropropane | | 0.007–0.23 | | 7125-84-0 |
| HCFC-234 ($C_3H_2F_4Cl_2$) | Dichlorotetrafluoropropane | | 0.01–0.28 | | 425-94-5 |
| HCFC-235 ($C_3H_2F_5Cl$) | Monochloropentafluoropropane | | 0.03–0.52 | | 460-92-4 |
| HCFC-241 ($C_3H_3FCl_4$) | Tetrachlorofluoropropane | | 0.004–0.09 | | 666-27-3 |
| HCFC-242 ($C_3H_3F_2Cl_3$) | Trichlorodifluoropropane | | 0.005–0.13 | | 460-63-9 |
| HCFC-243 ($C_3H_3F_3Cl_2$) | Dichlorotrifluoropropane | | 0.007–0.12 | | 460-69-5 |
| HCFC-244 ($C_3H_3F_4Cl$) | Monochlorotetrafluoropropane | | 0.009–0.14 | | |
| HCFC-251 ($C_3H_4FCl_3$) | Trichlorofluoropropane | | 0.001–0.01 | | 421-41-0 |
| HCFC-252 ($C_3H_4F_2Cl_2$) | Dichlorodifluoropropane | | 0.005–0.04 | | 819-0-1 |
| HCFC-253 ($C_3H_4F_3Cl$) | Monochlorotrifluoropropane | | 0.003–0.03 | | 460-35-5 |
| HCFC-261 ($C_3H_5FCl_2$) | Dichlorofluoropropane | | 0.002–0.02 | | 420-97-3 |
| HCFC-262 ($C_3H_5F_2Cl$) | Monochlorodifluoropropane | | +0.002–0.02 | | 421-2-3 |
| HCFC-271 ($C_3H_6FCl$) | Monochlorofluoropropane | | 0.001–0.03 | | 430-55-7 |

| Common name | 141b | 1,1,1-TCA | 113 |
|---|---|---|---|
| Scientific name | HCFC 141b | CFC 140a | CFC 113 |
| Atomic name | $C_2H_3FCl_2$ | $C_2H_3Cl_3$ | $C_2F_3Cl_3$ |
| Atom arrangement diagram | 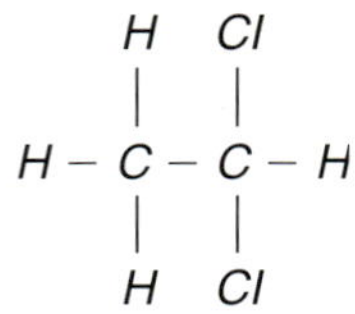 | 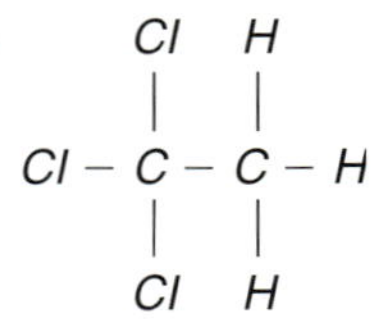 | 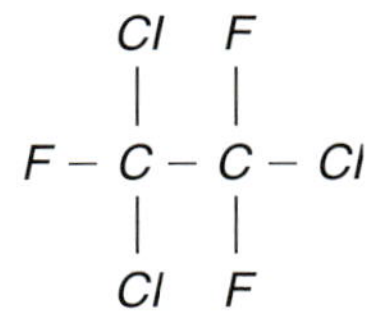 |
| Molecular Model | 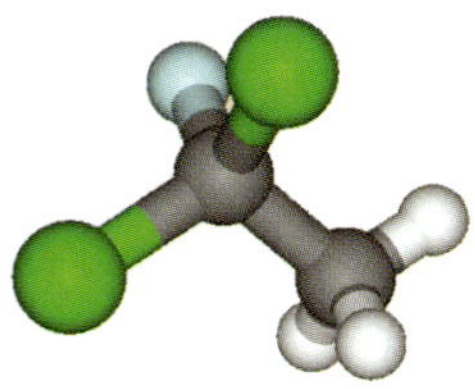 |  | 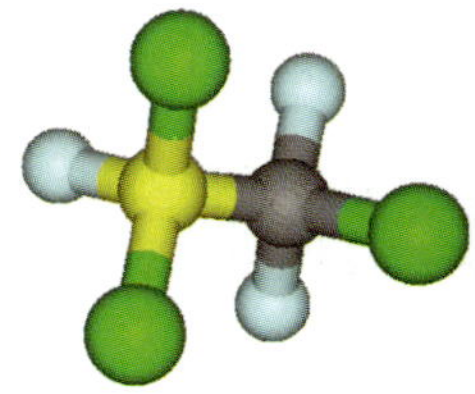 |
| | Hansen Solubility Parameter | | |
| Disperse | 6.0 | 16.8 | 14.7 |
| Polar | 12.9 | 4.3 | 1.6 |
| Hydrogen-bonding | 2.1 | 2.0 | 0.0 |

**Figure 2.3** Three similar, but different, molecules

**Table 2.4 Effect of Molecular Composition on ODP**

| Compound | Number of Chlorine Atoms | Number of Fluorine Atoms | Relative ODP | Action by US CAA |
|---|---|---|---|---|
| HCFC 141b | 2 | 1 | 0.17 | Use for cleaning allowed in US through 2003. Banned afterward |
| CFC-113 | 3 | 3 | 0.8 | Production banned in US after 1996 |
| 1,1,1-TCA | 3 | 0 | 0.1 | |

ODCs. US manufacture of all three is banned by Chapter 602 of the 1990 US CAA.

Table 2.4 shows unequal treatment by the US CAA for HCFC 141b and 1,1,-TCA. Yet their relative ODP is similar, but their solubility parameters are different.

Why is 1,1,1-TCA a Class I ODC, yet its ODP is less than the criteria value of 0.2? The reason is that, in the late 1980s, production of 1,1,1-TCA was far greater than that of HCFC 141b. So its effect on stratospheric ozone was more significant. Both emission volume and reactivity were considered.

#### 2.1.3.2 *The First Rule of Wrestling[23] (and Life)*

Why does 1,1,1-TCA have an ODP of only 0.1?

Why isn't perchloroethylene even listed as an ODC? After all, it contains four Chlorine atoms!

If Chlorine atoms react with ozone to destroy it, why aren't the molecules with the most Chlorine rated as the worst depleters of ozone?

The answer is that there is another factor.

The chemicals (solvents) have to survive in the Earth's atmosphere long enough to reach the

[23] The author is a former competitor and ardent fan of amateur wrestling.

stratosphere. This is where UV radiation is filtered from sunlight by ozone being repeatedly and reversibly converted to Oxygen.

That's the first rule of wrestling, and life. *You have to show up on the mat in order to win.*

Molecules without Fluorine atoms, and only Chlorine (or Bromine or Iodine) atoms, are too reactive in the Earth's troposphere. They decompose within the troposphere, and so don't survive to reach the stratosphere. The presence of Fluorine atoms within molecules confers neutrality upon the molecules.

So the worse depleters of the ozone layer are those chemicals which have a relatively unreactive halogen atoms (Fluorine) and an atom which is significantly more reactive (Chlorine or Bromine):

- Fluorine allows the chemical to survive the troposphere and arrive in the stratospheres.
- Chlorine reacts to destroy ozone when the chemical arrives within the stratosphere.

### 2.1.4 Atmospheric Environmental Chemistry

Ozone is made and destroyed in the stratosphere. Generally, ozone doesn't migrate there. Ozone ($O_3$) in the stratosphere (upper atmosphere) performs a valuable function for our planet.

Ozone intercepts UV radiation from the sun, and keeps it from impacting us on the Earth's surface. Without this chemical reaction, UV radiation will cause serious sunburn, skin cancer and eye disorders. UV radiation is that of wavelength less than about 380 nm.[24] The molecular model for ozone is Figure 2.4.

The ozone depletion process begins when CFCs or other ozone-depleting substances are released, usually near ground level. Winds inefficiently mix the troposphere (the closest vertical zone between the Earth's surface and an altitude of 10–15 km (6.2–9.3 miles) and evenly distribute the gases. CFCs are extremely stable, and they do not dissolve in rain.

After a period of several years, CFC molecules reach the stratosphere (the vertical zone between the troposphere and an altitude of about 50 km (31 miles).

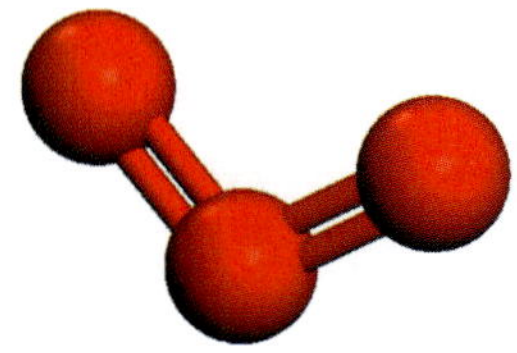

**Figure 2.4** Ozone

Loss of ozone by reaction with CFCs has produced what is popularly called the "ozone hole" (see Figure 2.1).

The chemical reaction for production of ozone involves two steps.[25] In the first step, an Oxygen molecule absorbs a photon of light (conventionally abbreviated as the Greek character $\nu$) with a wavelength shorter than 200 nm. The UV energy splits the Oxygen molecule into two Oxygen atoms.

The following is a simplified description of very complex chemical reactions:

$$O_2 + \nu < 200\,\text{nm} \rightarrow 2O^* \qquad (2.1)$$

$$2O_2 + 2O^* \rightarrow 2O_3 \qquad (2.2)$$

The * symbol on the Oxygen atom means that this species is not a complete molecule and able to react with another species. The produced Oxygen atom ($O^*$) is very reactive. In the second step of ozone formation, two Oxygen molecules and two Oxygen atoms react to form two ozone molecules.

The combined process, with UV as an active initiator, is:

$$3O_2 \rightarrow 2O_3 \qquad (2.3)$$

UV light of another wavelength is also involved in the destruction of ozone. UV light whose wavelength is between 280 and 320 nm splits an ozone molecule into an Oxygen molecule and an Oxygen atom:

$$O_3 + \nu\, 280 - 320\,\text{nm} \rightarrow O^* + O_2 \qquad (2.4)$$

The reactive Oxygen atom and ozone molecule combine to form two Oxygen molecules:

$$O^* + O_3 \rightarrow 2O_2 \qquad (2.5)$$

[24] 380 nm, or 380 billionths of a meter, or 380 angstroms (Å).

[25] http://www.nas.nasa.gov/About/Education/Ozone/chemistry.html

The combined process, with UV as an active initiator, is the inverse of Equation (2.3). Ozone is converted to Oxygen:

$$2O_3 \rightarrow 3O_2 \quad (2.6)$$

Up to 98% of the sun's high-energy UV light are consumed in these chemical reactions involving destruction and formation of atmospheric ozone. The global exchange between ozone and Oxygen is on the order of 300 million tons per day.[26]

Because ozone filters out harmful UV radiation, less ozone means higher UV levels at the Earth's surface. Ozone present in the stratosphere protects life on Earth by filtering out harmful UV rays from the sun. The more ozone depletion, the larger the increase in incoming UV.

Suppose the ozone were somehow eliminated in another chemical process. Then UV light would not be blocked from striking the Earth.

The process by which ozone is eliminated involves Chlorine or Bromine atoms from chemicals which include cleaning solvents. Fluorine atoms are too slow to react and don't participate in the ozone depleting reactions. Iodine atoms react too quickly and are consumed at altitudes lower than the stratosphere.

In the stratosphere, CFCs come into contact with short wavelength UV radiation which is able to split off Chlorine atoms from the CFC molecules.

The key reaction involves liberation of Chlorine or Bromine atoms from ODC molecules.[27] The equation is written for CFC-113:

$$CCl_3F_3 + \nu \rightarrow CCl_2F_3 + Cl^* \quad (2.7)$$

If there were no ODCs, the destruction of ozone by the following equations would not occur. The free Chlorine atom (Cl*) exists as a very reaction substance called a free radical:

$$2(Cl^* + O_3 \rightarrow ClO^* + O_2) \quad (2.8)$$

If this chemical reaction were not followed by others, the situation would not be so tragic. But three additional reactions regenerate the Cl radical:

$$ClO^* + ClO^* \rightarrow (ClO)_2 \quad (2.9)$$

$$(ClO)_2 + \nu \rightarrow Cl^* + ClOO^* \quad (2.10)$$

$$ClOO^* \rightarrow Cl^* + O_2 \quad (2.11)$$

The result of the last three equations is that:

1. Ozone is converted to Oxygen.
2. UV radiation is consumed.
3. The Chlorine radicals are regenerated so they can destroy more ozone. In this process, they are considered to be a catalyst because they are not consumed in the reaction scheme.

The above equations are not the complete set of all chemistry ongoing in the stratosphere. Actual processes are still being researched, and are more complicated than shown here. The effect of the Chlorine radicals can be mitigated via their conversion to hydrochloric acid. Also the $CCl_2F_3$ specie will also react with Chlorine radicals and regenerate the CF-113.

Because of the slow rate of air mixing between the lower and upper atmosphere it is theorized that stratospheric CFCs will stay at a significant level well into the next century. CFCs have a lifetime in the atmosphere of about 20–100 years, and consequently one free Chlorine atom from a CFC molecule can do a lot of damage, destroying ozone molecules for a long time.

### 2.1.5 The SNAP System in the US

In the US, after the CAA Amendments[28] were passed by the US Congress in 1990, the US EPA recognized that it needed a scheme to provide users with the tools needed to comply with the Montreal Protocol. Else, national and international compliance goals would not be achieved. The scheme needed to fulfill many needs:

- Users needed to have alternatives identified. Users needed alternatives which did not damage the planet's ozone layer (few users are expert in atmospheric chemistry). Users needed alternatives whose use would not be banned shortly

[26] Graedel, T.E. and Crutzen, P.J., "*Atmospheric Change: An Earth System Perspective*," (2nd ed.), Freeman, New York, 1993, p. 141.
[27] http://www.ausetute.com.au/cfcozone.html
[28] See Chapter 612 of the Clean Air Act (CAA).

after they adopted them. Users needed published lists (a public clearinghouse) of acceptable alternatives and prohibited alternatives so they could make plans and narrow choices.

- Suppliers of products and processes needed to know that their offering was not going to be restrained by regulatory decisions shortly after they spent resources to commercialize them. Suppliers needed to have their offering recognized in a published list, if it were acceptable.
- The US EPA needed to ascertain if use of proposed replacements for ozone-depleting solvents would actually reduce the emissions of ozone-depleting solvents.

All these needs were to be fulfilled by the US EPA's Significant New Alternatives Program (SNAP).[29] The US EPA did not, and has not, tested alternatives for efficacy or performance.

The first acceptability lists were for substitutes in these major industrial use sectors: refrigeration and air conditioning, solvents (including cleaning applications), foam blowing, fire suppression and explosion protection, sterilants, aerosols, adhesives, coatings, inks, and tobacco expansion.

All this is straightforward. The SNAP program was necessary to for the US to achieve its commitments to reduce emissions of ODCs under the Montreal Protocol.

Announcements and actions then showed the US EPA's interest in issues outside of ODP. Chiefly, suppliers (and users) became concerned that the US EPA was regulating conditions under which an alternative to ODCs could be used in the US. Exposure limits, not ODP classification, became paramount to users, suppliers, and the US EPA. No more significant example of this interest exists than the acceptance of *n*-propyl bromide (*n*-PB) under SNAP.[30]

Many questioned this focus. They claimed use conditions and exposure limits were not within the sphere of responsibility of the US EPA. However, those questions and claims produced no change in focus of the US EPA. They can,[31] and do determine what chemicals and conditions present a threat to human health and the environment.

Details of the SNAP program are simple:

1. Any supplier may petition the US EPA to add a chemical to either the accepted or prohibited list.
2. Petitions are granted or denied within 90 days.
3. Absence of denial within 90 days after petition means the proposed substitute chemical may be sold in the US without objection.

However, without acceptance, the US EPA may at any time deny the petition. Then the list of prohibited chemicals would be expanded by the chemical being petitioned. Thus users and other suppliers, generally don't rally to an alternative until is has been accepted. Such was the case with *n*-PB.[30]

A chronology of decisions made under SNAP is given in Table 2.5.[32] All individual decisions may be examined in detail through internet download.

### 2.1.6 Effects of the Montreal Protocol on Cleaning Work

The Montreal Protocol has influenced the cleaning industry,[33] but somewhat less so has influenced the Earth's ozone layer.[34]

---

[29] On March 18, 1994 EPA published the *Final Rulemaking* (FRM) (59 FR 13044).

[30] The initial SNAP petition was filed in 1995. Only in 2003, was a notice of proposed rulemaking published in the US Federal Register (68 FR 33284). Competitive suppliers, interested international users representing all points of view, regulators from other countries, and consultants provided comments. Identification of and completion of toxicological studies needed to justify an exposure limit consumed the most time. Many, including this author, believe the US EPA has not issued its final regulation (including an exposure limit) associated with *n*-PB.

[31] Chapter 612, Chapters C(1) and (2), of the Clean Air Act (CAA) Amendments provides that: "… the Administrator (of the US EPA) shall promulgate rules under this chapter providing that it shall be unlawful to replace any Class I or Class II substance with *any substitute substance* which the Administrator determines may present *adverse effects to human health or the environment*, where the Administrator has identified an alternative to such replacement that – (1) reduces the overall risk to human health and the environment; and (2) is currently or potentially available …"

[32] http://www.epa.gov/ozone/snap/chron.html

[33] An excellent summary of this influence is complied in the report prepared for the US EPA as "The US Solvent Cleaning Industry and the Transition to non-Ozone Depleting Substances," September 2004. It is available at http://www.epa.gov/Ozone/snap/solvents/EPASolventMarketReport.pdf. Data for Figure 2.5 is found in Tables 2-8 and 2-9, and for Figure 2.6 in Table 3-2. These data represent use by the sector identified in the report as precision cleaning. TCA is used in the larger quantity.

[34] Durkee, J.B. "The CFC Phase Out – How Did We Do?" *Metal Finishing Magazine*, 2005, Vol. 103, No. 6, pp. 60–62.

**Table 2.5 SNAP Decisions**

| Notices and Rulemakings Under the SNAP Program | Publication Date | Federal Register Citation | Effective Date |
|---|---|---|---|
| SNAP ANPRM | January 16, 1992 | 57 FR 1984 | |
| SNAP Proposal | May 12, 1993 | 58 FR 28094 | |
| SNAP Final Rule | March 18, 1994 | 59 FR 13044 | April 18, 1994 |
| Notice 1 | August 26, 1994 | 59 FR 44240 | August 26, 1994 |
| Subsequent Proposed Rule 1 | September 26, 1994 | 59 FR 49108 | |
| Notice 2 | January 13, 1995 | 60 FR 3318 | January 13, 1995 |
| Subsequent Final Rule 1 | June 13, 1995 | 60 FR 31092 | July 13, 1995 |
| Notice 3 | July 28, 1995 | 60 FR 38729 | July 28, 1995 |
| Proposed Settlement w/OZ Tech. | September 22, 1995 | 60 FR 49275 | |
| Denial of 1st OZ Petition | September 25, 1995 | 60 FR 49407 | |
| Subsequent Proposed Rule 2 | October 2, 1995 | 60 FR 51383 | |
| Notice 4 | February 8, 1996 | 61 FR 4736 | February 8, 1996 |
| Subsequent Final Rule 2 | May 22, 1996 | 61 FR 25585 | June 21, 1996 |
| Subsequent Proposed Rule 3 | May 22, 1996 | 61 FR 25604 | |
| Notice 5 | September 5, 1996 | 61 FR 47012 | September 5, 1996 |
| Denial of 2nd OZ Petition | September 30, 1996 | 61 FR 51018 | |
| Subsequent Final Rule 3 | October 16, 1996 | 61 FR 54029 | November 15, 1996 |
| Notice 6 | March 10, 1997 | 62 FR 10700 | March 10, 1997 |
| Subsequent Proposed Rule 4 | May 21, 1997 | 62 FR 27873 | |
| Notice 7 | June 3, 1997 | 62 FR 30275 | June 3, 1997 |
| Subsequent Proposed Rule 5 | February 3, 1998 | 63 FR 5491 | |
| Notice 8 | February 24, 1998 | 63 FR 9151 | February 24, 1998 |
| Notice 9 | May 22, 1998 | 63 FR 28251 | May 22, 1998 |
| Denial of 3rd OZ Petition | January 21, 1999 | 64 FR 3272 | |
| Interim Final Rule (MT-31; Subsequent Rule 7) | January 26, 1999 | 64 FR 3861 | January 26, 1999 |
| Interim Final Rule (HFP; Subsequent Rule 8) | January 26, 1999 | 64 FR 3865 | January 26, 1999 |
| Subsequent Proposed Rule 6 | February 18, 1999 | 64 FR 8038 | |
| Subsequent Proposed Rule 6 – Correction | March 25, 1999 | 64 FR 14417 | |
| Advance Notice of Proposed Rule 12 (*n*-PB) | February 18, 1999 | 64 FR 8043 | |
| Advance Notice of Proposed Rule 12 (*n*-PB) – Correction | March 25, 1999 | 64 FR 14417 | |
| Subsequent Final Rule 5 | March 3, 1999 | 64 FR 10373 | April 2, 1999 |
| Subsequent Final Rule 4 | April 28, 1999 | 64 FR 22981 | May 28, 1999 |
| Notice 10 | June 8, 1999 | 64 FR 30410 | June 8, 1999 |
| Notice 11 | December 6, 1999 | 64 FR 68039 | December 6, 1999 |
| Notice 12 | April 11, 2000 | 65 FR 19327 | April 11, 2000 |

(Continued)

**Table 2.5 SNAP Decisions (Continued)**

| Notices and Rulemakings Under the SNAP Program | Publication Date | Federal Register Citation | Effective Date |
|---|---|---|---|
| Subsequent Final Rule 6 | April 26, 2000 | 65 FR 24387 | May 26, 2000 |
| Notice 13 | June 19, 2000 | 65 FR 37900 | June 19, 2000 |
| Subsequent Proposed Rule 10 (Foams) | July 11, 2000 | 65 FR 42653 | |
| Notice 14 | December 18, 2000 | 65 FR 78977 | December 18, 2000 |
| Notice 14 Correction | March 7, 2001 | 66 FR 13655 | March 7, 2001 |
| Notice 15 | May 23, 2001 | 66 FR 28379 | May 23, 2001 |
| Notice of Data Availability (HCFC Use in Foams) | May 23, 2001 | 66 FR 28408 | |
| Subsequent Direct Final Rule 9 (Rescind Rule) | January 29, 2002 | 67 FR 4185 | April 1, 2002 |
| Notice 16 | March 22, 2002 | 67 FR 13272 | March 22, 2002 |
| Subsequent Final Rule 10 | July 22, 2002 | 67 FR 47703 | August 21, 2002 |
| Notice 17 | December 20, 2002 | 67 FR 77927 | December 20, 2002 |
| Subsequent Final Rule 11 | January 27, 2003 | 68 FR 4004 | March 28, 2003 |
| Notice 17 Correction | April 7, 2003 | 68 FR 16728 | April 7, 2003 |
| Subsequent Final Rule 11 Correction (Additional Information) | April 7, 2003 | 68 FR 16729 | April 7, 2003 |
| Subsequent Final Rule 11 Correction (Typographical errors) | April 7, 2003 | 68 FR 16749 | April 7, 2003 |
| Subsequent Proposed Rule 12 (*n*-PB) | June 3, 2003 | 68 FR 33284 | June 3, 2003 |
| Notice 18 | August 21, 2003 | 68 FR 50533 | August 21, 2003 |
| Subsequent Proposed Rule 12 (*n*-PB) Correction | October 2, 2003 | 68 FR 56809 | October 2, 2003 |
| Notice of Data Availability (reports on $CO_2$ total flooding fire extinguishing systems) | May 11, 2004 | 69 FR 26059 | |
| Final Rule on the Use of HFC 141 b in Foam | September 30, 2004 | 69 FR 58269 | November 29, 2004 |
| Notice of Data Availability (reports on $CO_2$ total flooding fire extinguishing systems) | October 1, 2004 | 69 FR 58903 | October 1, 2004 |

Anyone can track the progress of ozone concentration change. The US Department of Commerce's Climate Monitoring & Diagnostics Laboratory shares current data on their web site.[35]

Many countries have moved to reduce the use of CFCs. The impact of these chemical equations, and the political action they produced, is that CFC cleaning solvents are no longer produced in developed

[35] http://www.cmdl.noaa.gov/ozwv/ozsondes/spo/spoplots.html

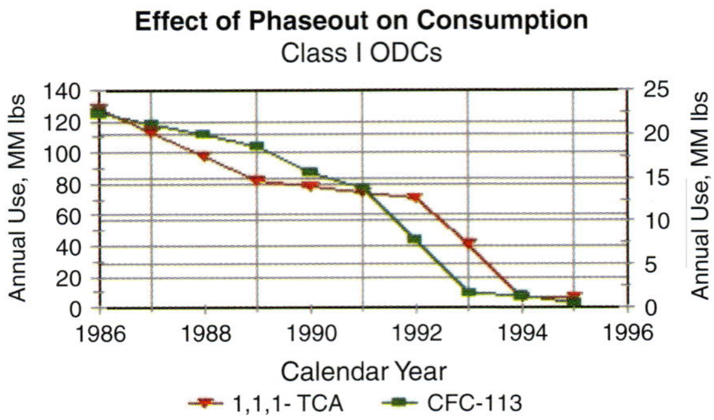


**Figure 2.5**

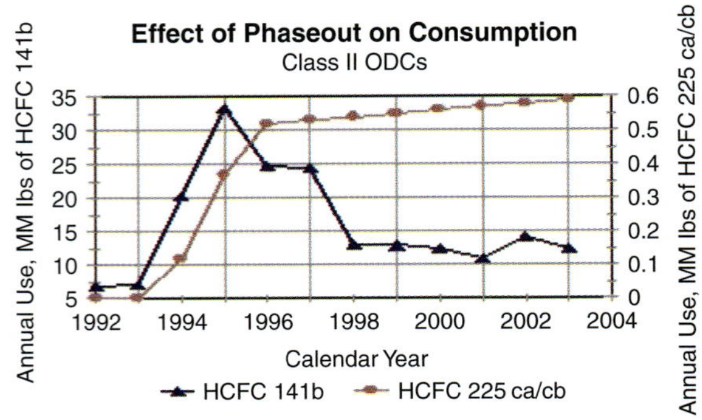


**Figure 2.6**

countries. Those doing solvent cleaning work were denied their use and have/will/must find an alternative.

It is interesting, 10 years after the fact, to examine how the pace of the phaseout was different among Class I and some Class II ODCs. Figure 2.5 shows the parallel action of managers using 1,1,1- TCA and CFC-113. Stockpiles were seldom retained, and only for essential uses generally accepted by the US EPA.

However, in 2003 and beyond a stockpile situation[36] existed for HCFC 141b – primarily used for aerosol cleaning operations.[37] This situation is unique in that the solvent cleaning agent had been sold for a price only about 10% of that required for its commonly-used replacement (HCFC 225 ca/cb[38]), *and* all use is emissive. In other words, the decisions of managers produced no momentum for phaseout when the main option was a greatly increased operation cost. As this is written, aerosol solvent cleaning is still in transition[39] (see Figure 2.6).

### 2.1.7 A Manager's Choices

More than a decade has passed since requirements of the Montreal Protocol were implemented by developed countries. It is highly unlikely that, as a manager in a developed country, you are managing a stockpile of ozone depleting solvents. Certainly, you have chosen to replace the ODC previously used with some alternative.

But as a manager in a developing country,[40] your initial deadlines are January 1, 2010 – for phaseout of CFC-113, or/and January 1, 2015 for phaseout of 1,1,1-TCA.

The decision for replacement cleaning solvent about is made to first match as nearly as possible the Hansen Solubility Parameters (HSP), and then to match the surface tension of the currently-used ODC (see Footnote 33).

For reference guidance, however, the decisions[41] made by managers in the use of cleaning agents to meet the 1996 phaseout deadline are listed in Table 2.6.

## 2.2 CLEANING CHEMICALS AS VOCs

The two-word phrases "low VOC" and "cleaning agent" have become as married as have the phrases "right-wing" and conservative, or "left-wing" and liberal.

Early in this decade, it seemed more important for a cleaning agent, solvent or aqueous, to be described by those two-word phrases, or the one most sought-after but hardest to obtain: *VOC exempt.*

---

[36] See Table 2.6 where it is noted that production and use of HCFC 141b is banned in the US after January 1, 2003.

[37] See Footnote 18, Table 2.1.

[38] Another Class II ODC, whose date for production phaseout in developing countries is 2020. See Table 2.1.

[39] These choices, and many others, are described in more detail in the forthcoming book by this author: *On Solvent Cleaning*, published in 2007 by Elsevier, ISBN 185617 4328.

[40] See Section 2.1.2 and Footnote 33.

[41] See Chapter 4, Section 4.4 of the reference noted in Footnote 33. The application segments are those of the reference: electronics cleaning, metal cleaning, precision cleaning, and aerosol cleaning.

**Table 2.6 Solvent Replacement Decisions**

| Metal cleaning | Precision Cleaning | Electronics Cleaning | Aerosol Cleaning |
|---|---|---|---|
| **ODC solvent used** | | | |
| 1,1,1 TCA | CFC-113 | CFC-113 | CFC-113 |
| **Primary replacement material or process** | | | |
| Trichloroethylene | HFCs, HCFCs, HFEs | No-Clean[42] | HCFC 141b |
| **Secondary replacement material or process** | | | |
| Aqueous cleaning | | HFCs, HFEs, and HCFCs | Various |

The reason is that the days of worry-free emissive cleaning are over – at least in most developed countries. Concerns about emissions from cleaning machines often dominate the methodology by which they are selected.

This section is about the chemistry of VOCs, and how that inherent chemistry affects how its use is regulated.

### 2.2.1 US VOC Definition

In the US, it is the EPA who defines which chemicals are and are not VOCs.

This definition[43,44] is continuously evolving. As of this writing, the definition of negligible photochemical reactivity is equivalence with that of ethane. A lower reactivity would be similarly classified as VOC exempt. A greater reactivity would mean no change from VOC status.

The definition is based on the molar-based reactivity rate, not weight-based rate.[45]

Solvents are, or are not, VOCs. Aqueous cleaning agents contain, or do not contain, VOCs. The level of VOC in a solvent can be less than that in an aqueous cleaning agent, or the reverse. In general, the reverse is more commonly true.

---

[42] See Chapter 1, Section 1.8.

[43] "... Per 40 CFR 51.100(s) Definition Volatile organic compounds (VOC)

(s) "Volatile organic compounds (VOC)" means any compound of carbon, excluding carbon monoxide, carbon dioxide, carbonic acid, metallic carbides or carbonates, and ammonium carbonate, which **participates in atmospheric photochemical reactions**.

(1) This includes any such organic compound other than the following, which have been determined to have negligible photochemical reactivity.

(2) For purposes of determining compliance with emissions limits, VOC will be measured by the test methods in the approved State Implementation Plan (SIP) or 40 CFR Part 60, Appendix A, as applicable. Where such a method also measures compounds with negligible photochemical reactivity, these negligibly-reactive compounds may be excluded as VOC if the amount of such compounds is accurately quantified, and such exclusion is approved by the enforcement authority.

(3) As a precondition to excluding these compounds as VOC or at any time thereafter, the enforcement authority may require an owner or operator to provide monitoring or testing methods and results demonstrating, to the satisfaction of the enforcement authority, the amount of negligibly reactive compounds in the source's emissions.

(4) For purposes of Federal enforcement for a specific source, the EPA shall use the test methods specified in the applicable EPA-approved SIP, in a permit issued pursuant to a program approved or promulgated under Title V of the Act, or under 40 CFR Part 51, Subpart I or Appendix S, or under 40 CFR Parts 52 or 60. **The EPA shall not be bound by any State determination as to appropriate methods for testing or monitoring negligibly reactive compounds if such determination is not reflected in any of the above provisions**...."

[44] Johnson, William, US EPA, Personal Communication to Durkee, J.B., June 3, 2003 and February 19, 2004.

[45] There has been considerable discussion within the US chemical industry and the US regulatory community about the appropriate basis for regulation. Some favor the idea that the reaction rate should be based on the unit with which emissions are regulated (mass units). Others favor the idea that regulation of VOC-exempt chemicals should be on the basis by which chemical reaction rates are typically measured (molar units).

Obviously, there are many uses for chemicals in addition to cleaning. VOC exemption of chemicals used in these other applications can be quite significant. For example, baking of bread causes emission of copious quantities of ethanol, which is a VOC.

Note that evaporation rate is not a defining issue about VOCs in the US. The characterization "volatile" is misapplied. Users and suppliers, often protest that their chosen solvent should be exempt from VOC status because its rate of evaporation is low. The US EPA assumes that no matter the evaporation rate, all the solvent will eventually reach the troposphere. Hence reactivity in the troposphere with sunlight is the defining issue.

VOCs, in the US, should be called PAPRs (Participates in Atmospheric Photochemical Reactions). That these compounds are volatile and are called organic has no direct relation to the reason for their regulation. Only in the US is the characterization VOC not relevant.

### 2.2.2 European VOC Definition

It's just the opposite in Europe. The definition of VOC in the US includes chemicals that are not classified as VOCs in Europe. The US requirements are more stringent than those of Europe. Ironically, the consequences of the less stringent European VOC definitions make it more difficult to practice solvent cleaning in Europe.

In Europe, a VOC is defined by its tendency to evaporate. This type of definition is aimed at reducing emissions from paints, coatings, furniture, architectural materials, etc. But those doing solvent cleaning are impacted as well.

The most usual definition in Europe is the one stated in the Solvent Emissions Directive[46,47] for the European Community (EC), which defines a VOC as:

> "... any organic compound having at 293.15 K (20°C or 68°F) a vapor pressure of 0.01 kPa[48] (0.075 mmHg) or more, or having a corresponding volatility under the particular conditions of use ..."

A number of European countries, however, have developed their own definition in specific contexts.[49] All are based on vapor pressure or boiling point, not reactivity:

- Austria defines, in the 1995 Solvent Ordinance[50] (in effect since August 25, 2001), that a VOC has a maximum boiling point of 200°C (392°F).

---

As this regulation has evolved during the 1990s, US EPA has taken both positions. All exempt chemicals, except acetone, were evaluated using molar units. Acetone was determined to be less reactive than ethane on a per-gram basis, but more reactive on a per-mole basis.

In the proposal to classify acetone as negligibly reactive, the Agency stated that it had "elected to adopt the grams ozone per-gram VOC basis, since grams (or tons), rather than moles, is the mass unit used in regulations dealing with VOC emissions" (59 FR 49878, September 30, 1994). There were no adverse comments on this proposed decision to use the per-gram basis.

This position was reversed in 1999 {40 CFR 51.100(s)} in the proposed recommendation about tertiary-butyl acetate. As of this writing, May 2005, the basis for VOC exemption in the US is molar reactivity relative to ethane. However, those chemicals for which application for VOC exemption has already been made but not decided (see Table 2.8) will be considered on a per-gram basis. See also *Federal Register*, 9/13/05, Vol. 70, No. 176, P5406–5451.

[46] http://europa.eu.int/eur-lex/en/consleg/pdf/1999/en_1999L0013_do_001.pdf see Chapter B.17, 1999.

[47] Council Directive 1999/13/EC of March 11, 1999 on the limitation of emissions of volatile organic compounds due to the use of organic solvents in certain activities and installations. *Official Journal* L 085, 29/03/1999 P. 0001–0022.

[48] To make vapor pressure data more useful, use the following table:

**Conversion of Units for Vapor Pressure**

| Pa | mbar | mmHg | Atm |
|---|---|---|---|
| **1** | .01 | $7.500617 \times 10^{-3}$ | $9.869233 \times 10^{-6}$ |
| 100 | **1** | 0.7500617 | $9.869233 \times 10^{-4}$ |
| 133.3224 | 1.33224 | **1** | $1.315789 \times 10^{-3}$ |
| 101325 | 1013.25 | 760 | **1** |

[49] http://www.esig.org/download/SED_FAQ.pdf

[50] http://www.umweltbundesamt.de/uba-info-presse-e/presse-informationen-e/p4102e.htm

Only high boiling (>200°C) solvents are VOC exempt.
- Switzerland, in its "VOC Ordinance,"[51] defines a VOC as an organic compound with a maximum boiling point of 240°C (464°F). Only high boiling (>240°C) solvents are VOC exempt.
- Germany, France, Italy, UK, and other countries share the EC definition above.[52,53]

The VOC definition in all European countries includes highly volatile solvents which are not VOCs in the US (as above, ethane, acetone, methyl acetate, and parachlorobenzotrifluoride).

Therefore the local legislation should always be consulted to ensure the appropriate definition is being used in the country of interest.

These VOC definitions are absolutely crucial for those who wish to clean with anything other than pure water and don't wish to emit regulated compounds.

### 2.2.3 Cleaning with "Oil" in Europe[54]

In Europe, such cleaning operations are limited to solvents which have very low vapor pressure (high boiling point). Consequently, these cleaning solvents are more like oils than traditional solvents:

- Must be heated to a high temperature when boiled in a vapor degreaser. High temperature may cause damage to parts.
- Will impose additional safety hazards because metal surfaces are significantly hotter (100+ °F versus 300+ °F).
- Will impose thermal stability problems to the solvent because of the increased temperature.
- Will be difficult to remove from parts because of a low evaporation rate.

It must be remembered the definition of which chemicals are VOCs in Europe was not chosen relative to any specific application. Painting, dyeing, oiling, and cleaning are all subject to the same requirement for VOC exemption. There should be no sense that extraordinary attention is being paid to cleaning work.

### 2.2.4 The Environment Is in the Eye of the Beholder

Differences among nations about definitions of VOC, or other environmental parameters, are not about to evaporate. This is most evident in the different environmental and product safety requirements that exist (or do not exist) in countries and regions throughout the world.

These requirements often stem from fundamentally different perceptions of acceptable risk to human health and the environment. And these in turn come from the more basic societal, cultural, economic, and even religious attributes of individual countries.

Such differences stubbornly resist easy reconciliation toward global uniformity. There are also legal/regulatory precedents. The stability of precedent is one of the most valuable aspects of any legal/regulatory system. It assures existing rules will not change readily so the public can confidently draft and sign contracts, agree to specifications, etc., on the basis of clear principles. Equally important, it allows for reasonable predictions of changes.

Another approach has been taken by the International Standards Organization (ISO) in their ISO 16000-6.[55] Here, vapor concentrations are measured via sampling in a work area or a test chamber. The sample is collected on a sorbent and analyzed with a gas chromatograph. This approach mainly applies to emissions not from surface cleaning machines but from products. The regulating agency[56] then specifies which solvents are of concern.

---

[51] http://www.swissmem.ch/eng/pdf/umweltpolitik-e.pdf

[52] http://dbe.invista.com/e-trolley/page_9166/

[53] Solvents favored by the phrase "low" risk in Germany are DBE (dimethyl glutarate + dimethyl succinate (CAS 106-65-0) + dimethyladipate (CAS 627-93-0)); NMP (N-methyl-2-pyrrolidone); and BGA (ethylene glycol monobutyl ether acetate/2-butoxyethylacetate).

[54] A good reference is the *European Commission National Emission Ceilings Directive* (NECD) and UN/ECE *Gothenburg Protocol*, which set national emission ceilings for pollutants for all signatories, to be achieved by 2010.

[55] http://www.iso.ch/iso/en/stdsdevelopment/techprog/workprog/TechnicalProgrammeSCDetailPage.TechnicalProgrammeSCDetail?COMMID=3660.

[56] http://www.umweltdaten.de/daten-e/agbb.pdf

**Table 2.7 Compounds Exempt From VOC Status in the US**

| Chemical Name | Industry Name | ODP | GWP |
|---|---|---|---|
| Methane | ME-thane | ~0 | ~0 |
| Ethane | E-thane | ~0 | ~0 |
| Methylene Chloride | Dichloromethane | ~0 | ~0 |
| 1,1,1-Trichloroethane | Methyl chloroform, TCA | .12 | 144 |
| 1,1,2-Trichloro-1,2,2-Trifluoroethane | CFC-113 | 1.0 | 6030 |
| Trichlorofluoromethane | CFC-11 | 1.0 | 4680 |
| Dichlorodifluoromethane | CFC-12 | 1.0 | 10720 |
| Chlorodifluoromethane | HCFC-22 | .05 | 1780 |
| Trifluoromethane | HFC-23 | | |
| 1,2-Dichloro-1,1,2,2-Tetrafluoroethane | CFC-114 | .94 | 9880 |
| Chloropentafluoroethane | CFC-115 | .44 | 7250 |
| 1,1,1-Trifluoro-2,2-Dichloroethane | HCFC-123 | .02 | 76 |
| 1,1,1,2-Tetrafluoroethane | HFC-134a | | |
| 1,1-Dichloro-1-Fluoroethane | HCFC-141b | .12 | 713 |
| 1-Chloro-1,1-Difluoroethane | HCFC-142b | .07 | 2270 |
| 2-Chloro-1,1,1,2-Tetrafluoroethane | HCFC-124 | .02 | 599 |
| Pentafluoroethane | HFC-125 | | |
| 1,1,2,2-Tetrafluoroethane | HFC-134 | | |
| 1,1,1-Trifluoroethane | HFC-143a | | |
| 1,1-Difluoroethane | HFC-152a | | |
| Parachlorobenzotrifluoride | PCBTF | | |
| Cyclic, branched, or linear completely methylated siloxanes | OS-10, OS-20, OS-30 | | |
| Acetone | Dimethyl ketone | ~0 | ~0 |
| Tetrachloroethylene | Perchloroethylene | .005 | |
| 3,3-Dichloro-1,1,1,2,2-Pentafluoropropane | HCFC-225 ca | .02 | 120 |
| 1,3-Dichloro-1,1,2,2,3-Pentafluoropropane | HCFC-225 cb | .03 | 586 |
| 1,1,1,2,3,4,4,5,5,5-Decafluoropentane | HFC 43-10mee | | |
| Difluoromethane | HFC-32 | | |
| Ethyl fluoride | HFC-161 | | |
| 1,1,1,2,3,3,3-Heptafluoropropane | HFC-227ea | | |
| 1,1,1,3,3,3-Hexafluoropropane | HFC-236 fa | | |
| 1,1,2,2,3-Pentafluoropropane | HFC-245 ca | | |
| 1,1,2,3,3-Pentafluoropropane | HFC-245 ea | | |
| 1,1,1,2,3-Pentafluoropropane | HFC-245 eb | | |
| 1,1,1,3,3-Pentafluoropropane | HFC-245 fa | | |
| 1,1,1,2,3,3-Hexafluoropropane | HFC-236 ea | | |
| 1,1,1,3,3-Pentafluorobutane | HFC-365 mfc | | |
| Chlorofluoromethane | HCFC-31 | | |
| 1-Chloro-1-Fluoroethane | HCFC-151 a | | |
| 1,2-Dichloro-1,1,2-Trifluoroethane | HCFC-123 a | .02 | 76 |

(Continued)

**Table 2.7 Compounds Exempt From VOC Status in the US (Continued)**

| Chemical Name | Industry Name | ODP | GWP |
|---|---|---|---|
| 1,1,1,2,2,3,3-heptafluoro-3-methoxy-propane | HFE-7000, or (n-$C_3F_7OCH_3$) | 0 | |
| 1,1,1,2,2,3,3,4,4-nonafluoro-4-methoxy-butane | HFE 7100 or $C_4F_9OCH_3$ | 0 | 320 |
| 1-ethoxy-1,1,2,2,3,3,4,4,4-nonafluorobutane | HFE 7200 or $C_4F_9OC_2H_5$ | 0 | 55 |
| 1,1,1,2,2,3,4,5,5,5-decafluoro-3-methoxy-4-trifluoromethyl-pentane | HFE-7300 or $C_2F_5CF(OCH_3)CF(CF_3)_2$ | 0 | |
| 3-ethoxy-1,1,1,2,3,4,4,5,5,6,6,6-dodeca-fluoro-2-(trifluoromethyl) hexane | HFE-7500, or | 0 | |
| 2-difluoromethoxymethyl)-1,1,1,2,3,3,3-heptafluoropropane | HFE 8200 or $(CF_3)_2CFCF_2OCH_3$ | 0 | 55 |
| 2-ethoxydifluoromethyl)-1,1,1,2,3,3,3-heptafluoropropane | $(CF_3)_2CFCF_2OC_2H_5$ | 0 | |
| Methyl Acetate | MeOAc, or $CH_3O(C = O)CH_3$ | 0 | 0 |
| Tertiary-butyl Acetate[57] | t-BAc, or $(CH_3)CO(C = O)CH_3$ | 0 | 0 |
| Methyl Formate | Methyl Formate, or $CH_3O(C = O)H$ | 0 | 0 |
| Perfluorocarbon compounds which fall into these classes:<br>I) Cyclic, branched, or linear, completely fluorinated alkanes.<br>II) Cyclic, branched, or linear, completely fluorinated ethers with no unsaturations. | | | |

### 2.2.5 VOC Exempt Chemicals in the US

The list of currently exempt chemicals,[58] in the US, is in Table 2.7.

### 2.2.6 Potential Future US VOC Exemptions

In addition to those chemicals already exempted, there are additional compounds for which their manufacturer, distributor, or trade association has filed for exempt status to the US EPA about being a VOC.[59,60] They are listed in Table 2.8. The purpose of this table is not to predict the future but to indicate the diversity of concern among manufacturers about providing VOC-free products.

#### 2.2.6.1 *A Personal Opinion*

In a sense the US EPA's policy of VOC exemption has not succeeded. The ideal policy would:

- **Classify solvents** as VOCs by their relative potential for smog formation. The current policy does not do this. The classification scheme is by negligible reactivity and nothing else.
- **Encourage substitution** of lower reactivity smog-formers for higher reactivity smog-formers. The current policy provides little incentive for solvent substitution. This is because it is a binary policy – either VOC or not VOC. There is no recognition of higher reactivity or lower reactivity.

[57] The exemption is not complete. "t-butyl acetate (TBAC) will not be VOC for purposes of VOC emissions limitations or VOC content requirements, but *will continue to be VOC for purposes of all recordkeeping, emissions reporting, and inventory requirements which apply to VOC*". Please see 69 FR 69298, effective November 18, 2004.

[58] Effective November 18, 2004.

[59] Received Petitions Requesting VOC Exempt Status and for which EPA has Published no Final Action (as of November 18, 2004).

[60] Listed in order of earliest application date. List is current as of May 5, 2006. Bold type indicates items about which EPA has published expected approval.

**Table 2.8 Chemicals for Which VOC Exempt Status Has Been Applied**

| Chemical Name | Petitioner |
|---|---|
| Chlorobromomethane | ICF Kaiser (SAI Division) |
| 1-Bromopropane (also known as n-propyl bromide) | Enviro Tech International, 5/10/1996; petition also submitted by Albemarle Corp. |
| Methyl Bromide | Chemical Manufacturers Association |
| n-Alkanes (C12–C18) | The Aluminum Association |
| Technical white oils | The Printing Industries of America and Pennzoil Products Co. |
| **Benzotrifluoride** | Occidental Chemical Company [now Islechem]; petition also submitted by Kowa America |
| Carbonyl Sulfide (COS) | E.I. du Pont de Nemours and Company; petition also submitted by Texas Mid-Continent Oil & Gas Association |
| trans-1,2-dichloroethylene | 3M Corporation |
| **Dimethyl succinate** | Dibasic Esters Group/Synthetic Organic Chemical Manufacturers Assn. |
| Dimethyl glutarate | Submitted in the same petition as for dimethyl succinate. |
| Carbon Disulfide | Texas Mid-Continent Oil & Gas Association |
| Acetonitrile | BP Chemicals and GNI Chemicals Corporation |
| Toluene Diisocyanate (TDI) | Chemical Manufacturers Association |
| Methylene Diphenyl Diisocyanate (MDI) | Chemical Manufacturers Association |
| **Propylene Carbonate** | Huntsman Corporation |
| Methyl Pivalate | Exxon Chemical Company |
| **Dimethyl Carbonate** | Kowa America Corp. |
| **Hydrofluoropolyethers (four HFPE compounds)** | Solvay-Solexis |

Today, there is considerable scientific discussion about a universal reactivity characteristic.[61] Thus, solvent substitution (except for that mandated by the Montreal Protocol) hasn't occurred in the cleaning industry to a significant degree.

In addition, use of the per-unit-weight basis is inconsistent with the selection of ethane as the reactivity benchmark. It creates a bias that causes reactive, higher molecular weight organics[62] to be classified as negligibly reactive.[63]

As in Section 2.2.2, legislation in the US and Europe are different and not connected. A product classified as VOC-exempt in the US is not automatically classified as not being a VOC in Europe. There are no exemptions in Europe from VOC status.

---

[61] There is a public task group (Reactivity Research Working Group), of which the author is a member, formed to study atmospheric reactivity data. The group will recommend, as of May 2005, that there is adequate scientific basis (mathematical airshed models based on kinetic data and validated mechanisms) to support definition of a new VOC Regulation Policy.

[62] An example is oligomeric components of coatings.

[63] Figure 4-1 of "The U.S. Solvent Cleaning Industry and the Transition to non Ozone Depleting Substances," September 2004 claims that about 60% of those using ozone-depleting solvents transitioned to it. The reference is available at http://www.epa.gov/Ozone/snap/solvents/EPASolventMarketReport.pdf.

Another point of view, that "...This study illustrates that products manufactured using a no-clean label are not a guarantee of long-term reliability...", can be found in the article "Analyzing the Debate of Clean vs. No-Clean," by Tosun, U., and Wack, H., in *SMT Magazine*, March 2006, pages 20 to 23.

**Table 2.9 Comparison of VOC Exempt (in US) Solvents**

| Solvent | Ozone Depleter? | Exposure Limit, ppm | Flash Point °F | Kb | Price (€/lb) |
|---|---|---|---|---|---|
| HFC 227ea | NO | 1000 (PEL) | NA | | >5 |
| Acetone | NO | 750 (ACGIH) | 1.4 | 0 | <1 |
| HFC 245 fa | NO | 300 (PEL) | NA | 0 | >5 |
| HFE 7100 | NO | 750 (ACGIH) | NA | 5 | >15 |
| HFE 7200 | NO | 200 (PEL) | NA | 5 | >15 |
| t-butyl acetate (t-BAC) | NO | 200 (PEL) | 33.8 | | <1 |
| HFC 43_10mee | NO | 200 (PEL) | NA | 9 | >15 |
| HFE 7100 | NO | 200 (PEL) | NA | 5 | >15 |
| OS_30 | NO | 200 (PEL) | 350 | 13.4 | >15 |
| OS_20 | NO | 200 (PEL) | 84 | 15.1 | >15 |
| OS_10 | NO | 200 (PEL) | 30 | 16.6 | >15 |
| Methyl acetate | NO | 200 (ACGIH) | 14 | | <1 |
| HFC 365 mfc | NO | 200 (PEL) | −13 | 13 | >10 |
| Perchloroethylene (Perc) | NO | 100 (OSHA) | NA | 90 | <1 |
| Methyl formate | NO | 100 (ACGIH) | −20 | | 1 |
| Parachlorobenzotrifluoride (PCBTF) | NO | 25 (PEL) | 109.4 | 64 | <1 |
| Methylene chloride (Meth) | NO | 25 (OSHA) | NA | 136 | <1 |
| CFC 113 | Class I | 1000 (PEL) | NA | 31 | >25 |
| HCFC 141b | Class II | 500 (PEL) | NA | 56 | <1 |
| 1,1,1-TCA | Class I | 350 (OSHA) | NA | 124 | <1 |
| HCFC 225 ca/cb | Class II | 100 (PEL) | NA | 31 | >15 |

## 2.2.7 VOC Exempt (US Only) Cleaning Solvents

Solvents on the US VOC exempt list, Table 2.7, which bring some value in cleaning applications,[64] are listed in Table 2.9. The list is sorted by exposure limit. These solvents bring significant value to users in compliance with national, state, and local environmental regulations.

Collectively, however, they do not bring significant value as a selection of solvents which can be used in a variety of applications. Hansen Solubility Parameters (HSPs) can define this value for each solvent.

A general, and unexpected, concern about US VOC exempt solvents is flammability. Those solvents with higher values of HSPs are flammable. Only two are classified as combustible (flash point above 140°F).

The polar and Hydrogen-bonding parameters are plotted in Figure 2.7. For comparison, HSP values for many common oxygenated solvents are plotted on the same scale in Figure 2.8.

Intermolecular forces, which produce increased values of polar and $H_2$-bonding HSPs, are considerable lessened in solvents which are VOC exempt compared those intermolecular forces within oxygenated solvents which are not VOC exempt in the US.

Comparison with halogenated solvents would support the same conclusion.

[64] The remainder of the chemicals on that list are either refrigerants or specialty chemicals.

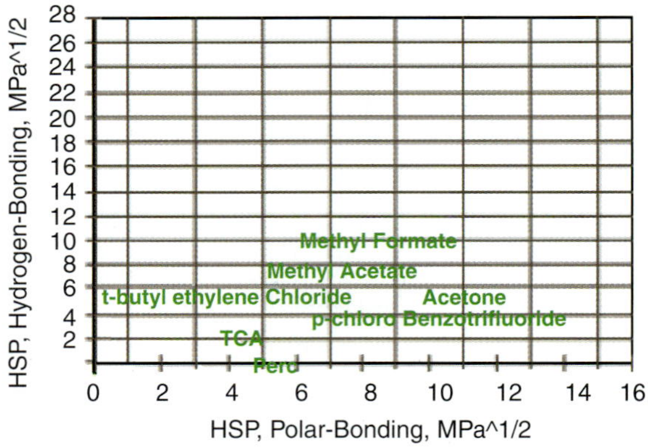


Figure 2.7

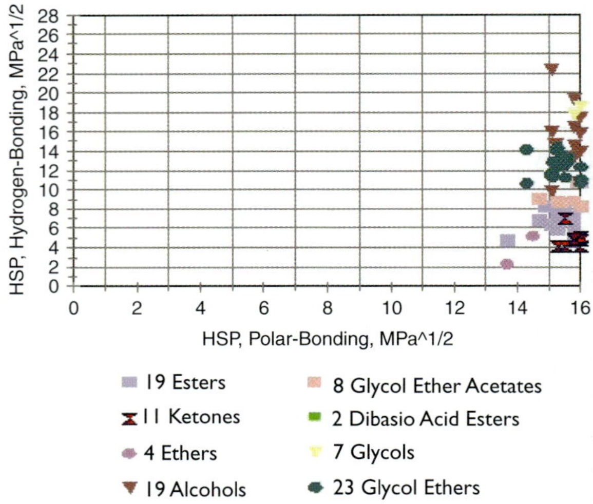


Figure 2.8

### 2.2.7.1 *Wither Solvent Substitution*

Cursory examination of these two figures shows one reason why the US EPA's binary policy about VOC exemption has not fostered solvent substitution.[65] The VOC-exempt solvents in Figure 2.7 don't have the same HSP values (solvency behavior) as do commonly used oxygenated solvents.

If solvent substitution becomes a cornerstone of US policy[66] toward management of VOC emissions, at least two elements are necessary:

1. A scale is needed by which reactivity leading to smog production can be evaluated.
2. A legal regulation in which differences in reactivity[67] on the scale are noted so that those who use chemicals less reactive toward smog production are given inducement to do so.

To date, US Federal policy acknowledges solvent replacement – not substitution. Smog-forming chemicals may be replaced with VOC-exempt chemicals. But difference in reactivity is not part of that decision.

### 2.2.7.2 *Be Careful for What You Wish!*

The current US binary system might be replaced with a scheme where relative reactivity with UV light determines the VOC "character" of a chemical. It would be a mistake to believe that action would be a "license to steal."

Mass, or volume of emissions, would matter:

- The current Federal binary regulation allows "unlimited" emission of compounds deemed VOC-exempt, and locally-determined emissions of VOCs.
- A Federal regulation embracing relative reactivity would likely limit "expected production of ozone or smog." This would be the product of emission rate times relative reactivity.

Consequently, one could not emit "unlimited" quantities of acetone, HFE-7100 (assuming one could afford same), or t-butyl acetate as is permissible

[65] Solvent substitution means replacing a solvent used to complete a function with another solvent which completes the same function. In this case, reduction of reactivity with UV light would be the intended purpose of the replacement.

[66] As of this writing, it is not clear that this is the preferred outcome. The existing binary system may or may not be retained.

[67] A small step has been taken in the direction of reactivity-based solvent substitution. In 2005, approval was granted of a new consumer products regulation as part of the California State Implementation Plan (SIP). The issue was managing volatile organic compounds (VOC) in architectural coating products. US EPA is allowing use of California's Tables of Maximum Incremental Reactivity (MIR) for determination of the contribution to formation of smog. See: http://www.epa.gov/ttn/oarpg/t1/fr_notices/15311finalcarb.pdf or http://www.epa.gov/ttn/oarpg/t1/fact_sheets/carbvocfinrulefs.html

But there is "no free lunch." Chemicals previously identified as negligibly reactive and exempt from EPA's regulatory definition of VOCs (See Table 2.7) now count towards a product's reactivity-based VOC limit for the purpose of California's aerosol coatings regulation.

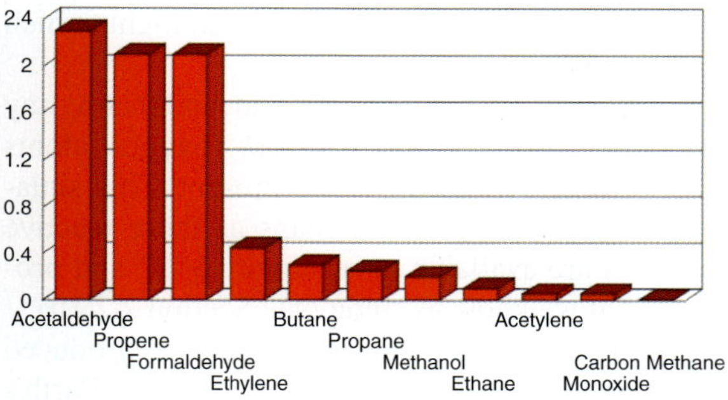


**Figure 2.9** Photochemical reactivity of various compounds

with the current Federal CAA. A mass or volume restriction would apply.

### 2.2.8 Reactions Leading to Smog Formation

VOCs can react with emissions from cars and diesel engines to cause air pollution problems in some areas.[68] VOCs are regulated because they react with sunlight and other chemicals in the atmosphere to produce what we know as photochemical smog.

Note that VOCs as emitted chemicals don't produce smog. They may add an odor, an texture, or a color to air. But they don't form smog without the presence of other pollutants.

Smog is produced by a complex photochemical reaction between hydrocarbons and nitrogen oxides, or just nitrogen oxides, in the presence of sunlight:

- Smog can be formed from just nitrogen oxides and sunlight – without presence of VOC. This smog is chiefly nitrogen oxide (NO) and ozone ($O_3$). It is also known as photochemical smog.
- Smog cannot be formed from just VOCs[69] and sunlight – oxides of nitrogen and an oxidizer (ozone) are required in the chemical reactions.

*Essentially the contribution of VOC emissions is to make existing levels of photochemical smog worse.*

The oxides of nitrogen generally come from combustion processes (Equations [2.12 and 2.13]) – chiefly gasoline-powered automobiles, forest fires, and fuel-burning power plants. That's why cars have catalytic converters and power plant stacks have scrubbers. The catalytic converter in automobile exhaust systems reduces air pollution by oxidizing hydrocarbons to $CO_2$ and $H_2O$ and, to a lesser extent, converting nitrogen oxides to $N_2$ and $O_2$. Oxides of nitrogen are often called $NO_X$, meaning that NO and $NO_2$ are included.

### 2.2.9 Photochemical Smog

The following equations are a simplified version of several very complicated processes. The first two equations are completed at ground level, by human-controlled activities (chiefly combustion processes):

$$N_2 + 2O_2 \rightarrow 2NO_2 \quad (2.12)$$

$$N_2 + O_2 \rightarrow 2NO \quad (2.13)$$

Production of nitrogen dioxide ($NO_2$) is the common outcome. Nitric oxide (NO) is relatively non-toxic at ambient concentrations. Oxidation of NO to $NO_2$ occurs naturally.

$NO_2$ persists in the atmosphere and is a potent respiratory tract toxin. $NO_2$ is not very water soluble and penetrates readily to the alveoli of our lungs where it forms nitrous acid ($HNO_2$) and nitric acid ($HNO_3$):[70]

$$NO_2 + \nu < 380\,\text{nm} \rightarrow O^* + NO \quad (2.14)$$

In the troposphere, $NO_2$ will decompose (disassociate) with energy supplied by UV light. Obviously, this reaction doesn't happen at night.

This reaction starts (or is part of) the smog-formation process. Why does it occur? Why doesn't it happen within smokestacks or exhaust pipes of automobiles? Then the subsequent reaction producing ozone wouldn't produce atmospheric smog! Why does the disassociation of $NO_2$ occur in the troposphere?

---

[68] http://chin.icm.ac.cn/database/mcmleeds.html

[69] This is one reason why the concept of VOC reactivity (see Section 2.2.1) is not simply a property of a specific chemical. It is because emission of VOCs does not produce smog without interaction with other factors – especially oxides of nitrogen (see Section 2.2.10). It is the complete environment (composition of all reactive components, frequency distribution of incident radiation, composition of auxiliary components such as water and particulate surface, and temperature) which affects smog generation. Some photochemical reactivity data are plotted in Figure 2.9.

[70] http://www.public-health.uiowa.edu/fuortes/Text/air_pol1.htm

The answer is that the sunlight striking the $NO_2$ molecule needs to have a certain amount of energy to instigate and maintain the disassociation reaction. That amount is about 72 kcal/g-mol at 25°C (129.6 K BTU/lb-mol at 77°F).[71] Sunlight in the visible (>380 nm) range of wavelengths and infrared radiation (>1000 nm) does not have energy to support the reaction in Equation (2.14). This is shown in Figure 2.10. Only sunlight in the UV range has enough energy to produce smog!

The disassociation reaction allows production of ozone, as in Equation (2.15). This happens faster than you can read about it because the Oxygen atom is *very* reactive:

$$O_2 + O^* \rightarrow O_3 \qquad (2.15)$$

But the NO produced in Equation (2.14) isn't stable. It reacts with ozone to regenerate $NO_2$ as in Equation (2.16). This reaction happens at night:

$$NO + O_3 \rightarrow NO_2 + O_2 \qquad (2.16)$$

The combination of these three Equations (2.14), (2.15), and (2.16), is a circular chemical process, as Equation (2.17):

$$NO_2 + O_2 + v < 380\,\text{nm} \rightarrow O^* + NO + O_2 \rightarrow O_3 + NO \rightarrow NO_2 + O_2 \qquad (2.17)$$

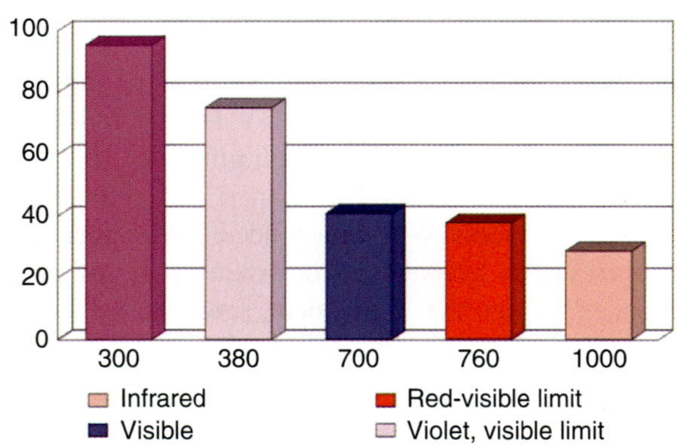


**Figure 2.10** Variation of radiation energy with frequency

In this process, UV is removed from sunlight (which is beneficial to humans).

But this circular process also means that at any time there are reactive oxidizers (oxygen atoms ($O^*$) or ozone molecules ($O_3$) present. This situation is not beneficial to humans as these reactive oxidizers are available to react with VOC and produce what we know as smog (see Section 2.2.10).

In summary, photochemical smog is produced from combustion products produced at the Earth's surface and UV light found in the troposphere:

- Nitrogen monoxide (NO) is a reddish-brown gas which is visible to the human eye at high concentrations (above about 1 ppm). NO is one component of smog.
- Ozone is a strong oxidizer – an somewhat unstable molecule. By itself it can cause eye and nasal irritation. Ozone has a harsh odor.

These are secondary pollutants – not directly emitted, but formed in the atmosphere. UV light (wavelength <380 nm, just below the visible range) is generally the catalyst or initiator. Secondary pollutants are damaging to plant life and lead to the formation of photochemical smog.

### 2.2.10 Smog Formed from VOCs

In bright sunlight, nitrogen oxides, hydrocarbons (VOC, also called HCs) and oxygen interact chemically to produce powerful oxidants like ozone ($O_3$), *and* peroxyacetyl nitrates (PANs). Ozone and PANs are temporarily created; but destroyed in subsequent reactions. PANs are primarily responsible for the eye irritation so characteristic of this type of smog. Both pollutants have significant effect on human health.

The path to smog formed from VOCs starts with the reactive oxidizers ($O_3$ and $O^*$) available from the chemistry associated with pollution byproducts from combustion (Equations [2.14], [2.15], and [2.16]):

$$O^* + H_2O \rightarrow 2HO^* \qquad (2.18)$$

$O^*$ reacts with water to produce another radical. The $HO^*$ species (called a hydroxy radical) is unstable and then quickly reacts with hydrocarbons.[72]

$$R\text{—}H + HO^* \rightarrow H_2O + R^* \qquad (2.19)$$

[71] http://naftp.nrcce.wvu.edu/techinfo/Smog/smog.html

[72] In this description, hydrocarbon solvents will be described as R—H, where "R" is the main solvent structure and "H" is a Hydrogen atom attached to that structure. This description applies to all-to-all solvents except those which don't have Hydrogen

The species R* is called a hydrocarbon radical. Like other radicals it is very reactive:

$$R^* + O_2 \rightarrow ROO^* \quad (2.20)$$

The species ROO* is called a peroxide radical. A peroxide is a pair of linked Oxygen atoms attached to a HC.[73] This species exists because there are free Oxygen atoms (O*) which produce its precursors R* and HO*. But it doesn't exist for very long.

The peroxide radical next reacts, in two steps, with pollution from combustion (NO) and Oxygen to make a stable compound, an aldehyde (CHO).

$$ROO^* + NO \rightarrow NO_2 + RO^* \quad (2.21)$$

Note that Equation (2.21) is the first in which pollution byproducts of combustion and VOCs interact. The hydrogen atoms come from decomposition of the hydrocarbon radical.

$$RO^* + O_2 \rightarrow R{-}CHO + H{-}O{-}O^* \quad (2.22)$$

An aldehyde is a compound which contains the group CHO. The group HOO is called a hydro peroxide. See Figures 2.11 and 2.12.

H–O–O–H

**Figure 2.11** Peroxide

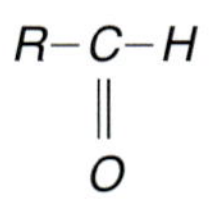

**Figure 2.12** Aldehyde

Aldehydes and PAN are common components in analyses of air above Mexico City, Santiago de Chile, Hong Kong, and other cities affected with pollution from combustion and hydrocarbons.[74]

The formation of PAN (peroxyacetyl nitrate) proceeds from aldehydes produced as above. Three rapid steps are involved:

$$R{-}CHO + HO^* \rightarrow R{-}CO^* + H_2O \quad (2.23)$$

$$R{-}C({=}O) + O_2 \rightarrow R{-}O{-}O{-}O^* \quad (2.24)$$

$$R{-}O{-}O{-}O^* + NO_2 \rightarrow R{-}C{-}O{-}O{-}O{-}NO_2 \quad (2.25)$$

For the case where R is $C_4$ (methane), the product in Equation (2.25) is PAN (Figure 2.13). PAN causes eye irritation, chest constriction, irritation of mucous membranes, and is also known to damage plants.

The hydro peroxide formed in Equation (2.22) is also a reactant which can produce byproducts hazardous to humans and the Earth:

$$H{-}O{-}O^* + NO \rightarrow HO^* + NO_2 \quad (2.26)$$

$$HO^* + NO_2 \rightarrow HNO_3 \quad (2.27)$$

In summary, the **simplified** sequence of reactions associated with smog formation is:

- Nitrogen oxides result from combustion processes (Equations [2.12] and [2.13]).

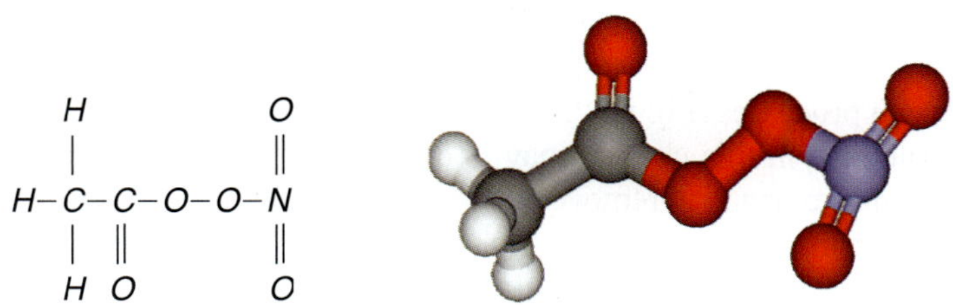

**Figure 2.13** PAN (Peroxyacetyl nitrate)

atoms. Perchloroethylene, trichloroethylene, and methylene chloride are such exceptions. Please recall that the unsaturated double bond between the two Carbon atoms is very reactive. Therefore, these compounds have a short atmospheric lifetime and don't significantly penetrate the upper atmosphere. In other words, R doesn't exist in the upper atmosphere; R—H does exist.

[73] "R" is the HC on to which the aldehyde is attached.

[74] http://www.atmos.anl.gov/ACP/Gaffney.pdf

**Table 2.10 Major Pollutants and Reaction Products**

| Pollutant | Description | Effects | Levels of Concern |
|---|---|---|---|
| NO | Nontoxic, colorless, odorless gas | Easy to oxidize to $NO_2$ | |
| $NO_2$ | Reactive gas | Gives polluted air a yellow to reddish brown color and pungent odor | Hundreds of ppb |
| $O_3$ | Active oxidant | Is most destructive to plants | Hundreds of ppb, or less |
| PAN | | Inflammation of air passages reducing lung capacity. Long-term damage to lungs and to ability to resist other lung and throat problems. Causes a plant disorder known as "silver leaf" | 10–20 ppb |

- Nitrogen oxides generate Oxygen atoms and ozone (Equations [2.14] and [2.15]).
- Oxygen atoms form hydroxyl radicals (Equation [2.18]).
- Hydroxyl radicals generate hydrocarbon radicals (Equation [2.19]).
- Hydrocarbon radicals form hydrocarbon peroxides (Equation [2.20]).
- Hydrocarbon peroxides form aldehydes (Equation [2.22]).
- Aldehydes form aldehyde peroxides and hydroperoxides (Equation [2.24]).
- Aldehyde peroxides form peroxyacylnitrates (Equation [2.24]).
- Hydro peroxides form nitric acid (Equation [2.27]).

The concern about these reactants, intermediates, and products is deadly serious (Table 2.10).

Hydrocarbon chemicals are not the prime cause of contamination of the Earth's air. There would be no formation of smog unless the influence of hydrocarbon solvent emissions were combined with the emissions from combustion processes (Figure 2.14).

The equations and structures in this chapter may appear complicated to a manager trying to remove oil and metal particles from machined injection valves.

Yet, these equations and structures are important. They are the scientific basis for regulations defining which solvents can be used to remove that grease, how the solvent cleaning equipment must be designed,

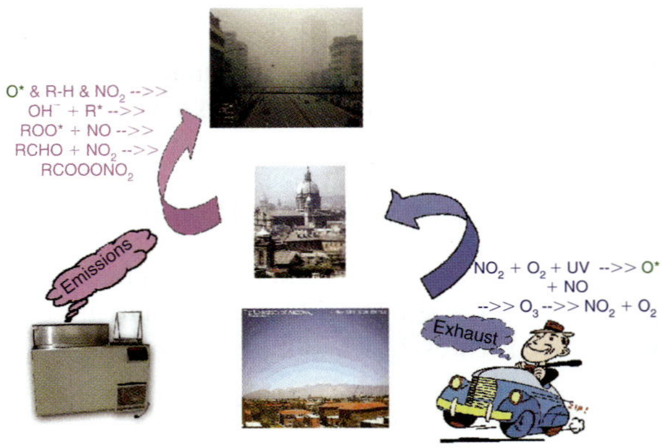


**Figure 2.14** Cartoon showing interaction of VOC with combustion products to produce smog

and how employees using that equipment must be protected.

## 2.3 CLEANING CHEMICALS AS AGENTS CAUSING GLOBAL WARMING

The pace of global warming is glacial. The nature of global warming is that the incremental changes of which it is composed are minuscule. That's why it is so difficult to recognize what it is, and what its causes are.

Essentially, the debate about global warming concerns the time by which certain countries should take what action. The existence of global warming is not under general scientific debate.[75]

[75] US Department of State, *The United States of America's Third National Communication Under the United Nations Framework Convention on Climate Change*, May 2002. See http://yosemite.epa gov/oar/globalwarming.nsf/content/ResourceCenter PublicationsUSClimateActionReport.html

## 2.3.1 What Global Warming Means

The phrase "global warming" is a synonym for massive climate change on our planet. The concern is that activities of man have affected the Earth's atmospheric environment to the extent that the Earth's climate will be affected.

These activities include, but are not limited to, agricultural and industrial changes associated with the industrial revolution. Today versus the past, our planet has more people, releases more byproducts to the atmosphere, and has fewer forests and wetlands.

The Earth naturally absorbs incoming solar radiation. It also reflects and emits longer wavelength terrestrial (thermal) radiation back into space. On average, the absorbed solar radiation is balanced by the outgoing terrestrial radiation emitted to space – atmospheric temperatures are stable.

A portion of this terrestrial radiation is absorbed by gases in the atmosphere. The energy from this absorbed terrestrial radiation warms the Earth's surface and atmosphere. This creates what is known as a "natural greenhouse effect."

The essence of global warming is that if gases in the upper atmosphere have a higher capacity to absorb energy, they will do so and the Earth's surface will become warmer.

Other factors, both natural and human, affect our planet's temperature. The atmosphere has always contained carbon dioxide, methane, and nitrous oxide. These gases – together with water vapor – create the natural greenhouse effect.

Emitted gases are called "greenhouse gases." This means they can absorb heat as does the atmosphere in a greenhouse. These gases have a high specific heat. Some naturally occurring ones are water vapor, carbon dioxide, methane, nitrous oxide, and ozone.

One definition of a greenhouse gas is a gas that transmits solar (short wavelength) radiation, but absorbs infrared (long wavelength) radiation.

Actually, global warming is not new, and not unwanted. A stored inventory of $CO_2$ and other energy-absorbing gases has accumulated over the several billion years of our planet's life. The Earth's temperature would be considerable colder without that heat sink. Citizens at the equator would dress more warmly. One estimate is that surface temperatures would be higher by >30°F!

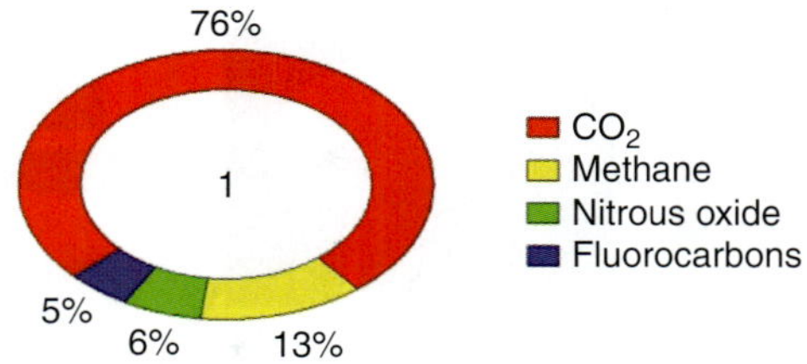


**Figure 2.15** Sources of greenhouse gases

There is a serious scientific controversy about both cause and effect. As about smog formation, scientists know which emitted chemicals can have a deleterious effect. In other words, we understand most of the chemistry.

About global warming, we know which gasses can absorb certain amounts of heat, because we can measure their physical properties. We do know that when the atmosphere becomes warmer the planet's surface will become warmer.

## 2.3.2 Doing What Comes Naturally

In fact, *natural sources* of greenhouse gases are by far the largest source of greenhouse gases[76] (Figure 2.15):

- Humans exhale carbon dioxide as they breathe. Plants absorb it from the air during photosynthesis. It is also absorbed in the oceans and, combined with other chemicals, stored as carbonate salts in sediments on the ocean floor.
- Methane is generated naturally by bacteria that break down organic matter in wetlands. It escapes from garbage landfills and open dumps. Methane gas also leaks out during mining, extraction and transportation of coal, oil, and natural gas.
- Each year we add 7–13 million tons of nitrous oxide to the atmosphere mainly by using Nitrogen-based fertilizers, disposing of human and animal wastes, and automobile exhausts.[77] When these Nitrogen-based fertilizers break down in the soil, nitrous oxide is released into the air.
- The world's oceans are a significant component of this situation.[78] See Figure 2.16. Carbon dioxide,

[76] http://globalwarming.enviroweb.org/ishappening/sources/
[77] http://globalwarming.enviroweb.org/ishappening/sources/sources_n2o_facts1.html
[78] http://yosemite.epa.gov/oar/globalwarming.nsf/content/emissions.html

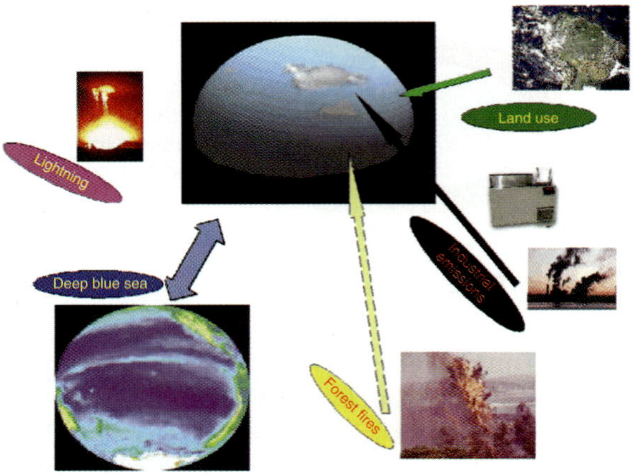


**Figure 2.16** Sources and sinks of global warming gases

nitrous oxide, and methane are all very soluble in salt water. Bodies of water are sinks in which natural greenhouse gases are stored.

A storage element is a major reason why emission of greenhouse gases lags in time their effect in the atmosphere. A sizable inventory in our oceans of energy-absorbing materials, whose volume is uncertain, is only one reason why it is difficult to model global warming effects.

### 2.3.3 Ignorance May Not be Bliss

Scientists think rising levels of greenhouse gases in the atmosphere are contributing to global warming, as would be expected; but to what extent is difficult to determine at the present time.[79] In other words, we don't know all the atmospheric chemistry and physics.[80]

What scientists don't know is:

- The approximate timing and extent of the effects on the atmosphere of emissions of all gases.
- The quantitative effect on the atmosphere of man-directed modifications to our planet's landscape.
- The geological/agricultural processes native to our planet which can be counted upon to moderate or accelerate the effects of the emissions of materials made by man.

### 2.3.4 Manmade Chemistry

Other greenhouse gases, synthesized by man, are HFCs, perfluorocarbons (PFCs), and sulfur hexafluoride ($SF_6$). They are generated in a variety of industrial processes.[81]

HFCs and PFCs have become commonly used as cleaning and drying agents. Oceanic uptake does not appear to be a significant sink for many of the HCFCs, PFCs, and HFCs.[82] So Figure 2.16 is incorrect in that regard.

Each greenhouse gas differs in its ability to absorb heat from the atmosphere. Each can participate in different reactions with radiation in the stratosphere. HFCs and PFCs are the most heat-absorbent (high specific heat).

### 2.3.5 The Debate About Global Warming

Nations which are the most industrialized are, to no surprise, the most significant emitters of greenhouse gases.

This is because all economies on this planet are based on combustion of fossil fuels which produce $CO_2$. At its heart, the portion of global warming controllable by man is about energy use, and production.

Investment in technology and infrastructure for energy use is one of the largest investments made by a developing or mature country. No country wants to (or probably can afford to) discard that investment, then replace it.

At least two crucial questions are being debated about global warming. The debate apparently must be settled before change occurs – countries are not likely to make significant voluntary[83] reductions in emissions. The 20% unilateral reduction proposed by the character of US President played by Michael Douglas in the entertaining 1993 film "The American

[79] http://yosemite.epa.gov/oar/globalwarming.nsf/content/ClimateUncertainties.html

[80] For an opposite points of view about the scientific evidence concerning global warming, see http://www.sitewave.net/news/ or www.oism.org/project

[81] http://www.fluorocarbons.org/frame.htm?chfamilies/HFCs/g_properties/g365mfc.htm

[82] http://www.cmdl.noaa.gov/publications/annrpt24/532.htm

[83] An excellent article about firms who have found it to be a good business practice to reduce emissions of greenhouse gases (chiefly through reducing energy consumption) is: Aston, A. and Helm, B., "The Race Against Climate Change," *Business Week Magazine*, December 12, 2005.

President" was a huge and to-date unreached commitment.

#### 2.3.5.1 *Delay, Linger, and Wait*

The first question being debated is **when is this action is needed**, so the cost of doing it may be better justified.

Putting a lid on all vapor degreasers which use fluorinated solvents now is not going to materially affect the stored inventory of greenhouse gases.

A manager's answer to that first question depends upon: (a) their understanding and acceptance of mathematical models of the planet's upper atmosphere, and (b) their estimate of the impact of predictions about how our planet's surface condition changes due to modifications in its upper atmosphere.

According to the National Academy of Sciences, the Earth's surface temperature has risen by about 1°F in the past century, with accelerated warming during the past two decades. Specific data are plotted in Figure 2.17.

The US EPA expects that "… Increasing concentrations of greenhouse gases are likely to accelerate the rate of climate change. Scientists expect that the average global surface temperature could rise 1–4.5°F (0.6–2.5°C) in the next 50 years, and 2.2–10°F (1.4–5.8°C) in the next century, with significant regional variation …"[84]

Undoubtedly, our planet is getting slightly warmer. Is this part of a long-term cycle of warming? Is this merely a short-term fluctuation? What is the practical meaning of an increase of ocean temperature by 1°F (or °C)?

Rising global temperatures are expected to raise sea level, change precipitation and other local/regional/global climate conditions. Changing regional climate could alter forests, crop yields, and water supplies. It could also affect human health, animals, and many types of ecosystems.[85] Deserts may expand into existing rangelands, and features of some of our National Parks may be permanently altered.

Unfortunately, many of the potentially most important impacts depend upon whether rainfall increases or decreases, which cannot be reliably projected for specific areas.

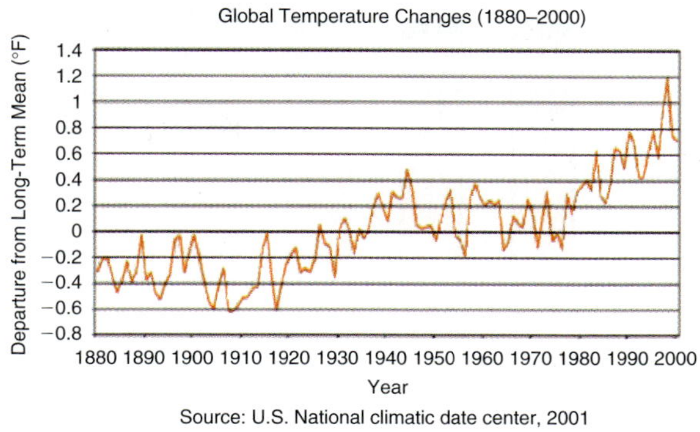


**Figure 2.17** Global temperature change

So the question about when action is needed may hinge on the nature of the consequences of no action. If, in the next millennium, local temperatures were to increase by 30°F, life for our descendants would not be as we know it!

In that case, we don't have to put down this book and take local action to reduce the inventory of greenhouse gases before finishing the book. But we must take positive action to do that in our lifetime.

The bottom line answer to the first debate question about global warming is that while we may delay positive action, *ultimate failure to act will be the ultimate cause of our planet's destruction as a site for human life.*

#### 2.3.5.2 *No! No! After You …*

Just which countries should take the first action?

See Figure 2.18. Estimates of greenhouse gas emission in the 1990s are presented in units of millions of metric tons of Carbon equivalents (MMTCE), which weights each gas by its GWP value or global warming potential.

The US, Japan, Germany, Canada, and the UK emit the largest volumes of greenhouse gases.

Another perspective is obtained, however, by dividing the emitted gas volume by a measure of economic activity (dollars of Gross Domestic Product [GDP]; see Figure 2.19). Here, the highest-ranking (least efficient with their emission of greenhouse gases) countries are Ukraine, Russia, Poland, South Africa, and China.

[84] http://yosemite.epa.gov/oar/globalwarming.nsf/content/climate.html

[85] Hurricane Katrina, which struck the southern US in 2005, is believed by many, without scientific cause-and-effect evidence, to have been produced by a change in global ocean temperatures.

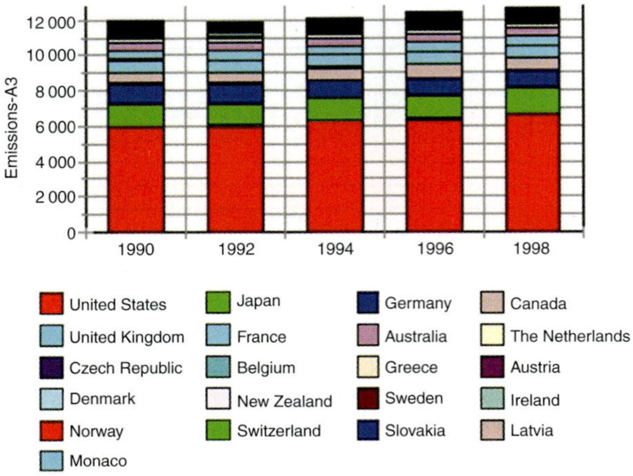


**Figure 2.18** Global emissions of greenhouse gases

Still another perspective is obtained by dividing the emitted gas volume by the country's population (see Figure 2.19b). Here the highest-ranking (where each person emits the most greenhouse gas) countries are the US, Canada, Australia, Netherlands, and Saudi Arabia.

The perspective of Figures 2.18, 2.19 (a) and (b) frames the second major question about global warming. *That is who should do what about it.*

If the needed action is to reduce emission volumes, which countries should make what reductions?

- Should the greatest reduction be made by the countries who emit the greatest volume? That's what Figure 2.18 suggests. After all, shouldn't the countries who are adding the most to the stored volume of greenhouse gases be responsible for making the largest reduction in the addition to that inventory?
- Should the greatest reduction be made by the countries who are least efficient in their use of energy and raw materials to produce economic benefit for their citizens. That's what Figure 2.19(a) suggests. After all, shouldn't the countries who have yet to implement the best technologies be charged with doing just that before other countries are asked to change?

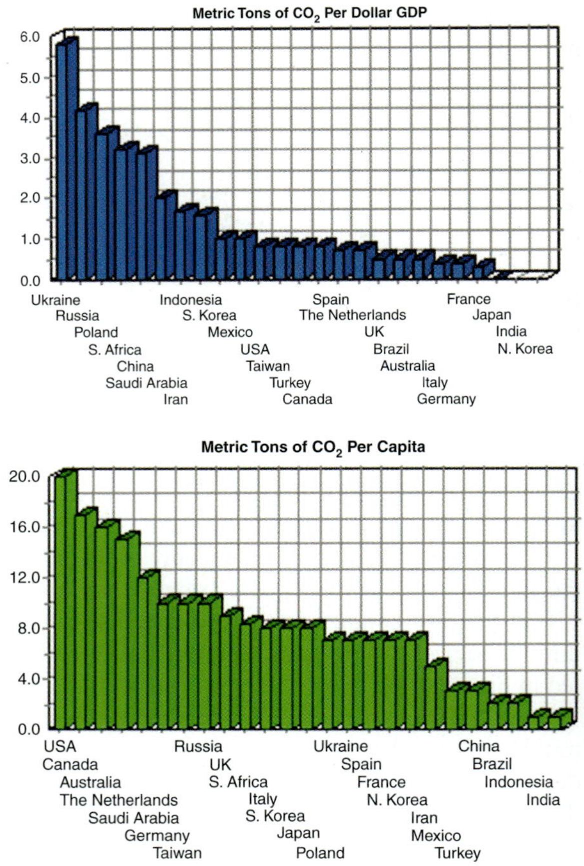


**Figure 2.19** Global emissions corrected for economic output

- Should the greatest reduction be made by the countries whose inhabitants each emit the greatest volume? That's what Figure 2.19b suggests. After all, shouldn't the citizens who are each most profligate with greenhouse gases be tasked with assuming more responsibility for control of emissions of greenhouse gases?

The Kyoto Treaty[86] was an attempt to forge a compromise (as was done in the Montreal Protocol) among the negotiating parties about the answer to these three questions.

[86] The Kyoto Treaty commits industrialized nations to reducing emissions of greenhouse gases, principally carbon dioxide, by around 5.2% below their 1990 levels in the decade beyond its approval. The proposed treaty would have required the US to reduce its emissions 31% below the level otherwise predicted for 2010. Put another way, the US under the Kyoto Treaty (effectively ratified when Russia's president approved it on November 5, 2004) would have had to cut 552 million metric tons of $CO_2$ per year by 2008–2012.

The treaty explicitly acknowledges as true that man-made emissions, principally from the use of fossil fuels, are causing global temperatures to rise, eventually to catastrophic levels. Treaty supporters believe that if countries of the Earth dramatically cut back, or even eliminate, fossil fuels, the climate system will respond by sending global temperatures back to "normal" levels. See Figure 2.15.

Proposed answers in the Kyoto Treaty to the three questions above (and others) have not been satisfactory to some governments. Chief among them is that of the US.[87]

As of this writing, the Kyoto Treaty hasn't received the support from enough nations, as did the Montreal Protocol. Some wonder why the Kyoto Treaty about global warming hasn't received the ultimately broad-based support that the Montreal Protocol about ozone-depletion did.[88]

This author's view is that the Kyoto Treaty requires decisions by all sovereign nations of the Earth about critical issues involving each nation's interests. Change more fundamental than that described by the Montreal Protocol is contemplated. The evidence is not as clearly defined. The consequences have broader ramifications.

Basically, the Kyoto Treaty asks questions whose answers pit:

- Established industrially mature countries against countries less well developed.
- One understanding of the consequences of experimental scientific observations versus another understanding.
- One analysis of the economic impacts of infrastructure change versus another analysis.
- One narrowly focused mind set versus another mindset equally narrowly focused but on another objective.

The answer to these questions will have both political and scientific components,[89] and hopefully will not be achieved due to climatological catastrophe or political upheaval.

### 2.3.6 Regulation of Solvent Cleaning[90] Because of Global Warming

Global Warming Potential (GWP)[91] has been developed as a metric to compare (relative to another gas) the ability of each greenhouse gas to trap heat in the atmosphere. Carbon dioxide ($CO_2$) was chosen as the reference gas to be consistent with the guidelines of the Intergovernmental Panel on Climate Change (IPCC[92]).

GWPs of common solvents which are important[93] to persons doing solvent cleaning are given in Table 2.11,[94,95,96] with that of other compounds of interest.

One of the points that should be taken from Table 2.11 is that global warming is an evolving science. That's why GWP values from the Third Assessment Report (TAR) are different from values published in the Second Assessment Report (SAR).

---

[87] The US government signed the Kyoto Treaty on November 12, 1998, but never submitted it to the US Senate for ratification (required by the US Constitution).

In 1997, the US Senate sent a powerful signal that the Kyoto Treaty was unacceptable. By a vote of 95 to 0, the Senate passed the Byrd–Hagel resolution, which stated that the Senate would not ratify the Kyoto Treaty if it caused substantial economic harm and if developing countries were not required to participate on the same timetable.

[88] For a good analysis, see Baumert, K. and Kete, N., "The US, Developing Countries, and Climate Protection: Leadership or Stalemate?", published by *World Resources Institute*, 2001.

[89] For an alternative point of view, see Michael Crichton's *State of Fear*, Harper Collins, 2004.

[90] The components of aqueous cleaning agents which are not water are not considered to have global warming potential.

[91] The Global Warming Potential (GWP) of a gaseous compound is a composite measure of its ability to absorb radiation in the infrared (IR) spectral region (typically $500^{-1}$–$200\,cm^{-1}$), **together with its expected atmospheric lifetime**. 10, 100 and 500 year lifetimes are considered. In this book, values (unless noted) are always calculated for the 100 year lifetime. Effectively the GWP of a chemical compares the amount of IR radiation absorbed by unit weight (e.g. 1 lb or Kg) of the chemical over a given time (taking into account its removal through degradation processes) with that absorbed by an equivalent weight of emitted $CO_2$. Because of atmospheric degradation of compounds (e.g. through reaction with OH radicals) the GWP decreases with time.

[92] The IPCC was created jointly by the World Meteorological Organization and the United Nations Environment Programme in 1988. The IPCC is responsible for compiling and synthesizing the growing body of scientific literature on climate change. The comprehensive assessments of IPCC form the scientific basis for climate change policies. They have produced a textbook titled *Climate Change 2001: The Scientific Basis*. It covers aspects about global climate change in additional detail. It can be purchased or found at this Internet-based address: http://www.grida.no/climate/ipcc_tar/wg1/index.htm

[93] http://www.grida.no/climate/ipcc_tar/wg1/248.htm#tab67

[94] Global Warming Potentials are calculated relative to the effect of carbon dioxide up to 100 years. All effects beyond 100 years are disregarded; this captures less than 40% of the total effect from $CO_2$.

[95] The Second Assessment Report (SAR) – *Climate Change 1995: The Science of Climate Change, ed.* Houghton *et al.*, Cambridge University Press, 1996. The values in the 1995 report have been adopted for use in the Kyoto Protocol. The Third Assessment Report (TAR) was published in 2000.

[96] Uncertainty in GWP is stated to be ±35% for both the Second and the Third Assessment Report values.

**Table 2.11 GWP at 100 Year Time Horizon**

| Compound Name | Formula | Source | Application | SAR | TAR |
|---|---|---|---|---|---|
| Carbon dioxide | $CO_2$ | | Human respiration | 1 | 1 |
| Methane | $CH_4$ | Natural | Bacterial activity | 21 | 23 |
| Nitrous oxide | $N_2O$ | | Fertilizers wastes | 310 | 296 |
| HFE-8200 | | | Cleaning agent | | 55 |
| HFE-7200 | $C_4H_9OCH_3$ | | Cleaning agent | | |
| HCFC-225ca/ HCFC-225cb | $CF_3CF_2CHCl_2$ $CClF_2CF_2CHClF$ | | Cleaning and drying agent | 181/620 | |
| HFC-23 | $CHF_3$ | | | 11 700 | 12 000 |
| HFC-32 | $CH_2F_2$ | | | 650 | 550 |
| HFC-125 | $CF_3CHF_2$ | | | 2800 | 3400 |
| HFC-134 a | $CF_3CH_2F$ | | Refrigerant | 1300 | 1300 |
| HFC-143 a | $CF_3CH_3$ | | | 3800 | 4300 |
| HFC-152 a | $CH_3CHF_2$ | | | 140 | 120 |
| HFC-227 ea | $CF_3CHFCF_3$ | | Fire extinguishant refrigerant | 2900 | 3500 |
| HFC-236 cb | $CF_3CF_2CH_2F$ | | | | 1300 |
| HFC-236 ea | $CF_3CHFCHF_2$ | | | | 1200 |
| HFC-236f a | $CF_3CH_2CF_3$ | Manmade | Fire extinguishant refrigerant | 6300 | 9400 |
| HFC-245 ca | $CH_2FCF_2CHF_2$ | | | 560 | 640 |
| HFC-245 fa | $CHF_2CH_2CF_3$ | | Blowing agent | | 950 |
| HFC-365 mfc | $CF_3CH_2CF_2CH_3$ | | Cleaning and drying agent | | 890 |
| HFC-43-10mee | $CF_3CHFCHFCF_2CF_3$ | | Cleaning and drying agent | 1300 | 1500 |
| Sulphur hexafluoride | $SF_6$ | | Enchant in manufacture of semiconductors | 23 900 | 22 200 |
| CFC-113 | $CCl_2FCClF_2$ | | Cleaning agent whose manufacture is banned because of ODP | | |
| Carbon tetrafluoride | $CF_4$ | | | 6500 | 5700 |
| Perfluoroethane | $C_2F_6$ | | | 9200 | 11 900 |
| Perfluoropropane | $C_3F_8$ | | Drying agent heat transfer fluid | 7000 | 8600 |
| Perfluorobutane | $C_4F_{10}$ | | | 7000 | 8600 |
| Perfluoro-cyclobutane | $c\text{-}C_4F_8$ | | | 8700 | 10 000 |
| Perfluoropentane | $C_5Fl_2$ | | | 7500 | 8900 |
| Perfluorohexane | $C_6Fl_4$ | | | 7400 | 9000 |

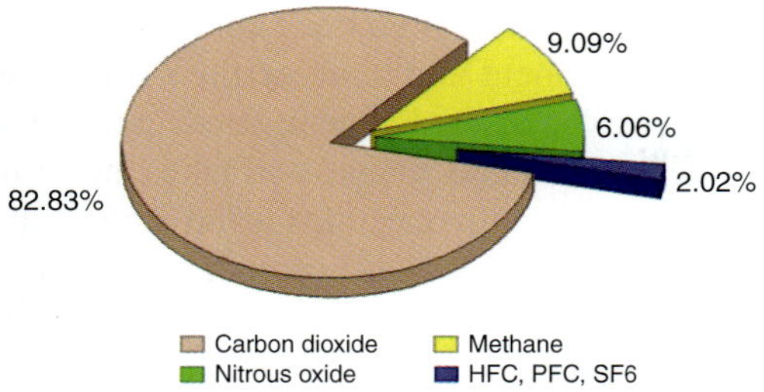


**Figure 2.20** Emission of greenhouse gases

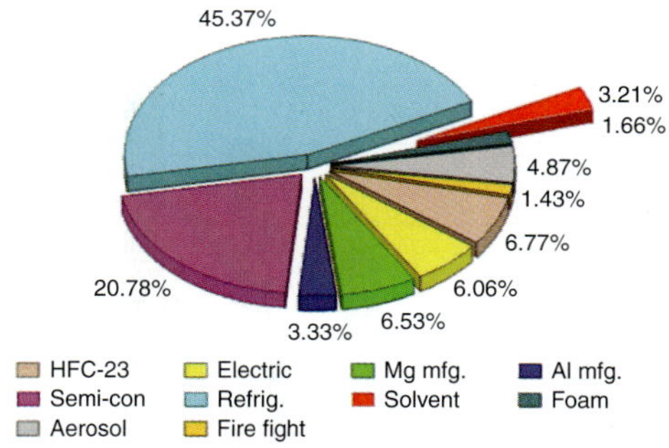


**Figure 2.21** Uses of fluorinated fluids

A second point, supporting Figure 2.20,[97] is that global warming would not be a problem where there is no human life about Planet Earth!

Said another way, both involuntary (non-discretionary) and voluntary (discretionary) human activities contribute to global warming.

There is only a portion of this serious problem on which humans can modify their activities to contribute to a solution.

### 2.3.6.1 *High Concern – High GWP Gases*

The gases of concern in Table 2.11 are the HFCs and PFCs. They are called high-GWP gases.

By application, most of these solvents are refrigerants.

Even on the basis of Carbon equivalents,[98] not actual volume emitted, these gases with high-GWP ratings are not a major contributor to the inventory of greenhouse gases. The blue exploded sector (2.02%) in Figure 2.20 shows actual emission (in Carbon equivalent units) in 1999 by source or type[99] of emission.

Given the distribution of applications for HFCs/PFCs/$SF_6$, solvent use is still a minor segment. Figure 2.21 (see Footnote 36) shows the distribution of estimated emissions of fluorinated fluids in 2010. The percentages are based on Carbon-equivalents, not volumes of gases.

Within that distribution, refrigerant use dominates 45%). Solvent use, not just for cleaning applications, is only around 3% of the Carbon-equivalent emission.

### 2.3.6.2 *"Why are THEY Picking on Us?"*

No single cleaning solvent is a major contributor to the problem of global warming – based on GWP rating and emission volume.

Use of HFCs, PFCs, and $SF_6$ draws concern, from environmental regulatory agencies, out of proportion to their volume (or Carbon-equivalent volume) of emission.

There are several reasons for a high level of concern by environmental regulatory agencies:

- Use is controllable with moderate regulatory effort. Contrast regulation of $CO_2$ emissions from human respiration or photosynthesis in plant life; regulation of $CH_4$ emissions from animals; regulation of $N_2O$ emissions from combustion of forests, with regulation of use of a chemical whose production can be restricted (or banned) by fiat, or whose use can be managed through choices of process equipment. With this perspective, focus on high-GWP gases is a sound regulatory strategy.
- Use of fluorinated gases is growing at an aggressive rate. That rate is expected to accelerate faster than general economic growth (see Figure 2.22 and Footnote 36). The HFC category is expected to grow from less than 10 MMTCE to more than 50 in 20 years. A major reason for growth of use of HFCs and PFCs is that they replace other substances which are CFCs (CFC-113 and HCFC 141b).

---

[97] The data is from 1999.

[98] Units of millions of metric tons of Carbon equivalents (MMTCE).

[99] http://www.epa.gov/ghginfo/pdfs/gwp_gas_emissions_6_01.pdf Exhibit ES-2.

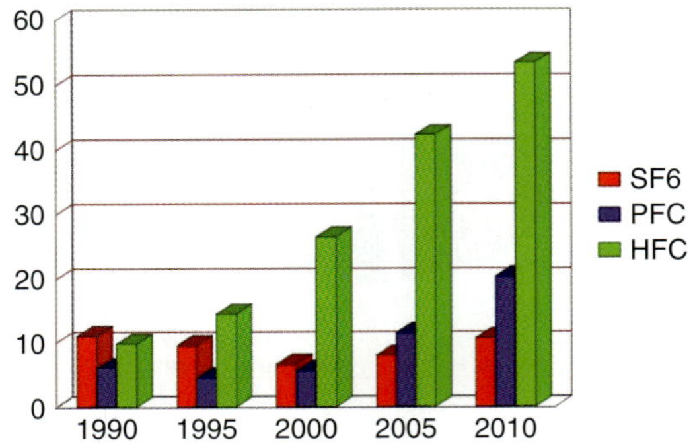


**Figure 2.22** Growth of use of fluorinated gases

- Regulation is consistent with the economic preferences of users. HFCs and PFC are quite expensive (€15–€20 per pound as of Fall 2003). Users have a real incentive not to emit them. Most applications, except for actions such as suppression of fires, are in equipment which contains emissions.
- Fluorinated gases have higher GWP values than other substances – see Table 2.11. Their use commands attention.

In summary, the focus on HFCs, PFCs, and $SF_6$ will continue to be out of proportion to their inventory in the atmosphere.

## 2.4 CLEANING AGENTS WHICH CAN BE BIOLOGICALLY OXIDIZED

How often does a Manufacturer Safety Data Sheet (MSDS) refer to an aqueous cleaning agent as biodegradable? How often does a manager read (or hear) this statement and imagine their spent cleaning agent can be discarded in any convenient waterway – believing some "bugs" will take care of it?

The answer to both questions is the same – all too often. "Biodegradable" has become, in the minds of many, a word meaning "I don't have those solvent disposal problems!"

An outcome of believing that answer is usually a problem which has technical, sanitary, environmental, or legal ramifications. This sub-chapter covers how to understand and avoid those problems.

### 2.4.1 Biological Oxidation

Waste water from cleaning operations represents complex mixtures of organic (Carbon-bearing) compounds and inorganic substances. Both the soil and the cleaning agent contribute to each type.

Biological oxidation is waste treatment – conversion of those organic materials to innocuous products. Living microbial systems (bacteria) effectively degrade these organic materials, with specific participation of Nitrogen, Phosphorous, and certain inorganic ions (e.g. "minerals").

There are two types of bio-oxidation – one that involves Oxygen as a feedstock,[100] and one which does not.[101] Most biological oxidation of cleaning wastes is done with Oxygen (air) as a feedstock.

Bacteria[102,103] of different species are the necessary ingredient in both types.

[100] Oxidation involves transfer of electrons between atoms. Oxidation is the removal of an electron from a molecule or atom; Reduction is the gain of an electron by a molecule or an atom. Oxygen, by "stealing" electrons from other atoms (or molecules) oxidizes those molecules.

[101] Other atoms or atomic groups can "steal" electrons. Some bacteria can use $NO_3^-$ (nitrate) or $SO_4^=$ (sulfate) as terminal electron acceptors instead of Oxygen. These organisms can carry-out anaerobic (without Oxygen) oxidation.

[102] Incidently, bacteria are very different from enzymes:

- Bacteria are single-celled organisms. Enzymes are proteins with specific amino acid sequences and three-dimensional structures that act as a catalysts.
- Bacteria are alive and reproduced by simple cellular division. Enzymes are chemicals and are not alive.
- Some bacteria require only minerals and a Carbon source such as sugar for growth. Enzymes don't grow.
- Bacteria show a wide range of nutrient requirements. Enzymes catalyze only one type of reaction.
- Bacteria break down organic material into carbon dioxide and water (usually). Enzymes can catalyze a wide range of reactions.
- Bacteria die when their food supply is exhausted. Enzymes are catalysts whose concentration doesn't change upon reaction.

[103] Usually, a biological oxidation process will be seeded with a population of specific bacteria sufficient to start oxidation at a useful rate, although smaller concentrations of capable bacteria may be present in ambient environment. Ground soil is full of free-living bacteria which help with biodegradation.

Figure 2.23

Figure 2.24

#### 2.4.1.1 *Aerobic Oxidation*

This most commonly used treatment involves direct reaction with Oxygen. The simplified reaction scheme (in words) is:

$$\text{Organic Matter} + \text{Oxygen} + \text{Bacteria A} \rightarrow \text{New Cells} + \text{Carbon Dioxide} + \text{Nitrates} + \text{Water} \quad (2.28)$$

There are several unique features of the process[104] which is Equation (2.28):

- Oxygen, from air, is the only feedstock (other than soil and cleaning agent components).
- Soluble nitrates can be produced if the organic matter (soil and cleaning agent components) contained Nitrogen. Soluble sulfates can also be produced if the mixture contains Sulfur. In this way, nutrients can be recycled.
- New cells of bacteria A are produced[105] as the supply or organic material increases. This makes biological oxidation a very flexible process.[106]
- But when the supply of oxidizable organic material declines, bacteria cells die. This produces a solid, often aromatic, waste product.[107]
- This is a low-cost process.[108] A major component of operating cost can be electricity to drive the blower which "pumps" air into the oxidation tank.
- Water and $CO_2$ are produced, and Oxygen is consumed. The process can't be completed in a sealed environment.

#### 2.4.1.2 *Anaerobic Oxidation*

This biological oxidation treatment, also common, does not involve reaction with Oxygen (atoms). Yet no Oxygen is fed to the treatment process (Figure 2.24).

---

[104] Figure 2.23 shows the size of the water pumping equipment necessary to operate a modest-sized community aerobic oxidation plant. Photo courtesy of Puyallup, WA, US.
[105] Cell growth rate is often exponential with time.
[106] It's interesting to note that aqueous cleaning is a very unforgiving process while treatment of the waste it produces is usually very flexible. Alternately, solvent cleaning is also a very flexible or forgiving process while treatment by distillation of the soil components produced by it can require very specific conditions.
[107] This material is referred to by the technical term sludge. Since some cells die even when their food supply is stable, production of sludge is normal and continuous.
[108] But is not a process with a small footprint – ground space requirement as shown in Figure 2.23. Photo courtesy of Puyallup, WA, US.

There are two sequential steps, involving different species of bacteria:[109]

1. First the organic material (soil and cleaning agent components) is oxidized to organic oxygenated products, Equation (2.29).
2. Then, those products are oxidized to $CO_2$ and other compounds, Equation (2.30). The two simplified reaction schemes (in words) are:

$$\text{Organic Material} + \text{Bacteria B} \rightarrow \text{New Cells} + \text{Organic Alcohols and Acids} \quad (2.29)$$

$$\text{Organic Alcohols and Acids} + \text{Bacteria C} \rightarrow \text{New Cells} + CO_2 + H_2S + CH_4 + NH_3 \quad (2.30)$$

As with aerobic oxidation, there are several unique features of the process which is Equation 2.30:

- Anaerobic oxidation does not produce soluble nutrients from Nitrogen and Sulfur as does aerobic oxidation.
- Completion of the anaerobic oxidation process requires both specie of bacteria to be present in the required amounts.
- Anaerobic oxidation is a process where the Oxygen is provided by the substance being oxidized – rather than being separately added.

The same organic feedstock may be oxidized by either process – with different bacteria.

In general, anaerobic oxidation is seldom practiced with wastes from industrial cleaning processes because the more simple aerobic oxidation is satisfactory.

## 2.4.2 Diversity in the Wash Tank

An expression in politics speaks about how laws and sausage are made and that it is often wise to not know too much about how either is done. The same can be said about how aqueous cleaning agents are formulated.

Basically, neutrally, and acidically, there is no such item as a single aqueous cleaning agent. There are almost as many aqueous cleaning agents as there problems in which their use is critical. Reasons for this diversity range include:

- Aqueous cleaning technology is less forgiving than solvent cleaning technology. Frequently, customization of a formulation is necessary for it to provide value.
- There is usually more than one way to solve a problem.
- Commercial competitiveness when a supplier has only their formulation and the quality of their service to differentiate themselves from their competition.

Diversity leads to a "good news, bad news" situation:

- It can be difficult to manage biological treatment of aqueous wastes because of the variety of components – all of which must be well-treated.
- Forgiveness is a hallmark of biological treatment.

Said another way, biological treatment and aqueous cleaning agents were made for one another – far beyond the fact that water is the largest volume ingredient in each.

### 2.4.2.1 *Composition of Aqueous Cleaning Agents*

Aqueous cleaning agents are composed of at least six general types of components.[110] Each must be compatible with the other, or the product would split into multiple phases. Each is present to complete a different function. *Each* must be oxidizable by the bacteria or this method of waste treatment won't be useful.

This diversity is why the inherent forgiveness of biological treatment is necessary, and valued.

[109] The bacteria labels A, B, and C in Equations 2.29 and 2.30 are used only for explanation.

[110] An excellent reference about bio-oxidation of components used in formulation of cleaning agents was published in 2001 by the Denmark Environmental Protection Agency. It is Environmental Project # 615, 2001, *Environmental and Health Assessment of Substances in Household Detergents and Cosmetic Detergent Products*. A PDF can be found at http://www.mst.dk/publications. Much of the detailed information in Table 2.12 about biodegradation was selected from research reported in this reference.

**Table 2.12 Bio-oxidation Characteristics of Components of Aqueous Cleaning Agents**

| Purpose | Nature | Type | Formed by Reactants as | Biodegradation Pathway | Biodegradability | Area of aquatic Concern |
|---|---|---|---|---|---|---|
| Surfactant | Non-ionic (Not Charged) | Alcohol Ethoxylate | Linear and Branched Fatty Alcohol, Ethylene or Propylene oxide | Aerobic | 80–100% (readily biodegradable), with shorter-chain oxides and unbranched alcohols being higher | Algae, Invertebrates, can be accumulated and toxic to fish |
| | | | | Anaerobic | 50–80% with linear alcohols being higher | |
| | | Alkyl Glycoside | Linear Fatty Alcohol, Glucose | Aerobic | 80 to 100% (readily biodegradable) | Algae, can be accumulated and somewhat toxic to fish |
| | | | | Anaerobic | Ultimately biodegradable | |
| | | Glucose Amide | Structure is R—$CH_2$—CO—N—$(CH_3)$—$CH_2$—$(CHOH)_4$—$CH_2$—OH | Aerobic | 80–100% (readily biodegradable) | Higher molecular weights are toxic to aquatic life |
| | | | | Anaerobic | Ultimately biodegradable, in absence of Oxygen | |
| | | Fatty Acid Amides | Linear Fatty Acid, Monoethanolamide or Diethanolamide | Aerobic | Ultimately biodegradable (30–80%) | Algae, can be accumulated and somewhat toxic to fish |
| | | | | Anaerobic | Ultimately biodegradable (<80%) | |
| | Cationic (Positively Charged) | Quaternary ammonium alkyls | Alkyltrimethyl-Ammonium Salts | Aerobic | Ultimately biodegradable (0 to 60%) | Algae are very sensitive. Acutely toxic to invertebrates and fish. |
| | | | | Anaerobic | | |
| | | | Dialkyldimethyl-ammonium salts | Aerobic | Ultimately biodegradable (0 to 60%) | Algae are very sensitive. Acutely toxic to invertebrates and fish |
| | | | | Anaerobic | | |
| | | | Alkyldimethylbenzyl-ammonium salts | Aerobic | Ultimately biodegradable (8–83%) | Very toxic to algae |
| | | | | Anaerobic | | |
| | | | Alkyl ester ammonium salts | Aerobic | Ultimately to readily biodegradable (76–94%) | Significantly less toxic than alkyl-ammonium salts |
| | | | | Anaerobic | Ultimately biodegradable | |

(Continued)

**Table 2.12 Bio-oxidation Characteristics of Components of Aqueous Cleaning Agents (Continued)**

| Purpose | Nature | Type | Formed by Reactants as | Biodegradation Pathway | Biodegradability | Area of aquatic Concern |
|---|---|---|---|---|---|---|
| | Anionic (Negatively Charged) | Alkyl Sulfates | Linear fatty alcohol, Sodium Sulfate | Aerobic | Rapid ultimate bio-degradation, unbranched alcohols being less so | Generally considered to have a low potential for bioconcentration in aquatic organisms. However, they can be toxic to algae and fish. |
| | | | | Anaerobic | Rapid ultimate bio-degradation, unbranched alcohols being less so | |
| | | Alkyl Ether Sulfates | Linear fatty ethoxolate, Sodium sulfate | Aerobic | Rapid ultimate bio-degradation, unbranched alcohols being less so | Less toxic to algae but more so to invertebrates and fish |
| | | | | Anaerobic | Somewhat ultimately biodegradable (23–88%) | |
| | | Linear Alkylbenzene Sulfonates | Linear alkyl benzene, Sodium sulfate | Aerobic | Rapid ultimate bio-degradation | Very commonly used, but low-to-moderate bioaccumulation potential toxic to invertebrates, and fish – especially increasing chain length |
| | | | | Anaerobic | | |
| | | Secondary Alkane Sulfonates | Paraffins, Sodium Sulfate | Aerobic | Rapid primary bio-degradation (56–99%) | Not especially, but increasing toxicity for algae, invertebrates, and fish with increasing chain length |
| | | | | Anaerobic | Not degraded | |
| | | α-olefin sulfonates | Linear and branched α-olefins, sodium sulfate | Aerobic | Rapid primary bio-degradation (95–100%) | Toxic to algae, invertebrates, and fish |
| | | | | Anaerobic | Not degraded | |
| | | Sulfo-succinates | Esters of succinic acid, Sodium Sulfate | Aerobic | Unsubstituted molecules normally biodegradable, substituted ones are less so | |
| | | | | Anaerobic | | |
| | | Fatty acid soaps | Vegetable oils, caustic soda | Aerobic | Ultimately biodegradable (63 to 100%) | Variable by type, to algae, and fish |

| | | | | | | |
|---|---|---|---|---|---|---|
| | | | | Anaerobic | Partially bio-oxidized ultimately biodegradable (79–99%) | |
| | Amphoteric (doubly chargable) | Betaines | The 'betaine' structure is (N,N,N-Trimethylglycine) | Aerobic | Readily biodegradable | Variable by type, to algae, crustaceans and fish |
| | | | | Anaerobic | | |
| | | Fatty alkyl Imidazolines | Fatty acids or esters, Alcohol Amines | Aerobic | Readily biodegradable | |
| | | | | Anaerobic | | |
| Builders | Phosphates | $-PO_3-H$ | Sodium Tripolyphosphate | None | Contains no biodegradable (Carbon-bearing) material | Not toxic, but a nutrient which can produce 500X its weight in algae |
| | Phos-phonates | $-C-PO_3-H_2$ | Phosphorous acid, formaldehyde, ammonia/amines | Aerobic | Little primary or ultimate bio-degradation because of stability of C–P bond | Little toxicity to fish and invertebrates |
| | | | | Anaerobic | | |
| | Poly-Carboxylates | Water-soluble polymers | Homopolymers of acrylic acid or copolymers of acrylic acid and maleic anhydride, a Sodium salt | Aerobic | Not readily biodegradable because of bond stability and higher molecular weight | Not generally toxic to aquatic life because they absorb on decayed bacteria sites (sludge) and are removed from the water |
| | | | | Anaerobic | | |
| | Citrates | Multiple-valence, non-phosphoric | Mono-, di-, and tri-sales of Sodium salts of citric acid | Aerobic | Rapidly and readily biodegradable | Low toxicity towards aquatic organisms |
| | | | | Anaerobic | | |
| | Zeolites | Insoluble | Aluminum silicate | Inert and inorganic | | Low toxicity towards aquatic organisms |
| Solvents | | Alcohols | Ethanol, isopropanol | Aerobic | Rapidly and readily biodegradable | Low toxicity towards aquatic organisms |

(Continued)

**Table 2.12 Bio-oxidation Characteristics of Components of Aqueous Cleaning Agents (Continued)**

| Purpose | Nature | Type | Formed by Reactants as | Biodegradation Pathway | Biodegradability | Area of aquatic Concern |
|---|---|---|---|---|---|---|
| | | | | Anaerobic | Rapidly and readily biodegradable | |
| | | Glycol ethers | 2-Butoxy ethanol (butyl glycol), Butoxy diglycol | Aerobic | Rapidly and readily biodegradable | Low toxicity towards aquatic organisms |
| | | | | Anaerobic | Rapidly and readily biodegradable | |
| | | Glycols | Propylene glycol, Glycerol, Dipropylene glycol | Aerobic | Rapidly and readily biodegradable | Low toxicity towards aquatic organisms |
| | | | | Anaerobic | Rapidly and readily biodegradable | |
| | | Alcohol amines | 2-Amino ethanol | Aerobic | Rapidly and readily biodegradable | Low toxicity towards aquatic organisms |
| | | | | Anaerobic | | |
| "Metal catchers" | "Ring-Formers" for Multi-Valent Ions | EDTA | Ethylenediamine tetraacetate, or its tetra-Sodium salt | Aerobic | Not biodegradable | Low toxicity towards aquatic organisms |
| | | Amino carboxylic acids | Nitrilotriacetate, or its Sodium salt | Anaerobic | | Low aquatic toxicity to algae, crustaceans and fish |
| | | | | Aerobic | Biodegradable by acclimated microorganisms | |
| | | | | Anaerobic | Variable and uncertain biodegradation | |
| Acids and bases | Organic acids | Acetic acid | $CH_3CH_2COOH$ | Aerobic | Ultimately biodegradable | Generally low toxicity to aquatic life |
| | | | | Anaerobic | | |
| | | Citric acid | $C(CH2 - COOH)_3OH$ | Aerobic | Readily biodegradable | Very low toxicity towards aquatic organisms |
| | | | | Anaerobic | | |
| | Inorganic acids | Sulfamic acid | $H_3NO_3S$ | Not organic. Not biodegradable | | Strong acid. Toxicity depends upon concentration |
| | | Phosphoric acid, sulfuric acid | $H_3PO_4$, $H_2SO_4$ | Not organic. Not biodegradable | | |

| | | | | | | |
|---|---|---|---|---|---|---|
| | Organic bases | Sodium Carbonate | $Na_2CO_3$ | Aerobic / Anaerobic | Biodegradable at low concentration (neutral pH) | Strong base; toxicity depends upon concentration |
| | | Sodium Bicarbonate | $NaHCO_3$ | Aerobic / Anaerobic | Biodegradable | Very low toxicity towards aquatic organisms |
| | Inorganic bases | Potassium hydroxide, caustic soda | KOH, NaOH | Not organic | Not Biodegradable | Strong base; toxicity depends upon concentration |
| Biocides | | Nitro-substituted compounds | 2-Bromo-2-nitropropane-1,3-Diol (Bronopol) | Not biodegradable, because it releases formaldehyde | | Very toxic to aquatic life |
| Corrosion inhibitors | | "Metal catchers" | 1,2,3-Benzotriazole | Not biodegradable | | Very toxic to aquatic life |
| Defoamers | Non-ionic | Block Copolymers | Ethylene Oxide and Propylene Oxide | Aerobic | Ultimately biodegradable (5–50%) | Persistence, but not toxic |
| Deodorants | Attacks bacteria | Alkyl ammonium chlorides | Hexamethyl-enetetramine chloroallyl chloride | Not biodegradable, because it releases formaldehyde | | Very toxic to aquatic life |
| Perfumes | Replace-ment odor | Orange | D-Limonene | Toxic to normal waste treatment bacteria, but is commonly used in aqueous cleaning agents | | Very toxic to aquatic life |

These components include the following:

2.4.2.1.1 *Aqueous surfactants* are generally grouped by their charge and may be non-ionic, cationic, anionic, or amphoteric. They are chosen by the formulator of aqueous cleaning agents based on the type and amount of soils expected:

- **Non-ionic surfactants** *don't ionize*[111] but do dissolve in water. They are not normally pure compounds. They can be mixtures of homologous structures composed of alkyl chains that differ in their number of Carbon atoms.[112] They contain at least one hydrophobic (water-insoluble) alkyl chain and a Oxygen-bearing group.[113]
- **Cationic[114] surfactants** *do ionize* after solution in water. They too are not normally pure compounds. They contain at least one hydrophobic alkyl chain and a hydrophilic (water-soluble) group carrying a *positive* charge.[115] In aqueous solution, they are positively charged. Thus they can absorb[116] onto negatively charged ionic materials, and simultaneously repel water via their alkyl end. Formulation in cleaning agents can be emulsifiers, wetting agents, and biocides.
- **Anionic[117] surfactants** also *do ionize* after solution in water. They too are not normally pure compounds. They too contain at least one hydrophobic alkyl chain and a hydrophilic (water-soluble) group carrying a *negative* charge.[118] In aqueous solution, they are negatively charged. Thus they too can absorb (see Footnote 116) onto positively-charged ionic materials, and simultaneously repel water via their alkyl end. Formulation in cleaning agents is usually as a detergent.
- **Amphoteric surfactants** are the transvestites of the surfactant family. They *do ionize* in water. They do exhibit *both* cationic and anionic behavior – but not at the same time![119] Molecular structure is very complex. They do tend to be pure compounds.

2.4.2.1.2 *Builders* are used for at least two purposes: (1) to contain the effects of water hardness on surfactants, and (2) to buffer pH in solution.[120] Builders are also called complexing agents. They are chosen by the formulator of aqueous cleaning agents based on cost and factors associated with general applications:

- **Phosphates** have not been commonly used to formulate[121] cleaning agents in the last decade or so because of eutrophication[122] of many fresh waters. But they are still in use.
- **Phosphonates** have become an effective replacement for phosphates as the Carbon-to-Phosphorous bond is stable and phosphorous is not released as a nutrient. But because of that bond stability, phosphonates are not readily biodegradable.

---

[111] Ionize, in this case, means to separate into charged species (ions). Non-ionic surfactants dissolve in water as intact molecules. Cationic, anionic, and amphoteric surfactants dissolve in water as two ions with opposite charges.

[112] They are often synthesized from varying numbers of hydrophilic ethylene oxide (EO), propylene oxide (PO), or butylene oxide (BO) units; and alcohols derived from vegetable oils (fatty alcohols). Their bifunctionality is derived from the internal ether (Oxygen) linkages, and the long hydrophobic (water-hating) hydrocarbon (fatty) chain.

[113] Hansen Solubility Parameters (HSP) would show substantial disperse values. The polar-bonding values would be similar to the Hydrogen-bonding values.

[114] A cation is a positively charged ion.

[115] Those constructed from quaternary (four) ammonium compounds are used in commercial products. The positive charge comes from a quaternary Nitrogen atom. Their bi-functionality comes from the long hydrophobic hydrocarbon (fatty) chain, and the positively charged quaternary ammonium ion.

[116] This attachment can make them very difficult to biologically oxidize because of interference by the species to which they are attached.

[117] An anion is a negatively charged ion.

[118] Commercial products are constructed to provide the negatively charged species from a sulfonate, sulfate, carboxylate or phosphate group.

[119] An anion (negatively charged) is produced in basic solution. A cation (positively charged) is produced in acid solution.

[120] The marketing phrase associated with Las Vegas might be restated as "… what starts in solution, stays in solution …". Builders earn their cost by keeping cleaning agents, water salts, and soils in solution and not redeposited on part surfaces.

[121] Sodium tripolyphosphate was the chief source of phosphates.

[122] One can have too much of a good thing. Eutrophication is the aging of a waterway. Excess nutrients (Phosphorous, Nitrogen) can support excess growth of algae and other aquatic organisms. As life continues and dies, silt and decayed aquatic life fill the bottom of the water way. It becomes a bog, and later land. Overstimulation of aquatic growth by excess phosphorous from components of cleaning products has lead to voluntary and statutory restrictions which limit the use of the detergent builder sodium tripolyphosphate.
Another successful approach has been to establish Phosphorus removal processes at major wastewater treatment plants.

- **Polycarboxylates**[123] have no phosphorous content so there is no concern about eutrophication. Like phosphonates, they are not readily biodegradable.
- **Sodium citrate** also contains no phosphorous. Unlike phosphonates and polycarboxylates, sodium citrate is easily biodegradable. Sodium citrate is used to complex just certain[124] metal ions which become water hardness – Calcium, Magnesium, Iron, etc.
- **Zeolites**[125] don't contain phosphorous, and aren't biodegradable. They are not commonly used to formulate aqueous cleaning agents in the US, but are elsewhere used.

2.4.2.1.3 *Solvents* For a solvent to be soluble in water, it must have some intermolecular forces or molecular characteristics similar to those of water. For those solvents used in formulated aqueous cleaning agents, a hydroxyl group ($OH^-$) is common to both solvent and water:

- **Alcohols, glycol ethers, glycols, and alcohol amines** are used depending upon the characteristics of the soils expected. All are readily biodegradable.

2.4.2.1.4 *"Metal Catchers"*[126] are "high-tech" builders.[127] While there is some application to control water hardness or replace phosphates, there is more intention to remove heavy metals from aqueous cleaning baths. Secure removal of metal ions allows: (1) ionizing surfactants to perform as intended, (2) metal corrosion to be controlled, and (3) collection of toxic heavy metals:

- **Ethylenediaminetetracetate (EDTA)** has been almost as long-used to scavenge metals from water as has soap been used to remove oil from it. It is absolutely not biodegradable.
- **Nitrotriacetate (NTA)** is a similar compound. It is only somewhat biodegradable.

Segregation of heavy metals (Chromium, Lead, Nickel, Zinc) with EDTA or NTA does not mean "... get out of jail free ...". It means that these metals have been removed from the cleaning bath, *but* have become concentrated in one place. Said another way, solid material collected after treatment of wastes from aqueous cleaning baths may become classified as hazardous because of presence of particles/scale/sludge which are rich in toxic metals.

2.4.2.1.5 *Acids and Bases* Three crucial functions are fulfilled by these components of aqueous cleaning agents, as: (1) controllers (buffers) of pH,[128] (2) reactants, and (3) corrodents.[129]

But not all acids or based can be treated as components of waste[130] from a cleaning machine via biological oxidation:

- **Organic acids**, such as acetic and citric acid, can also fulfill the function of a metal brightener

[123] The multiple Oxygen atoms in polycarboxylates exhibit intermolecular forces which allow absorption of these builders. They sorb onto particulate matter and sludge. Consequently, they are removed from biological treatment systems without being oxidized by bacteria.
Polycarboxylates are reported to inhibit the crystal growth of inorganic salts. They remain in suspension and do not precipitate.

[124] These are metal ions with multiple valence (+2 or +3). The Sodium ion remains soluble in the biological treatment system, and the Calcium/Magnesium/Chromium ions are precipitated and removed with the sludge. Sodium oxalate has also been used to chelate bi-valence metal ions. While organic, it is toxic to humans and aquatic life.

[125] Zeolites are commonly used in fixed bed ion exchange operations. As such, they would be an excellent choice to remove water hardness. However, they should not be confused with other builder components – Zeolites are insoluble and inert. Chemically, Zeolites are aluminum silicates.

[126] The technical term for these components is chelating/sequestering agents.

[127] Both EDTA and NTA are:
(1) Banned by the European Union for use in laundry detergents because of they are not efficiently biodegraded. See *Official Journal of the European Union*, L76/27, March 22, 2003.
(2) Commonly used to remove metal ions from boiler feed water streams.
(3) Used to sequester (capture) metal ions in ring structures.

[128] pH is a value on a measurement scale between being an acid, or a base. A pH value of 7.0 is neither (neutral). Higher values than 7.0 represent higher concentrations of soluble basic material. Aqueous cleaning is usually done between pH values of 8.5–12.0. Some aqueous cleaning is done with strong acids with pH values around 4.0. A single pH unit represents a 10-fold change in concentration of soluble acid or base.

[129] This is not an supplementary rat. A corrodent is a material which causes corrosion – usually of metal surfaces, possibly to enhance brightness.

[130] As soils shouldn't survive an aqueous cleaning process, so should not acids or bases. Each will react in a buffering step, and/or can react with a soil component. That residue is what must be bio-oxidized. Since the organic portion of that residue can be a solid, or absorbed on a solid, it may be difficult for it to be accessed by soluble bacteria.

(a corrodent) – as components of an aqueous rinsing agent, or single treatment chemicals. They are readily biodegradable.

- **Inorganic acids**, such as sulfamic ($H_3NO_3S$), phosphoric ($H_3PO_4$), and sulfuric acid ($H_2SO_4$) are less often reactants. Without any organic content (Carbon atoms), they are not biodegradable.
- **Organic bases** primarily buffer pH. Sodium carbonate ($Na_2CO_3$) and sodium bicarbonate ($NaHCO_3$) are common components of an aqueous cleaning agent. The carboxylic portion ($CO_3$) of these bases can be biodegradable.
- **Inorganic bases**, especially potassium hydroxide (KOH), are the feedstock for many alkaline aqueous cleaning systems. Nothing is biodegradable because there is no content of Carbon to oxidize.

Either $Na_2CO_3$/KOH or $H_3NO_3S/H_3PO_4/H_3NO_3S$ can be used to corrode a metal surface so that soil may be exposed for contact by detergents.

2.4.2.1.6 *Biocides/Corrosion Inhibitors/Defoamers/Deodorants/Fragrances*

- **Biocides**[131] are occasionally used to prevent biological oxidation in static cleaning tanks. Used at levels of 0.01–0.1%, they are usually present at concentrations less than those necessary for inclusion in an MSDS. Nitro-substituted compounds are one type. An example are one type, of which 1,2-Bromo-2-Nitropropane-1,3-Diol (Bronopol is an example).
- **Corrosion inhibitors** in aqueous cleaning agents are likely to also be "metal-catchers," acids, bases, pH-control compounds (buffers) or other chemicals. Yet if aqueous alkaline cleaning is intended, the corrosion inhibitor can't be an acid, base, or buffer! That's why one commonly used with Copper/brass/bronze is 1,2,3-Benzotriazole – a chelating agent for copper ions. Corrosion inhibitors are nearly always a proprietary additive, used at low concentrations, not identified on MSDSs, and may raise concerns about toxicity to human and aquatic life.
- **Defoamers** are additives which reduce the surface tension of water.[132] Also used at low concentrations because of their substantial effects on surface tension, they are seldom identified on MSDSs. De-foamers break existing foams; antifoams keep foams from forming. The same chemical may be used in different ways or concentrations for both situations. Block copolymers of ethylene and propylene oxide are often the backbone of antifoam agents.
- **Deodorants** can also be biocides, because often the odor is produced by bacteria. Many are designed to produce small amounts of formaldehyde, a potent toxin. Because of their low concentration in formulations, seldom will they be identified on an MSDS. One type is an alkyl ammonium chloride, called Quaternium-15.[133]
- **Fragrances/Perfumes/Odor** Maskants are attempts to conceal a problem. They overpower

---

[131] Obviously, this type of component is not compatible with biological oxidation of waste from an aqueous cleaning process. Yet, some applications require their presence and other methods of waste treatment must be found.

[132] A foam is a stable network of bubbles. If the bubbles of which a foam is comprised aren't stable, the foam won't be stable. Bubble stability a balance between buoyant forces (density differences between gas and fluid) seeking to expand the bubble, and surface tension forces seeking to contract its size. When the surface tension of a fluid is low, bubbles grow and collapse – foam dissipates. Bubbles formed in a fluid of zero surface tension would grow as fluid vapor or air was produced but soon burst as there were no surface forces to contain them as a cavity. Foams often form in water because of its high surface tension.

Defoaming agents (anti-foams) don't have to be soluble in the cleaning bath. All they must do is migrate to the vapor/air fluid interface. That's why hydrophobic (water-hating) non-polar silicone oils are useful as antifoam agents. Most silicone antifoam agents contain silica particles that are treated to make them hydrophobic so they will remain suspended, preferentially, in the silicone oils. It is thought that the tiny silica particles puncture the fluid-vapor film and allow the silicone oil to penetrate the interface surface. Obviously, this approach shouldn't be tolerated in surface cleaning applications – because of residual particles after rinsing.

Polar antifoam agents are non-ionic surfactants. Insolubility in aqueous solutions is the key property of these antifoam agents. Above their cloud points (crystallization temperatures), they produce insoluble droplets that are incorporated into the interfacial films as are other insoluble oils.

The best antifoam agent for cleaning operations is removal of the mechanical forces (usually agitation) which produce surface turnover. If bubbles aren't formed, an antifoam agent won't be needed, and traces (residue) of that agent won't be present on surfaces of rinsed parts. In some cases, the tendency to generate foams is enhanced by anionic soluble surfactants present in the formulated cleaning agent.

[133] Also used a preservative in personal care products.

existing odors with others felt to be more pleasing. Ironically natural D-limonene, reeking of the odor of oranges and a common cleaning agent, has been commonly used to mask other odors. Most odor maskants are natural products and aren't toxic but may biodegrade slowly.

Some additives are multi-functional. Carbonates, for example, contribute to detergency, soil holding or water-conditioning, and corrosion inhibition.

### 2.4.3 Find That Formulation!

You have to know what to manage. You have to know the general composition of what's used in the system which you manage:

- The aqueous cleaning agent
- The soil
- Any products of reaction between them

If you don't have this information, you might as well be "... seeking weapons of mass destruction ...". You won't know if *all*[134] the components of the aqueous waste produced by your aqueous cleaning system are treatable by biological oxidation. *You probably won't be able to get an environmental permit, or comply with it should you have one*!

If you have compositional and ingredient information, you can use Table 2.12 which lists the "environmental fate"[135] of many components of formulated aqueous cleaning agents.

#### 2.4.3.1 *Show Me the Money I*

Start with the MSDS. It's required to be provided with every unit of product (see Chapter 3, Section 3.9–3.9.3). Probably worthless for your purpose, but it's a starting point.[136]

Next, legally[137] obtain *all*[138] composition and ingredient information in the possession of the supplier's representative (or the supplier).[139] Sign any non-disclosure agreement (NDA) submitted by the supplier. Keep all commitments in that agreement.

#### 2.4.3.2 *Composition Counts*

"... Of course I'm not married ...", "... the check is in the mail ...", and "... sure, it's biodegradable ..." are among famous expressions honored more for their being incorrect or incomplete than the opposite.

Bio-oxidation is practical over certain ranges of composition of organics. Unfortunately, those composition ranges are seldom those of the cleaning agent in its packaged state, *or in its condition of use*:

- Aqueous cleaning agents are typically sold as concentrates to save the cost of shipping water. Twenty-five to fifty percent concentration is common.[140] But cleaning agents can't be biologically oxidized at that rich level of concentration.
- Aqueous cleaning agents are typically used at a concentration which is economically effective – for both buyer and seller. Dilution of the as-sold product at a level of 1 to 2 or 1 to 10 means that the concentration in the cleaning bath, and in the waste, is several percent.

---

[134] Remember, if a component isn't biologically oxidizable, it won't be oxidized. And untreated, it is part of the effluent from your site. Some other additional treatment step will have to be found, and added.

[135] This includes whether or not the component can be oxidized by bacteria, whether the oxidation is done by an aerobic (with Oxygen) or anaerobic (without Oxygen) process, and whether the component is toxic to aquatic life.

[136] An MSDS is a marketing tool. As such, information in it will describe positive aspects. Negative aspects will be minimized, or omitted through use of the ubiquitous "N/A". See Chapter 3, Section 3.19.

Second, an MSDS isn't written for the audience in which you as a manager find yourself. An MSDS is required to include information about hazardous components of the product. As a manager, responsible for waste treatment of the product and corollary materials, you need information about *all* components of the product. *All* components must be treated prior to disposal.

[137] Don't be afraid to search the internet about composition information, or the patent journals as well.

[138] This includes composition of ingredients whose concentration is quite low, and doesn't have to be reported on MSDS: corrosion inhibitors, biocides, antifoams, deodorants, and the like.

[139] Forage through the supplier's management tree until you reach a person who will provide that information – in return for appropriate commitments by your firm about confidentiality. The supplier has a perfect right to own and conceal the details of their formulation. You have a perfect right to know those details to use the supplier's product. But in return for that right you surrender the right to communicate those details to others. It's fair bargain: you're cleaning parts, you're not selling cleaning agents.

[140] Only occasionally is a neat (100%) product provided.

- Efficient biological oxidation typically occurs over the range of concentration from around 0.1% (1,000 ppm) to around 0.0025–0.0050% (25–50 ppm).

Consequently, the "biodegradable" waste from an aqueous cleaning bath may not be so for two excellent reasons:

- It is too rich in nutrients (organics), and must be diluted by with at least 10 and possibly 100 volumes of water so that efficient biological oxidation can be managed. Suppose the aqueous cleaning tank in the process you manage holds 100 gallons. That means you must provide at least 1,000 gallons of water (clean or dirty) in order to manage efficient biological oxidation.
- It contains soil components which may or may not be biodegradable. They, properly, are not listed on the MSDS for the cleaning agent.

It's not considered unethical to conceal information which doesn't have to be presented.

### 2.4.4 In the Heat of the Moment

Aqueous cleaning and biological oxidation[141] are processes which occur at temperatures best-suited for each. There is no reason to expect those temperatures to be similar, or even close to one another:

- Biological oxidation is normally conducted at ambient conditions, *ca.* 70 or 80°F (*ca.* 20 or 25°C).
- Aqueous cleaning is usually conducted at 120°F to 160°F[142] (*ca.* 50–80°C).

Both dilution and cooling can be accomplished with the same volume of water if water cooler than ambient and a mixing pond/tank are available. But that further increases treatment volume and cost.

### 2.4.5 OD-ing on Oxygen

Oxygen demand is the metric by which the strength of an aqueous waste is measured. A waste which is not biodegradable, obviously imposes no demand for Oxygen.

Waste strength is significant both before and after biological treatment:

- Aqueous waste strength before treatment is a measure which is part of the specification by which the biological treatment system is designed.[143]
- Aqueous waste strength after treatment represents the amount[144] of untreated pollution emitted to some waterway.

Waste strength is characterized in at least four ways, chiefly by the amount of Oxygen necessary to oxidize all material. The oxidation methods (and a significant one not involving oxidation) are compared in Table 2.13.

These parameters are useful in design[145] and monitoring of biological oxidation systems. Both off-line or on-line measurements can be made, and used. The units of these and similar parameters are ppm.[146]

### 2.4.6 Show Me the Money II

In Section 2.4.3.1, it was written that it was futile to use MSDSs to learn if it is practical to treat an aqueous cleaning agent by biological oxidation. In this chapter, it is written that it can be futile to use claims of biodegradability to learn the same thing.

The following statement is copied from an internet-based advertisement for a common aqueous cleaning product: "… is readily decomposed by naturally occurring microorganisms. … is biodegradable … The Biological Oxygen Demand (BOD),

---

[141] Moran, J.M., Morgan, M.D. and Wiersma, J.H. (1980). *Introduction to Environmental Science* (2nd ed.). W.H. Freeman and Company, New York, NY.

[142] Optimum of ultrasonic transducers to clean via cavitation is considered to be 160°F.

[143] Issues involved are type of bacteria, holdup time and volume, temperature, and input rate of air to be pumped and dissolved into the water.

[144] When multiplied by the volume of the effluent.

[145] It may be that these measured parameters suggest biological oxidation is *not* a suitable treatment method. For example, BOD or TOC values might be low while TOD values are high. This indicates there is little organic or biological oxidizable material within the waste, but the waste does have significant nonoxidizable components (such as Sulfur, Nitrogen, Phosphorous, or metals). An example might be where the ratio of TOD to BOD is *ca.* 100 to 1.

[146] Parts per million (ppm), by weight, is the standard for concentration in aqueous wastes. Recall that the density of water is 1 g/ml or 1000 g/l. Consequently, concentrations are also reported as milligrams (mg)/l which is an equivalent presentation.

**Table 2.13 Measures of Strength of Aqueous Wastes**

| Name | Oxidation Nature | Meaning | Oxidation Method |
|---|---|---|---|
| BOD | Bacteria | An estimate of the "food" available in the sample. Said another way, "food" is organic matter – Carbon and Hydrogen, but not Nitrogen or Phosphorous or Sulfur, or metals. | Oxygen, dissolved in water and fed to bacteria. Test duration is 5 days, though the normal time expected for nearly complete oxidation of "food" in a treatment plant is *ca.* 20–25 days. Test temperature is normally 20°C. |
| COD | Chemicals | An estimate of the total "non-metallic" content of the aqueous waste. Here, Carbon and Hydrogen as well as Nitrogen, Phosphorous, and Sulfur would be expected to be completely or partially oxidized. Metals, aromatic organics, and halides would not be oxidized. | Hot oxidation with dichromate solution. This is not a simulation of a treatment plant – it is a method of estimation. |
| TOD | Any means | An estimate of the total oxidizable content of the aqueous waste. Here, Carbon, Hydrogen, Nitrogen, Phosphorous, Sulfur, and metals would be expected to be oxidized. Halides would not be expected to be oxidized. | Combustion in a catalytic and controlled environment. This also is not a simulation of a waste treatment plant – it too is a method of estimation. |
| TOC | "Programmed oxidation" | An estimate of total Carbon content of the waste which is organic. Other Carbon may be present as carbonates (see Section 2.4.2.1). The differentiation is made because carbonates are relatively benign to aquatic life relative to most other components of aqueous wastes. | Oxidation is also accomplished by catalytic oxidation after carbonates are removed in a variety of ways. |

as a percentage of the Chemical Oxygen Demand (COD), after 4, 7, and 11 days was 56%, 60%, and 70%, respectively for … ".

No criticism is intended of this product. But this claim, which is intended to prove one point (biodegradability), actually proves another – that the product may not completely biodegradable. It also begs a question: what is the treatment method for the ~one-third of the material which is *not* biodegradable?[147]

Marketers of these useful products stress cleaning performance. But managers of aqueous cleaning agents are also responsible, in most countries, to manage their safe use *and* disposal, which includes:

- Disposal of *each* component of the cleaning agent, *and* soil.[148]
- Knowledge of the environmental fate of *each* component of the cleaning agent, and soil.
- A good process by which *each* environmental fate can be achieved.
- Management to assure that each environmental fate is achieved, as well as compliance with all regulations.

---

[147] An internationally-accepted test for "ready biodegradability" is promulgated by the Organization for Economic Cooperation and Development (OECD). It is known as the OECD Guidelines 301a–f, and can be found at http://www.oecd.org/dataoecd/17/16/1948209.pdf

[148] Commercial confidentiality will limit collection of information about components of the cleaning agent. Ignorance will limit collection of information about the soil. Yet, managers are responsible to obtain both because environmental regulators don't forgive fallibility about treatment of some waste components and not others.

The claim of biodegradability by a supplier must apply to all components – not just some. The user is responsible for disposal of all. Table 2.12 can be a resource in managing that responsibility.

### 2.4.7 US Regulations Concerning Biodegradable Cleaning Agents

Surprisingly enough, there are other process operations that produce wastewater. They include both domestic sewage and industrial waste sewage and industrial waste from other manufacturing sources. Metals, organic pollutants, sediment, bacteria, and viruses may all be found in wastewater.

In the US, the Clean Water Act (CWA) rules:

- The US EPA's Office of Wastewater Management (OWM), in cooperation with states, EPA offices, and other stakeholders, manages programs to achieve the goals of the CWA.
- The CWA requires that all point source wastewater dischargers obtain and comply with a National Pollution Discharge Elimination System (NPDES) permit. NPDES permits, introduced in 1972, regulate the discharges from Publicly-Owned Treatment Works (POTW) facilities, other wastewater treatment facilities, industrial facilities, concentrated animal feeding operations, aquiculture, and other "point source"[149] dischargers.
- The CWA provides that states may be empowered to operate their own NPDES programs, provided such programs meet minimum federal requirements. Waste water from aqueous cleaning operations is covered by NPDES permits; no specific operation must be identified. Only the specific parameters of the co-mingled waste are covered, not the component waste sources.

## 2.5 CLEANING AGENTS WHICH RAISE CONCERNS ABOUT TOXICITY

This sub-chapter is not necessarily about cleaning agents, nor is it a general warning against the use of chemicals.

But cleaning agents can and do play a role in harmful effects produced by some chemicals.

In 1970, the US Congress mandated that there should be a way to (at least) attract the attention of users and regulators to certain chemicals. These were chemicals whose use, the Congress believed, based on scientific data, exposed users and the environment to *significantly more risk* than did other chemicals.

The way found was to identify some chemicals as being in at least two categories:

1. *Criteria Pollutants* (*CPs*): US EPA refers to six "chemicals" that can have lethal or permanent effects on "sensitive" segments of the US population. The "chemicals" (pollutants) are: carbon monoxide (CO), nitrogen dioxide ($NO_2$), sulfur dioxide ($SO_2$), ozone ($O_3$), Lead (Pb), and particulate matter (abbreviated as PM).
2. *Hazardous Air Pollutants* (*HAPs*):[150] US EPA refers to189 chemicals that cause serious health and environmental hazards as HAPs or air toxics.

All states in the US are required to recognize in their local regulations that HAPs and CPs must be treated differently than other chemicals.

Countries outside the US have their individual classifications of chemicals which are usually different than the classification in the US.

Whether or not a cleaning chemical (solvent) is classified for special treatment in the country of your operation, inclusion on the US EPA's HAP or CP lists should mandate at least search for the primary reasons which produced the classification, alternative cleaning chemicals, and special techniques for use. All of that information is found in this book.

### 2.5.1 Criteria Pollutants

US EPA refers[151] to chemicals that can injure health, harm the environment, and cause property damage as CPs. Certain persons, called "sensitive"

[149] Point sources are discrete outfalls such as pipes or man-made ditches. Within an operating site, many operating devices may feed waste water into the site's single point source outfall.

[150] The term Hazardous Air Pollutants has been abbreviated by all as HAPs.

[151] The actual phrase was "… may reasonably be anticipated to result in an increase in mortality or an increase in serious irreversible or incapacitating reversible illness. …."

populations,[152] are more vulnerable to this damage and regulation of CPs is an approach to providing super-normal protection to them.

A US National Ambient Air Quality Standard (NAAQS) and a monitoring program has been developed for each CP. These standards are national in scope, not applicable to air emissions from specific sites, and do change with time.

The six critical pollutants are as follows:

1. *Carbon Monoxide* (*CO*): Motor vehicle exhaust contributes about 60% of all CO emissions.
2. *Lead* (*Pb*): Metal processing accounts for about one-half of the emission of lead to the atmosphere.
3. *Sulfur Dioxide* ($SO_2$): Sulfur oxides ($SO_2$ and $SO_3$) are formed when fuel containing Sulfur is burned. Actually, the concentration of $SO_2$ is taken as an indicator of the total amount of sulfur oxides present. Fuel combustion, from electrical generation or industrial plants, contributes about 75% of all $SO_2$ emissions.
4. *Nitrogen Dioxide* ($NO_2$): There are also multiple oxides of nitrogen, called $NO_X$. As with sulfur oxides, a single one is taken as an indicator of the total amount. The primary source of $NO_X$ is also fuel combustion. $NO_X$ is significant to managers of cleaning operations because VOCs emitted from those operations react with $NO_X$ to produce another CP, and smog.[153]
5. *Ozone* ($O_3$): Seldom emitted directly, and quite reactive, ozone is formed by chemical reaction in the atmosphere. The ingredients are various, but are most commonly thought to be VOCs, $NO_X$, and UV wavelengths of sunlight (see Section 2.2). Heat speeds the process. Ozone is harmful to humans, but as an active oxidant is very harmful to plants. It contributes to smog, but is not smog (see Table 2.12).
6. *Particulate Matter* (*PM*): This is a heterogeneous and uncontrolled mixture. Both solids and aerosols of liquid are included. Stability in the atmosphere defines particle size.[154] Larger particles are seen as smoke, dust, or soot. Smaller particles are unseen without magnification.[155] PM is emitted directly ("primary" particles) or formed in the atmosphere ("secondary" particles). Usually formation involves reaction with fuel combustion byproducts, sunlight, and humidity (water vapor).

What is unique about $O_3$ and PM is that they, and their formation, are mobile. There isn't a substantial concentration of either adjacent to an automobile parked next to a cleaning machine. A refinery emitting VOC in Ponca City, OK, US and rush-hour traffic around NYC, US can produce smog over Oxford, in the UK. This fact can make regulations about CPs difficult to accept, construct, and demand.

A manager of cleaning work might believe CPs are unimportant to their affairs when: (1) no specific permit regulating CPs is associated with their site, and/or (2) no cleaning system directly emits a CP. That belief would be totally incorrect. CPs tend to be regulated[156] indirectly because some ($O_3$, PM) are not directly emitted and $O_3$ and PM *are formed in the atmosphere from chemicals which are commonly emitted from cleaning systems.*

### 2.5.2 Hazardous Air Pollutants

Air toxics are released from sources throughout the country and from motor vehicles. As an example,

[152] These include children, older adults, and persons with asthma. Amendments to the 1970 Clean Air Act (CAA) also intended to provide protection from atmospheric pollutants to "public health" and "welfare." Within the framework of the CAA, "welfare" refers to the viability of agriculture and ecosystems (such as forests and wildlands).

A recent study demonstrates the seriousness of this type of differentiation. Consider Research Report 131 from the Health Effects Institute, "Characterization of Particulate and Gas Exposures of Sensitive Subpopulations Living in Baltimore and Boston," December 2005, available at http://www.healtheffects.org/Pubs/Koutrakis.pdf

[153] See Section 2.2.9, where formation of smog via $NO_X$, without VOC, is described.

[154] Atmospheric material large enough to settle in air is pollution, but not PM.

[155] PM standards are constructed based on differentiation by size. Particles above 10 μm in average size, usually visible, are regulated by volumetric concentration called $PM_{10}$. Particles above 2.5 μm in size, often not visible, are regulated by volumetric concentration called $PM_{2.50}$.

[156] As this book is being published (December 2005), the US EPA proposed strengthening – by nearly 50% of existing limits for PM. Public comment may affect this proposal.

gasoline contains such toxics as benzene and toluene, which have been identified by the US EPA as HAPs.

Use of gasoline without consideration of hazards it presents reflects a key characteristic of humans – we are deliberately willing to accept risk where we believe we can: (1) protect ourselves against its hazards, (2) benefit from its acceptance, and (3) mask it among other risks.

One might think that HAPs are never used in commerce or industry. The opposite is true. Some HAPs are major feedstocks for production of other chemicals, *and* are standalone products. Examples are methyl ethyl ketone (adhesives), acrylic acid (coatings), Chlorine (bleach and disinfectants), or ethylene oxide (glycols and surfactants).

Safe use is believed by many to be possible because exposure limits[157] have been developed through testing with animals and humans. Engineering schemes have been/can be in place to (1) control exposure to within those limits, and (2) monitor that exposure.

#### 2.5.2.1 *The Legal Stuff*

The US CAA, passed in 1970, required: "… for the purpose of establishing national primary and secondary ambient air quality standards, the Administrator shall within 30 days after December 31, 1970, publish, and shall from time to time thereafter revise, a list which includes each air pollutant:

> *"Emissions of which, in his judgment, cause or contribute to air pollution which may reasonably be anticipated to endanger public health or welfare;*
>
> *The presence of which in the ambient air results from numerous or diverse mobile or stationary sources; and*
>
> *For which air quality criteria had not been issued before December 31, 1970 but for which he plans to issue air quality criteria under this chapter.*[158] *…"*

The US Congress produced a list of 189 chemicals which were considered to meet the above requirements[159] and be considered HAPs. The list contained the 189 chemicals:

> *… pollutants which present, or may present, through inhalation or other routes of exposure, a threat of adverse human health effects (including, but not limited to, substances which are known to be, or may reasonably be anticipated to be, carcinogenic, mutagenic, teratogenic, neurotoxic, which cause reproductive dysfunction, or which are acutely or chronically toxic) or adverse environmental effects whether through ambient concentrations, bioaccumulation, deposition, or otherwise …*[160]

The list[161] can be expanded by fiat of the US EPA Administrator or contracted by review, including public comment, of a petition[162] to remove a specific chemical.

All states in the US are required to recognize in their local regulations that HAPs and CPs must be treated differently than other chemicals. Individual US states are allowed to choose to manage even more stringent regulation.[163]

---

For "inhalable coarse" particles, which are particles between 2.5 and 10 μm ($PM_{10-2.5}$), the proposed limit is a 24-hour standard of 70 μg per cubic meter (70 μg/$m^3$).

For fine particles which are particles 2.5 μm in diameter and smaller, EPA is also taking comment on a range of annual and 24-hour standards, including strengthening these standards as well as retaining the standards at their present levels – $PM_{10}$ as an annual value of 50 (50 μg/$m^3$) and a 24-hour average of 150 (150 μg/$m^3$); and $PM_{2.5}$ as an annual value of 15 (15 μg/$m^3$) and a 24-hour average of 65 (65 μg/$m^3$).

[157] The **excellent** *NIOSH Pocket Guide to Chemical Hazards*, can be downloaded for *free* at http://www.cdc.gov/niosh/npg/npg.html. It is a significant reference about exposure limits. See Chapter 3, Section 3.5.1.

[158] US Clean Air Act, Title 42, Chapter 85, Subchapter I, Part A, Section 7408.

[159] Through a clerical error, hydrogen sulfide was inadvertently included in this list. After petition, it was removed. Since then, caprolactam was removed from the list after petition, public review, and re-regulation.

[160] US Clean Air Act, Title 42, Chapter 85, Subchapter I, Part A, Section 7412 (b) (1).

[161] To locate the complete list of Hazardous Air Pollutants see http://www.epa.gov/ttn/atw/orig189.html or http://www.deq.state.ne.us/Publica.nsf/0/fa8b072045fedcf106256dc10060dd64/€FILE/03-150.PDF

[162] US Clean Air Act, Title 42, Chapter 85, Subchapter I, Part A, Sec. 7412 (b) (2,3).

[163] US Clean Air Act, Title 42, Chapter 85, Subchapter I, Part A, Sec. 7412 (d) (7).

#### 2.5.2.2 *Meaning of HAP*

In a reissue of the CAA (1990), the US Congress mandated that *identification* was not enough. There should be a US national *strategy*[164] to control emissions of those chemicals. The purpose of that strategy should be to achieve at least a 75% reduction in the incidence of cancer related to exposure to chemicals identified as HAPs.[165]

Chapter 112 of the CAA Amendments of 1990 requires the US EPA to evaluate and control emissions of the substances on the HAPs list and to identify source categories for which it must establish emissions standards. As such, the US EPA has issued proposed and final rules for various manufacturing sectors, referred to as the Maximum Achievable Control Technology (MACT) rules.

HAPs are also emitted from small stationary sources (e.g. dry cleaning machines) and large stationary sources (e.g. incinerators). Facilities that emit in a year more than 10 ton of any single HAP or 25 ton of total HAPs are likely to be subject to these rules.

#### 2.5.2.3 *Lists of HAPs*

Chemicals which are HAPs and which might reasonably be considered for cleaning operations are listed in Table 2.14.[166,167] Substitutes (alternatives) are suggested.[168]

The letter codes[170] in Table 2.14 refer to specific health-related issues.

---

[164] US Clean Air Act, Title 42, Chapter 85, Subchapter I, Part A, Section 7412 (k) (3) (a).

[165] US Clean Air Act, Title 42, Chapter 85, Subchapter I, Part A, Section 7412 (k) (2).

[166] To locate the complete list of Hazardous Air Pollutants. See http://www.epa.gov/ttn/atw/orig189.html

[167] For detailed information about each HAP listed in Table 2.14, see the following US EPA sites:
A. http://www.epa.gov/ttnatw01/hlthef/di-ethan.html
B. http://www.epa.gov/ttnatw01/hlthef/trichlor.html
C. http://www.epa.gov/ttnatw01/hlthef/tri-ethy.html
D. http://www.epa.gov/ttnatw01/hlthef/dioxane.html
E. http://www.epa.gov/ttnatw01/hlthef/acetonit.html
F. http://www.epa.gov/ttnatw01/hlthef/benzene.html
G. http://www.epa.gov/ttnatw01/hlthef/carbonte.html
H. http://www.epa.gov/ttnatw01/hlthef/chlorofo.html
I. http://www.epa.gov/ttnatw01/hlthef/cumene.html
J. http://www.epa.gov/ttnatw01/hlthef/di-forma.html
K. http://www.epa.gov/ttnatw01/hlthef/di-ethan.html
L. http://www.epa.gov/ttnatw01/hlthef/ethy-gly.html
M. http://www.epa.gov/ttnatw01/hlthef/glycolet.html
N. http://www.epa.gov/ttnatw01/hlthef/hexane.html
O. http://www.epa.gov/ttnatw01/hlthef/isophoro.html
P. http://www.epa.gov/ttn/atw/hlthef/methanol.html
Q. http://www.epa.gov/ttn/atw/hlthef/methylet.html
R. http://www.epa.gov/ttn/atw/hlthef/methyl-k.html
S. http://www.epa.gov/ttn/atw/hlthef/methylen.html
T. http://www.epa.gov/ttn/atw/hlthef/toluene.html
U. http://www.epa.gov/ttnatw01/hlthef/xylenes.html

[168] The general basis for substitution in Table 2.14 is: (1) absence from being the US EPA's HAP list, (2) relative equivalence of Hansen Solubility Parameters (HSP), (3) lower IRCHS rating, and (4) similarity of boiling point. Obviously, not all items can be fulfilled. Occasionally there is no similar chemical.

*It should not be expected that either air or water will be useful substitutes for a HAP.*

In some cases, a common conundrum evolves. It is about substitution of a chemical identified as a HAP for another not so identified, but which is flammable (see Table 2.14 for the substitutes for 1,1,1-Trichloroethane).

Use of a HAP should raise an issue other than substitution. That is replacement. In other words, consider replacing the use of a HAP in a process with another process – in which a chemical quite unlike the HAP is the central player.

[169] As this book is being published (December 2005), the US EPA has proposed to remove methyl ethyl ketone (MEK) from the Clean Air Act list of toxic air pollutants. Public comment may affect this proposal.

[170] For a supportive perspective on item L (glycol ethers), see the toxicity report and recommendations in "F036: Toxicity of 1,2-Dioxyethane (EGDEE) for Fertility Clarifications on the French Position for Classification," *ECBI/15/03 Add. 3*, January 14,

**Table 2.14 Cleaning Solvents Rated by US EPA as Hazardous Air Pollutants (HAPs)**

| Common Name | CAS # | Exposure Limit ppm | Primary Reason for Concern by US EPA[170] | Possible Alternative Cleaning Solvent |
|---|---|---|---|---|
| 1,1-Dichloroethane | 75-34-3 | 100 (OSHA & NIOSH) | A | 1,1-Dichloropropane, Dioctyl phthalate, Chloropropane, Lactic acid |
| 1,1,1-Trichloroethane (TCA) | 71-55-6 | 350 (OSHA & NIOSH) | B | Butyl chloride, Cyclopentene, Diethyl sulfide |
| 1,4-Dioxane | 123-91-1 | 100 (OSHA) | D | Thiophene |
| Acetonitrile | 107-13-1 | 2 (OSHA) | E | Acetyl acetone |
| Benzene | 71-43-2 | 10 (OSHA) | F | Mesitylene, Naphtha high-flash, Aromatic 150 |
| Carbon tetrachloride | 56-23-5 | 10 (OSHA) | G | Mesitylene, Naphtha high-flash, D-Limonene, Hexadecane |
| Chloroform | 67-66-3 | 50 (OSHA) | H | Anisole, Oleic acid |
| Cumene | 98-82-8 | 50 (OSHA) | I | |
| Dimethyl formamide | 68-12-2 | 10 (OSHA) | J | Trimethylphosphate |
| Ethylene dichloride | 107-06-2 | 50 (OSHA) | K | Dibutyl phthalate, Ethyl cinnamate, Benzaldehyde |
| Ethylene glycol | 107-21-1 | 50 (OSHA, ACGIH) | L | Glycerin |
| Glycol ethers derived From ethylene | Please see Table 2.15 | | M | Please see Section 2.5.2.4 |
| Hexane | 110-54-3 | 500 (OSHA) | N | Heptane |
| Isophorone | 78-59-1 | 25 (OSHA) | O | Ethyl formate, Methyl acetate |
| Methanol | 67-56-1 | 200 (OSHA) | P | 1,3-Butanediol, Ethanol, 1,4-Butanediol 1,3-Propylene glycol |
| Methyl ethyl ketone[169] | 78-93-3 | 200 (OSHA) | Q | 1,1-Dichloropropane, Vinyl propionate |
| Methyl isobutyl ketone | 108-10-1 | 100 (OSHA) | R | n-PB, Isobutyl acrylate |
| Methylene chloride | 75-09-2 | 25 (ACGIH) | S | Cyclohexanone, Pyrrole |
| Perchloroethylene | 127-18-4 | 50 (ACGIH) | B | Ditridecyl phthalate |
| Toluene | 108-88-3 | 200 (OSHA) | T | Naphtha high-flash, Aromatic 150, Mesitylene |
| Trichloroethylene | 79-01-6 | 10 (ACGIH) | C | Anisole, Oleic acid, Diethyl carbonate |
| Xylene (isomers) | 133-02-7 | 100 (ACGIH) | U | Naphtha high-flash, Aromatic 150, Mesitylene |

### 2.5.2.4 *Glycol Ethers – Special Cases*

Glycol ethers based on ethylene glycol were singled out as a special category in the list of HAPs because there are so many of them.[171]

The US EPA has based its original classification of glycol ethers on toxicological information for three such chemicals. The three glycol ether compounds[172] for which there is the most toxicological information are 2-Methoxyethanol (see Table 2.15), 2-Ethoxyethanol, and 2-Butoxyethanol.[173]

However, many high molecular weight glycol ethers were exempted from designation as HAPs.[174]

The members of the glycol ethers category of HAPs are derived from ethylene glycol, diethylene glycol, and triethylene glycol.

*The HAP category does not contain glycol ethers based on propylene glycol, dipropylene glycol, or tripropylene glycol. Propylene glycol ethers are not HAPs* (see Footnote 15).

2.5.2.4.1 *Ethylene is Propylene with a Carbon Atom Missing* Joe DiMaggio was a Hall-of-Fame baseball player for the New York Yankees. After his baseball career, he became an international spokesman for a major home product. His younger brother, Dominick (Dom) DiMaggio, was a baseball player for the Boston Red Sox for eleven seasons. After his baseball career, he founded a plastics manufacturing company and became wealthy. Could Dom have replaced Joe in center field? Absolutely – he was an All-Star seven times! Could Dom have sold coffee makers? No way – Dom *who*? Could Joe have built and managed his own business? Well, he didn't.

Slight variances in heredity can produce significant differences in success, outcome, or behavior. Or the differences in success, outcome, or behavior can be almost negligible.

In addition to the DiMaggio brothers, glycol ether solvents are another example where the effects of slight variances in heredity both do and do not matter. Glycol ethers based on propylene glycol are:

- *Superior to* glycol ethers based on ethylene glycol in every area where there is concern about environmental, health, or safety issues.
- *Significantly different* in a few ways which may have significant consequences, such as solubility parameters or flash point.
- *Indistinguishable* in most other physical or chemicals characteristic.

---

2004, ecb.jrc.it/classlab/1503a3_FR_diethoxyethane.doc. The recommended classification for ethylene glycol methyl, dimethyl, ethyl, and diethyl ethers was Category 2 [R62] {Possible Risk Of Impaired Fertility}.

But another report is less concerned about another ethylene glycol ether. See "French Proposals for the Environmental Classification of 8 Glycol Ethers," *ECBI/01/01 Add. 8*, September 28, 2001, ecb.jrc.it/classlab/0101a8_FR-prop-EG.doc. The recommendation was to not change the outcome produced in 1997 because of lack of new data. For ethylene glycol butyl ethyl ether, this classification in ISBN 92-828-8398-1 was as R10 – Xn {*hazardous to health because of flammability*}; R21 – Xi {*irritant to skin*}; R36 {*irritating to eyes*}. This outcome does not describe a potent toxin.

Similarly, see ECBI/01/01, on January 26, 2001, ecb.jrc.it/classlab/0101_FR_egpe026.doc. The recommended classification was R10 – Xn {*hazardous to health because of flammability*} – R21 {*Harmful By Inhalation And In Contact With Skin*} – R36 {*Irritating To Eyes*} – [R66 {*Repeated Exposure May Cause Skin Dryness Or Cracking*}]. This outcome also does not describe a potent toxin.

This information, and a trove of international toxicology data, can be found at the European Chemicals Bureau (ECB), http://ecb.jrc.it/. At this site the International Uniform ChemicaL Information Database (IUCLID) can be found. Two areas deserve attention: "Existing Chemicals" and "New Chemicals."

[171] http://www.epa.gov/ttn/atw/glycol2000.pdf US EPA 745-R-00-004

[172] See reference M of Footnote 167. Acute exposure to high levels of the glycol ethers in humans results in narcosis, pulmonary edema, and severe liver and kidney damage. Chronic exposure to the glycol ethers in humans results in fatigue, lethargy, nausea, anorexia, tremor, and anemia. Medical studies are reported at National Toxicology Program (NTP), "Toxicological Studies of Ethylene Glycol Ethers: 2-Methoxyethanol, 2-Ethoxyethanol, 2-Butoxyethanol (CAS Nos. 109-86-4, 110-80-5, 111-76-2) Administered in Drinking Water to F344/N Rats and B6C3F1 Mice", *TOX-26*. 1993.

[173] On November 29, 2004, the US EPA delisted 2-Butoxyethanol (ethylene glycol monobutyl ether, EGBE) from the HAP list. See 69 FR 692988: "… judge that the potential for human health and environmental effects (from EGBE) is sufficiently low to provide reasonable assurance that such adverse effects will not occur. …"

[174] The glycol ethers category is defined by the following formula: $R\text{–}(OCH_2CH_2)_n\text{–}OR'$, where: $n = 1, 2$, or 3; R = alkyl C7 or less, or phenyl or alkyl-substituted phenyl; $R' =$ H or alkyl C7 or less, OR′ consisting of a carboxylic acid ester, sulfate, phosphate, nitrate, or sulfonate. Chemicals that meet this category definition are reportable. See p. 6 of Footnote 171.

**Table 2.15 Comparison of Ethylene-Based and Propylene-Based Glycol Ethers**

| | Methyl Ether | | Methyl Ether Acetate | |
|---|---|---|---|---|
| | **Ethylene Glycol** | **Propylene Glycol** | **Ethylene Glycol** | **Propylene Glycol** |
| CAS # | 109-86-4 | 107-98-2 | 110-49-6 | 108-65-6 |
| | 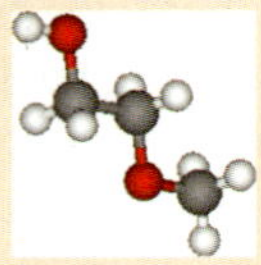 | 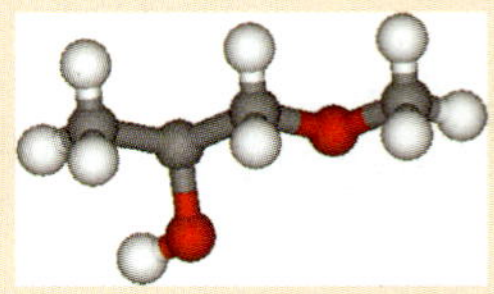 | 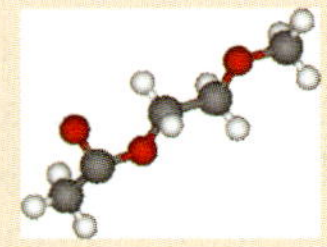 | 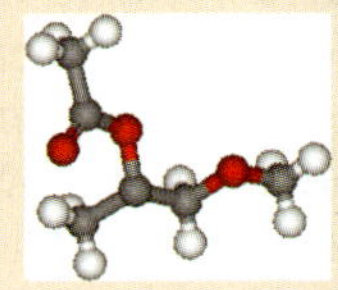 |
| Molecular formula | $CH_3OCH_2CH_2OH$ | $CH_3OCH_2CH_2$ $CH_2OH$ | $CH_3OCH_2CH_2O$ $(C{=}O)CH_3$ | $CH_3OCH_2CH_2$ $CH_2O(C{=}O)CH_3$ |
| Molecular weight | 76.1 | 90.1 | 118.1 | 132.2 |
| Density (g/ml) | 0.964 | 0.921 | 1.004 | 0.968 |
| Boiling point (°F, °C) | 255.2, 124.0 | 248.0, 120.0 | 293.0, 145.0 | 294.4, 145.8 |
| Flash point, closed cup (°F, °C) | **104.0, 40.0** | **89.6, 32.0** | 120.2, 49.0 | 108.0, 42.0 |
| Autoignition Temperature (°F, °C) | 285.0, 545.0 | 270.0, 518.0 | 410.0, 770.0 | 631.00, 333.00 |
| Evaporation rate, *n*-BA = 1 | .53 | 0.62 | 0.35 | 0.333 |
| HSP disperse | 16.2 | 15.3 | 15.9 | 15.6 |
| HSP, polar | 9.2 | 8.0 | 5.5 | 5.6 |
| HSP, $H_2$-bonding | 16.4 | 13.9 | 11.6 | 9.8 |
| Viscosity (cps) | 1.8 | 1.9 | 1.1 | 0.80 |
| Refractive index | 1.4020 | 1.4035 | 1.4030 | |
| Antoine equation constants, A & B, Log (Pv, °K) (mmHg & °C) | A = −2292.1, B = 8.6573 | A = −2104.8, B = 8.2554 | A = −2425.9, B = 8.7181 | A = −2376.3, B = 8.5531 |
| Surface tension (dyne/cm) | 30.8 | 27.7 | | 26.9 |
| HAP by US EPA | YES | NO | YES | NO |
| LEL (vol%) | 1.8 | 1.6 | 1.5 | 1.5 |
| UEL (vol%) | 14.0 | 13.8 | 12.3 | 7.0 |
| Exposure limit (ppm) | 25 (OSHA) | 100 (ACGIH) | 25 (OSHA) | 100 (AIHA) |
| NFPA, health | 1 | 0 | 1 | 0 |
| NFPA, flammability | 2 | 3 | 2 | 2 |
| NFPA, reactivity | 1 | 0 | 0 | 0 |
| Vapor hazard rating (HR), @ 20°C | 327.9 | 98.7 | 106.8 | 36.8 |
| Molar vapor concentration @ flash point (%) | 1.09 | 0.95 | 0.49 | 0.29 |

Two examples are in Table 2.15.

- Non-performance benefits can accrue from switching away from ethylene-derived to propylene-derived glycol ether solvents: higher exposure limits, reduced (almost) NFPA hazard ratings, reduced vapor hazard ratio, *and* avoidance of the US EPA HAP classification.
- Many significant performance properties are unchanged: density, viscosity, evaporation rate, and boiling point.
- The Hydrogen-bonding HSP for both methyl ether *and* methyl ether acetate solvents is consequently reduced (changed) for those which are propylene-derived versus those derived from ethylene. Significance of this change is application-specific.
- The flash point of the propylene-derived methyl glycol ether solvent is reduced below 100°F, while that for the ethylene-derived solvent is barely above 100°F. Since the "bright line" between the safety classification of flammable and combustible is at 100°F, the NFPA flammability rating[175] for the ethylene-derived methyl glycol ether solvent is superior to that of the propylene-derived one.

In summary, users can usually swap with impunity the ethylene-derived glycol ether solvent for the propylene-derived one and avoid the HAP classification by the US EPA. But as do parents of porcupines advise their young about mating, caution is advised.

2.5.2.4.2 *Management Guidance* As managers of cleaning or other operations, readers of this book should explain why their operations include ethylene-derived glycol ether solvents (HAPs) if propylene-derived glycol ether solvents (non-HAPs) will provide equivalent or similar performance.

It is acceptable for managers to empower or allow site use of chemicals deemed HAPs or similarly considered. Engineering controls, personal protective equipment, and peer-reviewed exposure limits are the tools by which staff are protected when hazardous chemicals are used.

But this author doesn't believe it acceptable to use ethylene-derived glycol ether solvents (HAPs) when propylene-derived ones (non-HAPs) are available without economic penalty. In other words, just because one can do something doesn't mean they should.

### 2.5.3 Concern Outside of the US

Pollution is pollution. Four of the six criteria pollutants which command interest of the US EPA also command attention by the World Health Organization (WHO).[176] This attention is manifested in the form of global air quality guidelines:[177]

- Updated air quality guidelines for Europe were to have been established by the end of 2005. They will replace global guidelines issued in 2000,[178] and be based on current medical knowledge and atmospheric sampling results. An interim report suggests the 2005 guidelines will not include Lead or CO. This was because it was believed that the toxicological mechanisms of CO and Lead were well established and were used as the basis of existing air quality guidelines.
- Final recommendations for guidelines will include PM, $O_3$, $NO_2$, and $SO_2$.
- VOCs are treated less as single chemicals but as symptoms of the need for improved control of evaporation.[179]

Past guidelines[180] offered specific limits on air concentrations for organic pollutants: acrylonitrile, benzene, butadiene, carbon disulfide, carbon monoxide, 1,2-Dichloroethane, dichloromethane (methylene chloride), formaldehyde, polycyclic aromatic hydrocarbons, polychlorinated biphenyls, polychlorinated dibenzodioxins and dibenzofurans, styrene, perchloroethylene, toluene, trichloroethylene, and vinyl

[175] See http://www.sefsc.noaa.gov/HTMLdocs/nfpahazard.htm, OSHA standard 1926.407, Hazardous (Classified) Locations, or NFPA 70 – the US National Electrical Code.

[176] *WHO Air Quality Guidelines – Global Update Planning Meeting*, 11 January 2005, London, United Kingdom. Available for free at http://www.euro.who.int

[177] These guidelines were scheduled to be available by the end of 2005. As of this writing, details were available at http://www.euro.who.int/air/activities/20050624_2. As of January 6, 2006, no update was available.

[178] *Air Quality Guidelines for Europe*, WHO Regional Publications, European Series, No. 9, 2000, ISBN 92 890 1358 3. Available for free at http://www.euro.who.int

[179] See Section 2.2.2 for the European perspective about management of VOCs, versus that of the US EPA.

[180] *WHO Regional Publications, European Series* No. 91, 2nd (ed.), ISBN 92 890 1358 3.

chloride. As well as Inorganic pollutants include: Arsenic, asbestos, Cadmium, Chromium, fluoride, hydrogen sulfide, Lead, Manganese, Mercury, Nickel, Platinum, and Vanadium. *But* also singled out for attention were the same materials generating concern five years later: PM, $O_3$, $NO_2$, and $SO_2$.

So it isn't that some countries, regions, or organizations see hazards others don't see. Differences in concentration limits are only different by degree of emphasis. Obviously local regulations dominate local affairs. But legislators and environmental regulators in every country recognize (1) the same hazards, and (2) that pollution from one country affects citizens from other countries, and interacts with pollution produced by other countries. What goes around comes around.

While the WHO has no legislative or enforcement standing, its recommendations herald future legislation and enforcement actions. Managers of cleaning systems are wise to be familiar with current local requirements, *and* expectations of future requirements via current WHO guidelines.

# 3

# Health and safety hazards associated with cleaning agents

**Chapter contents**



SHEA has become a common description of a senior (usually) management position in many operating organizations. Here SHEA, or variations of this jargon, stands for Safety, Health, and Environmental Administrator. In other words, safety and health issues have carried the same weight as have environmental issues in the minds of enterprise managers.

That's as it should be.

That's also easier said than done – which is why all the three issues are managed by the same administrator. All managers want to select cleaning agents and other chemicals with no safety, health, or environmental issues. With a few exceptions, that's unlikely.[1]

Consequently, the SHEA person must be able to manage *compromise*. An example is the interchange of parachlorobenzotrifluoride (PCBTF) for isopropranol, or the reverse. PCBTF has a good environmental characteristic (VOC exempt in the US) and an acceptable flammability characteristic (Class II) but its exposure limit is low (25 ppm). IPA is a VOC, is more flammable (Class IB), and its exposure limit (400 ppm) raises few concerns. Which would your prefer – *if* both would meet the application needs?

This chapter is about the SH of SHEA, and how to use available information to make those compromises. The E was covered in Chapter 2.

## 3.1 GENERAL HEALTH AND SAFETY HAZARDS

Cleaning agents are chemicals, or formulated mixtures of chemicals. Their hazards are essentially the characteristics of those chemicals[2] used in the formulation.

This is a fortuitous outcome. Hazards of a mixture can be assessed through a list of ingredients in

[1] A chemical would be unique if it was aggressive to soils, not to surfaces, and not to humans. Pure water and $CO_2$ are examples of such unique chemicals used as cleaning agents. Even so, use of either of these cleaning chemicals brings compromise about compatibility with latent soils or feasibility of a cleaning process.

[2] Almost without exception, the components of formulated mixtures are compatible – else they wouldn't be useful products. One wouldn't clean with a mixture of incompatible components. One of the incompatible components might be retained on the cleaned surface and act as a new soil! Hence, synergy or reaction between chemical components of formulated cleaning agents is seldom seen.

the mixture, and an analysis of the hazards associated with each ingredient. More than 30 years ago, the material safety data sheet (MSDS) was created to be a standardized ingredient list, and a description of hazards of each ingredient on the list:[3]

- The chief *safety* hazard associated with cleaning agents is that they may ignite (catch fire).
- The chief *health* hazard associated with cleaning agents is that contact with human tissue may damage that tissue.

### 3.1.1 General Methods for Assessing Hazards

Technologists use the same approach to assess both the safety and health hazards associated with a chemical.

That approach is to provide an exposure and observe if the exposure produces damage associated with the hazard:

- *Fire production* is assessed by exposing the chemical to a standard set of conditions, and observing if combustion associated with fire is produced. Two common sets of conditions are those involved in measurement of flash point and of explosivity.
- *Tissue damage* is assessed by either exposing non-human tissue, observing damage, and extrapolating to human tissue; or by examining damage where human tissue has been (usually inadvertently) exposed. The former is known as laboratory testing. The latter is known as epidemiological analysis.

Said another way, flash point data and lethal concentration ($LC_{50}$) for inhalation exposure data allow the same thing – relative ranking of a hazard to produce damage.

Neither flash point nor $LC_{50}$ data specifically refer to a situation you manage. Fire or respiratory damage from use of chemicals does result from how you manage the use of chemicals. If staff you manage uses chemicals to extinguish cigarettes or "get high," flash point or $LC_{50}$ data have no meaning to you.

## 3.2 FLASH POINT

Simply, flash point refers to ignition of the chemical to produce a fire. Fire is the first major hazard associated with the use of cleaning agents (chiefly solvents) in cleaning (or other) operations.

The flash point of a chemical has no direct effect on its capability as a cleaning chemical.

There is a paradox about flash point data. As data, it is meaningless (see Sections 3.1.1, 3.4, and 3.5) and often breeds confusion. But its impact can be dominant (see Section 3.16.1, Figure 3.28, and Table 3.45)! This is because of its direct effect on selection of the cleaning process, cleaning equipment, and cleaning procedures, as well as packaging, transportation, selection, and disposal of the cleaning solvent.

### 3.2.1 Definition of Flash Point

Flash point is the lowest temperature at which a solvent can form an ignitable mixture in air near the surface of the liquid. It is easier to ignite the liquid solvent the lower is its flash point. Specific details about how to measure flash point are found in Occupational Safety and Health Administrations (OSHA's) Standard 1910.106, part of which is excerpted below.[4] See also NFPA 30.

Measurement of flash point does not require production of a stable flame, it only requires ignition of

---

[3] See Section 3.19 for a detailed analysis of how and how well the MSDS has provided these two functions.

In most cases MSDSs present a list of ingredients, individual hazards of each ingredient within the mixture (product), and a formulation showing the proportion of each ingredient present. Ingredient proportions are usually stated in general terms or a range is given. This is done to keep competitive firms from producing the same product. Vagueness in composition can lead to a vagueness about the hazards of the formulated product.

But in some cases, enough is known about the formulated product, and its formulation is relatively unchanging, that the hazards of the formulated product are presented.

[4] (a)(14) "Flash point" means the minimum temperature at which a liquid gives off vapor within a test vessel in sufficient concentration to form an ignitable mixture with air near the surface of the liquid, and shall be determined as follows:

(a)(14)(i) For a liquid which has a viscosity of less than 45 SUS at 100°F (37.8°C), does not contain suspended solids, and does not have a tendency to form a surface film while under test, the procedure specified in the Standard Method of Test for Flash Point by Tag Closed Tester (ASTM D-56-70), which is incorporated by reference as specified in Sec. 1910.6, shall be used.

vapor and fuel.[5] The actual measurement which defines ignition is usually a temperature rise[6] produced by the combustion.

### 3.2.2 Practical Meaning of Flash Point

The practical use of flash point data is to rate the position of a chemical relative to local, state, national, and international regulations. This point is written in a common procedure for measurement of flash point:

> … Flash point measures the tendency of the specimen to form a flammable mixture with air under *controlled laboratory conditions. It is only one of a number of properties that shall be considered in assessing the overall flammability hazard of a material.*[7]…

A flash point measurement has no intrinsic value. The test, while extremely significant, is entirely artificial. It is used only in multiple hazard rating systems (see Section 3.16 and Table 3.25).

Measured results are intended to be comparative, because the test procedures require calibration against reference fluids.[8]

At least two misunderstandings have developed about flash point testing and results.

#### 3.2.2.1 *False Understanding 1*

Many users feel that "… if it can be ignited, it can burn …". That is, a solvent with a flash point will sustain combustion (Table 3.1).

**Table 3.1 False Understanding 1**

| False Understanding | The Truth |
|---|---|
| Existence of a flash point via a standard test means that the solvent will sustain combustion in "real world" operations. | A solvent which can be ignited in a flash point test **may or may not sustain** combustion and produce a stable flame outside these test conditions. |

A laboratory flash point tester produces an artificial and controlled environment. The environment is intended to be one of vapor (in a closed or open container) in equilibrium or saturated with the test solvent. An electrically produced ignition or a gas-fired flame is inserted for a period of about 1 second into this environment and a flash (*not* a stable flame) is or is not observed.[9]

Maintenance of a stable flame involves issues other than those involved with igniting a cloud of saturated vapor. A flame will be stable when:

1. Fuel is continually supplied by some mechanism.
2. Air is continually supplied by another mechanism.
3. Fuel and air remain within a relative composition range which supports combustion.
4. Heat is removed at a suitable rate (too high a rate of heat removal will reduce temperature and will quench the flame, too low a rate of heat removal will allow temperature to increase and results may not be comparable to other results).

---

(a)(14)(ii) For a liquid which has a viscosity of 45 SUS or more at 100°F. (37.8°C), or contains suspended solids, or has a tendency to form a surface film while under test, the Standard Method of Test for Flash Point by Pensky-Martens Closed Tester (ASTM D-93-71) shall be used, except that the methods specified in Note 1 to Section 1.1 of ASTM D-93-71 may be used for the respective materials specified in the Note. The preceding ASTM standards are incorporated by reference as specified in Sec. 1910.6.

(a)(14)(iii) For a liquid that is a mixture of compounds that have different volatilities and flash points, its flash point shall be determined by using the procedure specified in paragraph (a)(14) (i) or (ii) of this chapter on the liquid in the form it is shipped. If the flash point, as determined by this test, is 100°F. (37.8°C.) or higher, an additional flash point determination shall be run on a sample of the liquid evaporated to 90% of its original volume, and the lower value of the two tests shall be considered the flash point of the material.

(a)(14)(iv) Organic peroxides, which undergo auto-accelerating thermal decomposition, are excluded from any of the flash point determination methods specified in this subparagraph.

For more details, see http://www.osha-slc.gov/SLTC/smallbusiness/sec8.html

[5] Measurement of a flash point does not require the flame produced in solvent vapor by an ignition device to remain ignited (sustain burning).

[6] Fire point is a corollary term. It means the temperature at which a stable flame continues to burn. Normally, the fire point value (a temperature) is slightly higher than the flash point. Fire point requires burning of a flame for a duration of 5 seconds. See ASTM D-1310-86, 3.2.1. Flash point measurements do not require a stable flame. For low flash point liquids (Class I liquids – flash point <100_F) flash and fire points are generally so close as to be considered the same. For higher flash point liquids a distinct difference is noted. In general, one would expect the difference to be greater with increasing flash point.

[7] ASTM D56-02a, Section 5.1.

[8] See ASTM 56-02a, Chapter S2 Verification of Apparatus Performance.

[9] By either visual means, detection of a temperature rise, or detection of ionized materials present in a flame.

**Table 3.2 False Understanding 2**

| False Understanding | The Truth |
|---|---|
| Absence of the existence of a flash point via a standard test means that the solvent will not sustain combustion in "real world" operations. | A solvent which cannot be ignited in a flash point test **may or may not** be ignited at conditions other than those in a flash point test. |

5. An inert is not added which distorts the solvent/air composition ratio.
6. A flame front is not reflected from a containing surface back into the ignition source.

*None of these issues is managed in a flash point test apparatus* – because maintenance of a stable flame is not a goal of the flash point test.

### 3.2.2.2 *False Understanding 2*

Still more users feel that "… if it has no flash point, it can't burn …".

This point is also written in a common procedure for measurement of flash point (Table 3.2):

> … There are instances with pure materials where the absence of a flash point does not ensure freedom from flammability. Included in this category are materials that require large diameters for flash propagation, such as trichloroethylene. This material will not propagate a flame in apparatus the size of a flash point tester, however, its vapors are flammable and will burn when ignited in apparatus of adequate size.[10]

The same point is made in a literature article:

> … If an incorrect technique is used for the material in question, a result is obtained that is not valid. … All erroneously appear to have a flash point which the incorrect method is used.[11]

Here the authors are noting that cleaning solvents including *n*-propyl bromide, trichloroethylene, methylene chloride, HFE-72 DE, HFE-71 DE, and HFE-71 DA all appear to produce results which define a flash point when flash point testing is done outside of the ranges of American Society for Testing and Materials (ASTM) D93. Yet, when the testing is done within the test ranges of ASTM D93, no flash point is reported.[12] This reference is simply stating that some cleaning solvents which produce no flash point in the preferred test *can be ignited.*

**Table 3.3 Meaning of Flash Point Test Results**

| Result | Regulatory impact | Practical meaning |
|---|---|---|
| Measured value | **Solvent classification as per Section 3.16** | A stable flame may or may not be able to be sustained |
| No measured value | Solvent not classified | **The solvent may be able to be ignited** |

In summary, flash point results are often misunderstood. The numerical values are meaningless – relative to ignition potential.

But flash point values are priceless when these results are used in regulatory classification schemes about transportation, storage, allowable types of electrical equipment, and disposal methods. Results of the various regulatory classification schemes often dominate choice in a cleaning application. This situation is given in Table 3.3.

In other words, flash point measurements are like your weight – meaningless, until you can't fit into your clothes.

## 3.2.3 Two Factors Influencing Flash Point

Chemicals with the highest level of ignition hazard have the lowest flash points. For example, automotive gasoline has a flash point of −40°C (−40°F); 10–30 W motor oil has a flash point of +210°C (+410°F).

Note the phrase ignitable mixture in the definition of flash point. It is the vaporized liquid, not the liquid itself, that is ignited.

[10] See ASTM 56-02a, Chapter X2 Flash Point Test and Flammability of Mixtures.

[11] Kanegsberg, B.F. and Shubkin, R.L., "Solvent Flammability Basics," *Clean-Tech Magazine*, November/December 2003, p. 18.

[12] Here the "wrong test ranges" are the use of flash point test equipment outside the recommended range of use temperature.

Flash point is related to:

- *Vaporization*: In general higher levels of vapor generation, at the same temperature, produce increased flammability. In other words, solvents with high vapor pressures are characterized by low flash points.

See Figure 3.1 (or Figure 3.9) which shows this effect for oxygenated solvents:

- *Reactivity*: This is reactivity of the solvent vapor with Oxygen in air. Liquid Argon has no flash point, despite its rapid vaporization. Argon does not react with Oxygen in air. Liquid methane, which also rapidly vaporizes, does react with oxygen in air and has a low flash point (−306°F [−188°C]).

See Figure 3.2[13] which shows how smaller molecules, having lower molecular weights, are more difficult to ignite than are solvents with a higher molecular weight.

A more flammable solvent is one which can be ignited at a lower concentration. In other words, a more reactive solvent is more flammable.

For the same level of reactivity (atomic composition), chemicals which are more volatile have lower flash points. Other specific effects are difficult to discern from this figure.

At any rate of evaporation it is the chemical structure of the solvent molecule that determines reactivity with Oxygen – whether or not it has a flash point, and at what molar concentration.

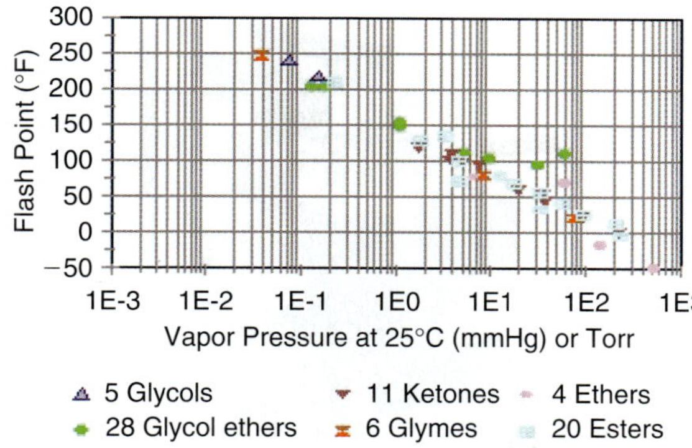


**Figure 3.1**

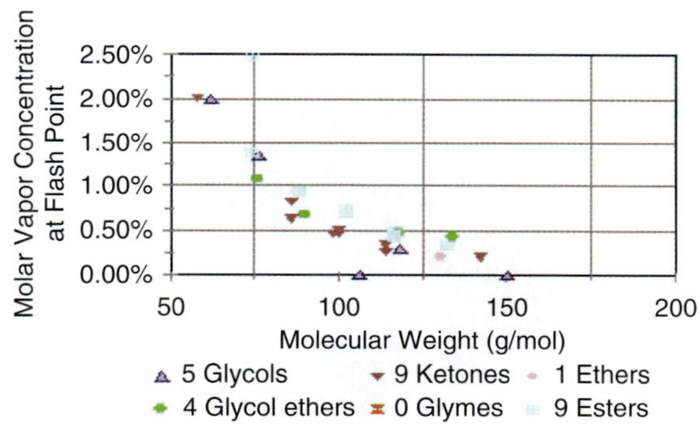


**Figure 3.2**

## 3.2.4 Test Equipment

Flash point test equipment has been designed to simulate the various ways in which solvents are used.

One model, the Pensky-Martens closed-cup tester, is used for testing viscous paints and coatings.

### 3.2.4.1 *General Types of Equipment*

The majority of solvent cleaning work, relative to flash point, is of two types. They are work done in an open tank or a closed tank:

1. An open tank would typically be used for cleaning at less than the boiling point (cold cleaning). The potential for ignition of solvents in cold cleaning, under shipment, or storage conditions is assessed by measuring flash point in an open-cup tester. One model of a flash point tester used for open-cup testing is shown in Figure 3.3.
2. A closed tank would include a storage or shipping container. A vapor degreaser is a closed tank, in this sense. The potential for ignition of solvents under cold cleaning conditions is assessed by measuring flash point in a closed-cup tester.

A closed-cup tester is intended to be operated with the vapor saturated with solvent held within the tester. One model of a flash point tester used for closed-cup testing is shown in Figure 3.4.

Obviously, the same solvent will give different absolute results, *and may give different results*

[13] The same oxygenated solvents form the data set for both Figures 3.1 and 3.2. The parameter plotted on the vertical axis in Figure 3.2 is the calculated amount of solvent vapor present at the point of ignition in a flash point test. The units are molar (volume), not weight.

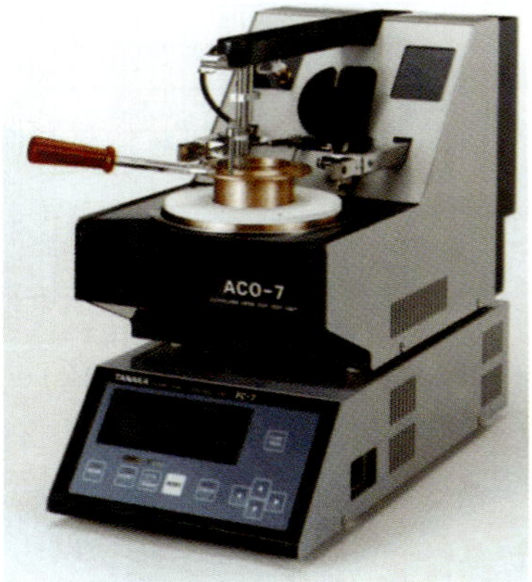


**Figure 3.3**

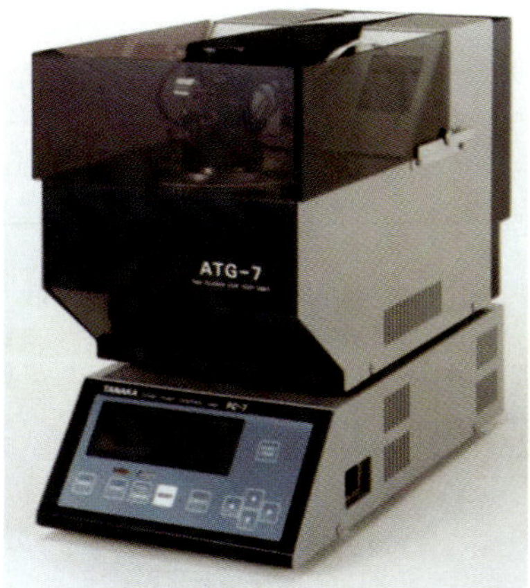


**Figure 3.4**

*relative to another solvent*, if tested in an open-cup or a closed-cup tester.

The closed-cup method prevents vapors from escaping. So, the open-cup tester will lose the most volatile components.

Thus open-cup flash points are higher than those for the same solvent measured in the closed-cup

**Figure 3.5**

tester. When several values are available, the lowest temperature is usually taken in order to assure safe operation of the cleaning process:

- Don't ever accept flash point information if the type of tester is not supplied as well. A manager must know if the test was done with an open-cup or a closed-cup tester.

#### 3.2.4.2 *Flash Point Equipment/ Procedures*

Flash points are determined experimentally by heating the liquid in a container (cup), and then introducing a small flame just above the liquid surface. The temperature at which there is a flash/ignition is recorded as the flash point.[14]

Test equipment and procedures are specified by the International Standards Organization (ISO).[15] Four commonly used sets of equipment and procedures are shown in Table 3.4.

---

[14] Per ASTM D56-02, Introduction "... Flash point values are a function of the apparatus design, the condition of the apparatus used, and the operational procedure carried out. Flash point can therefore only be defined in terms of a standard test method, and no general valid correlation can be guaranteed between results obtained by different test methods, or with test apparatus different from that specified. ..."

[15] Flash point test procedures strictly controlled by ASTM standards are:

D56 Tag Closed Tester
D92 Cleveland Open Cup
D93 Pensky-Martens Closed-Cup Tester

**Table 3.4 Comparison of Flash Point Test Equipment/Procedures**

| Name | Test Fluids | Temperature Range (°C) | Cup: Open/ Closed | Detection Method | ASTM test | Ignition Source |
|---|---|---|---|---|---|---|
| TAG | Cleaning solvents, etc. | −20 to +80 | Open/closed | Temperature rise | D1310 (open), D56-02 (closed) | Gas or electric ignitor |
| Setaflash | Diesel, kerosine, lacquers | 0 to 220 | Open/closed | Visual | D3278-82 (closed) | |
| Pensky-Martens | Paints, coatings | 60 to 360 | Closed | Temperature rise | D93 | |
| Cleveland | Oils, bitumens | >360 | Closed | Ionization, detection | D92 | |

Flash points for cleaning solvents are normally measured by the TAG[16] closed-cup test (TAG CC or CC).

## 3.3 EXPLOSIVE LIMITS

Both flammability and flash point relate to combustion of hydrocarbon-based (containing at least Hydrogen and Carbon Atoms) chemicals with Oxygen in the air. That's something managers of cleaning chemicals want to avoid in all operations!

Section 3.3 is a companion to Section 3.2 in that both cover a screening test to help manage chemical cleaning agents which can be ignited.

### 3.3.1 Combustion Basics

The combination of conditions that can allow an explosion[17] are known as flammability.

Flammability is not the same as flash point. But both are measurements about combustion. Both are important to users of chemicals in cleaning operations.

Three factors must be present in a situation for combustion to occur.[18] All three are present in both flash point and flammability testing. All three can be present in cleaning operations:

1. There must be a **fuel** present. This is the vaporized chemical, containing at least Hydrogen and Carbon atoms.
2. There must be a source of **Oxygen**. This is usually air. But it could be pure Oxygen in a closed system. Simply, this is why hydrocarbon solvents are not used to clean tubing used for National Aeronautics and Space Administration's (NASA's) liquid Oxygen rocket fuel or the US Navy's deep sea breathing mixtures.
3. There must be a **source of ignition**. This is usually a spark produced unintentionally by mechanical means. But it can be an electrical discharge produced intentionally, or an unexpected hot surface, or transfer of a large amount of static electricity.

---

D1310 Tag Open-Cup Apparatus
D3143 Cutback Asphalt with Tag Open-Cup Apparatus
D3278 Closed-Cup Apparatus
D3828 Small-Scale Closed Tester
D3941 Equilibrium Method with Closed-Cup Apparatus

[16] The abbreviation TAG refers to the person who developed the test (Tagliabue).

[17] For purposes of managers of cleaning operations, there is no difference between a fire and an explosion. BOTH are to be avoided with equal fervor.

But a technical difference is that a fire is a combustion process which produces heat and light while an explosion is any process which produces a rapid significant expansion of a large volume of gas, and possibly some heat and light as well.

[18] This is why the strategy for prevention of fires and explosions is to use equipment and procedures to assure that a *minimum* (and hopefully two) of one of these three factors *cannot be present* in managed situations.

### 3.3.2 It's the Concentration

The outcome in a flammability test, flash point test, or cleaning operations can be a combustion reaction. This occurs in the vapor phase[19] and can be generally described as below. It is the release of energy which can produce damage:

$$C_WH_XB_YO_Z + (W + \{½ \cdot ½ \cdot X\} - Z)\, O_2 - >>>$$
$$W \cdot CO_2 + X \cdot H_2O + Y \cdot B_2 + \textbf{Energy}$$

Here, *W* and *X* are the number of atoms of Carbon (C) and Hydrogen (H) in the hydrocarbon chemical molecule. *Y* and *Z* are the number of atoms of halogen (B) and Oxygen (O) if the molecule has other atoms substituted for Hydrogen:[20]

- A halogen atom will usually retard the combustion (see Section 3.4.1).
- An Oxygen atom will usually aid the combustion reaction.

A vapor mixture is flammable when there are the correct proportions of fuel vapor and Oxygen. If there is too much or too little of either fuel or Oxygen, a mixture won't be flammable.

The concept of concentration limits is a crucial one. This means a mixture which has too much fuel (rich mixture[21]) or too much Oxygen (lean mixture[22]). Combustion will only happen, given an ignition source, if the Oxygen/fuel ratio is within certain limits.[23]

### 3.3.3 Flammable Concentration Limits

For simplicity, those ratio limits are expressed in terms of fuel (chemical) concentration.

The minimum chemical (fuel) concentration in air is called the lower explosive[24] limit or LEL. The maximum fuel concentration in air is called the upper explosive limit or UEL:

- A match held over a chemical of low volatility *may not produce a flame* because enough chemical hasn't vaporized for the LEL to be exceeded.
- A match dropped in a liquid chemical *probably won't produce a flame* at least because the chemical concentration in air is well above the UEL.
- A match exposed to a chemical air mixture *will produce a flame* when the concentration in air[25] is above the LEL and below the UEL.

Any covered vessel containing liquid chemical produces a chemical (fuel) mixture in air when the liquid evaporates. The vessel can be a railroad tank car, a paint can, a storage tank, a vapor degreaser, transfer piping, or a cold cleaning tank. The location of this potentially flammable mixture in the vessel is called the headspace, or less commonly, ullage.

---

[19] Combustion in the liquid phase is extremely difficult to initiate because of the difficulty of achieving a substantial concentration of Oxygen dissolved into a liquid, and the relatively high heat dissipation in a liquid relative to a vapor. One can foolishly extinguish cigarettes in liquid gasoline. But one is betting that combustion does not take place in the vapor phase above the liquid before the cigarette is immersed in the liquid gasoline.

[20] The chemical described as $C_WH_XB_YO_Z$ is imaginary, and used only for general illustration, as is the letter B used to represent halogen atoms. For non-halogenated chemicals, B is zero. For non-oxygenated chemicals, Z is zero.

The necessary amount of Oxygen (not air) needed to completely react with a hydrocarbon chemical is $(W + \{½ \cdot ½ \cdot X\} - Z)$ per unit of hydrocarbon chemical. Here, the units are vapor volumes of each, or moles of each. The chemical engineering term for this amount is called the stoichiometric amount.

For pure hydrocarbon cleaning agents, such as hexane which is $C_6H_{14}$, $(W + \{½ \cdot ½ \cdot X\} - Z) = (6 + \{½ \cdot ½ \cdot 14\} - 0) = 9.5$ volumes of pure Oxygen per volume of hexane vapor. Since air is approximately 1/5 (20%) Oxygen by volume, the stoichiometric concentration of hexane in air is $(1/9.5)/5 = 2.1$ volume or mole %.

[21] A *rich mixture* (too much fuel) is one with less Oxygen than the stoichiometric amount needed for combustion. It is deficient in Oxygen relative to fuel. Such a mixture may well be ignited by a spark, but will not sustain combustion because the Oxygen present is depleted in the reaction below the level needed to support combustion.

[22] A *lean mixture* (too little fuel) is one with more Oxygen than the stoichiometric amount needed for combustion. It is deficient in fuel relative to Oxygen. Such a mixture may well be ignited by a spark, but will not sustain combustion because the fuel is depleted in the reaction below the level needed to support combustion.

[23] This is why a leak of a small amount of natural gas from a cooking stove normally does not immediately ignite. Good ventilation, via dilution of emissions with additional air, is a cheap and effective safety practice both home cooking and shop cleaning.

[24] Conventionally, the word explosive is used instead of the word combustion – probably for emphasis about the possible outcome. The terms "flammable limit" and "explosive limit" are used interchangeably.

[25] Obviously, this concentration is an increasing function of both temperature and pressure.

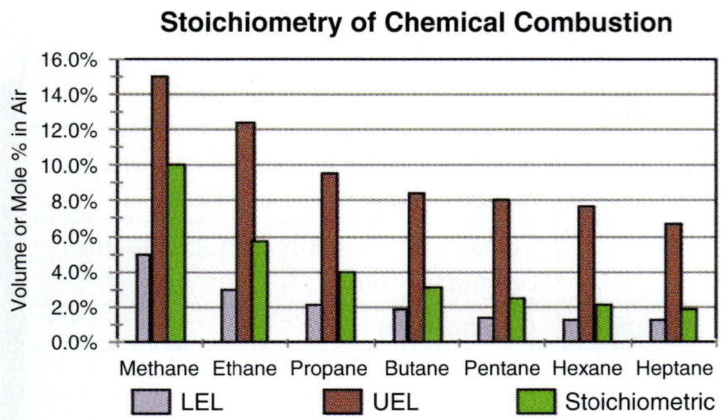


**Figure 3.6**

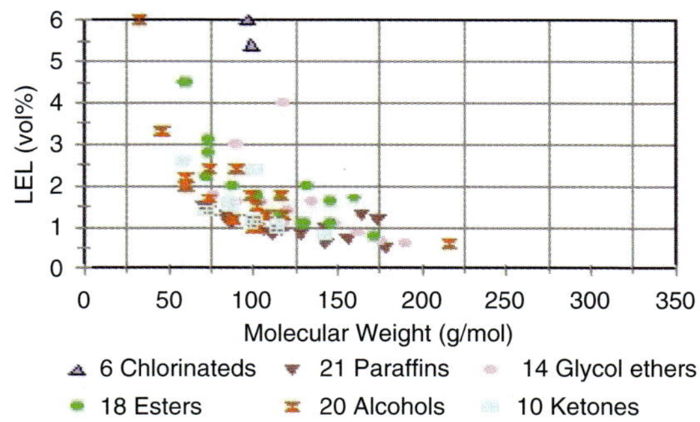


**Figure 3.7**

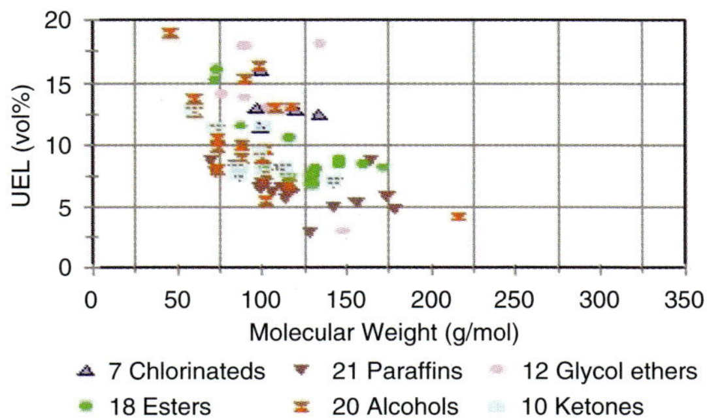


**Figure 3.8**

The stoichiometric concentration lies between the LEL and the UEL.[26] This is shown for common hydrocarbons in Figure 3.6. Note that the LEL is slightly below the stoichiometric concentration of fuel. This means that the combustion of hydrocarbon isn't complete, but is empirically stable and repeatable.

Flammability limits are a function of molecular weight as much as any other molecular characteristic.[27] This is shown in Figures 3.7 and 3.8.

### 3.3.4 Flammability Limits versus Flash Point

Table 3.5 compares these two measures of potential for ignition.

Note in this table the reason why *both* flash point and flammability limit are considered in safety evaluations:

- Flash point is intended to be a simulation of heating in an open or closed container, such as a vapor degreaser.
- Flammability (explosion) limit is intended to be a simulation of storage in a closed container.

Heating and storage compromise nearly all the ways in which a chemical might be used (see Section 3.3.6).

### 3.3.5 Flammability Test Equipment

The ASTM E-681 apparatus (see Figure 3.5) consists of a spherical glass flask of a specific volume containing a magnetic stirrer for mixing the materials and an ignition source, typically a match, spark, or hot wire. The top of the flask is sealed with a rubber stopper equipped with inlet tubes for air and solvent (fuel). The flask is enclosed in an insulated chamber and positioned above a magnetic stirrer drive.

For each test, the vessel is evacuated and precise amounts of test gases, measured by partial pressures, are added. Inlet tubes are then closed and the ignition source is triggered. "The upward and outward propagation of the flame away from the ignition source is noted by visual observation. The concentration of the flammable component is varied between trials until the composition which will just sustain propagation is determined.[28]"

The flammability limit is determined by finding the lowest concentration of fuel vapor that will result in flame propagation for a given spark. As the spark is made weaker, the concentration of fuel vapor has to be increased for the mixture to remain flammable. The strength of a spark igniter is measured

[26] An LEL and UEL are also known as lower flammability limits (LFL) and upper flammability limits (UFL).

[27] Flash point and LEL data are from the reference in Footnote 44. (Section 3.4.4) Also see Footnote 29 (Section 3.3.5.1).

[28] ASTM E 681-01, Standard Test Method for Concentration Limits of Flammability of Chemicals (Vapors and Gases).

**Table 3.5 Two Different Measures of Ignition Risk**

| Item | Flash Point | Flammability Limit |
|---|---|---|
| Situation under study | Heating of a solvent in an open or closed container | Storage with vaporized solvent in a closed container |
| Variables held constant in test | Strength of ignition source | Temperature (usually 25°C), strength of ignition source |
| Variables in test | Temperature directly and composition indirectly | Composition |
| Test temperature | As necessary to produce a flash | Usually 25°C |
| Test fixture volume | Small cup (~few milliliters) | Large spherical vessel (~5 l) |
| Detection method | Visual recognition of flame in manual testers, temperature or pressure rise in automatic testers | Visual recognition of flame |
| ASTM standards | D1310 (open) or D 56-02 (closed) for Taglibue tester | E-681 |
| Test outcome | Temperature at which a flame is first produced | Volumetric composition at which a stable flame front is produced |
| Reason for testing | Testing by solvent manufacturers to characterize product before it is released | Testing on site by users of headspace composition (ventilation requirements) versus LEL adjusted with safety factor |
| Use of outcome | Choice of shipping and packaging methods; design and specification of cleaning equipment and installation details; waste disposal method | Modification of site equipment and procedures to avoid unsafe condition which could produce explosion |

in terms of the stored electrical energy (joules) used to create the electrical discharge. The lower flammability limit is typically measured using a spark of energy 10–100 J. This is comparable to the arc created by a short circuit in household wiring.

#### 3.3.5.1 *Portable Flammability Test Equipment*

Traditionally, gas detectors such as explosimeter have been used to check that enclosed spaces are "gas free." They have also been used to measure the flammability of headspaces in terms of percentage of the lower flammable limit (LEL).[29]

Such detectors rely on a calibration carried out normally on a single hydrocarbon (e.g. methane) which may have LEL characteristics that are far removed from the hydrocarbons actually present in the headspace. When using an explosimeter to assess the degree of hazard in residual fuel oil tank headspaces, the instrument is usually calibrated with a pentane/air or hexane/air mixture. Calibration should be done as Al Capone reportedly advised citizens in Chicago to vote "early and often".

### 3.3.6 Screening Tests

The reason for these two measures is to simulate two different types of application which bracket how chemicals are commonly used (storage and heating). They are screening tests for ignition risk.[30]

[29] The US National Fire Protection Association (NFPA) normally recommends that measured concentrations of ignitable materials not exceed 25% of the LEL. This defines a safety factor of 4 to 1 for such measurements. (see Sections 3.16.1 and 3.16.2 for additional details).

[30] If a chemical has no flash point and no measured LEL or UEL values, that doesn't mean it can't react with Oxygen to produce energy. Under extreme levels of ignition energy, temperature, pressure, or configurations of equipment, reaction with Oxygen *can* occur at nearly all levels of composition.

Managers of fire safety have accepted that, for non-extreme conditions, a chemical won't react with Oxygen to produce energy or gas volume if it has no measured flash point (heating) or LEL/UEL[31] (storage).

Measurements of flash point or LEL/UEL have no inherent value. They are just numbers used as screening tests in conjunction with existing and proven systems for ranking relative hazard levels (see Section 3.7).

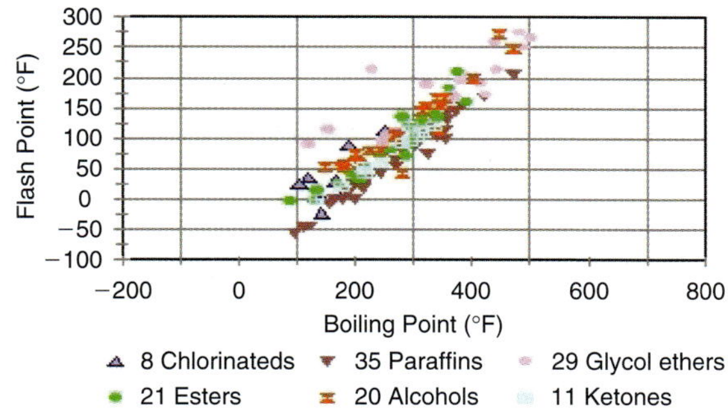


**Figure 3.9**

## 3.4 "FOOLING" THE FLASH POINT TEST

The phrase "fooling the test" means that some ordinarily reliable test is caused to produce an unexpected, and possibly unreliable, result.

The point of this chapter is not that flash point testing is not meaningful. It is the point is that flash point testing gives only a partial viewpoint of whether or not a chemical can catch fire. Some values of high (or no) flash point, which would normally lead a user to expect a low tendency to reach with Oxygen and catch fire, are not matched by values of explosive limit which would lead to the same expectation.

### 3.4.1 The Effect of Halogen Atoms

Halogen (Fluorine, Chlorine, Bromine, or Iodine) atoms are noted as being capable of reducing the reactivity of solvent vapors with Oxygen in air.[32]

Halogen atoms are incorporated within some organic chemicals to convert them into fire suppressants.[33] Nearly all commercial fire suppressants contain halogen atoms. Bromine is more commonly used than Chlorine – because the Bromine atom is more reactive and less stable.

### 3.4.2 Effects on Flash Point

The addition of halogen atoms to molecules containing Carbon and Hydrogen atoms depresses the tendency of liquids for ignition. In other words, halogen atoms raise flash points.

In Section 3.2.3, two factors were identified which greatly affect flash points. They are volatility and reactivity.

See Figure 3.9[34] which includes nearly 130 chemicals containing many different types of reactive atoms.

Data plotted in Figure 3.9 describe chemicals containing only Carbon, halogen, Oxygen and Hydrogen atoms.[35] A very broad range of volatility, and flash point, is included in this data set.

It is this data of volatility which must be used to assess the effect of halogen atoms on flash point. The reference data used is shown in Figure 3.10[36] and is repeated in Figures 3.11 and 3.12.

---

[31] In a sense the upper explosive limit (UEL) is a trap. While technically valid, UEL is not a useful parameter to screen for safe operations. A fuel-rich mixture, above the UEL, is not a safe situation – as can be a fuel-lean mixture, below the LEL. Inadvertent or uncontrolled dilution with air can convert a fuel-rich mixture into one which can be ignited. Don't use UEL as a screen test.

[32] Troitzsch, J., *International Plastics Flammability Handbook*, Hanser Gardner Publications, Munich, 1990.

[33] Burgess, D.R.F., Tsang, W., Westmoreland, P.R. and Zachariah, M.R., *Thermochemical and Chemical Kinetic Data for Fluorinated Hydrocarbons*, http://www.cstl.nist.gov/div836/836.03/papers/NistTNIntro.html

[34] Figure 3.9 (flash point versus temperature when vapor pressure is 760 mmHg) is an alternate presentation of the same information which is in Figure 3.1 (flash point versus vapor pressure at 25°C) both figures describe volatility.

Note that more volatile compounds boil at lower temperatures.

[35] Chemical structures are only paraffinic. There are no aromatic (ringed) compounds in this data set.

[36] One reason for scatter in Figure 3.10 is that branched or isomeric molecules have lower boiling and flash points than their paraffinic relatives at the same molecular weight (boiling point). Included are various organic chemicals which cover a wide variety of molecular weights and chemical structures.

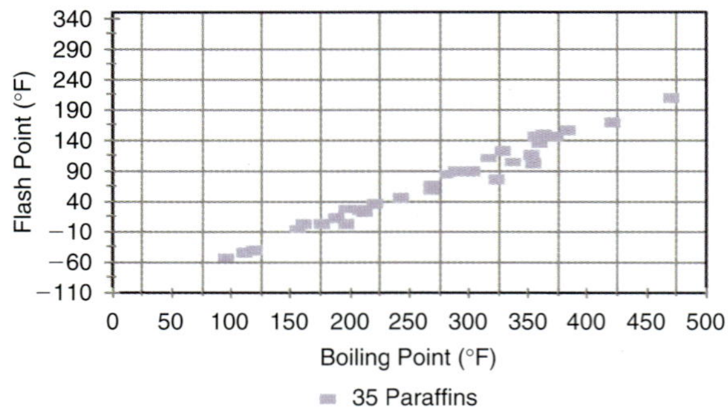


**Figure 3.10**

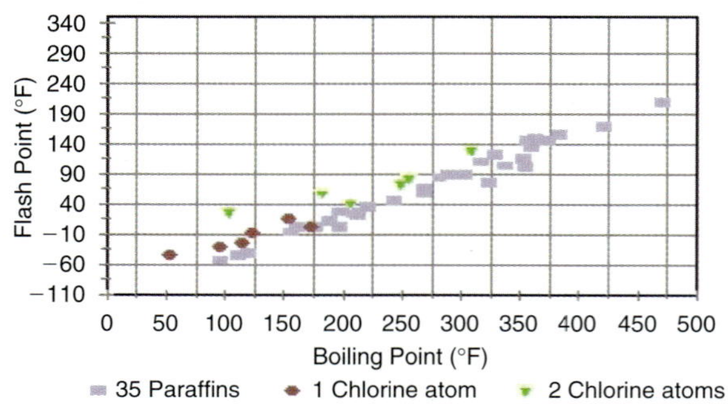


**Figure 3.11**

Figure 3.11 shows the effect of Chlorine atoms on flash point. Compounds with both one and two Chlorine atoms are shown:

1. Addition of a single Chlorine atom raises flash point.
2. Addition of a second Chlorine atom has the same effect – but it is not as pronounced. This is because the second Chlorine atom also increases volatility.

Similar suppression of ignition occurs when Bromine, Iodine, and Fluorine atoms are added. This is shown in Figure 3.12 (Bromine atoms), Figure 3.13 (Iodine atoms), and Figure 3.14 (Fluorine atoms[37]). Elevation of flash point, at the same boiling point, is unmistakable.

### 3.4.3 Relationships

Is there a relationship between flash point and explosion limits?

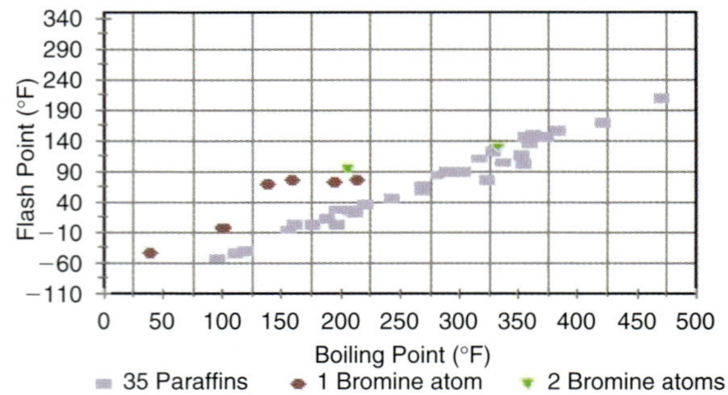


**Figure 3.12**

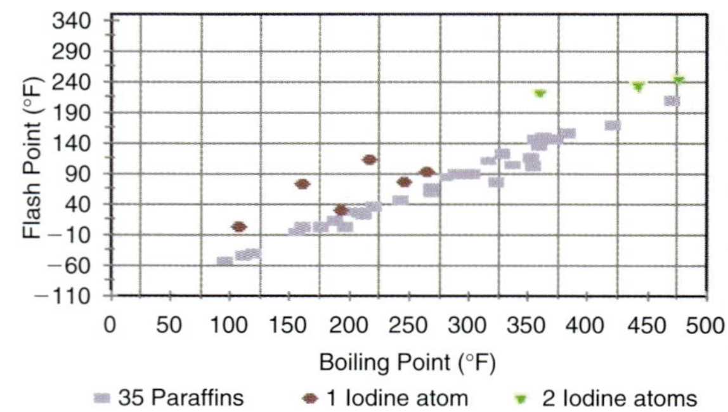


**Figure 3.13**

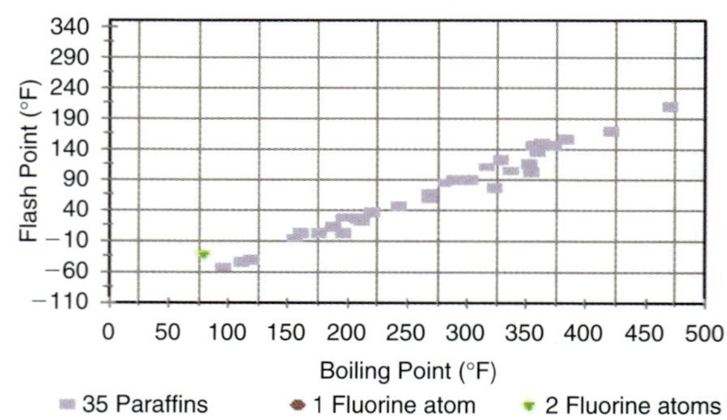


**Figure 3.14**

Both are measurements of the potential of a chemical for ignition – flash point as a simulation of heating and explosion limit as a simulation of storage. LEL and UEL are graphed in Figures 3.15 (LEL) and 3.16 (UEL) versus flash point.[38]

It's apparent that there is little or no relationship between the two types of measurement of ignition

[37] There is insufficient literature available to be visually useful. The chemicals are too volatile and flammable to be useful in cleaning operations.

[38] It's unfortunate that flash point data is usually subject to uncertainty. Seldom does data about flash point, reported in contemporary magazine articles, MSDSs, and databases, include a citation about the method by which it was measured. One should applaud when it is done. Open-cup and closed-cup data are seldom, but not always, differentiated. That uncertainty confuses understanding. Data in Figures 3.15 and 3.16 are from the same data set used in Figure 3.9.

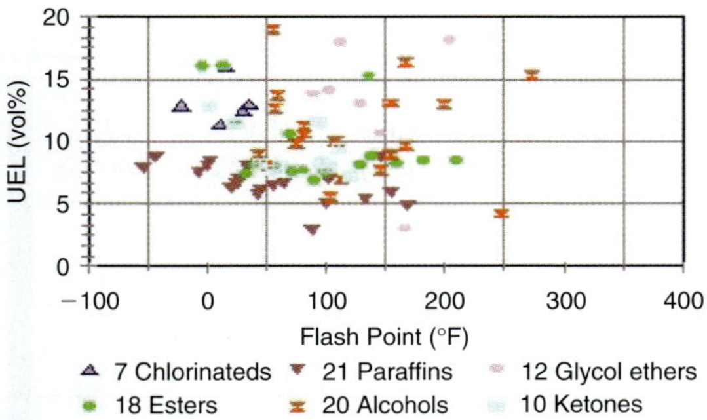


**Figure 3.15**

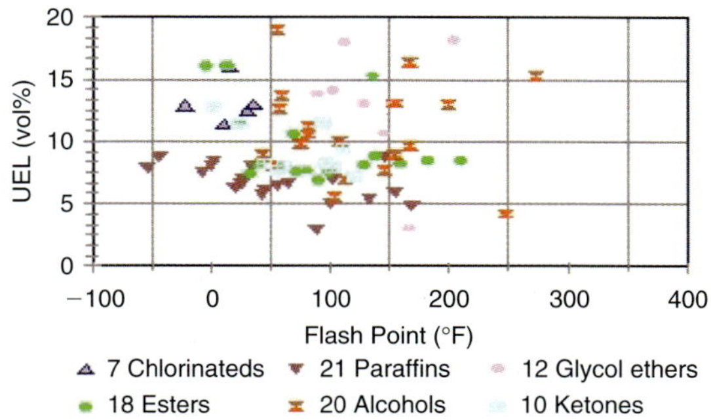


**Figure 3.16**

risk – for most chemicals. LEL and UEL values are typically within a modest range around the stoichiometric composition.[39]

Said another way:

- LEL and UEL values either exist or they don't. If they exist, they are similar to the stoichiometric concentration. If not, then they are not similar. *Managers should be more concerned about their existence and less concerned about their absolute value.*[40]
- Flash point values either exist or they don't. Volatility, reactivity with Oxygen, and content of halogen atoms determine their existence. *Managers should be more concerned with their absolute value*[41] *and less concerned about their existence.*

### 3.4.4 Life Outside the Flash Point Tester

Are users more safe when using halogenated solvents for cleaning operations because these chemicals have higher[42] flash points? Not necessarily.

Can halogenated solvents catch fire? Absolutely![43]

Flammability data[44] also describe combustion of chemicals with Oxygen. Consider the information in Table 3.6.

Data (see Footnote 19) in Table 3.6 show how the addition of single and dual halogen atoms to a chemical about which there is legitimate concern about fire safety have *dominant effects on flash points, but only modest effects on explosive limits.*

Perhaps this is more easily seen when this data is graphed, as in Figure 3.17 (ethane), Figure 3.18 (propane), Figure 3.19 (hexane), Figure 3.20 (benzene), and Figure 3.21 (ethylene).

Note how major change in flash point (red bars) occurs for ethane and propane when halogen atoms are added to hydrocarbons.

Only modest change in LEL (blue bars) and UEL (green bars) occurs.

For hexane, benzene, and ethylene as well, halogen atoms suppress combustion in flash point tests (raise values). There are lesser effects on flammability limits.

### 3.4.5 Ignition of Halogenated Chemicals

It's almost diabolical. Addition of halogen atoms to a hydrocarbon molecule raises the flash point of the chemical. That implies that use of it is less likely to produce combustion with Oxygen – a fire.

This is not so, if flammability (explosion limits) is (are) a measure of fire hazard – as they are.

---

[39] See Figure 3.6. LEL values can't be significantly less than the amount of fuel needed for the combustion reaction. An upper bound on UEL values is the fact that air is mostly not Oxygen.

[40] Because their absolute values are relatively unchanging.

[41] Because this determines the safety or hazard classification of the chemical.

[42] These flash points are not necessarily higher than those of other cleaning chemicals – just higher than the flash point of the unhalogenated chemicals (ethane, propane, ethylene, etc.).

[43] Material in this chapter is intended to support use of halogenated solvents where managers understand how they are different, and how to manage that difference versus other solvents.

[44] The major literature source for flash point and flammability data of halogenated chemicals is the database prepared by the Chemistry Department at the University of Akron (US). While individual references are not provided, the database is considered to an academic work. See http://ull.chemistry.uakron.edu/erd/

**Table 3.6 The Effect of Hologen Atoms on Ignitability**

| Base Chemical | Flash Point (°C) | LEL (vol%) | UEL (vol%) | Flash Point (°C) | LEL (vol%) | UEL (vol%) | Flash Point (°C) | LEL (vol%) | UEL (vol%) |
|---|---|---|---|---|---|---|---|---|---|
| | Ethane | | | 1-Chloroethane | | | 1,1-Dichloroethane | | |
| | −135 | 3.0 | 12.5 | −43 | 3.6 | 14.8 | −12 | 5.4 | 11.4 |
| | Propane | | | 1-Chloropropane | | | 1,3-Dichloropropane | | |
| | −104 | 2.1 | 9.5 | −32 | 2.6 | 11.0 | 4 | 3.4 | 14.5 |
| | Butane | | | 1-Chlorobutane | | | 1,4-Chlorobutane | | |
| | −60 | 1.8 | 8.5 | −17 | 1.8 | 10.1 | | | |
| | Pentane | | | 1-Chloropentane | | | 1,5-Chloropentane | | |
| | −49 | 1.4 | 8.0 | 3 | 1.4 | 8.6 | | | |
| | Hexane | | | 1-Chlorohexane | | | 1,6-Dichlorohexane | | |
| | −23 | 1.0 | 8.1 | 27 | 1.6 | 9.0 | 74 | 5.3 | 10.5 |
| | Cyclohexane | | | 1-Chlorocyclohexane | | | Dichlorocyclohexane | | |
| | −18 | 1.2 | 8.3 | 32 | 1.1 | 7.5 | | | |
| | Benzene | | | Chlorobenzene | | | *p*-Dichlorobenzene | | |
| | −1 | 1.4 | 8.0 | 24 | 1.3 | 11.0 | 66 | 1.8 | 7.8 |
| | Propane | | | 1-Bromopropane | | | 1,6-Bromopropane | | |
| | −104 | 2.1 | 9.5 | 24 | 4.6 | 13.8 | 54 | | |

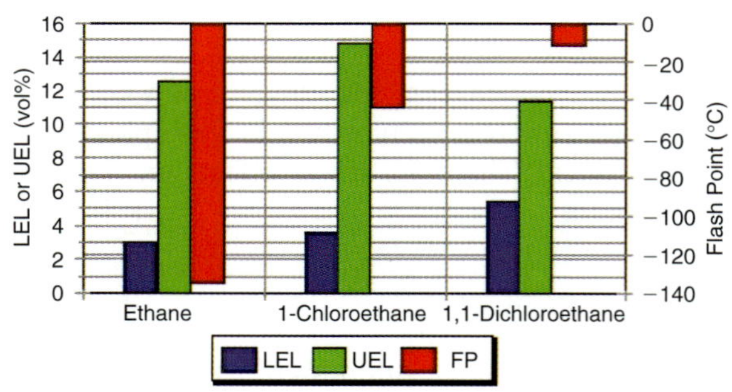


**Figure 3.17**

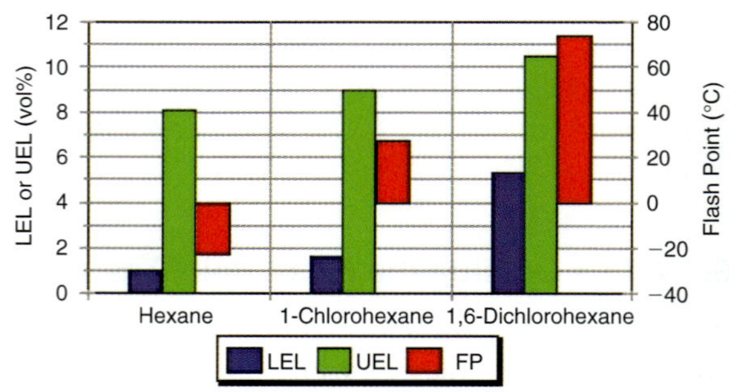


**Figure 3.19**

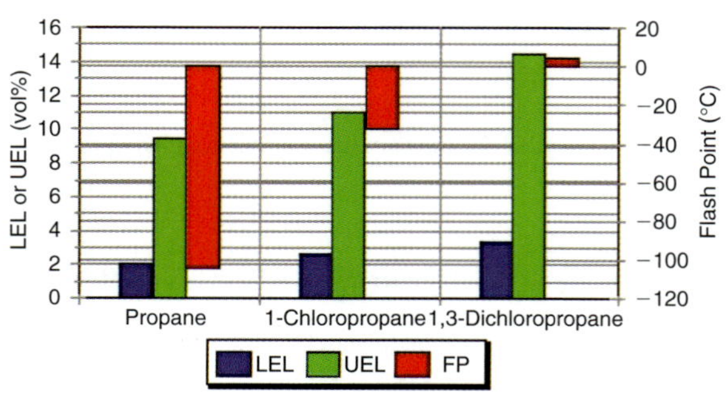


**Figure 3.18**

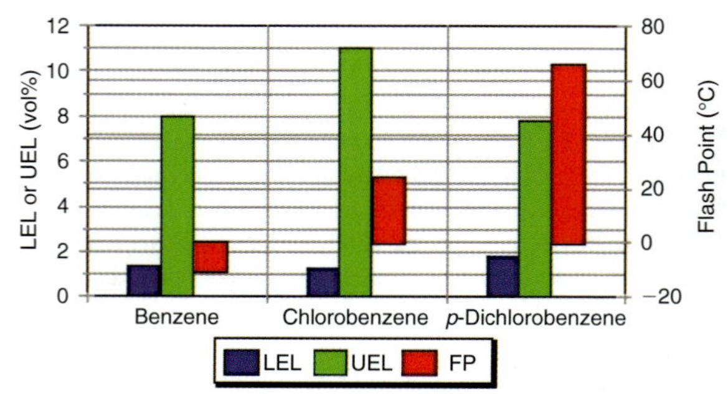


**Figure 3.20**

Consider the information[45] in Table 3.7. It shows that the vapor concentration in the headspace above a liquid-storage container exceeds the LEL for many halogenated chemicals.

Halogenated chemicals can catch fire.[46] Fumes, from a storage container, of all but one[47] of the chemicals listed in Table 3.7, can be ignited[48] as they display an LEL. This is because the concentration of chemical in fumes from a storage vessel is similar to or exceeds[49] the LEL.

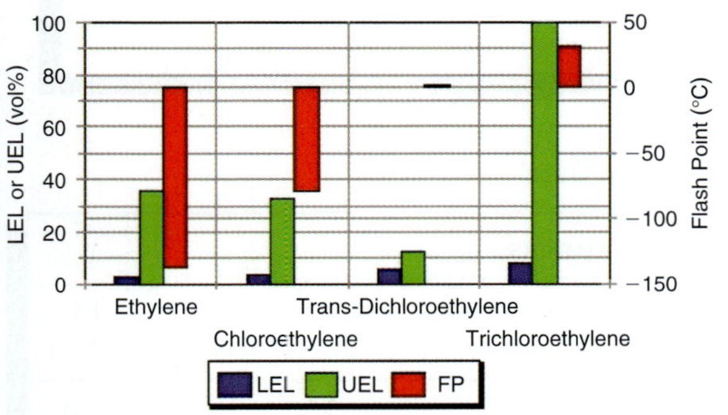


**Figure 3.21**

To be specific, equilibrium emissions from a storage tank containing any (except one) of the chemicals in Table 3.7 lack only a spark to become a fire. Further, dilution with some air may not improve safety with chemicals such as *n*-propyl bromide and methylene chloride. The diluted concentration in air may still be above the LEL.

### 3.4.6 Chemicals with No Measured Flash Points, and Measured LEL and UEL Values

Is this possible? Absolutely!

The flash point of a few halogenated chemicals can be increased[50] (suppressed) by the presence of the halogen atoms to where there is no measured flash point. But they can be ignited – measured LEL and UEL values.

Two examples are HCFC-141b and HFE-7200 (see Table 3.8).

It is beyond the scope of this volume why this is so. Of significance to managers is that absence of

**Table 3.7 Flammability of Common Halogenated Cleaning Chemicals**

| Chemical | Flash Point (°F) | LEL (vol%) | Calculated Vapor Concentration at Ambient Conditions (77°F), (vol%) |
|---|---|---|---|
| Methylene chloride | 25 | 13 | 41 |
| Trans-1,2-Dichloroethylene | 35 | 6 | 42 |
| 1,1,1-Trichloroethane | 30 | 7.5 | 13.2 |
| Trichloroethylene | 90 | 7.9 | 7.8 |
| Perchloroethylene | 113 | None | 1.3 |
| *n*-propyl bromide | 75 | 4.6 | 14.6 |
| Parachloro benzotrifluoride | 109 | 0.9 | 0.8 |
| Benzotrifluoride | 54 | 3.1 | 3.3 |

[45] Flash point and LEL data are from the reference in Footnote 44 (Section 3.4.4). Vapor concentrations at ambient conditions (77°F or 25°C) are calculated from calculated vapor pressures at that temperature divided by ambient pressure (760 mmHg). The Antoine equation is used to calculate vapor pressures as a function of temperature. And the two constants for the Antoine equation are obtained from literature data of vapor pressure.

[46] That's why two measures of ignition risk are employed in safety management. One simulates heating (flash point) and the other simulates storage (explosion limit).

[47] The chemical for which no LEL is noted in the literature is the one with the most halogen atoms (4).

[48] Given the presence of a spark.

[49] Exceedance of the upper explosive limit (UEL) provides no protection. A uncontrolled gust of air can dilute the concentration of fumes to a lower value but still above the lower explosive limit (LEL).

[50] The increase is from the flash point of the halogen-free hydrocarbon – ethylene, ethane, propane, etc.

**Table 3.8 Chemicals With Explosive Limits and No Flash Points**

| Chemical | Structure | Flash Point | LEL (vol%) | UEL (vol%) |
|---|---|---|---|---|
| HCFC-141b | 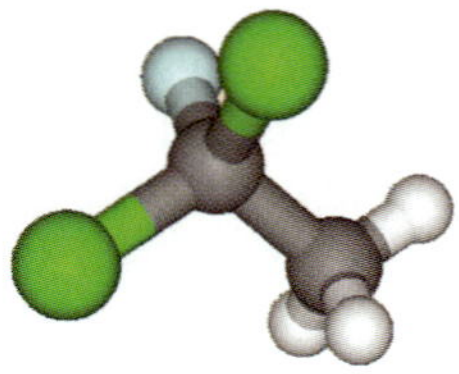 | None | 10.8 | 14.8 |
| HFE-7200 | 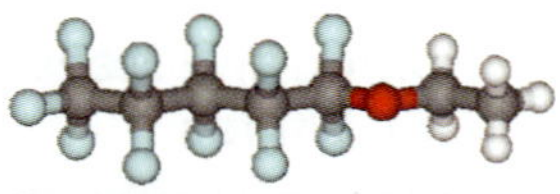 | None | 2.4 | 12.4 |

**Table 3.9 Chemicals About Which There is a Discontinuity In Data**

| Chemical | Flash Point, as Cited by Reference | | | LEL (vol%) | UEL (vol%) |
|---|---|---|---|---|---|
| | http://ull.chemistry.uakron.edu/erd/ | http://www.hsia.org | http://www.albemarle.com/brofrm.htm | | |
| Methyl bromide | −44°C | | None | 10.0 | 16.0 |
| Ethyl bromide | −20°C | | None | 6.7 | 11.3 |
| Methylene chloride | −4°C | None | | 13.0 | 22.0 |
| 1,1,1-Trichloroethane | −1°C | None | | 8.0 | 10.5 |
| Isopropyl bromide | 19°C | | None | 4.6 | 13.8 |
| *n*-propyl bromide | 24°C | | None | 4.6 | 13.8 |
| Trichloroethylene | 32°C | None | | 7.9 | 11.0 |
| Dibromomethane | 34°C | | None | | 27.2 |

flash point is not a free pass to avoid ignition risk. Of even more significance in the following section.

### 3.4.7 The Flash Point of *n*-Propyl Bromide[51]

There is legitimate scientific dispute about the characterization of some chemicals. All are halogenated solvents:

- One source of data is a public database developed by the department of chemistry at a US university (see Footnote 44).
- Two other sources of data are commercially related web sites – one being a trade organization and the other being a product manufacturer.

There is no attempt here to denigrate these commercial organizations or data supplied by them. Neither is there an attempt here to promote a public database.

There is an attempt to describe a discontinuity in technical information, and recommend how managers should deal with it. Data are listed in Table 3.9.[52]

Several items are apparent here:

- Flash point of halogenated chemicals can be difficult to measure reproducibly (see Footnote 33).

[51] This statement is no reflection on the feasibility of using *n*-propyl bromide. It is simply the one most familiar of those in Table 3.9.

[52] It is not the purpose of this volume to validate or invalidate publically available data sources. This author has studied several literature articles which support the data cited in Table 3.9, and is convinced they represent useful scientific work.

- Some chemicals can reproducibly ignite in one kind of exposure and not in another.
- Absence of a measured flash point does not mean a chemical can't catch fire.

The above discontinuity can't be dismissed. It is manifested in at least eight chemicals.

As managers considering use of halogenated solvents, you must consider both types of flammability testing and choose your equipment configuration, operating procedures, and protective equipment based on the worst-case scenario.

You cannot solely rely published flash point information. You must also consider explosion limits.

## 3.5 FLAMMABILITY AND ITS MEANINGS

The word flammability raises a legitimate concern, two misunderstandings, and possibly a misrepresentation.

### 3.5.1 The Concern

It is and should be real. It is that a *cleaning solvent will catch fire.*

Halogenated hydrocarbon solvents present a paradox to users:

- They contain Hydrogen and Carbon atoms – the stuff of which fires are made.
- But they also contain halogen atoms – the active ingredient in flame suppressants.

Which one dominates? Yes, that's the question!

This paradox has confused users of cleaning solvents such as HCFC-141b, 1,1,1-Trichloroethane, methylene chloride, trichloroethylene, HFE-7200, and more recently, *n*-propyl bromide. It is claimed that none of these common cleaning solvents has a flash point (see the information in Table 3.9).

### 3.5.2 The First Misunderstanding

It is that *a solvent with no measured flash point can't catch fire.*

That's wrong; some can catch fire.

All of the solvents mentioned above have no measured flash point. Yet each can be ignited. Each has a normally measured (ASTM E-681) value of an LEL and UEL. They range from 2.4 vol% to around 23 vol% in air at 25°C, respectively.

### 3.5.3 The Second Misunderstanding

It is that there are really *two meanings to the term "flammability."*

One is technical and the other is what most users think it means:

- The *technical* meaning is the basis for many regulations and industrial practices pertaining to cleaning solvents. Solvents are classified as flammable (flash point value <100°F) or combustible (100°F >flash point value <200°F) based solely on this data (classification published in CFR 1910.106 (a)(18), OSHA or NFPA30).
- The *commonly understood* meaning is that the described product will burn. Definitions for "flammability" and "flammable" in various non-technical dictionaries include: "… measure of the extent to which a material will support combustion …," "… any substance that is easily ignited, bums intensely, or has a rapid rate of flame spread …," "… capable of being easily ignited and of burning quickly …," etc.

### 3.5.4 An Opportunity for Misunderstanding (or Misrepresentation)

Misrepresentation might happen if solvent suppliers used phrases in advertisements like "… Flammability – None …".

In the case of *n*-propyl bromide, for example, that simple phrase points the user simultaneously in two directions:

1. *No*, a flash point is not reported for *n*-propyl bromide by the ASTM D-56-02 or D-1310 tests.
2. *Yes*, *n*-propyl bromide, for example "… will support combustion …".

Would a supplier advertise using such a phrase knowing some readers might more favorably value their product because they believe the product can't be ignited in one test when in fact it *can* be ignited in another? I hope not![53]

[53] But it's been commonly done. That's the reason for this chapter.

**Table 3.10 Different Measures of Ignition Risk**

| Item | Flash Point | Flammability Limit |
|---|---|---|
| Situation under study | Heating of a solvent in an open or closed container (Figure 3.3 or 3.4) | Storage with vaporized solvent in a closed container (Figure 3.5) |
| Variables held constant in test | Strength of ignition source | Temperature (usually 25°C), strength of ignition source |
| Variables in test | Temperature | Composition |
| Test fixture volume | Small cup (~few milliliters) | Large spherical vessel (~5 liter) |
| Detection method | Visual recognition of flame in manual testers, temperature or pressure rise in automatic testers | Visual recognition of flame |
| ASTM standards | D1310 (open) or D 56-02 (closed) for taglibue tester | E-681 |
| Test outcome | Temperature at which an ignition is first noticed | Volumetric composition at which a stable flame front is produced |
| Reason for testing | Testing by solvent manufacturers to characterize product before it is released | Testing on site by users of headspace composition versus LEL adjusted with safety factor |

As a manager, you must know that *the statement "... no flashpoint ..." (by itself) is meaningless.*

### 3.5.5 Le Denouement

To answer the question raised in Section 3.5.1: "mechanical circumstances" dictate whether the flame suppressant atoms in a solvent molecule will overpower the tendency of Hydrogen and Carbon atoms to act as fuel for combustion.

Said another way, the circumstances of the ignition test dictate whether ignition is or is not achieved. These circumstances are very different for the flash point and flammability tests (see Table 3.10).[54]

As you would expect, solvent molecules which have a higher ratio of flame suppressant to fuel atoms can't be ignited in either test (or in practice).

Common cleaning solvents such as HFE-7000, HFE-7100, HFE-7500, HFC-43 10mee, HCFC 225 ca/cb, perchloroethylene, and CFC-113 fail both test approaches – no flash point or LEL or UEL values. Each solvent has a high ratio of halogen atom to fuel atoms.

In summary the difference between flash point and flammability really doesn't matter in solvent selection – as long as *data from both tests* are considered.

## 3.6 STATIC DISCHARGE

Static electricity is the electricity trapped on the surface of a nonconductive body. Electricity on a conducting body, that is in contact only with nonconductors, is also prevented from escaping and is therefore not mobile and is "static."

Discharge of static electricity is a release of energy. Static electricity does not flow in a circuit. It does move when the charge is discharged. Sufficient static electricity can act as a spark, and cause ignition of a cleaning solvent.

### 3.6.1 Don't Give Me No Static

Where liquids flow through a pipe, static electricity is generated.

Conductive properties of the liquid[55] and the pipe network system affect the generation (charging) process:

- The network of pipe can be just the shipping or storage container.[56]

[54] This is an abbreviated form of Table 3.5.
[55] See Section 3.6.2.
[56] Rolling of filled, or partially-filled, drums will generate static electricity in the contained liquid. Both metal and plastic drums can produce the same effect if drums are rolled over wood or concrete surfaces – because those surfaces are non-conductive.

- When liquids flow through closed metal pipes, static electricity is not a hazard. This is because the liquid surface is already contacting the conductive metal pipe.
- Static electricity may become a hazard, however, when liquids are pumped into tanks. In some cases it is necessary to restrict[57] flow rates to control static generation. Charges produced in the liquid during pumping can accumulate on the surface of the liquid and cause sparking between the liquid surface and the tank or a projection in the tank.
- Filters in pipelines *greatly* increase the generation of static electricity. In one in aircraft fueling test[58] it was reported that the charge development was 10–200 times more with a filter than without one.

### 3.6.2 The Difference Between Air and Water

A solvent's dielectric constant (DC) is a measure of the relative effectiveness of that liquid as an electrical insulator. A good electrical insulator will allow static charge to build in a piping network, and not be dissipated. This is what we want to avoid, to avoid ignition:

- A solvent with a low DC is more of a safety hazard than one with a high DC. The former is a better insulator and can hold a static charge without dissipation. Ultimately the charge will be dissipated and that can produce a spark that can ignite an air/solvent mixture.

The perfect electrical insulator is a vacuum, which has a DC value of 1.00000. By comparison, air has a DC value of 1.00059, almost the same as a vacuum, and water has a DC value of 78.2. For a liquid, water is a poor insulator.[59] A solvent with a DC value of 10 will allow dissipation of 10 times the amount of static electricity as will a solvent with a DC value of 1.

A dielectric meter measures the relative DC of a solvent by measuring the difference in capacitance of the probe between a standard (usually cyclohexane, with true DC value of 2.025) and the solvent sample.

Recall that cyclohexane is a commonly used cleaning solvent. As an extremely poor dissipater[60] of static electricity, its use is raises significant safety concerns. This is true whether the cyclohexane is used alone or azeotroped with another chemical such as isopropyl alcohol.[61]

### 3.6.3 An Analogy

Think of static buildup, and discharge, as if it were anger.

Anger is normal. It must be dissipated at some point. We don't want anger to build up. Suppressed anger can result in rage or violence. We want to dissipate our anger as strongly held opinions or in exercise, and not rage.

The same is true with static electricity. A safer situation results when static electric charge is properly managed – dissipated at a low level rather than allowed to accumulate prior to discharge.

### 3.6.4 Bury Me Deep in That Old Ground

There is a basic requirement for the grounding (earthing) of process equipment to prevent ignition of flammable vapors by static discharge. The requirement is that grounding must be effective, which is not at all difficult to achieve. But failure to provide effective grounding may be fatal.

As managers of cleaning systems, we want the following to happen:

- Static charge generated by fluid movement should be transferred by conduction to an electrical ground system. This means only a small amount of static charge is stored in the liquid.

---

[57] This is a two-edged sword. Restriction of flow velocity can prevent generation of static electricity. BUT motion past the restricting device can also generate static electricity. Hence, generation of static electricity is best prevented by never allowing excessive velocities rather than restricting them.

[58] See http://www.environmental.usace.army.mil/library/faq/faqproeng/faqproeng.html

[59] Said another way, water is a good conductor.

[60] Water will dissipate static charge about 40 (78.2/2.025) times faster than will cyclohexane. There are no units to DC.

[61] Static discharge from cyclohexane is like any other problem to be managed. This author has worked at sites at which cyclohexane was manufactured. He has seen fires started by static discharge in small cyclohexane containers. He has also seen excellent management of grounding systems and procedures employed and believes that all fires involving static discharge from cyclohexane can be avoided.

As managers of cleaning systems, we want the following NOT to happen:

- A high concentration of static electricity to be accumulated within the liquid.[62]
- Accumulated static electricity to be discharged from the liquid to another surface. If the intensity of static transfer is great enough, the liquid can be ignited.

So what kind of equipment do managers want? Equipment that is well grounded.[63]

### 3.6.5 Real World Experience

The purpose of this subsection is not to discourage cleaning with any solvents. Rather, the purpose is to acquaint managers with the consequences of inadequate management of solvents and static electricity. Here are three past consequences:

1. Cyclohexane is an extremely common industrial chemical used in the manufacture of nylon and other polymers. However, on June 1, 1974 a vapor cloud explosion destroyed the Nypro cyclohexane oxidation plant at Flixborough, England, killing 28 people. Other plants on the site were seriously damaged or destroyed. The site presented a scene of utter devastation. The accident was traced to a poorly qualified design team that were asked to design and install temporary piping.
2. Static electricity-related fires at retail gasoline outlets are extremely unusual. However, the Petroleum Equipment Institute (PEI) and the American Petroleum Institute (API) say, between 2000 and 2002, there have been more than 150 cases of fires caused by sparks from static electricity igniting gasoline vapors.[64]
3. This author has personally seen small[65] fires started when tanks of cyclohexane were improperly sampled. Backup and extinguishing systems prevented injury.

In layman's terms, if Oxygen, fuel, and a spark are necessary to start a fire; cyclohexane flowing in a pipeline containing air is an incipient fire because it contains all three elements necessary to produce a fire. Only that the mixture might be too rich with fuel might avoid an ignition – if the pipeline is not well grounded.

### 3.6.6 A Preference

So what kind of solvents is safest to use?

- *Solvents which don't store a high concentration of static electricity.*

These solvents conduct the static electricity. These solvents act as good conductors or poor insulators.

The preferred solvents are good conductors. They have high DC. These preferred solvents are NOT poor insulators.

Solvents with low DC should be avoided.[66] We prefer, from the standpoint of safety, solvents which have high DC. From this standpoint *only*, we prefer water, and seek to avoid cyclohexane.[67]

### 3.6.7 DC United

You can compare the tendency to discharge static electricity with the DC data in Table 3.11. Remember, lower DC values means static charge is more easily discharged (drained or relaxed) as a spark and represent a solvent less safe to use.

---

[62] Cyclohexane can be safely used if grounding of all piping, fittings, vessels, and containers are well designed and maintained.
[63] Grounding and bonding connections can be made with pressure-type ground clamps; brazing, welding, battery-type clamps; or magnetic or other special clamps that provide good metal-to-metal contact. Surfaces to which grounding clamps are attached must be clean and free of paint, oil, grease, or other materials which would impede good contact. NFPA 77 includes some excellent diagrams of proper bonding/grounding equipment and techniques.
[64] Renkes, Robert, N., "Stop Static," *Tulsa Letter of the Petroleum Technology Institute*, January 25, 2000 and November 1, 2002. See http://www.pei.org/static/TLrefueling_fires.htm
[65] There is no such thing as a small fire. This statement was written in jest.
[66] Yet, for good cleaning performance, it may be necessary to use a solvent such as cyclohexane, acetone, limonene, or perchloroethylene. This is done every day by managers who establish and maintain effective grounding systems.
[67] Yet, if water doesn't produce adequate cleaning performance, there is no point expressing the preference and using it. The theme in this book is to use what works and manage it effectively.

**Table 3.11 DC Data**

| Solvent | Temperature (°F) | DC |
|---|---|---|
| Acetone | 77 | 1.4 |
| *n*-hexane | 68 | 1.9 |
| Heptane | 68 | 1.9 |
| Methylhexane | 68 | 1.9 |
| Cyclohexane | 68 | 2.0 |
| Cyclopentane | 68 | 2.0 |
| Toluene | 77 | 2.0 |
| Dichloroethane | 77 | 2.0 |
| Dichloromethane (methylene chloride) | 68 | 2.1 |
| Dichloroethylene | 68 | 2.2 |
| Limonene | 68 | 2.3 |
| Xylene | 68 | 2.4 |
| Tetrachloroethylene (perchloroethylene) | 70 | 2.5 |
| Glycerol | 77 | 2.5 |
| Methanol | 77 | 2.6 |
| Freon 113 | 70 | 2.6 |
| Isopropyl alcohol | 75 | 4.2 |
| Ethyl ether | 68 | 4.3 |
| Ethanol | 77 | 4.3 |
| Chloroform | 68 | 4.8 |
| Acetic acid | 68 | 6.2 |
| Methyl acetate | 77 | 6.7 |
| Propyl alcohol | 77 | 6.7 |
| Ethylene glycol | 68 | 7.0 |
| Cyclohexane | 68 | 8.3 |
| Trichlorethylene (TCE) | 68 | 8.5 |
| Trichloroethane (TCA) | 68 | 8.5 |
| Water | 68 | 78.2 |

## 3.6.8 Prevention of Discharge

The basic approach is to control two of the three factors necessary to start a fire.[68] They are:

- Oxygen – of which air about one-fifth.
- Fuel in a ratio between the LEL and the UEL.[69]
- A spark.

[68] See Section 3.3.1.
[69] See Section 3.3.3 and Figure 3.6.

### 3.6.8.1 *An Insurance Policy*

The principle of preventive fire (or explosion) protection comprises the reliable exclusion of just one of the requirements necessary for the development of an explosion.

Granted, perfect control of one of the above three factors will prevent an ignition or explosion. But

- *Control of a second factor is maintained for security – because perfection in the control of anything is seldom achieved, and because a mistake with a low DC solvent such as cyclohexane or acetone is equivalent to a fire.*

Some manufacturers of equipment, or operators of facilities, may take a less conservative approach than the above. Control of two factors is more expensive than control of one. This author can't recommend that approach.

### 3.6.8.2 *The Controlling Factor(s)*

The two controlled factors should be:

- Fuel level in the atmosphere (through measurement with a safety factor).
- Spark (through grounding of static electricity).

The general approach is given in Table 3.12.

The reason why this author can't and won't recommend use of just one of the above factors is security.

Do NOT bet the avoidance of a fire on the integrity of a ground connection, the continuity of an explosimeter, the actions of other human beings, or the position of an $N_2$ supply value.

## 3.6.9 Storage of Low DC Solvents

The following material is not a substitute for the details within references cited in Table 3.12, managers considering the storage of low DC solvents should know that:

- If storage tanks contain vapors which can be ignited, discharge of static electricity from within the liquid contents can ignite that fuel. Consequently,

**Table 3.12 Control of Static Electricity**

| Factor | Dual Requirements | Consequence of One Failure |
|---|---|---|
| Air exposure | Process control via continuous measurement of %LEL (in air):<br><br>• Alarms are set for levels such as 25% LEL (in air)<br>• The alarms produce positive action such as shutdown or flooding with Nitrogen | Should the fuel concentration approach the LEL because of failure of the process control system, static electricity can't produce a spark because it is dissipated from the liquid via a well-grounded storage tank. |
| Grounding of static electricity | This is a discipline with multiple details. A full textbook is necessary to cover all details. Designers, and managers, of well-grounded facilities will be intimately familiar with:<br><br>• OSHA 1910.106 (flammable and combustible liquids)<br>• OSHA 1910.107 (spray finishing using flammable and combustible materials)<br>• OSHA 1910.110 (storage and handling of liquefied petroleum gases)<br>• OSHA 1910.119 (process safety management)<br>• OSHA 1926.404(b)(1) (ground fault interrupters)<br>• 33 CFR 154 subpart E (vapor control systems (USCG))<br>• National Fire Protection Association's Flammable and Combustible Liquids Code, 2003 Edition (NFPA 30)<br>• National Fire Protection Association's Standard for Dipping and Coating Processes Using Flammable or Combustible Liquids, 2003 Edition (NFPA 34)<br>• National Fire Protection Association's Low-pressure Gas Code (NFPA 58)<br>• National Fire Protection Association's National Electrical Code (NFPA 70)<br>• National Fire Protection Association's Static Electricity Code (NFPA 77)<br>• American Petroleum Institute's standard on venting atmospheric and low-pressure storage tanks, non-refrigerated, and refrigerated (API 200)<br>• Process Safety (Perry's Handbook, Section 26) and/or<br>• The design standards of your firm | Should the ground system be poorly designed, incompletely installed, or not connected to the good Earth, a spark from static discharge can't produce ignition because the fuel concentration is below the LEL. |

flammable and combustible liquids should be introduced into storage tanks by means of fill pipes (also called dip pipes) which terminate within 6 in of the bottom of the tank (per NFPA 30). Said another way, splash[70] filling should be avoided.

- Internal structural members, and tank fittings should be and normally are electrically bonded to the tank to reduce the danger of internal sparks from lightning.
- When a tank is in contact with the Earth or if its connecting piping is grounded, special internal tank electrical grounding connections are not normally needed.
- If the network piping is ungrounded or nonconductive, grounding connections should be provided on tanks that are not in contact with the Earth.[71]

## 3.7 AUTOIGNITION TEMPERATURE

This parameter is not involved in classification of chemicals by risk of ignition as is flash point involved.

Rather, autoignition temperature (AIT) is involved in classification of how chemicals are used to minimize or avoid ignition risk. This is especially significant in cold cleaning.

### 3.7.1 Definition of AIT

The AIT of a chemical is the lowest temperature at which the chemical[72] will spontaneously ignite in the absence of an external ignition source, such as a spark or flame.

For a value of AIT to be a valid measurement, ignition must be initiated without external source of energy AND sustained without external source of energy.

### 3.7.2 Heat Transfer Rules

The general definition of an ignition source in Section 3.3.1 includes both external sources of energy (sparks or static electricity), and *hot surfaces*. All can produce a self-sustaining oxidation reaction between the Carbon and Hydrogen atoms in the chemical, and Oxygen atoms in the air.

As much as availability for reaction of Carbon and Hydrogen atoms in the chemical molecule is significant in controlling ignition, more so is heat transfer within the vapor mixture as well as between the vapor mixture and it surroundings.

A reaction can't be sustained if only a portion of the reactants is at a temperature below that necessary to sustain reaction.[73]

### 3.7.3 Measurement Methods

Most commonly, measurements of AIT are made in an apparatus similar to a closed-cup flash point tester using a procedure specified by the ASTM Method E659.

Differences between AIT and flash/fire points[74] are shown in Table 3.13.

### 3.7.4 Use of Autoignition Data

The phrase is "dipping and coating." While it may not seem to do so, it manifestly includes cleaning operations. "Dipping and coating" operations, which

[70] Splash filling is hazardous for two reasons: (1) it introduces a liquid which may have a considerable charge (acetone, cyclohexane, or trans-1,2-Dichloroethylene) into a flammable gaseous atmosphere, and (2) splash filling creates mists, which are much easier to ignite because mists of combustible liquids can be ignited at initial temperatures well below the flash point of the liquid.

[71] On concrete foundations.

[72] While the autoignition temperature is always cited as a property of a chemical, practically it is always the vapor phase of this chemical in which the combustion takes place.

[73] The hot surface and the spark both provide the same outcome – high local concentrations of energy. Each provides local heating of the chemical vapor and mixed air. Local means in the vicinity of the spark or the surface.

That local energy level may, or may not, be sufficient to locally activate the oxidation reaction. When it is sufficient, the reaction proceeds locally.

Since the oxidation reaction produces heat, surrounding volumes of chemical vapor and air are heated.

In 1936, J.E. Van't Hoff described autoignition in *Outlines of Chemical Dynamics*, "The temperature of self-ignition is that temperature at which the initial loss of heat caused by heat conduction, and so forth, is equal to the heat which at that instant is produced by the chemical change".

[74] Note that both flash and fire points are measured in a flash point tester. The difference in outcome is whether or not the "flash" is sustained.

**Table 3.13 Comparison of AIT With Flash and Fire Points**

| Issue | AIT | Flash point | Fire point |
|---|---|---|---|
| Relative value | Much higher than flash point | Much lower than AIT | Slightly higher than flash point |
| Measurement apparatus | Closed-cup tester | Closed-cup (usually) tester | Closed-cup tester |
| | No ignition source | Electric (usually) ignition source | |
| Variable in measurement | Temperature above static pool of liquid | | |
| Outcome of test | Ignition occurs and is sustained | Ignition occurs | Ignition occurs and is sustained |
| Test measurement | Temperature at which ignition occurs and is sustained | Temperature at which ignition occurs | Temperature at which ignition occurs and is sustained |
| Common procedure | ASTM E659 | ASTM D56-02 | ASTM D92 |

include parts cleaning, are limited by AIT in NFPA 34:[75]

- A user is not allowed to heat[76] fluid contents of an open tank[77] to within 100°F[78] of its AIT.
- Note that this temperature limit cannot exceed[79] the boiling point of the fluid, but *can exceed the flash point*.[80]
- Further, note that hazard classification (based on flash point) has no bearing on this temperature limit.[81]
- Open flames (cigarettes), spark-producing equipment, or hot surfaces which exceed the AIT are not permitted either.[82]

### 3.7.5 Relationships

An AIT is, somewhat surprisingly, not related to either boiling point (volatility) or reactivity with an ignition source (flash point). This is seen in Figures 3.22 (boiling point) and 3.23 (flash point).

Data used in these figures are that used for Figure 3.7 and the like.

## 3.8 MANAGING FLAMMABLE OR COMBUSTIBLE SOLVENTS

Technical information presented earlier in this chapter describes hazards associated with chemicals which can be ignited. Interesting and perhaps useful, but that doesn't speak to actions to be taken, or avoided. The guidance is provided in Section 3.17.1 along with general information about to manage all hazards (see Section 3.21).

But it is necessarily general because it can apply not only to industrial parts cleaning, but to commercial

[75] National Fire Protection Association's Standard for Dipping and Coating Processes Using Flammable or Combustible Liquids, 2003 Edition.

[76] Heating can be done directly or indirectly. Direct heating mean just that – a heating jacket or coil. Indirect heating means insertion of hot parts into the cooler fluid contents of the cleaning tank. This often occurs when machined, pressure-formed, treated parts are cleaned after that work. Here the cleaning tank is also used as a quench tank.

[77] A vapor degreaser with a close-able cover or top is not an open tank. Environmental regulations today essentially prohibit use of open tanks for vapor degreasing. Open tanks are today used for cold cleaning.

[78] Per NFPA 34, Section 5.10.1. This is 38°C.

[79] Per NFPA 34, Sections 5.10.1 and 5.10.2 require that adequate ventilation be implemented to remove produced vapors, and control instrumentation including sensors and shutoffs be used enforce this temperature limit.

[80] However, per NFPA 34, Section 6.2.1 through 6.2.1.2, when chemicals such as acetone, methyl acetate, or isopropanol (Class I Flammable) are used; or chemicals such as propylene glycol methyl ether acetate, parchlorobenzotrifluoride, or whitespirit (Class II Combustible); butyl lactate, isopar M, or propylene glycol monobutyl ether (Class IIIA Combustible) are heated, all electrical components and wiring must meet the articles 501, 505, and 516 of NFPA 70 (National Electrical Code).

[81] Yet, hazard classification, derived from flash point data, dominates other issues such as ventilation, electrical configuration, and storage volume.

[82] NFPA 34, Section 6.2.2. Note that there is no offset from the autoignition temperature.

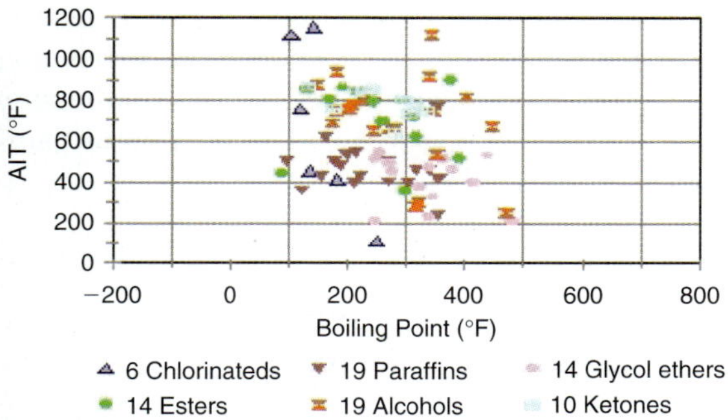


**Figure 3.22**

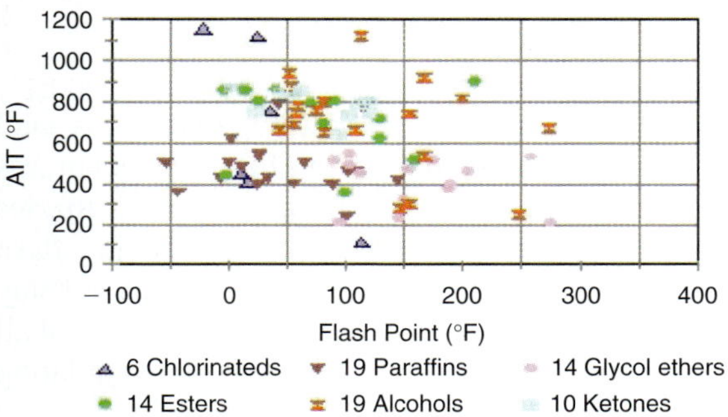


**Figure 3.23**

dry cleaning of clothes,[83] fueling of combustion engines in aircraft or automobiles, and storage of fuels.

## 3.9 HAZARDS OF AEROSOLS AND MISTS

Cold cleaning operations, in dip tanks, with solvents have been useful for many years. Automotive repair, part repainting, and cleaning of unassembled machines are commonly done in this way.

This is essentially immersion solvent cleaning. Spray cleaning (via impact) is not readily done with solvents:

- Solvent, when sprayed with or in air, produces mists or aerosols. And mists and aerosols are extremely flammable.

**Figure 3.24**

This is why the "sink-on-a-drum" facility is used to clean automotive parts. It has a nozzle and liquid supply system which splashes or wets liquid on the part surface rather than spraying the surface to gain cleaning by impact (see Figure 3.24).[84]

Here, aerosol generation with air and mineral spirits normally is avoided by design, and *must* be avoided, period.

### 3.9.1 Regulation of Aerosols

The flash point of the liquid does not govern its use in spray cleaning as is true with immersion cleaning.

Spray cleaning is covered in the US by OSHA 1910.123-126, as well as 1910.10, with strong assistance from the NFPA (Standard 33):

- A flammable aerosol is defined by the US OSHA as "... an aerosol which is required to be labeled flammable under the Federal Hazardous Substances Labeling Act (15 U.S.C. 1261). For the purposes of paragraph (d) of this chapter,

[83] Justifiable concerns raised about the toxicology of halogenated solvents has long spawned interest in dry cleaning with hydrocarbon solvents or $CO_2$. For example, see (among many other references) the US's Pennsylvania Title 34, Chapter 23 Laundering and Dyeing Establishments, Subchapter B. Dry-cleaning and Dyeing. The US's South Coast Air Quality Management District under Rule 1421 (Chapter d Part F) offers €20,000 grants to operators for purchase of equipment not using perchloroethylene because such operation will not be permitted after December 31, 2020.

[84] Figure 3.24 depicts the simplest design of such a "sink-on-a-drum" unit. Options, which add cost, include part transport, mechanical brushing equipment which is almost certain to be needed, liquid level controls and alarms, a pump driven by foot action, and inspection tools. Price for the unit in Figure 3.24 shouldn't exceed 500 Euro.

They can be used and are sold with aqueous-based cleaning agents as well as with solvents. Here, a quality issue is likely to be part dryness.

such aerosols are considered Class IA liquids.[85] …" Information in Figure 3.28 shows that a Class IA liquid is one with a flash point <73°F and a boiling point <100°F.

- The above applies to a liquid with a flash point of 380°F.[86] This means that if a solvent with a flash point of 380°F (which is classified as "unclassified" (essentially non-hazardous) by flash point information) is sprayed (without heating) so as to produce a flammable aerosol, the flash point of the air – solvent mixture shall be considered to be <73°F (classified flammable).

  *Said another way for emphasis: consider the flash point of an aerosol to be the ambient temperature where you are.*

In layman's terms, to be safe, treat all aerosols containing solvents as if they were flammable at whatever temperature they exist.

### 3.9.2 Creation of Aerosols and Mists

The condition which produces a flammable aerosol is a high velocity of air relative to liquid. Here tiny droplets of liquid are generated.[87] Lower air velocities relative to liquid produce larger liquid droplets which are less flammable.

This is because smaller droplets have significantly more surface area for evaporation, per unit mass or volume of fluid. Evaporation of the liquid produces local high concentrations of vapor. The outcome is that area near the spray zone becomes enriched in ignitable chemical.

To make guidance simple yet effective, considering the potential range of solvent properties and application equipment, US OSHA uses *application velocity* to define how to avoid production of flammable aerosols.[88,89]

### 3.9.3 US OSHA Requirements About Aerosols and Mists

Some specific requirements, by US OSHA, are good advice for managers in any country. To avoid production of or to contain flammable aerosols, managers should consider the following:

- The use of solvents for cleaning operations shall be restricted to those having flash points not less than 100°F; however, for cleaning spray nozzles and auxiliary equipment, solvents having flash points not less than those normally used in spray operations may be used.[90] Such cleaning shall be conducted inside spray booths and ventilating equipment operated during cleaning.[91]
- For a dip tank used in cold immersion cleaning, a direct low-pressure pumping system is necessary. A 10-gal (38-l), or smaller, gravity tank to supply the solvent for cleaning or paint must be used for flow coating. In case of fire, an approved heat-actuated device must shut down the pumping system.[92] This is especially important for spray cleaning of electronics components with sprays of isopropyl alcohol.

---

[85] See OSHA 1910.106(a). NFPA 30 B does not equate flammable aerosols to Class IA liquids. Class IA liquids and aerosol products can have very distinctly different behavior in a fire. Sizable fires involving the latter almost always involve rupturing and rocketing of the cans, which can spread the fire much faster than a sprinkler system's ability to respond. Also, high flash point liquids – even those with flash points in the 450°F – range, can be easily ignited when released under pressure in a very fine mist.

[86] White, Barry, J., *OSHA Regional Administrator, Standard Interpretation*, November 12, 1976.

[87] Static charge may also be generated (see Section 3.6.8).

[88] Note that the flammability (or inflammability) of an aerosol is not defined by test results (as flammability in liquids is defined by flash point testing). Note also that a flammable gas may be used as the propellant in the production of an aerosol product.

[89] And, yet, US OSHA and NFPA (per personal communication with R. Benedetti of NFPA, August 5, 2005) do not specify a velocity limit to be avoided. This author, who is not certified in fire protection technology, suggests that linear velocities in nozzles should be less than 1 foot/second.

[90] Note that the flash point of isopropyl alcohol is less than 100°F (57°F). Specifically, acetone (flash point 4°F) methyl ethyl ketone (flash point 23°F), and isopropyl alcohol should not be used to clean and flush spray nozzles or spray guns – unless they are used in the spray operation. Note also that if these solvents are diluted with water, their flash point can be measurably increased. *But* the acetone/methyl ethyl ketone/isopropyl alcohol will still evaporate in contact with high velocity air and can still form an aerosol. This is known as "fooling the test" (see Section 3.4).

[91] See OSHA 1910.107(g)(5). NFPA 33 requires that cleaning operations using flammable or combustible liquids must be conducted inside the spray booth or spray room, with ventilation operating.

[92] See OSHA 1910.126(b)(2). NFPA 33 does recognize an unenclosed spray area, such as spray painting structural steel members on the floor of a fabrication shop.

- The spraying operation must be enclosed.[93] Adequate mechanical ventilation must be provided so that there is enough inward air velocity to prevent the spray from leaving the vapor area.[94]
- Sound practice must be followed to prevent an ignition source. All metallic parts (including rotating parts) shall be bonded and grounded. Static collectors shall be installed.[95] Piping systems conveying flammable or combustible liquids shall be of steel or other material having comparable properties of resistance to heat and physical damage.[96] The tank should not be filled with air under pressure.[97]
- Containers supplying spray nozzles shall be of closed type or provided with metal covers kept closed.
- Containers not resting on floors shall be on metal supports or suspended by wire cables.
- Containers supplying spray nozzles by gravity flow shall not exceed 10-gal capacity.
- Original shipping containers shall not be subject to air pressure for supplying spray nozzles.
- Containers under air pressure supplying spray nozzles shall be of limited capacity:
    - not exceeding that necessary for 1-day operation;
    - shall be designed and approved for such use;
    - shall be provided with a visible pressure gage;
    - shall be provided with a relief valve.[98]
- Containers under air pressure supplying spray nozzles, air-storage tanks and coolers shall conform to the standards of the Code for Unfired Pressure Vessels, Chapter VIII of the ASME Boiler and Pressure Vessel Code – 1968 for construction, tests, and maintenance.[99]
- Finally, and obviously, there shall be no open flame or spark-producing equipment in any spraying area nor within 20-ft thereof, unless separated by a partition.[100]

### 3.9.4 Additional Considerations About Avoiding Aerosols and Mists

Three other significant restrictions affect use of solvents for cleaning in sprays:

1. There is no direct *de minimus* (small-size) exemption to the above requirements for spraying of solvents in air. A limit is implied in the definition of the word dangerous.[101] That is, in an aircraft maintenance shop, a worker removing nail polish via a small spray of acetone with air is not in a dangerous area. Another worker spraying D-limonene to remove painted insignia from aircraft wings is almost certainly in a dangerous area. Is a third worker at a maintenance bench using IPA to clean electronics or fuel injection components in a dangerous area? Apparently, that issue awaits a local decision with US OSHA guidance.
2. When a liquid with a flash point of 380°F is sprayed in air to produce an aerosol, the aerosol is to be treated as if it had a flash point of <73°F. By inference, a liquid without a flash point, but which does have an LEL and UEL, is similarly to be treated as if it had a flash point of <73°F (Class IA).[102] The inference comes from the fact that the solvent has an UEL/LEL:
    - That's why this author believes that *popular cleaning solvents, such as methylene chloride and n-propyl bromide, should be treated when*

---

93 See OSHA 1910.126(f)(1).
94 See OSHA 1910.126(g).
95 See OSHA 1910.126(c)(2).
96 See OSHA 1910.107(e)(7).
97 See OSHA 1910.126(b).
98 The relief valve should be set to operate in conformance with the requirements of the Code for Unfired Pressure Vessels, Chapter VIII of the ASME Boiler and Pressure Vessel Code – 1968, which is incorporated by reference as specified in Sec. 1910.6.
99 See OSHA 1910.107(e)(6).
100 OSHA 1910.107(c)(2). See also NFPA 410.
101 OSHA 1910.107(a)(2) defines the "spraying area" as any area in which dangerous quantities of flammable vapors or mists, or combustible residues, dusts, or deposits are present due to the operation of spraying processes. Personal communication between the author and Lisa Salters of OSHA, January 27, 2003 at (301)-515-6796.
102 OSHA 1910.107(d)(2) requires that all spraying areas shall be provided with mechanical ventilation adequate to remove flammable vapors, mists, or powders to a safe location and to confine and control combustible residues so that life is not endangered. Further, OSHA 1910.124(b)(1) requires ventilation adequate to prevent the vapor concentration from exceeding 25% of the lower flammable limit (LEL) of any flammable material. See also NFPA 77.

*sprayed in air as if they had a flash point and can catch fire.* The MSDS for each solvent states that it has no flash point (see Section 5.10.1, NFPA 34, and Section 3.7.4 of this book) but does report an LEL and a UEL. That means: it can catch fire!

3. US OSHA requires measurement of LEL when any areas in which dangerous quantities of flammable vapors or mists, or combustible residues, dusts, or deposits are present due to the operation of spraying processes.[103] Twenty five percent of the LEL is the allowable maximum.[104]

#### 3.9.4.1 *Mists versus Aerosols*

An aerosol is a suspension of solid or liquid particles in a gas. Dust, smoke, mists, fog, haze, and smog are various forms of common aerosols. Aerosol particles are found in different shapes (isometrics, platelets, and fibers) and different sizes. The range of diameters of common aerosol particles is between 0.01 and 100 mm (10–100,000 μm).

In general, the particles in an aerosol formed in solvent cleaning are liquid droplets, which are too small and stable to settle under the force of gravity. Particles in an aerosol can be kept aloft in air with a negligible velocity.

A mist is a failed aerosol. Here the particles of liquid are too large. They settle, slowly, as the buoyant forces aren't great enough to offset gravity.

Mists are almost always visible. Stable aerosols may not be so – because the particle size is too small.

But a *mist of liquid solvent should not be considered safer* than an aerosol containing the same solvent in air. Both contain rapidly evaporating solvent. Evaporation from an aerosol or mist is at a rate substantially greater than if the solvent were contained in a vessel because of the increased surface exposed to air.

#### 3.9.4.2 *Spray Cleaning*

Spray cleaning with pressurized solvents is usually effective, often convenient and can be harmful to the environment, and to users.

This author cannot recommend it unless the operation is managed as if it were a spray painting operation (see Sections 3.9.3 and 3.9.4).

If pressurized spray cleaning is essential, suitable aqueous-based materials (possibly containing particulate for a blast effect) should be used. That is deliberately beyond the scope of this book.

### 3.9.5 It's the Simplicity I Can't Figure Out

This complex guidance to managers appears to conflict with use of a simple "sink-on-a-drum" cleaning apparatus, or use of aerosol spray cans. It doesn't.

Many reputable suppliers have long understood these requirements. Their cleaning machines and products are designed to meet them. The machines are quite inexpensive, useful, and common. The products are quite safe if "used as directed."

The point of this section is for managers to understand the safety hazards of mists and aerosols, how/why cleaning machines and products are designed to avoid them, and how participation by them is necessary as well.

### 3.9.6 Fear and Loathing at the Wash Tank

Should users be afraid to do cold cleaning with solvents? Absolutely not! A properly designed "sink-on-a-drum" is quite safe. Probably more than 10,000 of these systems have been built and are being used in the US. They are safe to use because the US OSHA requirements are met. The US military has used them with mineral spirits cleaning agents (PD 680).

But don't even think about trying to "turbo charge" the unit by adding a high-pressure pump to increase cleaning power via a power spray, unless you enjoy watching fires.

### 3.9.7 Heating Liquids with Ignition Risk

There is a US regulation and an international standard which appear to conjoin:

- A user is not allowed to heat fluid contents of an open tank to within 100°F of its AIT (see Section 3.7.4, NFPA 34-2003, Section 5.10.1, and Section 3.74 of this book.).

---

[103] Weisenblatt, Irving, J., OSHA, Acting Director, Federal Compliance and State Programs, Standard Interpretation, March 20, 1979. see Also 1910.107.

[104] OSHA 1910.108(a)(2) or NFPA 34.

- When a combustible liquid (flash point >100°F) is heated to within 30°F (16.7°C) of its flash point, it shall be handled in accordance with the requirements for the next lower class of liquids.[105] This applies to open tanks only, not vapor degreasers, which are considered to be closed tanks.

Together, these three regulations:

1. Allow cold cleaning in open tanks with cleaning solvents whose flash points are between 100°F and 200°F (Classes II and IIIA).
2. Allow heating of these cleaning solvents to within 100°F of their AITs (above their flash points).[106]
3. Require the work to be done as if the solvent were classified as flammable (Class IB or IC if the boiling point is above 100°F and Class IA if the boiling point is below 100°F).

Said differently, cold cleaning work may be done with heated solvents, but they must be treated as if they are classified as flammable.

## 3.10 HOW CHEMICAL HAZARDS BECOME HUMAN DAMAGE

Bodily contact is the second (see Section 3.1) major hazard associated with the use of cleaning chemicals[107] in cleaning (or other) operations. This section is about what can happen if proper precautions are not taken to avoid improper human contact with chemicals.

This section is not a textbook about internal or external human medicine. A physician should be consulted when and as directed by the MSDS (see Section 3.19) for the chemical being used. This chapter describes the general contact hazards associated with use of chemicals, and how to prevent them.

The contact hazards of a chemical have no direct effect on its capability as a cleaning chemical. This is because the human body is much less able to retard effects of hazardous chemicals than are metals, plastics, glasses, or other substrates which are cleaned with chemicals.

However, contact hazards have a dominant indirect effect on selection of the cleaning process, cleaning equipment, and cleaning procedures, as well as packaging, transportation, selection, and disposal of the cleaning chemical.

In other words, in 1995 when corporate exposure limits (CEL) for *n*-propyl bromide were between 100 and 200 ppm, considerable cold cleaning work was planned; but in 2005, when the American Conference of Government Industrial Hygienists (ACGIH) threshold limit value (TLV) for *n*-propyl bromide is10 ppm, only cleaning work in enclosed machines is planned.

### 3.10.1 Routes of Entry

Simply, bodily contact refers to the three chief routes by which a chemical can enter a human body. Two are inadvertent: inhalation and skin contact. The third is usually intentional or inadvertent: ingestion. In the workplace the greatest risk is of skin damage, followed by skin absorption, inhalation, and ingestion of chemicals.

Some chemicals, such as strong acids and alkalis (e.g. chromic acid, sulfuric acid, nitric acid, sodium hydroxide) produce damage within a very short period of contact. In general, cleaning chemicals require prolonged, repeated contact before an effect is seen (e.g. liver damage and cancer by inhaled carbon tetrachloride, leukemia by inhaled benzene, allergic contact dermatitis from toluene and other chemicals). The effect on the user depends on the toxicity of the chemical, exposure time, amount, and individual susceptibility. Remember that the effects of long-term exposure to many chemicals are unknown.

Our bodies were not designed to protect us from hazardous chemicals, but from infecting agents (pathogens) and physical injuries. These features of our bodies are worth noting as they are our first line of defense against potential ravages of hazardous chemicals.

---

[105] See OSHA 1910.106(a)(18)(iii).

[106] If heating to within 100°F of the autoignition temperature is necessary for success of the cleaning operation, perhaps a piece of equipment other than an open tank should be used. This might be a vapor degreaser, which is designed to contain the significant level of evaporated solvent which will be produced in such a heating operation. Such a change will add immeasurably to safety and control of pollution.

[107] Information in this chapter applies to use of both chemicals and aqueous cleaning agents.

Finally, not all cleaning chemicals are hazardous to our bodies. We use acetone to clean some soils from parts and polish from our fingernails. We use isopropyl alcohol to remove some ionic residues from electronic parts and to disinfect our skin. We use limonene diluted in water to clean grease from kitchen appliances. And water is the most commonly used of all chemicals. But without any information, the safest practice for users is to assume that all chemicals are hazardous to their bodies.

The approach in this section will be to propose a first line of defense against the three main routes of bodily contact (inhalation, ingestion, and skin contact), and then cite the expected effects of that contact.

### 3.10.2 Damage by Inhalation

Inhalation is the most common route by which cleaning chemicals enter the body. That's why exposure limits, which don't speak to either skin contact or ingestion, are almost a universal parameter used to characterize the level of hazard presented by a chemical. This attitude, while understandable because of its simplicity, does not serve managers well because it exposes *them* to risks.

#### 3.10.2.1 *First Line of Defense*

Odor, a property of nearly all chemicals, is often the first indication of trouble. In general, humans can smell the presence of chemical at levels well below when the chemical can start to produce harm.

The cardinal rule is that if you can smell a chemical, use protective equipment (see Section 3.15.2) to prevent you and your staff from being exposed to the chemical.

Many examples of this rule are shown in Table 3.14,[108–111] where the odor threshold is considerably below the recommended TLV or exposure limit.

Note that there is no common reference for measurements of odor threshold.[112,113]

Unfortunately, the cardinal rule and the information in Table 3.14 is not all-inclusive. There are a few exceptions – some of which prove the rule:

- A significant one is formaldehyde. Formaldehyde is a potent irritant, a skin sensitizer and a carcinogen. The current permissible exposure limit (PEL) for formaldehyde is 0.75 ppm and the odor threshold for most people is 1 ppm. *Under no circumstances should formaldehyde be used as a cleaning chemical.*
- The exceptions include vinyl chloride, sulfur dioxide, quinone, phosgene, Chlorine, carbon tetrachloride, and carbon monoxide. Granted, only one of these chemicals (carbon tetrachloride) has been used as a cleaning chemical, but the odor threshold of each is below or around the recommended 8-hour average exposure limit. Here, one would be exposed without apparent recognition by smell.

### 3.10.3 Effects of Chemical Inhalation

Actual irritation of nasal passages may be the second line of defense from potential damage of inhaled

---

[108] US Coast Guard. *Chemical Hazards Response Information System (CHRIS) Manual.*

[109] Bureau of Explosives, American Association of Railroads (AAR). Emergency Action Guides, Washington, DC: AAR, 1996.

[110] American Industrial Hygiene Association (AIHA). *Odor Thresholds for Chemicals with Established Occupational Health Standards.* Akron, OH: AIHA, 1989. (see AIHA website, http://www.aiha.org).

[111] National Institute for Occupational Health and Safety (NIOSH), US Department of Health and Human Services (DHHS), *NIOSH Pocket Guide to Chemical Hazards.* This excellent, and free, reference contains lists of: TLVs , permissible exposure limits (PELs), and immediately dangerous to life and health (IDLH) values, as well as general industrial hygiene information for 398 chemical substances.

[112] This is because not all humans have the same sense of smell. To minimize this effect, odor threshold is determined by groups, called panels, of persons who differ in both in age and ethnicity. A odor panel, composed of persons who don't smoke and don't have a chronic allergy, will sniff the sample. They start with a very high dilution (small amount of sample to large amount of clean air). If the panelist cannot determine the difference in three presentations, the panel leader will then decrease the dilution by increasing the amount of clean air and the process will begin again. This will continue until the panelist can detect a difference in the three presentations. The concentration at that dilution is the detection limit for that panelist. Multiple outcomes are averaged.

[113] An excellent, though somewhat outdated in terms of exposure limits, reference is Amoore, J.E. and Hautla, E., "Odor as an Aid to Chemical Safety: Odor Thresholds Compared with Threshold Limit Value and Volatilities for 214 Industrial Chemicals in Air and Water Dilution," *Journal of Applied Toxicology*, 1983, Vol. 3, No. 6.

**Table 3.14 Odor Threshold and Exposure Limit of Chemicals**

| Compound | Odor threshold (ppm) | | | |
|---|---|---|---|---|
| Chemical | CHRIS | AAR | AIHA | TLV (ppm) |
| Acetaldehyde | 0.21 | 0.01–0.031 | 0.0028–1000 | 25 |
| Acetone | 100 | 0.66–320 | 0.037–0.15 | 750 |
| Ammonia (anhydrous) | 47 | 0.037–20 | 0.043–53 | 25 |
| Benzene | 4.7 | 0.16–320 | 0.78–160 | 10 |
| Carbon monoxide | Odorless | Odorless | Odorless | 25 |
| Carbon tetrachloride | >10 | 15–50 | 1.6–706 | 5 |
| Chlorine | 3.5 | 0.02–3.5 | 0.021–3.4 | 0.5 |
| Cumene | 1.2 | | 0.0051–1.3 | 50 |
| Cyclohexane | – | 0.41 | 0.52–784 | 300 |
| Dicyclopentadiene | 0.003 | 0.02 | 0.003–0.011 | 5 |
| Dioxin | – | – | 12 | 100 |
| Ethyl benzene | 140 | 0.25–2.3 | 0.092–0.60 | 100 |
| Ethyl mercaptan | – | – | 0.00035 | 0.5 (PEL) |
| *n*-hexane | – | – | 65–248 | 50 |
| Hydrogen sulfide | 0.0047 | 0.13 | 0.00007–1.4 | 10 |
| Methyl ethyl ketone | 10 | 11–27 | 0.25–85 | 200 |
| Naphthalene | n/a | 0.3–0.9 | 0.0095–0.64 | 10 |
| Phenol | 0.05 | – | 0.0045–1 | 5 |
| Phosgene | 0.5 | 0.125–1 | 0.12–5.7 | 0.1 |
| Phosphine | 0.14 | 0.02 | 0.01–5 | 0.3 |
| Quinone | – | – | 0.068 | 0.092 |
| Styrene monomer | 0.148 | 0.02–0.47 | 0.0047–61 | 50 |
| Sulfur dioxide | 3 | 3 | 0.33–5 | 2 |
| Toluene | 0.17 | 0.17–40 | 0.021–69 | 50 |
| Vinyl chloride | 260 | 260–25 000 | – | 5 |
| o, m, *p*-Xylene | 0.05 | 0.2–4 | 0.081–5.4 | 100 |

PEL, permissible exposure limit.

chemical vapors. Here the damage may have already started, depending upon the nature of the chemical and the affected person.

In this paradigm, irritancy (or other toxicity) generally occurs at a concentration somewhat higher (about 3 to 10 times higher) than the concentration at which odor is first detected (odor threshold).[114] Adherence to the cardinal rule prevents this irritation (Figure 3.25).

A person perceiving nasal irritation should remove themselves from the environment where they were

**Figure 3.25** UK irritancy symbol

[114] Walker, John, M., *et al.*, US EPA, *Potential Health Effects of Odor From Animal Operations, Wastewater Treatment, and Recycling of Byproducts*, A Workshop held at Duke University, Duke University on April 16–17, 1998.

effected. In this way, additional tissue damage may be prevented. In this way, nasal irritation should produce the same human reaction as should detection of odor – removal from the chemical-laden environment.

Nasal irritancy may be a harbinger of collateral damage. Chemical vapors which affect nasal passages often affect the tissue of the eye (see Footnote 1).

Irritation means that personal protective devices were not used as recommended or were ineffective. Headache, dizziness, and nausea resulting from inhalation may follow. If the condition persists, medical attention should be sought.

**Figure 3.26** UK very toxic symbol

But the effect may be more lethal. The likelihood of becoming sick from chemicals is increased as the amount of exposure increases. This is determined by the length of time and the amount of material to which someone is exposed – one can drink more than one can inhale.

### 3.10.4 Damage by Ingestion

Movies often don't illustrate reality. A character seen to drink poison quickly collapses, usually without extended demonstration of pain. In the real world, a person drinking a harmful or toxic chemical is in pain, demonstrates it, and usually continues to do so.

#### 3.10.4.1 *First Line of Defense*

Exposure through ingestion can occur by accident or by consuming contaminated food or drink or by eating food with contaminated hands or utensils. People have unwittingly drunk chemicals which have been kept in old, unlabeled drink containers.

The first line of defense is the taste of the chemical. That should cause one who has accidently, or intentionally, ingested some chemical to take notice and raise concern. Vomiting may be induced to get rid of most chemicals. In such cases dilution with water or milk may assist.

If a person knows that they have swallowed a substance they know or believe is toxic, they should immediately seek medical attention. Information on the MSDS or container label will prove invaluable (Figure 3.26).[115]

#### 3.10.4.2 *Effects of Chemical Ingestion*

Fortunately this contact is less frequent than the other two mechanisms of bodily contact. Further, the volume of ingested chemical may be low.

### 3.10.5 Skin Contact

For this author, the two major hazards of using chemicals intersect. In separate incidents in laboratories, this author was injured by both fire and repeated skin contact with chemicals (toluene). It's not obvious which is least pleasant. Impact by both does not quickly dissipate. Lessons from both incidents are in this volume.

#### 3.10.5.1 *First Line of Defense*

The human body possess number of physical and chemical barriers that prevent entry of pathogens or hazardous chemicals (Figure 3.27).

Of these, perhaps the most important physical barrier is the skin. The skin consists of two distinct layers: a relatively thin outer epidermis and a thicker layer, the dermis.

The epidermis consists of several layers of tightly packed epithelial cells that are dead and filled with a water-proof protein called keratin. Therefore, it acts as a physical barrier against entry of hazardous chemicals into the body.

The dermis contains a gland, called the sebaceous gland, that produces an oily secretion called sebum.

[115] While MSDSs have become marketing tools, versus information resources, where guidance in an MSDS is offered about action after human contact, for legal reasons that guidance is apt to be complete and sound.

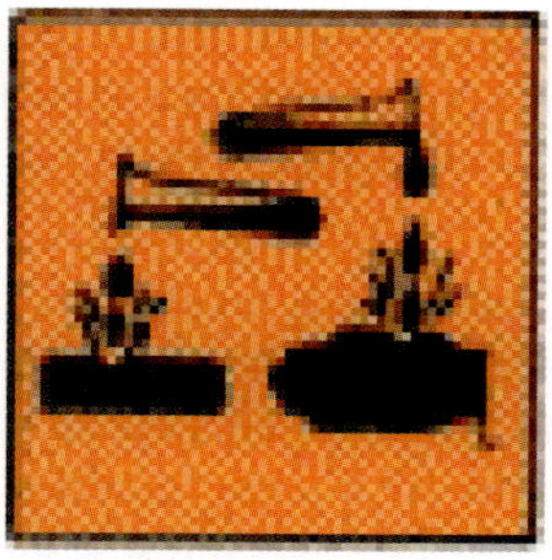

**Figure 3.27** UK corrosive symbol

Sebum consists of number of organic acids that maintain the pH of the skin between 3 and 5.[116]

Therefore, intact skin not only prevents entry of pathogens or hazardous chemicals but also inhibits the growth of most pathogenic bacteria due to its low pH.[117]

However, the skin does not cover the entire surface of the human body. Conjunctiva of the eye, alimentary, respiratory, and urinogenital tracts are not covered by dry, protective skin but by mucous membranes. Therefore, these places function as potential entry sites for pathogens or hazardous chemicals.

### 3.10.5.2 *Effect of Chemicals on Human Skin*

Skin exposure to organic chemicals can cause several problems. The three major ones are: irritant contact dermatitis, allergic contact dermatitis (see Section 3.10.5.2.1), and scleroderma.[118]

3.10.5.2.1 *Dermatitis* The most common problem is the former – irritant contact dermatitis. Industrial workers exposed to organic chemicals[119] are notable for demonstrating its effects.

Organic chemicals cause skin irritant contact dermatitis in two ways:

1. One is defatting.[120,121] Here the chemical dissolves (cleans) lipids, which are fats and oils, from beneath the outer surface (epidermis) of skin. This is easily seen as whitening after the skin is rubbed with some chemicals. This damage is reversible,[122] and usually not unduly painful.
2. The second type of dermatitis is skin irritation. Skin irritation is visualized as erythema (abnormal redness of the skin due to capillary congestion) and edema (abnormal infiltration and excess accumulation of serous fluid in connective tissue); the result of a local inflammatory process.[123] Generally, this damage is also reversible, (see Footnote 113) and while not generally painful can be extremely annoying.

Users of cleaning chemicals must know if skin contact with a specific chemical will produce irritant contact dermatitis. General studies[124] have shown

---

[116] See http://www.geocities.com/CollegePark/Quad/1267/indefense.htm

[117] Gerberick, G.F., PhD, Procter and Gamble Co., Cincinnati, OH, US, "The Importance of Exposure and Potency in the Assessment of Skin Sensitization Risk" Presented at the *International Conference on Occupational And Environmental Exposures of Skin to Chemicals: Science & Policy*, Hilton Crystal City, September 8–11, 2002. Or http://ns3.hgo.net/niosh_conf/s2t4.asp

[118] From the International Scleroderma Network (http://www.sclero.org/medical/about-sd/a-to-z.html). "... The systemic forms of scleroderma cause fibrosis (scar tissue) to be formed in the skin and/or internal organs. The fibrosis eventually causes the involved skin or organs to harden, which is why scleroderma is commonly known as the "disease that turns people into stone." ..."

[119] Wahlberg, J.E. and Boman, A. "Prevention of Contact Dermatitis from Solvents," *Current Problems in Dermatology, 1996*, Vol. 25, pp. 57–66.

[120] Goldsmith, L.B., Friberg, S.E. and Wahlberg, J.E., "The Effect of Solvent Extraction on the Lipids of the Stratum Corneum in Relation to Observed Immediate Whitening of the Skin," *Contact Dermatitis*, 1988, Vol. 19, pp. 348–350.

[121] Abrams, K., Harvell, J.D., Shriner, D., *et al.*, "Effect of Organic Solvents *In Vitro* Human Skin Water Barrier Function," *Journal of Investigative Dermatology*, 1993, Vol. 101, pp. 609–613.

[122] Wahlberg, J.E. "Erythema-Inducing Effects of Solvents Following Epicutaneous Administration to Man: Studied by Laser Doppler Flowmetry," *Scandinavian Journal of Work and Environmental Health*, 1984, Vol. 10, pp. 159–162.

[123] Wahlberg, J.E., "Erythema-Inducing Effects of Solvents Following Topical Administration," *Dermatosensitivity,* 1984, Vol. 3, pp. 91–94.

[124] Minako, I., Masayoshi, I., Jiusong, Z., Masafumi, E. and Katsumaro, T., "Evaluation of Skin Irritants Caused by Organic Solvents by Means of the Mouse Ear Thickness Measurement Method," *Journal of Occupational Health*, 2000, Vol. 42, pp. 44–46.

that this could be predicted before human exposure via laboratory experiments with animals versus epidemiological studies about past human exposure. The animals were guinea pigs, rats, and rabbits. Often the inside skin of the animal ear is exposed to chemical and observed to see if was swollen after a certain time interval.

A notable study (see Footnote 117) involved several common cleaning chemicals: toluene, *m*-xylene, trichloroethylene, 1,1,1-Trichloroethane, *n*-hexane, methyl ethyl ketone, butanol, and acetone. One conclusion of the study would have been expected by students of cleaning science. Skin lesions are more severe on skin which is exposed to lipophilic (oil-loving) organic chemicals (toluene, *m*-xylene, *n*-hexane, 1,1,1-Trichloroethane, and trichloroethylene) and less severe in more water-soluble (methyl ethyl ketone, acetone, and ethanol) chemicals.

3.10.5.2.2 *Dermatitis and Sensitization* Roughly three-quarters of all humans whose skin is irritated by exposure to chemicals recover within hours.[125] There is no permanent or semi-permanent effect from irritant contact dermatitis.

The other quartile is not as fortunate. These people are born with or develop a *hypersensitivity to chemicals*. When exposed to allergens (usually organic substances or chemicals), they experience an allergic reaction. To most people these allergens are harmless.

Their exposure to chemicals leaves a permanent consequence – allergic contact dermatitis. Allergic contact dermatitis is a reaction that occurs when the skin comes into contact with a substance to which the body is allergic (see Footnote 125).

Basically the difference between allergic contact dermatitis and irritant contact dermatitis is found in the stimulus/response ratio:

- Allergic contact dermatitis produces essentially the *same* skin irritation, allergic reaction, with every exposure.
- Irritant contact dermatitis produces a skin irritation with a *significantly less* exposure time or a lesser amount produced that effect. Alternately, the normal exposure produces a much more severe skin irritation which last longer.

Said another way, long-term effects can occur from repeated exposures to a chemical at levels not high enough to make one immediately sick.

The diagnosis of allergic contact dermatitis is established by positive patch test results and a thorough occupational history.

Fortunately most chemicals used for cleaning are not among those which provoke an allergic reaction. Some chemicals which *do* are: benzyl alcohol (CAS #100-51-6); cyclohexanone (CAS #108-94-1): 1,2-Dichlorobenzene (CAS #95-50-1); 1,3-Dichloropropene (CAS #542-75-6); diethylene dioxide (CAS #123-91-1); hexylene glycol (CAS #107-41-5); *D-Limonene* (CAS #138-86-3); propylene glycol (CAS #57-55-6); *turpentine* (CAS #8006-64-2); and of course toluene (CAS # 108-88-3).

Workers who have become sensitized to a particular agent may also exhibit cross-reactivity to other agents with similar chemical structures.[126]

A reduction in exposure to the sensitizer and its structural analogs may help to reduce the incidence of allergic reactions among sensitized individuals. For some sensitized individuals, however, complete avoidance in occupational and non-occupational settings provides the only means to prevent the immune responses to recognized sensitizing agents and their structural analogs.

Occupational allergic contact dermatitis can be avoided by personal hygiene, engineering control methods, good housekeeping, and personal protection:

- Personal hygiene, including hand washing, is very important to prevent contact dermatitis.
- Engineering control methods involve the enclosure of processes to separate workers from the harmful substances they work with. Local exhaust systems should be used where toxic substances may escape into the workroom. Non-hazardous substances should be substituted for hazardous substances.

[125] Brown, J.A., M.D., M.P.H., "Information on Hazardous Chemicals and Occupational Diseases," National Institute of Health, http://www.haz-map.com/workers.htm

[126] Benzene and toluene, for example.

- Good housekeeping includes proper storage of substances, frequent disposal of waste, prompt removal of spills, and maintenance of the equipment to keep it free of dust, dirt, and drippings. Articles heavily contaminated with chemical should be removed immediately when contamination occurs.

3.10.5.2.3 *Scleroderma* Fortunately rare, scleroderma is a connective tissue[127] disorder characterized by abnormal thickening of the skin. There are several types of scleroderma. Some types affect certain, specific parts of the body, while other types can affect the whole body and internal organs (systemic).

The cause of scleroderma is generally unknown. Environmental exposure to chemicals is one of the prime areas being investigated. Others include autoimmunity, genetics, and infections. There may be a genetic inclination along with exposure to a chemical or infection which triggers the illness.

## 3.11 HUMAN TOXICOLOGY AS AFFECTED BY CLEANING CHEMICALS

The following summary is not a substitute for a medical textbook. A qualified physician or industrial toxicologist should be consulted for any recommendations, or details.[128] This summary will focus on human organs or groups of organs providing valued functionality, and chemicals used in cleaning operations.

This summary is not an argument against chemical (specifically solvent) cleaning. Rather it is a realistic list of problems which are overcome by others who are safely doing cleaning work using chemicals.

Since most solvents are lipid (oil) soluble, solvents distribute to lipid-rich tissues like the central nervous system and to tissues with high blood flow like heart and liver.

Most chemicals have relatively short residence times in the body: from a few hours to a few days.

### 3.11.1 Central Nervous System

There is evidence[129] of varying quality that several cleaning chemicals can affect the central nervous system. Of concern as cleaning solvents are acetone, methyl ethyl ketone, methanol, xylenes, toluene, benzene, carbon tetrachloride, and trichloroethylene. Fortunately, most of these solvents are now used in cleaning work in machines which can and do meet the exposure limits thought to be necessary for safe exposure.

Solvents can have major acute (lasting a short time or requiring a short exposure) toxic effects on the central nervous system. This is because organic solvents usually act as anesthetics (substances which

---

[127] Connective tissue is composed of collagen, which supports and binds other body tissues.

[128] However, for a manager to do their own research, an absolutely vital internet reference is TOXINET at http://toxnet.nlm.nih.gov/

[129] There is a basic problem here. In general, there are few (fortunately) instances where statistically solid cause and effect relationships are established between an exposure in humans and damage to humans. A good example is the reference: Hageman, G., *et al.*, "Parkinsonism, Pyramidal Signs, Polyneuropathy, and Cognitive Decline after Long-term Occupational Solvent Exposure," *Journal of Neurology*, 1999, Vol. 246, No. 3, pp. 198–206. The reference describes three patients who had been exposed to various solvents for more than 20 years (25, 34, and 46 years). It concludes with the statement "… There is growing evidence that various organic solvents give rise to a Parkinsonism syndrome with pyramidal features in susceptible individuals. …"

Without question, the lives of these persons have been harmed. Was it the exposure to chemicals (methanol, toluene, carbon disulfide, and hexane) which caused the harm? The paper can't conclude that to the standard necessary for publication in a peer-reviewed scientific journal. Could a lawyer induce a jury to conclude that in a trial? The standard necessary for conviction may be different than that for publication.

So how should a manager protect people in his organization and use valued technology which involves chemicals? This author believes the wise approach is choose what works, AND rigorously enforce *exposure limits*. Don't fear use of chemicals. Don't ignore previous experience. Don't ignore science!

Rely on, and maintain compliance with, *exposure limits* – especially those which are established by *independent* agencies such as ACGIH (American Congress of Governmental Industrial Hygienists) or NIOSH (US National Institute for Occupational Safety and Health) or OSHA (US Occupational Safety and Health Administration). These limits are determined based on a statistical of an overall body of laboratory tests and human experience by a group of unbiased professionals without a vested interest in the outcome.

produces loss of sensation with or without loss of consciousness). Here the function of the central nervous system is depressed and the subject is greatly disoriented. Further, there may be intoxication to coma. Subjects have been noted to develop tolerance for this exposure to solvents.

Longer-term chronic (marked by long duration or frequent recurrence: not acute) effects often include neuropsychological dysfunction. Although neuropsychological dysfunction may be less severe than stroke, it can still be very disabling for patients and their families. Symptoms can include memory loss and changes in personality or in mental ability. Here there is a change in mood, personality, memory, cognition, etc. Other symptoms are appearance of being drunk, drowsiness (narcosis), tiredness, irritability, difficulty in concentrating, memory loss, and dementia.

### 3.11.2 Peripheral Nervous System

There is also evidence of varying quality that several cleaning solvents can affect the peripheral nervous system. The solvent of most concern is hexane. Also of concern are the cleaning solvents toluene, methyl butyl ketone, and xylene.[130]

While the central nervous system consists of the spinal cord and the brain, the peripheral nervous system is subdivided into the sensory-somatic nervous system and the autonomic nervous system. This includes control of other bodily functions or parts such as muscles, eyes, the heart, ears, taste, and lungs.

In some cases, a loss of control of these body parts has been associated in rats with inhalation exposure to propylene oxide.[131] Here, propylene oxide has been identified as a neurotoxin because the rats developed peripheral neuropathy,[132] a debilitating nerve disease.

Peripheral neuropathy causes numbness and tingling in the fingers and toes followed by progressive weakness and loss of feeling in the arms and legs. In severe cases, total loss of sensory perception in the hands and feet occurs, followed by muscle wasting. The disease progresses even after a person is no longer exposed to hexane, and it may take two or more years to recover, with no assurance of complete recovery. One person affected with peripheral neuropathy couldn't feel his arms or legs, and had lost so much motor control that he collapsed on the waiting room floor.

A *neurotoxin* produces disease of the nerve cells. Neurotoxins can involve both spinal and extremity nerves. Each nerve cell has one axon, which can be over a foot long. Axons connect nerve cells to the components of the peripheral nervous system. This disease is called an axonopathy and is a disorder characterized by axon swellings and secondary degeneration.

Over the past 25 years substantial research efforts have been devoted toward deciphering the molecular mechanisms of these presumed hallmark neuropathic features. However, recent studies suggest that axon swelling and degeneration are related to *subchronic low-dose neurotoxicant* exposure rates (i.e. mg toxicant/kg/day).[133]

Workers with hexane-related peripheral neuropathy have been reported in such workplaces as printing plants, sandal shops, and furniture factories throughout the world. In 2000, three workers using hexane-based brake cleaner were found to have peripheral neuropathy.[134,135]

Recognition that certain straight-chain (aliphatic) organic solvents have the potential to cause peripheral

[130] Sabri, M. and Spencer, P., *Research Brief 92: Biomarkers of Chromogenic Solvent Exposure and Neurodegeneration*, Oregon Health & Science University, August 16, 2002, and http://www-apps.niehs.nih.gov/sbrp/rb/rbs.cfm?Resbrfnum=92&view="
[131] Other solvents probably can contribute to neuropathy Ohnishi, A., Yamamoto, T., Murai, Y., Hayashida, Y. and Hori Tanaka, I., Propylene oxide causes central–peripheral distal axonopathy in rats. *Archives of Environmental Health,* 1988, Vol. 43, No. 5, pp. 353–356.
[132] Pronounced "Nur-Op-Ah-Thee."
[133] LoPachin, R.M., Lehning, E.J., Opanashuk, Lisa, A. and Jortner, B.S., "Rate of Neurotoxicant Exposure Determines Morphologic Manifestations of Distal Axonopathy," *Toxicology and Applied Pharmacology*, 2000, Vol. 167, pp. 75–86.
[134] Hexane-Induced Peripheral Neuropathy, Chronic Toxicity Summary – Hexane, http://www.oehha.ca.gov/air/chronic_rels/pdf/110543.pdf
[135] Wilson, M., "Phasing Out One Type of Health Hazard May Increase Another, Research Shows," BRIDGES – Center for Occupational and Environmental Health, December 2000, p. 1.

neuropathy is longstanding.[136] Recently it was established was that aromatic (ring-structure) solvents have this property.

### 3.11.3 Respiratory System

Solvents can have major acute (lasting a short time or requiring a short exposure) irritating effects on the respiratory tract.[137] Some solvents may produce chronic irritation and bronchitis/asthma.

It is water solubility that determines the level in the respiratory tract where the harmful effect occurs.

### 3.11.4 Cardiac System

Some chemical vapors, after being inhaled, are absorbed into the blood via the lungs. The absorbed chemical is then transported to the heart and other organs. The heart can be stimulated to abnormal vibrational activity levels by these chemicals.

This effect is called cardiac sensitization. It is so serious that there is an ASTM test for it.[138]

Halogenated alkanes (trichloroethylene, fluorocarbons, 1,1,1-Trichloroethane, perchloroethylene, and others) cause cardiac arrhythmias (an alteration in rhythm of the heartbeat either in time or force) and sudden death by altering cardiac sensitivity to endogenous catecholamines (drugs synthesized in the body that are released upon sympathetic nervous system activation).[139–141] Sudden death has been reported from inhaling typewriter correction fluid containing 1,1,1-Trichloroethane.[142]

Chemicals which stimulate cardiac activity make the heart more sensitive to normal human activity, which generates adrenaline, or to epinephrine (synthetic adrenaline). The result can be arrhythmias, which can be fatal.

### 3.11.5 The Liver

As well as storing vitamins and Iron, regulating blood sugar levels, and its role in digestion, the liver metabolizes foreign chemicals. Metabolism sometimes results in a chemical being changed into a more toxic specie (metabolite), and this is one way in which chemicals can damage the liver.[143]

As with other organs, liver toxins can be grouped together according to the kind of liver disease they cause, including acute hepatitis (inflammation of the liver) or chronic diseases such as cirrhosis and cancer. These diseases can also be caused by viruses such as hepatitis B, an important risk for health-care workers, as well as non-occupational factors.

Chemicals that cause acute hepatitis include carbon tetrachloride,[144] chloroform, dinitrophenol, dinitrobenzene, dioxin, polychlorobiphenyls, the pesticide dichlorodiphenyltrichloroethane (DDT), chlordecone, chlorobenzenes, the anesthetic halothane, the dye feedstock methylenedianiline, and the explosive trinitrotoluene (TNT).[145]

Also, exposure to more than one solvent at a time can cause synergistic enhancement of the hepatotoxicity.[146]

---

[136] It is interesting to note how this tendency is tightly focused in the straight-chain aliphatic compound *n*-hexane. This is seen by comparing the 8-hour exposure limits recommended byACGIH: butane ($C_4$) 500 ppm, pentane ($C_5$) 600 ppm, hexane ($C_6$) 50 ppm, cyclohexane (300 ppm), Heptane ($C_7$) 400 ppm, and octane ($C_8$) 300 ppm (see Table 3.24).

[137] London Hazards Centre Trust, *Chemical Hazards Handbook*, Interchange Studios, 1999, and http://www.lhc.org.uk/members/pubs/books/chem/chAAAAAA.htm

[138] ASTM E1674-99 *Standard Test Method for Cardiac Sensitization Study on Dogs*.

[139] Abedin Z., Cook R.C. and Milberg R.M., "Cardiac Toxicity of Perchloroethylene," *Southern Medical Journal*, 1980, Vol. 73, pp. 1081–1083.

[140] Reinhardt C., *et al.*, "Epinephrine-Induced Cardiac Arrhythmia's Potential of Some Common Industrial Solvents," *Journal of Occupational Medicine*, 1973, Vol. 15, pp. 953–955.

[141] Magos L., "The Effects of Industrial Chemicals on the Heart," In: T. Balazs (ed.), *Cardiac Toxicology*, CRC Press, 1981, pp. 206–207, ISBN 0849355559.

[142] King G.S., "Sudden Death in Adolescents Resulting from the Inhalation of Typewriter Correction Fluid," *Journal of the American Medical Association*, 1985, Vol. 253, pp. 1604–1606.

[143] Methylene chloride, metabolized to CO, may have acute cardiac effects. See http://www.pactox.com/organicsolvents.htm

[144] Folland D.S., *et al.*, "Carbon Tetrachloride Toxicity Potentiated by Isopropyl Alcohol, Investigation of an Industrial Outbreak," *Journal of the American Medical Association*, 1976, Vol. 236, pp. 1853–1856.

[145] Zimmerman H., *Environmental Hepatotoxicity*, Chapter III, Appleton-Century-Crofts/New York, 1978, pp. 279–345.

[146] Harris R.N., Harris J., Garry V.F. and Anders M.W., "Interactive Hepatotoxicity of Chloroform and Carbon Tetrachloride," *Toxicology and Applied Pharmacology*, 1982, Vol. 63, pp. 281–291.

Symptoms of acute hepatitis include headache, nausea, vomiting, dizziness, and drowsiness.

Only carbon tetrachloride and chloroform have been used as cleaning solvents, and are only seldom used today for the above reason.

### 3.11.6 The Kidneys and Urinary Tract

The kidneys, via urine, are the major route by which toxic chemicals are excreted from the body. Because of this, and the way the kidneys do their job, they are vulnerable to the toxic effects of chemicals. The damage caused is complicated, and in many cases still not well understood, but can result from acute and chronic exposure.

Proven or suspected kidney toxins (nephrotoxins) include toxic elements (Arsenic, Beryllium, Lead, Cadmium, Mercury, and Uranium), pesticides, and halogenated hydrocarbons. Aromatic amines are one group of chemicals known to cause bladder cancer. Human studies[147] have suggested glomerulonephritis[148] from exposure to halogenated hydrocarbons. Most other solvents have not been associated with kidney damage.

### 3.11.7 Blood

Benzene and chlorinated hydrocarbons[149,150] can cause aplastic anemia[151] and leukemia.[152]

### 3.11.8 Reproductive System

Teratogenicity[153] is the occurrence of structural malformations in a developing fetus when a substance is administered during pregnancy.

This terrible outcome is an excellent example of how toxicology science is learned today. Sanity dictates that no one would volunteer to participate in a possibly harmful exposure.

Another valid approach is to use anecdotical data where humans have been exposed in their normal course of living and working. This approach, in our industry, is called epidemiology.

An example is a recent study (see also Footnote 24);[154] 125 expectant mothers, in their first trimester, were identified who had been exposed in their work to "aliphatic and aromatic hydrocarbons, phenols, trichloroethylene, xylene, vinyl chloride, acetone, and related compounds" in their work environment. Major birth defects were seen in 13 children of mothers in the occupationally exposed group. In contrast, only one child with a birth defect was born to a woman in the control group (no occupational exposure) – 125 women matched to the study group. The authors conclude: "The main limitations of the study were its small size – only 125 occupationally exposed women and 13 children with birth defects – and the diverse exposures. It is not possible to link the cases to any particular solvents or occupations. Nor can one draw general conclusions about the reproductive hazards of solvents. Moreover, it would be expensive and impractical to conduct larger studies".

Consequently, tests with laboratory animals provide nearly all the information that toxicologists use to determine which exposures are safe and not. One of the many difficulties in developing this science is that laboratory animals are not human beings. Data found by animal testing is not always directly translatable to human exposure.

Nonetheless, animal experiments have shown that results such as teratogenicity have been found when laboratory animals are exposed to glycol ethers, chlorinated hydrocarbons, and ethanol. A long list of known and suspected teratogens has been published.[155]

---

[147] Ravnskov U., Forseberg B. and Skerfving S., "Glomerulonephritis in Exposure to Organic Solvents," *Acta Medica Scandinavica*, 1979, Vol. 205, pp. 575–579.

[148] A kidney disease. The kidneys' filters become inflamed and scarred and slowly lose their ability to remove wastes and excess water from the blood to make urine. The damage is irreversible.

[149] Greenberg, M.I. (ed.), *Occupational, Industrial, and Environmental Toxicology*, Mosby, St. Louis, 1997.

[150] Hathaway, G.L., *et al.*, *Chemical Hazards or the Workplace* (4th ed.), Von Nostrand Reinhold, New York, 1996.

[151] Aplastic anemia is a rare but extremely serious disorder that results from the unexplained failure of the bone marrow to produce blood cells.

[152] Leukemia is an acute or chronic disease characterized by an abnormal increase in the number of white blood cells in the tissues and often in the blood.

[153] Frazier, L.M. and Hage, G.M., *Reproductive Hazards of the Work Place*, Von Nostrand Reinhold, New York, 1998.

[154] *Journal of the American Medical Association*, 1999, Vol. 281, pp. 1106–1109, or http://www.ccohs.ca/headlines/text77.html

[155] Sax, N.I. and Richard, J., *Dangerous Properties of Industrial Materials* (9th ed.), Von Nostrand Reinhold, 1996, ISBN: 0442020252. See http://ptcl.chem.ox.ac.uk/MSDS/teratogens.html

In addition, a fetal solvent syndrome (similar to fetal alcohol syndrome) has been identified.[156] The best advice is the most common – expectant mothers should not work with solvents during their pregnancy.

## 3.12 CARCINOGENS

Carcinogens are agents that can cause cancer.

Carcinogenic effects are believed to be caused by cumulative exposure over all levels of dosage. From significant research, we believe a small number of molecular events can evoke changes in a single cell that can lead to uncontrolled cellular proliferation. This mechanism for carcinogenesis is referred to as "non-threshold," since there is theoretically no level of exposure for such a chemical that does not pose a small, but finite, probability of generating a carcinogenic response.

Non-carcinogenic effects, unlike carcinogenic effects, are believed to have a threshold; that is, a dose below which adverse effects will not occur.

A chemical is considered to be a carcinogen if:

- It has been evaluated by the International Agency for Research on Cancer (IARC) and found to be a carcinogen or potential carcinogen (see Table 3.15); or
- It is listed as a carcinogen or potential carcinogen in the Annual Report on Carcinogens published by the National Toxicology Program (Ref. [156], 9th ed.); or
- It is regulated by OSHA as a carcinogen.

Tests with animals and epidemiological analysis suggest that certain chemicals are (see Table 3.16), or might reasonably expected to be (see Table 3.17), human carcinogens.[157]

It should be noted that some manufacturers and some users do not agree with the selection of chemicals on these two lists.

The IARC classification system of risk to humans has four groups (see Table 3.15).

Most aren't cleaning solvents because users have been informed of this classification. Those chemicals which are or have been cleaning solvents and are "reasonably anticipated to be human carcinogens" are: carbon tetrachloride, chloroform, dichloromethane (methylene chloride), 1,2-Dichloroethane, tetrachloroethylene (perchloroethylene), tetrafluoroethylene, and trichloroethylene.

**Table 3.15 IARC Classification Scheme**

| | |
|---|---|
| Group 1 | The agent (mixture) is carcinogenic to humans. The exposure circumstance entails exposures that are carcinogenic to humans. |
| Group 2A | The agent (mixture) is probably carcinogenic to humans. The exposure circumstance entails exposures that are probably carcinogenic to humans. |
| Group 2B | The agent (mixture) is possibly carcinogenic to humans. The exposure circumstance entails exposures that are possibly carcinogenic to humans. |
| Group 3 | The agent (mixture or exposure circumstance) is not classifiable as to carcinogenicity in humans. |
| Group 4 | The agent (mixture, exposure circumstance) is probably not carcinogenic to humans. |

An indirect route to formation of cancerous cells is through *genotoxic chemicals*. They are those which are capable of causing damage to DNA. Such damage can potentially lead to the formation of a malignant tumor, but DNA damage does not lead inevitably to the creation of cancerous cells.

In Tables 3.16 and 3.17, chemicals reasonably expected to be used in cleaning appear in bold type.

## 3.13 UNEXPECTED HAZARDS

The discussion in Section 3.10 on bodily contact and Sections 3.2 through 3.8 on flammability is not meant in any way to dissuade users from use of cleaning chemicals, versus other technology. The choice to use solvent cleaning, versus aqueous or other cleaning technology not involving chemicals, should depend upon the application, not upon fear of the consequences of hazards.

[156] Coleman, C.N., Mason, T. and Robinson, S.E., "Developmental Effects of Intermittent Prenatal Exposure to 1,1,1-Trichloroethane in the Rat," *Neurotoxicology and Teratology*, 1999, Vol. 21, pp. 699–708.

[157] US Department of Health and Human Services, *Public Health Service, National Toxicology Program, 11th Report on Carcinogens*, Revised January 2001, Updated April 12, 2005. See http://ehp.niehs.nih.gov/roc/toc11.html

**Table 3.16 Agents, Substances, Mixtures or Exposure Circumstances *Known to be* Human Carcinogens**[157]

| | |
|---|---|
| Aflatoxins | Thorium dioxide |
| Alcoholic beverage consumption | Tobacco smoking |
| 4-Aminobiphenyl (4-aminodiphenyl) | Environmental tobacco smoke |
| Analgesic mixtures containing phenacetin | Erionite |
| Arsenic and certain arsenic compounds | Ethylene oxide |
| Asbestos | Melphalan |
| Azathioprine | Methoxsalen with ultraviolet A therapy (PUVA) |
| **Benzene** | Mustard gas |
| Benzidine | 2-Naphthylamine |
| bis(Chloromethyl) ether and technical-grade chlorom ethyl methyl ether | Radon |
| 1,3-Butadiene | Silica, crystalline (respirable size), e.g.: |
| 1,4-Butanediol dimethylsulfonate | Quartz |
| Cadmium and cadmium compounds | Cristobalite |
| Chlorambucil | Tridymite |
| 1-(2-Chloroethyl)-3-(4-methylcyclohexyl)-1-nitrosourea (MeCCNU) | Smokeless tobacco |
| Chromium hexavalent compounds | Solar radiation and exposure to sunlamps or sunbeds |
| Coke oven emissions | Soots |
| Conjugated estrogens | Strong inorganic acid mists containing sulfuric acid |
| Cyclophosphamide | Tamoxifen |
| Cyclosporin A | Tars and mineral oils |
| Diethylstilbestrol | 2,3,7,8-Tetrachlorodibenzo-*p*-dioxin (TCDD) |
| Dyes that metabolize to benzidine | Thiotepa |
| Direct blue 6 | |
| Direct black 38 | Vinyl chloride |

**Table 3.17 Agents, Substances, Mixtures or Exposure Circumstances *Reasonably Anticipated to be* Human Carcinogens**

| | |
|---|---|
| Acetaldehyde | *N*-methyl-*N'*-nitro-*N*-nitrosoguanidine |
| 2-Acetylaminofluorene | Metronidazole |
| Acrylamide | Michler's ketone |
| Acrylonitrile | Mirex |
| Adriamycin® | Nickel and certain Nickel compounds |
| 2-Aminoanthraquinone | Nitrilotriacetic acid |
| *o*-Aminoazotoluene | *o*-nitroanisole |
| 1-Amino-2-methylanthraquinone | Nitroarenes: |
| Amitrole | 1,6-Dinitropyrene |
| *o*-Anisidine hydrochloride | 1,8-Dinitropyrene |
| Azacitidine | 6-Nitrochrysene |
| Benzotrichloride | 1-Nitropyrene |
| Beryllium and certain beryllium compounds | 4-Nitropyrene |

(Continued)

**Table 3.17 Agents, Substances, Mixtures or Exposure Circumstances *Reasonably Anticipated to be* Human Carcinogens (Continued)**

| | |
|---|---|
| bis (Chloroethyl) nitrosourea | Nitrofen |
| Bromodichloromethane | Nitrogen mustard hydrochloride |
| Butylated hydroxyanisole | 2-Nitropropane |
| **Carbon tetrachloride** | *N*-Nitrosodi-*n*-butylamine |
| Ceramic fibers (respirable size) | *N*-nitrosodiethanolamine |
| Chlorendic acid | *N*-nitrosodiethylamine |
| Chlorinated paraffins (C12, 60% Chlorine) | *N*-nitrosodimethylamine |
| 1-(2-Chloroethyl)-3-cyclohexyl-1-nitrosourea | *N*-nitrosodi-*n*-propylamine |
| **Chloroform** | *N*-nitroso-*N*-ethylurea |
| 3-Chloro-2-methylpropene | 4-(*N*-nitrosomethylamino)-1-(3-pyridyl)-1-butanone (NNK) |
| 4-Chloro-*o*-phenylenediamine | *N*-nitroso-*N*-methylurea |
| Chloroprene | *N*-nitrosomethylvinylamine |
| *p*-Chloro-*o*-toluidine and *p*-chloro-*o*-toluidine hydrochloride | *N*-nitrosomorpholine |
| Chlorozotocin | *N*-nitrosonornicotine |
| C.I. basic red 9 monohydrochloride | *N*-nitrosopiperidine |
| Cisplatin | *N*-nitrosopyrrolidine |
| *p*-Cresidine | *N*-nitrososarcosine |
| Cupferron | Norethisterone |
| Dacarbazine | Ochratoxin A A |
| Danthron | 4,4′-Oxydianiline |
| Dichlorodiphenyltrichloroethane | Oxymetholone |
| 2,4-Diaminoanisole sulfate | Phenacetin |
| 2,4-Diaminotoluene | Phenazopyridine hydrochloride |
| 1,2-Dibromo-3-chloropropane | Phenolphthalein |
| 1,2-Dibromoethane (ethylene dibromide) | Phenoxybenzamine hydrochloride |
| 1,4-Dichlorobenzene | Phenytoin |
| 3,3′-Dichlorobenzidine and 3,3′-dichlorobenzidine dihydrochloride | Polybrominated biphenyls |
| **1,2-Dichloroethane** | Polychlorinated biphenyls |
| **Dichloromethane (methylene chloride)** | Polycyclic aromatic hydrocarbons: |
| 1,3-Dichloropropene (technical grade) | Benz[a]anthracene |
| Diepoxybutane | Benzo[b]fluoranthene |
| Diesel exhaust particulates | Benzo[j]fluoranthene |
| Di(2-ethylhexyl) phthalate | Benzo[k]fluoranthene |
| Diethyl sulfate | Benzo[a]pyrene |
| Diglycidyl resorcinol ether | Dibenz[a,h]acridine |
| 3,3′-Dimethoxybenzidine and 3,3′-Dimethoxybenzidine dihydrochloride | Dibenz[a,j]acridine<br>Dibenz[a,h]anthracene |
| 4-Dimethylaminoazobenzene | 7H-Dibenzo[c,g]carbazole |
| 3,3′-Dimethylbenzidine | Dibenzo[a,e]pyrene |
| Dimethylcarbamoyl chloride | Dibenzo[a,h]pyrene |
| 1,1-Dimethylhydrazine | Dibenzo[a,i]pyrene |
| Dimethyl sulfate | |

(Continued)

**Table 3.17 Profiles for Agents, Substances, Mixtures or Exposure Circumstances *Reasonably Anticipated to be* Human Carcinogens (Continued)**

| | |
|---|---|
| Dimethylvinyl chloride | Dibenzo[a,l]pyrene |
| 1,4-Dioxane | Indeno[1,2,3-cd]pyrene |
| Disperse blue 1 | 5-Methylchrysene |
| Epichlorohydrin | Procarbazine hydrochloride |
| Estrogens (not conjugated): estradiol-17β | Progesterone |
| Estrogens (not conjugated): estrone | 1,3-Propane sultone |
| Estrogens (not conjugated): ethinylestradiol | β-Propiolactone |
| Estrogens (not conjugated): mestranol | Propylene oxide |
| Ethylene thiourea | Propylthiouracil |
| Ethyl methanesulfonate | Reserpine |
| Formaldehyde (Gas) | Safrole |
| Furan | Selenium sulfide |
| Glasswool (respirable size) | Streptozotocin |
| Glycidol | Sulfallate |
| Hexachlorobenzene | **Tetrachloroethylene *(*perchloroethylene**) |
| Hexachloroethane | **Tetrafluoroethylene** |
| Hexamethylphosphoramide | Tetranitromethane |
| Hydrazine and hydrazine sulfate | Thioacetamide |
| Hydrazobenzene | Thiourea |
| Iron dextran complex | Toluene diisocyanate |
| Isoprene | *o*-Toluidine and *o*-toluidine hydrochloride |
| Kepone® | Toxaphene |
| Lead acetate and lead phosphate | **Trichloroethylene** |
| Lindane and other hexachlorocyclohexane isomers | 2,4,6-Trichlorophenol |
| 2-Methylaziridine (propylenimine) | 1,2,3-Trichloropropane |
| 4,4′-Methylenebis(2-chloroaniline) (MBOCA) | Tris(2,3-dibromopropyl) phosphate |
| 4-4′-Methylenebis(N,N-dimethylbenzenamine) | Urethane |
| 4,4′-Methylenedianiline and its dihydrochloride | 4-Vinyl-1-cyclohexene diepoxide |
| Methyl methanesulfonate | |

Rather, that information is meant to define the hazard so that appropriate precautions can be taken. Solvent cleaning, both cold cleaning and vapor degreasing, is being done safely in the US and around the globe.[158]

Bodily and flammability hazards are the two major types of general hazards presented to users of cleaning chemicals. In addition, there are at least two other circumstances which might be called hazards.

### 3.13.1 Legal or Regulatory Hazards

Only one chemical has been banned in the entire US based on use. That is HCFC 141b. The use of other chemicals such as 1,1,1-Trichloroethane and CFC-113 is not banned. Rather, in the US is it the *manufacture* of those chemicals which is banned by the 1990 Clean Air Act (CAA). These bans carry the weight of law in the US because they are regulations from the US Environmental Protection Agency (EPA) supported by an Act of Congress (the CAA).

[158] Gillman, A., Personal Communication, December 6, 2002. A reliable estimate is that there are at least 10,000 solvent cleaning units or machines being used in the US.

Outside the US, in developed countries, HCFC 141b is banned through the compliance with the Montreal Protocol (see Chapter 1, Section 1.2).

Locally or regionally, the use of certain chemicals in certain operations may also be banned for cleaning work.

In any of these cases, a potential or actual user of these solvents is under the hazard of legal sanction. While the author of this chapter occasionally recognizes numerous violations of these bans, he cannot recommend their violation in any circumstance.

Users or manufacturers of banned cleaning solvents should consider themselves being exposed to a hazard as significant as ignition or ingestion:

- This hazard is of being cited by a regulatory agency for non-compliance, or being sued by a worker or their agent for negligence (or similar fate).

Here, the technical protection is to seek an alternative or replacement chemical.

### 3.13.2 Economic Hazards

If legal sanction is a hazard, then economic loss is as well. This author is of the opinion that cleaning costs, if the system is well specified, designed, and used, should be almost negligible in the total of manufacturing cost. Users who choose to use a cleaning chemical with unique functionality or hazard classification have placed themselves exposed to another hazard – that of economic loss.

In the decade of the 1990s, this author saw many firms consciously or unconsciously expose themselves to the hazard of economic damage by mandating use of (or absence of use of) a uniquely functional cleaning chemical. Yet all but a few percent of those users remained under economic hazard because of that choice. Most accepted a change in chemical, process, or specification which increased their operating cost:

- This hazard which some managers have voluntarily chosen to accept under various rationales[159] is to add cost to their firm's balance sheet. That choice may or may not affect their personal employment security.

Again, technical protection is to seek an alternative or replacement solvent, process, or specification.

## 3.14 PROTECTION FROM HAZARDS

The two general strategies for containing flammability and limiting bodily contact are:

1. To avoid at least one (and preferably two) of the three factors necessary for ignition.
2. To assure any bodily contact is at levels (within exposure limits) which are not expected to cause significant harm.

The US OSHA defines their hierarchy of control as:

1. Engineering controls (see Section 3.22.1.1 for electrical regulations and Section 3.16 for hazard classification systems).
2. Work practice controls (see Section 3.15, on setting exposure limits).
3. Administrative controls (see Section 3.15.6, which covers meeting exposure limits).
4. Personal protective equipment (PPE; see Section 3.15.7, covers personal protective equipment to be used if exposure limits can't be met).

## 3.15 SETTING EXPOSURE LIMITS

In the US, as this book was being prepared, the US EPA agreed to accept *n*-propyl bromide as a replacement cleaning solvent for solvents which deplete the ozone layer. Authority and responsibility for this action comes from the US EPA's Significant New Alternative Program (SNAP).

A major component of that outcome is an exposure limit. This is a condition of use for the replacement solvent. We know these exposure limits under the "alphabet soup" or jargon of allowable exposure limit, corporate exposure limit, corporate guidance limit, permissible exposure limit, and threshold limit value (AEL, CEL, CGL, PEL, and TLV, respectively).

Users ask why there so many exposure limits (why not choose one of the above?), who creates them, upon what are they based, and why is the US EPA setting exposure limits instead of the US OSHA. These questions are answered in succeeding chapters.

[159] Some refer to these rationales as "political correctness," "social conscience," or "environmental security."

### 3.15.1 Definitions of Exposure Limits

Exposure limits are estimates. They are estimates of an amount of skin contact, ingested volume, and inhaled concentration. Almost always the exposure limit for solvents involves inhalation.

Exposure limits represent exposure which workers can sustain *without reversible damage*. This is a key point. It is the aim, when exposure limits are set, to avoid *any* permanent human damage.

These estimates of exposure:

- Cover all humans involved in specified work operations. There is no specific allowance in these estimates for known human differences of age, sex, weight, heredity, race, or sensitivity to chemicals.[160]
- Must cover both acute (short-term period between exposure and onset of symptoms) and chronic (long-time period between exposure to an agent and the onset of symptoms) types of exposure.
- Must cover consequences spanning nasal or dermal (skin) irritation to birth defects to kidney or heart failure and death.
- Must cover the possibly unknown slope of the dose–response curve.

They have the force of law in the US.[161]

In other words, exposure limits, however or by whom they are determined, represent the best-available technology for preventing injury. They are not guidance. They are not "best guesses." Compliance with them should be treated as workplace *requirements*.

### 3.15.2 Determination of Exposure Limits

Those working in industrial hygiene use a standard approach to developing exposure limits for non-carcinogenic effects.

Non-carcinogenic effects, unlike carcinogenic effects, are believed to have a threshold; that is, a dose below which adverse effects will not occur.[162] Carcinogenic effects are believed to be caused by cumulative exposure over all levels of dosage.[163]

This approach holds whether the method of bodily contact is dermal, oral, or by inhalation.

The approach for a untested chemical (solvent) is to:[164]

1. Obtain a complete body of scientific and industrial information where animals, and or humans, are exposed in one way (dermal, oral, or by inhalation) to this chemical. Information about humans is called epidemiological data. It represents experience of humans being exposed and suffering damage. Fortunately, epidemiological data is scarce.
2. Convene experienced toxicologists and industrial hygienists to examine this information and develop a health risk assessment. This includes derivation of at least two factors from the information:

   (a) The *no adverse effect level* (NOAEL). NOAEL is the maximum exposure or dose which

---

[160] This is seen in the charter from the US Congress to the Occupational Safety and Health (OSHA) Act of 1970: "The purpose of the OSHA Act is to "assure so far as possible every working man and woman in the nation safe and healthful working conditions and to preserve our human resources".

[161] 29 CFR Part 1910.1450 – Chapter of the Code of Federal Regulations: Occupational Exposures to Hazardous Chemicals in Laboratories.

[162] In the case of systemic toxicity, however, organic homeostatic, compensating, and adaptive mechanisms exist that must be overcome before a toxic endpoint is manifested. For example, there could be a large number of cells performing the same or similar function whose population must be significantly depleted before the effect is seen. The individual threshold hypothesis holds that a range of exposures from zero to some finite value can be tolerated by the organism with essentially no chance of expression of the toxic effect. See "*Reference Dose (RfD): Description and Use in Health Risk Assessments,* US EPA Background Document 1A, March 15, 1993, Section 1.2, or http://www.epa.gov/iris/rfd.htm

[163] In the case of carcinogens, the EPA assumes that a small number of molecular events can evoke changes in a single cell that can lead to uncontrolled cellular proliferation. This mechanism for carcinogenesis is referred to as "nonthreshold," since there is theoretically no level of exposure for such a chemical that does not pose a small, but finite, probability of generating a carcinogenic response. See *Reference Dose (RfD): Description and Use in Health Risk Assessments*, US EPA Background Document 1A, March 15, 1993, Section 1.2, or http://www.epa.gov/iris/rfd.htm

[164] Hertzberg, R.C. and Dourson, M.I., "Using Categorical Regression Instead of a NOAEL to Characterize a Toxicologist's Judgment in Non-Cancer Risk Assessment." In: *Toxicology of Chemical Mixtures: Case Studies, Mechanisms and Novel Approaches* (Ed. R.S.H. Yang), Academic Press, San Diego, CA, 1993. See http://www.tera.org/pubs/catreg1993.pdf. See also NFPA 704, about reactivity and compatibility of mixtures.

produced no repeatable deleterious effect. This is the maximum dose at which the response from the dose–response curve is zero.

(b) Uncertainty factor (UF). UF represents a relative method for characterizing the quality of the body of information, and the harm to be prevented.

Based on the quality of the body of information, UF is:

- Higher for information which is less reproducible.
- Higher if sensitive subpopulations are involved (children or the elderly).
- Higher for harms which have greater consequence (reproductive injury versus skin irritation).
- Higher for a higher degree of uncertainty believed to exist when experimental animal data are extrapolated to the general human population.

Based on the harm to be prevented, values of UF can be:

- ≈4 or 10 if the consequence is of lower impact such as respiratory or skin irritation.
- ≈100 for non-toxic effects such as reversible organ or tissue damage.
- 1,000 for reproductive damage.
- 10,000 or higher for the consequence being death.

The UF is sometimes known as a risk factor (RF). In any case, RF is always a judgment made by the convened experienced toxicologists and industrial hygienists.[165]

3. Set the exposure limit, or risk reference dose (RfD), as the ratio:

$$RfD = NOAEL/UF$$

RfD is the estimated daily contact with a chemical that is likely to not produce an appreciable risk of health effects over a human lifetime.

Note that a higher level of uncertainty in the body of information produces a higher UF and a lower (hopefully more safe) exposure limit.

Also note that the assessment is done by a group of experienced persons, not a single person. "Political correctness" is not involved. Group judgment is often unanimous.

### 3.15.3 Types of Data

Scientific and industrial toxicological information is very expensive of cost and time (hundreds of thousands to millions of dollars, and years of time). And cost and time become increasingly severe constraints as the degree of anticipated hazard increases. For example, a 28-day inhalation study (which takes about 90 days to complete) may suffice for some concerns. But a two-year feeding study may be required for teratogenic (reproductive) or nephrotoxic (kidney) concerns.

A major difficulty in toxilogical testing is uncertainty of results. An entire test program can be wasted if the dose levels are chosen too high or too low. In these cases, an NOAEL is not determined, or all data are NOAELs, respectively! Consequently, shorter (and cheaper) tests precede longer (and more expensive) ones. For example, the 28-day inhalation study may be preceded by a 7-day inhalation study. The two-year feeding study may be preceded by a six-month feeding study which may be preceded by a 28-day feeding study which may be preceded by a seven-day feeding study.

#### 3.15.3.1 *Other Types of Data*

Toxicologists[166,167] are studying statistical ways of enhancing the value of data already generated. For example, various curve-fit methods may be used with the dose–response data to estimate the dose at which 1% of the adverse effect is seen. Many will accept this dose as the NOAEL.[168]

[165] Uncertainty factor is dependent upon *both* the quality of the body of information, and the harm to be prevented. Usually factors of 10 are applied for each concern, with a value of 100 being common for many chemicals.

[166] Hertzberg, R. and Miller, M., "A Statistical Model for Species Extrapolation Using Categorical Response Data," *Toxicology and Industrial Health*, 1985, Vol. 1, No. 4, pp. 43–57.

[167] Hertzberg, R.C., "Fitting a Model to Categorical Response Data with Application to Species Extrapolation," *Journal of Health Physics*, 1989, Vol. 57, Suppl. 1, pp. 405–409.

[168] Lu, F.C. and Sielken Jr R.L., "Assessment of Safety/Risk of Chemicals: Inception and Evolution of the Adi- and Dose-Response Modeling Procedures," *Toxicology Letters*, 1991, Vol. 59, pp. 5–40.

#### 3.15.3.2 *Near Useless Types of Data*

Often an MSDS will not contain an exposure limit. Only a single data point from the dose–response curve will be provided.

For acute oral, dermal, or inhalation toxicity, the value may be presented as the $LD_{50}$ (or $LC_{50}$). This is the estimated dose that killed 50% of the specie being tested. $LD_{50}$s are expressed in terms of body mass (mg dose/kg mass), and identification of the animal specie (rats, mice, rabbits, etc.).

In general, $LD_{50}$ (or $LD_{10}$) values are nearly useless for users:

- This is because no one wants to operate at an undefined multiple of a dose which killed 50% of the exposed animals. Users want to operate at a experienced-based multiple of a dose which killed *no* exposed animals.

However, a comparison of $LD_{50}$s among solvents may provide some general guidance:

- *Remember, $LD_{50}$s are not exposure limits.* They are single points on a dose–response curve.

### 3.15.4 Costs of Toxicological Data

Occasionally, the body of information is not viewed as being complete by the toxicologists and industrial hygienists or the available information may lead toxicologists to suggest the need for additional testing of a different type. Both circumstances happened with *n*-propyl bromide.

Costs of toxicity testing to produce exposure limits is one cost users must pay (through price of the product). Maintenance of the personal environment below the exposure limit is the strategy by which we avoid bodily injury. In many cases that price is not paid by users – to the detriment of their employees.

In some cases, a solvent is never brought to market. The solvent manufacturer decides the cost of toxicological testing is not justified by the market potential for profit. At least one manufacturer of *n*-propyl bromide made this decision.

There is also a time cost of toxicological data. The need to plan, procure facilities, analyze data, and respond to peer review makes a two-year feeding study take six years.

### 3.15.5 Types of Exposure Limits

In general, there are two types of exposure limits developed per the equation above.

Both types can be present for either dermal, oral, or inhalation exposure.

The types differ by:

- WHO evaluated the information and developed the UF, and for
- WHAT PURPOSE the exposure limit is to be used.

The various possibilities are shown in Table 3.18 (WHO) and Table 3.19 (WHAT PURPOSE).

An implicit variable in Table 3.18 is time in the marketplace. CEL is the first determined exposure limit. In the US, EPA, OSHA, and NIOSH become involved when more toxicology data have been developed and the chemical has been in the marketplace for more time.

Like the US Supreme Court, the ACGIH is the evaluator of last resort. The ACGIH probably examines the largest body of data and use experience.[169] No corporation[170] can claim to have developed its own TLV; only the ACGIH can develop that estimate.

Exposure limits are not some magic thresholds that define the border between safe and dangerous. A PEL or STEL that was acceptable in 1950 may be recognized as dangerously high today. Alternately,

---

[169] A current example pertains to the cleaning solvent *n*-propyl bromide. In December 2004, the ACGIH determined the TLV to be 10 ppm.

[170] Corporations producing chemicals are whipsawed. If their proposed exposure limit (CEL, AEL, CGL) is quite low they can suffer the short-term and potentially fatal pain of low sales rate. If the proposed limit is quite high, they can suffer the long-term pain of harming people and the likely fatal pain of defending lawsuits about that. Juxtaposing these two issues should and usually does create a balanced attitude.

Some corporations have been known to avoid the discipline of achieving this balance by not proposing an exposure limit. The MSDS may be silent about an exposure limit. Managers considering use of this chemical for any purpose should seek another chemical from another manufacturer.

**Table 3.18 Types of Exposure Limits in the US – by Evaluation Body**

| WHO Evaluated the Information? | Name of Exposure Limit |
|---|---|
| The corporation (manufacturing the chemical) | Corporate Exposure Limit (CEL). This is occasionally identified as an Allowable Exposure Limit (AEL) or Corporate Guidance Limit (CGL). |
| Environmental Protection Agency (EPA) | Permissible Exposure Limit (PEL) |
| Occupational Safety and Health Administration (OSHA) | Permissible Exposure Limit (PEL) |
| National Institute of Occupational Safety and Health (NIOSH) | Permissible Exposure Limit (PEL) |
| American Conference of Government Industrial Hygienists | Threshold Limit Value (TLV) |

**Table 3.19 Types of Exposure Limits in the US – by Use**

| For What Use Will It Be Used? | Method of Use |
|---|---|
| Action level (AL) | A higher level at which some action should be taken to keep the level from going higher, or triggering certain compliance activities such as exposure monitoring, medical surveillance, or training. |
| Excursion level (EL) | A level which should not be exceeded for more than 15 minutes. |
| Time-weighted average (TWA) takes into account variable exposure through a full shift, 8-hour work day | The normal limit. For ethylene oxide, OSHA's AL is 0.5 ppm, EL is 10 ppm, and TWA is 1 ppm. |
| Short-term exposure limit (STEL) Limit of exposure during a short period, usually 15 minutes | STEL is defined by ACGIH. An STEL is not a separate independent exposure limit, but supplements time-weighted average limits where there are recognized acute effects from a substance whose toxic effects are generally chronic (long-term) in nature. |
| Absolute maximum level of exposure not to be exceeded (CEILING) | At no time should an employee be exposed to this concentration. |

recently-developed toxicological data about HCFC 225 ca/cb allowed the EPA to restate the exposure limit upward from 25 to 100 ppm.[171] There is a good overall summary of OSHA exposure limits.[172]

Finally, toxicologists[173,174] are considering supplementing tests where animals are sacrificed or harmed with some tests done "*in vitro*." This means using cells (or cell lines) instead of animals in acute toxicology tests.

#### 3.15.5.1 *Authorization to Set Emission Limits in the US*

Many (often those who didn't support the US EPA's SNAP decisions) have asked, "why is the EPA setting exposure limits? Isn't that OSHA's job? What is EPA's authority to do this?"

The answer is simple. The US EPA has authorization to do so under the CAA.[175]

[171] *Federal Register*, Vol. 67, No. 56, March 22, 2002, p. 13275.
[172] http://www.state.vt.us/labind/Vosha
[173] "Report of the International Workshop on *In Vitro* Methods for Assessing Acute Systemic Toxicity" (*NIH Publication 01-4499)*, 2001.
[174] "Guidance Document on Using *In Vitro* Data to Estimate *In Vivo* Starting Doses for Acute Toxicity" (*NIH Publication 01-4500*), 2001.
[175] In Section 612 of the CAA, the Agency is authorized to identify and restrict the use of substitutes for Class I and II ozone-depleting substances **where the Administrator has determined** that other alternatives exist that reduce overall risk to human health and the environment. See http://www.epa.gov/Ozone/snap/regs/59fr13044.html.

**Table 3.20 Comparison of Exposure Limits in US and UK**

| Cleaning Chemical | CAS # | WEL (ppm) | US Limit (ppm) |
|---|---|---|---|
| Trichloroethylene | 79-01-6 | 100 | 10 (ACGIH) |
| Toluene | 108-88-3 | 50 | 50 (ACGIH) |
| **Cyclohexane** | 110-82-7 | **100** | **300 (ACGIH)** |
| Isopropanol | 67-63-0 | 400 | 400 (ACGIH) |
| Methyl alcohol | 67-56-1 | 200 | 200 (ACGIH) |
| **Acetone** | 67-64-1 | **500** | **750 (ACGIH)** |
| Nitric acid | 7697-37-2 | 2 | 2 (OSHA) |
| Turpentine | 8006-64-2 | 100 | 100 (OSHA) |
| **Ethyl acetate** | 141-78-6 | **200** | **400 (ACGIH)** |
| **Perchloroethylene** | 127-18-4 | **50** | **100 (OSHA)** |
| Methyl acetate | 79-20-9 | 200 | 200 (ACGIH) |
| Propylene glycol methyl ether | 107-98-2 | 100 | 100 (ACGIH) |
| **Hexane** | 110-54-3 | **20** | **50 (ACGIH)** |
| **Diethyl ether** | 60-29-7 | **100** | **400 (OSHA)** |
| 1,2-Dichloroethylene | 540-59-0 | 200 | 200 (ACGIH) |
| Butyl lactate | 138-22-7 | 5 | 5 (NIOSH) |

Further, the US EPA's position is that exposure limits, or use conditions, that it sets are temporary.[176] Not everyone accepts this statement at face value.

### 3.15.5.2 *Authorization to Set Exposure Limits in the UK*

As of this writing (2005), the UK Control of Substances Hazardous to Health Regulations (COSHH) is being simplified to enable just a single type of exposure limit in the UK. Workplace Exposure Limits (WELs[177]) will replace Maximum Exposure Limits (MELs) and Occupational Exposure Standards (OESs).

These exposure limits have the force of law in the UK, and are not set by corporations or suppliers. Values for common cleaning chemicals, and those recognized in the US, are shown in Table 3.20.

Comparison with values recognized in the US is given to make this point:

- Exposure limits are judgment calls as per Section 3.15.5 and there is no reason to expect human judgment to be unanimous.

### 3.15.5.3 *Authorization to Set Occupational Exposure Limits*

The situation[178] in countries outside the US and the UK is given in Table 3.21.

Certainly, it is clear that a manager of a company with installations using chemicals in more than one country, or a manager selling a chemical into more than one country cannot practice the same technology in the same way. That commonality should not have been expected.

[176] "In imposing conditions on use, EPA does not intend to preempt other regulatory authorities, such as those exercised by the Occupational Safety and Health Administration (OSHA) or other government or industrial standard-setting bodies. Rather, EPA hopes to fill existing regulatory gaps during the interim period of substitution away from ozone-depleting compounds." See http://www.epa.gov/Ozone/snap/regs/59fr13044.html

[177] Current values of WELs can be found at http://www.hse.gov.uk/coshh/table1.pdf. Both 8-hour time-weighted average and 15-minute short-term exposure values are given.

[178] An excellent and current reference on this topic is provided by the European Agency for Safety and Health at Work from which the information in Table 3.21 is taken. See http://europe.osha.eu.int/good_practice/risks/ds/oel/

**Table 3.21 Exposure Limits Around the Globe**

| Country | Source of authorization | Force of law | Source for information |
|---|---|---|---|
| Austria | The Bundesministerium für Wirtschaft und Arbeit or BMWA (Ministry of the Economy and Labor) fixes these limit values. | Yes | http://www.bmwa.gv.at/ |
| Belgium | Royal decree of August 10, 1998 (Statute Paper of September 12, 1998). Most are ACGIH values. | Yes | http://www.emis.vito.be/ |
| Germany | The Ausschuss für Gefahrstoffe (AGS) or Committee on Hazardous substances establishes the Technical Rules. This Committee consists of members from all concerned groups. Maximum 8-hour values apply *only to healthy adults*. | Yes | http://www.hvbg.de/d/bia/pub/rep/rep.htm |
| Denmark | Arbejdstilsynet (Labor Inspectorate) publishes the occupational exposure limit (OEL) list and supervises its execution. The Arbejdstilsynet decides on the OELs after consulting with cmployer and employee representatives on the technical and economical feasibility of the proposed limit value level. | Yes. In the sphere of the Working Environment Act, employers are obliged to keep the exposure as low as reasonably possible and the limit values *should never be exceeded.* | http://www.arbejdstilsynet.dk/ |
| Spain | Developed by a working group of the Comisión Nacional de Seguridad y Salud en el Trabajo (National Commission of Safety and Health at Work).<br>There is a limit value, Valor Límite Biológico (VLB), for the content of the substance *in biological media* (i.e. blood, urine). VLBs have been defined for dozens of substances. | Yes. All limit values for hazardous substances are recommendations. They become legally binding when they are used by the Labor Inspectorate or other authorities in orders or instructions for companies. | http://www.mtas.es/insht/practice/vlas.htm |
| France | The OELs are established and revised by the Comité Scientifique pour la Surveillance des Atmosphères de Travail (Scientific Committee for Surveillance of the Workplace), working under the High Council for the Prevention of Occupational Hazards Environment. The committee is made up of experts, scientists, and industrialists involved with chemicals. | Yes. The OELs should be seen as a minimum requirement. | http://www.inrs.fr/dossiers/nd2098.pdf<br>http://www.inrs.fr/dossiers/vle.htm |
| Finland | The Ministry of Social Affairs and Health decides on OELs according to the proposal of the Advisory Committee for Occupational Health and Safety on Chemicals. The four subcommittee members represent the Ministry, the chemical industry, employers' organizations, and trade unions. | Yes. There are legally enforcing binding limit values (sitovat raja-arvot) and more orientating concentrations known to be harmful (haitalliseksi tunnetut pitoisuudet, HTP-arvot). Several other substances | http://www.occuphealth.fi/ttl/projekti/htp/english/index.htm |

(Continued)

**Table 3.21 Exposure Limits Around the Globe (Continued)**

| Country | Source of authorization | Force of law | Source for information |
|---|---|---|---|
| | | have also been set limit values for concentration in biological samples. | |
| Greece | The general law on "health and safety for workers" and several ensuing presidential decrees have achieved harmonization to European Union (EU) legislation. | Yes. Most of the OELs reported are equivalent to the published by the ACGIH. | http://gr.osha.eu.int |
| Italy | Determined by the National Institute of Occupational Safety and Prevention (ISPESL). Nearly all are ACGIH values. | Yes | http://www.ispesl.it |
| Ireland | Based on the Safety, Health and Welfare at Work (Chemical Agents) Regulations, 1994 (S.I. No. 445 of 1994). | Yes | http://www.hsa.ie/chemical.htm |
| Luxembourg | OELs used in Luxembourg are the same as used in Germany, unless specific OELs are provided. | Yes | http://www.itm.etat.lu/<br>http://www.mt.etat.lu/ |
| The Netherlands | • Legally binding OELs are based upon the health-based recommended occupational exposure level provided by the Dutch Expert Committee on Occupational Standards (DECOS) of the Health Council, but also *taking into account the socio-economic feasibility of that value.*<br>• Administrative OELs are not legally binding. They are regulated in a policy rule based on the working condition regulations. The Labor Inspectorate considers these values as levels that should in any case not be exceeded to protect workers. | Yes and No | http://www.dohsbase.nl<br>http://www.bsa-site.nl/Macwaarden.htm |
| Portugal | Values printed in the Portuguese Standard 1796 of 1988 are equivalent to the limit values published by ACGIH. | Yes, for Lead and vinyl chloride monomer. | http://www.ipq.pt/index.htm |
| Sweden | Scientific background documentation is prepared by the Kriterigruppen för hygieniska gränsvärde, Arbetslivsinstitutet (Criteria Group of the National Institute of Working Life). | Yes. According to the national regulations, the employer is obliged to keep the exposure level as far below the limit value as possible. | http://se.osha.eu.int/good_practice/ |

| | | | |
|---|---|---|---|
| Mexico | The Secretariat of Labor and Social Welfare (STPS) is responsible for setting standards regarding PEL. | Yes. The scientific basis is the use of ACGIH criteria documents. | http://www.naftaworks.org/ |
| Australia | Australian National Occupational Health and Safety Commission (NOHSC). | No. Advisory only. | http://www.nohsc.gov.au/OHS LegalObligations/National Standards/Expstd.htm |
| Canada | Canadian autonomous regions have different OSHA systems, which are applied according to the provincial regulations. | | http://www.ccohs.ca/oshanswers/legisl/intro.html |
| Canada – Quebec | Regulation of the Quality of the Workplace – includes both unions and operators | Yes | |
| Canada – Ontario | OEL task force – includes both unions, ministers, and operators. | Yes | http://www/ontariogazette.gov.on.can |
| Canada – Alberta | Technical Committee of OELs, established by Alberta's Occupational Health and Safety – includes both unions, ministers, and operators. | Yes | *Occupational Health and Safety Magazine* |
| Hong Kong | Basically, ACGIH values are used. | Not generally. Apply only to sites covered by the Factories and Industrial Undertakings Legislation | http://www.info.gov.hk/labour/ |
| Japan | The Japanese Association of Industrial Health recommends limits which are usually the ACGIH values. | No | http://joh.med.uoeh-u.ac.jp |
| New Zealand | An expert committee, consisting of representatives of the Occupational Safety and Health Service of the Department of Labor, toxicologists, and medical/scientific experts. | No, guidance only | http://www.osh.dol.govt.nz/ |
| Republic of South Africa | Department of Labour and the Department of Minerals and Energy. Do not apply to exposure to substances hazardous to health in mines. | No | http://www.saioh.org/index.htm<br>http://www.dme.gov.za/ |
| Switzerland | Decree of the Bundesrat and in conjunction with the Limit Value Commission of the Swiss Association for Occupational Medicine, Hygiene, and Safety. | It can be assumed that the health of the *vast majority of healthy employees* will not be harmed. | http://wwwitsp1.suva.ch/sap/its/mimes/waswo/99/pdf/01903-d.pdf |

It is interesting to note differences among countries around the world,[179] based on:

- Who is expected to be affected by the limits
  - All persons or the vast majority of persons
  - Healthy persons or healthy adults
  - Special groups such as the aged and children
- The meaning of the limits
  - Values never to be exceeded
  - Values which all operation will be less than
- Who defines the limits
  - All stakeholders – management, labor, government, etc.
  - Technical experts
- What can be considered in the determination of limits
  - Socio-economic feasibility or not
- How pervasive and valued is the scientific review process conducted by the ACGIH.

### 3.15.6 Meeting Exposure Limits

The fundamental lesson here is to not manage an environment where concentrations approach exposure limits.

For example, suppose the $PEL_{oral}$ for a new chemical is *X* g/kg. The work plan to avoid X is to use good hygiene and never drink from contaminated containers. That way the exposure is zero.

Again, suppose the $PEL_{dermal}$ for that chemical is *Y* g/kg. The work plan to avoid Y is to use good hygiene and never contact skin with liquid. Use tools, face shields, spray barriers, and tall freeboard to avoid contact. That way the exposure is zero:

- For dermal (skin) and oral exposure these goals are achievable.

Finally, suppose the $PEL_{inhalation}$ is *Z* ppm. Continue to use good hygiene. Use engineering controls such as covers, refrigeration coils, slow insertion rate, and tall freeboard to keep the chemical in the tank. Use plenty of ventilation with fresh air to dilute exposure below *Z* ppm concentration. That way the exposure is near zero:

- For inhalation exposure, ventilation doesn't dilute to zero concentration. And, it is almost never economically feasible to keep **all** the chemical in the tank.

The last resort to avoid two of the three types of exposure is to implement and Personal Protective Equipment (PPE) to avoid contact (see Table 3.22).

### 3.15.7 Protective Equipment

If exposure cannot be adequately controlled in any other way, workers should wear PPE. They may need to wear one or more of the following:

- Protective overalls. Aprons and overalls should be properly selected. Not all protective clothing

**Table 3.22 Meeting Exposure Limits**

| Type of Contact | Prime Methods for Meeting Limits | Backup Methods for Meeting Limits |
|---|---|---|
| Oral | Good hygiene. Care in cleaning of drinking containers. | Directed vomiting |
| Dermal | Good hygiene. Facilities to avoid skin contact. | PPE (gloves, aprons, and suits). |
| Inhalation | 1. Equipment designed to "keep solvent in the tank." 2. Dilution with good ventilation. | Secondary breathing apparatus (respiratory protective equipment), or personal protective equipment. |

[179] An interesting example is VOC exemption policy.

In France, Belgium, and Italy, *N*-methyl-2-pyrrolidone (NMP) has no exposure limit because of its low volatility, which causes it to be exempt from VOC status.

However, in the US, the exposure limit (not TLV) found on material safety data sheets is 10 ppm with the following two notes: (1) SARA 313: *n*-methyl-2-pyrrolidone is a regulated chemical under SARA Title III, Section 313; and (2) "This chemical is known to the State of California to cause developmental toxicity."

resists all substances. Overalls and contaminated personal clothing should be promptly discarded after use, or laundered and inspected before being re-worn (dermal damage).

- Appropriate gloves which have been specially selected to be resistant to the solvents-specific chemicals site. The MSDS, or the chemical manufacturer, will provide information about which glove materials are compatible with and will provide protection from their solvents or aqueous cleaning agents (dermal damage).
- Face shields or goggles (dermal damage).
- Respiratory protective equipment, where ventilation does not provide adequate control. Half-mask respirators fitted with the appropriate filter will often be sufficient in this instance, but compressed airline breathing apparatus may be necessary where spraying takes place, or where work is in a confined space (inhalation damage).

Those who need to wear PPE should be trained (see Section 4.18) and retrained in its proper use and in its limitations (see Section 3.21). PPE should be maintained and kept clean and fit for wear. Manufacturers' specifications should be followed. Store the equipment in clean, dry conditions away from chemicals – a locker would be suitable.

Barrier creams are used as substitutes for protective clothing, especially when gloves or sleeves cannot be used safely, but they do not shield as well as protective clothing.

## 3.16 HAZARD CLASSIFICATION SYSTEMS

It would be nice if there only one.[180]

That's not true. At least five have some degree of common acceptability in the US. They are from the:

- US Occupational Safety and Health Administration/Department of Transportation (OSHA/DOT) – see Section 3.16.1.
- National Fire Protection Association (NFPA) – see Section 3.16.2.
- National Paint and Coatings Association (HMIS) – see Section 3.16.3.
- American National Standards Institute (ANSI).
- US Environmental Protection Association.

The first three will be covered in this volume. As noted in Section 3.2.2 these are systems for classification of ignition hazards based on flash point data and other data. The first is supported by two US government agencies. The second one is quite similar, but different. The third one is supported by a private agency.

### 3.16.1 US OSHA/DOT Classification System for Ignition Risk

This system enables the force of law in the US.[181] The two agencies of the US government are the:

- US DOT
- US OSHA

[180] In addition to significantly more than one hazard classification system, the word classification is often replaced with the words ratings, rankings, evaluations, or assessments. While those latter four words may have different meanings in a dictionary, they are all taken in this volume to mean classifications.

[181] The connection between this US Federal Regulation and a legal citation against specific site in the US goes through what the NFPA identifies as a local "Authority Having Jurisdiction" (AHJ). Consulting engineers sometimes refer to this body as the "Questioning Authority."

The AHJ may be a identified as a commission, council, or board, and is empowered by local voters. The AHJ defines local plumbing, building, safety, and fire codes. They are local laws. Usually, but not always, these codes are based on standards independently developed by the NFPA or another like agency, AND local bias or needs. But the local AHJ can accept or reject any standard they choose. For example, on March 16, 2005, the California Building Standards Commission voted 8-2 to reverse its 2003 decision to adopt NFPA 5000 and NFPA 1. For local management of fire safety or ignition risk, NFPA 70 is most often chosen, and is referred to as the National Electrical Code. Most AHJs will nominate a fire chief or a fire marshal to enforce NFPA 70 – suitably fortified or weakened to meet local needs and bias.

NFPA 70 is a complex document, which in turn is supported by and linked to other NFPA codes. Definitions of flammable and combustible fluids used in NFPA 70 are those defined in Section 1.7 of NFPA 30 (Flammable and Combustible Liquids code) – which are those published by the US OSHA in 1910.106(a)(18). See Section 3.22 of this book.

In summary, the reason the US OSHA definitions of combustible and flammable chemicals have the force of law at your site is that your AHJ adopts the provisions of NFPA 30.

Their requirements are derived from the classification of a chemical based on its flash point (and to some extent on its boiling point). Note these two points: (1) the classification applies to all liquids, not just liquids used for cleaning and (2) the closed-cup method is used for all determinations of flash point. The classification system is published in 1910.106(a)(18) and is:[182]

- Class IA – Flash point less than 73°F; boiling point less than 100°F.
- Class IB – Flash point less than 73°F; boiling point equal to or greater than 100°F.
- Class IC – Flash point equal to or greater than 73°F, but less than 100°F.
- Class II – Flash point equal to or greater than 100°F, but less than 140°F.
- Class IIIA – Flash point equal to or greater than 140°F, but less than 200°F.
- Class IIIB – Flash Point equal to or greater than 200°F.

This information is collected in Figure 3.28.

Note that boiling point is only used to distinguish between Classes IA and IB.

Class IA liquids are extremely volatile, but there are few liquids that are so classed.[183]

Theoretically, there is no upper limit to Class IIIB, except that liquids with a closed-cup flash point above 200°F dry slowly and so are often poor choices for vapor degreasing. However, in Europe they may well be classified as VOC-exempt (see Chapter 2, Section 2.2.2) and so find application in cold (non-heated) cleaning work.

The US DOT has made a modification to this system. Because they are partners in a worldwide network of regulations about hazardous materials, DOT has changed its definition of "flammable liquid" by raising the upper limit to 141°F (60.5°C). However, DOT regulations include a so-called "domestic exemption" that allows a shipper to redesignate as a combustible liquid any chemical whose flash point is in the NFPA Class II range and which does not meet any other hazardous material definition.

### 3.16.2 NFPA Hazard Classification System

The National Fire Protection Association (NFPA) system uses[184] a diamond-shaped diagram. The diagram, reproduced as Figure 3.29, identifies four color-coded categories of hazard for each material.

These categories, also recognized in most countries outside the US, are as follows:

- Health hazard
- Fire hazard

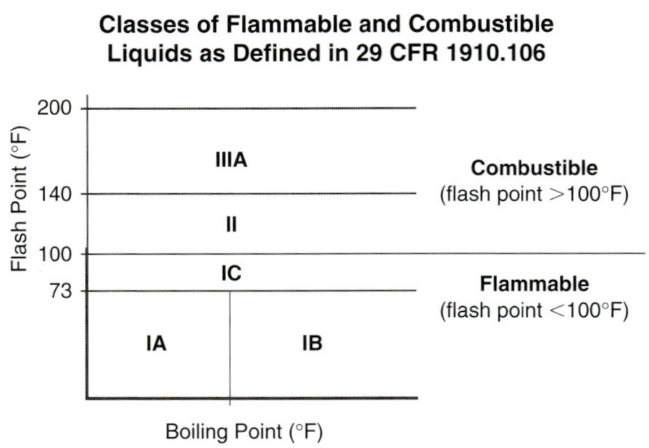


**Figure 3.28** OSHA/DOT flash point classification system

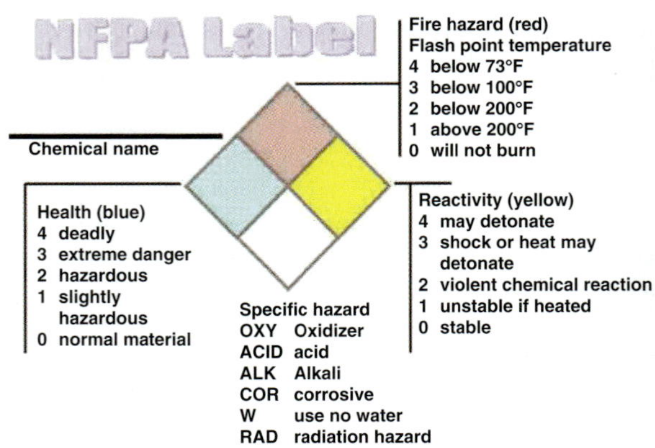


**Figure 3.29** NFPA hazard classification system

[182] See 29 CFR 1910.106, flammable and combustible liquids. In a sense, Class IA liquids distinguish all other flammable liquids from flammable gases.

[183] Some which are so, and have been used for cleaning work, are diethyl ether (IA), methyl formate (IA), ethyl chloride (IA), and pentane (IA). Of more current interest are methyl acetate (IB), acetone (IB), and benzotrifluoride (IB).

[184] Through their standard NFPA 704, Standard System for the Identification of the Hazards of Material for Emergency Response.

- Reactivity hazard
- Special hazard

Each of the categories is subdivided in levels of hazard potential with increasing numbers indicating increasing hazard.

The symbols and numbers indicate the degree of hazard associated with a particular chemical.

These colored diamond-shaped symbols are placed on containers of chemicals to identify the degree of hazard associated with the chemical.

The meaning of each degree of hazard in each of these categories is given as follows.[185]

### 3.16.2.1 *NFPA Health Hazard*

Both the form or condition of the chemical, as well as its inherent properties are considered. The chemical's degree of health hazard also indicates the degree of PPE required for working safety:

- **1** is for slightly hazardous (toxic) material which requires only minimal protection (e.g. safety glasses and gloves) in addition to normal work clothing to work with safely.
- **2** is for moderately toxic or a moderately hazardous material which requires additional PPE (e.g. chemical goggles, laboratory/work smock, local ventilation) in addition to that required for less toxic material. Consult the MSDS for specific health hazard and proper PPE to use with this material.
- **3** and **4** are for highly to extremely toxic (deadly) materials (and any carcinogen, mutagen, or teratogen). These materials will require specialized equipment (e.g. respirator or exhaust hood, full-face shield, rubber apron, specialized glove, handling tongs, etc.) beyond that required for moderately toxic material. Consult the MSDS and/or other safety information to determine the hazard (acute or chronic) and the proper PPE and engineering controls to safely use this chemical.

*It should go without saying that the choice of a cleaning chemical, which is rated 3 or 4 in the HEALTH category, should only be made when there are no other options, including non-chemical methods of cleaning.*

### 3.16.2.2 *NFPA Flammability or Fire Hazard*

The flammability (ignition risk) or fire hazards deal with the degree of susceptibility of the chemical to ignite and burn. Both the form or condition of the chemical, as well as its inherent properties are considered. Many hazardous materials such as acetone and gasoline, have a flash point (ignition temperature) far below freezing and will readily ignite with a spark if the vapor concentration is sufficient:

- **1** is for materials with a flash point above 200°F.
- **2** is for materials with a flash point below 200°F but above 100°F.
- **3** is for materials with a flash point below 100°F but above 73°F.
- **4** is for materials with a flash point below 73°F. Note that, as with the OSHA classification system, the NFPA system does not subdivide this classification by boiling point.

### 3.16.2.3 *NFPA Reactivity Hazard*

The reactivity hazards deal with the potential of a chemical to release energy. Some materials are capable of rapid energy release without any catalyst, while others can undergo violent eruptive or explosive reactions if they come in contact with water or other materials. Generally, this classification used to indicate the potential to react if the solvent is heated, jarred, or shocked:

- **1** indicates a chemical that may be reactive if heated and one that reacts with water.
- **2** indicates a chemical that may react violently without detonation.
- **3** indicates a chemical that may detonate or explode if subjected to a strong initiating force or heating under confinement.
- **4** indicates a chemical that readily detonates or explodes.

*It should go without saying that the choice of a cleaning chemical which is rated 2 or 3 or 4 in the category should only be made when there are no other options including non-chemical methods of cleaning.*

[185] The wording above, in many cases, of the degrees of hazard is reproduced exactly from NFPA literature (http://www.sefsc.noaa.gov/HTMLdocs/nfpahazard.htm) to avoid any miscommunication to the reader.

*A chemical with a Class 1 REACTIVITY classification should be used only when necessary for cleaning operations.*

#### 3.16.2.4 *Specific Hazard*

In addition to the three general categories of hazard, there are five specific hazards which are recognized and rated by the NFPA.

An open space at the bottom of the NFPA diagram is used to indicate additional information about the chemical or material. This information may include the chemical's radioactivity, proper fire extinguishing agent, skin hazard, its use in pressurized containers, protective equipment required, or unusual reactivity with water:

- OX or OXY indicates a material that is an oxidizer.
- W indicates a material that is water reactive.
- ALK indicates a material that is alkali.
- COR indicates a material that is corrosive.
- RAD indicates a material that is radioactive.

#### 3.16.2.5 *Special Labeling Requirements*

All containers that hold carcinogens, reproductive hazards, or acutely toxic chemicals must be properly labeled concerning the specific health hazard posed by the chemical.

This is generally done by the manufacturer. But OSHA makes it the responsibility of the *management of the site* to properly label older, mixed, or subdivided lots of such chemicals. When in doubt consult the MSDS (see Section 3.18).

#### 3.16.2.6 *NFPA Classifications of Cleaning Solvents*

The best source of this data for a single solvent should be the MSDS, which testifies to the general acceptance of this classification system. While there are always exceptions to generalizations,[186] seldom are the results of other hazard classification systems given on MSDSs.

As of this writing an excellent catalog of NFPA classifications for nearly 1,500 chemicals is at this web site.[187] The chemicals are listed alphabetically so one can easily compare classifications of competing chemicals.

Simplicity is the prime virtue of the NFPA and similar systems:

- One can identify without seeking a handbook that a health classification of 2 imposes more risks than a health classification of 1.
- One cannot learn why a value of 2 was given versus a value of 1, or what to do about from knowledge of the classification value.

This is shown by the entries in Table 3.23, where classifications using the NFPA system are given for some common cleaning chemicals.

The NFPA system provides recognition of hazards and ranking of them on a relative basis. However, it generally does not speak to the type of protection necessary to avoid them. Those decisions are left to local users.

### 3.16.3 Hazardous Material Identification System

The Hazardous Material Identification System (HMIS) was developed by a trade organization, the National Paint and Coatings Association (NPCA), as a tool in a comprehensive hazard communication (HAZCOM) program.

As such it is somewhat different than the OSHA/DOT or NFPA systems, which were created and are maintained to achieve different purposes:

- The OSHA/DOT system was created by a US Federal Law to be a benchmark for state laws.
- The NFPA system was created and is managed to provide a consistent comparative system for ranking the overall hazards of one chemical relative to another.
- The implementation for the HMIS is intended to fulfill the legal requirements of a HAZCOM program.[188]

In the HMIS system, numeric values for health, flammability, and physical hazards are developed

---

[186] This sentence is an observation. It is not an endorsement of the NFPA system. It is not a denigration other such systems.

[187] http://www.hazmat.msu.edu:591/nfpa/

[188] In the US, OSHA mandates through 29 CFR 1910 Chapter 1200 that workers be provided with hazard information and training about the chemicals they use (see Section 2.30.2).

**Table 3.23 NFPA Hazard Classification of Common Cleaning Chemicals**

| Solvent | CAS # | Health Classifications | Fire Classifications | Reactivity Classifications |
|---|---|---|---|---|
| 1,1,1-Trichloroethane | 71-55-6 | 2 | 1 | 0 |
| Trichloroethylene | 79-01-6 | 2 | 1 | 0 |
| Toluene | 108-88-3 | 2 | 3 | 0 |
| Cyclohexane | 110-82-7 | 1 | 3 | 0 |
| Isopropanol | 67-63-0 | 1 | 3 | 0 |
| Methyl alcohol | 67-56-1 | 1 | 3 | 0 |
| Acetone | 67-64-1 | 1 | 3 | 0 |
| Nitric acid | 7697-37-2 | 4 | 0 | 3 |
| Propylene glycol monomethyl ether acetate | 108-65-6 | 1 | 2 | 0 |
| Ethyl acetate | 141-78-6 | 1 | 3 | 0 |
| Water | 7732-18-5 | 0 | 0 | 0 |

by suppliers – not users, as with the South African Paint Manufacturers Association (SAPMA) system (see Section 3.16.5.1), not a collective panel of experienced technical persons as with NFPA, and not by regulators as was done with the OSHA/DOT system. The NPCA considers suppliers to be most knowledgeable about the inherent properties of the chemicals they sell.[189]

HMIS classification values must be purchased as part of a training program. They are not generally found on MSDSs or descriptive literature. And that's not necessarily a negative aspect because use of the HMIS provides useful information about *why* a chemical was given its classification.

The purchased HMIS system includes several workbooks, labels, placards, a "library," and instructional materials. The user has a manual with two pages for *each* chemical dedicated to explaining how the hazard information normally available on an MSDS is used to develop the HMIS classification for that chemical. The "library" contains roughly 450 chemicals already classified. Use of this system, as part of a training program, would allow workers to classify newly developed chemicals.[190]

---

[189] Another point of view about this opinion can be found in Section 3.18.3.

[190] **HMIS – Health hazard ratings**

The symbol (*) is attached to the classification value for health if it is believed that chronic (long-term) health effects may result from repeated overexposure:

0 *Minimal hazard*: No significant risk to health.
1 *Slight hazard*: Irritation or minor reversible injury possible.
2 *Moderate hazard*: Temporary or minor injury may occur.
3 *Serious hazard*: Major injury likely unless prompt action is taken and medical treatment is given.
4 *Severe hazard*: Life-threatening, major or permanent damage may result from single or repeated over exposures.

**HMIS – Flammability ratings**

0 *Minimal hazard materials* that will not burn.
1 Slight hazard materials that must be preheated before ignition will occur. Includes liquids, solids, and semi solids having a flash point above 200°F (OSHA/DOT Class IIIB).
2 *Moderate hazard materials* which must be moderately heated or exposed to high ambient temperatures before ignition will occur. Includes liquids having a flash point at or above 100°F but below 200°F. (OSHA/DOT Classes II & IIIA).
3 *Serious hazard materials* capable of ignition under almost all normal temperature conditions. This includes flammable liquids with flash points below 73°F and boiling points above 100°F as well as liquids with flash points between 73°F and 100°F (OSHA/DOT Classes IB & IC).
4 *Severe hazard flammable gases or very volatile flammable liquids* with flash points below 73°F, and boiling points below 100°F. Materials may ignite spontaneously with air (OSHA/DOT Class IA).

**Table 3.24 NFPA and HMIS Hazard Classification of Common Cleaning Chemicals**

| Solvent | CAS # | Health Classification | | Fire Classification | | Reactivity Classification | |
|---|---|---|---|---|---|---|---|
| | | NFPA | HMIS[190] | NFPA | HMIS | NFPA | HMIS |
| 1,1,1-Trichloroethane | 71-55-6 | 2 | 2* | 1 | 1 | 0 | 0 |
| **Trichloroethylene** | 79-01-6 | 2 | 2* | **1** | **0** | 0 | 1 |
| Toluene | 108-88-3 | 2 | 2* | 3 | 3 | 0 | 0 |
| Hexane:[136] | 108-10-1 | 1 | 1* | 3 | 3 | 0 | 0 |
| Cyclohexane | 110-82-7 | 1 | 2* | 3 | 3 | 0 | 0 |
| Isopropanol | 67-63-0 | 1 | 2* | 3 | 3 | 0 | 0 |
| Methyl alcohol | 67-56-1 | 1 | 2* | 3 | 3 | 0 | 0 |
| Acetone | 67-64-1 | 1 | 2* | 3 | 3 | 0 | 0 |
| Nitric acid | 7697-37-2 | 4 | 4 | 0 | 0 | 3 | 3 |
| Propylene glycol monomethyl ether acetate | 108-65-6 | 1 | 1 | 2 | 2 | 0 | 0 |
| **Ethyl acetate** | 141-78-6 | **1** | **2*** | 3 | 3 | 0 | 0 |
| Water | 7732-18-5 | 0 | 0 | 0 | 0 | 0 | 0 |

Because the NFPA and HMIS classification systems are developed by different staff for different purposes, it should be no surprise that there should be differences among the classifications for certain chemicals. Exact agreement, with occasional offsets, between the two systems are shown in Table 3.24 for the cleaning chemicals that were listed in Table 3.23.

Chemicals listed in boldface type have values of at least one hazard classification different between the NFPA and the HMIS systems.

This is not a flaw as the value of any classification system is in the action taken based upon it, and not in the absolute value of one specific classification.

### 3.16.4 Comparison of OSHA/DOT, NFPA, HMIS Classification Systems

Tables 3.25–3.27 clearly show the potential for confusion if classifications from these three systems are intermixed.

It isn't that any single system is inadequate. It is that there are three systems which produce three different results for the same chemical.

This author believes that the HMIS classification system must be viewed as part of an overall training program, which is what is sold by the NPCA. If any other training program is used, the HMIS system should be forgotten to avoid confusion.

---

**HMIS – Physical hazard ratings**

0 *Minimal hazard materials* that are normally stable, even under fire conditions, and will NOT react with water, polymerize, decompose, condense, or self-react; non-explosives.

1 *Slight hazard materials* that are normally stable but can become unstable (self-react) at high temperatures and pressures. Materials may react non-violently with water or undergo hazardous polymerization in the absence of inhibitors.

2 *Moderate hazard materials* that are unstable and may undergo violent chemical changes at normal temperature and pressure with low risk for explosion. Materials may react violently with water or form peroxides upon exposure to air.

3 *Serious hazard materials* that may form explosive mixtures with water and are capable of detonation or explosive reaction in the presence of a strong initiating source. Materials may polymerize, decompose, self-react, or undergo other chemical change at normal temperature and pressure with moderate risk of explosion.

4 *Severe hazard materials* that are readily capable of explosive water reaction, detonation or explosive decomposition, polymerization, or self-reaction at normal temperature and pressure.

**Table 3.25 Comparison of NFPA, HMIS, and OSHA/DOT Flammability Classification Systems**

| | NFPA | OSHA/DOT | HMIS |
|---|---|---|---|
| Materials with a flash point above 200°F | 1 | IIIB | 1 |
| Materials with a flash point below 200°F but above 100°F | 2 | IIIA for flash point between 140°F and 200°F | 2 |
| | | II for flash point between 100°F and 140°F | |
| Materials with a flash point below 100°F but above 73°F | 3 | IC | 3 |
| Materials with a flash point below 73°F | 4 | IB for boiling point >100°F | 4 |
| | | IA for boiling point <100°F | |

**Table 3.26 Comparison of NFPA and HMIS Classification Systems**

| NFPA | | HMIS | |
|---|---|---|---|
| | | 0 | *Minimal hazard*: No significant risk to health. |
| 1 | *Slightly hazardous* (toxic) material which requires only minimal protection (e.g. safety glasses and gloves) in addition to normal work clothing to work with safely. | 1 | *Slight hazard*: Irritation or minor reversible injury possible. |
| 2 | *Moderately toxic* or a moderately hazardous material which requires additional PPE (e.g. chemical goggles, laboratory/work smock, local ventilation) in addition to that required for less toxic material. | 2 | *Moderate hazard*: Temporary or minor injury may occur. |
| 3 | 3 and 4 are for highly to extremely toxic (deadly) materials (and any carcinogen, mutagen, or teratogen). These materials will require specialized equipment (e.g. respirator or exhaust hood, full-face shield, rubber apron, specialized gloves, handling tongs, etc.) beyond that required for moderately toxic material. | 3 | *Serious hazard*: Major injury likely unless prompt action is taken and medical treatment is given. |
| 4 | | 4 | *Severe hazard*: Life-threatening, major or permanent damage may result from single or repeated overexposures. |

Note that the HMIS classification differs from that of OSHA's standard 1910.106,[191] which is described in Section 3.16.1 – although it based on the same published information. Also, note that the flash point used is also a closed-cup flash point.

Note that for the health and physical hazard classification systems, the HMIS and NFPA have different numbers of parameters (4 versus 5).[192]

Note that the NFPA system does suggest within the definition of hazard classification the equipment necessary to protect against that classification of hazard.

To repeat, the real hazard of hazard classification systems is confusion. The use of one is mandated by US law (OSHA/DOT). Another is provided with every chemical (NFPA via the MSDS). Addition of

[191] For example, the NFPA uses larger numbers to classify chemicals with greater ignition risk. The OSHA system uses smaller numbers. For another example, the NFPA uses Arabic numerals and the OSHA system uses Roman numerals.
[192] Information in Tables 3.26 and 3.27 is directly quoted from the NFPA and HMIS sources.

**Table 3.27 Comparison of NFPA and HMIS Physical Hazard Classification Systems**

| NFPA | | HMIS | |
|---|---|---|---|
| | | 0 | *Minimal hazard*: materials that are normally stable, even under fire conditions, and will NOT react with water, polymerize, decompose, condense, self-react, and are not explosives. |
| 1 | A chemical that may be reactive if heated and one that reacts with water. | 1 | *Slight hazard*: materials that are normally stable but can become unstable (self-react) at high temperatures and pres sures. Materials may react non-violently with water or undergo hazardous polymerization in the absence of inhibitors. |
| 2 | A chemical that may react violently without detonation. | 2 | *Moderate hazard*: materials that are unstable and may undergo violent chemical changes at normal temperature and pressure with low risk for explosion. Materials may react violently with water or form peroxides upon exposure to air. |
| 3 | A chemical that may detonate or explode if subjected to a strong initiating force or heating under confinement. | 3 | *Serious hazard*: materials that may form explosive mixtures with water and are capable of detonation or explosive reaction in the presence of a strong initiating source. Materials may polymerize, decompose, self-react, or undergo other chemical change at normal temperature and pressure with moderate risk of explosion. |
| 4 | A chemical that readily detonates or explodes. | 4 | *Severe hazard*: materials that are readily capable of explosive water reaction, detonation or explosive decomposition, polymerization, or self-reaction at normal temperature and pressure. |

a third, or more, should be accepted only when the inherent confusion can be avoided by use of it and there is a clear unfulfilled need for it.

### 3.16.5 Non-US Hazard Classification Systems

Other systems are based on the needs of the local entities providing them. Some are more complicated than the NFPA system; some are less so.

Three non-US hazard classification systems will be covered in addition to the three US systems described above. They are as follows:

- South African Paint Manufacturers Association (SAPMA) – see Section 3.16.5.1
- Chemical Hazard Information and Packaging for Supply (CHIPS) – see Section 3.16.5.2
- European INventory of Existing Chemical Substances (EINECS) Hazard Classification – see Section 3.16.5.3

To no surprise, there are other significant international hazard classification systems, including:

- Workplace Hazardous Materials Information System (WHMIS) compiled by the Canadian government
- International Organization for Standardization (ISO)
- Japanese Industrial Standards (JIS)[193]

#### 3.16.5.1 *South African Paint Manufacturers Association*

This simple and intelligently thought out system speaks in terms of health hazards, and what protective equipment should be used to protect against them.

The SAPMA does not publish a list of chemicals with their classifications; users are expected to classify them themselves.

Health hazards are ranked in a numerical system. Protective equipment to combat them is listed in a

[193] For an example of how the variety of human experience leads to a variety of ways of completing the same task, consider JIS C 60695-2-10:2004 – *Fire Hazard Testing – Glow-Wire Apparatus and Common Test Procedure.*

**Table 3.28 Allocation of SAPMA Classifications**

| Hazard Group | Definition |
|---|---|
| Group 4: Minimal | *Minimally hazardous substances* are substances that are reasonably safe under all normal conditions of use. No residual effect is to be expected from accidental exposure even if no treatment is applied. |
| Group 3: Slight | *Slightly hazardous substances* are substances that require some sort of safeguards but are otherwise safe under normal conditions of use. Minor residual effect(s) could result from accidental exposure if no treatment is applied. |
| Group 2: Moderate | *Moderately hazardous substances* are substances that are to be handled with caution and careful regard for personal protection. Minor residual effect(s) could result despite prompt treatment and cumulative/chronic effects are possible. |
| Group 1: Serious | *Seriously hazardous substances* are substances that are to be treated with extreme caution, and either have to be handled with specializes equipment or personal protection has to be worn, or both. Major residual effect(s) could result despite prompt treatment, and severe chronic effects are possible. |

simple table. The two are combined into the SAPMA hazard ranking system. Both are shown in Tables 3.28 and 3.29.[194]

Combining the health hazard classification number with the letter indicating the PPE required makes up a classification for each chemical in the SAPMA database. These two symbols are separated by a letter "H" and the health hazard classification are shown in the form "3HC", "2HE", etc.

Examples of use of this hazard classification system are shown in Table 3.30.

This information, and other information, is found in MSDSs prepared to meet the requirements of the South African Occupational Health and Safety Act.

The appeal of the SAPMA system is that it communicates in a simple way a ranking of the hazards associated with chemicals, *and* the proper equipment to protect users from them.

[194] See http://www.ecochem.co.za/sapma.htm

**Table 3.29 SAPMA Protective Equipment Requirements**

| Hazard Classification | Required Protective Equipment |
|---|---|
| H A | No Specific Hazard |
| H B | Gloves and Dust Respirator |
| H C | Gloves and Safety Goggles |
| H D | Gloves and Safety Goggles and Dust Respirator |
| H E | Gloves and Safety Goggles and Gas Respirator |
| H F | Gloves and Safety Goggles and Foot Protection and Apron |
| H G | Gloves and Safety Goggles and Foot Protection and Apron and Gas Respirator |
| H J | Gloves and Foot Protection and Apron and Air-Fed Hood |
| H K | Consult Supervisor |

**Table 3.30 SAPMA Hazard Ranking of Common Cleaning Solvents**

| Solvent | CAS # | Classification |
|---|---|---|
| 1,1,1-Trichloroethane | 71-55-6 | 1HE |
| Trichloroethylene | 79-01-6 | 1HE |
| Diethyl ether | 60-29-7 | 2HE |
| Cyclohexane | 110-82-7 | 2HE |
| Isopropanol | 67-63-0 | 2HE |
| Ethyl alcohol | 64-17-5 | 2HE |
| Acetone | 67-64-1 | 2HE |
| Kerosene – straight run | 8008-20-6 | 2HE |
| Propylene glycol | 57-55-6 | 2HC |
| Ethyl lactate | 97-64-3 | 3HC |
| Water | 7732-18-5 | 4HA |

### 3.16.5.2 *In the CHIPS in the UK*

Quite the opposite is the CHIPS system. While the SAPMA classification system can be written on both sides of a postcard, the CHIPS system must be written in a small font on both sides of a sheet of legal-size paper. In actuality, it may have been.

This regulation, rewritten for the UK in 2002,[195] requires suppliers of dangerous chemicals (substances and preparations)[196] to classify them for their hazardous properties.

Suppliers must implement detailed criteria which are set out in the Regulations themselves and in the Approved Classification and Labeling Guide.[197]

Chemicals are classified differently by their hazards – depending upon whether they are being transported or supplied (sold). Chemicals may be classed as "Flammable" (Class 3) for transport, but "Harmful" or "Toxic" (equating to Class 6.1) for supply. "The reasons for this are the different types of risk exhibited in different situations. The end user, dealing with small quantities on a frequent basis over long time-periods, is at greater risk from any harmful/toxic effects due to contact with the product than from its flammable characteristics.[198]"

There are three parts to the classification system:[199]

1. A **single letter** corresponding to major description(s) of the hazard(s) associated with the chemical. The details are found in Table 3.31. Symbols, up to six, must be included on the package.
2. A **single letter and number combination** called a risk phrase (R-phrase). This information describes a hazard to persons, or to the environment. A chemical may have more than one R-phrase (see Table 3.32).
3. A **single letter and number combination** called a safety phrase (S-phrase). This information describes specific actions recommended to protect against the stated hazard(s). A chemical may have more than one S-phrase (see Table 3.33).

Compare the outcomes of the CHIPS and SAPMA hazard classification systems:

- Table 3.34 (CHIPS hazard classification system[200])
- Table 3.30 (SAPMA hazard classification system)

The same common cleaning solvents are described in each table.

It is interesting to note that a most detailed system for hazard classification apparently does not include three common chemicals which can and are used for cleaning work.

#### 3.16.5.3 *Hazard Classification in the European Community (EINECS)*

The UK is geographically proximate to Europe but is often politically displaced from other European countries. Both the European Community[201] and the UK have hazard classification systems which appear similar. Unfortunately, as with political opinions, the systems are based on very similar principles but produce somewhat different results.

A detailed list of safety phrases for the EINECS[202] system is given in Table 3.35.[203,204]

While these safety and risk phrases are quite similar to those in for the CHIPS program in Tables 3.32 and 3.33, they are not identical. Certainly, the general approaches are identical.

Risk phrases are given in Table 3.36.[205,206]

---

[195] Statutory Instrument, 2002, No. 1689 and ISBN 0 11 042419 0.

[196] Substances are individual chemicals. Preparations are formulated products.

[197] "Approved Guide to the Classification and Labelling of Dangerous Substances and Dangerous Preparations (1st ed.)" approved by the Health and Safety Commission on April 16, 2002.

[198] See http://www.the-ncec.com/hazchem/index.html

[199] See http://www.hse-databases.co.uk/chip/changes/environment/index.htm

[200] Results were obtained using the interactive system at http://www.the-ncec.com/cselite/

[201] Member countries are: Austria, Belgium, Denmark, Finland, France, Germany, Greece, Ireland, Italy, Luxembourg, The Netherlands, Portugal, Spain, Sweden, and the United Kingdom.

The EU seeks to eliminate the disparity among national laws that could pose barriers to the free movement of goods. EU law prevails over national laws.

[202] European Inventory of Existing Chemical Substances.

[203] Source of these safety phrases for the EC is the International Labor Organization at http://www.ilo.org/public/english/protection/safework/cis/products/icsc/dtasht/sftyphrs/index.htm

[204] Safety phrases ended with a * are those to which the following statement applies. "The phrase has been deleted by 28th Adaptation to the Technical Progress (ATP 28) (August 6, 2001), but may still appear in cards not modified since then."

[205] Source of these Risk Phrases for the EC is the International Labor Organization at http://www.ilo.org/public/english/protection/safework/cis/products/icsc/dtasht/riskphrs/index.htm

[206] Risk phrases ended with a * are those to which the following statement applies. "The phrase has been deleted by 28th Adaptation to the Technical Progress (ATP 28) (August 6, 2001), but may still appear in cards not modified since then."

**Table 3.31 Description of Major Hazards in CHIPS System**

| Abbreviation | Hazard | Symbol | Description of hazard |
|---|---|---|---|
| | | **Physicochemical** | |
| **E** | Explosive |  | Chemicals that explode |
| **O** | Oxidizing |  | Chemicals that react exothermically with other chemicals |
| **F+** | Extremely flammable |  | Chemicals that have an extremely low flash point and boiling point, and gases that catch fire in contact with air |
| **F** | Highly flammable |  | Chemicals that may catch fire in contact with air, only need brief contact with an ignition source, have a very low flash point or evolve highly flammable gases in contact with water |
| | | **Health** | |
| **T+** | Very toxic |  | Chemicals that at very low levels cause damage to health |
| **T** | Toxic |  | Chemicals that at low levels cause damage to health |
| **Carc Cat 1** | Category 1 carcinogens | | Chemicals that may cause cancer or increase its incidence |
| **Carc Cat 2** | | | Category 2 carcinogens |
| **Carc Cat 3** | | | Category 3 carcinogens |
| **Muta Cat 1** | Category 1 mutagens | | Chemicals that induce heritable genetic defects or increase their incidence |
| **Muta Cat 2** | | | Category 2 mutagens |
| **Muta Cat 3** | | | Category 3 mutagens |
| **Repr Cat 1** | Category 1 reproductive toxins | | Chemicals that produce or increase the incidence of non-heritable effects in progeny and/or an impairment in reproductive functions or capacity |
| **Repr Cat 2** | | | Category 2 Reproductive toxins |
| **Repr Cat 3** | | | Category 3 Reproductive toxins |
| **Xn** | Harmful |  | Chemicals that may cause damage to health |
| **C** | Corrosive |  | Chemicals that may destroy living tissue on contact |
| **Xi** | Irritant |  | Chemicals that may cause inflammation to the skin or other mucous membranes |
| **N** | Dangerous for the environment |  | Chemicals that may present an immediate or delayed danger to one or more components of the environment |

**Table 3.32 Risk Phrases in CHIPS System**

| Risk Phrases | Meaning of Individual Risk Phrase |
|---|---|
| R1 | Explosive when dry |
| R2 | Risk of explosion by shock, friction, fire, or other sources of ignition |
| R3 | Extreme risk of explosion by shock, friction, fire, or other sources of ignition |
| R4 | Forms very sensitive explosive metallic compounds |
| R5 | Heating may cause an explosion |
| R6 | Explosive with or without contact with air |
| R7 | May cause fire |
| R8 | Contact with combustible material may cause fire |
| R9 | Explosive when mixed with combustible material |
| R10 | Flammable |
| R11 | Highly flammable |
| R12 | Extremely flammable |
| R14 | Reacts violently with water |
| R14/15 | Reacts violently with water, liberating extremely flammable gases |
| R15 | Contact with water liberates extremely flammable gases |
| R15/29 | Contact with water liberates toxic, extremely flammable gases |
| R16 | Explosive when mixed with oxidizing substances |
| R17 | Spontaneously flammable in air |
| R18 | In use, may form flammable/explosive vapor–air mixture |
| R19 | May form explosive peroxides |
| R20 | Harmful by inhalation |
| R20/21 | Harmful by inhalation and in contact with skin |
| R20/21/22 | Harmful by inhalation, in contact with skin and if swallowed |
| R20/22 | Harmful by inhalation and if swallowed |
| R21 | Harmful in contact with skin |
| R21/22 | Harmful in contact with skin and if swallowed |
| R22 | Harmful if swallowed |
| R23 | Toxic by inhalation |
| R23/24 | Toxic by inhalation and in contact with skin |
| R23/24/25 | Toxic by inhalation, in contact with skin and if swallowed |
| R23/25 | Toxic by inhalation and if swallowed |
| R24 | Toxic in contact with skin |
| R24/25 | Toxic in contact with skin and if swallowed |
| R25 | Toxic if swallowed |
| R26 | Very toxic by inhalation |
| R26/27 | Very toxic by inhalation and in contact with skin |
| R26/27/28 | Very toxic by inhalation, in contact with skin and if swallowed |
| R26/28 | Very toxic by inhalation and if swallowed |
| R27 | Very toxic in contact with skin |
| R27/28 | Very toxic in contact with skin and if swallowed |
| R28 | Very toxic if swallowed |

(Continued)

**Table 3.32 Risk Phrases in CHIPS System (Continued)**

| Risk Phrases | Meaning of Individual Risk Phrase |
|---|---|
| R30 | Can become highly flammable in use |
| R31 | Contact with acids liberates toxic gas |
| R32 | Contact with acids liberates very toxic gas |
| R33 | Danger of cumulative effects |
| R34 | Causes burns |
| R35 | Causes severe burns |
| R36 | Irritating to eyes |
| R36/37 | Irritating to eyes and respiratory system |
| R36/37/38 | Irritating to eyes, respiratory system, and skin |
| R36/38 | Irritating to eyes and skin |
| R37 | Irritating to respiratory system |
| R37/38 | Irritating to respiratory system and skin |
| R38 | Irritating to skin |
| R39 | Danger of very serious irreversible effects |
| R39/23 | Toxic: danger of very serious irreversible effects through inhalation |
| R39/23/24 | Toxic: danger of very serious irreversible effects through inhalation and in contact with skin |
| R39/23/24/25 | Toxic: danger of very serious irreversible effects through inhalation, in contact with skin, and if swallowed |
| R39/23/25 | Toxic: danger of very serious irreversible effects through inhalation and if swallowed |
| R39/24 | Toxic: danger of very serious irreversible effects in contact with skin |
| R39/24/25 | Toxic: danger of very serious irreversible effects in contact with skin and if swallowed |
| R39/25 | Toxic: danger of very serious irreversible effects if swallowed |
| R39/26 | Very toxic: danger of very serious irreversible effects through inhalation |
| R39/26/27 | Very toxic: danger of very serious irreversible effects through inhalation and in contact with skin |
| R39/26/27/28 | Very toxic: danger of very serious irreversible effects through inhalation, in contact with skin and if swallowed |
| R39/26/28 | Very toxic: danger of very serious irreversible effects through inhalation and if swallowed |
| R39/27 | Very toxic: danger of very serious irreversible effects in contact with skin |
| R39/27/28 | Very toxic: danger of very serious irreversible effects in contact with skin and if swallowed |
| R39/28 | Very toxic: danger of very serious irreversible effects if swallowed |
| R40 | Limited evidence of a carcinogenic effect |
| R41 | Risk of serious damage to eyes |
| R42 | May cause sensitization by inhalation |
| R43 | May cause sensitization by skin contact |
| R42/43 | May cause sensitization by inhalation and skin contact |
| R44 | Risk of explosion if heated under confinement |
| R45 | May cause cancer |
| R46 | May cause heritable genetic damage |
| R48 | Danger of serious damage to health by prolonged exposure |
| R48/20 | Harmful: danger of serious damage to health by prolonged exposure through inhalation |

(Continued)

**Table 3.32 Risk Phrases in CHIPS System (Continued)**

| Risk Phrases | Meaning of Individual Risk Phrase |
|---|---|
| R48/20/21 | Harmful: danger of serious damage to health by prolonged exposure through inhalation and in contact with skin |
| R48/20/21/22 | Harmful: danger of serious damage to health by prolonged exposure through inhalation, in contact with skin and if swallowed |
| R48/20/22 | Harmful: danger of serious damage to health by prolonged exposure through inhalation and if swallowed |
| R48/21 | Harmful: danger of serious damage to health by prolonged exposure in contact with skin |
| R48/21/22 | Harmful: danger of serious damage to health by prolonged exposure in contact with skin and if swallowed |
| R48/22 | Harmful: danger of serious damage to health by prolonged exposure if swallowed |
| R48/23 | Toxic: danger of serious damage to health by prolonged exposure through inhalation |
| R48/23/24 | Toxic: danger of serious damage to health by prolonged exposure through inhalation and in contact with skin |
| R48/23/24/25 | Toxic: danger of serious damage to health by prolonged exposure through inhalation, in contact with skin and if swallowed |
| R48/23/25 | Toxic: danger of serious damage to health by prolonged exposure through inhalation and if swallowed |
| R48/24 | Toxic: danger of serious damage to health by prolonged exposure in contact with skin |
| R48/24/25 | Toxic: danger of serious damage to health by prolonged exposure in contact with skin and if swallowed |
| R48/25 | Toxic: danger of serious damage to health by prolonged exposure if swallowed |
| R49 | May cause cancer by inhalation |
| R50 | Very toxic to aquatic organisms |
| R50/53 | Very toxic to aquatic organisms, may cause long-term adverse effects in the aquatic environment |
| R51 | Toxic to aquatic organisms |
| R51/53 | Toxic to aquatic organisms, may cause long-term adverse effects in the aquatic environment |
| R52 | Harmful to aquatic organisms |
| R52/53 | Harmful to aquatic organisms, may cause long-term adverse effects in the aquatic environment |
| R53 | May cause long-term adverse effects in the aquatic environment |
| R54 | Toxic to flora |
| R55 | Toxic to fauna |
| R56 | Toxic to soil organisms |
| R57 | Toxic to bees |
| R58 | May cause long-term adverse effects in the environment |
| R59 | Dangerous for the ozone layer |
| R60 | May impair fertility |
| R61 | May cause harm to the unborn child |
| R62 | Possible risk of impaired fertility |
| R63 | Possible risk of harm to the unborn child |
| R64 | May cause harm to breast-fed babies |
| R65 | Harmful: may cause lung damage if swallowed |
| R66 | Repeated exposure may cause skin dryness or cracking |
| R67 | Vapors may cause drowsiness and dizziness |
| R68 | Possible risk of irreversible effects |
| R68/20 | Harmful: possible risk of irreversible effects through inhalation |

(Continued)

**Table 3.32 Risk Phrases in CHIPS System (Continued)**

| Risk Phrases | Meaning of Individual Risk Phrase |
|---|---|
| R68/20/21 | Harmful: possible risk of irreversible effects through inhalation and in contact with skin |
| R68/20/21/22 | Harmful: possible risk of irreversible effects through inhalation, in contact with skin and if swallowed |
| R68/20/22 | Harmful: possible risk of irreversible effects through inhalation and if swallowed |
| R68/21 | Harmful: possible risk of irreversible effects in contact with skin |
| R68/21/22 | Harmful: possible risk of irreversible effects in contact with skin and if swallowed |
| R68/22 | Harmful: possible risk of irreversible effects if swallowed |

**Table 3.33 Safety Phrases in CHIPS System**

| Safety phrases | Meaning of individual safety phrase |
|---|---|
| S1 | Keep locked up |
| S1/2 | Keep locked up and out of the reach of children |
| S2 | Keep out of the reach of children |
| S3 | Keep in a cool place |
| S3/7 | Keep container tightly closed in a cool place |
| S3/7/9 | Keep container tightly closed in a cool, well-ventilated place |
| S3/9/14 | Keep in a cool, well-ventilated place away from … (incompatible materials to be indicated by the manufacturer) |
| S3/9/14/49 | Keep only in the original container in a cool, well-ventilated place away from … (incompatible materials to be indicated by the manufacturer) |
| S3/9/49 | Keep only in the original container in a cool, well-ventilated place |
| S3/14 | Keep in a cool place away from … (incompatible materials to be indicated by the manufacturer) |
| S4 | Keep away from living quarters |
| S5 | Keep contents under … (appropriate liquid to be specified by the manufacturer) |
| S6 | Keep under … (inert gas to be specified by the manufacturer) |
| S7 | Keep container tightly closed |
| S7/8 | Keep container tightly closed and dry |
| S7/9 | Keep container tightly closed and in a well-ventilated place |
| S7/47 | Keep container tightly closed and at temperature not exceeding … °C (to be specified by the manufacturer) |
| S8 | Keep container dry |
| S9 | Keep container in a well-ventilated place |
| S12 | Do not keep the container sealed |
| S13 | Keep away from food, drink and animal feedingstuffs |
| S14 | Keep away from … (incompatible materials to be indicated by the manufacturer) |
| S15 | Keep away from heat |
| S16 | Keep away from sources of ignition – no smoking |
| S17 | Keep away from combustible material |
| S18 | Handle and open container with care |

(Continued)

**Table 3.33 Safety Phrases in CHIPS System (Continued)**

| Safety phrases | Meaning of individual safety phrase |
|---|---|
| S20 | When using do not eat or drink |
| S20/21 | When using do not eat, drink or smoke |
| S21 | When using do not smoke |
| S22 | Do not breathe dust |
| S23 | Do not breathe gas/fumes/vapor/spray (appropriate wording to be specified by the manufacturer) |
| S24 | Avoid contact with skin |
| S24/25 | Avoid contact with skin and eyes |
| S25 | Avoid contact with eyes |
| S26 | In case of contact with eyes, rinse immediately with plenty of water and seek medical advice |
| S27 | Take off immediately all contaminated clothing |
| S27/28 | After contact with skin, take off immediately all contaminated clothing, and wash immediately with plenty of … (to be specified by the manufacturer) |
| S28 | After contact with skin, wash immediately with plenty of … (to be specified by the manufacturer) |
| S29 | Do not empty into drains |
| S29/35 | Do not empty into drains, dispose of this material and its container in a safe way |
| S29/56 | Do not empty into drains, dispose of this material and its container at hazardous or special waste collection point |
| S30 | Never add water to this product |
| S33 | Take precautionary measures against static discharges |
| S35 | This material and its container must be disposed of in a safe way |
| S36 | Wear suitable protective clothing |
| S36/37 | Wear suitable protective clothing and gloves |
| S36/37/39 | Wear suitable protective clothing, gloves and eye/face protection |
| S36/39 | Wear suitable protective clothing and eye/face protection |
| S37 | Wear suitable gloves |
| S37/39 | Wear suitable gloves and eye/face protection |
| S38 | In case of insufficient ventilation wear suitable respiratory equipment |
| S39 | Wear eye/face protection |
| S40 | To clean the floor and all objects contaminated by this material use … (to be specified by the manufacturer) |
| S41 | In case of fire and/or explosion do not breathe fumes |
| S42 | During fumigation/spraying wear suitable respiratory equipment (appropriate wording to be specified by the manufacturer) |
| S43 | In case of fire use … (indicate in the space the precise type of fire-fighting equipment. If water increases the risk add – never use water) |
| S45 | In case of accident or if you feel unwell seek medical advice immediately (show the label where possible) |
| S46 | If swallowed, seek medical advice immediately and show this container or label |
| S47 | Keep at temperature not exceeding … °C (to be specified by the manufacturer) |
| S47/49 | Keep only in the original container at temperature not exceeding … °C (to be specified by the manufacturer) |
| S48 | Keep wet with … (appropriate material to be specified by the manufacturer) |
| S49 | Keep only in the original container |

(Continued)

**Table 3.33 Safety Phrases in CHIPS System (Continued)**

| Safety phrases | Meaning of individual safety phrase |
|---|---|
| S50 | Do not mix with … (to be specified by the manufacturer) |
| S51 | Use only in well-ventilated areas |
| S52 | Not recommended for interior use on large surface areas |
| S53 | Avoid exposure – obtain special instructions before use |
| S56 | Dispose of this material and its container at hazardous or special waste collection point |
| S57 | Use appropriate containment to avoid environmental contamination |
| S59 | Refer to manufacturer/supplier for information on recovery/recycling |
| S60 | This material and its container must be disposed of as hazardous waste |
| S61 | Avoid release to the environment. refer to special instructions/safety data sheet |
| S62 | If swallowed, do not induce vomiting: seek medical advice immediately and show this container or label to medical staff |
| S63 | In case of accident by inhalation: remove casualty to fresh air and keep at rest |
| S64 | If swallowed, rinse mouth with water (only if the person is conscious) |

**Table 3.34 CHIPS Hazard Ranking of Common Cleaning Solvents**

| Solvent | EINECS # | Label Information | Risk Phrases | Safety Phrases |
|---|---|---|---|---|
| 1,1,1-Trichloroethane | 200-756-3 |  | Xn; N; R20-59 | S(2-)24/25-59-61 |
| Trichloroethylene | 201-167-4 |  | R40-52/53 | S(2-)53-45-61 |
| Diethyl ether | 200-467-2 | 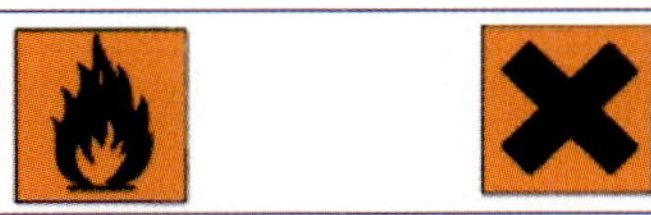 | $F^+$; Xn; R12-19-22-66-67 | S(2-)9-16-29-33 |
| Cyclohexane | 203-806-2 |  | F; Xn; N; R11-38-50/ 53-65-67 | S(2-)9-16-33-60-61-62 |
| Isopropanol | 67-63-0 |  | F; Xi; R11-36-67 | S(2-)7-16-24/ 25-26 |
| Ethyl alcohol | 200-578-6 |  | F; R11 | S(2-)7-16 |
| Acetone | 200-662-2 |  | F; Xi; R11-36-66-67 | S(2-)9-16-26 |
| Kerosene – straight run | 232-366-4 | | | |
| Propylene glycol | 200-338-0 | | | |
| Ethyl lactate | 202-598-0 |  | Xi; R10-37-41 | S(2-)24-26-39 |
| Water | | | | |

**Table 3.35 Safety Phrases in EINECS Hazard Classification System**

| Safety Phrase | |
|---|---|
| S1/2 | Keep locked up and out of the reach of children |
| S2 | Keep out of the reach of children |
| S3 | Keep in a cool place |
| S3/7 | Keep container tightly closed in a cool place |
| S3/7/9 | Keep container tightly closed in a cool, well-ventilated place* |
| S3/9 | Keep in a cool, well-ventilated place* |
| S3/9/14 | Keep in a cool, well-ventilated place away from (incompatible materials to be indicated by the manufacturer) |
| S3/9/14/49 | Keep only in the original container in a cool, well-ventilated place away from … (incompatible materials) |
| S3/9/49 | Keep only in original container in a cool, well-ventilated place |
| S3/14 | Keep in a cool place away from … (incompatible materials to be specified by the manufacturer) |
| S4 | Keep away from living quarters |
| S5 | Keep contents under … (appropriate liquid to be specified by the manufacturer) |
| S6 | Keep under … (inert gas to be specified by the manufacturer) |
| S7 | Keep container tightly closed |
| S7/8 | Keep container tightly closed and dry |
| S7/9 | Keep container tightly closed and in a well-ventilated place |
| S7/47 | Keep container tightly closed and at a temperature not exceeding … °C (to be specified by the manufacturer) |
| S8 | Keep container dry |
| S9 | Keep container in a well-ventilated place |
| S12 | Do not keep the container sealed |
| S13 | Keep away from food, drink and animal feedingstuffs |
| S14 | Keep away from … (incompatible materials to be indicated by the manufacturer) |
| S15 | Keep away from heat |
| S16 | Keep away from sources of ignition – No smoking |
| S17 | Keep away from combustible material |
| S18 | Handle and open container with care |
| S20 | When using do not eat or drink |
| S20/21 | When using do not eat, drink, or smoke |
| S21 | When using do not smoke |
| S22 | Do not breathe dust |
| S23 | Do not breathe gas/fumes/vapor/spray (appropriate wording to be specified by the manufacturer) |
| S24 | Avoid contact with the skin |
| S24/25 | Avoid contact with skin and eyes |
| S25 | Avoid contact with eyes |
| S26 | In case of contact with eyes, rinse immediately with plenty of water and seek medical advice |
| S27 | Take off immediately all contaminated clothing |
| S27/28 | After contact with skin, take off immediately all contaminated clothing, and wash immediately with plenty of … (to be specified by the manufacturer) |
| S28 | After contact with skin, wash immediately with plenty of … (to be specified by the manufacturer) |

(Continued)

**Table 3.35 Safety Phrases in EINECS Hazard Classification System (Continued)**

| Safety Phrase | |
|---|---|
| S29 | Do not empty into drains |
| S29/35 | Do not empty into drains, dispose of this material and its container in a safe way |
| S29/56 | Do not empty into drains, dispose of this material and its container at hazardous or special waste |
| S30 | Never add water to this product |
| S33 | Take precautionary measures against static discharges |
| S34 | Avoid shock and friction* |
| S35 | This material and its container must be disposed of in a safe way |
| S36 | Wear suitable protective clothing |
| S36/37 | Wear suitable protective clothing and gloves |
| S36/37/39 | Wear suitable protective clothing, gloves, and eye/face protection |
| S36/39 | Wear suitable protective clothing and eye/face protection |
| S37 | Wear suitable gloves |
| S37/39 | Wear suitable gloves and eye/face protection |
| S38 | In case of insufficient ventilation, wear suitable respiratory equipment |
| S39 | Wear eye/face protection |
| S40 | To clean the floor and all objects contaminated by this material, use … (to be specified by the manufacturer) |
| S41 | In case of fire and/or explosion do not breathe fumes |
| S42 | During fumigation/spraying wear suitable respiratory equipment (appropriate wording to be specified by the manufacturer) |
| S43 | In case of fire, use … (indicate in the space the precise type of fire-fighting equipment) |
| S44 | If you feel unwell, seek medical advice (show label where possible)* |
| S45 | In case of accident or if you feel unwell, seek medical advice immediately (show the label where possible) |
| S46 | If swallowed, seek medical advice immediately and show container or label |
| S47 | Keep at temperature not exceeding … °C (to be specified by the manufacturer) |
| S47/49 | Keep only in the original container at a temperature not exceeding … °C (to be specified by the manufacturer) |
| S48 | Keep wet with … (appropriate material to be specified by the manufacturer) |
| S49 | Keep only in the original container |
| S50 | Do not mix with … (to be specified by the manufacturer) |
| S51 | Use only in well-ventilated areas |
| S52 | Not recommended for interior use on large surface areas |
| S53 | Avoid exposure – obtain special instructions before use |
| S56 | Dispose of this material and its container to hazardous or special waste collection point |
| S57 | Use appropriate container to avoid environmental contamination |
| S59 | Refer to manufacturer/supplier for information on recovery/recycling |
| S60 | This material and its container must be disposed of as hazardous waste |
| S61 | Avoid release to the environment. Refer to special instructions/safety data sheets |
| S62 | If swallowed, do not induce vomiting: seek medical advice immediately and show this container or label to medical staff |
| S63 | In case of accident by inhalation: remove casualty to fresh air and keep at rest |
| S64 | If swallowed, rinse mouth with water (only if the person is conscious) |

**Table 3.36 Risk Phrases in EINECS Hazard Classification System**

| Risk Phrase | |
|---|---|
| R1 | Explosive when dry |
| R2 | Risk of explosion by shock, friction, fire, or other sources of ignition |
| R3 | Extreme risk of explosion by shock, friction, fire, or other sources of ignition |
| R4 | Forms very sensitive explosive metallic compounds |
| R5 | Heating may cause an explosion |
| R6 | Explosive with or without contact with air |
| R7 | May cause fire |
| R8 | Contact with combustible material may cause fire |
| R9 | Explosive when mixed with combustible material |
| R10 | Flammable |
| R11 | Highly flammable |
| R12 | Extremely flammable |
| R13 | Extremely flammable liquefied gas* |
| R14 | Reacts violently with water |
| R14/15 | Reacts violently with water, liberating extremely flammable gases |
| R15 | Contact with water liberates extremely flammable gases |
| R15/29 | Contact with water liberates toxic, extremely flammable gas |
| R16 | Explosive when mixed with oxidizing substances |
| R17 | Spontaneously flammable in air |
| R18 | In use, may form flammable/explosive vapor–air mixture |
| R19 | May form explosive peroxides |
| R20 | Harmful by inhalation |
| R20/21 | Harmful by inhalation and in contact with skin |
| R20/21/22 | Harmful by inhalation, in contact with skin and if swallowed |
| R20/22 | Harmful by inhalation and if swallowed |
| R21 | Harmful in contact with skin |
| R21/22 | Harmful in contact with skin and if swallowed |
| R22 | Harmful if swallowed |
| R23 | Toxic by inhalation |
| R23/24 | Toxic by inhalation and in contact with skin |
| R23/24/25 | Toxic by inhalation, in contact with skin and if swallowed |
| R23/25 | Toxic by inhalation and if swallowed |
| R24 | Toxic in contact with skin |
| R24/25 | Toxic in contact with skin and if swallowed |
| R25 | Toxic if swallowed |
| R26 | Very toxic by inhalation |
| R26/27 | Very toxic by inhalation and in contact with skin |
| R26/27/28 | Very toxic by inhalation, in contact with skin and if swallowed |
| R26/28 | Very toxic by inhalation and if swallowed |
| R27 | Very toxic in contact with skin |
| R27/28 | Very toxic in contact with skin and if swallowed |
| R28 | Very toxic if swallowed |
| R29 | Contact with water liberates toxic gas |

(Continued)

**Table 3.36 Risk Phrases in EINECS Hazard Classification System (Continued)**

| Risk Phrase | |
|---|---|
| R30 | Can become highly flammable in use |
| R31 | Contact with acids liberates toxic gas |
| R32 | Contact with acids liberates very toxic gas |
| R33 | Danger of cumulative effects |
| R34 | Causes burns |
| R35 | Causes severe burns |
| R36 | Irritating to eyes |
| R36/37 | Irritating to eyes and respiratory system |
| R36/37/38 | Irritating to eyes, respiratory system, and skin |
| R36/38 | Irritating to eyes and skin |
| R37 | Irritating to respiratory system |
| R37/38 | Irritating to respiratory system and skin |
| R38 | Irritating to skin |
| R39 | Danger of very serious irreversible effects |
| R39/23 | Toxic: danger of very serious irreversible effects through inhalation |
| R39/23/24 | Toxic: danger of very serious irreversible effects through inhalation and in contact with skin |
| R39/23/24/25 | Toxic: danger of very serious irreversible effects through inhalation, in contact with skin and if swallowed |
| R39/23/25 | Toxic: danger of very serious irreversible effects through inhalation and if swallowed |
| R39/24 | Toxic: danger of very serious irreversible effects in contact with skin |
| R39/24/25 | Toxic: danger of very serious irreversible effects in contact with skin and if swallowed |
| R39/25 | Toxic: danger of very serious irreversible effects if swallowed |
| R39/26 | Very toxic: danger of very serious irreversible effects through inhalation |
| R39/26/27 | Very toxic: danger of very serious irreversible effects through inhalation and in contact with skin |
| R39/26/27/28 | Very toxic: danger of very serious irreversible effects through inhalation, in contact with skin and eyes |
| R39/26/28 | Very toxic: danger of very serious irreversible effects through inhalation and if swallowed |
| R39/27 | Very toxic: danger of very serious irreversible effects in contact with skin |
| R39/27/28 | Very toxic: danger of very serious irreversible effects in contact with skin and if swallowed |
| R39/28 | Very toxic: danger of very serious irreversible effects if swallowed |
| R40 | Limited evidence of a carcinogenic effect* |
| R40/20 | Harmful: possible risk of irreversible effects through inhalation* |
| R40/20/21 | Harmful: possible risk of irreversible effects through inhalation and in contact with skin* |
| R40/20/21/22 | Harmful: possible risk of irreversible effects through inhalation, in contact with skin and if swallowed* |
| R40/20/22 | Harmful: possible risk of irreversible effects through inhalation and if swallowed* |
| R40/21 | Harmful: possible risk of irreversible effects in contact with skin* |
| R40/21/22 | Harmful: possible risk of irreversible effects in contact with skin and if swallowed* |
| R40/22 | Harmful: possible risk of irreversible effects if swallowed* |
| R41 | Risk of serious damage to eyes |
| R42 | May cause sensitization by inhalation |
| R42/43 | May cause sensitization by inhalation and skin contact |
| R43 | May cause sensitization by skin contact |
| R44 | Risk of explosion if heated under confinement |
| R45 | May cause cancer |

(Continued)

**Table 3.36 Risk Phrases in EINECS Hazard Classification System (Continued)**

| Risk Phrase | |
|---|---|
| R46 | May cause heritable genetic damage |
| R47 | May cause birth defects* |
| R48 | Danger of serious damage to health by prolonged exposure |
| R48/20 | Harmful: danger of serious damage to health by prolonged exposure through inhalation |
| R48/20/21 | Harmful: danger of serious damage to health by prolonged exposure through inhalation and in contact with skin |
| R48/20/21/22 | Harmful: danger of serious damage to health by prolonged exposure through inhalation, in contact with skin and if swallowed |
| R48/20/22 | Harmful: danger of serious damage to health by prolonged exposure through inhalation and if swallowed |
| R48/21 | Harmful: danger of serious damage to health by prolonged exposure in contact with skin |
| R48/21/22 | Harmful: danger of serious damage to health by prolonged exposure in contact with skin and if swallowed |
| R48/22 | Harmful: danger of serious damage to health by prolonged exposure if swallowed |
| R48/23 | Toxic: danger of serious damage to health by prolonged exposure through inhalation |
| R48/23/24 | Toxic: danger of serious damage to health by prolonged exposure through inhalation and in contact |
| R48/23/24/25 | Toxic: danger of serious damage to health by prolonged exposure through inhalation, in contact with skin and if swallowed |
| R48/23/25 | Toxic: danger of serious damage to health by prolonged exposure through inhalation and if swallowed |
| R48/24 | Toxic: danger of serious damage to health by prolonged exposure in contact with skin |
| R48/24/25 | Toxic: danger of serious damage to health by prolonged exposure in contact with skin and if swallowed |
| R48/25 | Toxic: danger of serious damage to health by prolonged exposure if swallowed |
| R49 | May cause cancer by inhalation |
| R50 | Very toxic to aquatic organisms |
| R50/53 | Very toxic to aquatic organisms, may cause long-term adverse effects in the aquatic environment |
| R51 | Toxic to aquatic organisms |
| R51/53 | Toxic to aquatic organisms, may cause long-term adverse effects in the aquatic environment |
| R52 | Harmful to aquatic organisms |
| R52/53 | Harmful to aquatic organisms, may cause long-term adverse effects in the aquatic environment |
| R53 | May cause long-term adverse effects in the aquatic environment |
| R54 | Toxic to flora |
| R55 | Toxic to fauna |
| R56 | Toxic to soil organisms |
| R57 | Toxic to bees |
| R58 | May cause long-term adverse effects in the environment |
| R59 | Dangerous for the ozone layer |
| R60 | May impair fertility |
| R61 | May cause harm to the unborn child |
| R62 | Possible risk of impaired fertility |
| R63 | Possible risk of harm to the unborn child |
| R64 | May cause harm to breast-fed babies |
| R65 | Harmful: may cause lung damage if swallowed |
| R66 | Repeated exposure may cause skin dryness or cracking |

(Continued)

**Table 3.36 Risk Phrases in EINECS Hazard Classification System (Continued)**

| Risk Phrase | |
|---|---|
| R67 | Vapors may cause drowsiness and dizziness |
| R68 | Possible risks of irreversible effects |
| R68/20 | Harmful: possible risk of irreversible effects through inhalation |
| R68/20/21 | Harmful: possible risk of irreversible effects through inhalation and in contact with skin |
| R68/20/21/22 | Harmful: possible risk of irreversible effects through inhalation, in contact with skin and if swallowed |
| R68/20/22 | Harmful: possible risk of irreversible effects through inhalation and if swallowed |
| R68/21 | Harmful: possible risk of irreversible effects in contact with skin |
| R68/21/22 | Harmful: possible risk of irreversible effects in contact with skin and if swallowed |
| R68/22 | Harmful: possible risk of irreversible effects if swallowed |

Compare Table 3.34 (CHIPS hazard classification system[207]) with the results of the Table 3.37 (EINECS hazard classification system[208]); and with the results of the Table 3.30 (SAPMA hazard classification system). The same common cleaning solvents are described in each Table.

Differences among the CHIPS and EINECS systems are probably related to the fact that different persons within different organizations[209] produced the two classifications. It is pointed out in Section 3.16.4 how these differences can be ameliorated if users are involved in the classifications of hazards. In this way, local differences with classification done by NFPA or other agency can be discussed and dissolved.

Most MSDS sheets now contain codes such as R23 or R45 which correspond to certain "risk phrases," and S17 or S24 which correspond to certain "safety phrases."

## 3.16.6 How Much Hazard Classification Is Needed?

There is probably a consensus answer which no one will accept. It is that users need less than they have now – but to use some of it more.

A second part of that answer is that all which is needed is to faithfully, rigorously, and accurately use any one system. If managers adopt a single recommendation from using this volume, it would be just that:

- Know the hazards of chemicals used in operations you manage – via any credible classification system.
- Protect persons, property, and enterprises you manage from those hazards by making informed choices.

In principle, there is no wrong in developing your own hazard classification system, and using it. Wrong can come in confusing classification outcomes about any chemical from multiple classification systems.

### 3.16.6.1 *Hazard Classification on a Local Scale*

It is not just governments, trade organizations, and interest-specific firms who develop and support hazard classification systems. Universities,[210] laboratories, (see Footnote 210) and chemical manufacturers[211]

---

[207] Results were obtained using the interactive system as http://www.the-ncec.com/cselite/index.html

[208] Results were obtained from the International Labor Organization's site as http://www.ilo.org/public/english/protection/safework/cis/products/safetytm/clasann4.htm

[209] The CHIPS classification was produced by the UK's The National Chemical Emergency Centre. The EINECS classification was produced by the International Labor Organization.

[210] For example, see http://www.pp.okstate.edu/ehs/KOPYKIT/HAZCOM.PPT, http://www.stanford.edu/dept/EHS/prod/aboutus/documents/safetyman/hazmatsystem.html, etc.

[211] The author participated in these activities at many plants owned by both Conoco and Du Pont.

**Table 3.37 EINECS Hazard Ranking of Common Cleaning Solvents**

| Solvent | EINECS # | Label information | Risk phrases | Safety phrases |
|---|---|---|---|---|
| 1,1,1-Trichloroethane | 200-756-3 |   | Xn; N; R20-59 | S(2-)24/25-59-61 Stabilizer content may affect phrases |
| Trichloroethylene | 201-167-4 | Carcinogen Category 2; R45 Mutagen Category 3; R68 R67 Xi; R 36/38 R52-53 | R40-52/53 | S(2-)53-45-61 |
| Diethyl ether | 200-467-2 |  | R12-19 | S(2-)9-16-29-33 |
| Cyclohexane | 203-806-2 |  | R11 | S(2-)9-16-33 |
| Propan-2-ol, isopropanol | 67-63-0 |  | R11 | S(2-)7-16 There may be isomers of this compound |
| Ethyl alcohol pure | 200-578-6 |  | R11 | S(2-)7-16 |
| Acetone | 200-662-2 |  | R11 | S(2-)9-16-23-33 |
| Kerosene – straight run | 232-366-4 | | | |
| Propylene glycol | 200-338-0 | | | |
| Ethyl lactate | 202-598-0 | | R10 | S(2-)23 |
| Water | | | | |

all maintain and support their own hazard classification systems.

This author believes that the six major hazard classification systems above bring value, but should be supplemented in many cases by locally supported systems. Here, those with a voice can be the most affected by decisions – the actual users of chemicals. That's why this author is so fond of the SAPMA system.

## 3.17 HAZARD MANAGEMENT – WITH INFORMATION

This approach to control of chemical hazards is shown in Figure 3.30.

Information is the key fuel energizing every stage of this progress. There are at least four major players in this work – each is providing and receiving information to the others:

- *Suppliers* notify governments of the existence of their single-chemical and formulated products.
- Government *regulators* reflect knowledge of products of other governmental agencies and private firms.
- *Agencies* and firms do risk assessments and hazard classifications about these products.
- *Users* receive those work products.

No country has a patent on this management approach – all industrialized countries practice hazard management in this manner. Yet, the system is voluntary.

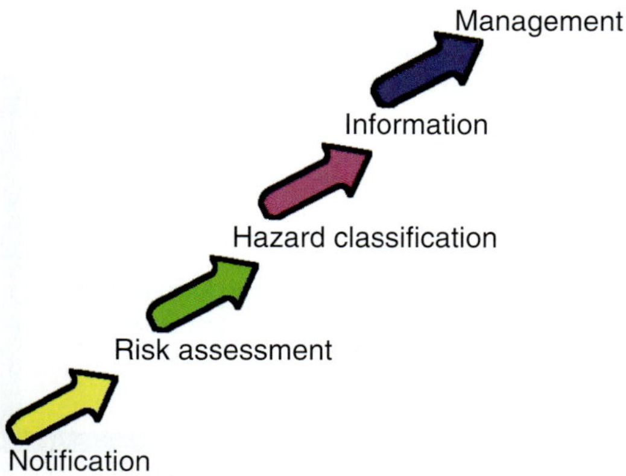


**Figure 3.30**

### 3.17.1 Notification: The Starting Point of Hazard Management

In 1967, the European Union (EU) introduced legislation on chemicals applicable to products manufactured in or imported into the territory of the EU. The objective was and is to protect the public health as well as the environment.

This is Directive 67/548/EEC.[212] It outlines the procedures for the classification, packaging and labeling of dangerous substances. This 1967 legislation has been amended directly 8 times as of this writing and adapted to technical progress 28 times.[213]

Directive 67/548/EEC is complex. The basic goal is to "classify" hazards of chemicals so that hazards can be communicated and managed. There are two distinct categories:

1. Manufacturers and importers of chemicals must classify all substances according to the law's description of "dangerous".[214] Once classified, chemicals are then required to be packaged and labeled accordingly.
2. "New" products must be "notified"[216] to the Member State Competent Authorities. Products considered to be "new" are those which were not included in the EINECS.[215] New chemicals, introduced subsequently, form a cumulative index, European LIst of Notified Chemical Substances (ELINCS) periodically updated as an Official Journal.

The EU is not the only authority which uses "notification" as a key tool for users in their managing of chemical hazards. Other authorities and their notification lists are as follows:

- Canadian Domestic Substances and Non-Domestic Substances Lists (DSL/NDSL)
- Australian Inventory of Chemical Substances (AICS)
- Philippines Inventory of Chemicals and Chemical Substances (PICCS)
- Korean Existing Chemicals List (ECL)
- Existing and New Chemical Substances (ENCS) promulgated by the Kashin Act of Japan
- Chemical Substance Inventory of the Toxic Substances Control Act (TOSCA) in the US[217]
- Israel Hazardous Substances Law[218]

Notification allows risk assessment. This provides a qualitative or quantitative evaluation of the threat posed to human health and the environment by the actual or potential presence of chemicals (solvents).

---

[212] *Official Journal P* 196, 16/08/1967, pp. 0001–0098.

[213] See http://www.eurunion.org/legislat/chemical.htm

[214] As defined in Article 2(2) of 67/548/EEC (as amended). These classifications range from explosive to toxic to carcinogenic to flammable.

[215] European Inventory of Existing Chemical Substances, a closed list of products on the EU market as of 9/18/81.

[216] Notification of a new product entails provision of the results of specified testing and the subsequent compilation of a technical dossier. A substance is subject to notification if it is not covered by one of the exemptions granted by Directive 92/32/EEC.

[217] The Toxic Substances Control Act (TOSCA) of 1976 was enacted by the US Congress to give EPA the ability to track the 75 000 industrial chemicals currently produced or imported into the United States. EPA repeatedly screens these chemicals and can require reporting or testing of those that may pose an environmental or human-health hazard. EPA can be the manufacture and import of those chemicals that pose an unreasonable risk. See http://www.epa.gov/opptintr/newchems/invntory.htm/

[218] Israel's Hazardous Substances Law prohibits the sale of hazardous substances by anyone holding a "poisons permit" to anyone not holding such a permit. This measure enables the authority to prevent the distribution of hazardous substances to anyone not authorized to deal with them.

**Table 3.38 Dichotomy in Risk Analysis**

| Most Significant Risk | | Least Significant Risk | |
|---|---|---|---|
| **US EPA** | **US Public** | **US EPA** | **US Public** |
| Global warming | Chemical waste disposal | Hazardous waste sites – active and inactive<br>Underground storage tanks | Indoor air pollution |
| Indoor air pollution, including Radon (high health risk) | Water pollution | | Exposure to chemicals in consumer products |
| Surface water pollution (high ecological risk) | Chemical plant accidents | | Global warming |

## 3.17.2 Risk Assessment

Risk assessments about the use of chemicals are conducted for a number of reasons, including to:

- Establish whether an ecological risk exists or not.
- Identify the need for additional data collection.
- Focus on the dangers of a specific chemical or the risks posed to a specific site.
- Help develop contingency plans and other responses to chemical releases.

The virtue, and problem, of risk assessment is that the outcome is based on the preferences of the person(s) doing the assessment:

- Risk assessment should be done by those involved because they must cope with the outcome.
- Yet risk assessment should also be done by persons equipped by training, experience, and serious interest. That combination is often difficult to coordinate.

An excellent example of this dichotomy is found in Table 3.38.[219] Here the contractor agency (US EPA) and their client (the US Public) have very different understandings of what constitutes risk which they wish to mitigate.

A risk assessment should be performed in four distinct steps:

1. *Data compilation and evaluation*: Here the objective is to verify that the data (from which assessments will be made) are appropriate for use and are considered to be representative of current conditions.
2. *Exposure assessment*: In this task, the objective is to estimate the type and magnitude of exposures from the chemicals of potential concern that are present at or migrating from a process/site/facility. Involved are the exposure setting, the physical environment, and the exposure pathways. For example, vapor degreasing and cold cleaning present very different exposure settings, and pathways for involved workers. A solvent which produced reversible respiratory irritation might be quite suitable for vapor degreasing and much less so for cold cleaning.
3. *Toxicity assessment*:[220] Here there are two aims – identification of hazards and assessment of dose–response relationships. The former determines whether exposure to a chemical can increase the incidence of a particular adverse health effect and determines the likelihood of occurrence in associated workers. The latter is more difficult to quantify: the relationship between the magnitude of exposure and the known adverse effects.
4. *Risk characterization*:[221] The above three steps are combined, analyzed, and reported upon. The work product of this endeavor is an opinion(s) – not an exposure limit, a hazard classification, or risk phrase. No matter how risks are defined or quantified, they are usually expressed as a probability of adverse effects associated with a particular

[219] US General Accounting Office Report to Congress, *Environmental Protection: Meeting Public Expectations with Limited Resources* (GAO/RCED-91-97, June 1991). Obviously, public perceptions about any topic will change with time and current events.
[220] Exposure assessment and toxicity assessment are usually conducted simultaneously.
[221] An excellent tutorial on risk management is given at http://risk.lsd.ornl.gov/homepage/tutorials.shtml

activity. This is because risk is the chance of injury, damage, or loss.

The US EPA maintains an excellent library[222] of information about risk for more than 550 individual chemicals. Many are cleaning chemicals.

### 3.17.3 Classification of Hazards

Risk assessment leads to classification of hazards, and the chemicals associated with them.

If a tree falls in the wilderness, and no one is there, is a sound made? If information about risks is used to classify the hazards of chemicals, and the people who will use the chemicals aren't part of the classification scheme, will the classification of hazards be useful?

The answer to these questions, respectively, are: that's beyond the scope of this book, and probably not, but that's the way its normally done.

The six hazard classification schemes in Section 13.16 (OSHA/DOT, NFPA, SAPMA, UK, and EINECS) are not the only ones available. But they are the ones most commonly used. Each shares the virtue of being produced by experienced professionals. All but one share the drawback of not being produced by any of the large numbers of people who use the results of these schemes. The situation is similar to that of risk assessment.

In most cases, the classification of hazards is done by a central authority. South Africa's SAPMA is an exception to this generality.[223]

### 3.17.4 Free Flow of Information

Hazard classification produces information. And information can be used to manage affairs, if it is shared.

Figure 3.31[224] shows how various entities produce and share hazard-related information.

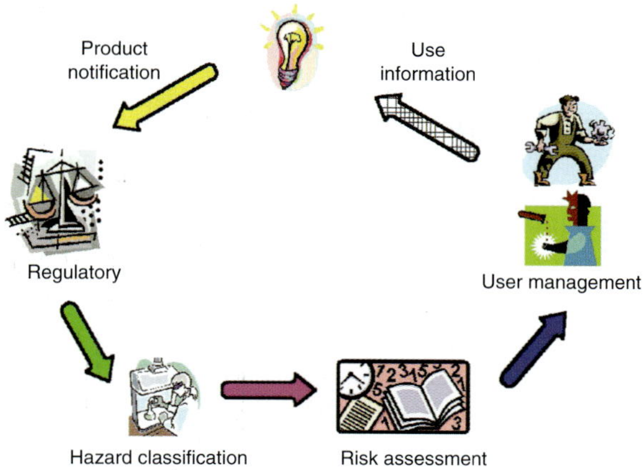


**Figure 3.31**

Not shown in this diagram is the interconnection without barriers between these various entities.

It is the free sharing of safety, health, and environmental information which is the hallmark of most firms who manufacture chemicals (solvents). Competition will limit exchange about pricing, quality control, and applications of solvents. But information about hazards is generally exchanged freely.

Suppliers and users share toxicity, safety, and hazard information, jointly fund laboratory studies with animals to learn about the effect of chemicals, jointly set exposure (corporate) limits for products which are sold competitively, and often provide to others epidemiological data about human exposure in their operations.

## 3.18 NUMERICAL HAZARD CLASSIFICATION SYSTEMS

Hazard management is, and often should be, arbitrary. Some sites or laboratories won't use cleaning chemicals with a 1 (or 2) classification in any NFPA category, certain risk phrases, or recommendations

[222] The library is known as the Integrated Risk Information System (IRIS), and can be found at http://www.epa.gov/iris/index.html

[223] The SAPMA does not publish a list of chemicals with their classifications; users are expected to classify them themselves. This generally requires someone versed in the subject. It is up to the person classifying to make a final choice, using the tools in Tables 3.28 and 3.29 and their experience. See http://www.sapma.org.za/SAPMAHHR.html

Classification is not a free pass to avoid responsibility in safety and environmental affairs. A firm that classified the hazards of benzene as if they were equivalent to the hazards of water would soon find their legal staff overworked. Lawsuits for having fraudulently caused injury through willful misidentification of chemical hazards would soon destroy the financial and commercial credibility of that firm.

[224] A companion to the diagram in Figure 3.30.

for use of certain protective equipment. Those locally-made decisions may make excellent sense if they are supported by those involved.

Other sites are interested in replacing chemicals with specific hazards with other cleaning chemicals with different (presumably lesser) hazards.

Involved persons then ask questions about which level of hazard is lesser or greater:

- For example, is an NFPA classification of 2 about health of more consequence than an NFPA classification value of 1 about flammability?
- Would the answer be different if both chemicals had the same numeric NFPA classification value?

The answer should depend upon the details of the cleaning process and capabilities of the operating staff.

As sites strive to effectively use and safely dispose of cleaning solvents, they seek tools to provide them with further insight into the ramifications of the chemistries they choose. For many, numerical hazard classification systems or algorithms are the answer.

### 3.18.1 Indiana Relative Chemical Hazard Score

In the 1990s, the US EPA was seeking improved methods to measure progress in pollution prevention. At that time, more than 50 numeric algorithms had been developed and publicized.[225] Yet, there was not then (and is not now) a scientific consensus on methods of ranking risk.

The Indiana Relative Chemical Hazard Score (IRCHS)[226] was the work product provided to the US EPA.[227] It is probably the best available algorithm for comparing hazards in use of various solvents. A database has been developed of rankings (scores) for more than 1,250 solvents (chemicals). They are sorted by Chemical Abstract (CAS) number and value.

The algorithm is complicated.[228] There are many components, but there are two general elements:

1. A normalized *worker exposure hazard value*. The workplace hazard value has three components: the health effects, the routes of exposure, and a safety hazard value.
2. A normalized *environmental hazard value*. The environmental hazard value has four components: water, air, land, and a global hazard value.

Each of these seven components has sub-components.

The classifications are no better than the data input from which they are calculated. Not all of the input are absolute measurements. Some are judgments. Others are results of toxicology testing which are often updated when additional studies are done. Such is the case with *n*-propyl bromide, methylene chloride, and HCFC 225 ca/cb.

Values of the IRCHS numerical ranking range from 60: to 48.0 (benzene), to 37.5 (perchloroethylene), to 31.0 (hexane), to 25.9 (xylene), to 18.4 (*n*-butyl acetate), to 12.7 (HFC-43 10mee), to 7.8 (D-limonene), to 4.1 (carbon dioxide), and to 0 (pure water[229]).

The IRCHS, then, is a mechanical calculation scheme for removing human bias and judgment from evaluations of hazards associated with chemicals. For some, that's a good thing. For others, it is the opposite.

#### 3.18.1.1 *Use of the IRCHS*

Classification values for common cleaning chemicals are given in Table 3.39.[230] These are the same chemicals as rated by the schemes previously mentioned.

In the view of many, an unexpected result of the IRCHS classification scheme is that flammable cleaning solvents, which are an anathema to some users, have distinctly different classifications than do chlorinated solvents.

---

[225] University of Tennessee Center for Clean Products and Clean Technologies, "Comparative Evaluation of Chemical Ranking and Scoring Methodologies," *EPA Order No. 3N*-3545-NAEX, April 7, 1994.
[226] Simpson, C., "Solvents by the Numbers," *Clean-Tech Magazine*, June 2002.
[227] Two US EPA Pollution Prevention Incentives for States (PPIS) grants were awarded in 1994 and 1996 to the Indiana Clean Manufacturing Technology and Safe Materials Institute to develop a method for ranking chemicals by their environmental and workplace hazards.
[228] The evaluation process cannot be completed via a mental calculation.
[229] Recall that the hazard is both in the chemical and the way it is used – one can drown in pure water.
[230] The value for kerosene is the average of the values for mineral spirits and for Stoddard solvent.

**Table 3.39 IRCHS Hazard Ranking of Common Cleaning Chemicals**

| Solvent | CAS # | Classification |
|---|---|---|
| 1,1,1-Trichloroethane | 71-55-6 | 36.6 |
| Trichloroethylene | 79-01-6 | 41.6 |
| Diethyl ether | 60-29-7 | 35.1 |
| Cyclohexane | 110-82-7 | 23.5 |
| Isopropanol | 67-63-0 | 14.2 |
| Ethyl alcohol | 64-17-5 | 13.2 |
| Acetone | 67-64-1 | 15.9 |
| Kerosene – straight run | 8008-20-6 | ~15 |
| Propylene glycol | 57-55-6 | 14.2 |
| Ethyl lactate | 97-64-3 | Not included |
| Water | 7732-18-5 | 0 |

This author believes that this type of system is valuable, but incomplete.

IRCHS speaks to the "what," but not to the "how."

While the rankings do provide perspective and the ability to compare among choices, they do not include any information about how the chemicals are to be used. After all, cold cleaning and vapor degreasing produce different spectra of hazards for any solvent.

To make a point with a question: how should the use of benzene in an enclosed vacuum machine be compared to hot alkaline aqueous cleaning agents in an open tank?

The answer is probably not a number. But the author's experience with clients who use the IRCHS is that they ban the use of chemicals with IRCHS values above *X*.

### 3.18.2 Vapor Hazard Ratio

Vapor hazard ratio (VHR) is basically a tool for establishing the chance that an evaporated chemical can reach its exposure limit via normal evaporation.

The cornerstone idea is that chemicals which have relative low exposure limits for entry into the body by inhalation, and are quite non-volatile, are more safe to use than are chemicals with relatively higher exposure limits which are considerable more volatile. Said another way, it does not matter if it is hazardous if it does not evaporate.

VHR is calculated from three pieces of information: (1) the use temperature, (2) vapor pressure versus temperature data (or equations), and (3) the exposure limit.

The procedure is simple:

1. Estimate the vapor pressure of the solvent at the use temperature via calculation or data.
2. Divide result in 1 by the ambient pressure, in the same units. The result is the partial pressure of the solvent in the exposure atmosphere. The units of the quotient are ppm.
3. Divide the result in 2, in ppm, by the exposure limit[231] in the same units. This quotient is the VHR:

$$\text{VHR} = (P_V/P_T)/(\text{TLV, CEL, PEL}) \qquad (3.1)$$

The result is normally a number above zero and less than several hundred (or several thousand). Obviously, lower VHR values represent a classification representing a solvent more safe to use.

VHR is useless as absolute information.[232] Like the IRCHS value, the VHR value has meaning only when compared to a like value for other chemicals. Classification values, at 25°C, for common cleaning solvents are given in Table 3.40. These are the same solvents as rated by the schemes previously mentioned.

The VHR system is favored by those who manufacture, or prefer by use, cleaning chemicals which are relatively non-volatile and have low exposure limits for whatever reason.

An excellent example of the perspective supplied by the VHR system is the classification of dibasic acid ester solvents (DBE, etc.).

The VHR of DBE is 87.

The exposure limit is 1.5 ppm – based on *reversible* nasal irritation produced at low concentrations.

---

[231] This can be the TLV recommended by the ACGIH, a CEL recommended by a supplier or distributor, or a PEL recommended by a regulatory agency.

[232] The CHIPS, EINECS, and SAPMA systems all produce absolute information. In other words, the rankings, recommendations, and results are self-supporting. The NFPA parameter and the HMIS parameter are also absolute information. They do not require comparison with other solvents to provide a perspective about a single solvent.

**Table 3.40 VHR Hazard Ranking of Common Cleaning Solvents**

| Solvent | CAS # | Classification |
|---|---|---|
| 1,1,1-Trichloroethane | 71-55-6 | 375 |
| Trichloroethylene | 79-01-6 | 774 |
| Diethyl ether | 60-29-7 | 1451 |
| Cyclohexane | 110-82-7 | 338 |
| Propan-2-ol, isopropanol | 67-63-0 | 103 |
| Ethyl alcohol pure | 64-17-5 | 58 |
| Acetone | 67-64-1 | 325 |
| Kerosene – straight run | 8008-20-6 | ~29 |
| Propylene glycol | 57-55-6 | 5E-6 |
| Ethyl lactate | 97-64-3 | No exposure limit established |
| Water | 7732-18-5 | ~0 |

This exposure limit, taken alone, would suggest DBE is as lethal as benzene.[233] The two are not comparable – benzene is a carcinogen, DBE makes one's nose itch!

In summary, VHR provides a valuable perspective often hidden to an investigation limited to exposure limits. But it is not a comprehensive classification system because specific hazards (flammability, health, etc.) aren't considered and no recommendations are made.

### 3.18.3 Other Hazard Classification Systems

There are a variety of other hazard classification systems. Each is related to a specific purpose or a specific regulatory agency. Some examples are as follows:

- The Globally Harmonized System of Classification and Labeling of Chemicals (CHS) developed by the International Maritime Organization (IMO) for the purpose of classifying environmentally hazardous chemicals relative to the aquatic environment.[234]
- The National Paint and Coatings Association (US) adapted a quick numeric system[235] for classification of three key potential hazards that can be associated with the chemical components found in most paints.
- The system developed for the International Association of Firefighters, which focuses on chemical reactivity.[236]

### 3.18.4 Modification of the VHR

Although the VHR correctly describes the risk potential based on exposure limits, it fails to describe specific hazards.

A common concern when using solvents is the risk of fire. The VHR system has been supplemented[237] via the addition of a parameter which speaks to a specific hazard: flammability:

- The solvent hazard parameter (SHP) is the VHR multiplied by the NFPA classification for flammability.
- A *more inclusive approach* would be to multiply the VHR by the sum of the three NFPA classifications for flammability, health, and reactivity.

The drawback to either the SHP parameter, or this more inclusive approach, is that it doesn't allow individual judgments[238] about weights of flammability, health, and reactivity according to either the constraints of a cleaning process equipment, or individual interests.

Flammability or health or reactivity may be more significant to some users or in some applications.

For completeness, both the SHP and the more inclusive approach are shown in Table 3.41. The values for DBE are 175 and 350, respectively.

[233] ACGIH recommends a TLV of 1 ppm.
[234] See http://hazmat.dot.gov/globharm.htm
[235] See http://www.paint.org/ind_issue/ib_march01.pdf
[236] Barton, J. and Rogers, R., *Chemical Reaction Hazards* (2nd ed.), Institution of Chemical Engineers, UK, 1997.
[237] See Kob, N.B., "Systematic Evaluation of the Hazard Potential of Solvents," available at http://dbe.invista.com/doc/files/
[238] These are precisely the questions asked in Section 3.18.

**Table 3.41 Hazard Ranking of Common Cleaning Solvents by Other Approaches**

| Solvent | CAS # | SHP classification | "More Inclusive Approach" |
|---|---|---|---|
| 1,1,1-Trichloroethane | 71-55-6 | 375 | 1125 |
| Trichloroethylene | 79-01-6 | 774 | 2324 |
| Diethyl ether | 60-29-7 | 5803 | 8704 |
| Cyclohexane | 110-82-7 | 1016 | 1355 |
| Isopropanol | 67-63-0 | 308 | 411 |
| Ethyl alcohol pure | 64-17-5 | 175 | 175 |
| Acetone | 67-64-1 | 977 | 1302 |
| Kerosene – straight run | 8008-20-6 | ~60 | ~60 |
| Propylene glycol | 57-55-6 | ~1E-4 | |
| Ethyl lactate | 97-64-3 | No exposure limit established | |
| Water | 7732-18-5 | ≈0 | |

### 3.18.5 Which Would You Prefer?

The existence of so many systems illuminates the dissatisfaction with each. None is perfect[239] by any standard. All have flaws. Each has widespread acceptance, and rejection, among some experienced users.

For many users, the question is irrelevant. Their government, state, region, or firm requires a certain approach to management of chemical solvents. The value of the information in this chapter is then to allow comparison among systems and provide the basis for suggestions about improvement.

A complex system, such as the CHIPS or EINECS, will be difficult to memorize and communicate. Labels will require a written code for translation. A system with many details is one which may foster mistakes by those who use it.

This author has considerable experience with the OSHA/DOT, NFPA, and HMIS systems. However, were one creating a new laboratory management system, the SAPMA system must be considered because of its simplicity and power.

## 3.19 THE MSDS

Use of cleaning solvents, or aqueous cleaning agents, absolutely requires familiarity with and use of hazard classification systems. It is the MSDS where the results of those systems are communicated.

Too often the interests and needs of users are frustrated because this MSDS also serves as a marketing vehicle for the supplier. This chapter is about MSDSs – how they often don't contain the information users need, and what are better sources of information.

### 3.19.1 The MSDS Lecture[240]

A significant source of information for users are MSDSs about chemical products used in cleaning.

In general MSDSs are worth what they cost.

US OSHA, the UK, the EC, and many other countries require one to be included with each initial product sample.[241,242]

### 3.19.2 What's Wrong with MSDSs

More than a generation ago, US OSHA intended through regulation to produce informed purchasers of chemical products or informed workers who use them. That hasn't happened. Why?

MSDSs don't contain all the information users need to select products, make good decisions, and

[239] A good site for information about many hazard classification systems is http://stneasy.cas.org. The site requires a fee.

[240] This author is known to be overly concerned about the completeness and certainty of information contained in many MSDSs. Past writings have been considered to be lectures.

[241] See 29 CFR-1910.1200(g).

[242] See 29 CFR-1910.1200(g)(6)(I).

inform workers of all hazards. Significant health, safety, and composition information are often not current or present.

OSHA requires MSDSs to be updated with new scientific studies within 3 months.[243] This requirement is honored mainly in the breach.

Flash points and explosive limits are often given as "N/A" when data can be found by a thorough literature search.

More importantly, very often the ingredient list is given as "proprietary."

### 3.19.3 The Situation for Manufacturers I

Manufacturers have three imperatives:

1. *Identify and evaluate* hazards as above.
2. Promote their products through their MSDSs.
3. Thwart competitive products as described in *their* MSDSs.

In general, manufacturers do this by "*selective identification or evaluation*" (words of the author) of hazards.

Many reputable manufacturers comply with these imperatives by providing MSDSs which contain incomplete information about the products they represent. The reader can ascertain this for themselves by reading, on the Internet, an MSDS and the sources of information recommended noted in Section 3.19.5.

A manager can get useful information from MSDSs, but it's likely not to be all the information they need to comply with OSHA's HAZCOM standard.[244]

Don't think for a minute that MSDSs aren't marketing tools. They most definitely are that. That's how you should view them.[245]

### 3.19.4 The Garden State in the US

New Jersey gets a bad rap about environmental matters – often deservedly so. But they are the undisputed leader in providing quality information about hazards of chemicals. Their "Right to Know" program deserves praise from this author.

MSDSs issued by manufacturers for sale of the chemicals in New Jersey must comply with a more stringent requirement. In fact, the information sheets are called hazardous substance data sheets (HSDSs). A specific format is required. HSDSs are frequently more thorough and accurate.

The most significant difference between MSDSs and HSDSs is that MSDSs are prepared by manufacturers but HSDSs are prepared by a professional panel including a physician, a toxicologist, and a industrial hygienist. The last numbered HSDS is 2053.

For more details, see http://www.state.nj.us/health/eoh/rtkweb/index.html, especially when you are scouting for a new product.

Don't mistake a MSDS for an information source. It is the strong recommendation of this author that HSDSs be used for any evaluation of chemical products. You can get them for free at the above site! If there isn't an HSDS for a product in which you are interested, ask the supplier for the MSDS they would provide for customers in New Jersey.

### 3.19.5 Other Better Sources

Use your search engine on the Internet. Sources of reference information about pure chemicals are widely available. As of this writing (2005), good ones are:

- http://chemfinder.cambridgesoft.com/ http://toxnet.nlm.nih.gov/
- http://www.indiana.edu/~cheminfo/
- http://ull.chemistry.uakron.edu/erd/
- http://www.sis.nlm.nih.gov/Chem/ChemWebLinks.html
- http://hackberry.chem.trinity.edu/cheminf.html

Given the rate of change and diversity found on the Internet, these named sources may not be available. But others of at least this value will be available. Also, see Chapter 6, Section 6.13.

### 3.19.6 The Situation for Manufacturers II

Manufacturers of formulated cleaning chemicals usually aren't "bad guys."

---

[243] See 29 CFR-1910.1200(g)(5).

[244] See http://www.osha.gov/pls/oshaweb/owadisp.show_document?p_table=STANDARDS&p_id=10103

[245] Should an organization you manage be required to prepare an MSDS, a useful guide to doing so is ANSI Z400.1-2004, *Hazardous Industrial Chemicals – Material Safety Data Sheets – Preparation*. It is available for purchase at http://www.ansi.org

They are selling users two things:

1. Their formulation
2. Service to correctly use it

If they publish their formulation recipe on their MSDS any competitor can copy it, for free, add some "wiffle dust," reduce the price by a pfennig, and compete for your business. That's why a complete ingredient list is seldom available.

Both OSHA[246] and New Jersey[247] have provisions for handling information called a trade secret. The provisions are seldom used because manufacturers don't trust them. They believe, often correctly, that if competitors learn their recipe they will lose their business.

But OSHA (in the US), and New Jersey, require you, the manager, to inform your employees of ALL hazards associated with products they use.[248] If the MSDS is silent about ingredient lists and short about health and safety hazards, how is this serious requirement to be met?

### 3.19.7 The Situation for Managers

Talk tough to the salesperson. Tell them that you won't purchase their product without this information. Mean that! Otherwise, don't purchase their chemical product.

Reputable suppliers continuously meet this legitimate need with secrecy agreements and other business methods. As a professional consultant, this author won't allow a client to use a product where they don't have and can't check this information. As a manager, you shouldn't either.

This guidance cuts two ways. If a supplier provides you under a written agreement with recipe information, you must guard it as if it were your own. You must explain the need to keep this information confidential to the employees in the shop who use the product. When a competitive salesperson visits they can't be allowed to learn by asking your operators "what's in this stuff?"

Some countries do a better and others do a poorer job of mandating the content of MSDSs. But as long as the MSDS serves the dual purposes of HAZCOM and marketing, its fidelity to completion of either purpose will be flawed. This is not likely to change.

## 3.20 LABELS

Every cleaning material should be stored in and used from a container with a label. This includes the purchased cleaning agents, soil samples, and part samples:[249]

- Labels are provided with all purchased cleaning agents. The label will have information based on the hazard classification system (see Section 3.18) used by the supplier (or location), *and* information to allow the cleaning to be recognized by managers and their staff.
- Local authorities (federal, provincial, town, corporation) will have their own requirements.[250]

Labeling should not be an end to itself but part of an overall HAZCOM program (see Section 3.21.2).

Managers should understand that a label is more than a sticky sheet applied to a drum. Labels can and will be signs, placards, process sheets, batch tickets, etc.

Space limitations usually limit the amount and type of information available on a *package label. Consequently, managers should use labels for identification, and not for reference.* A label should identify. Description of hazards should come from other sources of information.

## 3.21 USES OF HAZARD INFORMATION

Information exists to inform. If not used, it is wasted.

US OSHA requires information about chemical hazards to be provided in MSDS (Section 3.19.2) and labels (Section 3.20). These sources are intended to be tools by which hazards of using solvents (and all chemicals) can be identified by users, communicated to workers, and *used to protect workers*

[246] See http://www.osha.gov/pls/oshaweb/owadisp.show_document?p_table=STANDARDS&p_id

[247] See http://www.state.nj.us/health/eoh/rtkweb/rtkact.pdf

[248] 29 CFR-1910.1200(a)(1) requires this to be accomplished by means of comprehensive hazard communication programs.

[249] The latter two collections should be labeled so that the materials can be recognized and used.

[250] These are too numerous to be within the scope of this book.

*from those hazards*. Unless it is so used, this information has no value.

### 3.21.1 Sources of Information

Hazard information comes from more than MSDSs and labels. These sources may not be complete. This is because:

- MSDSs are both communication and marketing tools.
- Labels don't have multiple pages.

Consequently, managers shouldn't limit their accumulation of information to that provided with the chemical.

Today managers have another method of communication of information – the Internet. A manager can use their favorite search engine and type something similar to the following phrase:

> "*n*-methyl pyrrolidone" health safety environment EPA OSHA.

You will learn something you needed to know about *n*-methyl pyrrolidone (e.g.). You may learn:

- That there is a better choice of chemicals (less hazard to humans or the environment, better performance, or lower cost). This search phrase will undoubtedly uncover alternative solvents or cleaning processes which may be new to your workers.
- Of experiences from users such as your firm before you share that experience. You may learn that "... was OK. We proved for 2½ years that it could be used. But it was a stepping stone. As with any product, you need to be looking for something better..."
- Of a forthcoming regulation which will affect how you use that solvent.
- How to recycle it so costs are reduced.

More to the point, encourage your workers to do this search on the Internet as part of the communication process.

An excellent source of information about solvent (chemical) hazards has been developed by NIOSH.[251] It is superior to any MSDS. It is also free!

### 3.21.2 Communication of Information

There are some governmental requirements in the US.[252] No cleaning work is exempt from these requirements.[253] Here are the basics of what you are required to do:

- **Communicate** information about chemical hazards.[254–257] This communication must be **written**. In that way information can be accessible in an emergency.
- **Train** those affected in how to protect themselves from those hazards.[258]

---

[251] See http://www.cdc.gov/niosh/npg/

[252] There is an **informal** globally harmonized system (GHS) for the classification and labeling of hazardous chemicals. It is the goal of an effort by the US and other countries to promote common, consistent criteria for classifying chemicals according to their health, physical, and environmental hazards, and to develop compatible labeling, material safety data sheets for workers, and other information based on the resulting classifications. See http://osha.gov/SLTC/hazardcommunications/global.html

[253] *Fed Register No.*: 59:6126-6184. "... there were still some comments submitted which suggested that certain industrial sectors should be exempted from the rule, or only covered by limited provisions. The majority of these were from representatives of the construction industry, and from distributors of hazardous chemicals. The arguments generally involved the degree of risk encountered in the industry, and the feasibility of the requirements. OSHA has not found the arguments regarding infeasibility to be persuasive, nor is there any justification for lessening the protections afforded employees in the industries in question..."

[254] CFR-1910.1200(h)(1). Employers shall provide employees with effective information and training on hazardous chemicals in their work area at the time of their initial assignment, and whenever a new physical or health hazard the employees have not previously been trained about is introduced into their work area.

[255] *Fed Register No.*: 59:6126-6184. All employees have the "right-to-know" the hazards and identities of the chemicals they work with.

[256] *Fed Register No.*: 59:6126-6184. "... To the extent that the hazards of these materials are biological hazards, the Hazard Communication Standard would not apply in any event..."

[257] *Fed Register No.*: 59:6126-6184. "... The existing Hazard Communication Standard includes a total exemption for hazardous waste when regulated by EPA under the Resource Conservation and Recovery Act (RCRA) ... regulated by EPA under the Comprehensive Environmental Response, Compensation, and Liability Act (CERCLA) ..."

[258] *Fed Register No.*: 59:6126-618. "... training ... whether it needs to be done specifically on each chemical, or whether employers can train regarding categories of hazards. Either method would be acceptable ..." "... Employers shall provide employees with information and training on hazardous chemicals *in their work area at the time of their initial assignment, and whenever a new hazard is introduced into their work area* ...." "... The re-training required by the rule is when a new hazard is brought into the workplace, not a new chemical. If a new chemical is flammable, and the employer has already trained regarding flammability, there is no re-training required..."

- **Maintain files** of the communicated information.[259–261] Failure to meet this requirement is an excellent way to obtain a citation from an OSHA inspection!
- **Maintain on-site records** of those communications and training.

In addition this author strongly recommends that you:

- **Repeat the training** on an annual basis.

Any method of HAZCOM training can be used. But the language of the worker must be used.

It is not the responsibility of the employee to train himself. The requirements are not satisfied solely by giving employee the data sheets to read. The employer maintains the responsibility to ensure that their employees are adequately trained, and are equipped with the knowledge and information necessary to conduct their jobs safely.

The written program must reflect what employees are doing in a particular workplace. For example, the written plan must list the chemicals present at the site, indicate who is responsible for the various aspects of the program in that facility and where written materials will be made available to employees.

The written program must describe how the requirements for labels and other forms of warning, material safety data sheets, and employee information and training are going to be met in the facility.[262]

A manager doesn't have to develop their own written program. Several commercial sources, among many, are as follows:

- OSHA's Mine Safety Health Administration (MSHA) has developed a course for miners that offers more general coverage.[263] This course has 11 lessons. Each lesson covers a different topic related to the HAZCOM rule. The course can be completed in one sitting or one lesson at a time.
- The HMIS system (see Section 3.16.3).

This US OSHA HAZCOM requirement preempts all state or local regulations.

## 3.22 ELECTRICAL CLASSIFICATIONS

A manager, or his designate, must be familiar with NFPA 70[264] – the National Electrical Code (NEC).[265]

It relates the flammability classifications made in Section 3.16 (Figure 3.28) to something significant:

- Specific requirements for electrical equipment that are used in locations with chemicals of various levels of flammability hazard.

The basic idea is that a liquid or vapor of any level of flammability can be safely used – if the appropriate

[259] CFR-1910.1200(h)(3)(iv). The details of the hazard communication program developed by the employer, including an explanation of the labeling system and the material safety data sheet, and how employees can obtain and use the appropriate hazard information.

[260] CFR-1910.1200(g)(8). The employer shall maintain in the workplace copies of the required material safety data sheets for each hazardous chemical, and shall ensure that they are readily accessible during each work shift to employees when they are in their work area(s). (Electronic access, microfiche, and other alternatives to maintaining paper copies of the material safety data sheets are permitted as long as no barriers to immediate employee access in each workplace are created by such options.)

[261] MSDSs that represent non-hazardous chemicals are not covered by the Hazard Communication Standard. OSHA does not require nor encourage employers to maintain MSDSs for non-hazardous chemicals. See http://osha.gov/html/faq-hazcom.html

[262] OSHA Fact Sheet (January 1, 1993).

[263] See http://www.msha.gov/Hazcom/Buttons/index.htm. There is also an online training course at http://www.cbs.state.or.us/external/osha/educate/training/pages/205.htm

[264] Managers can read, for free, every National Fire Prevention Association (NFPA) standard at http://www.nfpa.org. Click on Codes and Standards, click on Document List and Code Cycle Information, click on the code you want, at the bottom click on Preview this Document, click on "I agree," and at the bottom click on "Open." in a new window, the Java applet rings up the TOC, WAIT the first page of the code comes up. Key the index icon at the bottom, WAIT. Click on your choice from the TOC, and WAIT. Your choice will appear in the second window.

[265] This work has stature in countries outside the US, and it used as a reference for electrical classification systems in many countries (see Section 3.22.3).

electrical (and other) equipment and procedures are used. The NEC specifies what that equipment and those procedures must be as a function of the details of the operation.

### 3.22.1 Classifications of Locations in the US

A two-tier system is employed throughout:

> "... Locations shall be classified based on the *properties* of the flammable vapors liquids or gases ... that may be present, and the *likelihood* that a flammable or combustible concentration is present ...".[266]

Said another way: there are six possibilities because there are three levels of flammability classification (flammable, combustible, and neither) and two levels of likelihood of a ignitable concentration (present and not present):

- The three possibilities about ignition risk are identified as Class I[267] (ignitable fluids), Class II[268] (ignitable dusts), and Class III[269] (ignitable fibers)[270] Table 3.42 demonstrates this classification.
- The two levels of likelihood are identified as Division 1 (normally present) and Division 2 (normally not present[272]). The difference is based on intent – whether or not the presence of flammable or combustible materials is present through an unintentional release (Division 2) or through expected use (Division 1).

Within this two-tier system, there is a sub-categorization of Class I atmospheres, based on the specific type of hazardous material present. These

**Table 3.42 NFPA 70 Classification of Electrical Requirements**

| Boiling Point (°F) | Flash Point (°F) | Ignition Risk with Fluids | Classification | | |
|---|---|---|---|---|---|
| | | | NFPA Fire | OSHA/DOT | NEC |
| <100 | <73 | Flammable | 4 | IA | Class I[271] |
| >100 | | | | IB | |
| All | >73 and <100 | | 3 | IC | |
| | >100 and <140 | Combustible | 2 | II | |
| | >140 and <200 | | | IIIA | |
| | >200 | Ignitable | 1 | Not | |

[266] NFPA 70, *National Electrical Code*, 2005 Edition, Chapter 500.5 (A). See also US OSHA Standard 1910.307. The Class designation basically refers to the type of fuel.

[267] NFPA 70, *National Electrical Code*, 2005 Edition, Chapter 501. Class I is ignitable glass and vapors (from flammable/combustible liquids).

[268] NFPA 70, *National Electrical Code*, 2005 Edition, Chapter 502. Class II is combustible dusts.

[269] NFPA 70, *National Electrical Code*, 2005 Edition, Chapter 503. Class III combustible fibers and "flyings" such as waste from a fabric handling operation.

[270] Recall that flash point and boiling point are the two criteria used to identify the level of ignition risk presented by a chemical.

[271] Equipment usable in Class I installations must be "intrinsically safe." This equipment is defined as "equipment and wiring which is incapable of releasing sufficient electrical or thermal energy under normal or abnormal conditions to cause ignition of a specific hazardous atmospheric mixture in its most easily ignited concentration" (ISA-RP12.6).

[272] Here, the hazardous liquids, vapors, or gases will normally be confined within closed containers or closed systems. That's Division 1. An example might be a working process area.

Hazardous materials can escape only in case of accidental rupture or breakdown of these containers or systems, or in case of abnormal operation. It is the escape or release situation which defines Division 2. An example might be a storage area in which there is normal exposure to hazardous materials but in which there can be exposure if a storage container ruptures or there is a spill.

**Table 3.43 Sub-Categorization of Class I Flammable Locations**

| Group | Class | Description of Atmosphere |
|---|---|---|
| A | I | Containing acetylene |
| B | I | Containing Hydrogen, gases, or vapors of equivalent hazard such as manufactured gas |
| C | I | Containing ethyl-ether vapors, ethylene, or cyclo-propane |
| **D** | **I** | **Containing gasoline, hexane, naphtha, benzene, butane, propane, alcohol, acetone, benzol, lacquer solvent vapors, or natural gas** |
| E | II | Containing combustible metal dust – including Aluminum, Magnesium, and their commercial alloys, and other metals of similarly hazardous characteristics |
| F | II | Containing combustible carbon black, coal, or coke dust[273] |
| G | II | Containing combustible flour, starch, or grain dusts[274] |

**Table 3.44 Summary of NFPA Electrical Classification System as Applied to Ignitable (Class I) Cleaning Chemicals**

| Ignition Risk – via Flash Point | Likelihood of Occurrence | Specific Atmosphere |
|---|---|---|
| NFPA Fire Class 3 or 4 – Flammable fluid | Division 1: Normally present | Groups labeled A through G D *is the one most applicable to chemicals used in cleaning* |
| NFPA Fire Class 2 – Combustible fluid | Division 2: Normally NOT present | |
| NFPA Fire Class 1 – Ignitable fluid | | |

atmospheres are identified as groups. The identification is shown in Table 3.43.

For managers involved in solvent cleaning work, Group D is the only one of potential concern. The NFPA electrical continuous for use of acetone in cleaning operations would be:

- Class I (ignitable), Division 1 (normally present), and Group D (acetone).

This classification system is often perceived to be complex by some managers. A simple summary is presented as Table 3.44.

#### 3.22.1.1 *Outcomes of NFPA 70 Electrical Classification*

The two most common electrical classifications which apply to cleaning solvents are the same:

1. Class I, Division 1, Group D – flammable solvents normally present.
2. Class I, Division 1, Group D – combustible solvents normally present.

The practical effect of all these classifications is given in Table 3.45.

---

[273] READER BEWARE: The word "Class" is used by two different organizations to mean two different things. Both are shown in Table 3.42; all are listed by NFPA70 as Class I because the fluids can be ignited:

- OSHA/DOT classifies liquids by flash point/boiling point and use the designation Class IA, Class IB, Class II, etc. (Note Roman numerals.)
- NFPA classifies liquids by point/boiling point and by type and uses the designation Class 4/3 (flammable fluids), Class 2 (combustible fluids), and Class 1 (ignitable fibers). (Note Arabic numbers.)

Division 2 means the ignitable atmosphere is normally prevented by containment of the material or by ventilation or some other control such as inerting. Said another way, Division 2 means two things have to go wrong: simultaneous fault in the electrical system causing an arc or spark AND breakdown of whatever controls the fuel (accidental release from a pipe, failure of the ventilation system).

Division 1 means the ignitable atmosphere is or is likely to be present. Said another way, Division 1 means only one thing is needed for a fire to occur: a source of ignition from the electrical system, either one that's there (sparking commutator brushes) or from breakdown (short). The fuel is considered present or likely to be so.

[274] This can also be a location in which easily ignitable fibers are stored or handled, except in process of manufacture.

**Table 3.45 Outcome of Electrical Classification for Various Fluids which Pose Ignitable Risk (NFPA Class I)**

| OSHA/DOT Classification | Example Solvent | Division | Wiring Methods | Comments |
|---|---|---|---|---|
| Flammable, Flash Point <100°F (Class IA, IB, IC) | Acetone, Isopar C, Isopar E | 1 (Present) | Threaded rigid metal conduit | Pressurized enclosures allowed So-called explosion-proof fixtures |
| | Isopropanol, Methyl Acetate, Diethyl Ketone | 2[277] (Not normally present) | Threaded rigid Enclosed gasketed Flexible metal Non-incendive field arcing devices such | |
| Combustible, Flash Point <200°F >100°F (Class II, IIIA) | Parchloro-benzotrifluoride, Whitespirit, Isopar H | 1 (Present) | Construction is basically "enclosed and gasketed" Dust-ignition-proof means dusts are excluded by seal and the enclosure will not permit interior arcs, sparks or generated heat | Class II enclosures need not "contain" an explosion as in Class I enclosures |
| | | 2 (Not normally present) | | |
| Neither, Flash Point >200°F (Not classified) | Isopar L, *n*-Methyl Pyrrolidone, Propylene Glycol Monobutyl Ether | 1 (Present) | Conduit bodies and junction boxes should have dust-tight gaskets | Dust-ignition-proof equipment is not required |
| | | 2 (Not normally present) | | |

Users and suppliers of equipments have a common interest. It is for there to be an unbiased way to demonstrate compliance with NFPA (and other) requirements.[275] That need is fulfilled by independent agencies[276] whose evaluation, if positive, is found in the application of their mark to the equipment.

### 3.22.2 Electrical Classification Outside of the US

As with fashion, the politics of national defense, favored cuisine, and VOC-exemption policies, electrical classification is not uniformly accepted or managed in non-US countries.

[275] That same need is felt by local fire prevention agencies responsible for enforcing local codes (see Footnote 180, Section 3.16.1).

[276] In North America, at least three agencies have jurisdiction over the acceptance or approval for the installation of specific equipment as being in compliance with NFPA (and other) codes. They are as follows:

1. Underwriters Laboratories, Inc. (UL) is probably the best-known agency. UL is a for-profit agency that serves the insurance industry by determining that products meet certain minimum safety standards. *UL does not approve any item for use in an application*. They list items that meet their minimum safety standards. This is an important distinction. There is no UL approval – including of this book.
2. Factory Mutual Research Corporation (FM) is also an insurance rating company (for profit). FM rates and approves heater systems, and components for use in industrial applications insured by the Factory Mutual Insurance Company. Unlike UL, *FM approves products for use in facilities insured by their parent company*.
3. The Canadian Standards Association (CSA) is a quasi-governmental agency that determines the suitability of products for use in Canada. It has some of the characteristics of both UL and FM. CSA provides an approval and follow-up service for other commercial and industrial products. Like UL, CSA reviews the safety aspects of a product – but relative to the Canadian Electric Code (CEC).

[277] Equipment that has been approved for a Division 1 location may be installed in a Division 2 location of the same class and group.

While NFPA 70 (NEC) has respected standing in nearly all countries, a manager of cleaning technology cannot specify the same equipment to be used with cleaning solvents in every country in which their firm has operations.[278]

There are three additional major and several additional minor electrical classification systems whose acceptance by a country defines how flammable and combustible cleaning solvents are to be used within that country.

As of this writing,[279] the following situation is prevalent:

- In the US, the NEC system is compulsory.[280]
- In Canada, the CEC[281] is compulsory.
- In Europe, the CENELEC[282] system is compulsory.
- In all other countries, there are a number of admissible standards which do include the NEC,[283] CEC, IEC,[284] and CENELEC.[285]

The basic difference between all other systems and the NEC is that the division system is replaced with a zone system. The zone system allows more detailed classification about the *likelihood of occurrence* of a flammable (Class I) or combustible (Class II) vapor mixture being present. Table 3.46 shows how the three zones and the two divisions are interrelated.

Issues relevant to zone classification are: (1) the emission level of gas/vapor, (2) the type of openings which currently exist, (3) the method (natural circulation or forced) and level of ventilation.

For a solvent vapor cleaning system (vapor degreaser), the zone system might[286] be applied as follows:

**Table 3.46 Comparison of Zone and Division Electrical Classification Systems**

| Explosive Atmosphere is Continuously Present | Explosive Atmosphere is Often Present | Explosive Atmosphere May Accidentally Be Present |
|---|---|---|
| Zone 0 | Zone 1 | Zone 2 |
| Division 1 | | Division 2 |

- *Zone 0*: The volume within the vapor degreaser, the distillation system, the storage vessel.
- *Zone 1*: The piping connecting the solvent-storage vessel to the degreaser, the piping connecting the degreaser to the distillation system, sample lines, etc.
- *Zone 2*: The volume outside the vapor degreaser or solvent-storage area, piping through which pressure upsets might be relieved, etc.

### 3.22.3 A Manager's Local Interests

National codes are often amended by regional or municipal authorities. These amendments may be to "grandfather" existing practices.[287] Occasionally, these amendments will cover local issues not addressed by the national code.

A manager, and their staff, must be knowledgeable about their local needs and represent them to their local authorities. If they don't, no one else will do so. But if they do, they are still responsible for their own operation.

[278] Babiarz, P.S., Schwarz, G., and Wehinger, H., "The Outlook for Global Unity for Hazardous Area Equipment," *Petroleum and Chemical Industry Conference*, 1998, Record of Conference Papers. The Institute of Electrical and Electronics Engineers Incorporated Industry Applications Society 45th Annual on Pages: 105–109. Carbon disulfide is a unique chemical, that has been used as a cleaning solvent, for which there are no commercially available electrical devices that are considered safe. This is because the autoignition temperature (90 °C) is both exceptionally low and close to the boiling point (46 °C).

[279] Fall 2005.

[280] World standards for the classification of hazardous areas are moving toward harmonization – though progress is slow because the class/division system has such broad and deep acceptance in the US. Parts of the zone system were incorporated into the US NEC as Article 505 in the 1999 NEC.

[281] Canadian Electric Code.

[282] *European Committee for Electrical Standardization*. This is supported by a group of 19 European countries and 11 affiliated countries that have CENELEC standards based on the parallel working IEC/CENELEC.

[283] The symbol "AEx" designates equipment built to NEC® standards for use in NEC zone designated areas.

[284] International Electrotechnical Commission (see http://www.iec.ch/). Their standards, as are those of ASTM and other agencies, are developed by independent experts. Each National Committee of the IEC handles the participation of experts from its country. It is this author's opinion that the NFPA still provides electrical classification of more equipment because of its technical credibility, but the IEC has greater global reach (Asia, Africa, South America, North America, and the Middle East).

[285] The symbol "EEx" designates equipment built to comply with CENELEC standards.

[286] Obviously, the specific classifications would depend upon the design of the specific vapor degreaser.

[287] To make them acceptable for local use.

4

# Control of industrial cleaning processes

**Chapter contents**



## 4.0 INTRODUCTION

This chapter applies to management of cleaning with solvents, to management of cleaning with aqueous technology, or management of non-cleaning processes.

Management techniques can overcome mistakes made in selection of process or cleaning agent. That's not the preferred plan for operation, but it is a possibility.

Yet management of cleaning processes normally receives the level of attention in technical magazines normally afforded email-based spam. Readers of this book will find serious guidance here about management of cleaning processes, presented in a humorous vein. The aim is to attract attention to a valuable and occasionally complex subject (statistics[1]) via use of (hopefully) simple text and humorous sayings produced by others.[2]

[1] Reader beware! These sections are not a substitute for course in statistics. These sections teach how to manage cleaning work using designed experiments and statistical methods.

The author's industrial experience is that these technologies aren't utilized in cleaning work – chiefly because: (a) they aren't understood, and/or (b) their value in use isn't appreciated. It is the aim of this book to contribute to the elimination of both concerns.

[2] This author has chosen to use humor to attract the reader's attention because his writing skills are unable to simplify the science of statistics. Some quotations are taken from J.E.H. Shaw's *Some Quotable Quotes for Statistics*, http://www.ewartshaw.co.uk/

## 4.1 ISSUES AROUND MANAGEMENT OF CLEANING SYSTEMS

*Just think of all the billions of coincidences that don't happen.*[3]

One manages something to produce success and avoid failure – however one defines those.

Some definitions of success and failure for the case of management of cleaning systems are given in Table 4.1. Many other manufacturing and maintenance processes would have similar outcomes. Also noted are the sections of this book in which each outcome of management is covered.

All five aspects of performance (quality, consistency, cost, safety/health, and environmental) are the responsibility of the system manager, whose performance will be judged, at least, upon whether these aspects are viewed as successfully managed.

**Table 4.1 Success and Failure in Management of Cleaning Systems**

| Outcome to be managed | Success | Failure |
|---|---|---|
| **Quality** of cleaned parts | Allow satisfactory completion of the next process stage (which may be packaging or actual use). | The next process stage is not accomplished in a satisfactory manner, or worse still – managers don't know whether or not the flaw is or is not with the cleaning system. |
| | See Chapter 5 and Appendix 2 for information on the selection and use of cleanliness/validation tests. | |
| **Cost** of operation | Cost meets or is less than expectations. | Cost exceeds expectations, or worse still – for either unknown reasons or reasons outside managers' span of control. |
| | See Chapter 6, Section 6.4 for information on expected costs and tactics for control of operating cost. | |
| **Consistency** of operation | No correlation between operation and quality of cleaning, that is there are no *special causes* and *common causes* do not cause variation to exceed control limits. | Quality of cleaning dependent upon operation, and worse still – for unknown reasons. |
| | See Sections 4.3–4.14 for information about control strategies (useful for cleaning and other operations) and about managing data about cleaning quality. | |
| **Compliance** with safety and health policy and regulations | No incidents. | Incidents affecting personnel whether or not covered by regulation. |
| | See Chapter 3, Sections 3.2–3.9 for information about flammability hazards associated with use of cleaning chemicals, and how to avoid same.<br>See Chapter 3, Sections 3.10–3.12 for information about health hazards associated with use of cleaning chemicals, and how to avoid same. | |
| **Compliance** with environmental policy and regulations | No incidents. | Incidents affecting the environment whether or not covered by regulation. |
| | See Chapter 3, Section 3.15 for information about setting and meeting exposure limits.<br>See Chapter 3, Sections 3.16–3.21 for information about risks to the environment associated with use of cleaning chemicals, and how to avoid those risks. | |

[3] Dick Cavett.

### 4.1.1 Personal Definitions of Success and Failure

*I was thrown out of NYU my freshman year ... for cheating on my metaphysics final ... I looked within the soul of the boy sitting next to me.*[4]

The definitions of success and failure in Table 4.1 probably represent those of the enterprise being managed, but may not fully represent the personal views of the manager.

Human managers may not subscribe to just the aspects of Table 4.1. They may have supplementary or separate visions of success and failure.

Certainly, compensation and promotion play a role. That's as it should be. Many enterprises have synchronized their organizational goals with the compensation and promotion goals of their managers. Many have not – perhaps to their detriment.

The normal time lag between management action and personal reinforcement because of that action can cause personal problems for managers, and their enterprises. Cleaning work does not generally provide instantaneous feedback between managerial decision and outcome. For example:

- Reduction of staff training probably won't immediately produce defects in cleaning quality.
- Reduction of environmental controls or monitoring won't immediately produce a citation from a regulating authority (fortunately).
- Reduction of ventilation around flammable chemicals won't immediately produce a fire (fortunately).
- Reduction of availability of personal protective equipment (probably) won't cause cases of contact dermatitis to develop (fortunately).

These reductions in use of resources, if implemented, can temporarily boost the status of a manager within an enterprise because their consequences haven't accrued.

Only individual character and integrity can impede managerial actions designed to provide short-term success at the long-term expense of the enterprise. While personal character is outside the scope of this volume, no topic in this volume is more significant to the manager, and the enterprise.

## 4.2 NEED FOR AND MEANING OF STATISTICS

*'Would you like me to quote you some statistics?'*
*'Er, well...'*
*'Please, I would like to. They, too, are quite sensationally dull.'*[5]

In a Utopian universe, we wouldn't need statistics. There, every sample would perfectly represent the true population from which it was taken. There, every measurement would have no error between it and the true value.

Managers wouldn't need the science of statistics. But managers don't live in that universe, and do need statistics.

Simple statistical methods allow managers to:

- Make judgements with imperfect measurements – those which don't always represent the true value.
- Understand the effect on their decision-making capability of using imperfect measurements.
- Determine what sampling procedures should be used when sampling and measurements are imperfect.

That's why users put up with the sometimes complex nomenclature, tools, and procedures of which the science of statistics is comprised.[6]

---

[4] Woody Allen (Allen Stewart Konigsberg).

[5] Adams, Douglas, *Life, the Universe, and Everything*, Ballantine Books, New York, 1996.

[6] This volume is not a substitute for a treatise on the science of statistics. That science is necessary for efficient operation of any enterprise which depends upon data for its success.

The approach used in this chapter is twofold: (1) use functions provided in spreadsheet programs to manage operating data, and (2) present background information which describes the concepts incorporated in the programming of those functions. There are many examples.

### 4.2.1 Nature of Statistics

*Statistics are no substitute for judgment.*[7]

Statistics can be thought of as a method of extrapolation – from a few to all, or from one time to another. Two basic terms in statistics are **population** and **sample**:

1. If one wanted to choose an administrative officer, one might hold an election of all those eligible to choose the officer. *A **population** is all those eligible to choose.*
2. If one wanted to predict in advance who would be elected, one might ask a few of those eligible to choose about their preference. The sample is those chosen to be asked about their preference, at a time before prediction. *A **sample** is a few chosen from the population.*

In a sense, the science of statistics allows one to have the best chance of managing the prediction so that it will reflect the election outcome.

In cleaning work, if a system produces one thousand units of something in an operating period, managers don't want to have to test all one thousand units via one of the cleaning tests mentioned in Chapter 5.

Normally, one can afford the resources to test for cleanliness only a small fraction of those one thousand units. Naturally, managers want the cleaning test results from a few units to represent the condition of the entire one thousand units.

The nature of statistics is the methodology by which we do just that – extrapolate from the few to the many.

### 4.2.2 Nature of Data

*If you torture the data long enough, it will confess.*[8]

What can you say about the population when all you've got is a sample? The sample is only an approximation for the population.

If you drew a different sample, you'd get a different value. Even worse, if you took the same sample and measured its value with a different instrument, a different value would result. Still worse, if you took the same sample tomorrow and made the same measurement using the same instrument, a still different value would result.

In other words, measured information (data) can change with:

- The sample of units which are tested.
- The facilities (equipment, staff, and procedures) used to do the testing.
- The time at which the testing is done.

The nature of data is that it represents a non-uniform population – so it's not uniform.

### 4.2.3 Nature of Information

*Everybody gets so much information all day long that they lose their common sense.*[9]

An odd aspect of quantum mechanics (and life) is contained in the Heisenberg Uncertainty Principle[10] which it states that it is impossible to simultaneously know both the position and velocity (momentum) of a particle. The idea is that when one uses tools small enough to exactly locate the position of a particle, one has to "touch" it and affect its velocity, and the reverse.

If one uses tools less precise, both the position and velocity of a particle can be known – but the results only approximate the true values because of the imprecision of the tools used. Events in tiny atoms are subject to quantum mechanics, yet people deal with larger objects in the laboratory, where the "classical" physics of Newton prevails.

[7] Henry Clay.
[8] Ronald Coase.
[9] Gertrude Stein.
[10] Named for Werner Heisenberg, who was one of the greatest physicists of the twentieth century. He wrote in 1927, "The more precisely the position is determined, the less precisely the momentum is known."

In other words, the closer one looks at something, the more they see.

Yet, sometimes more is less. An excellent example is the art of twentieth century French impressionist painters, such as Claude Monet. In paintings such as *Field of Poppies at Giverny, The Waterlily Pond and Bridge*, and *The Red Road* the close viewer sees streaks, segments, and dots of brightly-colored paint. Yet the observer who stands back a distance is rewarded with a totally different view – a scene of perspective, depth, and rich color as the individual paint elements coalesce into a full image.

In other words, one can look too closely, see too much, and miss the overall point.

The nature of information is that the meaning of it can change based on how it is obtained, and how it is judged.

### 4.2.4 Nature of Users

> *To be positive: To be mistaken at the top of one's voice.*[11]

In cleaning work, different users place different values on information derived from cleaning tests:

- The site manager values only one result – the overall average quality yield produced throughout a reporting period.
- The operations manager values the knowledge, derived from data and analysis, that quality as measured by the cleaning test is controlled, and is validated in a further test.
- The technical engineer values all data produced, and the statistical analysis of it.
- The shop foreman values that there are procedural actions he can take when the cleaning test result is not within prescribed limits.
- The laboratory technician who conducts the cleaning test values not having to use instrument "X" which he "knows" will always give a higher value than the true value.

The nature of users is that they have different needs for information, and place different values on it.

### 4.2.5 Nature of Samples

> *By a small sample we may judge the whole piece.*[12]

Which sample is necessary to describe a population? Consider the above example of an election:

- Do we have to test every member of a population? Then there would be no need for an election – the sampling would be the election.
- Can we test just one member every sampling interval? Then the opinions of different voters would be masked because we sampled only one voter.
- Can we test several members just once during operation? Then the opinions of voters would be assumed not to change during the period prior to the election.
- Can we test only left-handed voters, or those with any other definable characteristic? Then that characteristic would be assumed to be related to voter interest.

The nature of sampling is that it can have more influence on the test outcome than can the test itself – if it's not properly done.

### 4.2.6 Nature of Testing

> *I have not failed. I've just found 10,000 ways that won't work.*[13]

If you repeat a test, will the same result always be produced? There is a simple answer – of course it will be repeated, it's the same test!

The correct answer is the opposite.

---

[11] Ambrose Gwinett Bierce, *The Devil's Dictionary*, 1911.
[12] Cervantes (Miguel de Cervantes Saavedra).
[13] Thomas Alva Edison.

Consider a determination of Non-volatile Residue Analysis (NVR, as described in Appendix 2). There are five stages to this cleaning test. Each stage can and usually will produce a different effect each time it is repeated:

1. Select a sample of cleaned (supposedly) surface. If the residue isn't perfectly distributed on the surface in a uniform manner, the selected sample of surface will contain different amounts of residue to be collected during the cleaning test. But managers know that the distribution of soil is never perfectly uniform on any part surface. *What force would make it so?*
2. Weigh the part whose surface has been selected. The reported weight will likely be different if the measurement is repeated. Possible reasons are, but are not limited to:
    - The part was not placed on the same position on the weighing pan.
    - An air current deflected the weighing pan.
    - A small amount of humidity (water) was deposited on the part surface, or the weighing pan.
    - External vibration affected the "settling" of the balance to an equilibrium value.
    - The weight measurement was not taken after the same amount of "settling" time as the previous one.
    - The analytical balance was not zeroed (tared) to the identical value as with previous measurements.
    - Fluctuation in line AC voltage may cause and affect operation, including detection of position equilibrium of the balance pan.
    - The internal machinery of the analytical balance may not operate as previously due to wear.
    - The electronic circuitry of the analytical balance may not operate as previously because it is not at the same temperature, as there is heat buildup with additional use.
    - The internal machinery of the analytical balance may not operate as previously due to unknown causes.
3. Extract the part surface in a boiling solvent. The extraction process may not always produce the same result. Possible reasons are, but are not limited to:
    - Unexpected impurities in the extraction solvent, which affect its solvency properties or boiling temperature.
    - Inconsistencies in how the soil is chemically or mechanically bonded to the part surface.
    - Inconsistencies in soil composition[14] so that more or less is dissolved in the extraction solvent.
    - Temperature gradients produced in the extraction zone because the condensing temperature of refluxed solvent is changed because the coolant temperature[15] is not controlled.
    - The soil material being located in the same position as the part is placed in the extraction apparatus. Consequently, the soiled areas aren't exposed to the same temperature, temperature gradients, and fluid motion as are other soiled parts.
    - Additional impurities (mineral deposits) on the part surface cause local boiling on the part surface, which increases the fluid force acting upon the parts.
4. Re-weigh the part whose surface has been selected. The same reasons why the measurement of weight may vary (even if the true weight is exactly the same as in the previous determination) are given above.
5. Calculate the density of soil (NVR) as mass removed per unit area. Procedural errors can cause the calculated NVR value to be different than the true value. While nearly all procedural errors represent mistakes, human beings do make mistakes, such as:
    - Transposing numbers.
    - Making an incorrect calculation.
    - Mis-identifying a sample.
    - Dropping (mishandling) a sample piece, without aborting the testing of this piece.

While these testing errors may each seem trivial (and individually they may well be so), they can be (and are) combined.

The nature of testing can be described by the conversational phrase "stuff happens."

[14] Remember, soil is a material without specification for composition.

[15] This might be the temperature of tap water.

### 4.2.7 Nature of Error

*An error doesn't become a mistake until you refuse to correct it.*[16]

Accuracy is a general term denoting the absence of error of all kinds. Measurement error renders all measurements inaccurate.

Sources of error in measurement are classified as either **random** or **systematic**.[17]

Sources of both random and systematic error can be human-derived, or not. These two sources of error are not wholly separate.

The nature of measurement error is that it is all these: partially controllable, known, unknown, and identified through an aggregation of measurements.

#### 4.2.7.1 *Random Error*

*The generation of random numbers is too important to be left to chance.*[17]

Authors[18] define random error as "that part of our experience that we cannot predict." From a statistical perspective, random error can also be thought of as sampling variability:

- Where there is no random error, there is precision. That is a desirable attribute of measurement and estimation.[19]

The major strategies for reducing the role of random error are:

1. *Take more samples*: A larger sample, other things being equal, will yield more precise estimates of population parameters. It is this single strategy which usually has the greatest beneficial effect.
2. *Improve sampling procedures*: This can include increased care to avoid contamination, or the opposite – inadvertent removal of soil components. It can also include changes in sample size (number of parts tested), sampling frequency, and timing when the sample is taken. This strategy can reduce sampling variability compared to simple random sampling.
3. *Reduce measurement variability*: This means enforcing strict measurement protocols. It can also mean better instrumentation, or taking averages of multiple measurements.

#### 4.2.7.2 *Systematic Error*

*Reality is merely an illusion, albeit a very persistent one.*[20]

Systematic error, or bias, is a difference between an observed value and the true value due to all causes other than sampling variability.[21] Systematic error can arise from innumerable sources, including factors involved in the choice of sampling plan and factors involved in the definition and measurement of study variables (see above section):

- Where there is no systematic error or bias, there is validity. That is also a desirable attribute.

Systematic error or bias is by definition not affected by sample size.[22] There is only one strategy for reduction or elimination of systematic error:

- *Review, criticize, and modify testing procedures.*

A user can do this and expect to produce measurements with less bias afterward:

- Some testing has been done to elucidate the extent to which the total error is or is not composed of random error.

---

[16] Orlando A. Battista.
[17] Robert R. Coveyou.
[18] Rothman, K.J. and Greenland, S., *Epidemiology in Medicine*, Lippincott–Raven, 1998, ISBN 0-316-75780-2, p. 78.
[19] The presence of random variation must always be kept in mind in designing studies and in interpreting data. Generally speaking, numbers whose magnitude is small lead to imprecise estimates. Therefore, small differences based on small numbers must be regarded with caution since these differences are as likely the product of random variation as of something interpretable.
[20] Albert Einstein.
[21] Mausner, S. and Bahn, S., *Epidemiology: An Introductory Text*, (2nd ed.), W.B. Saunders Company, 1985, p. 139.
[22] Systematic error or bias is the extent to which an estimate differs from the true value of the parameter being estimated, even after sample size is increased to the point where random variation is negligible. An example of systematic error is when a butcher (inadvertently or not) puts his thumb on the weigh scale when weighing your purchase.

- Validation of the cleaning test has been done to independently suggest what is the true value (see Chapter 5, Section 5.12).

It is a mistake to believe that systematic errors produce a constant bias or offset between the true value and the measured value. Bias can be inconsistent, especially when there are multiple sources of it.

#### 4.2.7.3 *Overall Error*

> *There is nothing so useless as doing efficiently that which should not be done at all.*[23]

Overall error is the sum of random and systematic errors, many of which are unidentified. The sum of random and systematic errors is not a constant – found in every test.

Components of overall error can be identified and their impact can be estimated in advance. This allows selection of desirable equipment and proper procedures.[24]

But actual errors can't be known without data of measured (from a cleaning test) and true (from validation) values.

Every identified component of total error is not present in every measurement. And where present in multiple measurements, there is no certainty that any identified error component is always present to the same extent.

Even if the true test value were unchanging, the measured value would change because of the variable nature of error:

- Where there is no variation in actual errors, there is repeatability. That is also a desirable attribute.

The nature of overall error is nothing more than that it is the variable difference between measured value and true value.

### 4.2.8 Nature of Chance

> *If there is a 50-50 chance that something can go wrong, then 9 times out of ten it will.*[25]

Chance is what's left after human effort to improve.

Whatever performance is produced in a cleaning test after random and systematic errors have been removed to the extent deemed practical, optimal, or affordable is related only to the true cleaning outcome and chance.[26]

It is sheer chance which explains why measurements aren't perfectly repeatable (constant error). In other words, since we have chosen not to control some factors which affect our measurements (or haven't identified them), the effect those factors have is governed by chance, not by us.

The nature of chance is that it controls that which we have chosen not to control or of that which we are unaware.

### 4.2.9 Nature of Confidence

> *Be suspicious of anything that works perfectly – it's probably because two errors are canceling each other out.*[27]

Confidence is a belief.

It is a belief that the known and unknown uncontrolled factors which affect measurements will not combine so that by chance the difference between measured (from a cleaning test) and true (from validation) values won't be more extreme than a certain amount. Extreme can mean both large and small differences.

This belief, confidence, is fundamental. It means that we expect the cleaning test to provide value to

[23] Peter F. Drucker.
[24] An error analysis should be an ongoing activity when any testing method is employed.
[25] Paul Harvey News, 1979.
[26] This statement is in no way intended to mean that the result of cleaning (or other) tests is driven by chance. Users should continually seek to improve cleaning tests (and other work) via detection and elimination of random and systematic errors. Let smaller errors be affected by chance rather than larger ones.
[27] Dave Bartley.

us more than the cost of doing it. A belief otherwise would lead to another choice of cleaning tests or a decision not to use them.

Typically confidence, a belief, is described as a probability that the error between the true and measured values won't be more than a certain amount.

Confidence can't be estimated without data of measured and true values. These data allow calculation of confidence (*a probability*).[28] *Confidence* is typically not expressed as the amount of error.

The nature of confidence is that it is a quantitative belief that certain actions (cleanliness testing) will be worthwhile.

## 4.3 MANAGEMENT OF CLEANING PROCESSES FOR CLEANLINESS

*The only statistics you can trust are those you falsified yourself.*[29]

The underlying issue around process control is information. Control is the means by which information affects management of cleaning operations. Get your statistics book or spreadsheet. Look up the section on "*t*"-tests. We're going to make this simple!

Answers to specific questions are what are covered in this and subsequent sections.

The approach in this chapter will be to ask and answer questions which the author believes it is significant to answer, and which managers probably have already asked. Readers can substitute their own information and use the methods shown (with a spreadsheet) to answer these questions as they apply to their cleaning systems.

### 4.3.1 Data[30]: the Staff of Life

*I checked it very thoroughly, said the computer, and that quite definitely is the answer. I think the problem, to be quite honest with you, is that you've never actually known what the question is.*[31]

*Q (question)1: If all your parts look and feel the same, how do you tell which (if any) are clean?*

The simple answer is to test them for cleanliness via some test[32] which is believed (and can be proven by validation) to measure the amount of soil remaining on them after being processed through the cleaning system.

The complicated answer is to test them via completing the normal stage of their next use. The complication is that the next use stage may simultaneously not be performing properly. In that situation you can't differentiate part cleanliness from integrity of the next use stage.

### 4.3.2 Timing of Testing

*The major difference between a thing that might go wrong and a thing that cannot possibly go wrong is that when a thing that cannot possibly go wrong goes wrong it usually turns out to be impossible to get at and repair.*[33]

*Q2: If I know how to perform a cleanliness test, when should I use it? Should it be when I think the cleaning process isn't working, or when it is working?*

---

[28] The reverse can be done, one can specify a probability (confidence) that they are willing to accept that errors will be within some range, and then estimate that range. But only actual data will show if the true data do lie within that range.

[29] Winston Churchill.

[30] While Lt. Commander Data was a staff of life on the starship Enterprise, in this context, we are actually referring to numerical information.

[31] Adams, Douglas, *The Hitchhiker's Guide to the Galaxy*, 1979.

[32] See Chapter 5, Section 5.5.

[33] Adams, Douglas, *Mostly Harmless*, 1992.

As in Table 4.1, the commitment to cleanliness testing is a means of assuring that the next use stage is performed satisfactorily. If that stage is periodically performed, cleanliness testing can be organized around the timing of that performance.

But that's a reactive strategy. It will only identify goods which don't meet needs. This strategy doesn't identify when the cleaning process is not properly functioning at a time when there is opportunity to make adjustments to it – before failure is produced.

The recommended strategy is to:

- Test while the cleaning process is operating so that resources spent on operating it can be seen to be well spent, or inadequate performance can be kept from becoming worse.

Testing for cleanliness should be done as a normal part of process operation.

### 4.3.3 Frequency of Testing

*[On Statistics]: Here we have a game that combines the charm of a Pentagon briefing with the excitement of double-entry bookkeeping.*[34]

*Q3: If I know how to perform a cleanliness test, how often do I need to perform that test to be sure of getting the right result?*

The answer is that test frequency depends upon:

- **How often** you want to know if a certain amount of cleaning work has been done well. This refers to the work schedule. Choices might be once per hour, shift, day, grade or type of material being cleaned, or significant change in operation of the cleaning process. For continuous operation, once per day or shift is often adequate. Where the cleaning work is valued more highly, a manager would want to know about performance more often.
- **How well** the measurements of cleanliness correctly represent the true cleanliness condition of the parts. This refers to the error of measurement.[35]
- **How sure** you want to be that the measured results correctly represent the true cleanliness condition of the parts. This refers to confidence limits.[36]

The number of samples to be taken and tested is given in Table 4.2,[37] for any choice of how often it is desired to know about the performance of the cleaning machine.

For example, a manager might decide that their cleaning operation involved high-value goods. Accordingly, they might choose to test cleanliness every shift of operation.

Further, this manager wants to be at least 99% certain that the true value of the cleanliness measurement lies within 4% (0.04/1.00) of 1.00[38] when the standard deviation (error) of the cleanliness measurement is 5% (0.05/1.00).[39]

Accordingly, this user would complete 14 tests for cleanliness every shift. See the numbers in **red** in Table 4.2.

For operations involving lower value goods, another user might choose to complete six cleanliness tests every day – to be at least 95% certain that the true value of the cleanliness measurement lies within 5% of 1.00 for the same quality of test performance (5% measurement error). See the numbers in **blue** in Table 4.2.

[34] Adams, Cecil, *The Straight Dope*, http://www.straightdope.com/

[35] See Section 4.2.7, etc.

[36] See Section 4.2.8 and Appendix 1. Confidence limits define the range around the population true value within which measurements are likely to occur. In other words, the confidence limit is the percent of the time the true mean will lie in the interval estimate given.

[37] Details of this calculation, with various levels of measurement error, are given in Appendix 1.

[38] This value is chosen for convenience as any numeric data can be normalized to produce values relative to 1.00 by dividing each by the average value.

[39] Any other values of standard deviation of cleanliness measurements are possible. Table 4.2 can always be used. All values in that table are proportional to standard deviation. Suppose your standard deviation of your cleanliness test measurements is 10% (0.10 in Table 4.2), you want the expected error from the true value to be 5% (0.05 in Table 4.2), and you want this to happen 97.5% of all measurements. To use Table 4.2, *mentally* double all values of expected error to account for the fact that your measurement error is twice that of 0.05 upon which the table is based. Then enter the table at 97.5% confidence, and read down to an expected error value of 0.025 (your goal). Read to the left to **23** cleanliness tests. See the numbers in **purple** in Table 4.2.

**Table 4.2 Selection of Sample Size for Measurement Error = 0.05 and True Value = 1.00**

| | Degree of confidence | | | | | |
|---|---|---|---|---|---|---|
| | 80.0% | 85.0% | 90.0% | 95.0% | 97.5% | 99.0% |
| Number of samples | Expected error of collection of measurements from true value | | | | | |
| 1 | 0.154 | 0.208 | 0.316 | 0.635 | 1.273 | 3.183 |
| 2 | 0.067 | 0.081 | 0.103 | 0.152 | 0.219 | 0.351 |
| 3 | 0.047 | 0.056 | 0.068 | 0.092 | 0.121 | 0.169 |
| 4 | 0.038 | 0.044 | 0.053 | 0.069 | 0.087 | 0.115 |
| 5 | 0.033 | 0.038 | 0.045 | 0.057 | 0.071 | 0.090 |
| 6 | 0.029 | 0.034 | 0.040 | 0.050 | 0.061 | 0.076 |
| 7 | 0.027 | 0.031 | 0.036 | 0.045 | 0.054 | 0.066 |
| 8 | 0.025 | 0.028 | 0.033 | 0.041 | 0.049 | 0.059 |
| 9 | 0.023 | 0.026 | 0.031 | 0.038 | 0.045 | 0.054 |
| 10 | 0.022 | 0.025 | 0.029 | 0.035 | 0.042 | 0.050 |
| 11 | 0.021 | 0.023 | 0.027 | 0.033 | 0.039 | 0.047 |
| 12 | 0.020 | 0.022 | 0.026 | 0.031 | 0.037 | 0.044 |
| 13 | 0.019 | 0.021 | 0.025 | 0.030 | 0.035 | 0.042 |
| 14 | 0.018 | 0.020 | 0.024 | 0.029 | 0.034 | 0.040 |
| 15 | 0.017 | 0.020 | 0.023 | 0.028 | 0.032 | 0.038 |
| 16 | 0.017 | 0.019 | 0.022 | 0.026 | 0.031 | 0.037 |
| 17 | 0.016 | 0.018 | 0.021 | 0.026 | 0.030 | 0.035 |
| 18 | 0.016 | 0.018 | 0.020 | 0.025 | 0.029 | 0.034 |
| 19 | 0.015 | 0.017 | 0.020 | 0.024 | 0.028 | 0.033 |
| 20 | 0.015 | 0.017 | 0.019 | 0.023 | 0.027 | 0.032 |
| 21 | 0.014 | 0.016 | 0.019 | 0.023 | 0.026 | 0.031 |
| 22 | 0.014 | 0.016 | 0.018 | 0.022 | 0.026 | 0.030 |
| 23 | 0.014 | 0.016 | 0.018 | 0.022 | 0.025 | 0.029 |
| 24 | 0.013 | 0.015 | 0.017 | 0.021 | 0.024 | 0.029 |
| 25 | 0.013 | 0.015 | 0.017 | 0.021 | 0.024 | 0.028 |
| 26 | 0.013 | 0.015 | 0.017 | 0.020 | 0.023 | 0.027 |
| 27 | 0.013 | 0.014 | 0.016 | 0.020 | 0.023 | 0.027 |
| 28 | 0.012 | 0.014 | 0.016 | 0.019 | 0.022 | 0.026 |
| 29 | 0.012 | 0.014 | 0.016 | 0.019 | 0.022 | 0.026 |
| 30 | 0.012 | 0.013 | 0.015 | 0.019 | 0.022 | 0.025 |
| 31 | 0.012 | 0.013 | 0.015 | 0.018 | 0.021 | 0.025 |
| 32 | 0.012 | 0.013 | 0.015 | 0.018 | 0.021 | 0.024 |
| 33 | 0.011 | 0.013 | 0.015 | 0.018 | 0.020 | 0.024 |
| 34 | 0.011 | 0.013 | 0.014 | 0.017 | 0.020 | 0.023 |
| 35 | 0.011 | 0.012 | 0.014 | 0.017 | 0.020 | 0.023 |
| 36 | 0.011 | 0.012 | 0.014 | 0.017 | 0.019 | 0.023 |
| 37 | 0.011 | 0.012 | 0.014 | 0.017 | 0.019 | 0.022 |
| 38 | 0.011 | 0.012 | 0.014 | 0.016 | 0.019 | 0.022 |
| 39 | 0.010 | 0.012 | 0.013 | 0.016 | 0.019 | 0.022 |
| 40 | 0.010 | 0.012 | 0.013 | 0.016 | 0.018 | 0.021 |

(Continued)

**Table 4.2 Selection of Sample Size for Measurement Error = 0.05 and True Value = 1.00 (Continued)**

| | Degree of confidence | | | | | |
|---|---|---|---|---|---|---|
| | 80.0% | 85.0% | 90.0% | 95.0% | 97.5% | 99.0% |
| Number of samples | Expected error of collection of measurements from true value | | | | | |
| 41 | 0.010 | 0.011 | 0.013 | 0.016 | 0.018 | 0.021 |
| 42 | 0.010 | 0.011 | 0.013 | 0.016 | 0.018 | 0.021 |
| 43 | 0.010 | 0.011 | 0.013 | 0.015 | 0.018 | 0.021 |
| 44 | 0.010 | 0.011 | 0.013 | 0.015 | 0.017 | 0.020 |
| 45 | 0.010 | 0.011 | 0.013 | 0.015 | 0.017 | 0.020 |
| 46 | 0.010 | 0.011 | 0.012 | 0.015 | 0.017 | 0.020 |
| 47 | 0.009 | 0.011 | 0.012 | 0.015 | 0.017 | 0.020 |
| 48 | 0.009 | 0.011 | 0.012 | 0.015 | 0.017 | 0.019 |
| 49 | 0.009 | 0.010 | 0.012 | 0.014 | 0.017 | 0.019 |
| 50 | 0.009 | 0.010 | 0.012 | 0.014 | 0.016 | 0.019 |

Several general conclusions can be drawn from Table 4.2:

- A single sample may not represent the population.[40]
- A single measurement provides little value – unless the standard of confidence is exceedingly low or the expected error from the true value can be quite large.
- Significantly more tests are required to be more sure (higher confidence level) that results are within a certain range of a certain value. For the 5% errors assumed in Table 4.2, the number of samples must be increased from 6 to 10 to achieve an increase in confidence from 95% to 99% that the true value is within 5% of 1.00. See numbers in **dark green** in Table 4.2.
- In industrial practice, the 95% confidence limits are commonly used, and values below 95% are seldom chosen. Recent emphasis on quality have raised that chance to 97.5–98% and in some cases to beyond 99%.[41]
- There is a point of diminishing returns – additional cleaning tests beyond ~15–20 produce little incremental value for their increased cost.
- But a few measurements add considerable value beyond the value of one.
- It is possible to have an expected error of a collection of measurements be less than the standard error of a single measurement. This is done via many measurements.

### 4.3.3.1 *No Effect of Population Size*

> *Everything should be made as simple as possible, but not simpler.*[42]

Note that the size of the population is not a factor in choosing sample frequency.

If you need to conduct six cleaning tests to be 95% certain that the parts are clean, when the error of cleanliness measurement is 5% of the true value, that's the testing plan for cleaning 100, 1,000, or 10,000 parts in the chosen work period.[43]

The defining issues are the variation in the measured data and the population data, and how certain you must be of your conclusions, not the size of the population.

---

[40] See Section 4.2.1. An analogy would be to forecast an election outcome by polling a single voter.

[41] The trademarked term Six Sigma refers not to confidence limits, but to an error or defect level of not 5% as above, which is 50,000 defects per million cleaned parts, but just 3.4 defects per million cleaned parts.

[42] Albert Einstein.

[43] See Appendix 1 for an explanation of why this is so.

### 4.3.4 Sampling Methods

*The government are very keen on amassing statistics. They collect them, add them, raise them to the nth power, take the cube root and prepare wonderful diagrams. But you must never forget that every one of these figures comes in the first instance from the village watchman, who just puts down what he damn pleases.*[44]

*Q4: If I choose to perform cleanliness tests on six cleaned parts every day,[45] just which six parts should be chosen?*

There are several common strategies used for sampling.[46] Two are recommended for cleaning work. They are:

1. *Random sampling*: Here a user identifies each part by a number. Then six part numbers are drawn each day at random,[47] the cleanliness of these six parts is measured by a chosen test, and the collective result is taken to represent the cleanliness of the work done in that day. If the parts can't be numbered in advance, assign a number to sequential production times during the day (minutes or hours), randomly choose a different set of six numbers each day, and test a part made at those six times.
2. *Stratified sampling*: Here a user makes sampling coincident with some event: after and before change-out of solvent charges, when downstream operation is changed, when a certain type of part is cleaned, when a certain compliment of staff is assigned to the operation, or when someone remembers to do it.

Of these two choices, random sampling is *greatly* preferred because one doesn't generally know which events correlate with cleaning performance.

One approach which should *not* be used is convenience sampling. Here a user tests the next six parts which they encounter. This approach will omit learning of any changes in performance occurring when they didn't encounter parts.

## 4.4 THE "GOLDEN LOT" BENCHMARK

*Errors using inadequate data are much less than those using no data at all.*[48]

*Q5: To what should good cleaning test data be compared (or now that I have it, just what do I do with it)?*

In cleaning work, associated judgements usually aren't absolutes. Nearly all judgements about cleanliness are made relative to something else.

This arises because the value to an enterprise of cleaning work is that it allows successful completion of the next process stage.[49] The level of cleanliness needed is whatever necessary to complete subsequent operations.

While that level may be defined or treated by the enterprise as an absolute "bright line," the necessary cleanliness was initially determined by whatever level allowed successful downstream processing.

If the acceptable level of non-volatile residue (NVR) is 8.4 mg/SF,[50] it might have been defined by data (assumed) which showed that *individual* lots with NVR levels of 6.2, 5.1, 8.4, and 2.9 mg/SF processed acceptably and *individual* lots with NVR levels of 11.5, 13.6, 10.0, 15.7, and 19.9 mg/SF did not. Would an NVR level of 8.5[51] allow acceptable

---

[44] Anonymous English Judge, quoted by Sir Josiah Stamp in *Some Economic Matters in Modern Life* (1929).

[45] To be at least 95% certain that the true value of the cleanliness measurement lies within 5% of 1.00 for the same quality of test performance (5% measurement error).

[46] See Appendix 1. Here, sampling means choosing which parts to test for cleanliness, or any other action.

[47] This is easily done using a random number generator present in spreadsheet programs.

[48] Charles Babbage.

[49] That could range from sale to a customer to additional machining. So, the level of needed cleanliness from an operation depends upon what is to be done next with the cleaned article.

[50] Numerically higher values refer to more soil on the parts – dirtier parts.

[51] In this hypothetical example, the 8.5 is one unit of measurement (0.1 in this case) above the highest level of residue which was not too much residue (8.4).

operation? It might. Would an NVR level of 9.9[52] allow acceptable operation? It might, as well. One just doesn't know.

When the enterprise adopts the "Golden Lot"[53] strategy of cleanliness management, it means that the aim point for process control becomes:

- "Make it like that #, it worked", or
- "Don't make it like that #, it didn't work."

The "Golden Lot" strategy is too commonly used to define limits of cleanliness acceptability. But there are better ones which are derived from statistical procedures (see Section 4.6.1).

Two problems with the "Golden Lot" strategy are that:

1. The goal becomes a single value – 8.4 in the example above. In Section 4.3.3 and Appendix 1, it is shown how poorly a cleanliness measurement with a single lot can predict the cleanliness of the entire volume of cleaning work. The value of 8.4 could be well above the average (mean) measurement expected for this single sample as in Figure 4.1. Or it could be well below the average (mean) measurement expected for this single sample as in Figure 4.2.

There is no way to tell with a single measurement. The value characterizing the "Golden Lot" could be meaningless!

2. There is certain to be lost production. This is cleaned product, to use the example above, whose NVR measurements are between 8.4 (the maximum acceptable value) and 10.0 (the minimum unacceptable value). This is shown in Figure 4.3. If a lot of product is cleaned and a single value of NVR is measured between 8.4 and 10.0, it might produce acceptable downstream operation, or not. There is no way to tell with a single measurement.

The analysis in Table 4.2 suggests that measurements with six parts can support a superior and excellent position[54] for use of the "Golden Lot" benchmark. Example data for NVR measurement are shown in Table 4.3.

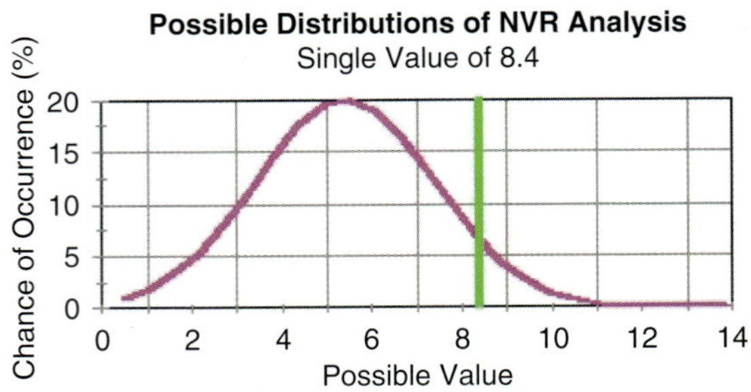


Figure 4.1

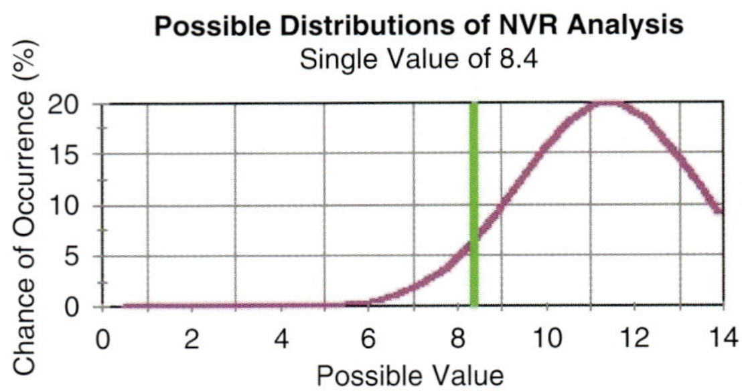


Figure 4.2

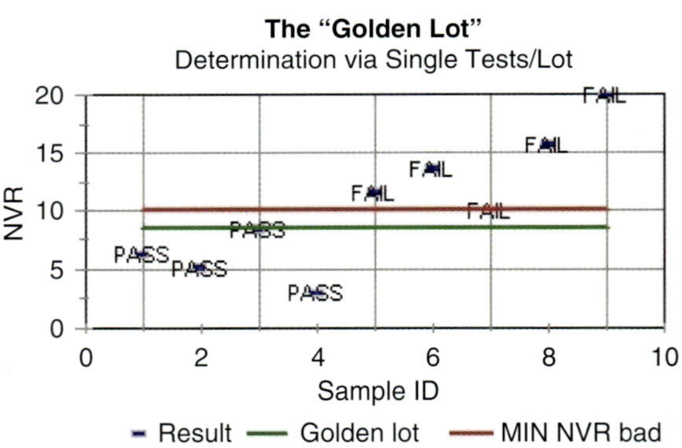


Figure 4.3

And that's the point: if the "Golden Lot" strategy is to be used, make the goal *not* to be the single measurement at which downstream operation was successful. Instead, choose the goal as the individual measurements of cleanliness for that "Golden Lot."

[52] In this hypothetical example, the 9.9 is one unit of measurement (0.1 in this case) below the lowest level of residue which was too much residue (10.0). Values between 8.5 and 9.9 lead to uncertainty in this example.

[53] Here the term "lot" refers to an identifiable amount of parts to be cleaned. This could be one from the production of one type made in 1 hour, 100 parts in a box, or a single part.

[54] So that a user is at least 95% certain (superior) that the true value of the cleanliness measurement lies within 5% of 1.00 with 5% measurement error (excellent).

**Table 4.3 "Golden Lot" NVR Cleanliness Measurements**

| Sample ID | NVR value, mg/SF |
|---|---|
| 1 | 6.3 |
| 2 | 11.0 |
| 3 | 4.2 |
| 4 | 9.2 |
| 5 | 15.0 |
| 6 | 4.7 |
| Mean (average) value = 8.4 mg/SF | |

Note that of the data in Table 4.1, some of the values are greater than the average (mean)! That's the nature of an average. That's OK!

*The benchmark becomes not 8.4 but **the data** which produced the 8.4.*

## 4.5 DECISIONS ABOUT CLEANLINESS QUALITY

*Statistics means never having to say you're certain.*[55]

*Q6: If the data coming from your cleaning test show the desired performance, how do you know if any of those numbers mean anything?*

The cleanliness evaluation for subsequently cleaned goods is to compare the new set of cleanliness measurements to those associated with the "Golden Lot" and make a statistically based judgement about whether or not those data sets are equivalent.

Data of a subsequent set of NVR cleanliness measurements are listed in Table 4.4. If they are from the same population of measurements as is the "Golden Lot," then the subsequent cleanliness evaluation shows no change from that of the "Golden Lot."

**Table 4.4 Off-specification NVR Measurements**

| Sample ID | NVR value, mg/SF |
|---|---|
| 1 | 8.8 |
| 2 | 8.2 |
| 3 | 8.4 |
| 4 | 7.7 |
| 5 | 7.7 |
| 6 | 8.0 |
| Mean (average) value = 8.13 mg/SF | |

That's the question to which users require an answer. Are the subsequently cleaned goods *equivalent* to the goods identified as the "Golden Lot?"

The question about better is a separate question. The question about poorer is moot if the goods are shown to be equivalent.

Equivalency of two data sets is best and commonly evaluated through the use of the "*t*"-test.

## 4.6 USE AND POWER OF THE "*t*"-TEST

*Truth comes out of error more readily than out of confusion.*[56]

The "*t*"-test is normally used to compare one set of measurements to another – when the requirement is to learn if both come from the same population of measurements. In this case, one set of measurements is from the "Golden Lot."

Spreadsheets can do the "*t*"-test calculations.[57] Both Microsoft's Excel and Corel's Quattro-Pro have the same function: Corel uses the name @TTEST and Microsoft uses the name =TTEST. The parameters for use are the same for both spreadsheets.[58] The output is a fraction:

=TTEST(Data Set #1, Data Set #2, 2, 2) for Excel

[55] Anonymous.

[56] Francis Bacon, *Novum Organum*, 1620.

[57] The *t*-distribution has a mean at x = 0, has the familiar "bell" shape, is symmetrical about that mean, is continuous, and never reaches the x-axis – as does the normal distribution.

[58] The number of data points must be the same in both data sets, or the routine will return ERR. The first value **2** means that both plus and minus deviation from the mean should be considered (a "two-tailed test"). The second value 2 means that the two data sets are to be treated as if they do have the same variability.

@TTEST(Data Set #1, Data Set #2, 2, 2) for Quattro-Pro

- When converted to a percent, the output of @TTEST is the % chance that Data Set #1 came from the same population as did Data Set #2.
- When converted to a percent the output of (1-@TTEST) is the % chance that Data Set #1 came from a different population than did Data Set #2.

In other words, this calculation identifies the probability that the second data set came from the same population which produced the first data set. Neither data set represent the entire population, both are samples.

For the NVR data in Tables 4.3 (Data Set #1) and 4.4 (Data Set #2), @TTEST is 0.879. This means that there is a 87.9% probability that the subsequent data set came from the general population of data from which the "Golden Lot" was produced. There is a 12.1% chance (100–87.9%) that both lots are *not* from the same population of cleaned parts.

Since 87.9% probability is a low value (or 12.1% is a high value), relative to 95% probability, one should assume that Data Set #2 (Table 4.4) is NOT from the same population as is Data Set #1 (Table 4.3).

Said another way, you should decide that Data Set #2 is not equivalent to Data Set #1 – *even though the average NVR is better (lower) for Data Set #2 than for Data Set #1*, if you insist (as most do) that your decisions be made with at least 95% confidence of being right.

### 4.6.1 A Better Benchmark than the "Golden Lot"

*Statistics are like a bikini. What they reveal is suggestive, but what they conceal is vital.*[59]

There are three institutional hazards in using a "Golden Lot" to represent the least clean your parts must be to produce acceptable performance in downstream operations. These hazards are:

1. There is only one "Golden Lot." It may be unique for many reasons other than the measured amount of soil remaining on it. For example, the soil might be preferentially located in zones where it has less effect upon downstream operation. Or, it may be that it isn't the "Golden Lot" which is unique – it is the downstream operation which was operating in a unique way when the "Golden Lot" was processed.
2. Likely, the "Golden Lot" will be defined by only a few measurements (the number you normally use).
3. By definition, the "Golden Lot" does not represent your entire population of cleaned parts. It represents only some which are more well cleaned, and with which downstream operations were not compromised.

Thus the "Golden Lot" is not a robust benchmark. A more secure strategy is to expand the information content of the "Golden Lot."

You do this by adding more data. Choose results of additional cleaning work and add the measured cleanliness values associated with it. Accumulate data from the "Golden Lot" and the additional data into a master database. Use that database as the reference benchmark to which the cleanliness of newly produced is compared.

The data which should be added is that which has relatively high, *and low*, levels of uncleaned soil and (*with which downstream operations are as expected)*. In other words, use all cleanliness data in which downstream operations are as expected.[60]

Don't screen data to identify work which has relatively low levels of uncleaned soil. While those observations are a portion of the normal population of cleanliness measurements (and machine performance), they are not representative of all performance from the cleaning machine.

*The basic idea is that the benchmark represents good operation, not the maximum level of residue which can be tolerated.*

All three of the above hazards are surmounted in this way.

[59] Aaron Levenstein.

[60] One might ask why results aren't included in the accumulation of data which describe unacceptable downstream performance. The answer is that these data, of *unacceptable* downstream performance, are by definition *different* than those which describe acceptable performance. So they don't belong in the reference accumulation of data.

## 4.7 DISCRETE CLEANLINESS DATA

*The solid wealth of insurance companies and the success of those who organize gambling are some indication of the profits to be derived from the efficient use of chance.*[61]

The preceding analysis is based on an assumed ability to define cleanliness via some test as a continuous spectrum of performance. Cleanliness data is not always continuous.[62] Cleanliness data can be discrete.[63]

Discrete cleanliness data usually occurs when the next processing stage is used by customers. They may only report dissatisfaction (*NO GO*) or speak in general terms (*fair, good, better*).

The information content of data is related to the number of possible outcomes. Discrete cleanliness data is almost barren of information, which can be correlated to the operations which produced it. Management of cleaning processes can be difficult where the only measurement of performance is discrete data.

With discrete data, accuracy is limited by observational accuracy and precision is non-existent.

At least three strategies have been successfully used in these situations. They are as follows:

1. Convert the discrete data to continuous data by a change in methodology:
   - Use a measurement scale with more than two outcomes. One example is a Likert scale[64] which is often used in surveys to increase the information content of responses. Another example is to replace measurement of a diameter or thickness by a "GO/NO GO" plug gage with use of a micrometer.
   - Communicate better with customers so that the information content of their responses is enriched.
2. Convert the discrete data to a numerical frequency, and manage that via the above or other statistical methods. The common approach is to compute DPMO.[65]

   Smaller numbers of opportunities can obviously be used as a basis for comparison. Or the basis could be defects per period of operation (months, hopefully).

   This conversion adds a variable (opportunities, not time) which is not likely to be a factor that controls cleanliness. This approach won't add information to the basic data because no new data has been added. But this approach does have two virtues as it does allow management of (1) discrete data and (2) infrequent failures.[66]

   A consultant specializing in statistical management may be helpful to verify that the conversion is properly done.
3. Convert the information about test failures (NO GO) into graphs versus time, and use "telephone theory." The basic aim is to identify the frequency of failure during an interval of cleaning operation and compare it to past behavior.

   Ever wonder why inappropriate, unfortunate, or "bad things" seem to happen at the same time? Shouldn't random failures occur at times roughly equally spaced throughout a time interval?

   No, they shouldn't. Statistical theory predicts that a *small number* of truly random events will not occur at equally spaced times through an interval.

   NO GO results from cleaning tests will be distributed through a time interval in a manner

[61] Edward de Bono.

[62] Continuous data is information that can be measured on a continuum or scale. Continuous data can have almost any numeric value and can be meaningfully subdivided into finer and finer increments, depending upon the precision of the measurement system.

[63] Discrete data (also called attribute data) is that in which only a few values are permitted. An example is "GO/NO GO" data, or binary (0, 1) data, in which only two outcomes are possible. Only a finite number of values is possible, and the values cannot be subdivided meaningfully.

[64] A Likert scale (pronounced "lick-ert") is an often used questionnaire format. It requests respondents to specify their level of agreement to each of a list of statements. The Likert scale is named after Rensis Likert, who invented the scale in 1932. A typical question using a five-point Likert scale might make a statement, then ask the respondent to indicate whether they strongly disagree, disagree, neither agree nor disagree, agree, or strongly agree.

[65] Defects per million opportunities (DPMO) is the average number of defects per unit observed during an average production run divided by the number of opportunities to make a defect on the product under study during that run normalized to one million.

[66] The frequency or DPMO approach is frequently used in the semiconductor and other industries where the failure occurs less.

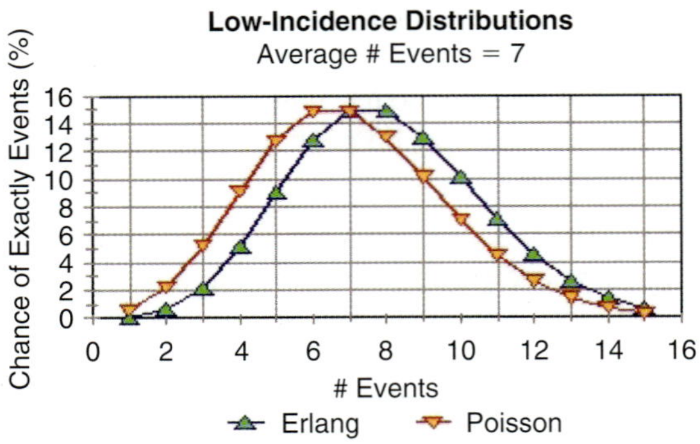


Figure 4.4

described by either the Poisson or Erlang distributions.[67] Both are shown in Figure 4.4.

Details of an example analysis are given in Appendix 2.

In summary, information-lean discrete cleanliness measurements should be avoided wherever possible – but they can be used to manage cleaning operations.

## 4.8 CONTROL OF A CLEANING PROCESS

### 4.8.1 Control for the Cleaning Process

*By a small sample we may judge the whole piece.*[68]

*Q7: If your process is running today apparently as it was running yesterday, has anything changed? And, do you care?*

A cleaning process is different from other operations, but not dissimilar. The techniques described above, and in the Appendix 1, can be used to control performance of cleaning processes and other operations. More to the point, techniques normally used to control refinery, biological, and semiconductor operations can be used with cleaning processes.

### 4.8.2 Control for the Customer

*If you live to be one hundred, you've got it made. Very few people die past that age.*[69]

Just because a process is in control does not necessarily mean that it is a useful process.

The capability of a process is measured by its ability to meet the needs of its customers. That means comparing the output of the process with the requirements of the customer. The customer speaks in terms of specifications or requirements. The process speaks via the data collected from it.

Measuring how well a stable distribution of data (i.e. a process in control) matches up with the specifications is the essence of process capability.

It matters not how uniform and invariable are the data produced by a cleaning (or other) process. Those data must directly translate (or be equivalent) to a requirement or specification of the next user of the cleaned materials. If not, the cleaning process is likely a waste of resources. Said another way, the final stage of cleaning must meet the needs of the next user or there is no point in completing it.

[67] These are the probability distributions of the number of occurrences of an event that happens rarely but has very many opportunities to happen. These distributions are used to model occurrence of events such as "wrong numbers" or "phone busy" situations with communication systems, "server busy" with network systems, or child suicides. The distributions are of an integer number of NO GO events – there is no interest in MAY GO results.

The Poisson distribution is the probability distribution of the number of events that would occur within a preset time. The Poisson distribution has only one parameter – the mean. The standard deviation is the square root of the mean. This distribution can be generated by the spreadsheet function @ POISSON (or =POISSON). The arguments are @ POISSON (mean failures expected over a time interval, the exact number of failures within a portion of the time interval, 1).

The Erlang distribution is the probability distribution of the amount of time until a specified number of events has occurred. Both distributions are similar. Only the Poisson distribution can be generated via a spreadsheet function.

[68] Cervantes (Miguel de Cervantes Saavedra).

[69] George Burns.

### 4.8.3 The Risks and Rewards of Control

*Fate laughs at probabilities.*[70]

The goal of a process control system is to enable economically sound decisions about actions affecting the process:

- This means balancing the risks of taking action when action is *not necessary*[71] versus failing to take action when action *is necessary*.[72] Some process operators take action *when* it was necessary, and have continued to do so. All deviations produced "correction." But continual process-adjustment in reaction to non-conformance actually increases variation and degraded quality.
- Useful process control also means taking the right actions for the right reasons, in the *right amounts*. Over-reaction and timidity are also sins in process control. The amount of action taken is often called gain.[73] The human belief that "more is better" has led some to make poor situations worse by applying an over-large correction.
- Timing, as in negotiations, sport contests, and love, can also betray the best of intentions. A control action taken for the right reason and in the right amount, but taken only after management approval, is likely a recipe for a double disaster. The expected process benefit may not be achieved because the process situation may have changed. And so management may believe those in control aren't!

### 4.8.4 Reactive versus Preventive Control

*Prediction is very difficult, especially of the future.*[74]

An overly simple evaluation is to declare there are two strategies for process control – reactive and preventive:

1. A *reactive* strategy is one which uses negative results (NO GO for example) of cleaning tests to justify and often calibrate modification to an operating cleaning process. This is the conventional approach.
2. A *preventive* strategy is one which uses ongoing information about process operation to do the same. When a preventive strategy is well-conceived and implemented, cleaning tests are not needed to provoke action. They are used only as required by customers or to provide "organizational comfort."

Moving from a product control strategy (product inspection to guarantee quality) to a process control strategy (process control to manage quality) can be a major paradigm shift from conventional operation. Nevertheless, it is the approach strongly recommended, and described in Section 4.10.

## 4.9 PROCESS VARIATION: COMMON AND SPECIAL

*I don't have any solution, but I certainly admire the problem.*[75]

As there are two types of measurement error,[76] not surprisingly, there are two analogous sources of process variation:

1. There is *systematic measurement error*, and there are *special causes* of process variation. Behind both is a reason. Because there is a reason for the

[70] Edward George Bulwer-Lytton, *Eugene Aram*, Book I, Chapter 10.
[71] Specialists call this a Type I error. One might call this a "nervous process".
[72] Specialists call this a Type II error. One might call this a "negligent process".
[73] Usually this is taken to mean the ratio of control action to process stimulus. Also usually, process stimulus is the difference between the process current performance and the desired process performance.
[74] Bohr, Niels Henrik David – a mentor of Werner Heisenberg.
[75] Ashleigh Brilliant.
[76] See Section 4.2.7.

process variation, a user can seek to find that reason (cause) and eliminate it:

- Accumulation of oil (soil) in the cleaning solvent[77] is a special cause which *gradually* harms cleaning performance. Insertion of cold parts into a vapor degreaser is another special cause which *immediately* harms cleaning performance.[78]
- Only those parts cleaned after the oil level becomes excessive or dried after the vapor blanket is disabled are affected. Not all parts are affected.

2. There is *random measurement error*, and there are *common causes* of process variation. Behind both is no known reason, just chance. Because there is no known reason for any specific observed variation, one can't take action to eliminate that variation:

- It is the interaction of all unchanged[79] system parameters[80] which are the common causes of process variation.
- This interaction affects all parts. But not all are affected to the same degree.

### 4.9.1 Examples of Common versus Special Causes

> *The manipulation of statistical formulas is no substitute for knowing what one is doing.*[81]

A distinction between special and common causes is not of merely academic interest. Rather, it is crucial to successful management of any process. Users respond by taking different action if they suspect either type of cause is affecting cleaning performance:

- If a special cause is suspected of harming performance, the user first seeks to identify it, then take specific action to remove it. An example would be where cleaning performance is poor on only one shift of operation. A user might observe more closely and find that the operator on that shift uses manual control on the power hoist rather than the installed control program. An action to remove this cause of poor performance would be to retrain this specific operator to not bypass the hoist control program to speed up action.
- If a user seeks to improve overall performance, they practice the strategy of *continuous improvement*. This means continually upgrading the means by how tasks are done – whether or not there is specific justification to do so (a special cause). An example would be to periodically retrain operators – with no particular emphasis on any phase of their work. Other examples would be to:
  - Recalibrate all thermocouples and the refractometer.
  - Clean out the bottom of both cleaning and rinsing sumps.
  - Drain the water separator.
  - Time the actions of the power hoist.
  - Have the supplier do a full assay analysis of a sample of distilled solvent (including metals).
  - Change the filter cartridges in the solvent recirculation lines.
  - Review the techniques used to array parts within parts baskets, and retrain operators to implement any chosen changes.

---

[77] Most of the specific examples used in this section will be associated with solvent cleaning technology. This is because solvent cleaning is more easily described mathematically than aqueous cleaning technology. So realistic, and hopefully useful, examples are easier for this author to create. Where possible, reference for both common cleaning technologies will be made.
[78] Cooling the vapor blanket above a heated liquid sump causes the vapor to condense and the blanket to collapse.
[79] Unchanged by deliberate action.

Occasionally there is not a "bright line" between a specific or special cause and a cause not generally activated. An example is the environment in which the solvent degreaser is operating. An excessive ventilation rate may excessively cool the freeboard area – causing the vapor blanket to collapse. Rinsing and drying of all parts will be affected. Here, personnel operating the cleaning machine cannot identify and remove the special cause because it is outside of their span of control. They can change no parameters of the cleaning system. Yet the root cause (inadequate environmental control) does affect all operation, and can be identified and removed.

The distinction should be made not on how the cause was activated, but whether or not it affects all parts.
[80] System parameters include but are not limited to: physical and thermal mass of parts being cleaned; arrangement, position, and contact of parts in rack (basket); speed and timing of movements by hoist; interaction of sump thermostat and heater as well as interaction of the headspace thermostat and chiller coil; temporary pressure or volume fluctuations in output from spray nozzles, etc.
[81] Blalock Jr., Hubert M. *Social Statistics (2nd ed.)*, 1972.

- Measure emission losses and compare values to those expected.
- During operation, inspect the defrost system for the cooling coils.
- Have an uninvolved but experienced person inspect the machine.
- If a standard load of parts has been developed (good idea), use it.

A process is said to be operating in a state of statistical control when common causes are the only source of variation.

### 4.9.2 Identification of Common and Special Causes

> *It's no exaggeration to say the undecideds could go one way or another.*[82]

Special causes are relatively easy to identify. Common causes are less so. A different strategy is generally used to identify each type of cause of process variation. See Section 4.13 on control charts.

- By definition, *special causes don't affect all operation*. So, affected operation may be identified through some differences versus normal operation.[83] Likely, this will isolate the timing of when the cause is implemented. From knowing "the when," a user can often use logic[84] to identify "the what."
- By definition, *common causes do affect all operation*. Consequently, their total impact may be judged by computing the overall variance[85] of the cleaning process. But they are most effectively identified through a method variously known as continuous improvement, the Deming cycle, the Shewhart cycle, or the PDSA cycle.[86]

Said another way, data contain "noise" and "signal." To be able to extract the information from any data set one must separate the "noise" from the "signal." The "signal" is likely produced by the special cause. The "noise" is likely produced by the common causes. Control charts, described in Section 4.13, are the major tools used to separate the "signal" (special causes) from the "noise" (common causes).

Operating staff can usually determine special causes for manufacturing defects by consulting the workforce, but dealing with common causes is a management responsibility.

---

[82] George H.W. Bush.

[83] See Section 4.13 where control charts are discussed.

[84] See Section 4.20 for advice about troubleshooting and problem solving.

[85] See Appendix 2 for information about how to evaluate the uniformity (variance) of a process by using control charts.

[86] Fundamentally, this is a strategy for management of any activity – including life. PDSA is a continuous quality improvement model consisting out of a logical sequence of four repetitive stages for continuous improvement and learning: Plan, Do, Study (Check) and Act:

- *Plan*: plan ahead for change. Analyze and predict the results.
- *Do*: execute the plan, taking small stages in controlled circumstances.
- *Study*: check, study the results.
- *Act*: take action to standardize or improve the process.

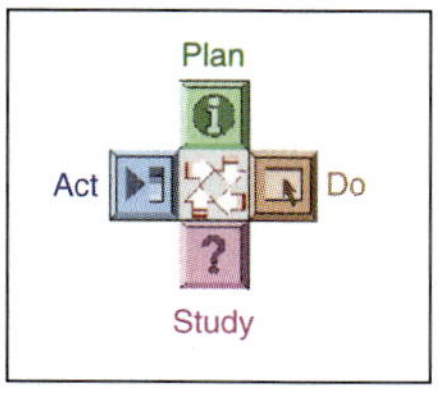


The PDSA model is cyclical. One continues to grow/improve via planning, executing the plan, reviewing the results of the plan, and acting based on the results. It was pioneered by Walter A. Shewhart in the 1920s and taken up by W. Edwards Deming with significant effect by Americans during the World War II to improve aircraft production.

Application of PDSA to the situation described in Section 4.9.1.1 might produce the following sequence of events:

- Recognition (PLAN) that the programmable power hoist is the means by which timing of the cleaning cycle is established.
- Observation (DO) that one can bypass the hoist program and operate it by manual control.
- Examination (STUDY) of cleaning performance (NVR results) for all operating shifts and observation that performance is only poor on one shift.
- Inquiry (ACT) as to the actual procedures used on that shift.
- Understanding (PLAN) that the actual procedures used aren't those as scheduled.
- Retraining (DO) of operating staff on that shift about the importance of cleaning quality versus cleaning production rate.
- Continued examination (STUDY) of cleaning performance and observation that performance level has not changed on that shift.
- Installation (ACT) of a new control program for the hoist to avoid manual operation.
- Recognition (PLAN) that operations on all shifts are similar, but not adequate …

#### 4.9.2.1 *Recognition of Common and Special Causes*

*Consistency requires you to be as ignorant today as you were a year ago.*[87]

No one wants a cleaning operation to fail. Yet failure to identify and correct both special and common causes can produce operational failure.

These causes don't shout at you saying "… here I am, come fix me." They have to be identified – often by painstaking attention to detail, and more often by trial and error.

Causes of failure which others who have used solvent cleaning or some aqueous machines have identified and removed are listed in Table 4.5:

- In some cases, special causes (which don't affect all parts) contaminate the entire stock of cleaning solvent and become a common cause.
- In other cases, the same root cause is not always practiced. Hence the same flaw can be both a special and common cause.

Both are also noted in Table 4.5.

Identification that both special and common causes are at work is covered in the section about control charts (see Section 4.13).

## 4.10 "PRODUCT-BY-PROCESS" MANAGEMENT

*An expert is a man who has made all the mistakes, which can be made, in a very narrow field.*[88]

*An expert is someone who knows some of the worst mistakes that can be made in his subject, and who manages to avoid them.*[89]

A preventive (front-end) strategy of process operation is completely different than a reactive (back-end) strategy.

The former is referred to as "product-by-process" (PBP). The latter is more conventionally practiced.

PBP means that cleaning quality for the next part use is assured by how the cleaning process is operated, not by completion of cleaning tests on supposedly cleaned parts. This is basically the approach recommended (or required) under various quality management programs such as ISO 9001 2000.[90]

Cleaning performance or validation tests should (see Chapter 5) be used not as arbiters of cleaning quality, but as (a) recognition that the expected quality was in fact produced, or (b) tools used in troubleshooting when the process is not "in control."

## 4.11 INPUTS AND OUTPUTS

*"'What's one and one and one and one and one and one and one and one and one and one and one and one?'*
*'I don't know' said Alice. 'I lost count.'*
*'She can't do addition.' said the Red Queen."*[91]

It's all about the inputs and outputs:

- Inputs are what are to be controlled. Outputs are the way in which users justify the resources spent on managing the cleaning process.
- Inputs are the complete set of variables by which the cleaning process is managed. Outputs are how users recognize the effects of their management decisions.
- In other words, inputs are the levers by which change is initiated. Outputs are the change.

Selection of inputs requires understanding of how a process works, and what environmental or safety

[87] Berenson, Bernard, *Notebook*, 1892.
[88] Niels Henrik David Bohr.
[89] Werner Heisenberg, *Der Teil und das Ganze*, 1969 – Student of Niels Bohr.
[90] Clause 4.1 of ISO 9001 2000 makes it very clear that organizations must identify and manage the processes that make up their quality management systems. An ISO 9001 2000 quality management system is made up of many processes. In the terms of ISO 9001 2000, a process is an integrated set of activities that uses resources to transform inputs into outputs. These processes are glued together by means of many input–output relationships. For details and contact information see http://www.iso.org
[91] Lewis Carroll (Charles Lutwidge Dodgson), *Through the Looking Glass*, 1872.

**Table 4.5 Examples of Special and Common Causes**

| Example of a Special Cause | Example of a Common Cause |
|---|---|
| **Easy to Identify** | |
| Solvent changeout is delayed for reasons of economy, and parts cleaned with that material are not well-cleaned, whereas parts cleaned with refined solvent are well-treated. Cleaning quality varies with time. | The wrong cleaning process was specified. |
| The machine cover is left off by one shift of operators. In addition to environmental issues raised, parts are poorly dried when the cover is removed (the vapor temperature drops). | The wrong cleaning cycle was specified. |
| Parts with cupped bottoms retain liquid and aren't well-dried. | Ultrasonic transducers are not turned on, so removal of particles from all parts is poor. |
| Parts with a peculiar feature (a blind hole or complex internals) aren't well-cleaned. | Compressed air, used for drying, is dirty and infects parts with particles. |
| Over-tall parts are inserted into the machine. They are not completely immersed. | Room air (the environment) is dirty and infects parts with particles. |
| The level transmitter fails low and parts aren't completely immersed. | The next user changes their requirements and no cleaned parts meet the new criteria. |
| **Difficult to Identify** | |
| Parts racked in the corners of baskets don't receive full rinsing action and retain residues. Other parts are well-treated. | The bath thermocouple develops an offset, and parts are treated with subcooled solvent, or the bath is boiled dry. |
| Light parts, with a low thermal mass, heat quickly aren't washed or rinsed with adequate condensing solvent. | Solvent purification/distillation is not properly working. Soil is not removed but is recycled and parts are reinfected. |
| Heavy parts, with a high thermal mass, remove too much heat and collapse the vapor blanket. | A single component is preferentially removed from an azeotropic mixture. |
| The programmable hoist is mis-programmed and one stage is shortened. | The cleaning test fails, for any consistent reason |
| One rinsing or drying nozzle is mis-aimed. Parts located in some zones of the bath are untreated. | Stabilizer is not added, or is depleted. Parts are corroded by acid. |
| **Where a Special Cause Becomes a Common Cause** | |
| Some parts processed upstream bear extraordinarily high levels of soil. This makes them more difficult to clean, and fouls the bath. | All parts become difficult to clean because the bath becomes fouled with excessive soil. |
| An intractable soil is removed from parts, but is retained within the cleaning machine. | All parts become reinfected with soil not purged from the cleaning machine. |
| Some parts processed upstream bear an unusual soil component. It is not removed in purification/distillation. | An unusual soil component is occasionally added to the bath, but not removed in purification/distillation and all parts are re-soiled by it. |
| Small parts are not retained within the rack. They fall to the bottom of the cleaning sump. | The parts basket is not fully immersed because of debris on the bottom of the cleaning sump. |
| **Where the Same Flaw Produces Both a Special and Common Cause** | |
| **Some** packaging is dirty, and infects parts not tested for cleanliness with particles. | **All** packaging is dirty, and infects parts not tested for cleanliness with particles. |
| Parts aren't well rinsed and dried because **operators on one shift** shorten the performance cycle. | Parts aren't well rinsed and dried because **all** operators of the basket hoist shorten the performance cycle. |
| The wrong solvent is used, but is **useful for some** soils or soil components. | The wrong solvent is used. |

requirements must be fulfilled. A desire to be complete often leads users to define everything about a process (except the output) as an input. But that approach misses the point and leads to unneeded complexity and cost.

When selecting just which process inputs which should be monitored and managed in a control scheme, one should be mindful of Ockham's Razor.[92] In more current language, this is known as the KISS[93] principle.

### 4.11.1 Selection of Process Inputs

*There is nothing to it. All one has to do is hit the right keys at the right time and the instrument plays itself.*[94]

The quality of solvent cleaning/rinsing/drying is *determined*[95] by the parameters given in Table 4.6 – for a properly designed solvent[96] cleaning process.

**Table 4.6 Process Inputs for Solvent Cleaning Process**

| General input | Item to be controlled/managed while NOT operating | Item to be controlled/managed while operating |
|---|---|---|
| The soil level | • Consistency of upstream operation | • Total weight or volume of soil on a collection of uncleaned parts |
| The solvent | • Selection, and quality<br>• Appearance<br>• Changeout on schedule | • **Temperature(s)** in cleaning/rinsing sump(s)<br>• **Temperature(s)** in and above vapor blanket (for an open-top machine) or level of vacuum (for a vacuum machine)<br>• **Refractive index** (soil level)<br>• *Stabilizer content*<br>• *Impurity buildup* (acidity, etc.) |
| The cleaning cycle | • Operator training about conforming to cycle specifications versus productivity<br>• Hoist control program | • **Dwell time** within cleaning (sumps), rinsing (vapor freeboard), and drying (freeboard) zones (open-top machine) or periods (vacuum machine)<br>• *Number of cleaning cycles* since previous solvent changeout or purge to distillation |
| The way in which the solvent is applied | • Arrangement and blockage of spray nozzles<br>• Overflow weirs not fouled | • *Level and appearance* of fluid in water separator |
| The way in which the parts are "presented" | • Placement and inventory of parts in parts baskets | • Audit about work flow |
| General operation | • Audit about conformance to maintenance standards | • Audit about conformance to operating instructions |

[92] Ockham's Razor (also Occam's Razor) is a principle attributed to the fourteenth century English logician and Franciscan friar, William of Ockham. In its simplest form, Ockham's Razor states that one should not make more assumptions than needed (*Essentia non sunt multiplicanda praeter necessitatem*). When multiple explanations are available for a phenomenon, the simplest version is preferred.

[93] Keep it simple.

[94] Johann Sebastian Bach.

[95] Items in Table 4.6 are not the only items which can affect the quality of cleaning performance. "Acts of God," upstream discontinuity, malicious behavior, and the like are not included here.

[96] In an aqueous cleaning process, dynamic surface tension would be used in place of refractive index to measure soil level.

### 4.11.2 Selection of Process Outputs[97]

> *If a man will begin with certainties, he shall end in doubts; but if he will be content to begin with doubts, he shall end in certainties.*[98]

The quality of solvent cleaning/rinsing/drying is *recognized* by the parameters in Table 4.7 – for a properly designed solvent cleaning process.

### 4.11.3 Scorekeeping: Monitoring Input/Output Information

> *If a man look sharply and attentively, he shall see Fortune; for though she is blind, she is not invisible.*[99]
>
> *The harder I work, the luckier I get.*[100]

The flood of information which could be produced by diligently recording everything listed in Tables 4.6 and 4.7 can be stemmed by vigilant supervision

**Table 4.7 Process Output for Solvent Cleaning Process**

| General output | Item to be audited while NOT operating | Item to be audited while operating |
|---|---|---|
| No safety incidents | • Employees trained with standard operating instruction<br>• Safety placards displayed<br>• Employees familiar with Manufacturer's Safety Data Sheet (MSDS) information (Occupational Safety and Health Administration (OSHA) Hazmat) | • Staff following operating instructions (especially pertaining to flammable solvents)<br>• Staff using personal protective equipment<br>• Area ventilation in use |
| No environmental incidents | • Degreaser cover in place<br>• Staff stores and transfers solvent properly | • No odor in operating area<br>• Degreaser cover in place and use |
| Compliance with environmental regulations | • Periodic emission checks or measurements within limits | |
| General part appearance | • Upstream operation with parts as expected | • No harm to part |
| Part cleanliness | • Operating environment doesn't contaminate cleaned parts | • Satisfactory results of cleaning tests, and downstream use |
| Consistency of operation | • Appropriate (not all) employees trained about use of control charts | • Control charts show operation to be within preset control limits |
| Cost of operation | • Create cost sheet for cleaning operation, based on measurable parameters and allocated cost assignments from enterprise accounts<br>• Prepare forecast, or budget, as stated expectations | • Fulfillment of expectations<br>• Analysis of variance (versus expectations)<br>• Outcome of PSDA (see Section 4.92) |

[97] Note that these process outputs are coincident with the characteristics of success and failure in management of cleaning systems as described in Table 4.1.

[98] Francis Bacon, *The Advancement of Learning*, 1605, I v 8.

[99] Francis Bacon, *Essays On Fortune*, 1625.

[100] Darrell Royal.

and operating staff. Ockham's Razor also applies to management of records.

However, some information deserves special consideration:

- Four items, noted using *italic* typeface in Table 4.6, should be recorded using methods suitable to the operating area. A simple checklist would suffice. Frequency of recording should also be suitable to the operating area. But recording could be done coincident with testing for cleanliness as suggested in Section 4.3.3. The four items are:
  - (a) *Stabilizer content* (results of periodic measurements as directed by solvent manufacturer).
  - (b) *Impurity buildup* (results of periodic measurements (e.g. acidity or moisture) as directed by solvent manufacturer).
  - (c) *Number of cleaning cycles* since previous solvent changeout or purge to distillation.
  - (d) *Level and appearance of fluid* in water separator.
- Four items, noted using **bold** typeface in Table 4.6, should be recorded using continuous instrumental methods:
  - (a) **Solvent temperature** in cleaning sump (normal increases reflect accumulation of soil and will indicate at some point that solvent changeout, purge, or purification should be done).
  - (b) **Solvent refractive index**[101] in cleaning sump, (normal changes reflect accumulation of soil, and existence of additional soil types. This will indicate at some point that solvent changeout, purge, or purification should be done).
  - (c) **Solvent temperature** in vapor blanket[102] (this will indicate chiller coils are restraining hot solvent vapors from escaping as air pollution).
  - (d) **Dwell time** of various stages during cleaning/rinsing/drying cycle (this will indicate that the proposed cleaning cycle is being completed as scheduled, and is a part of the *PBP* methodology).

Three of the four quantities should be measurable as part of the normal equipment provided with a vapor degreaser. If sump and head-space thermocouples aren't provided, don't purchase the cleaning apparatus – buy something else. A programmable hoist should also provide electronic output describing its actions.

Solvent refractive index can now be reliably measured on-line for an investment which is both reasonable (<€400 for a hand-held instrument ~€1,000 for a basic on-line unit, or ~€2,500 for a unit with more accessories) and justified by improved cleaning operation.

## 4.12 CONTROL TARGETS

*I wish to God these calculations had been executed by steam.*[103]

It is written here that cleaning is soil management. That begs the question, what happens to the soil after it is removed from the parts?

The expected and hoped-for answer is that it stays with the cleaning agent – and then it is removed from the cleaning machine with the cleaning agent. Accumulation of soil within the machine represents a system design which has failed.[104]

Soil accumulation within the cleaning agent can become "too much of a good thing." Parts can be reinfected with soil removed from previously

[101] This on-line measurement is an indirect measurement of soil concentration. Refractive index *per se* has no inherent meaning in cleaning work. In fact, soil concentration can be inferred from other kinds of measurements such as solvent surface tension, opacity, or density.

Because aqueous cleaning technology does not transport soil in a single phase, on-line measurement of soil content is problematical. The issue is distribution of the soil materials in a multiple-phase mixture.

Further, soil concentration may be directly measured off-line in some cases. See methods in Appendix 2.

[102] This may be required by the US EPA's NESHAP covering solvent cleaning machines. See http://www.epa.gov/ttnatw01/degrea/haloforms.pdf

[103] Charles Babbage.

[104] The same is also true for aqueous and solvent cleaning. A failed system design for aqueous cleaning is one which separates soil from the parts with mechanical force, heat, and detergents, but doesn't remove it from the aqueous cleaning system.

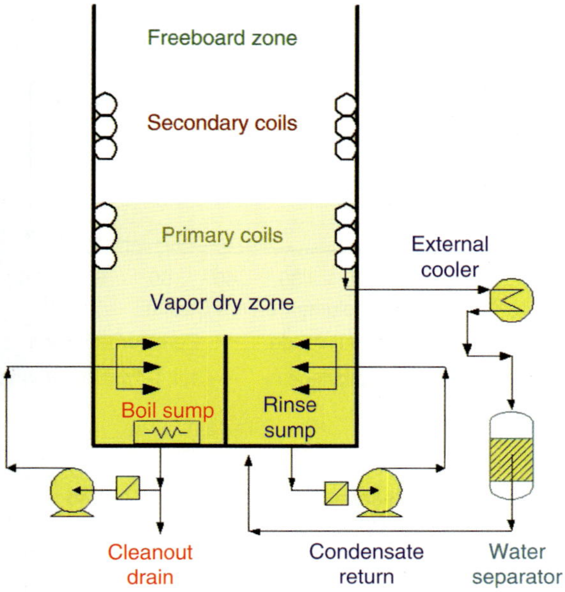


**Figure 4.5**

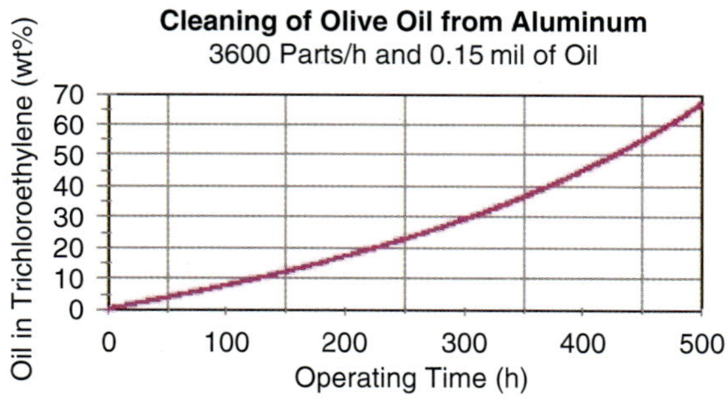


**Figure 4.6**

cleaned parts. The extent to which this happens is really dependent upon how and how often the soil-laden cleaning agent is removed from the cleaning or boil sump.[105] That's a management issue.

Poor soil management can be seen with a calculated example using solvent cleaning technology.[106] A generalized drawing of a solvent cleaning machine (vapor degreaser) is shown in Figure 4.5.

Consider use of this vapor degreaser[107] to clean a thin stamping oil (similar to olive oil) from stamped Aluminum parts[108] using trichloroethylene.

For an assumed 0.15 mil (0.00015 in or 0.00038 cm) film thickness of olive oil, and a stamping rate of one part per second. Figure 4.6 shows that the contents of the degreaser's cleaning sump becomes more than one-half olive oil before a month's operation is completed!

In other words, the cleaning sump in the degreaser becomes an applicator for the stamping lubricant!

This is not an extreme example. The assumed soil loading is actually an abnormally thin coating of lubricant, though the production rate is somewhat high for that size degreaser machine.

Part cleanliness deteriorates with accumulation of oil. In this calculated example, the cleaning agent (trichloroethylene) becomes a rich mixture of the soil to be cleaned and the cleaning solvent.

Normal dragout[109] contaminates the rinse sump because the liquid adhering to the parts inserted into the rinse sump is the oil-rich liquid from the cleaning sump.

Thus the rinse sump, which started as pure trichloroethylene, also becomes contaminated with oil.

The effect, without corrective action, is shown in Figures 4.7 (calculated boiling point) and 4.8 (calculated[110] cleaning performance).

---

[105] This is because the soil isn't separated from the solvent in the degreaser sumps, but in the external distillation column (or elsewhere). These facilities are not shown in Figure 4.5.

[106] Here parts are cleaned via immersion into a volume (boil sump) of solvent containing previously removed soil (oil). They are then removed and rinsed in a volume (rinse sump) of solvent which has been more recently purified of retained soil. Liquid clinging to parts on removal from the boil sump will be mixed within the contents of the rinse sump, that is oil is transferred from one sump to another.

[107] For this example, the sump volumes are 15 gal each. Each sump is 18 in deep. Top area is 12 in by 15 in. These values describe a commonly used small unit.

[108] The parts are Aluminum stamped pieces of size 1 W × 3″ L × ¼ in thick, and coated on all six faces with a stamping oil. The parts are racked in a basket. Cycle time for cleaning is 2.5 minutes. Production rate is 3600 parts/hour.

[109] Liquid adhering to the "cleaned" parts on removal from the cleaning sump. See Chapter 1, Sections 1.12.5 and 1.12.6, and Chapter 7, Section 7.6.

[110] The calculation is based on: (a) dilution of the cleaning solvent in the cleaning sump by oil removed from the parts and loss of cleaning solvent to emissive losses (90% of that allowed by the US EPA); (b) estimated dragout film thickness (1.0 mil) on parts leaving the cleaning sump, (c) dilution by the cleaning solvent in the rinsing sump of the dragout film clinging to the parts, and (d) the relationship between solution density of solvent with oil and boiling point displayed in Figures 4.9 and 4.10.

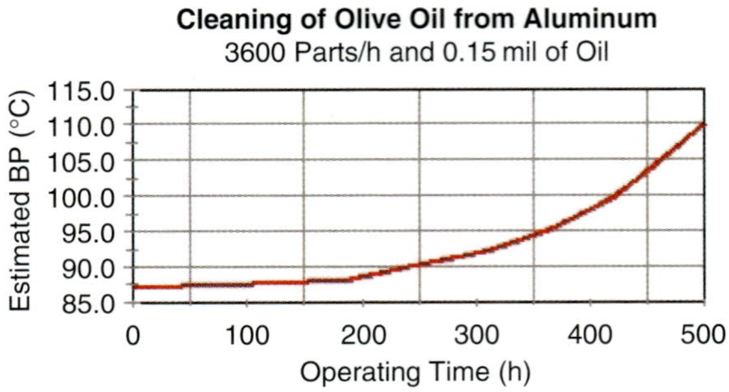


Figure 4.7

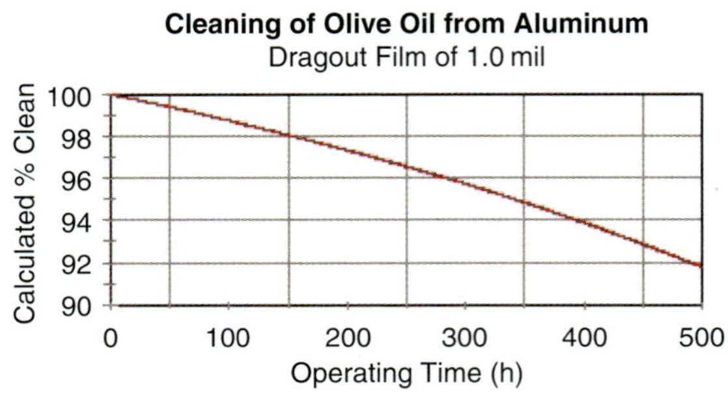


Figure 4.8

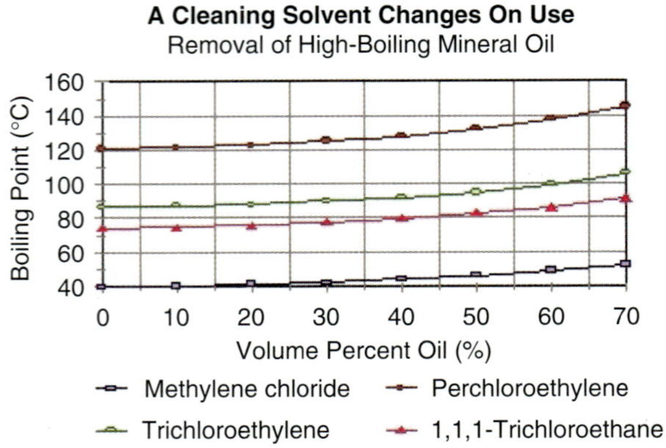


Figure 4.9

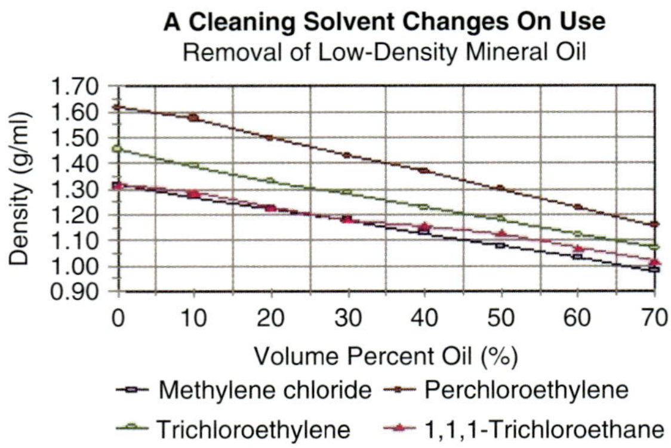


Figure 4.10

After one month's operation, part cleanliness has long ceased to be satisfactory. Basically, the cleaning solvent is being used to transport soil from one part to another and one sump to another.

And the solution boiling point bears no resemblance to that of pure solvent.

This is plainly the ultimate in poor soil management!

This effect is well-known. Measured effects on boiling point and solution density are presented in Figures 4.9 and 4.10.

Our commonly used chlorinated solvents and mineral oil were used for the data displayed in these two figures:[111] methylene chloride, perchloroethylene, trichloroethylene, and 1,1,1-Trichloroethane.

The graph titles tell all – soil accumulation in a cleaning solvent changes its character and may render is useless for cleaning work.

### 4.12.1 The Principle of "On-Aim" Control

*Those are my principles, and if you don't like them...well, I have others.*[112]
*You can't learn too soon that the most useful thing about a principle is that it can always be sacrificed to expediency.*[113]

[111] Data courtesy of DOW Chemical's manual on vapor degreasing which can be found at http://www.dow.com

[112] Groucho Marx.

[113] William Somerset Maugham, *The Circle*, 1921, Act 3.

There are two remedies for the poor operation shown in Figures 4.6–4.10. Both involve removing the oil-laden solvent from the cleaning bath, followed by separation of the olive oil soil from the trichloroethylene cleaning solvent. The remedies differ by whether or not the oil is continuously removed or removed (purged) periodically.

The monotonic behavior described in Figures 4.6–4.8 is a smooth passage to catastrophe.

### 4.12.2 Periodic Control Action

*The most useful piece of learning for the uses of life is to unlearn what is untrue.*[114]

The periodic behavior described in Figures 4.11 and 4.12 is a rough passage to poor quality cleaning.

Here, every 120 hours (5 days), the complete contents of the cleaning sump is purged to a distillation column for separation of oil, or to a waste dumpster for disposal. Then the cleaning sump is refilled with oil-free solvent. Weekly pumpout is a very common practice.

While the maximum percent concentration of oil achieved in the solvent is now a single-digit number (versus 70%), cleaning quality on the sixth day of operation may be unfamiliar to those who process parts cleaned on the fifth day.

This common method of control is called periodic purging. It produces a level of cleaning quality which can be guaranteed not to be constant. In order for cleaning tests to be meaningful, they should be done with parts produced at the same time in the periodic purging cycle. That seems unlikely!

Periodic purging is used because it is thought to be simple to do, and other solutions are thought to be considerably more difficult to implement. The reader can judge that for themselves, based on the material in this chapter.

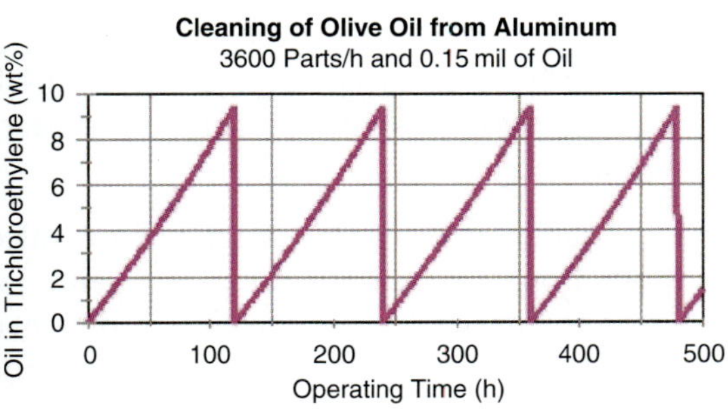


**Figure 4.11**

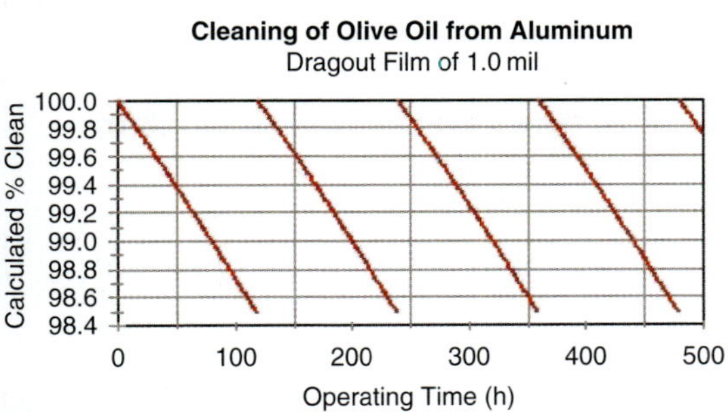


**Figure 4.12**

### 4.12.3 Continuous Control Action

*The weaker the data available upon which to base one's conclusion, the greater the precision which should be quoted in order to give the data authenticity.*[115]

The operating behavior in Figures 4.11 and 4.12 is well-intentioned, though mis-aimed. Periodic solvent cleanup (by purging to distillation or disposal) is a poor choice. Nothing is ever held constant – soil composition or cleaning quality! It isn't that 98.5% is or isn't a satisfactory level of cleaning. It is that the level isn't constant.

The second general remedy is to continuously remove some material from the cleaning sump – in order to maintain the level of oil at a constant value. This means that oil is allowed to accumulate, and then a continuous purge is started based on the measured (or estimated) entry rate of oil when the oil was allowed to accumulate. The purge keeps the oil level, and cleaning quality constant at 98.5%.

---

[114] Antisthenes.

[115] Norman R. Augustine.

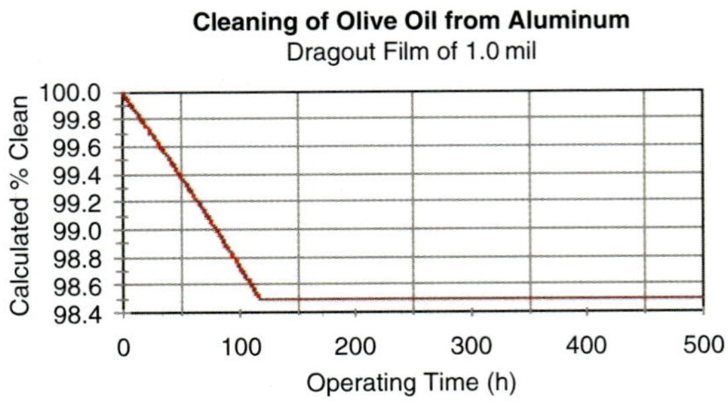


**Figure 4.13**

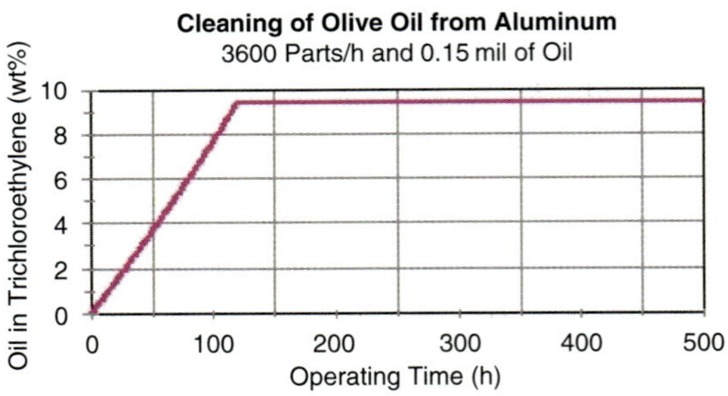


**Figure 4.14**

This continuous approach is shown in Figures 4.13 and 4.14.

But this also well-intentioned approach manages to maintain the cleaning quality at the wrong value – something different than 100%. There are two reasons for this:

1. Neither the entry rate of oil nor the oil concentration in the cleaning sump are known. Hence the purge rate can't be correctly set.
2. Purging doesn't start with system startup. It starts afterward – when the cleaning quality has already deteriorated.

### 4.12.4 On-Aim Control

*I am returning this otherwise good typing paper to you because someone has printed gibberish all over your examination on statistics and put your name at the top.*[116]

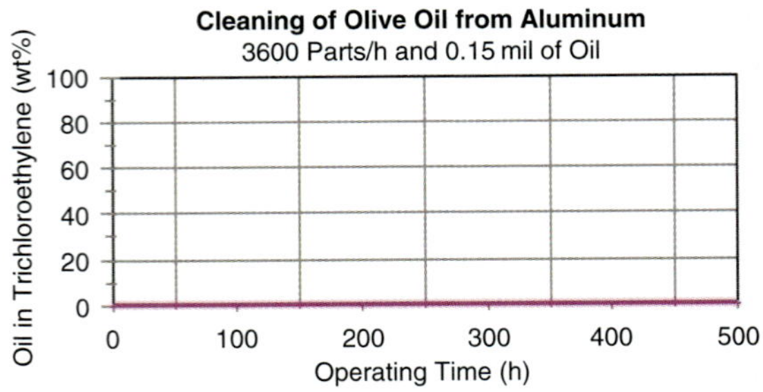


**Figure 4.15**

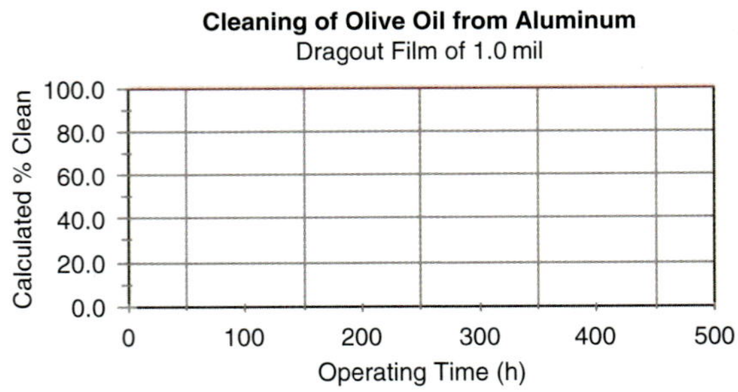


**Figure 4.16**

The idealized performance in Figures 4.15 and 4.16 is a smooth passage to the proper port – nearly 100% cleanliness:[117]

- This is done by setting the purge rate of soiled-solvent from the cleaning sump at a value low enough so that the dragout contains an amount of oil not detected on the stamped parts by the chosen cleaning test.
- The purge rate is set at the start of operation. Oil is never given an opportunity to accumulate. This is done by methods mentioned in Section 4.12.4.1 and described more fully in Section 4.13.

#### 4.12.4.1 *Two Approaches to On-Aim Control*

*Never practice two vices at once.*[118]

The behavior in Figures 4.15 and 4.16 is idealized – oil entry rate is neither fixed, known, or controllable.

[116] Anonymous Professor, Ohio University.
[117] The only way to guarantee 100% cleanliness would be to remove all the oil from all the solvent as the oil is removed from parts.
[118] Tallulah Bankhead.

*Consequently, a nearly oil-free cleaning bath isn't practical.*

What is practical is to:

- Set and maintain a purge rate of oil-laden solvent from the cleaning bath. The purge is started when the unit is started. This purge rate is one expected to maintain the soluble oil level at a low enough level and the cleaning quality at a high enough level to simultaneously meet the organization's needs for cost and quality. This approach is caused "fixed purge" or *manual control* of cleanliness.
- Measure the oil content of the oil-laden solvent, and adjust the purge rate to maintain oil content at a suitable and chosen level. This approach is called "smart purge" or *automatic control* of cleanliness.

This second approach requires measurement of a new parameter – refractive index (see Table 4.6 and Section 4.11.3). Either a batch-use, hand-held instrument or a continuously reading on-line instrument can be used. Both are affordable and justifiable – depending upon manpower availability and cleanliness standards.

"Smart" oil control provides compensation for the variable, unknown, and uncontrollable flow of oil into the cleaning bath as soil on valued parts.

Examples of both approaches are shown in Figures 4.17–4.20. The same methodology which produced Figures 4.11 and 4.12 is used.[119]

It isn't the differences of numerical values that are the significant learning from this calculated example. Rather, the key learning is that:

- If a manager is serious enough about managing part cleanliness to acquire and read this book, they must be serious enough to value the improved cleanliness produced by automatic ("smart") control of the solvent cleaning bath![120]

**Manual Control of Oil Buildup in Cleaning Sump**

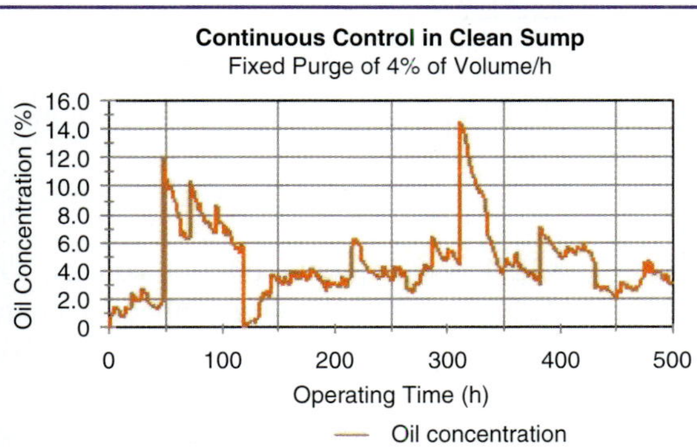


**Figure 4.17** Oil concentration

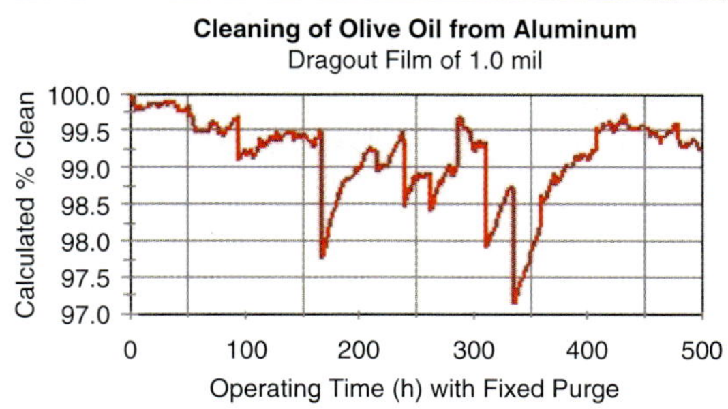


**Figure 4.18** Cleanliness

For fixed purging of oil to produce on-aim control of cleanliness, the operator is unaware of and not controlling changes in oil loading. Average percent cleanliness is around 98.5% with a minimum of nearly 97%.

[119] The calculation assumes the same equipment, parts, work flow, solvent, soil, and overall film thickness of soil. However, operation is made more difficult by adding the real-world complexity of (a) random variation in the film thickness of soil (oil loading) over time, and (b) a temporary 25% increase of oil loading which occurs for 1 hour each day. Thus, inlet oil level is both "noisy" and periodically spikes. The calculation starts with a pristine cleaning tank.

[120] Note the subtitle in Figures 4.17–4.20. The controlled purge rate is such that the complete solvent volume in the cleaning sump must be purged (turned over) to distillation or other solvent recovery process each day. That represents 4% per hour × 24 hour per day. Let not this level of solvent processing be surprising. The assumptions behind these calculations are completely normal. Cleaning quality will be no better than solvent quality. Solvent quality will be better than how well (frequently) it is purified (cleaned). The same soil removal problem is present with aqueous cleaning technology – it's just solved in a different way.

**Automatic Control of Oil Buildup in Cleaning Sump**

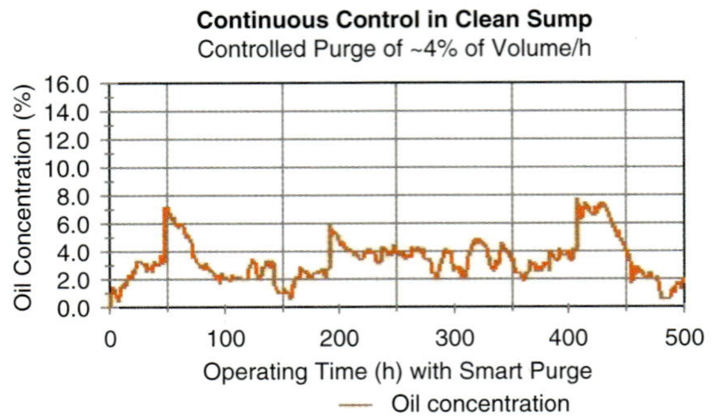


**Figure 4.19** Oil concentration

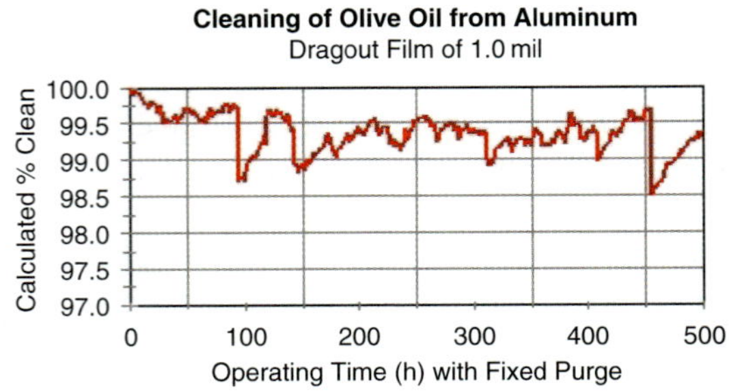


**Figure 4.20** Cleanliness

For "smart" purging of oil to produce on-aim control of cleanliness, the operator does not care about changes in oil loading. Average percent cleanliness is above 99%, and the minimum is above 98.5%.

### 4.12.5 Summary About Control Action

*It is true that you have missed everything of importance, but you have hit upon the method.*[121]

On-aim control is a basic phrase from the vocabulary of statistical process control (SPC). It means that the user should aim to achieve that which they desire.

Users of cleaning (or other) equipment don't value parts exiting the cleaning machine which don't pass the required cleaning test. Users desire cleaned parts which pass their test.

So on-aim control is simply a strategy for achieving that. On-aim control specifically means that the user should maintain in their degreaser constant conditions (as well as practicable) which are expected to produce parts which are 100% clean.

There are always two approaches to doing this – with and without instrumentation and automatic (*smart*) control.

## 4.13 CONTROL CHARTS

*That which we must learn to do, we learn by doing.*[122]

Section 4.12.4 speaks about on-aim control. One can't implement on-aim control without a control chart. These charts allow users to understand what their process is doing. These charts should allow judgements and action about if that is what is wanted.

Figure 4.7 is a primitive control chart. It clearly shows increase of boiling temperature in the cleaning sump over about one month's operation. One can't miss the continuous passage over a month's time to catastrophe.

But Figure 4.7 is not a useful control chart because it doesn't visually shout out "DO SOMETHING!"

- A good control chart visually speaks to the need for action when action is necessary.
- An excellent control chart visually speaks to the need for action *before* action is necessary.

[121] Sir Arthur Conan Doyle, *The Adventures of Sherlock Holmes: A Case of Identity,* 1892.
[122] Aristotle, *Nicomachean Ethics II.*

In this book, three types of control charts will be recommended, based on experience, for use in cleaning work. They are: R-bar ("R") charts, X-bar charts, and CUSUM charts.

### 4.13.1 Use of Control Charts

*If you think it's expensive to hire a professional to do the job, wait until you hire an amateur.*[123]

Process variation can be partitioned into two components.[124] Special cause variation is typically caused by some problem or extraordinary occurrence in the system. Natural process variation, frequently called *common cause* or system variation, is the naturally occurring fluctuation or variation inherent in all processes.

It is the purpose of control charts to allow detection of special cause variation and natural process variation.[125]

The two basic ideas behind detection of *special causes* and *common causes* are:

1. A *special cause* is likely to be acting upon the process when some process variable suddenly shifts from an established value to a different one. Said another way, a *special cause* will cause the *range* of the process variable to change.

   Remember, a *special cause* acts on operations at just certain times, with certain parts, or with certain procedures. Consequently, the range of individual measurements about operation is examined to detect *special causes*. Users seek to reduce specific non-conforming operation by eliminating *special causes*.
2. A *common cause* is likely to be acting upon the process when the average of some process variable suddenly shifts from an established value to a different one. Said another way, a *common cause* will cause the average of the process variable to change.

   Remember, a *common cause* acts on all operation. Consequently, averages of operation are examined to detect *common causes*. Users seek to reduce average variability by eliminating *common causes*.

In summary, *special causes* are detected by observing changes of ranges of performance, and *common causes* are detected by observing changes of average performance.[126]

### 4.13.2 Content of Control Charts

*It is commonly believed that anyone who tabulates numbers is a statistician. This is like believing that anyone who owns a scalpel is a surgeon.*[127]

A control chart can be constructed for nearly any parameter. Both *input* and output variables listed in Tables 4.6 (inputs) and 4.7 (outputs) can be considered in control charts.

Adoption of the "PBP" strategy described in Section 4.10 requires emphasis chiefly on those control charts which examine *input* variables. If charting the known input variables isn't a powerful tool in management of cleaning quality, then the input variables haven't been correctly or completely identified.

Control charts in which measurements of cleaning quality (an output variable) are displayed may be necessarily for good organizational reasons. But if they are useful in actual management of cleaning quality, the cleaning process has yet to be correctly and fully understood.

### 4.13.3 Limits on Control Charts

*Don't be afraid to ask dumb questions. They're more easily handled than dumb mistakes.*[128]

The horizontal axis of control chart graphs is operating time, or additional units cleaned/processed. The vertical axis of control charts is a statistical parameter and the limits associated with it.

---

[123] Red Adair, on his fee for extinguishing oil well fires after the Gulf War in 1991.
[124] Shewhart, W., *Economic Control of Quality of Manufactured Product*, Van Nostrand, 1931.
[125] See Table 4.5.
[126] See Section 4.9.2.1 for examples of each.
[127] R. Hooke, *How to tell the Liars from the Statisticians*, 1983.
[128] William Wister Haines.

There are basically three kinds of control lines (called limit lines) in a control chart. They are:

1. The central line (the chosen statistical parameter).
2. The upper control limit (UCL) for the statistical parameter.
3. The lower control limit (LCL) for same.

Use of the limit lines is that when the statistical parameter falls outside the UCL or LCL, some form of change must occur in the cleaning operation.[129]

What distinguishes the value of a control chart from the lesser value of run charts (see Figure 4.6 or 4.7) is the control limits. Run charts display data from operation. Control charts allow conclusions about action to be drawn from that data.

The LCL and UCL are determined from the statistical principles described above and employed in the Appendix 1.

In summary, it is the intersection of the statistical parameters from the process with the control limits which visually shout out "DO SOMETHING" (versus Figure 4.7).

### 4.13.4 "R" (R-bar) Control Charts

*The wireless telegraph is not difficult to understand. The ordinary telegraph is like a very long cat. You pull the tail in New York, and it meows in Los Angeles. The wireless is the same, only without the cat.*[130]

The character "R" stands for range – the absolute arithmetic difference between consecutive values of the statistical parameter.

The word range has meaning to those experienced in fire of artillery weapons. The meaning is that the weapon can't do its intended work until variation in the range of fire is negligible, and the weapon is on aim.

The analogous meaning is that the cleaning process isn't doing its intended work until variation in the range of statistical parameters is negligible, and the process is on aim.

Reduction in the range of operating statistical parameters should be the goal of persons managing cleaning machines:

- It is *special causes*, not acting continuously, which produce excess range in operating statistical parameters.

Examination of control charts allows unprejudiced recognition of the impact which *special causes* can and do produce. That's why managers of processes have prepared and examined them for more than fifty years.

As there is no rest for the weary, the preparation and examination of "R" control charts is a continuing affair. One can either root out and suppress *special causes* of variation in cleaning processes, or have it done by the next user of the cleaned part.

Three examples of the evolution of a stable cleaning process are shown in Figures a, b, and c in Table 4.8.[131]

In the case presented in Figure (c) in Table 4.8 there is essentially only one control limit line. The LCL is superfluous because the oil concentrations below zero have no meaning.

Details of the construction of a "R" control chart are found in the Appendix 1.[132]

---

[129] A major flaw in the use of LCL and UCL lines is the human tendency to note that a trend has developed which indicates the process is leading away from the center line; and then to take corrective action *prior* to a time when one of the control limits is exceeded.

The whole point of a control limit line is that data between it and the center line is not known to be different than that representing the center line – within the level of confidence upon which the control limit is based. More simply, data between control limits is noise and should be ignored. Only data outside of control limits should suggest/require/mandate action. See Appendix 1.

[130] Albert Einstein.

[131] The basis for these three calculated figures is the same as that in Section 4.12. The differences in them are produced by arbitrarily reducing the periodic entry of pieces with excessively high levels of oil soil.

Here, the *special cause* is not located in the cleaning machine. The *special cause* is located upstream. It is assumed in these examples that the dispenser of stamping oil in the stamping machine periodically leaks excessive oil into the stamping machine. Parts formed at that time are neither better nor worse than parts produced with the specified amount of stamping lubricant. But cleaning of them AND THOSE WHICH FOLLOW will be not done as well as specified because of the entry of excessive oil into the cleaning machine.

The source of this *special cause* is not an anomaly. Cleaning machines, in a sense, act as erasers for the effects of abnormal upstream operation.

[132] Carefully note the units of a "R" control chart. The horizontal axis has the units of number of groups of samples taken (basically time or amount of cleaning work done). The vertical axis has the units of the measured variable.

**Table 4.8 Evolution of a Cleaning Process**

<table>
<tr><th>"R" control chart</th><th>Comments</th></tr>
<tr><td>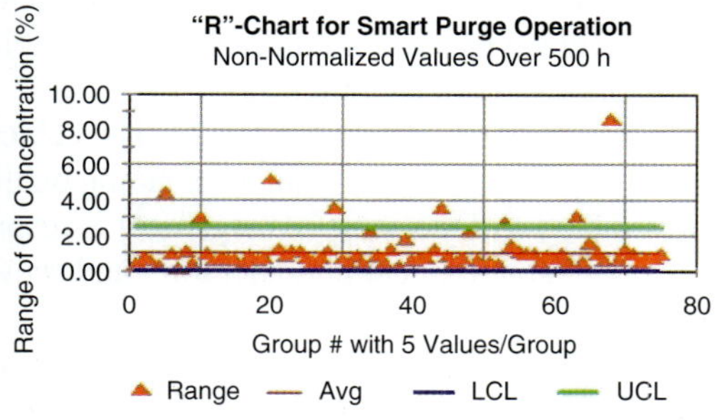

(a)</td><td>Measured concentrations of oil soil in the cleaning sump are the statistical parameters (vertical axis).<br>The horizontal axis is more production (time).<br>The horizontal control limit lines enclose most operating data. But there are more than a half-dozen instances of major shifts in the range of oil concentration. Each was produced by one or more special causes.</td></tr>
<tr><td>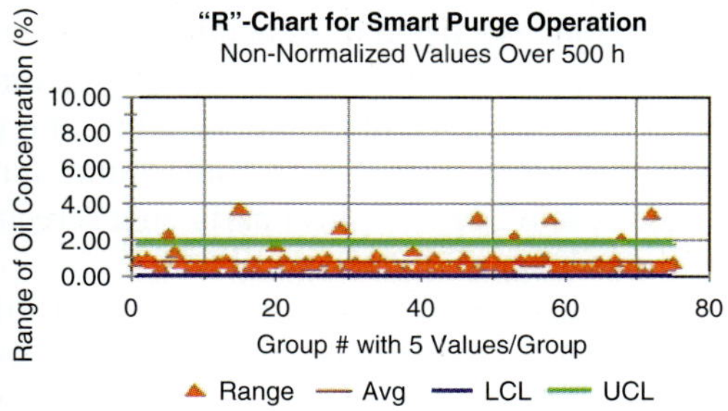

(b)</td><td>Some progress has been made here through elimination of some special causes.<br>The UCL line is closer to the center line – indicating that the standard deviation of measured oil concentration has been reduced.<br>Yet, there are still around a half-dozen major shifts in the range of oil concentration.</td></tr>
<tr><td>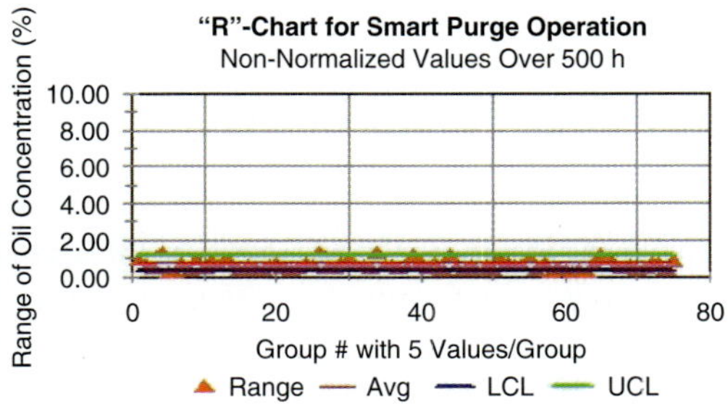

(c)</td><td>Considerable additional progress has been made in routing out special causes.<br>The UCL line is very close to the center line. There are no instances where the range of oil concentration exceeds the UCL line.<br>Examination can now start of this process to identify if there are common causes which affect cleaning operation.</td></tr>
</table>

### 4.13.4.1 *Identification of Special Causes*

> *It is a capital mistake to theorize before one has data. Insensibly one begins to twist facts to suit theories, instead of theories to suit facts.*[133]

There is no season for *special causes*. One is always hunting them. One is always finding them – hopefully.

The recipe for pinpointing *special causes* includes observation, understanding, and thinking. In management seminars it's called logical analysis. Your boss may call it "getting your head out of your posterior regions."

[133] Sir Arthur Conan Doyle, *The Adventures of Sherlock Holmes, A Scandal in Bohemia,* 1892.

There are at least two powerful tools to aid users here:

1. *Time coincidence*: The horizontal axis in "R" control charts is basically time (or amount of cleaning work done). The noted change of range outside of the UCL begs a question about what was happening during or just before the time at which the change in range was noticed.
2. *Logic or inference*: The vertical axis in "R" control charts is basically change in some statistical parameter derived from operating data. The noted change in range begs a question about what can influence that parameter. The power of logic must be balanced against the realization that logic is only a proposal and not a guarantee.

*Special causes* are often inconsistent human behavior that is not compliant with operating procedures. Here, interpersonal skills outside the scope of this book may be required to eliminate these *special causes*.[134]

### 4.13.4.2 *Standards for Special Causes*

> *The nice thing about standards is that there are so many of them to choose from.*[135]

Because there are no specifications for dirty parts, special causes are often acting upstream of the cleaning machine – as well as within it. Elimination of them can involve political as well as technical forces – if the same management doesn't "own" both the machine which produces the dirty parts and the machine which cleans them.

Adoption of the strategy of Product by Process (PBP) (see Section 4.10) requires that the issue of standardization of soil level be faced.

To some, with good reason, this is as serious as an attempt to standardize the contents of a garbage can.[136] To those operating cleaning machines, it is perfectly serious because they are responsible for an outcome (clean parts) without being responsible for the factors which define their task.[137]

Can standardization of input soil level be achieved? Probably not in this world – unless cleaning quality is highly valued and the same management is responsible for both soil application/generation and removal/cleaning.

If standardization or control of input soil level is a "fool's errand," identification of it is not.

Soil level can be identified or estimated by a variety of simple means – many of the cleanliness tests or validation tests listed in Appendix 2 can be used. A common practice is to boil a set of parts in the normal cleaning solvent, while weighing the parts before and after treatment. Another, and simpler practice is to place wet (oily) parts on a screen and note the volume of drained soil.[138]

When a history of such information is available for comparison, current similar information can be examined. Then the effect of a *special cause* can be seen as having originated upstream of the cleaning machine, or not.

### 4.13.5 X-Bar Control Charts

> *Certitude is not the test of certainty. We have been cocksure of many things that are not so.*[139]

X-bar charts are also called Shewhart charts[140] to recognize the contributions of a pioneer in statistical process control.

The term X-bar refers to the symbol $\bar{x}$ which is the statistical symbol for an average (mean). This is

---

[134] A single person can curb success whereas a team is usually necessary to produce it.

[135] Grace Murray Hopper.

[136] Yet, concerns about recycling of waste have already changed thinking and actions here.

[137] One action that is commonly taken is pre-cleaning. Goods are brushed or flushed before entry into the cleaning machine in order to remove gross levels of soil. While this can serve a leveling or standardizing function, it usually is done to save cost by reducing the duty on the cleaning machine.

[138] This is easy and necessary to do in operations such as cold header or machining operations where forming fluids, coolants, and lubricants are often applied as if they were free.

[139] Oliver Wendell Holmes.

[140] During the 1920s, Dr. Walter A. Shewhart proposed a general model for control charts. Basically, Shewhart charts are plots with running averages of the measured variable with the control limits from the center line, expressed in terms of standard deviation units.

because it is the current average of an operating statistical parameter which is plotted – along with both LCL and UCL.

X-bar control charts are normally constructed and examined in sequence with "R" control charts:[141]

- The "R" control chart is first constructed and used to eliminate *special causes*.
- Then the X-bar control chart is prepared to enable elimination of *common causes*.

Details of the construction of an X-bar control chart are found in the Appendix 1.[142]

The utility of an X-bar control chart is often seen when process operation is examined from startup through continuing operation. Examples of the startup of a stable solvent cleaning process are shown in Figures (a)–(c) in Table 4.9.[143]

In this example, stable means that there are no significant *special causes* and that the level of *common causes* is barely within the control limits.

To the extent that oil content in the cleaning sump affects cleaning quality, cleanliness of parts produced during this example should be very satisfactory.

### 4.13.5.1 *Identification of Common Causes*

> *Windows NT 5.0 is an evolutionary, not revolutionary, release of the Windows NT operating system. While there are important new features in this release, version 5.0 will build on a proven system architecture and incorporate tens of thousands of bug fixes from version 4.0.*[144]

The recipe for pinpointing *common causes* is **also** found in Section 4.9.2 – continuous improvement. But it's not as easy to implement:

- Since common causes affect all operation, there is no clue provided by time coincidence because there isn't any.
- Logical change is interwoven with random events in a fabric which conceals cause and effect. When one tests a deliberate change, which is suggested by a logical analysis of operation, one has to allow the outcome to be seen through a designed experiment which measures the effects of both random variation and directed effect.

### 4.13.5.2 *She Thought She was Special, But I Thought She was Common*

> *"Data! Data! Data!" he cried impatiently. "I can't make bricks without clay."*[145]

The power of R and X-bar control charts to separate *special cause* and *common cause* variation deserves further recognition. Two cases illustrate:

1. That X-bar control charts display random variation (*common causes*) while R control charts don't.
2. How "R" control charts display directed variation (*special causes*) and X-bar control charts don't.

Calculated results[146] are shown in Table 4.10 – with only extreme random variation (*common cause*) and no directed variation (*special cause*).

---

The main drawback of Shewhart charts is that they only use the information about the process contained in the last data point. The second drawback is they best monitor large changes in the measured variable. The latter drawback can be most significant for those doing cleaning work because cleaning development, management, and validation involve **small** differences among parameters difficult to measure (NVR, particle or defect count, solution concentration, surface thickness, misalignment, and optical quality).

[141] Usually, one chooses not to prepare an X-bar control chart until special causes have been eliminated after their effect has been seen via an "R" control chart. The reason for this choice would be that *common causes* usually have less impact and are more difficult to isolate than are *special causes*. However, this author has seen cases where the opposite is true.

[142] Note that the units of an X-bar control chart are not those of an "R" control chart. The horizontal axis has the units of number of groups of samples taken (basically time or amount of cleaning work done). The vertical axis of an X-bar control chart is dimensionless as all values are divided by the overall group average.

[143] The basis for these three calculated figures is the same as that in Section 4.13.4 (Table 4.8). However, operation here is stable by elimination of the periodic entry of pieces with excessively high levels of oil soil, **and** by arbitrarily reducing the level of random variation in all values of oil concentration.

[144] Microsoft, from http://www.microsoft.com/ntserver/community/

[145] Sir Arthur Conan Doyle, *The Adventures of Sherlock Holmes: The Adventure of the Copper Beeches*, 1892.

[146] On-aim operation (see Section 4.12.4) was assumed in these calculations.

**Table 4.9 Startup of Stable Solvent Cleaning Process**

| Control chart | Comments |
|---|---|
| 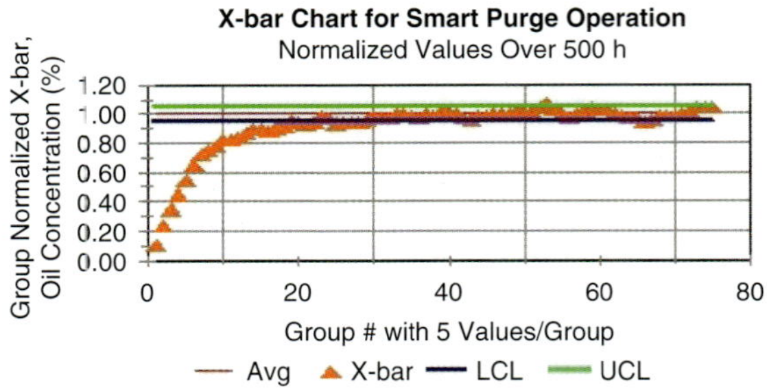<br>(a) | This X-bar control chart shows startup of a solvent cleaning machine.<br>Note that the period of startup operation (until the *overall* average value has been reached) is longer than for stable operation (until average values are within control limits).<br>Oil level during groups ~30 to ~50 is lower than the average value – so parts produced during this time should be cleaner than will normally be expected. |
| 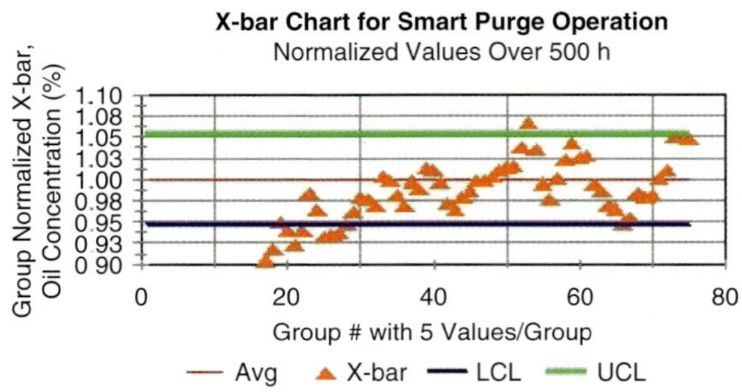<br>(b) | This is the same control chart as Figure (a), except that the span has been truncated to eliminate the period of startup operation.<br>This represents real-world good-quality operation of a solvent cleaning machine.<br>*Common causes* can never be completely eliminated. There will always be some random process variation that is within control limits and considered acceptable. |
| 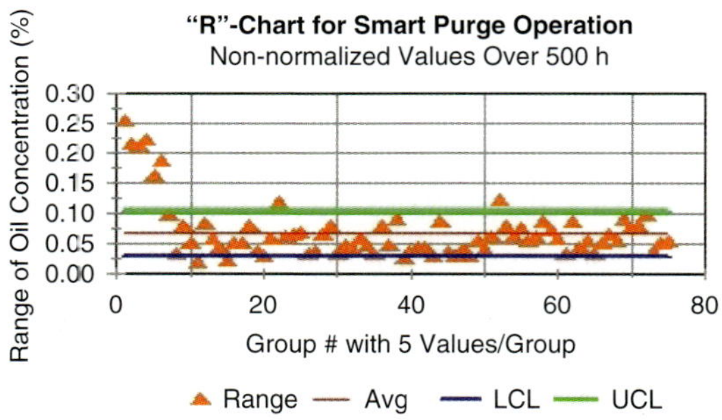<br>(c) | This is the complimentary "R" chart to Figure (a). The startup period includes a special cause – the startup. |

Results are also shown in Table 4.11 – with only extreme directed variation (*special cause*) and no random variation (*common cause*).

The calculated figures in Tables 4.9 and 4.10 illustrate how simple and well-used technology such as "R" and X-bar control charts can illuminate both special and common causes of process variation, and make them visible within a fog bank of operating data.

#### 4.13.5.3 *Overlap Between Common Causes and Special Causes*

*Still, it is an error to argue in front of your data. You find yourself insensibly twisting them round to fit your theories.*[147]

---

[147] Sir Arthur Conan Doyle, *His Last Bow: The Adventure of Wisteria Lodge*, 1917.

**Table 4.10 When Common Causes Become Too Common**

| 10X Random variation | 1X Random variation |
|---|---|

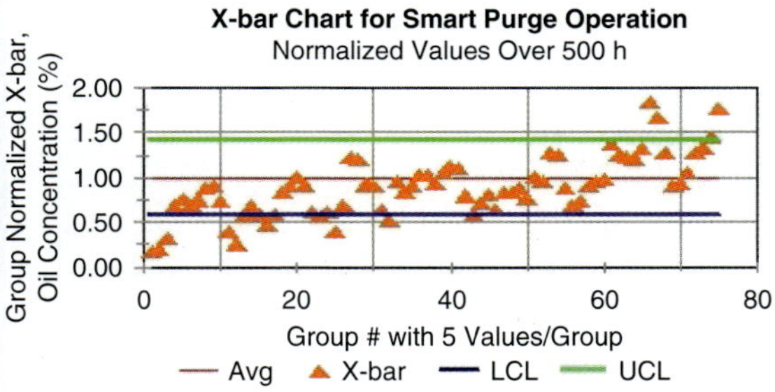


(a)

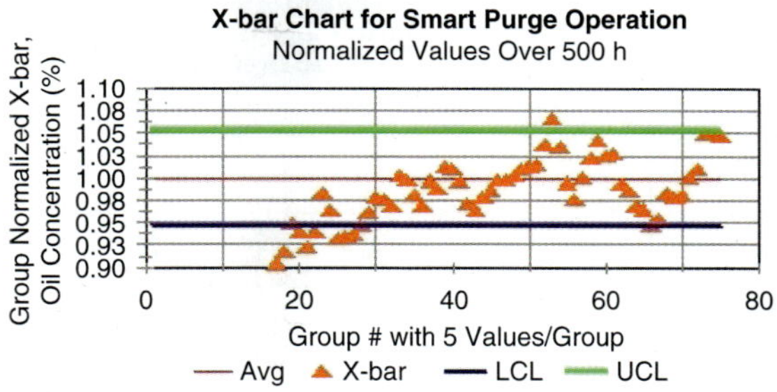


(b)

These two X-bar control charts show essentially the same behavior – a cleaning process with only a modest level of experience exceeding the average after startup.

BUT the span between the LCL and the UCL is visually different – it's 10X larger as it was intended to be.

The lesson is that one must pay attention to:

- Excess of average behavior beyond the LCL and UCL, but also to
- The span between them. **It is the gap between the LCL and UCL which speaks to the level of random variation, which is the level of *common cause*.**

The latter point is the reason for constructing X-bar control charts – to be able to estimate the level of *common causes* acting on the solvent cleaning machine through the span between control limits, and to notice of those limits are exceeded.

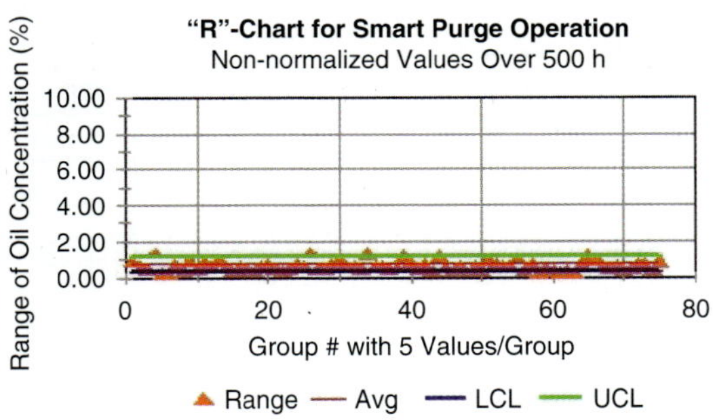


(c)

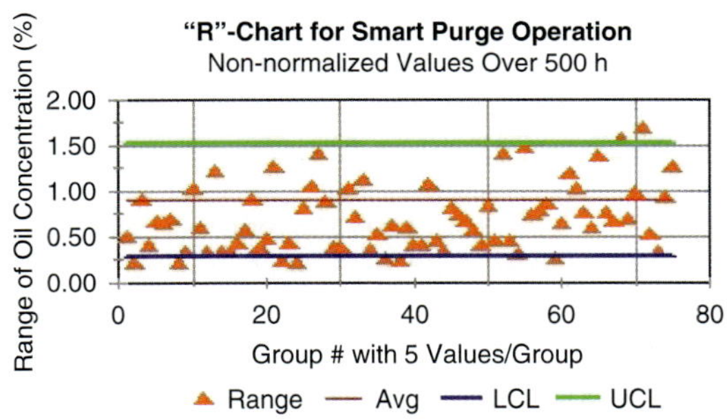


(d)

These two "R" control charts also show essentially the same behavior – a process with essentially no special causes having an effect. The range of oil concentration is quite consistent – within control limits.

This is as expected because of the assumptions used in calculations. Granted the span between the LCL and the UCL has increased by 10X, but that is of no consequence in an R control chart which only speaks to *special cause* variation.

The lesson is that one should **not** pay attention to:

- "R" control charts to learn about random or *common cause* variation.

**Table 4.11 When Special Causes Become Too Special**

| 10X Directed variation | 1X Directed variation |
|---|---|

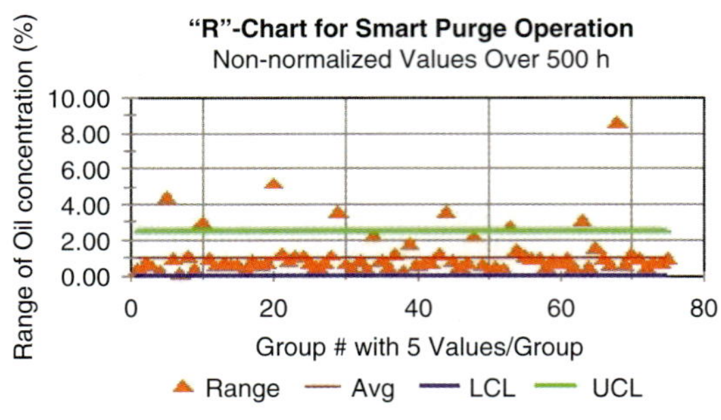


(a)

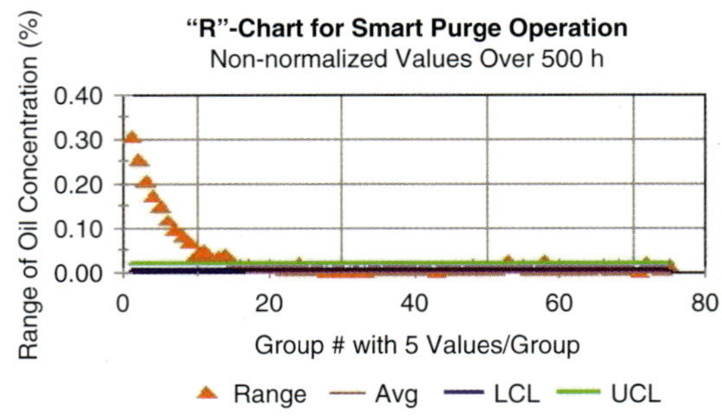


(b)

There are more than one-half-dozen occurrences (in Figure (a)) where the *special cause* of periodic oil intrusion forces the range of oil content to be displaced from between the control limit lines.

The difference in these two "R" control charts isn't the span between the two control limits.

The difference is that the range:

- Exceeds the control limits when oil leakage (*special cause*) feeds from the stamping machine to the cleaning machine.
- Does not exceed the control limits when there is no periodic oil intrusion.

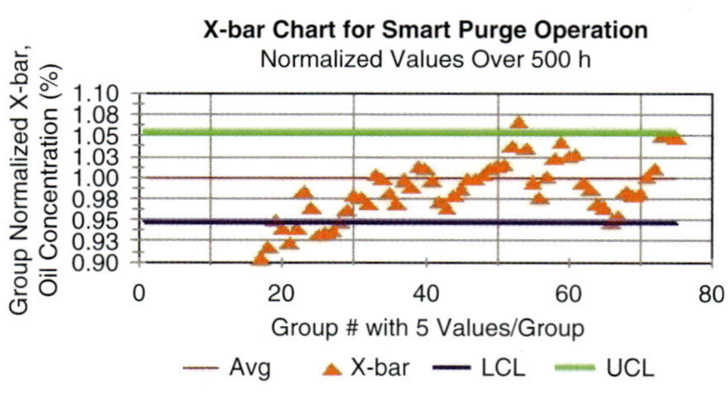


(c)

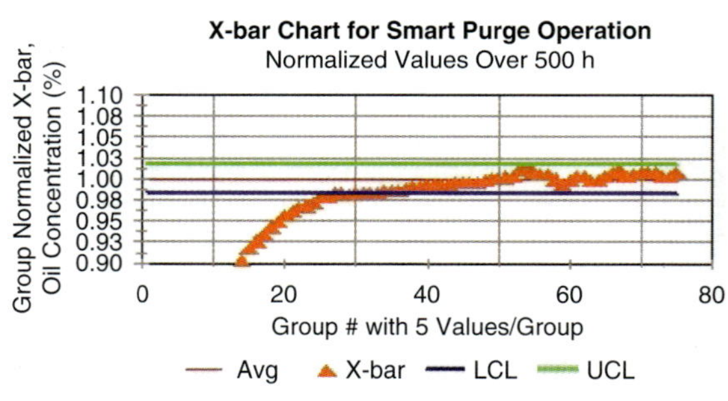


(d)

These X-bar control charts for the same operation clearly show (especially Figure [c]) that no *common causes* are acting on the cleaning system. All operation is within the control limit lines. This is despite the existence of a significant periodic application of a special cause – an oil leakage that feeds from the stamping machine to the cleaning machine.

Random events can sometimes be seen as both *special causes* and *common causes*.

As pointed out in the Appendix 1, inappropriate, unfortunate, or "bad things"[148] seem to happen at the same time. Random failures **don't** occur at times roughly equally spaced throughout a time interval! Random failures tend to occur in clusters.

[148] This is true of "good things" as well.

This behavior can be mistaken for *special cause* variation. If an excursion beyond a control limit line is simultaneously noted in *both* "R" and X-bar control chart, this is likely the case.

## 4.14 PROCESS CONTROL

*You see, but you do not observe.*[149]

Control charts were created by Dr. Shewhart (see Footnotes 124 and 140). They were modified and used by Dr. Deming[150] and others for the purpose of improving the quality of nearly anything by removing the sources of variation involved in them.

But process control is more than a search for special and common causes of variation.

Process control can be a search for:

- *Stability*: One uses algorithms to adjust parameters to avoid variation without knowing its sources. A satisfactory outcome is expected if the value of all parameters are non-variant.
- *Compliance*: On-aim control, such as that defined by the protocols of ISO 9001 2000, seeks to adjust parameters to specified values and maintain them at those values. A satisfactory outcome is expected if all inputs are within specifications.
- *Performance*: Algorithms are also used to produce an output condition. This may be an output parameter, an event, or the avoidance of either. A satisfactory outcome is expected if the output is fulfilled or avoided.

In the control of cleaning processes, managers need additional tools to allow them to find and achieve those aims.

While "R" and X-bar control charts are exceptionally valuable in operating cleaning machines because they allow identification and elimination of causes of variation, they are less useful in managing stability, compliance, or performance. Additional tools are needed.

There are three reasons for this, when managing any type of cleaning machine.[151] They are as follows:

1. Change of sump or bath temperature, soil content, stabilizer or detergent content, etc., in cleaning machines occurs less often and is smaller in magnitude than change in other operations.
2. "R" and X-bar control charts aren't focused on specific defects – only variation in some measurement. Control of a specific defect requires tools sensitive only to it.
3. "R" and X-bar control charts only allow illumination of variation. They don't directly speak to the source which is necessary to demonstrate stability, compliance, or performance.

### 4.14.1 Tools for Process Control

*I don't see the logic of rejecting data just because they seem incredible.*[152]

Most textbooks about statistical process control (SPC) call them the "Magnificent Seven." While this reference may compromise the reputation of an otherwise classic movie, most of the techniques described by this sobriquet are useful in managing cleaning processes.

Collectively, the focus of all is to identify and enable actions which enhance process stability, compliance, and performance.

The seven SPC techniques are:

1. *Histogram*: To plot overall cleaning performance by its frequency of occurrence.
2. *Check sheet*: To collect historical cleaning data.
3. *Pareto chart*: To analyze past cleaning defects by their frequency of occurrence.
4. *Cause-and-effect diagram*: What you think it is.

[149] Sir Arthur Conan Doyle, A Scandal in Bohemia, in *The Adventures of Sherlock Holmes*, 1892.

[150] Deming, W.E. "On a Classification of the Problems of Statistical Inference," *Journal of the American Statistical Association,* No. 37, Vol 218, pp. 173–185, 1942. In this paper, Deming advocated the application of statistical process control to disease surveillance and adverse healthcare events.

[151] For reasons given earlier in this chapter, most examples refer to operation with a solvent cleaning machine.

[152] Sir Fred Hoyle.

5. *Defect concentration diagram*: An image of the parts showing the locations where cleaning is poor.
6. *Scatter diagram*: Any X versus Y relationship between cleaning performance and potential cause.
7. *CUSUM control chart*: To identify the need for action in a timely and positive manner.

All are covered in more detail in Appendix 1. But because of its value, the CUSUM control chart will be described in Section 4.14.2. Complete directions for use of a spreadsheet in preparation of a CUSUM chart are also given in Appendix 1.

### 4.14.2 The CUSUM Control Chart

*If the facts don't fit the theory, change the facts.*[153]

If you use only one of the seven tools for process control of your cleaning system, it should be this one.

There are four reasons. CUSUM charts are: (1) easy to customize, (2) very responsive, (3) illuminating of problem situations, and (4) easy to implement with a spreadsheet.

#### 4.14.2.1 *Shooting Pool (of Variance) with a CU(SUM)*

*People see only what they are prepared to see.*[154]

CUSUM is an acronym for cumulative sum. Statisticians call the cumulative sum "… a random walk with mean zero …."[155]

The CUSUM recognizes deviant behavior.[156] It is of the deviations of the current value of some measurement from the overall average value of that measurement.

The deviations are algebraically added. In this way the CUSUM is responsive to changes of the measurement from above the average to below the average, or the reverse. The CUSUM approximates zero for a measurement which varies aimlessly (randomly) around the overall average.

Where the CUSUM brings value is in recognition of deviation that is not aimless but is systematic. Persistent deviation of a measurement above or below the overall average brings magnitude to the CUSUM. In fact, if that persistent deviation switches from above to below the overall average (or the reverse), the CUSUM can also be persistently offset from zero.

CUSUM, "R", and $\overline{x}$ control charts all use UCLs and LCLs. But there are three points of differentiation between the CUSUM control chart and the "R" and $\overline{x}$ control charts:

1. CUSUM is based on all the data collected after an overall starting point. Individual "R" and $\overline{x}$ values are just that – averaged over a few recent (perhaps hourly) measurements.[157]
2. The CUSUM can be defined to NOT respond to changes the user deems to be insignificant,[158] that is to ignore changes less than a specified amount.
3. There is a single value for range in "R" control charts and a single value for average in $\overline{x}$ control charts. But there are two values for CUSUM in CUSUM control charts:
   - One is of the positive deviations. The other is of the negative deviations. If the deviations are in only one direction, only one CUSUM value gains value different from zero.[159] The other CUSUM value stays at zero.

---

[153] Albert Einstein.
[154] Ralph Waldo Emerson, *Journals*, 1863.
[155] Montgomery, D.C., *Introduction to Statistical Quality Control*, (4th ed.), Wiley, 2001, ISBN 0-47-31648-2, p. 408.
[156] The manager need have no personal fear here.
[157] The CUSUM can be defined so as to place greater weight on more current measurements and less on those from previous operation. Please see Montgomery, D.C., *Introduction to Statistical Quality Control*, (4th ed.), Wiley, 2001, ISBN 0-47-31648-2.
[158] Though it isn't useful, and may be counterproductive in cleaning work, the CUSUM can also be defined so that it is given a "head start" toward exceedance of a control limit. See the above reference.
[159] The CUSUMs are calculated so that this is true for either positive or negative deviations. See Figure 4.21.

- Two CUSUMs exist so that if deviations have been persistently positive (or negative), and then over the short term become negative (or positive), that change can be immediately recognized.

### 4.14.2.2 *The Aim of CUSUM Control Charts*

*Measure what is measurable, and make measurable what is not so.*[160]

Although CUSUM control charts can be used to identify special and common causes of variability, they are better used for process control. This is because:

- $\overline{x}$ and "R" control charts were designed for identification of special and common causes.
- CUSUM control charts were designed to allow timely and justified response to process changes.

If a measurement is worth the resources needed to produce it, examine it with regard to past behavior and your concerns about what its value might mean to you. If it is worth those resources, construct a CUSUM control chart using those values to do just that. If it's not worth those resources to construct the chart, don't take the measurement.

### 4.14.2.3 *A Tale of Two Sums*

*One does not discover new lands without consenting to lose sight of the shore for a very long time.*[161]

Effective use of two sums (called $C^+$ and $C^-$ for the sum of positive and negative sums, respectively) is shown in Figure 4.21.

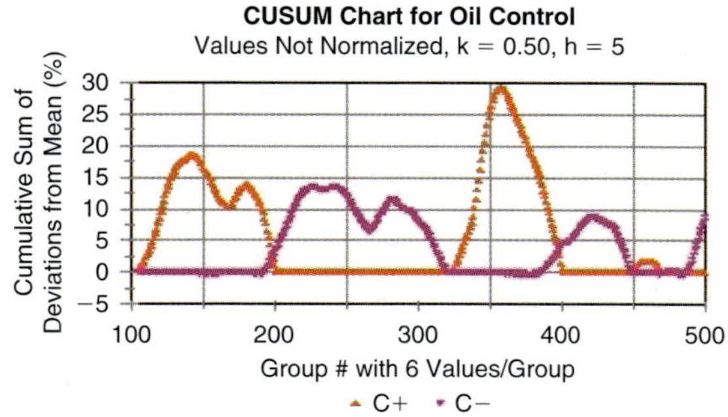


**Figure 4.21**

Examples where change in direction of deviation of measurement from average occur around group numbers 200, 330, 400, 450, and 475.

In each case, the measurement swings from one side of the overall average value to the other side. Note how the sum valued at zero rapidly increases before the positively valued sum declines to zero. Illumination and recognition of behavior around groups #200 and #400 is the reason CUSUM technology is so powerful.

That's why two sums are used – because you can't wait until a single sum of deviations declines to zero.[162]

### 4.14.2.4 *CUSUM Control Charts Produced by a Spreadsheet*

*640K ought to be enough for anybody.*[163]

Honesty requires declaration of a strong preference. This author finds spreadsheets quite useful. All of the numerical examples in this volume were created with a spreadsheet.

One good use is to construct control charts. The reason is that as you accumulate data within a table, the spreadsheet makes it easy to produce a graphical representation of that data.[164]

---

[160] Galileo Galilei.

[161] André Gide.

[162] CUSUM is a graphical technique. It was developed around five decades ago to be used with a piece of graph paper. More significantly, it was developed to allow *visual* recognition of deviations from aim. Using a spreadsheet, the technique can be used to provide any response including automatic action.

[163] Bill Gates, 1981.

[164] The Section on scatter plots in Appendix 1 illustrates the hazard of plotting data from a situation where there is no underlying meaning. While the plot image may appear as if it were designed by an advertising agency, the data on which the image is based may have been produced through the use of random numbers. The outcome can be confusion and mistakes.

A second reason is that all the statistical functions are available for analysis of past and current measurements. While these functions can be used without understanding of the basis behind them, hopefully the information in this volume will prevent that occurrence.

Construction of CUSUM, "R," and $\overline{x}$ control charts is taught in Appendix 1. The specific formulas necessary to construct a CUSUM control chart are shown for entry in each cell of the spreadsheet.

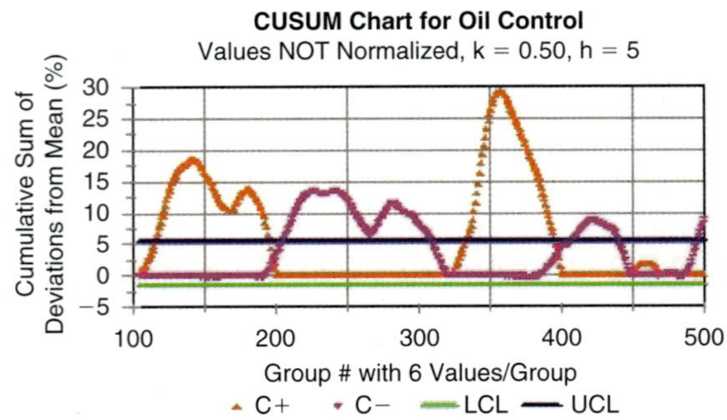


**Figure 4.22**

### 4.14.2.5 *In Control*

*Mathematicians are like Frenchmen: whatever you say to them they translate into their own language, and forthwith it is something entirely different.*[165]

As with "R" and $\overline{x}$ control charts, users want to know when to be concerned. One answer, from Section 4.13.3, is in use of control limits.

UCL and LCL are estimated from a statistical evaluation of the normal overall variation in the measurement.

Users select control limits by specifying the extent to which they want variation to exceed normal overall variation before they are notified. Exceedance of the control limit lines is that notification.

Usually the specification is made as a multiple of the normal overall variation (standard deviation). A number around 3–5 is commonly used for the multiplier.

The same information in Figure 4.21 is plotted with both control limit lines as Figure 4.22. This is a typical CUSUM control chart.[166]

### 4.14.2.6 *Delay, Linger, and Wait*

*Either I've been missing something or nothing has been going on.*[167]

Material in the previous section described how CUSUM control charts can be customized to recognize certain levels of variation through the use of control limit lines.

CUSUM control charts can also be customized to *ignore* smaller levels of variation.

This is done by subtracting a preset quantity from the individual deviations. If there is a remainder, that amount is the deviation used in the summation calculations. If there is no positive remainder, zero is used in the summation calculations.

Typically the preset quantity is a multiple of the normal overall variation (standard deviation). A common value for the multiple is one-half. That is, the deviation between measurement and overall average must exceed one-half of the normal variation (standard deviation) or it is neglected. Values for the multiple seldom exceed 1.

### 4.14.2.7 *Customized Process Control*

*My mother's menu consisted of two choices: Take it or leave it.*[168]

- Do not underestimate the power to ignore variation which we wish to ignore and to respond to that we deem significant.

---

[165] Johann Wolfgang von Goethe.

[166] Note that the units of the horizontal axis are those used in control charts above. Those of the vertical axis are those of the native measurement as CUSUM control charts are not normally normalized.

[167] Karen Elizabeth Gordon.

[168] Buddy Hackett.

- Do not also fail to recognize that poor choices about these amounts of variation can produce silence until failure or unrelieved noise.

That's why CUSUM control charts should be constructed using a spreadsheet.[169] Then the values of variation we wish to ignore and to which we wish to respond can be re-selected at any time. The CUSUM plot can be redrawn, and the effect of our choices re-examined in real time.

In another sense CUSUM control charts are excellent tools for communication and identification of areas of dispute. Members of a group can see the effects of their opinions about sensitivity of process control when the CUSUM control chart is plotted on a spreadsheet.[170]

#### 4.14.2.8 *Nothing Common or Systematic*

> *The great general is the man who stakes everything on his hypothesis while realizing that it is only a hypothesis and must be modified from moment to moment.*[171]

Note that CUSUM control charts are not used to identify and eliminate either common or system causes. That is presumed being done simultaneously with "R" and $\overline{x}$ control charts using the same measurement data within the same spreadsheet program.

CUSUM control charts are used to identify, justify, validate the need for action – not necessarily to identify what that action should be. Replacing opinion with statistical reasoning makes it difficult for those who speak most loudly to dominate discussion.

As to what should be controlled, see Tables 4.6 and 4.7, in which process inputs and outputs are described, respectively. Also see Section 4.26 about troubleshooting.

## 4.15 PROCESS MANAGEMENT

> *Depend upon it, there is nothing so unnatural as the commonplace.*[172]

Process management is the use of process control, and other technologies, to achieve the aims of the enterprise. And the chief aim of any enterprise is to avoid problems so that profit can be made.

Process management is not process control with a fancy suit and higher price tag. It is more the opposite. It is using your nose, eyes, touch, and mind to avoid and solve problems. If you manage a solvent or aqueous cleaning machine, you'll meet the aims of your enterprise if you can anticipate problems in it or which it might cause, and keep them from happening.

### 4.15.1 Mission of Management

> *Computers make it easier to do a lot of things, but most of the things they make it easier to do don't need to be done.*[173]

Cleaning operations contribute to an enterprise in two ways:

1. **By allowing product to meet both the needs of end users and their expectations**. Some products won't properly function if their surfaces are not clean. And no end-use customer will be satisfied with a product which contains the residue of how it was produced.
2. **By eliminating some problems inherent in the production system**. Many products can't be properly formed, milled, inspected, assembled, etc., if their surfaces are not clean.

---

[169] Raw measurements should be recorded on paper, and keyed into appropriate cells in the spreadsheet. The paper record allows audit for a short-term period. An electronic backup of the spreadsheet should be the copy for reference and long-term audit.

[170] Yet, the basic values for $h$ and $k$ recommended in the Appendix 1 should be discarded only when there is sufficient reason for doing so.

[171] John Burdon Sanderson Haldane, *Possible Worlds: The Duty of Doubt*, 1927.

[172] Sir Arthur Conan Doyle, *The Adventures of Sherlock Holmes: A Case of Identity*, 1892.

[173] Andy Rooney.

In other words, the reason management of an enterprise supports the investment in and use of a cleaning system is to directly produce profit and to avoid problems which affect profit. Operation of that system must contribute to profit and eliminate problems or it will not exist within the enterprise.

Clean parts bring little or no inherent value to an enterprise. Clean parts bring value only when cleanliness allows further function or relieves difficulty.

Never assume a cleaning machine is valued because of what it is. It is valued because it produces or allows profit and avoids problems.

## 4.16 MAINTENANCE

> *All parts should go together without forcing. You must remember that the parts you are reassembling were disassembled by you. Therefore, if you can't get them together again, there must be a reason. By all means, do not use a hammer.*[174]

Do not discard the maintenance manual provided by the machine's manufacturer. It may be your only chance to know how the manufacturer intended the machine to look and function.

Manufacturer's literature to the contrary, a cleaning machine is often more a collection of components than an integrated system.[175] This means that your attention should be periodically directed toward the half-dozen or so components crucial to the machine's function rather than the machine itself. For a solvent cleaning machine, they include:

- The hoist and its programmable controller (see Section 7.8).
- The heating element and its programmable controller (see Chapter 7, Section 7.7).
- The refrigeration package.
- The distillation package.
- The hardware for temperature measurement (thermocouples).
- Level indicators and switches, particularly the low-level alarm for liquid in the boiling sump.

The components will likely be made by different manufacturers and be useful when integrated into products other than cleaning machines. You should collect the maintenance information (manual), contacts for local support, and parts lists for each of these items. This collection should be saved, used, and relied upon.

### 4.16.1 Clean Maintenance

> *Sirs, I have tested your machine. It adds a new terror to life and makes death a long-felt want.*[176]

The advice in the previous section is straightforward and you did not need to purchase this book to learn it.

From this author, the best advice about maintenance for your aqueous or solvent cleaning machine is to not treat it like a garbage can. This too commonly happens.

Cleaning machines of indeterminate age often appear rusted even when made of stainless steel, when off-line have an layer of oil on top of the black-colored cleaning agent and a heap of parts below it, have a water separator full of water, reek of something vile, have frosted cooling coils, are topped with an ill-fitting cover, and contain a bent parts basket. Occasionally, one has to step over something to access the machine.

The reason is all too simple – a bad attitude. Soil enters the machine without limit. The machine's owners treat it as if it were a repository for that soil – as long as enough clean parts are produced.

The best maintenance strategy is keep your cleaning machine[177] looking clean. Make it look as if clean parts will emanate from it:

- When the first rust appears, check the cleaning agent pH. Check the stabilizer level. Investigate

---

[174] *IBM Maintenance Manual*, 1925. This approbation flies in the face of advice tended to me at an early age by my father – if it won't go, get a bigger hammer!

[175] See Chapter 7 for a lengthy coverage of which components (pumps, nozzles, tanks, etc.) are found in top quality cleaning machines.

[176] Sir Herbert Beerbohm Tree (to a gramophone company who asked him for a testimonial).

[177] Whether it be based on aqueous or solvent technology.

why stainless steel is corroding. It shouldn't – that's why you purchased one made of stainless steel.

- Examine the cleaning agent when the machine is shut down. If it's oily, investigate why the distillation/decanting system isn't being used or isn't working.
- Check the color of the cleaning agent. If it has one, analyze it to identify the colored specie. Investigate why that specie isn't being removed by distillation, filtration, or some other way. If you don't remove the colored specie, it will stay there forever.
- Go fishing. Dredge the bottom of all sumps for parts. You're sure to find some because parts baskets are always being overfilled. If debris keeps your parts baskets from being fully immersed with cleaning agent, there is no reason to operate the machine.
- Look at and into the water–solvent separator or oil–water separator. It should be partially full of water when performing well – but not every day after drainage. For a solvent cleaning machine, the likely source of water is humidity, but parts can also be wet with water. If the solvent–water separator is solvent-laden, you probably have an unexpected soil component which is acting as a surfactant (making solvent compatible with water).
- An uncharacteristic odor should motivate you to identify it and eliminate it rather than tolerate it.
- If the cooling/condensing coils are white with ice, your defrost cycle isn't working and the cover is allowing too much moist air to enter.
- Observe the cover when the machine is used. Is it a seal or an obstruction?
- Observe the parts basket. If it looks as if it had been in a train wreck, you should make that happen and get another.
- If the condition of the cleaning machine is too shabby for it to be demonstrated to the plant manager, you're not a candidate for his job.

In summary, place the same value on the condition of your cleaning machine as you do about the cleaned parts it produces. That's a simple and effective maintenance program.

## 4.17 OPERATIONS

> *We trained hard ... but it seemed that every time we were beginning to form up into teams we were reorganized. I was to learn after ... that we tend to meet any new situation by reorganizing, and what a wonderful method it can be for creating the illusion of progress while actually producing confusion, inefficiency, and demoralization.*[178]

It is an oversimplification to write that a solvent or aqueous cleaning machine with a programmable hoist should almost "manage itself" – just place the parts into their carrier(s) and remove them when cleaning is complete.

The truth involves preparation and training – especially training. You and your staff shouldn't operate a vapor degreaser or an aqueous degreaser until you have thoroughly studied the manufacturer's instructions. Then copy them, store the original, and build a effective training program.

## 4.18 STAFF TRAINING

> *The best way to have a good idea is to have a lot of ideas.*[179]

But that's just the starting point. Learning is most effective when it is:

1. Self-taught
2. Put in to practice

A manager's training program should be built around those two principles.

First, write the operating manual – including maintenance information and safety information specific to the working solvent. Here are some guidelines:

- Preparation of the manual should be a team project.[180]

[178] While often elloneously attributed to Petronius Arbiter Circa 27-66 A.D., it earliest reference dates only to 1970. Its true author is unknown. See http://en.wikipedia.org/Petronius Arbiter (and others).

[179] Linus Pauling.

[180] An operating manual should never be written by one person (who has the perspective of a single experience), or the person who designed the item described in the manual (who frequently underestimates the difficulties contemplated by an inexperienced user). Software manuals often seem to have been produced this way.

- Rephrase the manufacturer's general and specific guidance into specific guidance for you and your staff.
- Use methods of illustration and jargon familiar to all.
- Amplify areas where some have concern.
- Set specific limits and parameters versus general suggestions.
- Images are easy to produce with a digital camera and graphics software. Use them early and often.
- Condense pertinent material into a single page which can be copied repeatedly, sealed in plastic, and located everywhere.

Then, assemble a "How Manual"[181] which should include:

- The single page.
- The comprehensive operating instructions.
- The comprehensive maintenance instructions.
- A copy of materials provided by the manufacturer.
- Reference information about the components from which the machine is assembled (see Chapter 7).
- The MSDS for all chemicals used in the area.
- A list of specific hazards associated with use of this machine.
- Documentation of when and where each person associated with the machine was trained using these materials.[182]
- The troubleshooting and problem solving materials you develop from Section 4.26.
- The cleaning test procedures and specifications associated with your machine and parts (see Chapter 5).

### 4.18.1 "Dummy Running"

*Multiplication is vexation,*
*Division is as bad;*
*The 'Rule of Three' doth puzzle me,*
*and Practice drives me mad.*[183]

Second, "dummy run"[184] the operating instructions by using them and the cleaning machine. This is especially important when the machine is first used after purchase or off-site maintenance.

You should spend a modest amount of time using the training and the manual to operate the machine. This is done in stages. "Dummy running" might require only one-half to a full day of commitment, but will pay dividends when mistakes aren't made. For a solvent cleaning machine, the sequential stages should include operation:

1. **Without parts or fluid**: This involves checkout using a primary control – the programmable hoist – and the parts transport system. Problems here, with a machine full of hot boiling liquid and valuable parts, can be significant problems. Also, adjustment of some programmable hoists can be confusing. Do not skip this stage!
2. **With parts and water**: This involves the heating and cooling systems, including distillation. It is likely that the heating and cooling capabilities are matched to the cleaning solvent and not water. This is not a concern if operation is not stable with water. The aim is to develop personal familiarity/comfort and checkout equipment functionality.[185]
3. **With cold solvent**: This involves safe solvent handling and transfer, AND initial checkout of personal protective equipment (goggles, gloves,

[181] This is jargon for a portable and changing volume that is a compendium of information about *how* to do every task associated with a device.

[182] In the US, an excellent way for a firm to fail a site inspection by the Occupational Safety and Health Administration (OSHA) is to not have an available file of MSDS information, not have done a process hazards review, or not have documentation of training.

[183] Anonmyous *Elizabethan Manuscript*, 1570.

[184] This again is jargon. It refers to startup training. The phrase is not a characterization of the persons operating the equipment. It means that the equipment is to be tested using simulated conditions. For example, new operators will normally "dummy run" a refinery using water to verify equipment functionality, and their training.

[185] A major requirement of, and drawback to, operation with water is that a solvent cleaning system must be thoroughly dried of water before production cleaning work is started. This is sequentially done by draining water and allowing time for evaporation, and then using the normal distillation and water separator equipment.

aprons, etc.). If not previously done, the fire alarm and extinguishing system, the ventilation system, and the evacuation system should all be operated though they shouldn't (hopefully) be needed.

4. **With boiling solvent and no parts**: Basically, the above three stages of training are first put into practice by the trainees, though with an empty parts basket. Confidence and familiarity are the obvious goals here. Operation of the water separator should be evaluated as the last traces of water from Stage 2 are removed.
5. **With boiling solvent and distillation**: For the first time, operators get an opportunity to implement the distillation system. The only soil should be "system dirt."
6. **With all systems on and clean parts**: Here an inexperienced staff has the opportunity to witness what is more easily shown than explained – the impact of cold parts upon a hot cleaning and drying environment. Process stability becomes earned through action and reaction rather than learned through study. All must participate here and convert the unstable to stable.
7. **With all systems on and normal parts**: At this point, nothing significant should happen. Staff should be at least familiar and somewhat comfortable with equipment and procedures. Further, significant soil accumulation requires between a shift and several days of operation.

### 4.18.2 The Difference Between Training and Participation

*I often say that when you can measure what you are speaking about, and express it in numbers, you know something about it; but when you cannot measure it, when you cannot express it in numbers, your knowledge is of a meagre and unsatisfactory kind.*[186]

There is a third principle of training: audit. While OSHA only requires operating staff to be exposed to information,[187] you should insist on a higher standard: participation.

Set some metrics (minimums) for each of the above seven stages of "dummy running." All operating staff must meet these requirements or repeat the training.

Here are some suggestions for these metrics,[188] for the stages listed in Section 4.18.1:

| | |
|---|---|
| Stage A | Timing of hoist operation stages versus a goal after adjustment from unprogrammed settings |
| Stage B | Stability of measured temperatures |
| Stage C | No spills |
| Stage D | No odor |
| Stage E | Stability of measured temperatures |
| Stage F | Stability of measured temperatures |
| Stage G | Cleanliness of parts versus cleanliness specifications |

In summary, as with maintenance, place the same value on training your staff as on the cleaned parts your machine produces. Poorly trained staff can ruin the performance of any machine and, even worse, can cause a safety or environmental incident.

## 4.19 STARTUP OF A CLEANING MACHINE

*There's no sense in being precise when you don't even know what you're talking about.*[189]

Whether for the first time or after a weekend's down time, startup of a vapor degreaser, or other cleaning machines should be a non-event.[190]

[186] William Thomson, Lord Kelvin, *Lecture to the Institution of Civil Engineers*, May 3, 1883.
[187] Via their Hazard Communication Program.
[188] Obviously, more than operator skill is being tested here. Machine design is also being evaluated. Metrics should be pertinent to local needs and those for multiple training stages should be combined where justified. The point isn't that the metrics above are wrong for your operation, it is that you should develop and use those right for your operation.
[189] John von Neumann.
[190] If it's not, consult the manufacturer's representative ASAP!

The following 10 steps are general and should apply to nearly all machines:[191]

1. *Use your senses*: Inspect visually. Notice any unusual aroma type or level in the area. Recognize hot or chilled surfaces which shouldn't be so. "Dummy run" the programmable hoist. Check the alignment and seal of the cover.
2. *Check the solvent*: It is what's doing the work. Recognize color change, suspended material, or uncommon aroma. Fill all compartments as necessary.
3. *Be in control*: Power all control devices.[192] Be sure all safety controls (interlocks or over-rides) are active.[193]
4. *First, chill out*: Turn on water flow to cooling coils. Turn on self-contained or external refrigerant package systems and initiate flow to condensing coils. Verify measured temperatures and adjust controllers as necessary to attain goal temperature values.
5. *Then, get hot*: Activate heat supply (gas, hot water, electricity, etc.).
6. *Rise to the clouds*: Periodically open the cover and observe generation of the vapor blanket above the boil sump(s). It should rise above the sumps. Adjust (trim) heat supply so that the top level of the vapor zone is around the midpoint of the condenser coils. It should never be close to the top of the condenser coils.[194]
7. *Stay in control*: Check all instruments. Adjust any values not in proper ranges.
8. *Go*: Fill parts baskets and initiate basket transports. Start parts cleaning work. Among other outcomes, this will properly cause the vapor blanket to condense[195] and shrink downward towards the boiling zones.
9. *Equilibrate*: Increase heat supply to boil sump(s) so that the top level of the vapor zone is around the midpoint of the condenser coils. Normally this is a manual adjustment.[196]
10. *Inspect or audit*.

Some users may adopt a simpler startup procedure. Turn on the condensing/cooling and heating supplies, turn on the part transport apparatus, and depart. Certainly, we automobile drivers don't complete a startup checklist before we depart for work in the morning. But pilots of airplanes are required to complete a checklist before takeoff. Truth for users lies within those extremes and is determined by how well clean parts are valued by the enterprise, and the experience of the operating staff.

---

[191] While these instruction are written specifically for solvent cleaning machines, managers of aqueous cleaning machines will find the same 10 stages necessary as well. For example, the aim of Stage 6 is to observe that the interface of the cleaning agent with the equipment is in proper condition and stable. For an aqueous cleaning machine, verify that the top surface of the wash liquid is clean and free of supernatant soil.

[192] This is not a trivial suggestion. A past client applied electric heat to a vapor degreaser to initiate boiling without having first pressurized the pneumatic-powered automatic control system. Naturally the system boiled dry – because the coolant valves couldn't be powered open. Fortunately there was no fire.

[193] With some electrically heated systems, safety controls aren't activated until the main power switch to heat supply is engaged. This is not a desirable feature but is common.

[194] Uncondensed vapor will be entrained by upward movement of the parts basket, and possibly buoyancy versus air, and can escape the machine when the cover is opened for parts removal. The local fire marshal may call this situation a fire hazard. The local environmental permit writer may call this situation an exceedance.

[195] That's why it's called vapor degreasing.

[196] A properly designed solvent cleaning machine will have an adequate imbalance between heat supply and condensing capacity so that the top level of the vapor zone is always within the condensing coils. This minimizes the possibility for solvent emission and maximizes energy use by the cleaning machine. The imbalance is necessary so that the vapor blanket may be sustained when cold parts enter and properly cause vapor to condense.

This equilibrium can be destroyed if unusual parts are cleaned – those whose thermal capacity (mass times specific heat) is larger than that upon which the design is based. In this case, the top of the vapor blanket will shrink below the condensing coils. Parts will not likely be well-cleaned as there will be no vapor degreasing:

- If heat supply is already at maximum, these parts can't be cleaned in this machine.
- If heat supply is not at maximum but condensing capacity is at maximum, there may be an emission of solvent vapor from the machine.

A trivial example is that a machine designed to clean thin metal stampings probably won't clean engine blocks of the same soil. Another trivial example is the reverse.

## 4.20 FIXTURING (RACKING) PARTS FOR CLEANING

*I like work; it fascinates me. I can sit and look at it for hours.*[197]

Some parts can be "thrown" into a parts basket and the parts will be well treated when the parts basket is inserted and removed from the cleaning machine. Cannonballs may be an example of such parts.

However, most parts must be arranged when exposed to the cleaning process so that all surfaces are well treated by both the liquid cleaning agent and the solvent vapor, or the aqueous cleaning agent. Coffee cups are certainly an example of parts which must be fixtured.[198] For a solvent cleaning machine:

- If the parts aren't fully contacted by the liquid (not vaporized) solvent, they won't be fully cleaned.
- If the parts aren't fully contacted by the vaporized (not liquid) solvent, they won't be fully dried.

Part arrangement is always a compromise between productivity and practicality. The perfect arrangement of parts would be one part per basket in a basket with no surfaces contacting the parts. The compromise should, to the extent practicable:

- Assure complete contact with the liquid cleaning agent when parts are immersed or sprayed. Parts, including cannonballs, shouldn't contact one another on any surface. Minimum contact with the fixturing device is preferred, but not always possible.
- Allow full drainage of the liquid solvent from all surfaces of the parts when they are removed from the immersion sump. This includes internal recesses and external crevices. That's why coffee cups should not be arranged open side up.
- Forbid soil-laden solvent to contact parts sited lower within the parts basket.

Retained (trapped) cleaning agent will increase dragout, which reduces the efficiency of the rinsing process. Soluble soil within retained cleaning agent will likely become a cleaning defect. See Chapter 1, Sections 1.12.5 and 1.12.6.

## 4.21 OPERATING A CLEANING MACHINE

*For every complex problem, there is a solution that is simple, neat, and wrong.*[199]

After startup, a properly designed and maintained cleaning machine will almost "manage itself" or behave normally unless part or soil character is drastically changed.

That's true. *Except that sometimes it won't.*

Normal operating behavior is described in Sections 4.18.1 and 4.19. Behavior outside normal should be identified. The cause of it should be identified as special or common, and appropriate action should be taken to eliminate or moderate the cause. The purpose of Sections 4.9–4.13 is to provide skills with which to recognize aberrant behavior, and take action to correct it.

Self-management won't be effective when some factor upstream or within the cleaning machine changes excessively due to a common or system cause. That's why control charts (see Section 4.13) are useful. They are easy to produce with a spreadsheet (see Appendix 1), and should be a normal operating behavior.

## 4.22 INSTRUMENTATION NEEDS

*A computer lets you make more mistakes faster than any invention in human history, with the possible exceptions of handguns and tequila.*[200]

---

[197] Jerome K. Jerome.
[198] The word fixture is used here as a verb. One fixtures parts within a parts basket for maximum effectiveness. The word fixture is also commonly used (ungrammatically) as a participle as above. A synonym for fixture is "rack". Both are used as verbs and adjectives to mean to arrange for maximum effectiveness.
[199] Henry Louis Mencken.
[200] Mitch Ratcliffe.

Instrumentation (electrical or mechanical) for cleaning machines is of three general types:

1. ***Safety Action***: used to protect operators, or the environment.

Normal behavior for a solvent cleaning machine is characterized by the following interlocks[201] being energized (not bypassed) and effective:

- A safety cutoff switch on heat flow – activated by rise in vapor temperature immediately above the chiller zone.
- A safety cutoff switch on heat flow – activated by rise in temperature in the boil sump.
- A safety switch which allows full coolant flow – activated by rise in temperature in the boil sump.
- A safety switch on heat flow – activated by decline in liquid level in the boil sump.
- Appropriate safety switches if the heat supply is not electricity (a pressure regulator on steam supply, or a detector on the gas pilot light).

Each of these devices should have a manual reset. Operating instructions should identify the few persons who allowed to implement the reset and when they may do so. Manual resets for interlocks are not "*Get out of Jail Free*" cards.

2. ***Information***: used to demonstrate that the machine is operating properly.

Normal behavior is characterized by the following controlled process outcomes:[202]

- Work insertion rate controlled via programmable (or manual) hoist at a rate not greater than 11 feet/minutes (5.6 cm/s).[203]
- Cooling water temperature is normal for the area and season[204] (probably between 80°F (27°C) and 110°F (43°C).
- Vapor temperature immediately above the chiller zone is no more than 30% of the normal boiling point of the solvent.[205]
- Vapor temperature immediately below the condensing zone is equivalent to the solvent boiling point.
- Temperature of the boiling solvent in the cleaning sump does not exceed the normal boiling point by more than about 8°F (5°C).[206]

3. ***Quality or Production Control***: used to demonstrate that parts cleaning is being done properly.

Normal behavior is characterized by the following controlled process outcomes:

- Timing (dwell) for the various actions of which the cleaning cycle is composed: vapor degreasing, immersion, rinsing, and drying as per operating instructions. These time values are normally controlled by the programmable hoist or by an operator's action.

---

[201] These devices are conventionally referred to as interlocks, safeties, or safety switches. Occasionally, they are called some unprintable name when they impede an operator from taking precipitous action which is highly desired at the moment but not within operating procedures.

[202] For a solvent vapor degreaser.

[203] This rate is a limit required for emission control by the US EPA's NESHAP (US CFR Vol 65, No. 197, September 8, 2000, or Guidance Document EPA-453/R-94-081) for vapor cleaning equipment. The basic idea is not to entrain upward or displace downward solvent vapor by moving the parts basket at a high rate of travel.

[204] Remember, the temperature of cooling water from a cooling tower is related to the outside air temperature. This is also true for cooling water supplied from and returned to a river.

[205] This limit is also specified in the US EPA NESHAP for halogenated solvents (Guidance Document EPA- 453/R-94-081, pp. 2–23). But the approach is recommended by this author for all vapor degreasing solvents as a way of minimizing emissions.

For example, if IPA is the cleaning solvent, the temperature above the top refrigeration coil should not exceed 54°F (12°C). This recommendation implies, and strongly suggests, that a refrigerated freeboard coil (RFC) be used as the upper condensation device to minimize emissions (versus a coil with cooling water as the heat sink).

This measurement can be easily made by attaching a thermocouple probe with flexible leads to the parts basket when the machine is operating but not processing parts.

[206] This limit is artificial, and basically specifies the maximum soil level allowed in working solvent. You should select your own limit, based on the quality of cleaning your value. See Sections 4.8–4.14 for guidance about controlling soil level and cleaning quality.

- Temperature of the non-boiling solvent in the rinse sump is as per operating instructions. This is probably lower than that within the boil sump.
- Ultrasonic transducers are powered. This includes setting of power level as per operating instructions and setting of sweep frequency as per manufacturer's recommendation.

### 4.22.1 Human Instrumentation

*Discovery consists of seeing what everybody has seen and thinking what nobody has thought.*[207]

Not all information comes through a wire, pneumatic tube or as a radio signal. Some are provided by those associated with the cleaning machine.[208] Normal behavior (for a vapor degreaser) is observations of:

- Water flowing (dripping) from the refrigeration coils into the water separator.
- Clear effluent overflowing on the top of the water separator, and no material trapped between the water and the solvent layers.
- Full level in the condensate tank or clean rinse fluid storage compartment.
- No unanticipated (hopefully none) odor in work area.
- No missing parts – the parts basket or hooks always appear on exiting the machine as they do on entering it.
- Overflow weirs not fouled.
- Spray nozzles properly arranged for specific parts and not blocked.
- Condensed solvent is flowing to the water separator, and through it to the proper degreaser compartments.[209] Water overflow is continuous and not blocked.
- No observation of droplet sprays[210] being emitted when the cover is opened for work entry and removal.
- Not all cleaning tests are described in Appendix 2. An operator, or innocent bystander, may notice parts being not well cleaned on exit from the machine. That observation should be taken as equivalent to an analytical test result until the contrary is proven.
- No unexpected incidents which imply either that operation is not fully understood or that a special or common cause must be detected and eliminated/moderated.

## 4.23 USE OF INFORMATION

*"Wisdom consists in being able to distinguish among dangers and make a choice of the least harmful."*[211]

*"Nothing has really happened until it has been recorded."*[212]

This author has made a deliberate decision not to recommend a specific data sheet for aqueous or solvent cleaning operations. There are two reasons for this omission: (1) a desire for simplicity as expressed by Occam's Razor or the KISS principle in Section 4.11, and (2) diversity among applications[213] means "one size does *not* fit all."

However, a fine starting point would be a data sheet (or check sheet as described in Appendix 1) based on the generalized process inputs given in

[207] Albert Szent-Gyorgyi, in I.J. Good, *The Scientist Speculates,* 1962.

[208] Occasionally, needed information is provided by others who are not associated with the cleaning machine, but happen to be unintentionally and negatively affected by it. When this happens, plainly a special or common cause has acted and should be removed/moderated before operation is allowed to continue.

[209] With some machines, distilled solvent flows directly to storage compartments and not first to the water separator.

[210] This is because all spraying should be done low within the vapor zone, or within the liquid immersion tanks.

[211] Niccolo Machiavelli, *The Prince*, 1513.

[212] Virginia Woolf.

[213] Batch atmospheric or vacuum vapor degreasers, continuous cross-rod vapor degreasers, or batch cold cleaning degreasers would all have different requirements about information.

Table 4.6 and the generalized process outputs given in Table 4.7. To supplement that you should consider Section 4.22:

- Safety/environmental information about interlock status. There should be a requirement for an examination of interlock functionality at semi-annual or annual maintenance shutdowns.
- Human-recognized information.
- Instrumental information.

However, there is a clear recommendation: don't record information which is not worth entering in a spreadsheet program and comparing to a previously established metric. This consultant has seen too many users keep nearly all information, or essentially no information – and then not manage their affairs with the use of that information. In other words, use it or lose it!

## 4.24 PART TRANSPORT

*Hard work never killed anybody, but why take a chance?*[214]

The component most likely to fail in a cleaning system is the part transportation (transfer) system.

The reason is that it has the most moving parts. That means it may also be the most expensive subsystem in your cleaning process.[215]

While cleaning quality is determined by events occurring within your aqueous or solvent cleaning machine, productivity is likely determined by events occurring outside it. Both quality and productivity are necessary for your enterprise to profit.

## 4.25 IDLING MODE VERSUS SHUTDOWN

*I think we consider too much the good luck of the early bird, and not enough the bad luck of the early worm.*[216]

Degreasers, whether aqueous or solvent powered, are more like lovers than automobiles when activation is considered. Automobiles can (usually) be started and shut off on demand. Degreasers, like lovers, require support prior to and after contact.

### 4.25.1 Vapor Degreasers

*Start by doing what's necessary; then do what's possible; and suddenly you are doing the impossible.*[217]

Since solvent vapor degreasers use boiling solvent, time must be provided during the startup phase for the solvent to be heated to boiling. After use, the situation is reversed. Heat-up and cool-down time are normally measured in hours.

Vapor degreasers are often placed in standby or idling mode[218] when not needed, rather than being fully shutdown between scheduled works. To conserve energy, idling mode normally involves heat input sufficient to overcome heat losses while maintaining sump temperature at the boiling point. This saves energy two ways: nothing is boiled and there is little boiled material to condense.

A solvent vapor degreaser in shutdown mode has no heat energy input and no removal of energy via coolant.

---

[214] Charlie McCarthy (Edgar Bergen).

[215] One can add 25–100% of the price of new cleaning process by the choices one makes around automation of functions, especially part transport.

This author is not convinced that this level of expense is necessary in order to insure reliable operation or reduce labor costs. Recent improvements in the price and capability of sensors, robotic placement devices, miniaturization, and wireless data transfer have made it possible for users to retrofit automated part transfer capability for a cost considerably less than the above. It's not clear that these changes have been embodied in current pricing of new automated cleaning processes.

[216] Franklin D. Roosevelt.

[217] St. Francis of Assisi.

[218] The US EPA's NESHAP covering solvent cleaning machines recognizes different emission rates from solvent degreasers operating in idling or working mode. See http://www.epa.gov/ttnatw01/degrea/haloforms.pdf

### 4.25.2 Aqueous Degreasers

*There is no abstract art. You must always start with something.*[219]

Aqueous degreasers also have a heat-up cycle. Similar times (hours) must be allotted, though for a different reason. Aqueous degreasers don't operate near the boiling point of water; however the volume of fluid which must be heated is typically a decade larger than that for a solvent degreaser.

But the biggest change affecting an aqueous degreaser, which is not found in solvent degreasers, is in the integrity of the detergents and other additives.[220] These chemicals are the lifeblood of an aqueous degreaser. Their performance is strongly dependent upon temperature. The issues around integrity are:

- Solubility in water.
- Formation of micelles between hydrophobic and hydrophillic species.
- Evaporation of water which produces enrichment of detergents and other additives.[221]

Depending upon the level of retained soil, shutdown of an aqueous degreaser can leave the detergents and additives commingled with soils on the water surface. If this debris removed as it should be, so also are the detergents and additives.

When the degreaser is restarted, and heat is applied, the surface goods won't magically materialize into complete solution. Circulation with a pump will be necessary. And so also will be replenishment of the detergents and other additives.

A key difference between aqueous and solvent degreasers is that a composition adjustment is likely necessary upon startup of an aqueous degreaser.

## 4.26 SOLVING OF PROBLEMS

*Machines should work. People should think.*[222]

While Appendix 1 describes recognition (via control charts) and identification (by observation and logic) of special and common causes, little mention is made of what to do about them. The former is troubleshooting. The latter is problem solving.

Actions found practical to solve problems with solvent degreasers are given in Table 4.12.[223]

Despite the guidance in Table 4.12, the best method for solving problems with cleaning equipment, politics, or relationships is to return to what has worked.

In the case of cleaning equipment, operating the process with "on-aim" control using the process inputs in Table 4.6 is preferred to all other methods of troubleshooting or problem solving.

## 4.27 DWELL TIMES

*To compose music, all you have to do is remember a tune that nobody else has thought of.*[224]

There is a rhythm or pace to the operation of degreasers – whether aqueous or solvent. What they do is done in stages. Each stage is a unit of the entire cleaning process. The sequence of stages doesn't change with urgency or identity of the process operator (or it shouldn't). The stages are scheduled to each require a certain time – because of a certain reason. Elapsed time for each stage normally changes as does the nature of the part.

---

[219] Pablo Picasso.
[220] Subcooled solvents are still useful cleaning agents. The reason solvents are used in boiling service is so that vapor can be produced and condensed as pristine rinsing agents.
[221] Enrichment is not necessarily a positive. Detergents hold hydrophobic soils in water only over a range of composition.
[222] Larry Lorenzoni.
[223] The assumption behind Table 4.12 is that the proper solvent has been selected and that the cleaning machine has been properly designed, but may not be properly operated or maintained.
[224] Robert Schumann.

**Table 4.12 Actions to Solve Common Problems with Solvent Degreasers**

| Observation or Deduction | Proposed Action |
|---|---|
| Emissions must be too high because measured concentrations in the local atmosphere always exceed limitations, or operators complain of solvent odor. | 1. Check exit and entry rate of parts basket versus NESHAP limit of 11 feet/ minute.<br>2. Is parts basket area more than 50% of degreaser area?<br>• Either would cause solvent vapors to be entrained or forced from the machine.<br>3. Check temperature above refrigeration coils. It should be no more than 30% of the solvent normal boiling pont. Boilup and condensation may not be in balance.<br>4. Check cover of degreaser. Is it well sealed? Do operators close it after entry and removal of parts baskets?<br>5. Check solvent properties for formation of a low-boiling azeotrope with water. |
| Cleaning (versus rinsing) quality is poor. This involves a judgment.<br>Poor cleaning and poor rinsing have different symptoms.<br>With poor cleaning, debris is *usually* located throughout all parts or in specific sections of some parts. | It is likely that either the solvent is not properly contacting the parts, or the parts are being reinfected after cleaning is complete.<br>1. Check boil sump temperature:<br>• Too low means solvent is not boiling and heater output is limited.<br>• Too high means excessive soil accumulation. If so, change solvent or check operation of distillation.<br>2. Check vapor blanket temperature (just below cooling coils) after entry of cold parts:<br>• Too low before immersion of basket means cleaning with condensate is reduced. The cause is imbalance between thermal mass of cold parts and heat supply during boilup. Try a reduced part loading in basket and check for heater limitation.<br>3. Extend dwell times – both in the vapor rinse and possibly liquid immersion zones. Parts should remain in the vapor rinse zone until all condensation stops. See Section 4.27.<br>4. Input soil loading may have changed in mass, type, or method of attachment.<br>5. Investigate solvent quality – acid may have been generated via reaction with water, or a new soil component may be an impurity. |
| Rinsing (versus cleaning) quality is poor.<br>With poor rinsing, debris is *usually* randomly located spots. | 1. Check rinse sump for soil accumulation via dragout.<br>2. Check agitation level via either/both liquid sprays or ultrasonic transducers.<br>3. Check temperature in rinse sump. Lower temperatures mean more vapor can condense on parts as the final solvent rinse in the vapor rinse zone.<br>4. Extend rinse dwell times in rinse sump and vapor degreasing zone. |
| Solvent dragout too high, causing rinsing to be poor or require too much time/solvent.<br>Investigate to learn if cupped sections or crevices trap solvent. | 1. Review how parts are racked or fixtured in baskets.<br>2. Can rinse nozzles be redirected?<br>3. Can parts be disassembled or redesigned? |
| Sprays of solvent observed outside machine. | 1. It is likely that the cover is not providing a seal.<br>2. Vapor spray nozzles are probably located too high in the degreaser. They should be at least 6 in below the vapor level.<br>3. Is excessive pump pressure being used? |

(Continued)

**Table 4.12 Actions to Solve Common Problems with Solvent Degreasers (Continued)**

| Observation or deduction | Proposed action |
|---|---|
| Parts aren't well-dried.<br>This probably means liquid droplets are seen as thin remaining surface films of solvent that will readily disappear via vaporization. | 1. Check vapor temperature as it may be below the normal solvent boiling point due to reduced heat input.<br>2. Extend dwell time in vapor zone. However, drying rate will only change asymptotically with increased time.<br>3. Review situation above with solvent dragout. |
| Water separator is flooded with water, and remains so. | Though water can enter machine in a variety of ways, a major one can be parts whose sections trap water. |
| Rusting of equipment (condensate trough) or parts.<br>Water spots on parts.<br>Degreaser has generated acid ("gone acid"). | Check stabilizer level.<br>Add fresh solvent which contains virgin stabilizer.<br>Reduce water input. |

If the rhythm of solvent vapor degreasing is that of a tango or of Irish step-dancing, the rhythm of aqueous degreasing is that of slow waltz.

### 4.27.1 Solvent Vapor Degreasers

*Thus, the task is not so much to see what no one has yet seen, but to think what nobody has yet thought, about that which everybody sees.*[225]

There are seven steps (stages) in a vapor degreasing cycle. Elapsed time for any or each is referred to as dwell time for that steps (stages).

The machine's manufacturer will have made general recommendations which may or may not have been based on experience of testing your parts with your soils. Absent such test experience, this author can do no better, but can explain the reasons behind this guidance.

The steps are shown in order of occurrence in Table 4.13.

Note that if the immersion rinse is completed at the same temperature as immersion cleaning, there will be no final vapor-produced rinse with condensate. This is because the parts will be at the solvent boiling point and there will be no temperature difference to cause condensation.

Finally, an increased cycle time is not always better than a shorter one. If problems described in Table 4.12 aren't solved, an infinite cycle time won't help matters.

### 4.27.2 Aqueous Degreasers

*Get a bunch of those 3-D glasses and wear them all at the same time. Use enough to get it up to a good, say, 10- or 12-D.*[226]

The steps (stages) are shown in order of occurrence in Table 4.14. Note that there are two types of aqueous cleaning process: immersion and spray. The comments in Section 4.27.1 about manufacturer's recommendations apply here as well.

[225] Erwin Schrödinger.
[226] Rod Schmidt.

**Table 4.13 General Dwell Times for Solvent Vapour Degreasers**

| Process stage | Factor(s) limiting Dwell Time | Reason for Factor | Range of Dwell Times, (Second) |
|---|---|---|---|
| Entry | Linear vertical speed of 11 feet/minute (5.6 cm/s) per US EPA NESHAP. | Emission control | 10–30 |
| Vapor Degreasing | Heatup of parts from entry temperature to normal solvent boiling point. | Soil removal | 15–45 |
| Immersion Cleaning (single stage) | 1. Access of solvent to all areas of parts. 2. Solution of soil in solvent. | Cleaning quality | 30–120 |
| Immersion Rinsing | 1. Access of solvent to all areas of parts. 2. Mixing of dirty solvent with clean solvent. | | 15–60 |
| Final Vapor Degreasing Rinse with Condensate | Temperature in immersion rinse sump. | Final rinse with pristine solvent condensate to produce cleanest parts | 0–30 |
| Drying | Evaporation rate of liquid. | Solvent removal | 30–120+ |
| Removal | Linear vertical speed of 11 feet/minute (5.6 cm/s) per US EPA NESHAP. | Emission control | 10–30 |
| Total cycle time | | Seconds | 110–435 |
| | | Minutes | ~2–7+ |

There are at least two conclusions from the Tables 4.13 and 4.14:[227]

1. The experience with aqueous cleaning technology will last significantly longer than that with solvent cleaning technology.
2. Elapsed time for spray and immersion aqueous cleaning operations is similar, although the processes themselves are not.

## 4.28 COPING WITH A SOLVENT DEGREASER ON ACID

*The most misleading assumptions are the ones you don't even know you're making.*[228]

An acidic cleaning solvent, at its boiling point for many hours can be quite corrosive. Those are the conditions in a vapor degreaser.

This is a serious problem for three reasons:

1. Parts can be damaged, often irreparably. Rust spots while wet or gray oxide deposits while dry are produced by acid attack on metal.
2. Even stainless steel equipment can rust over months (or less) of exposure. Components of other materials will fare even more poorly.
3. Operating staff may be harmed by contact with acid materials.

An acidic degreaser usually doesn't evolve overnight.[229] Damage gradually occurs over a period of

[227] The estimates of dwell (cycle) in Table 4.13 share some common assumptions. The most important ones are: (1) that the machine design is reasonably appropriate for the soil load, (2) the parts have some complex configuration, (3) that cleaning quality is on the order of 98% soil removal, and (4) that drying quality is "to the touch."

Can these estimates of time be shortened? Sure. But be careful if you plan operation of an enterprise based on significantly shorter dwell times. Laboratory data, particularly for an aqueous degreaser, usually are taken with a clean system. Ask for a demonstration of cleaning quality versus dwell times when the system is loaded with soil – which will be its condition in operation by your enterprise.

[228] Adams, Douglas and Carwardine, Mark, *Last Chance to See*, Ballantine Books, New York, 1966.

[229] But that's how the damage has been recognized when users weren't closely monitoring operation of their degreaser. A common reaction is: "*[Expletive]! Our degreaser went acid last night!*."

**Table 4.14 General Dwell Times for Aqueous Degreasers**

| Process stage | Factor(s) limiting dwell time | Reason for factor | Range of dwell times (second) |
|---|---|---|---|
| Entry for immersion or spray | Linear vertical speed of ≈22 feet/minute (11.2 cm/s) | Operational integrity | 5–15 |
| Part heatup to immersion bath temperature | Heatup of parts from entry temperature to bath temperature of 120–160°F | Detergent performance | 15–45 |
| Immersion cleaning (first stage) | 1. Access of solution to all areas of parts<br>2. Removal of soil via emulsion<br>3. Drainage of dragout (see Chapter 1, Section 1.12.5) | Cleaning quality | 75–180 |
| Immersion rinsing (first stage) | 1. Access of solution to all areas of parts<br>2. Dilution of dirty cleaning agent with clean | Rinsing quality | 15–30 |
| Immersion cleaning (second stage) | 1. Access of solution to all areas of parts<br>2. Removal of soil via emulsion<br>3. Drainage of dragout (see Chapter 1, Section 1.12.5) | Cleaning quality | 75–180 |
| Immersion rinsing (second stage) | 1. Access of solution to all areas of parts<br>2. Dilution of dirty cleaning agent with clean<br>3. Drainage of dragout | Rinsing quality | 30–45 |
| **Total immersion cycle** | | | 210–495 |
| Part heatup via spray contact | Heatup of parts from entry temperature to chamber temperature of 120–160°F | Detergent performance | 45–150 |
| Spray cleaning (first stage) | 1. Access of solution to all areas of parts<br>2. Drainage of dragout | Cleaning quality | 90–180 |
| Spray rinsing (first stage) | 1. Access of solution to all areas of parts | Rinsing quality | 30–60 |
| Spray cleaning (second stage) | 1. Access of solution to all areas of parts<br>2. Drainage of dragout | Cleaning quality | 90–180 |
| Spray rinsing (second stage) | 1. Access of solution to all areas of parts<br>2. Drainage of dragout | Rinsing quality | 45–75 |
| **Total spray cycle** | | | 300–645 |
| Transport | Relocation to dryer | Process stage | 0–30 |
| Drying | Evaporation rate of liquid | Water removal | 120–600+ |
| Removal | Linear vertical speed of 22 feet/minute (11.2 cm/s) per US EPA NESHAP. | Emission control | 5–15 |
| | Total cycle time | Immersion–Seconds | 340–1155+ (~6.5–20 minutes) |
| | | Spray–Seconds | 430–1305+ (~7–22 minutes) |

days to weeks once the (im)proper conditions are established and not corrected.

### 4.28.1 Water, Water Everywhere

*Technical skill is mastery of complexity while creativity is mastery of simplicity.*[230]

It's all about the water. Some popular cleaning solvents react with water to produce acid.[231] Halogenated solvents are those best known for this behavior.[232] A degreasing solvent can't decompose to produce acid if there is no water present.[233] That's a major reason why vapor degreasers have water separators.

The chemical reaction, for *n*-propyl bromide solvent, is given as Equation (4.1).

$$CH_3CH_2CH_2Br + HOH \rightarrow CH_3CH_2CH_2OH + HBr \quad (4.1)$$

The acid produced is hydrobromic acid (HBr). Chlorinated solvents would produce hydrochloric acid (HCl). Fluorinated solvents would produce hydrofluoric acid (HF).

There are another two factors – reactivity of the halogen atom, and temperature.

1. Fluorine atoms are more stable than are Chlorine atoms, which are more stable than Bromine atoms, which are more stable than Iodine atoms. This is why there are no commercial cleaning solvents containing Iodine atoms, and cleaning solvents containing Fluorine atoms are too stable to decompose to produce acid.[234]
2. Cleaning solvents are generally used at their boiling points, and chemical reactions generally proceed at higher rates when the temperature is elevated. So formation of acid with chlorinated or brominated cleaning solvents is of less concern when they are used in cold cleaning.

### 4.28.2 Dealing with an Acid Degreaser

*A simple rule of thumb: never trap for an error you don't know how to handle.*[235]

Shut it down. Neutralize the acid. Get rid, safely, of the contents. Don't do it again!

An acid degreaser can't be "fixed." Even if the water inflow can be stopped, there is no practical way to collect only the acid and safely manage it.

First, take the degreaser off-line and prepare it for maintenance – stop work flow, turn off the heat, and later refrigeration.

The second step is to neutralize the contents of the degreaser with a water solution of sodium carbonate (soda ash[236]). A modest amount of heat will be generated. Agitate the solvent–water solution to enhance heat transfer, and reaction. When the temperature drops to ambient, remove all contents. Similarly, treat the boil sump, the rinse sump, the contents of the distillation system, and all storage.[237]

The third stage is to discard the fluid contents – using the normal handling and environmental procedures associated with this degreasing solvent. Just because the utility of this particular batch of solvent has disappeared doesn't mean that its hazards have as well.

---

[230] E. Christopher Zeeman, *Catastrophe Theory*, Selected Papers 1972–1977.

[231] For example, methyl acetate can slowly react with water to produce acetic acid and methanol. But acetic acid is not nearly as corrosive to metal or as effective a catalyst for other reactions as are halogen acids.

[232] The 1,1,1-Trichloroethane, whose manufacture is banned because of high ozone depletion potential, was particularly reactive with water to produce hydrochloric acid.

[233] Under extreme conditions ester solvents can decompose to produce acid and an alcohol. This is not a concern with cleaning solvents as other reactions are more probable.

[234] By the Pauling Scale, used to rate the reactivity of halogen atoms, Fluorine is rated 4.0, Chlorine is 3.0, Bromine is 2.8, Iodine is 2.5, Carbon is 2.5, and Hydrogen is 2.1. The difficulty of breaking a Carbon–Chlorine bond is 3.0–2.1 or 0.9 Pauling Units. That for removing a Bromine atom is a similar 0.7. But Iodine is 0.4 and Fluorine is 1.9.

[235] Ed Wilts. The author's corollary to this statement is to never ask a question to which you may not want to know the answer.

[236] Soda ash is a relatively safe, low-priced chemical that is commonly available. It will react with (titrate) the acid. Obviously, it is basic material to be able to react with acid. This means soda ash is ***also*** corrosive to human eyes, digestive systems, and skin. Soda ash should be added to the acidic solvent in about a 1–2% solution by weight – say ½ kg in 5 gal of water.

[237] Assume all solvent is infected with water unless you have the capability to conduct a specific analysis (Karl Fischer) for water. Halogenated solvents are low in value relative to your parts, and to the cleaning machine.

### 4.28.3 Repairing an Acid Degreaser

*I heard that in relativity theory space and time are the same thing. Einstein discovered this when he kept showing up three miles late for meetings.*[238]

Four actions must be completed before the cleaning machine is restarted for cleaning work:

1. First, use additional soda ash solution in water to thoroughly flush **all** wetted areas. Retained acid will continue the corrosion, albeit at a reduced rate. Boil the basic solution within the degreaser if possible. Finally, flush the soda ash solution several times with water to avoid any deposits of soda on the parts.
2. Second, clean the machine using the procedures of Section 4.29.
3. Third, repair the damage done by corrosion. It is likely damage will be concentrated at points of mechanical weakness – bend lines, corners, or weld lines. It would not be embarrassing to consult the machine's manufacturer, especially for advice about welding metal patches over corroded areas.
4. Finally, inspect the vapor degreaser as if it were a new machine. Make a photographic record of the repaired damage.

Fill with fresh solvent. Test operation with written procedures.

### 4.28.4 Stopping Intrusion of Water

*The primary purpose of the DATA statement is to give names to constants; instead of referring to pi as 3.141592653589793 at every appearance, the variable PI can be given that value with a DATA statement and used instead of the longer form of the constant. This also simplifies modifying the program, should the value of pi change.*[239]

Before restart as was originally done, the root cause of the corrosion (intrusion of water) must be detected and eliminated. If the root cause is not eliminated, this exercise will become habit forming.

The likely source of water is **not** humidity – though that certainly can be a parallel source. It is **also** not likely that the part surfaces were wet with water, as suggested in Section 4.16 (Maintenance). The water separator should be able to eliminate those levels of water intrusion, and the stabilizer package able to neutralize that amount of acid.

Consider the following items as potential source(s) of large amounts of water. The source(s) might be:

- Leakage from a cooling line.
- Drainage of condensed frost from the refrigeration coils during a defrost cycle.
- Transport of liquid water within the configuration of parts if they are not drained before or during insertion into the parts basket.
- An unexpected soil component which is acting as a surfactant (making solvent compatible with water).

Carefully observe operation of the water separator during startup. Some water will always enter a degreaser (as humidity and "drag-in"), and it should always be generally immiscible with halogenated cleaning solvents. Thus, the water separator should be needed, and normal rejection of water as a top layer should be observed.

If rejection of a top layer is not observed, then that difficult and unusual situation needs to be understood. Either water is not reaching the separator or water has become compatible with the solvent. Here, the manufacturer or a cleaning professional should be consulted.

## 4.29 CLEANING A CLEANING MACHINE

"Furious activity is no substitute for understanding."[240]

---

[238] Steven Wright.
[239] *FORTRAN Manual for Xerox Computers.*
[240] H.H. Williams.

How can it be possible to have to cleaning a cleaning machine? Shouldn't a cleaning machine clean itself?

Well, no. Cleaning machines remove soil from surfaces. That's cleaning. In this section, the removal of soil from the surfaces of a cleaning machine will be covered. That's cleaning a cleaning machine.

### 4.29.1 Cleaning A Solvent Vapor Degreaser

> *LAMARR: Be still, Taggart. My mind is aglow with whirling transient nodes of thought careening through a cosmic vapor of invention.*
>
> *TAGGART: Ditto.*[241]

Four basic steps (stages) are required: (1) removal of all the solvent; (2) preparation of the machine for maintenance; (3) removal of all the soil(s), and (4) preparation for re-use.

**1. Solvent removal**

- Stop work flow and remove all parts baskets.
- Turn off the heat supply.[242]
- Flush sumps and distillation to remove residue not bonded to surfaces.
- When sump cools, turn off cooling system(s).
- Drain all solvent, either through an internal sump or through the distillation system. Collect solvent in containers for distillation prior to reuse, or disposal.

**2. Preparation for maintenance**

- Ventilate all system internals with fresh air, taking care not to infect surrounding areas with solvent-laden exhaust.
- Isolate system from site utilities:
  - Disconnect and lock out[243] connections associated with heat supply (electrical, gas, steam, hot water, etc.).
  - Disconnect and lock out all fluid connections.
  - Inspect for completeness of isolation prior to maintenance work.
- Remove all attached components which could be damaged during maintenance or need inspection/repair during this work. These include safety devices, instruments, heating coils, etc.
- There should be no zones where material (solvent, air, etc.) can be contained or where pressure or vacuum can be retained.

**3. Soil removal**

- Examine system, and develop some written work plan (however simple). Decide what tasks are and are not to be done; who is to do which task and not to do others; and how well individual tasks are to be done. Certain tasks are:
  - Take digital photographs showing parts accumulation and soil buildup (or lack of it) every time the machine is shutdown for inspection.
  - Note and remove any parts that were unintentionally retained within the system.
  - Most soil (metal chips, sludge, shop dirt, etc.) should be removable by hand or via flushing with water.
  - Intractable soil (cured or baked) will probably be removed via mechanical action with a hand tool or possibly with a motorized tool. A small quantity of solvent may be applied to soften the intractable soil prior to mechanical action.
- Understand, and commit to writing, how internal operation allowed soil to be accumulated. This is especially important if, as expected, retained soil exists only in certain locations. The reason for this item, and the next two, is that retained soil inside a cleaning machine is not under your control and may reinfect already cleaned parts.
- Decide how to prevent the observed accumulation in the future and complete that action.
- Choose a time for the next inspection for soil accumulation.

**4. Preparation for reuse**

- Inspect for corrosion damage and take appropriate action with the manufacturer's guidance. Take

---

[241] Norman Steinberg, Mel Brooks, Andrew Bergman, Richard Pryor and Alan Unger, *Blazing Saddles*, 1974.

[242] This is a good example of why a cleaning machine should have both controls for individual functions (heatup, refrigeration, part transport, etc.) and an overall control for immediate shutdown.

[243] The phrase "lock out" is one fundamental to good safety practice. It means that one is to not just break apart a connection but: (1) make it impossible for someone to inadvertently reconnect it, via a seal which is locked to the machine side of the connection, and (2) maintain the key to that lock only in the possession of the staff actually doing the maintenance work.

digital photographs of affected areas as well as areas where corrosion might be expected to occur.[244]

- Repair mechanical devices (pumps, heat exchangers, etc.), instruments, or metal surfaces as needed.
- Conduct any inspections required by local codes or national pressure vessel codes.
- Assess condition of cleaning system and make a conscious decision to return it to active service, or not.

This work should be done through a written procedure. There should be written work orders associated with the actions involved with isolating the degreaser from site utilities. Safety is the reason for emphasizing written versus *ad hoc* action. Retention of digital photographs is useful for understanding any future unexpected situations. Note the date this work was done and any unusual operating conditions which preceded it or followed it.

### 4.29.2 Aqueous Degreaser

*LAMARR: Be still! My mind is a raging torrent flooded with rivulets of thought cascading in a waterfall of creative alternatives.*[245]

The procedure for cleaning work with an aqueous degreaser is quite similar, and may be simpler. The same four stages are to be done, in the same order.

The crucial difference is that the hazards presented by the aqueous cleaning agent are not those of the solvent cleaning agent.

## 4.30 PREPARATION FOR ENTRY OF SOLVENT OR AQUEOUS CLEANING MACHINES

*Time is what prevents everything from happening at once.*[246]

Entry happens for a variety of good and normal reasons.[247] Entry into the confined space of *either type* of degreaser can be done safely and should not be avoided *when necessary*. Vessel entry is common with continuous (cross-rod) solvent degreasers or those used to clean large parts.

The maintenance work described in Section 4.16 is necessary, but not sufficient.

Additionally, *prior to entry by any person*, it is necessary for the following to be completed:[248]

- An experienced person(s) must measure the air quality *inside the degreaser*, via *several* independent measurements using a sampling tube:
  - **Oxygen content:** with a portable calibrated oxygen analyzer. Content should never be less than 19.5 volume % or greater than 23 volume %.
  - **Explosivity:** relative to the Lower Explosive Limit (LEL), using a portable calibrated analyzer. Level should never approach or exceed 10% of the published LEL.
  - **Chemical exposure:** with a Drager tube or like sampling device. Level should never exceed the short-term exposure limit noted on the MSDS.
  - **Temperature:** value should never exceed 120°F.
  - **Airborne Dust:** should never obscure visibility at a distance of 5 ft or less.

If test results allow conclusion that the atmospheric condition of the confined space is unacceptable, entry is prohibited until such conditions are brought into acceptable limits. A written notice about the acceptability of entry should be posted at all points of possible entry.

- Appropriate personal protective equipment must be used by all who enter the degreaser:
  - The person entering the degreaser must wear a harness and lifeline.
  - The lifeline must be tethered to a solid point external to the degreaser.
  - Protective equipment should be at least that used by staff supporting the degreaser or handling solvent: gloves, face shields, goggles, aprons, etc.

[244] These are points of mechanical weakness – weld or bend lines or cove corners.
[245] Norman Steinberg, Mel Brooks, Andrew Bergman, Richard Pryor and Alan Unger, *Blazing Saddles*, 1974.
[246] John Archibald Wheeler. *American Journal of Physics*, 1978 vol. 46, p. 23.
[247] Workers normally enter either type of cleaning machine because its size (usually height) makes it impossible or difficult to complete some tasks, or because a closeup inspection of a surface is required by the enterprise or a local authority. Workers may also have to enter for removal of retained soil materials or removal of retained parts.
[248] Excellent specific references can be found at the US Occupational Safety and Health Organization (OSHA) web site, http://www.osha.gov/SLTC/confinedspaces/index.html

  - A locally approved self-contained breathing apparatus should be available to be worn by any person entering the degreaser. It should be of the pressure-demand type with full facepiece.
- Appropriate mechanical devices must be available:
  - If a vertical rescue would be required from the confined space, and the depth of the space is more than 3–4 ft (as normal with a vapor degreaser), a mechanical lifting device is needed. This device should have a lifting advantage of at least 4:1.
  - Steps, ladders, ropes, etc.
  - Ventilation equipment (fans, blowers, etc.).
- A rescue person must be on standby at the entry to the degreaser at all times:
  - The rescue person should be as familiar with the degreaser as is the person who has entered it. In addition, they should be trained in the basics of medical rescue.
  - The rescue person must have access, or be connected to, the lifeline connecting to the person within the degreaser.
  - There should be a positive and open communication link between the person within the degreaser and the rescue person, **and** a second communication link between the rescue person and the external support.
  - The rescue person should have and wear the same facilities and PPE which support the person entering the degreaser. But they do not enter the degreaser unless required or requested by the person who first entered the degreaser.
- The rescue person should have external support of a third person, who is to be notified if it is necessary for the rescue person to enter the degreaser.

Written procedures should be available for all foreseeable actions. Associated staff should be trained in those procedures.

A component of those procedures must be a written permit system where access to and entry into either type of degreaser must be approved by site supervision and accepted by the person(s) entering the machine.

## 4.31 MULTISTAGE CLEANING OPERATIONS

> *If you are doing things the same way as two years ago, you are almost certainly doing them wrong.*[249]

It is important to understand multiple-stage cleaning, rinsing, and drying operations. They are common and managers should be able to recognize those units which are well designed.

While appearing complex, multiple-stage operations are simply repetitions of a few process stages. Those for a two-stage aqueous degreaser are listed in Table 4.14. Those for a single-stage solvent degreaser are found in Table 4.13.

### 4.31.1 A Single Stage of Any Cleaning Technology

> *OTTO: I stand on my record. Fifteen crashes and not a single fatality.*[250]

A single stage[251] is defined as a wash contact followed by a rinse contact.[252]

Single-stage cleaning is shown *functionally* in Figure 4.23. Note:

- Fresh cleaning agent is always used.
- Rinsing and washing are not done in the same chamber.[253]
- Rinsed parts are never sprayed or immersed with the cleaning agent used in the current stage.[254]

[249] Sir John Harvey-Jones.

[250] Otto, *The Simpsons*.

[251] In Figure 4.23, the part is represented by the star symbol, whose color lightens with increased cleanliness.

[252] Cleaning work is nearly always organized as a wash and rinse – never in the same vessel. This is because the soil removed in the wash contact remains within the cleaning agent (solvent or aqueous) and is still in contact with the goods being washed. Cleaning isn't considered complete within any stage until that soil-laden cleaning agent is rinsed away.

[253] Or when they occasionally are, the chamber is first emptied of the fluid used for cleaning in that stage.

[254] That would produce no improvement. One wouldn't rinse oneself in one's bathwater, would one (see Chapter 7, Section 7.6)? Another charge of cleaning agent is used – presumably less soil-laden. This additional charge of cleaning agent may or may not be pristine (soil-free) and may or may not be pure water (in aqueous technology).

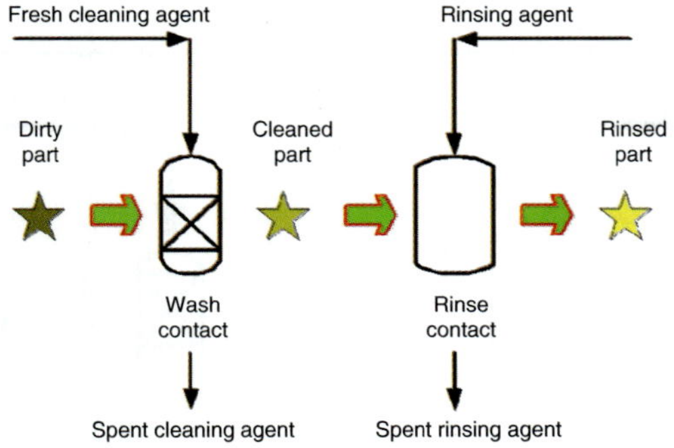


**Figure 4.23**

- Since washing and rinsing are done in different chambers, there can be different time schedules for each (see Tables 4.13 and 4.14).

The single stage can be of any technology, aqueous, solvent, mechanical, plasma, or whatever else.

### 4.31.2 Rate Limitations Within a Single Stage

> *It's a nitwit idea. Nitwit ideas are for emergencies. The rest of the time you go by the Book, which is mostly a collection of nitwit ideas that worked.*[255]

The driving force for cleaning in a single stage is the difference between the concentration of "cleanable soil"[256] at the part surface and the same in the cleaning agent. When both concentrations are equivalent, there is no driving force for cleaning in that stage, and the cleaning rate is negligible:

- Said another way, the rate of soil removal for parts which are nearly clean is only a small fraction of the rate of soil removal for parts which are quite dirty.

That's the limitation of any single stage of a cleaning process – when there is no driving force for additional cleaning work.

A second (or third or additional) stage of cleaning is then needed if the level of cleaning is thought not to be adequate:

- The succeeding cleaning stage will use fresh cleaning agent, as per Figure 4.23. If soil-laden cleaning agent were used, versus fresh cleaning agent, that driving force would be reduced and the additional cleaning stage would add lesser (or no) value.
- These additional stages may be scheduled for longer elapsed times because the rate of cleaning in them is slower than in initial cleaning stages.

### 4.31.3 Quality Limitations Within a Single Stage

> *CALVIN: It's not denial. I'm just very selective about what I accept as reality.*[257]

If a perfectly clean surface is contacted with a cleaning agent containing one molecule of soil, that soil molecule can be transferred (at some slow rate) to the clean surface and infect it. Then the driving force is reversed, and the soil molecule can be transferred back (at some slow rate) to the cleaning agent:

- So a perfectly soil-free surface is theoretically impossible. If all the soil on it were transferred (at some slow rate) to a cleaning agent, then there would be a driving force to transfer the soil back to the surface.

The concept of a limiting driving force explains why perfect cleaning cannot be obtained in any single stage of cleaning work. By analogy to Heisenberg's uncertainty principle:[258]

- A surface can be no more clean that the cleanliness of the cleaning agents in contact with it.

This is not an abstract concept. It's why vapor degreasing was developed so that surfaces could be rinsed with pure liquid solvent produced on the part surface from condensed vapor. It's why the final

[255] Larry Niven and Jerry Pournelle, *The Mote in God's Eye*, 1974.
[256] This term, introduced here, is the soil surrounded by the cleaning agent which removes it from parts. For solvent cleaning technology, "cleanable soil" is soil dissolved in solvent. For aqueous cleaning technology, "cleanable soil" is the organization of micelles which join detergent, soil, and water.
[257] Bill Watterson, *Calvin and Hobbs*.
[258] See Section 4.2.3, and Chapter 1, Section 12.6.

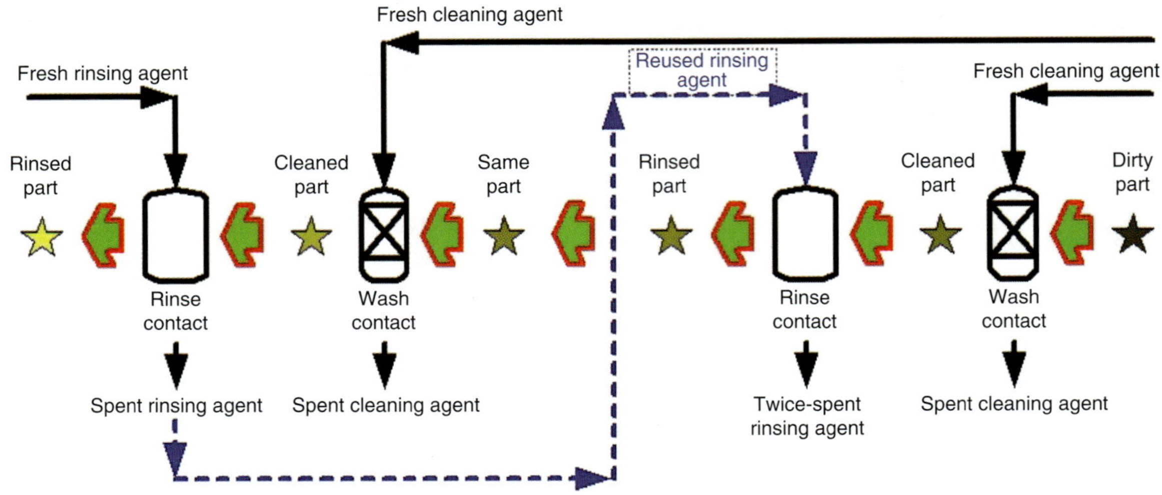


**Figure 4.24**

contact on critical[259] surfaces in aqueous cleaning is with the purest and most mineral-free water, deionized (DI) water.

### 4.31.4 Multiple Stage Cleaning with any Cleaning Technology

> *This 'telephone' has too many shortcomings to be seriously considered as a means of communication. The device is inherently of no value to us.*[260]

Un-rinsed parts from any cleaning stage may be rinsed with rinse fluid previously used. Here, the level of soil is less than that in the stage from which the parts have just been removed. This is called feed-forward rinsing.

Figure 4.24 functionally shows these relationships. In it, Figure 4.23 is repeated to show:

- The transported part is cleaned and rinsed in Stage 1 on the right. It is then transported to the left for cleaning and rinsing in Stage 2.[261]
- Fresh cleaning agent is used in both cleaning stages.
- Fresh rinsing agent is *not* used in both rinsing stages. The rinse agent used in Stage 2 (where soil level is low) is reused in Stage 1 (where soil level is high). Said another way, the cleanest rinse fluid is used in final contact with the parts.[262]
- This is also called "counterflow rinsing" – where the flow direction of parts is counter to the flow direction of rinse fluid.

Only cost and floorspace limit the number of wash and rinse stages which can be used.

The operations described in Figure 4.24 are often simplified in critical cleaning work to omit the cleaning stage. In this case, after the level of cleaning necessary is competed, the parts are repeatedly rinsed. Three and more sequential rinses with DI water are commonly used in a counterflow program.

Note that no relationship with time is implied in Figure 4.24. The cleaning and rinsing operations in Stage 2 can be completed a week after those within Stage 1.

[259] The phrase "critical cleaning" means that the value of the cleaning stage is critical to the success of the enterprise. With semiconductor products, for example, soils left on the surface render the product useless.
[260] Western Union, *Internal Memo,* 1876.
[261] This is indicated by the lightening of color of the part symbol.
[262] As described in Section 4.31.3.

# 5

# Testing for cleanliness

**Chapter contents**



*Cleaning is soil management*. That's all.[1]

There are at least three management steps:

1. **Removal of soil from parts**: This management step includes what is popularly known as cleaning, and also what is similarly known as rinsing, and what is also similarly known as drying.
2. **Removal of soil from the cleaning machine(s)**: This management step should be included as a part of any useful cleaning process. If this isn't done, the parts are likely be reinfected with soil or soil components, and considered not clean.
3. **Removal of the soil from the cleaning machine and collection of the soil materials into forms suitable for efficient disposal or economic recycle/reuse**: This management step should also be included as a part of any useful cleaning process. If this isn't done, the cleaning process won't be economically viable, environmentally permitted, or in existence.

If *all* of these steps aren't completed and well done, there is no point in testing for cleanliness. This is because soil components and materials will locate somewhere they are not supposed to be. They will likely reinfect parts at some time in a discontinuous and inconsistent manner. So measurements of cleanliness will characterize how ineffectively such a cleaning process is being managed instead of characterizing how well it's being managed. Said another way, managers will be sorting the chaff or weighing the dross instead of grading the wheat or assaying the Gold.

## 5.1 TO KNOW CLEANLINESS YOU HAVE TO KNOW SOIL

Soil is not what your lawn is growing in your flower garden. Soil is whatever managers don't want to be on or within their parts.[2]

Three aspects of soil are:

1. Its *chemical character* (see Section 5.1.1).
2. From *where it came* (see Section 5.1.2).
3. How it's *recognized* (see Section 5.1.3).

### 5.1.1 Chemical Characterization of Soil

Chemically, soil can be:

1. *Specific (single component) and chemically known*[3] *materials* such as brand "X" silicone

[1] Recall the insightful expression favored by the author "if something gets clean, something else gets dirty."

[2] In this sense, the term "parts" can be quite general. Parts can be widgets, the machine which produced the widgets, or the assembled/packaged widget.

[3] Here the word known means that the chemical identity of the single-component soil material is known. In this way a favorable choice of cleaning agents can be made. The soil is unlikely to have a single component.

grease, cooking lard or peanut oil, brand "Y" automotive antifreeze (generally ethylene glycol), linolenic acid (used as an antioxidant), isopropyl myristate or stearic acid (used in cosmetics as a thickening agent and emollient), chlorinated paraffins (used as extrusion aids), or mineral oil.

2. *Various (multi-component) mixtures*[4] such as drawing fluids, extrusion aids, greases, inks, protective coatings, food products such as catsup, coolants, stamping oils, jeweler's rouge, lubricants for cutting or machining, bodily fluids, or synthetic motor oil. There is a special case of various (multi-components) processing compounds having been *added beyond the amount needed*[5] for processing. Examples are:

    - Incompletely reacted components of paints or coatings. These are usually low-molecular-weight materials which have been only partially cured or polymerized.
    - Excess solder materials (paste or solder metal) not included in the finished joint.
    - Excess pressure-forming fluid (used in cold-header machines) or coolant (used in screw-cutting machines).

3. *General*[6] *materials* of unspecified composition such as mill scale, pyrogens,[7] particulate, "tar", "smut" or "smuge," shop dirt, fingerprints, wax, or haze/sheen. There is another special case of atmospheric organic or biological contamination. Usually, this contamination adsorbs on glass, metal, or semiconductor surfaces when the surface is exposed to air of ambient quality. This is why high-value operations are conducted in controlled environments (cleanrooms).

4. The *result of a measurement* which represents something never unidentified on a surface before or after the cleaning work is completed such as:

    - Non-volatile residue (NVR) on optic materials.
    - Results of a test on flat surfaces, such as the American Society for Testing and Materials (ASTM) F22-02 "water break" test, which involves indirect measurements of surface forces.
    - Total organic carbon (TOC) on tubing used for Oxygen service.
    - An instrumental reading of optically stimulated electron emission (OSEE[8]) from a scan of a surface for organic films.
    - Particle count as measured in a solution used to wash a surface.

5. *Part of the base substrate*: Oxide layers form on the surfaces of many metals (Aluminum, Iron, Titanium, etc.), as well as alloys (stainless steel). These layers are responsible for the inertness to corrosion displayed by these metals. But these layers, which are only a dozen angstroms in thickness (or less), can also be soils in some applications.[9]

For only the first of these five types of soil is their chemical identity well enough known so their presence or absence can be established by a specific and formal analytical method.

In other words, specific procedures developed from analytical chemistry to identify single chemicals are seldom useful as cleaning tests. Soils seldom are single compounds and their identity is not generally known.

[4] Almost all of these multi-component soils are commercial products and thus have Manufacturer's Safety Data Sheet (MSDS) (or similar) information which can be known. This should include identify of all components present in significant amounts. The point to remember is that *all* components must be removed in the cleaning operation.

[5] While it may be true that these materials have some pedigree (see Footnote 4, Section 5.1.1), they can't be assumed to be identical to the same materials which have only participated in one cycle of use in the upstream process. It is between possible and likely that oxidation and thermal decay will have occurred.

[6] In this case, though the name of the chemical family of this soil is known, the specific chemical identity may or may not known. The best information about mill scale might be that it is mainly composed of metal oxides and rich in iron oxides. If the right cleaning process is chosen, these soils can be managed just as well as those where the specific soil is chemically identified.

[7] Walls of dead cells.

[8] OSEE – an instrumental technique for characterizing organic films by amount present, and perhaps type (see Appendix 2).

[9] In recent decades, bonding (joining) technology has begun to migrate from construction of bonds (joints) made where a metal surface is melted and then fused (e.g. welding, soldering, or brazing). An evolving technology is construction of bonds made by adhesion between non-melted surfaces. For example, aircraft structures are now, and will be more so, made from metals or composite materials bonded by some polymeric adhesive. In these applications, metal oxide in the first few angstroms of depth is a significant soil on a metal surface. The high liability associated with aircraft structures, for example, makes cleanliness testing (inspection) vital prior to bonding operations.

**Table 5.1 Comparison of Cleaning Agents and Soils**

| Cleaning Agents | Soils |
|---|---|
| Single-component products (usually) | Multi-component mixtures |
| • Blends of chemicals, each selected to provide a different function, and assembled via a recipe chosen for some directed reason<br>• Azeotropes of single components with a fixed and known recipe[10] | Composed of usually unknown ingredients<br>Prepared with unknown and variable proportions<br>Not homogenized<br>May include tramp[11] materials, or contain components not always present |
| Qualified by some specifications<br>Evaluated to some standards | Recognized, at best |
| Purchased via contract | Often produced inadvertently |
| Backed to some extent by a provider, who at least can provide some related information | Produced by *you* |

### 5.1.2 The Pathology of Soils

If baking of pastry requires fixed formulas for ingredients and specific conditions as procedures, and if cooking of food requires recipes which may be modified to suit taste in addition to criteria for mixing or cooking time that involve matching an observed result produced by other chefs, then cleaning of soils is a time-sensitive pie-eating contest where both hands can be used and no napkins are provided.

Those managing cleaning work must know at least as much about soils as they do about cleaning agents – if their cleaning work is to meet their requirements. After all, the soil is the enemy to be conquered.

Soils are not like cleaning agents. For one evaluation of that difference, see Table 5.1.

Yet, soils are composed of the same, or similar, chemicals as are cleaning agents. In fact, often the best choice of cleaning agent may be the one in which the components in the soil were formulated, or one similar to it.

Soils are not "dirt," "junk," "scum," or "stuff." They are collections of chemicals which need to be removed from their present site (the parts) and relocated in a concentrated form to somewhere else (the waste bin). Most contain multiple components.

That's what cleaning work does. To do that efficiently and repeatedly, one needs to know something about soils:

- At *least*, one hopes to know the list of ingredients from which they are produced.
- At *most*, one may need to know that and:
  - If heat or pressure have caused some transformation via chemical reactions among components or decomposition.
  - If foreign substances have been added.
  - If these soils have become bonded to the parts in a way other than with van der Waals forces.
  - If there is particulate material attached.

---

[10] Except for refined hydrocarbons produced to performance specifications and not to assay specifications, soy-based solvents which are refined from multi-component vegetable oils, dibasic acid-based solvents which are produced as byproducts from chemical reactions, and the like.

[11] The word tramp is commonly used to mean unexpected or uncontrolled. A tramp soil is one which isn't always present. Tramp soils are similar to hobos – along for the ride, like a remora on a shark.

A tramp soil may be oil spilled periodically into a coolant sump, sediment which has accumulated and is only flushed onto the equipment when levels in some containers are low, material only produced during off-aim operation such as startup or shutdown, or the result of an operational mistake.

Here is a practical example. The operator of a machine producing screw machine (tool formed) or cold header (pressure formed) parts will collect them in a bin. When full, the bin will be fed to the cleaning machine. Parts in the top of the bin will contain little lubricant used in the forming process as it will have drained to the bottom of the bin. When the last parts are fed to the cleaning machine, a charge of lubricant is added to a cleaning machine which hasn't previously received any. That lubricant is a tramp soil and the cleaning machine is likely to choke on it!

**Table 5.2 Unit Operations**

| Unit Operation | Meaning |
|---|---|
| Cleaning | Removal of unwanted materials |
| Rinsing | Removal of unwanted materials diluted in cleaning agents |
| Drying | Removal of dilute or pure cleaning agent |

To some it may see foolish to apply any level of science and engineering to making the best choice for dealing with soils.[12] To some it may seem proper that soils should be removed with acid, a blowtorch, or a belt sander because their "integrity" is inconsistent with controlled manufacturing. This author believes that point of view is a understandable, but is a short-sighted viewpoint because it ignores consequences to the product/parts, workers, and the environment.

This author recommends that selection of cleaning agents to clean soils should be done with some relevant science, appropriate diligence, and realistic expectations. In this and other publications, he has endeavored to do just that.

#### 5.1.2.1 *Soil isn't Just Dirt*

Soil can also be the outcome of incomplete cleaning work, as so:

- *Dilute soil materials dissolved in cleaning agent.* This may be clinging to parts or contained within their configuration. Removal of this material is done in the *rinsing* stage (see Chapter 1, Section 1.12).
- *Pure, or very clean, cleaning agent* clinging to parts or contained within their configuration. Removal of this material is done in the *drying* stage (see Chapter 1, Section 1.13).

Table 5.2 shows how the three unit operations normally combined into the phrase cleaning are related.

### 5.1.3 Recognition of Soils

These general definitions make the point that any contamination which degrades the value of use of an article is a soil. If users don't want a material on a surface, it's soil. Pure water is a soil on parts to be adhesively bonded.

Correspondingly, if materials present on a surface don't compromise the next use of an article, they are of little or no interest. Examples are water films on parts to be electroplated or electropolished in water solution, compatible cleaning agent retained on surfaces to be painted or coated, or wax retained for support on powder structures prior to firing (powder metallurgy).

In other words, if managers recognize surface contamination by whatever means, they should only use the unit operations of cleaning, rinsing, or drying to remove it if the cost of allowing it to remain exceeds the cost of removing it. These operations aren't free.

## 5.2 CLEANLINESS TESTING

In Chapter 1, Section 1.10, clean parts were identified as one of five characteristics of a successful cleaning operation. This was taken to mean that the next use of the parts could be completed without regard to contamination. While that statement is concise and correct, it's not very useful and it may encourage unwarranted risk. After all, no user wants to invest the cost of the next operation without some assurance that contamination has been adequately removed.

[12] Many have attempted to define precision cleaning versus critical cleaning versus what is called industrial (metal) cleaning. The standard approach is to focus on level or the value of cleaning quality (Table 5.3). Another approach is to focus on the soil:

- Industrial (metal) cleaning and similar operations involve soils somewhat consistent with the descriptions in Table 5.1.
- Precision cleaning and similar operations usually involve soils whose components are fairly well defined, and much more consistent with the descriptions in Table 5.1.
- Critical cleaning is precision cleaning done in a cleanroom so the environment may be managed and parts aren't reinfected with soils. That and similar high-value operations usually involve soils about which much is known because significant effort is spent in keeping parts free from infection by these soils.

Concerns such as these have produced a most commonly asked question – how clean is clean?

The specific[13] answer to that question has produced three technical tools:

1. **Cleanliness tests**: These are examinations – usually of a surface.[14] The examination produces a result.
2. **Cleanliness metrics**: These allow a judgment call – about whether or not the examination result justifies a pass or fail grade about future use.
3. **Validation of cleanliness tests**: These are independent examinations – occasionally conducted to prove that the cleaning test hasn't been unintentionally compromised.

## 5.2.1 Cleanliness Tests

Probably more questions come to a consultant about cleaning tests than on any other topic. This author believes the reason for this is that there is not a single answer to these questions. Cleanliness is both subjective and objective!

### 5.2.1.1 *What Cleanliness Is (And Is Not) – Subjective View*

Measures of cleanliness with industrial parts have to do with things in common with our human experiences. For example:

1. *Cleanliness is generally observed, not measured* (by ourselves and our customers). This means that cleanliness is not established by the process of cleaning. It didn't matter how hard or long you rubbed those chrome bumpers, the street rod didn't shine until the guys on the street said it shined.
2. *Cleanliness is established by whether or not we can do what we want to do next* (e.g. have our customers buy and use our product). This means that downstream success or failure defines cleanliness. My desk may look neat and clean, but if I need cash and can't find my checkbook on my desk, the desk is not neat and clean.
3. *Cleanliness is the result of an opinion*. This means that cleanliness is subjective. You and the customer may have different opinions about the cleanliness of the same parts. In the marketplace, only the customer's opinion counts; yours doesn't.

### 5.2.1.2 *What Cleanliness Is (And Is Not) – Objective View*

And yet cleanliness can be, often is, and should be defined by the result of some test and comparison of that test result to a metric.

The reason for this duality is that metrics are needed to conduct business. Opinions may drive business decisions, but business opinions are (or should be) based on objective information. Consider another interpretation about the above examples:

1. *Cleanliness is generally observed, not measured.* Your customer doesn't care to witness operation of your cleaning machine or to know how many stages of rinsing are in it or to learn why you have used aqueous or solvent cleaning technology. Your customer, which can be yourself, cares only that the outcome of cleaning does or doesn't meet an accepted objective specification. The form that specification takes may be a number, an opinion about a photograph, a trial use of the cleaned materials, or just an opinion based on experience.
2. *Cleanliness is established by whether or not we can do what we want to do next.* Granted, visual appearance may be the only reason parts are cleaned. That simply means a component of the next use is a judgment about appearance.
3. *Cleanliness is the result of an opinion.* The opinion about cleanliness may be that the parts don't meet the customer's numerical specification for cleanliness. In that case, cleaning tests you conduct don't matter. The defining opinion is that of the customer (or end user, which may be you). It is only *their* tests which matter. In this case, the disagreement is not about cleanliness but about implementation of cleaning tests. Parts aren't clean until the customer's own tests (or use) convince them that the parts are clean.

In summary, cleanliness of parts (after performance of a cleaning/rinsing/drying operation) is analogous to the taste of prepared food. Both cleanliness and taste:

- Are opinions based experience.
- Are dependent upon how they were produced.
- Can be quantified on some objective scale.
- Ultimately judged by the user.

[13] The general answer is "I don't know!"

[14] Kuhn, A.T., "Starting Off with a Clean Slate: Using Dyne Liquids Is One of the Easiest and Most Cost-Effective Means of Assessing Surface Cleanliness," *Metal Finishing Magazine*, May 2005, Vol. 103, No. 5, pp. 72–79.

More simply, if you select a cleaning test from the following information and don't involve the next user in the definition of performance, you've wasted your time!

#### 5.2.1.3 *Tip of the Iceberg*

This chapter and this book are about management of cleaning – of things made, maintained, or used. Some call that parts cleaning, and it's a multi-billion dollar industry with highly valued applications around the globe. Technology described here can and has and will be used for decontamination of items brought back from and sent into space, preparation of prosthetics for implant into humans, and getting cutting oil off screws.

What's not covered in this chapter and this volume, is the majority of cleaning applications – cleaning prior to or during our activities. These include janitorial or home maintenance cleaning, restaurant operations, surgical preparations (both medical and dental), archaeological methodology, cleaning of clothing, packaging of raw food and drink, personal or pet sanitation, forensic examination, creation of catalytic environments, metrology, restoration of artifacts, and many other aspects of life.

Yet there are a few principles common to all these endeavors:

- Imbesi's Law of the Conservation of Filth (Footnote 1): Whenever something becomes clean, something else becomes dirty.
- The First Law of "cleaning thermodynamics": soil is like entropy – always created, never destroyed.
- The Second Law of "cleaning thermodynamics": you can never get something completely clean.
- The application of Heisenberg's uncertainty principle to cleaning operations – to get a particle off a surface, first you have to find both.

This book is limited to areas where this author can contribute. Cleaning is not limited to this author's experience.

## 5.3 SPECIFIC CLEANLINESS TESTS

Suppose you don't manage a laboratory with X-ray spectral equipment, high-salaried chemists who can do elemental analysis of metal surfaces, or well-trained technicians who can operate particle counters based on scattering of laser-produced light. In other words, suppose the firm you manage is like most others.

How can you do valued and sophisticated cleanliness testing? That's described in Sections 5.4–5.6 and Appendix 2.

### 5.3.1 Separation by Level of Cleaning

Sections 5.4–5.6 are separate because each contains information about measuring cleanliness at three different levels: **industrial**, **precision**, and **critical**. Readers, and this author, should beware that some cleanliness verification techniques can be used in different ways or situations to provide information about all three levels of cleanliness. In other words, cleanliness measurement is organized in this book by level of cleaning quality desired, not by application (because there are too many applications to count), and not by method.[15]

The term industrial cleaning may sound gross. It's not. See Table 5.3 and Footnote 12 for general differentiation among levels of cleaning performance. In this book:

- **Industrial (metal or gross) cleaning** describes cleaning work done to *gain acceptance* of the material, surface, part, or work being cleaned. Acceptance usually means that the preceding actions aren't a bar to those next planned.
- **Precision cleaning** describes cleaning work done to *meet a standard* of precision set by a customer, standards organization, or a manager. Here, the word precision means faithfulness to that standard.
- **Critical cleaning** describes cleaning work done *to avoid* legal/regulatory liability, human/societal casualty, or economic/infrastructure dislocation. Consequences of this dislocation are by definition critical.

All cleanliness tests are summarized in Section 5.7. See Section 5.9 for information about selecting/collecting tools for cleaning inspection and testing. Additional specificity about some cleanliness tests is provided in Appendix 2.[16]

[15] This choice was made because most managers know more about what they want to achieve (management of a certain level of cleaning) than how to do it (via use of specific methods).
[16] The purpose of Appendix 2 is to provide additional information about how to accomplish certain cleanliness tests.

**Table 5.3 Comparison of Cleanliness Levels**

| | Industrial | Precision | Critical |
|---|---|---|---|
| Chapter of this Book | 5.4 | 5.5 | 5.6 |
| Acceptance Is a Key Issue | **YES** | Probably not | Probably |
| Standard Involved? | Probably not | **YES** | Probably |
| Health, Legal, Economic Dislocation Involved | No | Probably not | **YES** |
| Final Level of Soil (Generally) | Variable | Small | Minuscule |
| Final Level of Soil (Specifically), Mg/SF | >25 | 1 to 25 | Immeasurable to <1 |
| Particle Sizes of Concern | 10 microns (10,000 nm) | >0.5 microns (500 nm) | <<0.5 microns (500 nm) |
| Operator Training | Modest | Probably significant | At least as much as is required in production of the item being cleaned |
| Investment Testing Facilities | Less than the price of a coach airline ticket | Could be similar to the price of a first-class trans-oceanic airline ticket | Could be similar to the cost of a private airplane |
| Site Location | Anywhere | Dedicated facility | Cleanroom |
| Cleanliness Verification | Probably | Certainly | Absolutely |
| Cleanliness Validation | Probably not | | |
| Example Applications | • Flushing mud from tires so tread can be evaluated<br>• Pre-cleaning to reduce the load on a downstream cleaning machine | • Surface treatments to allow good performance by paints and coatings<br>• Products for sale to military organizations | • Preparing stents or implants for human use<br>• Bonding composites to metals on aircraft frames<br>• Removal of CMP or other preparation waste from semiconductor structures |

## 5.4 TESTS PROVIDING INFORMATION ABOUT AN INDUSTRIAL LEVEL OF CLEANING

These approaches may be all that is required, as long as the limitations are understood.

### 5.4.1 Visual Inspection

This is a traditional non-critical method of verification for cleanliness. A chef takes a taste. A sommelier takes a sniff. A sculptor reaches out and touches. A manager of cleaning facilities looks closely.

Visual inspection should be the *first cleanliness test done in every situation.* It's the cheapest. It's completed the most quickly. It involves the most experienced knowledge (yours).

At least three actions should be taken when a plan of visual inspection is implemented:

1. Have currently produced references available for comparison of output – both acceptable and not acceptable.[17,18]
2. Observe from more than one perspective. Said another way, adopt the attitude that the item being inspected may not be acceptable and that the

[17] A report that something "looks OK" is a waste of time. The report should speak in terms of what others understand – previous performance or standards.

[18] Metrics (see Section 5.10) for visual observation can and should be developed. A set of five should suffice – with two of them being a completely cleaned piece and the other being one not yet cleaned. That leaves three intermediate ones. Digital photographs of suitable production will provide good dividends.

inspector's role is to identify how this is so.[19] Be thorough, but efficient.[20]
3. Document something[21] that was observed – even if it was nothing.

Microscopic examination may be useful – but only after visual inspection without magnification is complete.[22] This author's experience is that inspection for quality control with magnification often wastes resources on evaluation of discontinuities which aren't defects at this level of cleanliness.

Generally, visual examination should be non-destructive.

**Figure 5.1**

### 5.4.2 Observation with UltraViolet ("Black") Light[23]

This is another form of visual inspection. It is chiefly focused at identifying the presence of hydrocarbon soils on metal or glass (non-polymeric) surfaces.

Many organic[24] materials, *including polymers, cleaning agents, and soils*, will fluoresce when exposed to "black" light. See Figure 5.1 for a whimsical demonstration, and Figure 5.2[25] to see a simple low-cost source of black light.

The energy in ultraviolet (UV) light stimulates some – but not all – chemicals. A few electrons in their atoms are stimulated after absorption of a photon of UV light. These electrons become excited and are boosted to a higher energy state. When these electrons "calm down," they return to their home position in the atom's structure and visible light is normally (but not always[26]) emitted:

- Basically, light of one frequency (below visible) is absorbed and then light of another frequency (visible) is emitted.

[19] This should be a key training principle for inspectors. Their role is not to PASS goods, but to identify the ones which FAIL.
[20] Recall the advice of basketball coach John Wooden: "Be quick, but don't hurry."
[21] If it isn't noted, it's not worth doing. Digital cameras with automatic date and time notation and low-cost removable digital storage media should render all objections to doing this moot.
[22] See the forest first, then the trees. Then, see Section 5.5.3.
*Microscopic examination can be very useful* – when a defect or pattern of defects has been identified. Here the magnification is not used for quality control but as an aid to understanding the mechanism of defect formation (see Section 5.6.1).
[23] Light just beyond the violet edge of the visible spectrum is called ultraviolet (UV) light. The best UV light sources produce both long-wavelength (300–400 nm "black" light) and short-wavelength (less than 300 nm) light. The appellation "black" means that the light can't be seen by humans.
Commonly, "black" light units consist of partially evacuated glass tubes containing a small amount of Mercury. In a "black" light unit, the glass is tinted to allow the UV light to pass through, but not the visible light. Fluorescent black light lamps can be purchased. When buying fluorescent black light lamps, check the part number stamped on the glass near the end. The part number should end in a "BLB" suffix. The glass should be a dark purple. Expect to spend around 15 euro or less.
Dark colored objects do not "glow" under a "black" light because they absorb the UV radiation. Light-colored objects, especially those washed with detergent package containing optical brighteners, appear to glow when illuminated with "black" light. Optical brighteners aren't cleaning agents. They don't remove soil. They camouflage it. Optical brighteners absorb UV light. Then they fluoresce, emitting the photons of visible light, and the fabric surface appears brighter.
[24] Fluorescence was first noted with minerals. In 1852, George Stokes noted fluorescence in the mineral fluorite. Other minerals which fluoresce are diamonds, calcite, gypsum, ruby, talc, opal, agate, quartz, and amber. Fluorescent molecules tend to have rigid structures and delocalized electrons.
[25] Image courtesy of Amberica West. The unit shown sells for less than 30 euro.
[26] Moog, R. *et al.*, *Physical Chemistry: A Guided Inquiry*, Houghton Mifflin, Boston, MA, US, 2004.

**Figure 5.2**

Chemicals which fluoresce when radiated with "black" light include chlorophyll, quinine, eosin (a dye used in medical testing), "day-glow" paints, blood, urine, semen, Vitamin A and the B vitamins (thiamine, niacin, and riboflavin), many inks including those used on postage stamps, and common components of paints, fabric, and plastics such as stabilizers and antioxidants.

Management of cleanliness via observation under "black" light involves the following steps:

- Establish that the soil[27] being used does fluoresce, and that it can be seen by you when illuminated under "black" illumination[28] both neat and when diluted in the cleaning agent.
- Establish that the native part does not fluoresce in the same way.
- Develop some reference standards.[29] Here parts are wetted in synthetic mixtures of pure soil at known and dilute concentration in cleaning agent, and digital images are made. These parts should then be processed using the next operating step in the normal manner. The standards are the images and the processing outcomes.
- Train staff in use of these reference standards, and in evaluating cleaning quality of actual parts.

The "black" light examination is also non-destructive. Visual examination, with any light source, should be done in an environment at least as clean as is intended for the parts. Examination of parts larger than a coffee cup can be inefficient.

### 5.4.3 Tape Sampling Tests

Your mother used this technique to get dog hair off the couch or to remove lint from your father's[30] suit. The collection and preservation of microtraces, such as fibers, using flexible tape is generally accepted as being very practical and efficient. It is also used in forensic and security investigations.

Traditionally, a technician showed the manager a piece of "mending" tape[31] which they had applied to a cleaned (and possibly dirty) part. The adhesive on the tape acted as a solid-based cleaning agent and removed some (or all) loose particles, fibers, skins, and similar other solid materials. When the tape appeared identical (or similar) to virgin tape, the parts were considered to be clean.

"Cleaning" with "mending" tape (for the purpose of cleanliness evaluation) is probably not useful for soil pieces which are too small to be seen, or liquid soils.

However, a particular tape is useful. It is used for masking in wave soldering of circuit boards and collection[32] of forensic evidence. It is not two-component product – being produced as a extruded

[27] This is why it was noted in Section 5.1 that in order to manage cleanliness one has to know something about soils. "Black light" techniques don't have value if the soils don't fluoresce. The observation of no fluorescence has little meaning if the soiled surface is steel with mill scale – which doesn't fluoresce.

In fact, soils which don't fluoresce can be made to do so via addition of a tracer compound which does so. For details, see http://www.angstromtechnologies.com/. Today, forensic and security investigations of printed documents, paper currency, fabric, postal inks, and other items involve use of this technology. Why shouldn't it be used in management of cleaning work?

[28] As shown in Figure 5.1, this is probably done in an environment without visible illumination.

[29] Without this step, managers won't know if absence of fluorescence under black light means the parts are adequately clean, or not.

[30] Your father may have even used it to remove blonde hairs from that suit.

[31] These products are usually two components – a polymeric film with an adhesive coated on one side.

[32] Chable, J., Roux, C. and Lennard, C. "Collection of Fiber Evidence Using Water-Soluble Cellophane Tape," *Journal of Forensic Sciences*, 1994, Vol. 39, No. 6, pp. 1520–1527.

film of polyvinyl alcohol which is structurally self-supporting, an excellent adhesive, fully water-soluble, and biodegradable.[33] The virtue obtained in cleaning evaluations is that everything recovered by this tape can be examined by filtration or analysis of the tape-water solution.

An assumption around its use is that the surface sampled is a representative of the character of the entire surface including sections which can't be reached by the tape.

Sampling with tape is low-cost, non-destructive, and probably underutilized.

### 5.4.4 "White Glove" Treatment

For many years, non-commissioned officers (NCOs) would examine the barracks of military trainees for order and cleanliness. Both were indicators for discipline. A surface contacted with the NCO's white glove would tell the tale. The least amount of dust on the glove could produce the greatest amount of punishment.

The "white glove" (white cloth) test is still very useful and easy to employ in management of discipline in operating of parts cleaning facilities. It is also absolutely non-quantitative.

Obviously, it can only be used on parts which have been thoroughly dried – unless the object of the test is to find moisture on the glove (cloth).

It is most effectively used with oily or greasy soils, colored films, or large ($\gg$10 μm) particles. As with tape sampling, if the wearer of the "white glove" doesn't wipe the dirty sections, the wrong decision will be made about discipline.

"White gloves" can be simply fabric: any clean, dry piece of white fabric, a small wad of cotton that might be used to swab the barrel of a gun, or even a kitchen paper towel. Beware of commercial personal care products (tissues and wipes), however; though they may feel dry to the touch, they may contain mixtures of glycol-based lubricants and isopropanol – known cleaning solvents. Use of such products may provide both testing and cleaning functionality, rendering the former invalid.

"White gloves" are used in critical cleaning, but not to test for cleanliness. In these applications, woven and non-woven wipers are used to trap particles and absorb liquid.[34]

Clients of this author have had more success with the tape sampling tests as economics force reuse of the "white gloves" and loss of identity of successive outcomes. Perhaps that's a discipline issue.

### 5.4.5 The "Water Break" Test

Measurement of the contact angle of a droplet on a surface is used to determine the wettability of the surface:

> *Wettability, with water, is taken as "proxy" (being synonymous) for cleanliness. While that is sometimes or often true, it cannot be taken as an article of faith.*

Liquids that wet a surface (or spread out upon it) have a low contact angle; liquids that do not wet, but rather form a bubble or drop over the surface, have a high contact angle.[35]

This is the basis for the "water break" test. In the water break test:

- If water "beads," the surface is considered to be contaminated with a hydrophobic substance (oil/grease) (see Figure 5.3).[36]
- If the distilled purified water "breaks" or "sheets" off, the surface is considered clean (see Figure 5.4).[37]

---

[33] See http://www.3m.com/intl/kr/img/single/pdf/5414.pdf

[34] For a comprehensive discussion of the use of wipers in critical cleaning applications, see Howard Siegerman's *Wiping Surfaces Clean*, ISBN 0-9748753-5-X, Vicon Publishing, Amherst, NH, 2004.

[35] See Section 5.4.6.1. The term "hydrophilic" is used to refer to surfaces that are wettable by aqueous fluids (cleaning agents, soils, or bodily fluids).

[36] Figure 5.3 is of a demonstration where a single drop of water was placed on an synthetic hydrophobic (water-hating) surface. This image would be of the "ultimate" water bead. Imagine, the mass of water is being totally supported by surface forces existing at the small area of contact between the drop and the surface. Note the perfectly spherical shape where the surface energy is minimized.

[37] Figure 5.4 is of another demonstration. A single drop of water was collected on a leaf. Since the leaf structure contains significant water, the surface is hydrophilic (water-loving). This image would be of the "ultimate" water sheet. The mass of water is supported over a broad area of the leaf because surface forces don't repel the water drop.

Figure 5.3

Figure 5.4

The "water break" test is very subjective and difficult to reproduce:

- What constitutes "breaking" may be seen as different among observers. *There is no control state where cleanliness is known.*
- The test is difficult to interpret at best with flat surfaces, but nearly impossible to interpret if the surface is convoluted.
- Very light or scattered contamination may not be discernable.
- Usually, only a small portion of the surface is evaluated at one time.
- Some contaminants are water-soluble – especially residues from aqueous-based cleaners.

The "water break" test *can* be used with solvent cleaning agents. Solvent remaining after drying will likely be hydrophobic and cause water droplets to be seen as beads. Since poor quality drying is a flaw of the overall cleaning process, any useful cleaning test should illuminate that flaw.

The outcome of a "water break" test is usually binary – GO/NO GO (sheets/beads).

The ASTM test method[38] covers the detection of the presence of hydrophobic (non-wetting) films on surfaces and the presence of hydrophobic organic materials. If properly conducted, the test may enable detection of molecular layers of hydrophobic organic contaminants. On very rough or porous surfaces, the sensitivity of the test may be significantly decreased.

This author/consultant cannot recommend use of any form of the "water break" test, except:

- *To identify soiled parts, parts which are not clean.*[39] Beaded droplets are fairly easy to recognize and they form reproducibly. Cleanliness, however, is less well reproducibly recognized – *and that's about what managers most want to know.*

The simplicity of the "water break" test is not generally outweighed by its variability.

#### 5.4.5.1 *An Quantitative Version of the "Water Break" Test*

If the "water break" test presents a macro view of the action of liquids on surfaces, this test[40] is based on a micro view.

Referred to by some as the sessile[41] drop test (SDT), it is plainly old[42] technology. For some, the SDT can be valuable technology.

[38] See ASTM F22-02 Standard Test Method for Hydrophobic Surface Films by the Water-Break Test.

[39] *Please beware of the trap of identifying as clean a part surface on which water does not bead!* That observation only means that there may not be hydrophobic materials (oily soils) present – which may be only true in a relative sense. Or there may be non-oily soils present – such as un-rinsed aqueous cleaning agents, aqueous food residues, or aqueous bodily fluids.

[40] Kuhn, A.T., "Determining Whether a Metal Surface is Really Clean," *Metal Finishing Magazine*, September 2005. Also see Section 5.4.6.2, Figure 5.10, and Appendix 2.

[41] The technical term is "sessile drop". Derived form Latin, the word sessile means "without stem". In zoology, a permanently attached or fixed barnacle which is not free-moving is referred to as a sessile barnacle. A sessile drop is one which can be isolated, usually by placement with a micro syringe.

[42] Bikerman, J.J., "A Method of Measuring Contact Angles," *Industrial and Engineering Chemistry*, June, 15, 1941, pp. 443–444.

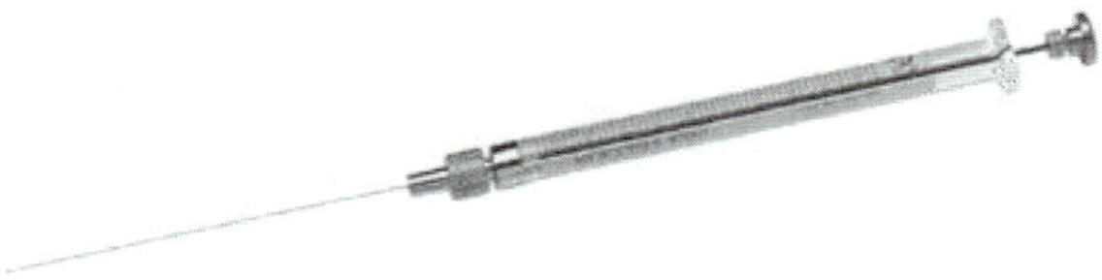

**Figure 5.5**

**Figure 5.6**

**Figure 5.7**

**Figure 5.8**

As opposed to ASTM F22-02, where the applied volume of water is measured in cubic centimeters or larger volumes, the volume of water used in the SDT is controlled to be a few microliters.[43]

Dispensing can be done with a micro-pipette (micro-syringe) when a tiny single drop is applied. This can be the same equipment which is used to inject liquid samples in gas chromatography (GC) (see Figure 5.5).[44]

The basis of the SDT test can be seen in Figures 5.3 and 5.4 where volumes of water on surfaces exhibit different diameters depending on how they wet that surface:

- Water on a clean surface will form sheets whose horizontal dimension (diameter) for the same applied volume is large (see Figure 5.6).
- Water on a soiled surface which will form drops whose horizontal dimension (diameter) for the same applied volume is small (see Figure 5.7).

In other words, measure the diameter of a water drop whose volume is known and you can estimate surface cleanliness. The parameter representing surface cleanliness is contact angle (see Section 5.4.5).

This technique, published in 1940,[45] involves estimation of contact angle of a tiny and known volume of liquid (usually water) via measurement of its diameter. The equation relating controlled drop volume, measured diameter, and calculated contact angle ($\theta$) is:

$$\frac{(\text{Drop diameter})^3}{\text{Drop volume}} = \frac{24 \times [\text{Sin}(\theta)]^2}{\pi \times \{2 - [3 \times \text{Cos}(\theta)] + [\text{Cos}(\theta)]^3\}} \quad (5.1)$$

The best way to measure drop size is with magnification. A digital camera (see Section 5.9) can record the image within 1 or 2 seconds (see Section 5.4.6.2) after the drop is deposited on the surface by the syringe. Then the digital image can be examined by whatever software is available, and the diameter estimated (see Figure 5.8).

While not descriptive of quantum mechanics, Equation (5.1) is not solved by visual inspection. Graphical solutions are shown in Figures 5.9 and 5.10 for drop sizes of 2, 5, 10 and 20 μl.

This author doesn't recommend conducting the SDT by solving Equation (5.1). Although this equation is easily programmed in a spreadsheet, there is little reason to do so.

Rather, *use the measurements of drop diameter as the result of the SDT.*

Use measurements of drop diameter as indicative of surface cleanliness. When the diameter has

[43] 1 μl = 0.001 ml.

[44] Such a syringe can be easily purchased. Useful ones have been seen at the auction web site, eBay. Bid no more than 25 euro or much less.

[45] Bikerman, J.J., *Transactions Faraday Society*, 1940, Vol. 36, p. 412.

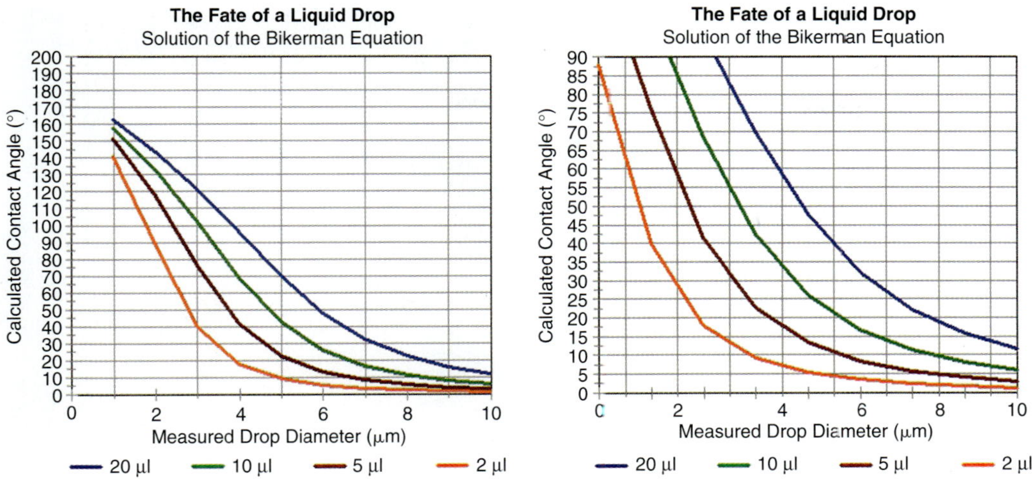


**Figure 5.9**

increased for the same syringed volume, cleanliness has improved – at the spot where that drop was applied. A control state can be developed (see Section 5.4.6).

The SDT offers resolution of some of the issues concerning precision around the "water break" test. However, the issues around variability remain as only a tiny fraction of a surface is examined. Also, remember this test does not directly measure surface cleanliness, it measures surface energy (see Section 5.4.6.1, Footnote 48).

## 5.4.6 A Controlled Approach[46] to the "Water Break" Test

This means that differentiation between beads and sheets can be done via something other than a "calibrated" opinion as in ASTM F22-02.

Use of a control state can be a substantial advantage, if the "water break" test does bring value. The price to obtain this gain is a loss of simplicity and increase of cost.

While a manager can always require completion of the "water break" test on a surface specimen known to be clean[47] (having met requirements for downstream processing), that manager would also prefer to make the outcome more quantitative.

Both aims can be achieved.

### 5.4.6.1 *It's the Tension*[48]

A manager can replace the "water break" test, which involves a GO/NO GO determination[49] with a test where one dispenses a prepared liquid (not

[46] This title may appear counterintuitive. After all, based on Section 5.4.5, why would a manager wish to make quantitative measurements around a phenomenon considered to have significant variability? Quantization will not have impact on the basic five concerns about the "water break" test mentioned in Section 5.4.5. Nevertheless, repeatable numerical output that is representative of the visual difference between beads and sheets of water is valued by some managers because of the paucity of information contained in a result from the "water break" test.

[47] This specimen is known as the control or reference specimen. Its condition of cleanliness is known, and satisfactory.

[48] Surface tension results from intermolecular forces between molecules. At a surface, molecules of a liquid undergo a net force which pulls them inward to the bulk volume of liquid.

Surface energy is not the same as surface tension. Surface energy describes reactivity of solid surfaces. The units of surface energy (free energy per unit area) and surface tension (force per unit length) do happen to be the same, mN/mm (dyne/cm).

In general for a soil to adhere to a surface, it must wet the surface. That means surface energy must overpower soil surface tension. In other words, the surface must exert a stronger attraction to the soil than the inward pull of the soil to itself. If soil surface tension overpowers surface energy, the soil is not bonded to the surface and forms drops. That would make cleaning easy! Soils adhere to surfaces, as ink does to paper, when the surface energy overpowers soil surface tension by at least 10 mN/mm (dyne/cm). Normally but not always, water-based soils are likely to be less well-bonded to surfaces than are solvent-based soils because surface tension of the carrier water is higher than that of a carrier solvent

[49] About whether or not water flushed on a test surface either forms sheets (low contact angle, non-oily hydrophilic surface) or beads (high contact angle, oily hydrophobic surface).

pure water) of known surface tension on to a test surface:

- If the prepared liquid doesn't wet the test surface, it will be seen as forming beads. That means the surface tension of the prepared liquid does not exceed the surface energy[50] of the surface.
- If the prepared liquid does wet the test surface, it will be seen as forming sheets. That means the surface tension of the prepared liquid at least exceeds the surface energy of the test surface.

The test is repeated with another prepared liquid on another surface area thought to have the same surface quality:

> *The outcome is identification of the prepared liquid whose surface tension at least exceeds the surface energy of the test surface.*

Basically, surface energy of the test surface is independently and directly estimated (not measured). The outcome is numerical, not GO/NO GO.

Nothing else between a drop of wetting liquid and a flat surface is measured. No equations need to be solved. No expensive equipment is needed.

The control state is a complete repetition of the above procedure on a surface known to be clean via having met requirements for downstream processing:

- It is the estimate of surface energy,[51] when compared to that measured on surface considered to be clean, which characterizes cleanliness.

There are at least two techniques for implementation of this technology: dyne liquids and the Nordtest method. See Sections 5.4.6.2 and 5.4.6.3.

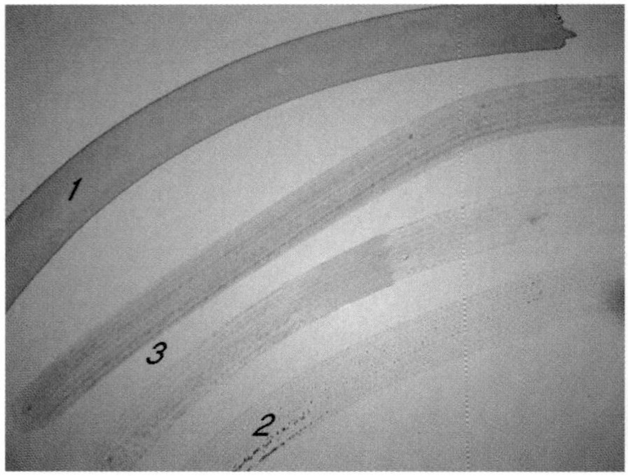


**Figure 5.10**

### 5.4.6.2 *Surface Tension Test Fluids*

Printers,[52,53] coaters, and other operators are very concerned about coverage of liquids on solid surfaces. After all, if ink or paint doesn't wet a surface, it can't adhere to it via adhesion or saturate it via absorption.

That's why surface tension test fluids[54] were developed. They are binary prepared (formulated) mixtures whose surface tension is known. They are portable surface tension standards.

They can be used to establish surface character – cleanliness relative to a standard.[55]

Dyne liquids are applied as a continuous film to about one square inch (6.5 cm$^2$) of surface under study by a brush, swab, wick, or felt-tipped pen.[56] The operator carefully observes when and if the continuous film *retracts* and breaks up into droplets. Experience has shown that wetting is normally adequate when the continuous film of dyne liquid

---

[50] See Section 5.6.3.2.

[51] This outcome is also known as the "critical surface energy." See also Section 5.4.6.3.

[52] See http://www.pillartech.com/corona_tech3a.html for additional information.

[53] Boyle, E., "Taking the Measure of Surface Treatment is a Learning Process," *Paper, Film, & Foil Converters*, September 1, 1996. See http://pffc-online.co/mag/paper taking measure surface/index.html.

[54] Also known as dyne liquids or test inks. They are most frequently used to bracket the surface energy level of polymer or paper substrates.

[55] See Footnote 14. Also see Section 5.6.3.2 and Appendix 2.

[56] As a product, they are produced in 1 or 2 mN/mm (dyne/cm) increments. A container of them resembles a box of crayons. Three suppliers of these products can be found at the http://www.shermantreaters.co.uk, http://www.tigres.de/ and http://www.softal3dt.com.

Care must be taken in their application so that mechanical cleaning work isn't done by moving or removing soil by contact with the applicator.

*remains intact* for at least 2 seconds. There are at least two possible outcomes (see Figure 5.10):[57]

1. The continuous film quickly retracts or decays into droplets – for at least 2 seconds – or doesn't decay and remains continuous. This means that the surface tension of the test ink is *not less* than the surface energy of the surface.
2. The continuous film quickly decays into droplets, almost immediately as applied. This means that the surface tension of the test ink is *less* than the surface energy of the surface.

Normally a user *starts* with a test liquid believed to have a *higher* surface tension than the surface energy of the surface. Here they see a continuous film.[58] Then test liquids with incrementally lower surface tensions are tried on different areas. The desired outcome is when a test liquid is identified which doesn't form a continuous film and does form discrete drops. It is conventional to say the surface has a "dyne value" equal to the surface tension of the solution which maintained a continuous film.

In this way, the surface energy of the test surface is estimated, or bracketed.[59]

In other words, the level of cleanliness of the test surface has been estimated or bracketed. Comparison should then be made with the level of surface energy measured using the surface of a cleaned piece (part) known to perform properly in downstream operations.

There are at least four international standards which cover this determination, chiefly for printing applications on fibers, papers, or films. They are ASTM D2578-04a,[60] ISO 8296,[61] JIS K6768 (1971, polyethylene and polypropylene films), and DIN 53,364.

### 5.4.6.3 *The Nordtest[62] Method*

This method is singled out for consideration because it is a quantitative method which was developed for cleaning evaluations, and because of its parent.[63] Its proper use includes a control state.

Nordtest Poly 176 uses fluids which are normally prepared by the user. They are applied as droplets from a pipette (not the micro-pipette of Figure 5.5):

- A droplet of a liquid of low surface tension will *spread* upon a surface whose energy level is higher. This lowers the energy of the combined system. An observer using the Nordtest method will see the droplet as spontaneously wetting this surface – a small film or sheet.
- Another liquid droplet, whose surface tension is higher than that of the surface, will *not spread*. An observer using the Nordtest method will see the droplet as not spontaneously[64] wetting this surface: a small bead.

A surface with relatively high energy will preferentially be spontaneously spread upon by a liquid with a relatively lower energy, thus decreasing the surface energy of the system.

---

[57] Figure 5.10 courtesy of Sherman Treaters (http://www.shermantreaters.co.uk). The three "stripes" of surface tension test fluid represent the following three outcomes:

1. **Good coverage**. The dyne liquid lies evenly on the material in a continuous line. There is no reticulation (breakup into a network) of the dyne liquid. The surface tension of the material is *at, or higher than*, the dyne level of the test fluid.
2. **No coverage**. The dyne liquid reticulates (breaks up) into droplets. The surface tension of the material is *well below* the dyne level of this test fluid.
3. **Intermediate coverage**. The dyne liquid line is defined but there is partial reticulation from the edges. The surface tension of the material is *just below* the dyne level of this test fluid.

[58] Hansen, C.M., "Characterization of Surfaces by Spreading Liquids," *Journal of Paint Technology*, 1970, Vol. 42, No. 550, pp. 660–664.

[59] See Chapter 4, Section 4.4–4.6 for methodology to establish the reference values ("Golden Lots") of cleanliness, or critical surface energy.

[60] "Standard Test Method for Wetting Tension of Polyethylene and Polypropylene Films." This method uses mixtures of ethylene glycol monoethyl ether and foramide which are not readily volatile and may be good solvents for some soils. The latter is decidedly not wanted in a cleanliness test fluid.

[61] "Plastics – Film and sheeting – Determination of wetting tension."

[62] Hansen, C.M., "New Nordtest Method Easily Shows Contamination on Surfaces," *Pigment and Resin Technology*, 1998, Vol. 27, No. 5, pp. 304–307. The method, NT Poly 176, can be obtained at http //www.nordicinnovation.net/nordtestfiler/poly176.pdf.

[63] Charles Hansen is the developer of solubility parameters based on three intermolecular forces. The parameters allow matching the character of cleaning agents to soils.

[64] Here spontaneously means that the drop will either spread or not within 2 seconds. First impressions should be considered final.

**Table 5.4 Spreading Surface Tensions for Clean Surfaces**

| Material Surface | Surface Tension, mN/mm (dyne/cm) |
|---|---|
| Hot rolled steel | 32–34 |
| Cold rolled steel | 30–32 |
| Stainless steel | 26–28 |
| Aluminum | 24–26 |
| Bronze | 24–26 |
| Copper | 22–24 |
| Polyethylene | 22–24 |

The test fluid used in Nordtest Poly 176 is a mixture of ethanol and water,[65] though other combinations can be used.

Information from Footnotes 105 and 109, about how surface energy differs among materials, is given in Table 5.4 (see Section 5.4.6.4).

Note the values are ranges because the determination is made via a "bracketing" process.

Also note the values in Table 5.4 are provided only for illustration, to show how different materials have different surface energy. *The values in Table 5.4 should not be used as reference values for clean surfaces (control states).* That should be determined by *your* staff using each material common to *your* operations which has been identified as clean in *your* downstream operations.

### 5.4.6.4 *Compatibility and Incompatibility of Methods*

Results of the Nordtest method, the dyne liquid tests and the "water break" test are not compatible. They shouldn't be interchanged. One must make a selection:

- *Different criteria are used.* The defining criteria in the Nordtest method is spontaneous *spreading* of applied liquid. This refers to the transition between soiled and cleaned – *when the surface is just not soiled.* The defining criterial in use of dyne liquids is *retraction* of applied liquid. This refers to the transition between cleaned and soiled – *when the surface is just not clean.* While similar, and sometimes the same, these two transitions are not theoretically identical.[66] Values of surface energy from the Nordtest are often lower than those from tests with dyne liquids.
- *The nature of the interface between the test drop and the soiled surface matters.* Since different fluids are used among the two tests, the molecular interaction between the soil and the test fluid will be different. If the test fluid happens to be a preferred solvent for the soil, the test fluid will more easily and fully wet the surface; and conversely.
- The *test fluids used are different.*

This author prefers the dyne liquid test, though not for a technical reason. Dyne liquids have been long and commonly used. Evaluations with them may be more easily accepted by end-use customers.

Yet, there is a technical reason upon which to make a choice. The Nordtest Poly 197 can be used with any set of liquid mixtures. The test fluids in the dyne liquid pens are only those commercially useful in certain applications such as printing. This author prefers the Nordtest Poly 197 when it is necessary to choose the test liquid (see Section 5.6.5.3).

### 5.4.6.5 *Evaluation of Surface Tension Test Fluids for Cleanliness Testing*

In a sense, the dyne liquid or Nordtest technology is perfect:

- One can't spend €100 on a reasonable supply of test fluids.
- There is no investment needed for facilities.
- The range of outcomes due to differences in operator judgment is reduced from GO/NO GO to one or two increments of surface tension.
- The outcome is not binary as is the "water break" test. A range of values of critical surface energy are possible as cleanliness varies. Yet, a GO/NO GO value of surface tension can be chosen if necessary.[67]

[65] Not Scotch over ice, these are a series of calibrated mixtures of ethanol and water with surface tensions varying by two units between mixtures. Ethanol and water both naturally disappear from the surface under test (evaporate), are inexpensive, and are harmless to persons, the environment, and most substrates.

[66] Hansen, C.M. and Pierce, P.E., "Surface Effects in Coating Processes," *Industrial Engineering Chemistry – Product Research & Development*, 1974, Vol. 13, No.4, pp. 218–225.

[67] See Chapter 4, Section 4.7 about the statistical hazards of using discrete cleanliness data.

- The outcome requires less judgment than does the "water break" test, and a control state is provided.

Yet, in another sense, both are useless:

- If the soil is soluble in the dyne liquid or contains particles.
- The tests can't be used with metals whose surfaces are "live" – reacting with Oxygen and forming a protective oxide layer in the first dozen angstroms of surface thickness. This is because the surface energy level is not stable[68] over time. There will be no control state.
- If the cleaned surface is still wet with cleaning agent as the test liquid will be diluted.

The details of your application will separate a perfect method from one of negligible value.

### 5.4.7 Gravimetric Methods

If contractors in Europe can assay their success in cleaning HVAC ducts via gravimetric means,[69] and scientists at the US NASA can do so in assembly of spacecraft,[70] surely one can do so in management of general cleaning work.

This is a clear example, as mentioned in Section 5.3.1, where the same technique can be and is being used to quantify quite different levels of soil on quite different parts.

Gravimetric methods refer to the practice of weighing a part (a cleaning duct or a piece of Oxygen piping) before and after a cleaning process stage. The difference in weight is presumed to be what was removed in the cleaning stage.

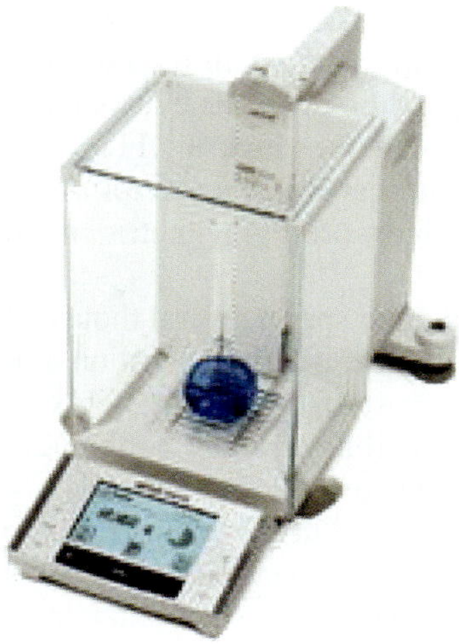

**Figure 5.11**

Gravimetric methods are practiced in two general ways:

1. Weighing the part (which includes the weight of the remaining soil).
2. Weighing an extract of fluid[71] which was produced by treating the part after it was cleaned.[72] This can be:
   - A filter through which the extract was passed. The material weighed was the particulate matter remaining on the parts after cleaning.[73]
   - The residue remaining after the extracting fluid is removed by evaporation.[74]

An analytical balance with superior resolution is required to recognize a change of <1 mg/SF residue on a part whose size covers 1 SI of area (~0.08 mg).

[68] If one starts with a pristine metal configuration (no oxides), surface energy is ~400 mN/mm (dyne/cm). As some metal ions interact with Oxygen, oxide sites develop and begin to cover the surface. When, the first dozen or so angstroms of surface thickness become oxide sites, no more can be formed as native metal ions are protected (passivated). Now the surface energy is at some minimum for this configuration.

The surface energy outcome is both asymptotic with time and surface specific.

[69] Broms, S. and Cramer, S., "A Gravimetric Measurement Method for Ventilation Cleanliness, Before and after Cleaning. Description and Application," available through the Asociación Técnica Española de Climatización y Refrigeración http://www.atecyr.org. The required level of cleaning was ~100 mg/SF residue.

[70] ASTM E1235, "Test Method for Gravimetric Determination of Nonvolatile Residue (NVR) in Environmentally Controlled Areas for Spacecraft." The required level of cleaning was <1 mg/SF residue.

[71] Deionized (DI) water is often used as the extract fluid, even when the soils are minuscule amounts of oil, fingerprints, or grease. Here the extraction is not done by solution.

[72] This extract is not the actual cleaning solution. Rather it is part of a cleanliness verification scheme.

[73] This is the method commonly used to assay the quality of cleanliness in tubing used by the US Navy in Oxygen service (see MIL-STD-1330D and MIL-STD-1622B).

[74] If some of the part were to be removed in the cleaning stage, that wouldn't be noticed in this cleanliness test (see Section 5.4.1).

This balance must be located in an environment where humidity, temperature, and air currents are controlled, and be operated by someone other than the local butcher (see Figure 5.11).

Absent such capability, even for assaying residue in HVAC ductwork, gravimetric methods aren't practical.

Be prepared to spend several thousand Euro for a suitable analytical balance. Insist on a training program for your staff, and calibration standards. The controlled environment may be the largest component of capital cost.

Cleaning work in a century past could have included use of an analytical balance suitable for gravimetric analysis of soil remaining on cleaned parts. Most of the technology listed in Appendix 2 could not have been used because it hadn't been invented. In this century, gravimetric technology has less of a role to play because of its lack of specificity and high cost to obtain precision. If you manage facilities with a suitable analytical balance and controlled environment, use them. If not, choose another technology.

## 5.5 TESTS PROVIDING INFORMATION ABOUT A PRECISION LEVEL OF CLEANING

The best answer to the question, "Just what *is precision* cleaning?" came from someone who remarked offhand, "Well, it's just as good as it has to be!"

*Precision cleaning* is an often-used and probably meaningless[75] technical phrase. In this book it means cleaning work done to a standard which speaks to how to good it has to be.

### 5.5.1 ASTM Methods as Cleaning Tests

A very efficient way to conduct and manage cleaning tests is to use what the ASTM has done: develop open, unbiased standards about performance.

The use of ASTM standards[76] is efficient because the standards useful to you are probably already developed, internationally available, can be performed at your site or by any commercial testing laboratory, have been openly evaluated by experts, and are chiefly developed by volunteer users[77] for other users.

Most ASTM standards cover two aspects:

1. How to do something (standard practices).
2. How to know if that something was satisfactorily done (standard test methods).

Many standards contain validation procedures.

### 5.5.2 Specific ASTM Procedures

In order to establish simple performance tests for cleanliness in your operations, you need to know:

- The material from which your parts are made.
- What will be done next with the cleaned parts. A key issue is whether that next step involves bonding or some other operation where elimination of particles is the main objective.

If particulate removal is NOT crucial to the success of the next operation, use Table 5.5.[78] You will also need to choose whether your cleanliness test is to be general (measures the apparent surface quality) or end-use (measures surface bonding quality).

For all operations where particle removal is of concern use Table 5.6.

Obviously, the ASTM is not the only organization producing standards which bring value when used in cleaning tests. Others who offer similar technology are:

- American National Standards Institute (ANSI), http://www.ansi.org
- National Resource for Global Standards (NSSN), http://www.nssn.org
- National Institute of Standards and Technology (NIST), http://www.nist/gov

[75] The author's web site is http://www.precisioncleaning.com.

[76] Covering more operations than one wants to know about, there are more than 10,000 ASTM standards published annually in 70 volumes, including those for the manufacture of polymers, industrial lighting, fire extinguishing agents, use of liquid penetrants, waste management, expansion joints on bridges, lithographic imaging, and industrial hygiene. Standards purchased from the ASTM are not expensive, and your library may even have copies. You can contact the ASTM at 100 Barr Harbor Drive, West Conshohocken, PA 19428-2959, US; or (610)-832-9500; or service@local.astm.org.

[77] In the interest of full disclosure, this author is a practicing ASTM member. ASTM standard subcommittees are currently developing and reviewing specific standards for various cleaning methods.

[78] Durkee, J.B. and Gawenis, C., "Industrial-Strength Cleanliness Tests," *Parts Cleaning Magazine*, July, 1997, pp. 10–12.

**Table 5.5 Tests for Cleaning Operations Where Particulate is *Not* of Concern**

| Material Type | ASTM Test |
|---|---|
| General materials | D6361-98(2004)e1 Standard guide for selecting cleaning agents and processes<br>G122-96(2002) Standard test method for evaluating the effectiveness of cleaning agents |
| General metals | D1280-00 Standard test method for total immersion corrosion test for soak tank metal cleaners |
| Carbon and low-alloy steels<br>Cobalt and Nickel alloys<br>300 and 400-series stainless steels<br>Cast Iron epoxy matrices (with metals)<br>Titanium alloys<br>Aluminum, Magnesium, brass, bronze<br>Copper and Copper alloys<br>Carbon and low-alloy steels<br>Cobalt and Nickel alloys<br>300 and 400-series stainless steels<br>Cast Iron epoxy matrices (with metals) | F483-98(2002)e1 Standard test method for total immersion corrosion test for aircraft maintenance chemicals<br>D1280-00 Standard test method for total immersion corrosion test for soak tank metal cleaners<br>F485-02 Standard test method for effects of cleaners on unpainted aircraft surfaces<br>F1110-02 Standard test method for sandwich corrosion test<br>ARP 1512 Sandwich corrosion test<br>ASTM E21.05, "Standard Method for Measurement of Non-volatile Residue (NVR) on Surfaces"<br>F519-97e2 Standard test method for mechanical Hydrogen embrittlement evaluation of plating processes and service environments<br>F1111-02 Standard test method for corrosion of low-embrittling cadmium plate by aircraft maintenance chemicals |
| Titanium alloys | F945-01 Standard test method for stress-corrosion of Titanium alloys by aircraft engine cleaning materials<br>ARP 1795 Stress-corrosion of Titanium alloys |
| Rubber compounds | AMS 3209/3209 Test for rubber compatibility<br>D2240-04 Standard test method for rubber property – durometer hardness |
| Acrylic plastics<br>Unspecified thermoset plastics | F484-02 Standard test method for stress crazing of acrylic plastics in contact with liquid or semi-liquid compounds<br>F791-96(2002) Standard test method for stress crazing of transparent plastics |
| Painted surfaces | F502-02 Standard test method for effects of cleaning and chemical maintenance materials on painted aircraft surfaces |

- Canadian Standards Association (CSA), http://www.csa.ca
- International Organization for Standardization (ISO), http://www.iso.ch

Other standards organizations involving aerospace or military operations include:

- Aerospace Material Specification (AMS)
- All-Russia Institute of Aviation Materials (VIAM)
- All-Russia Institute of Light Alloys Joint-stock company (VILS)
- Central Aerohydrodynamic Institute (TSAGI)
- European Aerospace Association (AECMA)
- US Military (MIL[79])

### 5.5.3 Using ASTM Tests as Cleaning Tests

In practice, the selection of cleaning tests is easier than the appearance of Tables 5.5 and 5.6 might

[79] As this is written general MIL standards are being replaced with performance-based standards meeting the needs of individual services. But a well-used general standard is MIL-STD-1246 (now IEST-STD-1246D, "Product Cleanliness Levels and Contamination Control Program." It can be downloaded from http://cadigweb.ew.usna.edu/~midstar/downloads/main/references/milstd1246c.pdf.

**Table 5.6 Tests for Cleaning Operations Where Particulate *is* of Concern**

| Material type | General Cleaning Tests Where Apparent Surface Quality Can Be Evaluated | End-use Cleaning Tests Where the Surface Will Be Bonded to Another |
|---|---|---|
| Carbon and low-alloy steels<br>Cobalt and Nickel alloys<br>300 and 400-series stainless steels<br>Cast Iron epoxy matrices (with metals)<br>Titanium alloys<br>Aluminum, Magnesium<br>Brass, bronze<br>Copper and Copper alloys | F483-98(2002)e1 Standard test method for total immersion corrosion test for aircraft maintenance chemicals<br>D1280-00 Standard test method for total immersion corrosion test for soak tank metal cleaners<br>F485-02 Standard test method for effects of cleaners on unpainted aircraft surfaces<br>F519-97e2 Standard test method for mechanical Hydrogen embrittlement evaluation of plating processes and service environments<br>F1111-02 Standard test method for corrosion of low-embrittling cadmium plate by aircraft maintenance chemicals | D3167-03a(2004) Standard test method for floating roller peel resistance of adhesives<br>D1876-01 Standard test method for peel resistance of adhesives (T-Peel test)<br>D1781-98(2004) Standard test method for climbing drum peel for adhesives<br>D1002-01 Standard test method for apparent shear strength of single-lap-joint adhesively bonded metal specimens by tension loading (metal-to-metal)<br>D2919-01 Standard test method for determining durability of adhesive joints stressed in shear by tension loading<br>D903-98(2004) Standard test method for peel or stripping strength of adhesive bonds<br>ASTM E21.05, "Standard method for measurement of non-volatile residue (NVR) on surfaces" |
| Titanium alloys | F945-01 Standard test method for stress-corrosion of Titanium alloys by aircraft engine cleaning materials<br>ARP 1795 Stress-corrosion of Titanium alloys | |
| Aluminum, Magnesium | | D3762-03 Standard test method for adhesive-bonded surface durability of Aluminum (wedge test) |
| Rubber | AMS 3209 (MS28900) | |

make it appear. A broad variety[80] of ASTM tests have been included to show the power of this approach.

Remember, these ASTM tests were designed for other purposes. These are compatibility tests, not cleaning effectiveness tests.

Yet for a surface to be compatible with an adhesive, it must be clean. That's what these ASTM tests do – *they all provide quantification of the character of a surface so it may be used in downstream operations*. That's what one wants their cleaning test to provide.

Here are some examples:

- **F483-98(2002)e1** measures corrosion of aircraft metals with time under conditions of total immersion by a combination of weight change measurements and visual qualitative determination of change. If a surface is considered to be corroded, it probably isn't clean.
- **D1876-01** measures the relative peel resistance of adhesive bonds between flexible surfaces. If an adhesive can't stick to a surface, the surface probably isn't clean.
- **F484-02** measures a crazing effect on a plastic surface. If a surface is characterized as being crazed, it probably isn't clean.

You may need to employ only one, or maybe two, of these tests to determine if these parts are clean. So, how do you choose which to adopt as your cleanliness test?

Choose the cheapest and simplest one that most closely matches the actions of your next process step. For example:

- For Titanium parts having particulate, where paint adhesion must be tested, choose peel resistance (ASTM D-1876-72) or possibly peel/stripping strength (ASTM D-903-49). Both tests are commonly used in coatings work.
- Or for Titanium parts having particulate, where the chemical condition of the surface must be tested, choose *effects on unpainted surfaces* (ASTM F-485).

In summary, the use of ASTM (*or other*) standards for surface characterization as cleaning tests allows users to indirectly test for cleanliness with proven procedures which are easily accepted by both international suppliers and users. Further, they have negligible cost.

## 5.6 TESTS PROVIDING INFORMATION ABOUT A CRITICAL LEVEL OF CLEANING

Critical cleaning is all about cleanliness testing. The phrase refers more to the consequences of inadequate performance than to the amounts of soil removed or the physical size of individual soil fragments.[81]

Managers of critical cleaning insist upon an extraordinary level of cleanliness testing, perhaps every item tested at every step of processing. They do that when critical is the consequence to their enterprise if they're wrong about the cleanliness of some surface. They do that when the cleanliness of the assembly is of issue, not the cleanliness of its components.

Here the standard of performance (versus Section 5.5) is at or below the detection limit.

### 5.6.1 A Core Principle

Managers responsible for producing superconducting tape, vast quantities of semiconductor switches, micro-machined objects, composite materials for aircraft structures, etc. have learned that:

- Sometimes the consequences of use of their work product is related to items they can't detect or quantify.

[80] ASTM E1548-03 "Standard Practice for Preparation of Aerospace Contamination Control Plans" is a resource for developing a cleanliness management system when allowable residue levels are exceedingly low – BUT it is also an excellent educational guide for developing a cleanliness management system for any purpose.

[81] In a past life, this author was responsible for development of technology to remove large (10–100 μm) fragments of over-polymerized material from polymer films. These defects were called "gel." The films were used in manufacture of laminated glass fused in automotive windshields. The business consequence of any measurable defects in any assembled windshield was that all windshields would have to be purchased and recalled by the author's enterprise. Though not understood at the time, this was critical cleaning, even though the defect size was not minuscule.

That's one reason why these works are conducted and completed in controlled environments:[82]

- In other words, cleanliness testing may prove nothing, because the soils are at levels which are undetectable.
- In still other words, cleanliness testing may prove nothing, because the soils have not been chemically or physically identified, or even located.
- In yet other words, outcomes of cleanliness testing (and operations where cleanliness is produced) will be no better than the quality of the ambient environment. Garbage in, garbage out!

A manager's enterprise avoids negative critical outcomes by restriction of soil transfer rather than removal or detection of it.

### 5.6.2 Change Happens

Information about detection of minuscule quantities of residues published here is likely to be outdated. Some of the forces driving this change are not generally related to quality improvement, but are:

- US Homeland Security issues, which foster increased levels of research and development in the detection and analysis of surfaces contaminated with explosives, hazardous biological materials, DNA, illegal drugs, or other materials of interest:
  - Forensic examination of surfaces thought to be contaminated with DNA or drugs.
  - Recognition of specific personnel via software algorithms.[83]
- Legal liability in medical, dental, and food service[84] operations.
- Growth of applications in nanotechnology.

Evolution of high-capacity computers is an enabling factor in this change. Computer vision enables the integration of views from many cameras into a single, consistent "superimage."

### 5.6.3 Visual (Human) Inspection with Magnification

Clients with critical (high consequence of failure) cleaning operations often perform cleanliness tests by training and empowering a person to observe, identify, and count particles (or other defects) on a surface. These surfaces have included components to be inserted in the human body (stents and screws) as well as components to be bonded into aircraft frames, and metal tape for superconductors.

Liability is the reason for this. Some of these clients require acceptance by the US Food and Drug Administration's (FDA) through protocols in their Quality Systems Manual (QSM). Personal inspection is often thought to be the most secure way of providing compliance. Many clients have believed it is easier to validate[85] a human-based cleanliness detector than a mechanical or chemical instrument.

Training and auditing are the two key activities necessary to success (avoidance of failure). Auditing is probably the more important of the two activities (see Chapter 4, Section 4.1.8.2).

### 5.6.4 Detection of Biological Residue in Food Service

Not all cleaning work is involved with removing grease from dirty screws or particles from optic components. Some managers are responsible for cleanliness outside of parts cleaning.

---

[82] This topic is outside the scope of this book, but managers should consider *Controlled Environments Magazine*. Full disclosure requires mention that this author writes a monthly column on critical cleaning for this magazine.
[83] Picardi, M. and Jan, T., "Recent Advances in Computer Vision," *The Industrial Physicist Magazine*, February/March, 2003.
[84] Food service includes significantly more than restaurant service. Fermentation of beer or bread, bottling of wine and dairy products, and packaging of meat, poultry, and eggs are just a few of the food service applications where clean surfaces are critical to managers.

The U.S. Food Safety and Inspection Service (FSIS) sponsors a program called Pathogen Reduction/Hazard Analysis and Critical Control Point (PR/HACCP – pronounced "hass up"). HACCP is a system that identifies and monitors potential biological, chemical, or physical food-borne hazards which can adversely affect the safety of food products. Compliance with HACCP is beyond the scope of this book.
[85] The view of this author is that the issue is less of validation and more of responsibility. Managers believe it is easier to assign fault and replace a human person versus a mechanical/chemical device in the maelstrom of activity following critical consequences.

Cleanliness is critical in food service industries. As customers, that's what we expect. Cleaning processes in food service are probably more simple than those processes used in other operations where failure is critical. Cleaning processes in food service typically involve contact with biodegradable aqueous detergents, hot and or superheated water, mechanical abrasion, and possibly ultrasonic-driven agitation. There is usually only a single stage of washing and one of rinsing.

Versus critical cleaning work in nanotechnology, there is often little inspection or purification of cleaning ingredients in most food service such as water. Verification (inspection) of cleanliness has become more a role for local regulators and less for operating managers.

A major reason for this is that inspection for retained biological (food residue) contamination is expensive, time-consuming, and can be cumbersome. There are three basic approaches used to assess retained biological contamination in food service:

1. ATP bioluminescence.
2. Detection of specific residues, usually bacteria.
3. Conventional surface analysis for general contamination.

#### 5.6.4.1 *ATP Bioluminescence*

ATP is not three doors down from where one gets cash from a slot in a wall. ATP is Adenosine Triphosphate. It is a chemical compound found in all living cells or cells that were once living. This include bacteria, food debris, yeast, and mold.

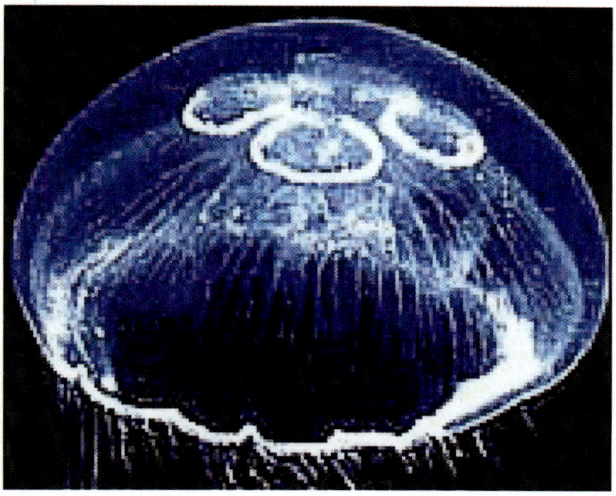

**Figure 5.12**

ATP is, or is a proxy for, food contamination.[86] Food residues, both living and dead, can be detected via detection of ATP.

That can be done in an unusual way by bioluminescence.[87] This is emission of visible light by living organisms such as the firefly[88] and various fish (see Figure 5.12), fungi, and bacteria. Bioluminescence occurs when the enzyme[89] luciferase[90] comes in contact and reacts with ATP.

The amount of light emitted in this reaction is directly proportional to the amount of ATP detected (food residue) in a sample. Light output from the reaction that is measured by a luminometer, which can be a handheld device. Calibration data are shown in Figure 5.13.[91] Note that the output quantity is an amount of ATP not a concentration of ATP.

[86] ATP serves as the major energy source within the cell to drive a number of biological processes such as photosynthesis, muscle contraction, and the synthesis of proteins. The complicated structure of ATP is shown at right.

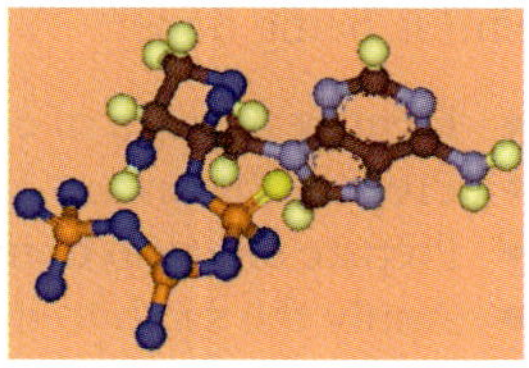

It is a highly stable compound that persists long after a cell has died. So, it's an excellent substance to validate cleanliness. Unclean food contact surfaces have should identifiable amounts of ATP from residue and perhaps microbial cells. That's the core of the cleanliness verification method (see Figure 5.13).

[87] Specifics of the main chemical reaction producing bioluminescence is beyond scope of this book, but a general form of the chemical reaction is:

$$\text{ATP} + d\text{-Luciferin} + O_2 \rightarrow \text{Oxyluciferin} + \text{AMP} + \text{Pyrophosphate} + CO_2 + \text{Light (at 560 nm)}$$

[88] The firefly uses this enzyme to emit flashes of light to attract its mate. See: DeLuca, M.A. and McElroy, W.D., "Purification and Properties of Firefly Luciferase," *Methods in Enzymology*, 1978, Vol. 57, pp. 3–15.

[89] Enzymes are catalysts. They aren't normally consumed in chemical reactions, but sometimes they are modified. Many firms sell test kits with these enzymes for use in cleanliness testing of food service and other operations. Examples can be found at http://www.neogen.com and http://www.promega.com/

[90] Luciferase is a generic name for the many enzymes which produce bioluminescence. The name Luciferase does not refer to part of the devil. It is derived from Lucifer, which is Latin for light-bearer.

Luciferin is an another generic name. It refers to light-emitting pigments found in organisms capable of bioluminescence.

[91] See *Promega Technical bulletin #268*. The Luciferin reactant must be stored under refrigeration, and carefully used to avoid contamination with other ATP-bearing species.

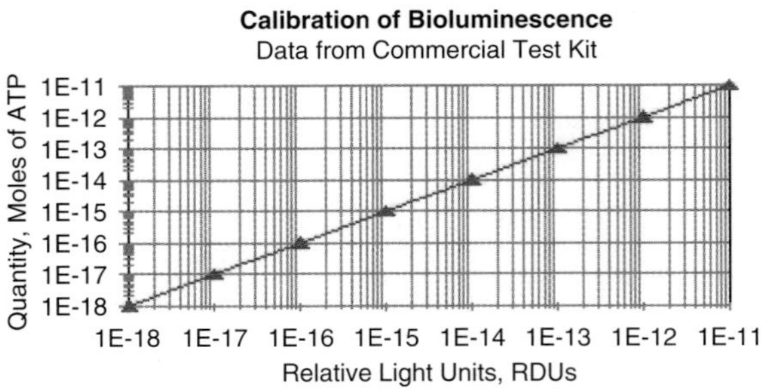


**Figure 5.13**

Results take only a matter of minutes after setup. This allows immediate response. ATP bioluminescence is not an analysis for a specific microbial species. Although measuring total biological residue is a better indicator of cleanliness, that test takes more time and is commonly performed off-site. ATP bioluminescence speaks only to total surface cleanliness, and can be easily done on-site. That's almost always adequate for this industry.

There is a cost and commitment. Investment in facilities can range from 500 to more than 5,000 euro depending upon the degree of automation required. Training for technique and avoidance of contamination is essential to valued results. But most retail food service establishments don't participate in use of ATP. That speaks to their perceived value of surface cleanliness and difficulty of justifying the capital for investment.

### 5.6.4.2 *Detection of Specific Bacteria Residues*

Here is how it had been done for many years and still is being done:

- Identify the area of concern.
- Prepare the area.
- Prepare the swab from the package to take a sample.
- Apply the swab to the area.
- Seal the swab in the sample container.
- Forward the container to a local lab that prepares specific cultures to test for specific bacterial growth.
- Receive the results in several days or so.

Tests can be done for *Listeria* and *Listeria monocytogenes*, *Salmonella*, *E. coli* O157:H7, *Staphylococcus aureus*, Generic *E. coli* and total coliform, and other bacteria of concern.

Today, managers can purchase test kits[92] from operating budgets and do these (and other) tests in-house. This provides a tradeoff versus justification of investment capital for a luminometer to measure ATP bioluminescence. Basically the trade is classical – investment versus operating cost, specificity versus generality:

- The test kits avoid capital investment and provide specific identification of significant bacteria.
- Luminometers require investment capital and measure total food residue.

That's the type of choice managers are paid to make!

### 5.6.4.3 *Conventional Surface Analysis*

Managers who use ATP bioluminescence technology are concerned about total contamination with food residue. Managers who use the test kits are concerned about specific contamination. The analytical techniques listed in the appendix to this chapter are a conventional approach to monitor specific chemical contamination.

That can be other than food residue or associated bacteria – it can be spent cleaning, disinfecting, or other treatment chemicals. All can represent infringement of the Food Safety and Inspection Service (FSISs) Hazard Analysis and Critical Control Point (HACCP) protocols.

## 5.6.5 No Sheet(s), The Contact Angle is High!

Sheets of water (meaning hydrophilic or a "clean" surface) can be differentiated more fully from beads of water (meaning hydrophobic or a "dirty" surface).

The approach avoids the methods of ASTM F22-02 – attempting judgment from non-quantitative observations of the supernatant (floating) water on the surface whose cleanliness is in question.

The approach also avoids the methods associated with dyne liquids or the Nordtest Poly 197, which are

[92] Each test kit contains antibody-coated wells which are marvelously complex layered flow systems. The sample moves through the wells via wicking action. The antibodies are specific to the kit's target bacteria substance. The sample moves through the system and binds to a conjugate enzyme. A substrate is added to produce a color change – usually from red to blue. The more conjugate, the more blue color, and the more target substance that is detected.

described in Section 5.4.6.2, improving judgment by using a graduated scale of cleanliness outcomes.

This differentiation involves measuring the angle of contact between the water bead/sheet and the flat surface.

The contact angle for a idealized water *bead* is shown in Figure 5.14. Taken from the horizontal and tangent to the bead surface, the angle in this figure is well more than 90°.

The contact angle for a idealized water *sheet* is shown in Figure 5.15. Here, the contact angle is well less than 90°.[93]

The basic idea is that:

- A high contact angle (beads of water; oily hydrophobic surface) means the surface is more dirty. A contact angle of almost 180° represents almost complete detachment of the water from the surface (almost no wetting).
- A low contact angle (sheets of water; non-oily hydrophilic surface) means the surface is more clean. *A zero contact angle represents complete wetting.*

### 5.6.5.1 *The Energetics of Surfaces*

Contact angle is a measure of interfacial energy[94] – between the cleaned or soiled surface, and a liquid – presumably water.

Surface energy[95] is a term used to describe the reactivity of the surface liquid to a solid surface.

Contact angle and surface energy are conveniently related through an semi-empirical model. The model, like success, has many fathers.[96] It is referred to as the Girifalco – Good – Fowkes – Young equation,[97,98] and Equation (5.2)

$$\text{Cos}(\theta) = -1 + 2x\sqrt{\frac{\gamma_{\text{Interfacial solid}}}{\gamma_{\text{Liquid}}}} \quad (5.2)$$

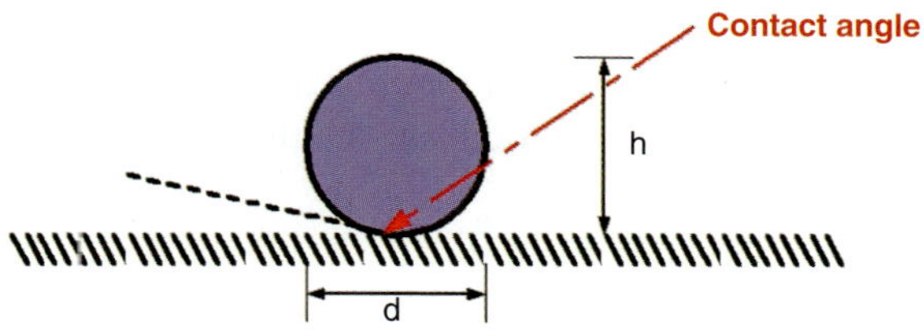


**Figure 5.14**

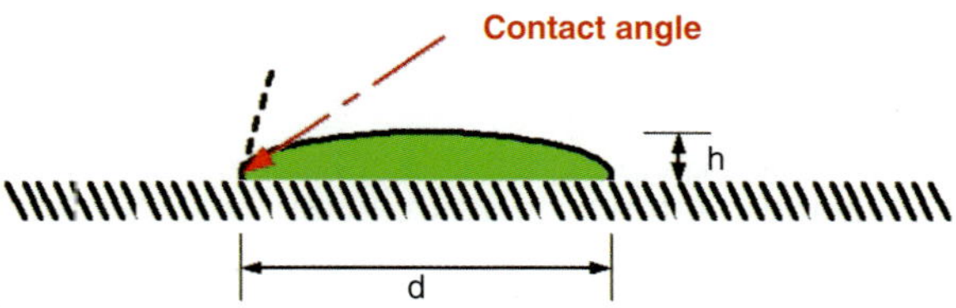


**Figure 5.15**

where: $\theta$ = Contact angle, in degrees.
$\gamma_{\text{Interfacial solid}}$ = Critical surface tension of the clean or contaminated solid at the interface between liquid and the solid surface, in mN/mm (dyne/cm).
$\gamma_{\text{Liquid}}$ = Liquid surface tension, in mN/mm (dyne/cm).

Based on the surface composition, $\gamma_{\text{Interfacial solid}}$ is a property of that surface. Measured values of non-metals and some metals can be found in reference 109, page 365, and Table 5.7.

Equation (5.2) describes how contact angle ($\theta$) is related to the:

- Nature of the solid surface (clean or contaminated) for a given liquid.
- Nature of various liquids for a given solid surface (clean or contaminated).

---

[93] When the contact angle is zero (which is physically impossible unless the liquid surface tension is also zero, which it can be under supercritical conditions), the surface will be completely wetted by a liquid. With a contact angle of 15°, there will be good wetting of a surface. When the contact angle is more than 90°, good wetting isn't going to happen, beads of liquid will be seen.

[94] Packham, D.E., "Work of Adhesion: Contact Angles and Contact Mechanics," *International Journal of Adhesion and Adhesives*, 1996, Vol. 16, pp. 121–128. See http://www.bath.ac.uk/~mssdep/paper5.htm.

[95] Surface energy has the same units as surface tension, dyne/cm or mN/m. While contact angle is dependent upon the fluid used, surface energies have the benefit of theoretically being independent of choice of fluid. See Section 5.4.6.1 where this term is used with surface tension test fluids.

[96] Bikerman, J.J., "Surface Energy of Solids", *Physica Status Solidi*, 1965, Vol. 10, No. 3.

[97] See page 376, Equation X-46 within the reference in Footnote 109.

[98] Girifalco, L.A. and Good, R.J., *J Phys Chem*, 1957, Vol. 61, No. 094.

Equation (5.2) is plotted[99] in Figures 5.16 (water on various surfaces) and 5.17 (various liquids on PTFE). Both sets of information are plotted together as Figure 5.18.

### 5.6.5.2 *Quantitative Identification of Surface Cleanliness*

For a clean surface,[100] $\gamma_{\text{Interfacial solid}}$ can be estimated by:

- Using a clean liquid whose surface tension is known ($\gamma_{\text{Liquid}}$).
- Making multiple measurements (see Section 5.6.5.3) of contact angle ($\theta$) using various parts of the surface.[101]
- Solving Equation (5.2) for $\gamma_{\text{Interfacial solid}}$ for each measurement.

Equation (5.2) is restated as Equation (5.3) in terms of $\gamma_{\text{Interfacial solid}}$:

$$\gamma_{\text{Interfacial solid}} = \left[\sqrt{\frac{(\text{Cos}(\theta) + 1)}{2}}\right] \times \gamma_{\text{Liquid}} \quad (5.3)$$

Average of measured values of contact angle, *using a single liquid*, of $\gamma_{\text{Interfacial solid}}$ on any surface can be used to define the energy state representative of that clean surface:

- It is this value of $\gamma_{\text{Interfacial solid}}$ which should be the aim point[102] for cleanliness testing. One should use contact angle ($\theta$) as the measured parameter, a clean liquid whose surface tension is known ($\gamma_{\text{Liquid}}$), and Equation 5.3.

In summary, cleanliness can be established relative to a reference condition[103] for any solid surface by measuring the contact angle of a drop of liquid placed upon the surface, and by use of Equation (5.3). This approach converts the qualitative "water break" test to a quantitative one.

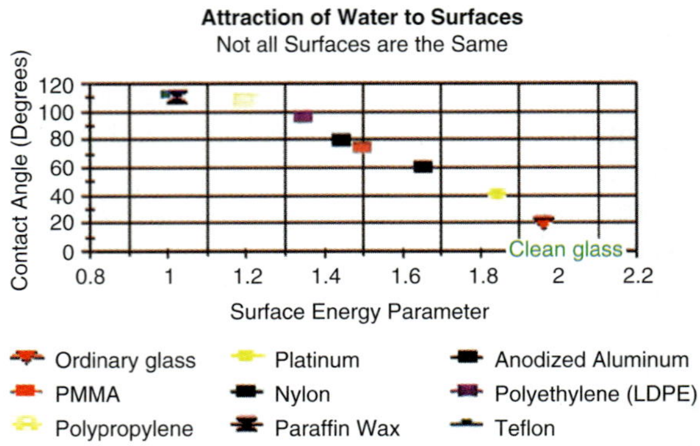


**Figure 5.16**

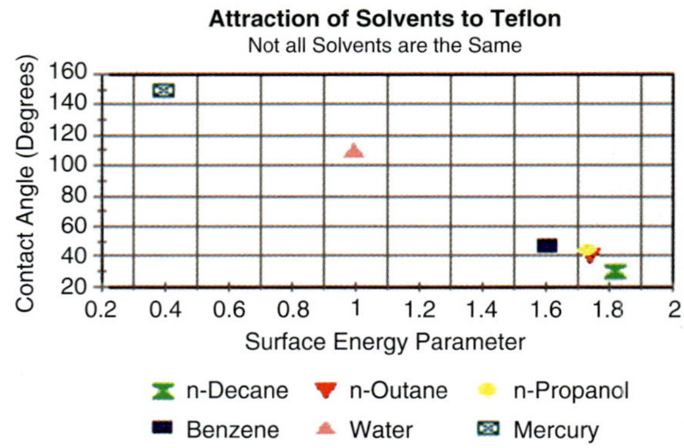


**Figure 5.17**

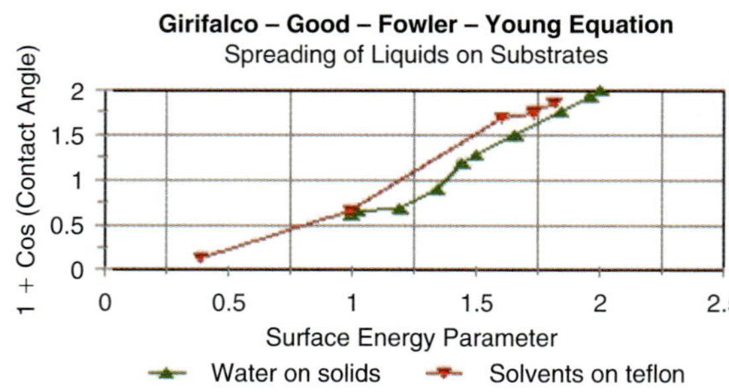


**Figure 5.18**

[99] Data for both plots is from Tables 5.7 and 5.8. The surface energy parameter is the right side of Equation (5.3).

[100] Obviously, a clean surface is one which has enabled satisfactory downstream processing or operations.

[101] Obviously, the surface tension of liquid water (or whatever liquid is used) at use temperature must be obtained from any handbook (72.8 dyne is the value at 20°C).

[102] This aim point is a surface energy whose units are mN/mm (dyne/cm).

[103] See Chapter 4, Sections 4.4–4.6 for methodology to establish the reference values ("Golden Lots") of $\gamma_{\text{Interfacial solid}}$, and Section 5.6.5.3.

**Table 5.7**

| Liquid Applied to Polystyrene | Liquid Surface Tension, Dyne/cm | Contact Angle (°) θ | (1+Cos θ)/2 | $\gamma_{\text{Interfacial Solid}}$ (via Eq. 5.3), Dyne/cm |
|---|---|---|---|---|
| Water | 72.6 | 93 | 0.474 | 34.40 |
| Glycerol | 63.4 | 82 | 0.570 | 36.11 |
| Formamide | 58.2 | 76 | 0.621 | 36.14 |
| Trichlorobiphenyl | 45.3 | 18 | 0.976 | 44.19 |
| 1-Bromonapthalene | 44.6 | 18 | 0.976 | 43.51 |
| Tricresyl Phosphate | 40.9 | 14 | 0.985 | 40.29 |
| Hexachloropropylene | 38.1 | 11 | 0.991 | 37.75 |
| 1,1-Diphenylethane | 37.7 | 12 | 0.989 | 37.29 |
| *t*-butyl Naphthalene | 33.7 | 5 | 0.998 | 33.64 |
| Dicyclohexyl | 32.8 | Spread | | |
| bis (2-ethylhexyl)-Orthopthalate | 31.3 | 7 | 0.996 | 31.18 |
| Squalane | 29.5 | Spread | | |
| Hexadecane | 27.6 | Spread | | |
| Tetradecane | 26.7 | Spread | | |

### 5.6.5.3 *An Effect of Liquid Surfaces*

Relative to Equation (5.3), which liquid should be used for testing to establish contamination, and the reference of cleanliness? Should water always be used?

The choice of liquid does matter.[104] See Table 5.7 where various liquids are applied to polystyrene.[105]

The point of the information in Table 5.7 isn't with what chemical to clean polystyrene. The point is how contact angles are dependent upon the interaction between the liquid and the surface which it contacts. Notice how the contact angle varies from:

- Very high when the liquid is quite unlike polystyrene (water).
- Very low when the liquid is much more like a hydrocarbon (hexadecane).

The test liquid for *all* cleanliness testing (from "water break"[106] to dyne liquid to Nordtest Poly 197) should be chosen based on the nature of the soil which is contaminating the surface.

The choice should *always favor a liquid which is repelled by the soil*:

- For an aqueous soil such as an water-based cutting fluid, choose a hydrocarbon test fluid such as any paraffin solvent.
- For an oil or oil-like soil such as a lubricating fluid, choose water, ethanol or isopropanol.

Choose the liquid for measuring cleanliness on the opposite basis that you would use to choose the agent with which to clean it![107]

---

[104] Per Roger Woodward, FirstTenAngstroms, Inc., "*Contact angle* has no meaning without specifying the test liquid. Also, omitting the test fluid would be like taking the labels off the dyne pens. You really would not know much."

[105] Jarvis, N.L., Fox, R.B. and Zisman, A., "Surface Activity at Organic Liquid-Air Interfaces," published in Advances in Chemistry Series, *Contact Angle Wettability and Adhesion*, American Chemical Society, 1964, pp. 316–331. See Table II, p. 321. The word "spread" means that the liquid fully wet the surface, and the contact angle was very low.

[106] That is correct! If the simplicity of the "water break" test has appeal, avoid a major drawback to its use and choose the test liquid as something other than water, using the rule in Section 5.6.5.3. Water was chosen because of its availability, and because most soils were oil-based.

[107] By this choice, when the surface is completely free of contaminant which repels the test liquid, the test liquid will fully wet the cleaned surface. The low contact angle will be easy to notice and measure.

**Table 5.8 Contact Angles for Various Wetted Surfaces**

| Liquid | Surface Tension, MN/mm (dyne/cm) | Surface Substrate | Surface Energy, MN/mm (dyne/cm) | Contact Angle (°) |
|---|---|---|---|---|
| Water | 72 | Clean glass | 73 | 0 |
| | | Ordinary glass | 70 | 20 |
| | | Platinum | 62 | 40 |
| | | Anodized Aluminum | 50 | 60 |
| | | PMMA | 41 | 74 |
| | | Nylon | 38 | 79 |
| | | Polyethylene (LDPE) | 33 | 96 |
| | | Polypropylene | 26 | 108 |
| | | Paraffin wax | 19 | 110 |
| | | PTFE | 18 | 110 |
| | | Fluorad FC-721 | | 119 |
| *n*-Decane | 21.8 | PTFE | 18 | 30 |
| *n*-Octane | 23.9 | | | 40 |
| *n*-Propanol | 24 | | | 43 |
| Benzene | 28 | | | 46 |
| Water | 72 | | | 110 |
| Mercury | 470 | | | 150 |

### 5.6.5.4 *Measurement of Contact Angle*

Contact angle is defined geometrically as the angle formed by a liquid at the three-phase boundary where a liquid, gas, and solid intersect. Measurements are less difficult when the liquid suspended upon a surface is static.

Unfortunately, the balance between viscous forces, surface tension forces, and buoyant forces is metastable.[108] Contact angles are seen to advance and recede, which is known as contact angle hysteresis. A prime cause of hysteresis is inconsistent cleanliness character of the surface at the micro level. A second cause is surface roughness. In other words, water moves on surfaces.

Contact angles depend upon both the fluid applied to a surface and the surface. Some values for water on various surfaces as well as other liquids are in Table 5.8.[109,110]

Review Table 5.8. You know that water "sheets up" (wets) on glass and "beads up" (doesn't wet) on plastics, especially fluorinated ones such as PTFE. You also know that Mercury "beads up" (doesn't wet) on any surface.

The basic point is that measured contact angle characterizes the intersection of a liquid with a surface.[111]

---

[108] Interactions at the boundary surface can cause the contact angle to change considerably with time. Some are: (1) change of surface tension as surfactants or dissolved solutes migrate to the surface of the drop, (2) loss of liquid through evaporation, or (3) change in the solid through swelling with liquid. An excellent reference is at http://www.kruss.info/.

[109] Adamson, A.W. and Gast, A.P., *Physical Chemistry of Surfaces* (6th ed.), Wiley-Interscience, 1997, Table X-2, p. 365, ISBN: 0471148733.

[110] http://www.firsttenangstroms.com/pdfdocs/Cleanliness.pdf.

[111] For printing applications, one wants the ink to wet the surface so that the contact angle is as low as possible. The ink drop will spread out until the liquid's cohesion is balanced by its adhesion to the surface.

This criterion applies to more applications than just printing. A liquid will wet a solid surface completely when the work of adhesion between the solid surface and the liquid is greater than work of cohesion within the liquid. The difference is meaningful. The more this difference exceeds zero, the more the liquid will wet the solid.

If the same liquid is used, differences in surface texture/character can be inferred from changes in contact angle. Those differences in surface texture/character can be changes in surface cleanliness:

- Contaminants will change the contact angle if their surface energy is different from that of the clean surface.[112] Since contact angle is a measure of surface interfacial energy, most contaminants will have a different chemistry and different surface energy (see Section 5.6.5).

Contact angle is commonly measured in two ways, for surfaces which aren't porous.[113]

#### 5.6.5.5 *Goniometry*

One observes a single "element" of a common fluid on the test surface. This is called goniometry. It is an analysis of the shape of an "element" of test liquid placed on a solid surface.[114] The basic components of a goniometer include a light source, sample stage, lens, and image capture. Both static and dynamic (moving) measurements are made.

The "piece" of fluid can be tiny, but it must generally be placed on a flat surface to achieve static measurements.

The surface, of course, is the small section of the cleaned part whose cleanliness is to be established via measurement of contact angle. Curved surfaces can be examined via a movable stage.

While goniometry is not inherently a destructive test, sample preparation for it is usually so. The sample is the part whose cleanliness is being evaluated (see Figure 5.19).[115]

Obviously, only a small piece of the surface is evaluated with a single "piece" of liquid (usually water).

Computer analysis of the image of the shape of the "piece" is done to generate consistent contact angle data.

*Note: contact angle is observed and measured, versus tensiometry.*

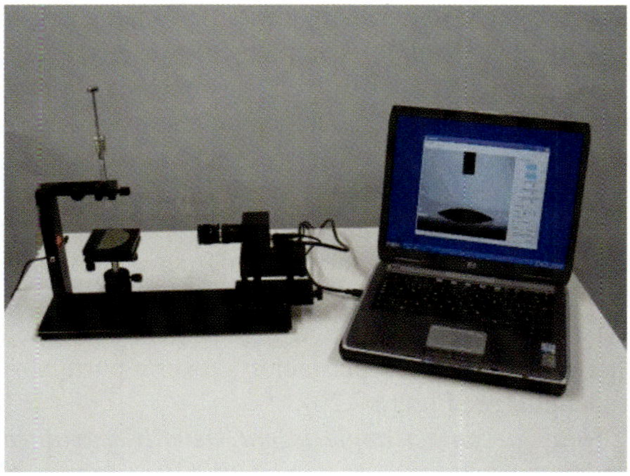

**Figure 5.19**

#### 5.6.5.6 *Tensiometry*

A tensiometer is an instrument which measures the effect of forces present when a sample of cleaned solid surface is brought into contact with a test liquid (usually water). If the forces of interaction, geometry of the solid, and surface tension of the liquid are known, the contact angle may be calculated.

The actual measurement is of the forces necessary to immerse the solid sample to various depths in the test liquid. Contact angle is computed from the net immersion force. The equation used is Equation (5.4). Note that the cosine of an angle is unity when the angle is 0°. This occurs when the test liquid perfectly wets the solid, when the wetting force is only necessary to overcome the surface tension of the liquid. The contact angle will be high, and the cosine of it low, when the liquid doesn't wet the solid – presumably because of contamination.

$$\left[\text{Measured wetting force}\right] = \left[\begin{matrix}\text{Part}\\\text{perimeter}\end{matrix}\right] \times \left[\begin{matrix}\text{Measured}\\\text{surface tension}\end{matrix}\right] \times \text{Cos.}\left(\begin{matrix}\text{Contact}\\\text{angle}\end{matrix}\right) \tag{5.4}$$

---

[112] Shieh, S., "An Analysis of Contact Angle Measurement," *Parts Cleaning Magazine*, p. S-8, March, 2001. A copy can be found at http://www.p2pays.org/ref/13/12920.htm. See Section 5.4.6.4.

[113] Contact angle can be measured if the porous structure can be saturated (filled) with the test liquid.

[114] The surface can be metal, glass, or plastic. Water is commonly used, because of its high surface tension. HFE-7200, which thoroughly wets clean metal surfaces, wouldn't be used. It has a surface tension of one-fifth that of water, beads would not be as easily differentiated from sheets. Liquids with low surface tension wet any surface more easily than those with higher surface tension.

[115] Other commercial facilities can be found at http://www.esrf.fr/exp_facilities/BM32/gmt/gmt.htm, http://www.kruss.info/, http://www.firstttenangstoms.com, and http://resonance.on.ca/.

Contact angle is normally a function of the extent to which the sample is immersed. So, immersion depth is usually standardized. In some tests, a profile of contact angle versus immersion depth is generated[116] so that examination of additional surface is accomplished. This adds to the value of the test.

Tensiometry has valued applications for measuring the surface character of regular structures such as fibers, rods, needles, or plates. But the sample must be formed or cut in a regular geometry such that it has a constant perimeter[117] over a portion of its length (see Figure 5.20).[118]

*Note: a contact angle is not visually examined with a tensiometer as is done with a goniometer.*

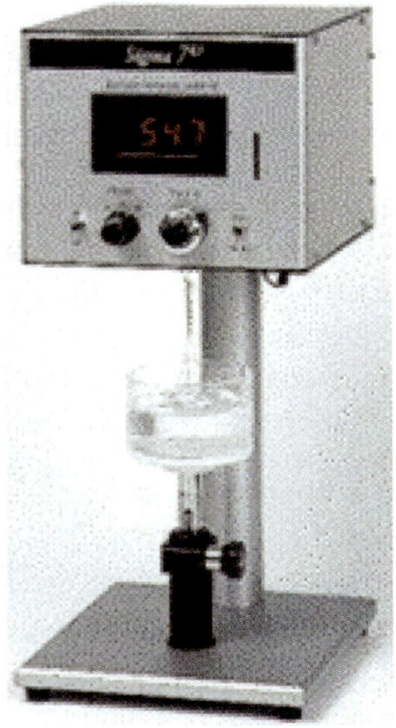

**Figure 5.20**

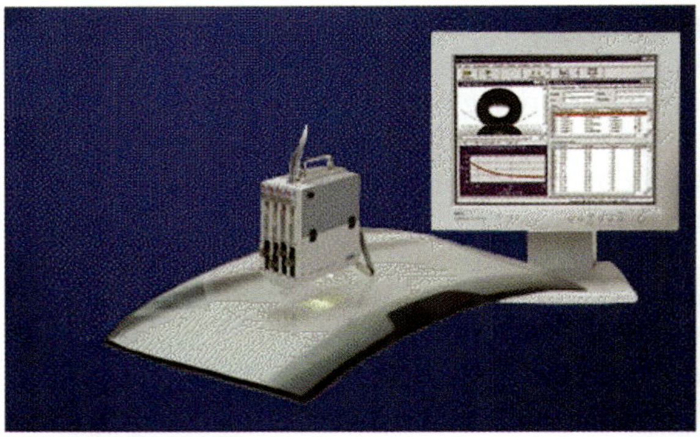

**Figure 5.21**

### 5.6.5.7 *Uses of Contact Angle Information*

Contact angle does not define cleanliness. At worst, it may describe un-cleanliness. At best, it may be correlated with downstream processing. But contact angle provides no information about the amount or identity of surface contamination.

Managers can use data of contact angle to avoid qualitative judgments about "beading" (not wetting) or "sheeting" (wetting) of water added to the surface of a cleaned part. In other words, the judgmental aspects of the "water break" test can be eliminated.[119,120]

In addition, the conventional binary (GO/NO GO) output of the "water break" test can be replaced with a spectrum of quantitative information (e.g. contact angle versus immersion depth).

Some curved parts may be evaluated via tensiometry which could not be evaluated via the conventional water break test because of small size or extreme curvature. But neither instrument displayed in Figure 5.19 or 5.20 is capable of examining sheets of stamped metal, or parts of any large size.

However, the instrument displayed in the commercial photograph as Figure 5.21 is claimed[121] to be capable of examining parts of large size and curved shape as a robotic tool.

Limitations do remain (see Section 5.4.5): (1) low levels[122] of contamination may or may not be recognized, (2) only a small portion of the part surface is

[116] Since surfaces aren't perfectly symmetrical, it does make a difference whether the profile is done as the sample piece is inserted or removed from the container of test liquid.

[117] Note that it is perimeter and not wetted area which is used. Surface forces exist at the interface between a liquid, a solid, and a gas. So it is the length of the interface, the wetted perimeter, and not the wetted area which is significant. Frictional forces are exerted over an area; surface forces are exerted at an interface.

[118] Other commercial facilities can be found at http://www.ksvltd.com/content/index/surfacechemistry.

[119] Chao, D.F., "New Method Developed to Measure Contact Angles of a Sessile Drop," *NASA Research and Technology*, July 16, 2002. See http://www.lerc.nasa.gov/WWW/RT2001/6000/6712chao.

[120] Contact angle measurements can reliably be made in industrial applications with ±2° accuracy.

[121] The author has no experience with this product. Image courtesy of KRUSS US. For details, see http://www.kruss.info/. See Figures 5.9 and 5.10.

[122] Measurements of contact angle can only assess the effect of contamination on surface energy upon or quite near the surface. This is because contact angles are determined by the upper two or three monolayers of molecules. So what is beneath this thin layer is hidden from the contact angle measurement.

being examined, and (3) water-soluble contamination (if water is used as the test fluid) or particulate will not likely be recognized.

## 5.7 COMPARISON OF CLEANING TESTS

For applications that relate to semiconductors, printed circuit boards, military specifications, pharmaceuticals, and any other instances where high-precision cleaning is required, instrumental methods are probably necessary.

Cleaning tests developed and used by client firms have included simple examinations such as the measuring and comparing the weight of clean and uncleaned parts, conventional determination of NVR or measurement of soil components in a extract by gas chromatography/mass spectrometer (GCMS), and more sophisticated ones in which surfaces are directly examined such as Electron Spectroscopy for Chemical Analysis (ESCA) or Fourier Transform Infra-red Spectroscopy (FTIR).

These and many other analytical chemical methods which are commonly used as cleaning tests are described in general terms in Appendix 2.

### 5.7.1 Comparison of Testing Methods Based on Surface Forces

In Sections 5.4 and 5.5, several techniques have been described, and are summarized in Table 5.9.

These tests are given significant space in this volume because they are numerous, easy to use, require little investment, and are sensitive to changes in surface quality.

## 5.8 AVOID THE END-USE TEST

An enterprise can, has, and will, develop proprietary cleaning tests. Often these tests are simply completion of the next processing step. This practice brings risk, which is not present in use of ASTM or other standards as cleaning tests. The risk is of a false negative:[123]

- This means that if the next processing step is not operating properly, the conclusion will be that the parts are not clean, when in fact the flaw is not of part cleanliness but of something else.

A good cleaning test should evaluate surface character in a way which is meaningful relative to the next use of the parts, and is not a test of that next use:

- But it's through examination of the next use of the parts that one establishes the reference point (benchmark, aim point, goal, standard, etc.) for all cleaning test methods.

## 5.9 TOOLS FOR CLEANLINESS TESTS

Those whose interest in cleaning is significant enough to purchase this book should maintain a kit of useful tools for use in evaluating part cleanliness. Depending on the level of cleaning required, the kit should include at least the following:

- A clean, well illuminated, *well ventilated*, unobstructed work area not contaminated by airborne dust or particulate.[124]
- A ~15 power loupe (magnifier used to inspect the test piece for visible particles).
- Squares of white cloth (used to wipe test pieces for dirt which can be seen). Commercial products include both wet and dry wipers, as well as non-woven fabrics which can be excellent for collecting particles.[125] Residual particulate, oil, and grease can often be detected by wiping the test piece with dry, white cloth. Swabs may also be useful.
- Several probes for assisting in surface inspection. One should be a toothbrush or non-shedding

[123] This serious risk can be mitigated if it is convenient to provide parts which are known by some independent method (see Section 5.10 about validation) to be clean. If these parts are processed, and the results **un**satisfactory, it can be reasonably assumed that it is the process which contains the flaw and not the cleaning work.

[124] This is the most significant item needed, and the one most often omitted. Remember Imbesi's Law of the conservation of Filth, Section 5.2.1.3.

[125] See ASTM D6650-01 Standard Test Method for Determining the Dynamic Wiping Efficiency, Wet Particle Removal Ability, and Fabric Particle Contribution of Nonwoven Fabrics Used in Cleanrooms. See Section 5.44 and Reference 34.

**Table 5.9 Cleanliness Tests Involving Surface Forces**

| Test | Section | Parameter Being Measured | Instrument | Useful on | Size of Surface Examined | Reproducibility | Outcome |
|---|---|---|---|---|---|---|---|
| Water break | 5.4.5 | Visual evaluation of liquid as being present in sheets or drops | Human eye | Metals | Up to a square foot | Poor | GO/NO GO for oil-like contamination |
| Goniometry | 5.6.5.5 | Contact angle | Goniometer | Metals, polymers | One square inch or so | Excellent | Comparison to same of clean surface |
| Tensiometry | 5.6.5.6 | Wetting force | Tensiometer | | | | Surface energy of contaminated surface versus clean surface |
| $\gamma_{\text{Interfacial solid}}$ | 5.6.5.1 | Contact angle | Various static contact angle and surface tension analyzers | | | | |
| SDT | 5.4.5.1 | Contact angle (or drop diameter) | Digital camera and magnifying view | | | Good | Comparison to same of clean surface |
| Nordtest Poly 197 | 5.4.6.3 | Visual examination of the fate of a liquid drop | Human eye | | | | Surface tension of fluid which just wets surface |
| Dyne liquid | 5.4.6.2 | Visual comparison of two liquid streams, one of which is continuous and the other is discontinuous | | Polymers (films, papers, fabrics, etc.) | Several square inches | | Surface tension of fluid which just doesn't wet surface |

paintbrush with soft bristles. Another should be a pair of tweezers.

- A can of pressurized air to blow liquids or other material off parts so surfaces can be inspected.
- Small (one-ounce) bottles of solvents expected to be suitable for cleaning the soils expected to be present. Also, provide a catch pan for waste and spills.
- A black light (ultraviolet) used to visually inspect for residual hydrocarbon-based material, and an unlighted place to use it.
- Numerous paper envelopes, plastic bags, Aluminum foil for packaging both clean and soiled parts, and labels and a marker. No unidentified sample should be retained. Depending upon the application, those goods should be purchased from a supplier of goods for use in a cleanroom.
- A location where packaged samples, records, reference parts, and these tools can be stored without being contaminated.

The total bill for a basic part cleanliness testing kit should not exceed 50 euro.

Additional items which will prove of value include a digital camera with zoom macro lens,[126] a 20X lighted viewer, an analytical balance capable of weighing to four decimal places if parts weigh less than ~500 g, a vise to hold parts for inspection of internal sections, an ultrasonic-powered jewelry cleaner, and a computer into which image files and process data can be stored.

### 5.9.1 Use of Tools Around Cleaning Tests

*The purpose of the above tools is not to conduct specific cleaning tests.* Rather, the purpose is to allow investigation of the interaction between a cleaning process and the parts which feed it and come from it.

There is no substitute for tactile learning. Wet a soiled part with solvent. Wipe the soiled area. Attempt to brush off the soil. Immerse it and apply ultrasonic force, etc. Touch the wet surface. Examine it. Think about it. Learn about the difficulty of moving and removing soil components. Use that knowledge in designing and operating your solvent or aqueous technology cleaning process.

## 5.10 CLEANLINESS METRICS

These allow a judgment call about whether or not the examination result justifies a pass or fail grade about future use. In addition to having a cleaning test, you have to know what to do with the results. Is a peel strength of 28 descriptive of a clean or a dirty part?

Guidance about selection and use of metrics are found in Chapter 4, Section 4.4 (The "Golden Lot" Benchmark) and Section 4.13 (Control Charts).

### 5.10.1 Why Metrics?

Specific metrics must be relative because the cleanliness requirements of the next processing step are different in every application. Metrics are the answer to the question, "How clean is clean?"

The answer, of course, is "Just well enough." As part tolerances are specified to allow an expected use, as services are structured to fulfill the values of customers, as scientists know their work will be scrutinized by other scientists, so should those doing cleaning work know that their work must be done only well enough.

Metal finishers often clean to allow the next finishing step. Later they may clean the same article to prepare it for inspection. Still later they will clean the same article prior to packaging. The goals and standards for each cleaning step are probably different. Granted the cleaning for each need may be done in the same process. *But the level of cleaning performance does not necessarily have to be the same for each operation*:

- There is no point to cleaning oily screws from a cutting line to the standard to which one would clean a metallic mirror.
- There is no point in cleaning an Aluminum piece which will become a tripod leg for a camera as well as one would clean a Titanium piece to be implanted for support in a person's leg.

[126] Emphasis cannot be placed too highly about use of digitally-produced images and the Internet to communicate understanding about cleaning (and other) operations. This author can (and does) witness cleaning operations and inspect parts without making expensive site visits. Since image storage can be so efficient, it is possible to easily retain evidence about past operations.

- There is no point to cleaning in preparation for surface grinding or mechanical deburring to the standard to which one would clean in preparation for plating or electropolishing.

The phrase "just well enough" means that resources aren't to be wasted in providing a level of quality not needed in sequential operations or actual use.[127]

Metrics allow identification and specification of "just well enough."

## 5.11 SPECIFIC RECOMMENDATIONS BASED ON APPLICATIONS

Information in Tables 5.5 and 5.6 speaks to how and when ASTM standards can be used as tests for cleanliness – with and without particulate, respectively. These standards contain guidance for selection of metrics.

The first choice for any cleanliness metric is that of the customer. This volume isn't thick enough to list all cleanliness metrics in common use.[128] However, some are:

- AAMI TIR30 specification of processes, materials, test methods, and acceptance criteria for cleaning reusable medical devices.
- ASTM B322-04 specification of cleaning metals prior to electroplating.
- DIN EN 12300 specification of the cleanliness of all surfaces of equipment containing cryogenic fluids.
- ISO 8504 or ISO 11126 specification of blast cleaning of steel surfaces to be painted.
- ISO 8501 specification of visual assessment of surface cleanliness.
- JIS H 8300 (Japanese Standard) or BS 2569 (British Standard) for specification for sprayed coatings.

While every application is different, two approaches are commonly used:

1. **Particle counting** is one of the most common fluid analysis tests. It can be used to determine the cleanliness of new oil, identify dirt ingress, verify filter performance, or indicate the onset of active machine wear.[129] Consider a metric (benchmark) of no more than 4,000 particles/ml whose size is greater than 5 μm.[130]
2. **Residue of Solvent Extract (ROSE)** is another common method of characterizing the cleanliness of assemblies. Here, a manager believes there are no zones of high concentration of residues that all surfaces on the circuit board, multiport injection valve, gear assembly, or conduit network are equally unclean. Consider a metric for printed circuit boards of: no material removed from the bare board,[131] 1–5 μg chloride[132]/square inch (μmg/SI), 10–15 μmg/SI bromide, and 0.004% residue.[133]

The point of this chapter is not that any of these recommendations apply to the operation you manage. They probably don't because of the diversity of situations involving cleaning.

The point of this chapter is that you should be using metrics. If you haven't chosen useful ones, this chapter should demonstrate that they exist, you can identify them, others are using them, and you should implement them.

## 5.12 VALIDATION OF CLEANLINESS

This exercise is an insurance policy. Users conduct validation to prove that their normal cleaning test is currently valid (hasn't been "fooled"[134]). Validation represents additional work and isn't justified for

[127] If the goal of cleaning work is to produce output of a quality suitable for the next task, then *continuous improvement* is the process by which the standards around that next task may be strengthened.

The goal of just well enough is not compatible with slipshod work. Just well enough means that the level of excellence in cleaning work matches that needed for subsequent operations, today and tomorrow.

[128] A good source for location and purchase of international and national standards is found at: http://webstore.ansi.org/ansidocstore.

[129] Like most analysis tests, obtaining a representative sample is paramount to the accuracy of particle count data. Particles are often a precursor to catalytic or other harmful events.

[130] This is equivalent to National Institute of Science and Technology (NIST) Standard Reference Material (SRM) #2806a.

[131] *Per IPC 650 TM (2.4.1)*. See http://www.ipc.org.

[132] Measured by ion chromatography.

[133] *Per IPC 650 TM (2.3.27)*. See http://www.ipc.org.

[134] To "fool the test" is a phrase that describes a situation when test results don't reflect the true conditions, when the test has been conducted according to standard procedure (e.g. see Chapter 3, Section 3.4).

every operation. It is justified when, as a manager of cleaning operations, you can't afford to be "fooled."

### 5.12.1 "Fooling" of Cleaning Tests

Cleanliness (and nearly all other) tests can be "fooled." Usually this involves the presence of an unexpected type of contamination. It may involve contamination located in an unexpected place. Here are some examples. If the normal cleaning test is:

- **A visual inspection in which particle contamination is counted:** The unexpected presence of an organic film probably won't be detected and may produce failure in the next step of use.
- **An evaluation of surface energy, or contact angle:** If there is a non-uniform distribution of cleanliness (or soil) among part surfaces with different locations or character, those differences won't be detected and may produce failure when those surfaces are engaged in the next step of use.
- **A measurement of NVR:** The unexpected presence of a volatile residue probably won't be detected and may produce failure in the next step of use before the residue is fully volatilized.
- **A measurement of TOC:** The unexpected presence of an inorganic residue probably won't be detected and may produce failure in the next step of use.

In each case above, the reverse could also be true.

A frustrating situation in which validation is essential occurs when there is no flaw in the cleaning test. Yet performance in the next step of use fails due to a cause unrelated to the cleanliness of the parts. Often the defect is incorrectly attributed to the cleaning process. Effort is wasted in troubleshooting a process which is performing flawlessly while the true cause of failure remains undetected.

### 5.12.2 Too Many Not to Choose One

Three general approaches toward validation of cleanliness are described below:

1. **Overall validation** where parts are treated in a way certain to remove all soil. Then a gravimetric comparison is made between a part shown clean by the normal test and a part from which it is certain all soil has been removed (see Section 5.12.4).
2. **Procedural validation** where each step of the cleaning test is independently validated (see Section 5.12.5).
3. **Specific validation** where the normal cleaning test is supplemented with another based on a different principle of analysis (see Section 5.12.6).

A fourth general approach can be found in the "product by process" methodology for process control in Chapter 4, Section 4.10.

One role of a manager is to recognize and choose. If validation of the situation you manage can't be done with one of the four general approaches above, then your challenge is to identify which approach will contribute and take that action. Review the material about "on-aim" control on Section Chapter 4, 4.12.1 or about problem-solving in Chapter 4, Section 4.26. Ask what the competition does. Ask for professional help.

### 5.12.3 What is Not Validation

A quotation attributed to many is "fanaticism is the practice of redoubling your efforts when all is lost." Validation does not mean that the normal test should be rerun many times under increased scrutiny.

More of similar data adds no new insight. More eyes looking in the wrong direction won't see approaching disaster. More thought with the same point of view won't produce inspiration.

One validates a result by examining it from another point of view.

This means something different must be done.

### 5.12.4 An Overall Approach Toward Validation

There is no ASTM, DIN, or ISO standard of or for validation of cleanliness. While many problems have the same cause, their solution is obscure because the symptoms are different.

This author often recommends a "total cleanliness" approach to validation of cleanliness tests. Others call this approach "brute force." Said another way, one seeks to measure the amount of soil removed by processes all would accept as providing complete cleaning.

Instead of measuring NVR, TOC, using an OSEE instrument, viewing liquid drops, or counting particles, do all of the following:

- *Boil* a part in a suitable material(s) to extract everything which contaminates its surface Contact time might be 30–120 minutes. Weigh the part before and after. Calculate the weight of soil removed.
- *Fire* (pyrolize) another part to a suitable temperature to vaporize everything which contaminates its surface.[135] Contact time might be 15–30 minutes. Weigh the part before and after. Calculate the weight of soil removed.
- Clean another part in the normal manner. Weigh the part before and after. Calculate the weight of soil removed. Execute the normal cleaning test with the cleaned part.

Then compare the total weight of contamination removed by extraction and pyrolysis with the weight of contamination removed by the cleaning process. Answer these questions. What should be the result of the cleaning test? Is it that which was measured? Is the cleaning process effective?

Convergence is expected here. The focus should not be placed on the *similarity* of the two measurements of total contamination to the specific contamination detected in the cleanliness test. *Equivalence* of total and specific contamination shows the cleaning process is removing all soil and the normal cleaning process is detecting that condition.

This general approach can be very useful if it can be applied when operation is inconsistent while results of the normal cleaning test are consistent.

Obviously, this approach isn't the one for all circumstances, such as when there are multiple kinds of contamination, where parts are irreplaceable, or where the method of removing all soil would also remove some the part.

### 5.12.5 A Procedural Approach Toward Validation

Process validation by procedure is commonly done in the manufacture of pharmaceuticals, in the manufacture and testing of medical devices, in the operation of computer and information management systems, in genetic research, and other activities where it is believed to be justified. Validation is necessary to receive ISO 9001 certification.

There are roadmaps, computer programs, guidebooks and manuals, and written procedures to implement validation in general and specific[136] situations. Providing an answer to one question is the basis for all:

- How can (could) this happen (be prevented)?[137]

Two alternatives, based on the same ideas but based on different experiences, should be considered:

- The discipline and methodology associated with validation are practiced in the assessment of safety risks (see Chapter 3, Section 3.17.2). Here the Failure Mode and Effect (FME) and Fault Tree (FTA) approaches are described.
- The rigor and responsibility associated with validation are practiced by those seeking certification by the US FDA. Several excellent resources are available without cost.[138,139]

Validation of testing can be complicated and expensive, involve professional consultants, and be lengthy. That doesn't mean it should be avoided where it is justified. *That's* a decision for a manager.

---

[135] Without harming the part. If this is impossible, choose another approach!

[136] There is even a peer-reviewed technical reference – the *Journal of Validation Technology*. Subscriptions and information can be obtained at http://www.ivthome.com.

[137] In the case of cleanliness testing, the situation to be avoided is being "fooled" by the cleanliness test.

[138] *Guideline on General Principles of Process Validation*, May 1987, FDA/CDRH/CDER. See http://www.fda.gov. The entire FDA *Medical Device Quality Systems Manual* can be viewed and downloaded without charge at http://www.fda.gov/cdrh/qsr/contnt.html. Scanning this material would not be a waste of time for a manager whose responsibility involved critical consequences of cleaning work.

[139] *Guide to Inspections Validation of Cleaning Processes*, which can be viewed and downloaded at http://www.fda.gov/ora/inspect_ref/igs/valid.html.

**Table 5.10 Specific Approaches to Validation of Cleanliness Tests**

| Parameter Being Measured | Normal Cleaning Test | Validation Cleaning Test |
|---|---|---|
| Residue(s) | NVR | • Visual examination using the tools mentioned in Section 5.4.1<br>• ASTM E21.05, "Standard method for measurement of NVR on surfaces"<br>• The extraction method described in Section 5.12.4 |
| Organic film(s) | OSEE | • Visual examination using the tools of mentioned in Section 5.4.1<br>• Ultraviolet Photoelectric ("black light" UV) Emission<br>• The pyrolysis method described in Section 5.12.4<br>• TOC/TOD analysis |
| Particulate(s) | ASTM D1876-01 Standard test method for peel resistance of adhesives (T-peel test) | • The *indirect* (optical examination) method for particle counting as described in Appendix 2<br>• The *direct* (examination of cleaning fluid) method for particle counting as described in Appendix 2<br>• D903-98(2004) Standard test method for peel or stripping strength of adhesive bonds |

### 5.12.6 A Specific Approach Toward Validation

Instead of the general (total cleanliness) approach toward validation in the above section, consider an approach toward validation focused on the nature of the cleaning test under scrutiny.

This means to measure the parameter being measured in the normal cleaning test using another approach sensitive to the same parameter (see Table 5.10).

With this approach the validation test is conducted at low frequency (perhaps every 20–100 tests), and the results are compared with those from the normal test. Table 5.10 shows how important the eyes of an educated observer are in solving problems associated with parts cleaning – they should be the first source of validation information.

### 5.12.7 Revalidation

"As long as the process operates in a state of control and no changes have been made to the process or output product, the process does not have to be revalidated".[140]

This guidance, provided for those manufacturing medical devices, should be adequate for those managing other critical cleaning operations.

This guidance also links the material about cleanliness testing (Chapter 5) to the material on process control (Chapter 4).

[140] US FDA *Medical Device Quality Systems Manual*, Chapter 4, available per Footnote 137.

6

# Challenging situations in critical, precision, and industrial cleaning

**Chapter contents**



This chapter is about situations. They include:

- Use of unusual cleaning technology (without chemistry).
- Business management issues such as how to choose a supplier, attend a trade show, choose a contract cleaning firm, or hire a consultant.
- Removal of pyrogens and particles.
- An unusual and effective strategy for managing a cleanroom.

## 6.1 TRADEOFFS: CLEANING WITHOUT CHEMISTRY

In Chapter 1, Section 1.2 and elsewhere, three actions were noted as being common to every cleaning process: mechanical action, thermal action, and chemical action.

Cleaning can be done without chemical action. It has been done, is being done, and will be done. It is not necessary to purchase, control the use of, and dispose of cleaning chemicals to successfully and fully complete valued cleaning work.

If that be the case, why hasn't every manager done just that – to save purchase and disposal cost, to eliminate safety and environmental concerns, and to reduce the number of factors to be managed?

The reason is simple – the use of cleaning chemicals generally brings net value above their true cost of use.

This chapter will cover how and when some of that value might be captured by managers willing to change from use of cleaning chemicals.

### 6.1.1 The Role of Chemistry in Cleaning

Seven momentum-based (mechanical) methods of cleaning described in Sections 6.13 through 6.19 don't benefit from the value which chemicals can bring to cleaning operations.

That role is to capture soils.

Chemical solvents dissolve soils. Chemical detergents form micelles with soils. These soils normally are "out of action" (captured) relative to the surfaces of parts. The remaining process steps (e.g. rinsing, skimming, distillation) are used to remove the captured soil from the cleaning machine (and cleaning chemicals).

When soils aren't captured, they can reinfect cleaned parts (see Section 6.1.5.1).

### 6.1.2 Methods of Cleaning Without Chemistry

Cleaning without chemistry is about mechanical force, and occasionally heat – but chiefly about application of mechanical force to soiled surfaces.

After all, if the chemical action is removed from the three basic actions of a cleaning process, that's what's remaining.

It's also about trading one set of problems for another. That's why a manager chooses to discard the most common tool for removing soil – they believe the "grass is greener," and sometimes it is!

### 6.1.3 Blast Cleaning

If you can make a particle out of it, you can use it for blast cleaning. Particles of water (liquid droplets or solid ice), $CO_2$[1] (dry ice), wheat or rice hulls, plastic or glass beads, metal shot, salt[2], corn cob grit (see Figure 6.1), silica sand, smelter slag, and other solids are entrained in high-velocity streams of liquids and gases, and shot at soiled surfaces.

The solid particles must be clean and dry or else they may clog the dispensing apparatus which feeds the blast nozzle – a common problem.

Blast cleaning is being done in applications where the level of cleanliness is critical to some operation,[3] where a protective grease is being removed from stored cable, or where old labels are being removed from containers that are being reused, or where large particles are removed from small machined parts by impact with jets of compressed air (no particle involved). Where blast cleaning is *not* being done is with fragile parts.

Soil removal is purely by mechanical action. The blast media impacts the soil. Momentum is transferred to the soil. That produces a net force, which can overcome the adhesive forces that bond soil to the surface.

Equipment for blast cleaning is not expensive or complicated. Like plasma cleaning, a significant amount of work is done by contract service providers.

Abrasive blasting systems typically include three essential components, an abrasive container (i.e. blasting pot), a propelling device (i.e. compressed air or pressurized water), and a set of blasting nozzles.

But the most significant facility is the one required to contain the action – the noise, the debris, the dust,[4] and the scrap blast cleaning agent. Like painting, most blast cleaning is done in a containment booth (see Figure 6.2).

**Figure 6.1**

**Figure 6.2**

---

[1] Dry ice ($CO_2$) is commonly used for remediation of wood by removal of mold spores versus removal with mechanical brushing.

[2] Salt is a general term. Chemically it can be NaCl, $NaCO_3$, $NaHCO_3$, or another solid chemical.

[3] Caimi, R.E.B., Lin, F-N. and Thaxton, E.A., "Gas-Liquid Supersonic Cleaning and Cleaning Verification Spray System," US Patent 5,730,806, March 24, 1998. See Section 6.6.3 for an analogous technology.

[4] Silica sand is frangible and brings significant concern about dust. It is used and directly disposed because of low price. A substantial amount, perhaps as much as 25%, of small diameter blast media is lost as particulate pollution (see South Coast Air Quality Management District, Chapter 2: *Unconfined Abrasive Blasting*, Draft Document, El Monte, CA, September 8, 1988).

Drawbacks can be significant. They are environmental[5] and personal,[6] in addition to the potential for part damage. Yet, blast cleaning is given significant consideration as an alternative to solvent and aqueous cleaning because it gives significant opportunities for pollution prevention.[7]

### 6.1.4 Pressure Washing

This can be the simplest method of cleaning many items, including houses, driveways, siding, fences, decks, roofs, concrete, and industrial parts. Pressure washing, however, is an underused technique for parts.

In its simplest form, pressure washing is application of mechanical force to dislodge soil from surfaces. The force is delivered by a jet of high-pressure water[8] (see Figure 6.3).

Velocity is not necessarily high as flows of only 1–5 gal/minute are common. But piston plunger pumps routinely discharge at 1,000 psi (or significantly more).

Generally, the work is not parts cleaning. It is cleaning of structures, piping, vehicle underbodies, or plate stock; goods taken from or being prepared for storage; or goods to be later cleaned to a higher level of cleanliness (pre-cleaning[9]).

Equipment is not expensive, usually portable, and simple to operate (see Figure 6.4).[10] Cycle time is either very short (few seconds) or very long (minutes), depending on the amount of soil and its adhesion to the surface.

**Figure 6.3**

**Figure 6.4**

[5]Most countries have applicable regulations because blast cleaning affects the environment in at least three ways: noise pollution, dust hazards, and disposal of waste. For example, see the regulation for Australia – EPA 108/03, at http://www.environment.sa.gov.au/epa/pdfs/guide_abrasive.pdf.

[6]Particulate matter (PM) and particulate Hazardous Air Pollutants (HAP) are major concerns relative to abrasive blasting. The US EPA's regulations about $PM_{2.5}$ and $PM_{10}$ are particularly significant because these particulate matter are *criteria pollutants*. The subscripts denote particles equal to or smaller than 2.5 and 10 μm in aerodynamic diameter, respectively (see Chapter 2, Section 2.5.1). Respirators are essential for all associated staff. Ventilation requirements are covered, for the US, in CRF 1910.94 (US Occupational Safety Health Administration (OSHA)).

[7]*EPA/625/R-96/003, Manual: Pollution Prevention in the Paints and Coatings Industry*, September, 1996.

[8]Occasionally, additives are injected to the water. Dilute detergents may be injected into the supply tank. These formulations must contain additives which minimize foam. Particles (sand, salt, etc.) may be entrained into the moving stream. Not oily soil, but metal can be cut or pierced by pressurized jets of sand-laden water. Finally, a phosphate-based or other coating material may be applied to the cleaned surface to prevent rust or other surface deterioration.

[9]Pre-cleaning such as this is nearly always a good idea. Any soil which can be removed upstream of a cleaning machine doesn't have to be later collected and removed from the cleaning machine. Yet, it's a tradeoff between the cost of labor and floorspace versus the cost of a more complex cleaning machine.

[10]Not shown in Figure 6.4 is a fluid supply tank, a containment system, and personal protective equipment for operators.

However, a manager must recognize the serious issues involved when blast cleaning is done. They include, at least, staff safety, environmental effects, and product quality:

- An acute[11] safety hazard is present. The cleaning wand is not a spray nozzle as children use in play. *It is a weapon*. The nozzle is aimed at surfaces for the purpose of dislodging soil. Its discharge will do that. Imagine what that discharge will do to human flesh! A manager's job is to prevent that.

  Somewhat less serious is overspray or rebound. Jets of pressurized water lose their impact within a few inches of their nozzle face. Yet good practice dictates that previously expanded fluid not impacting a surface or expanded rebounding from a surface be directed away from adjacent workers.
- Containment and waste disposal are the environmental issues. Blast cleaning produces dozens to hundreds of gallons of waste water.[12] Normally directed to sewers without concern, this dirty water is owned by the manager's firm, which is responsible for its safe containment, proper treatment, and lawful disposal.
- Quality control can be poor. It is easy to over-clean (remove surface as well as soil), which doesn't usually happen with an aqueous or solvent-based cleaning process. It can also be easy to remove soils from a location but move them to another location on the parts. Generally, standards are low in pressure washing so the latter situation isn't of concern.

Pressure washing is not critical cleaning – it can be effective cleaning, it can be efficient cleaning, and it can be low cost pre-cleaning. But perhaps more than other situations, management attention is required.

### 6.1.5 Cleaning with $CO_2$ Snow

One might not expect[13] to clean with "snowflakes." But the momentum of a "storm" (also referred to an aerosol) of snowflakes moving at hundreds of feet per second velocity is significant compared to the adhesion force holding a sub-micron sized particle to an otherwise clean surface.[14]

These snowflakes aren't the ones found on Christmas trees. They are generated by expansion of pressurized liquid $CO_2$ across a special nozzle, which is aimed at the parts. The nozzle is special because it is an asymmetric venturi which produces supersonic flow conditions at a constant level of enthalpy. Parts must be fixtured (supported) because of the momentum applied by the $CO_2$ stream.

$CO_2$ snow can also remove some hydrocarbon-based deposits and films[15] through combination of the solvent action of liquid $CO_2$ which temporarily exists, and momentum transfer. Even some silicone-based contamination can be removed.

This technology has been used with Argon and Nitrogen[16] as well. $CO_2$ is commonly used because of its cost, availability, inertness, and because its use raises no environmental concerns.

Obviously, applications are restricted to high-value parts where the level of contamination is minuscule. But equally obviously, the value of removal of small residues[17] is also high. Aerospace components and Silicon strata for semiconductors have been prime applications.

---

[11] Acute means lasting a short time or requiring a short exposure. Chronic means marked by long duration or frequent recurrence: not acute (see Chapter 3, Section 3.11.1).

[12] Some firms sell portable containment areas, called "islands," where runoff can be collected and overspray can be restrained (see http://www.pressureisland.com). A clever manager could design and build an effective island using internal resources.

[13] However, its founder, Dr. Stuart Hoening of the University of Arizona, did so and published the initial finding as: Hoening, S.A., *Compressed Gas Magazine*, August 22, 1986. Prof. Hoening also developed OSEE measurement technology (see Appendix 2).

[14] Sherman, R. Hirt, D. and Vane, R., *Journal of Vacuum Science and Technology,* 1994, Vol. 12A, pp. 1876–1881. Robert Sherman, as much as anyone, has fostered development of this technology.

[15] OSEE is commonly paired with $CO_2$ snow cleaning of organic films as a test for cleanliness (see appendix 2).

[16] This is not the impingement technology for removal of particles described in Chapter 1, Section 1.6.5. That technology involves *entrainment* of solid particles of Argon in a moving gas stream. Technology described in Chapter 6, Section 6.1.5 involves production of a particle (snowflake) via *expansion* of a liquid across a nozzle.

[17] See Chapter 1, Section 1.6 on the general topic of removal of particles from surfaces.

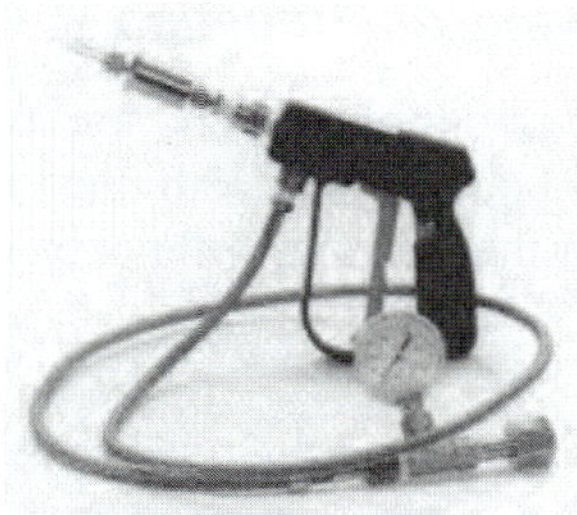

Figure 6.5

Systems are not expensive – one can purchase a complete one for ~2,000 euro. A simple nozzle is shown in Figure 6.5.

### 6.1.5.1 *Reinfection*

That's of major concern when cleaning technology that doesn't involve chemicals is used (see Section 6.1.1). This is especially true with $CO_2$ snow cleaning, because it is used in situations where the soil level is the lowest and the form of soil can be tiny mobile particles.

Several tactics should be considered to minimize reinfection of parts by material removed with any of the above momentum-based cleaning methods:

- Capture and contain (*the normal role of cleaning chemicals*) the effluent from the cleaning area – with sumps or catch basins, screens or nets, HEPA filters, or whatever is appropriate, given the nature of the outfall.
- Direction all flow from a dirty area to a clean area. No uncleaned parts should ever pass through the clean area.
- Inspect for reinfection. Understand and remove its cause.

## 6.1.6 Abrasive Cleaning (Mass Finishing)

This process combines parts cleaning, done in a somewhat abnormal way with another unit operation – surface smoothing[18] (deburring). It is more known for the latter function. The parts cleaning function (deoiling, and perhaps drying as well) is viewed as being provided at no net cost.

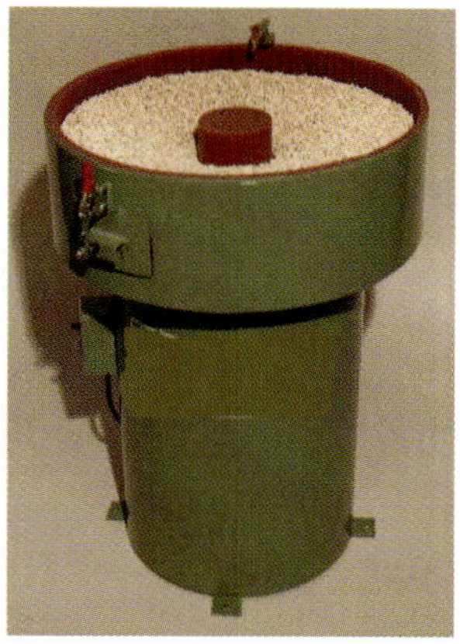

Figure 6.6

The parts to be finished are placed in a vibratory machine – a bowl, drum, barrel, bed, etc. (see Figure 6.6). Abrasive media are added. So also may be water and a detergent compound. The abrasive media also fulfills the role of a sorbent.[19]

The bowl is only partially fully of media and parts. Space is necessary for turnover of the contents so all surfaces are repeatedly contacted.

Action or movement of the vibratory machine causes the media to press and rub against part surfaces, edges, and corners to alter surface characteristics. The compound emulsifies the soils and carries them away from the parts.[20]

Usually, there is a large volumetric ratio of media to parts so that the liquid (oily soils, water solutions, etc.) are sorbed. Generally, dry parts are produced.

There are many kinds of media. This author has done effective deoiling of steel fasteners using the compound normally sold to recover spilled liquid from floors. Other choices are natural waste products

[18] Surfaces can be smoothed for plating or roughened for printing – depending on the choice of media. Multi-stage operation is only seldom seen as this technology is perceived as low cost.

[19] Mentioned in this chapter about blast cleaning, granulated corn cob grit or wheat hull is a common abrasive cleaning material. It is sold dust-free. It is biodegradable as compost. And if the soils are oils and greases, it can incinerated to recover fuel values. *Adsorption* and *absorption* are different, like dating and marriage. The former is a surface attraction. The latter is a volumetric saturation.

[20] Parts commonly cleaned and finished in this way are: Copper pipe fittings, ball bearings, springs, heads of golf clubs, jewelry and buttons, brass cartridge casings, and rivets/fasteners.

from refining of food products (ground corn cob[21]), finely ground ceramic materials, and polymeric granules onto which oil can adsorb. Whatever media is used, it should be evenly sized particles. The choice is likely to be made by iteration with the two parameters varied being softness/hardness of the granule and adsorptive capacity of its surface.

### 6.1.7 Cleaning with Ultrasonic Power

Immersion cleaning of oil and grease soils without chemistry is technically feasible, has been patented,[22] and there are some commercial applications.[23]

Only mechanical force is involved.

The force is generated by ultrasonic pressure transducers which normally produce cavitation bubbles. Hence, one might assume that the nature of the force is cavitation-induced implosion of tiny bubbles.

This assumption is incorrect. Cavitation plays no role in this cleaning process.

In fact, conditions are deliberately controlled to be opposite of those which normally produce cavitation bubbles. These conditions are given in Table 6.1.

**Table 6.1 No-Chemistry Cleaning is Not Similar to Cleaning Via Cavitation**

| Item | Conventional Cavitation | No-Chemistry Cleaning |
|---|---|---|
| Water degassing | Essential | Not done |
| Optimum temperature | ~71°C (160°F) | Any temperature |
| Hydraulic conditions | Static flow conditions | Moving flow conditions |
| Standing waves | Avoided via frequency sweep | No frequency sweep |
| Bubbles | Observed | Not observed |
| Cavitation | Yes | No |

#### 6.1.7.1 *Making An Emulsion*

The process is simple. There are two steps are involved (see Footnote 9). Both must be performed completely, or the cleaning will not be effective. The steps are to:

1. Produce an emulsion via ultrasonic pressure waves.
2. Rinse the emulsion from the parts.

There are three conditions necessary (per Footnote 22) to produce a water-soil emulsion:

1. The transducers must be located close to the parts. The distance of 1.27–7.62 cm (0.5–3 in) is recommended.
2. The volumetric power intensity must be high versus applications involving ultrasonic-produced cavitation (>~25 W/l [~100 W/gal]).
3. There must be motion in the fluid (to complete the rinsing step).

A single transducer can be located above or below the parts. A pair of transducers may be located above and below the parts. Line of sight access is not required, as nests of small parts may be cleaned.

If multiple transducers above and below the parts are used, the same frequency is used for each.[24] The purpose of the multiple transducers is to expose the surface which would not ordinarily be exposed to ultrasonic waves. Flat parts, which have two sides, are a common example.

The process is self-limiting. It is complete when there is no soil left to emulsify. Photographs of the generation of the water–soil emulsion[25] are shown in Figure 6.7.[26]

---

[21] Cob grit is a material selectively ground from the woody part of a corn cob. The grit particles are sized between 10 and 60 mesh, and come in shapes as elongated rounds or chopped kernels. Depending on the application, cob grit can probably hold about 50% of its weight in absorbed liquid without reinfecting parts (see Figure 6.1).

[22] Johnson, W.J., "Washing Parts with Ultrasonic Energy," US Patent 6,368,414, April 20, 2002.

[23] It has been called "No-Chemistry" cleaning.

[24] Swainbank, H.B. *et al.*, "Ultrasonic Strip Cleaning Apparatus," Versus US Patent 4,788,992, December 6, 1988, which requires a difference in frequency.

[25] The final gel-emulsion was stable under boiling conditions for at least 1 week nonstop and contained approximately 90% water.

[26] The photographs are shown in time sequence (from Figure 6.7). The images were part of a paper presented by this author at the *Phoenix Solvent Substitution Conference*, as "The Ultimate in Solvent Substitution-Cleaning without Chemistry," December 4, 1999.

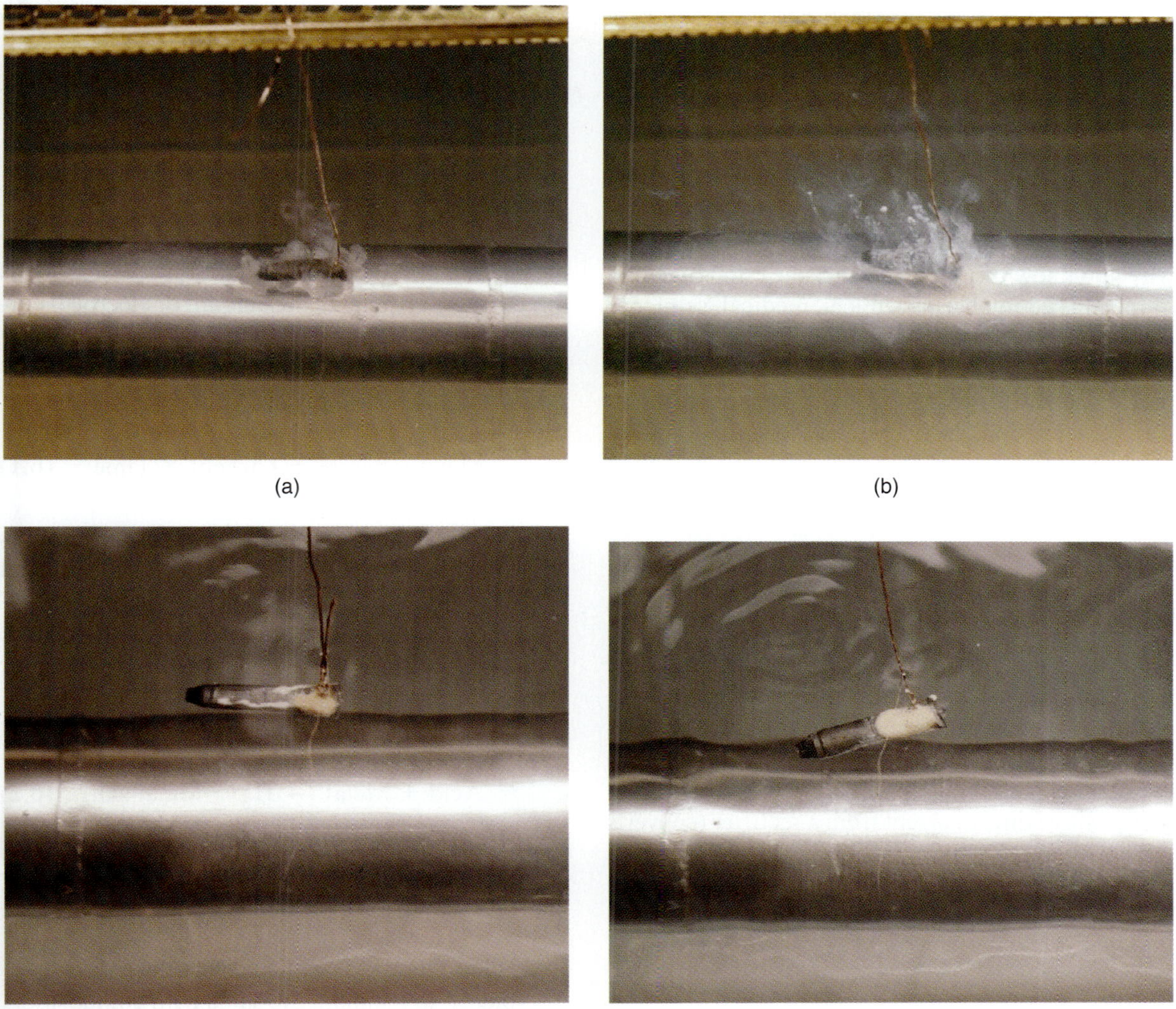

(a) (b) (c) (d)

**Figure 6.7**

### 6.1.7.2 *Process Issues*[27]

The rinse step is absolutely crucial. The rinsing nozzles can be located under water with the transducers, or the parts can be raised above the water and rinsed there. Obviously, a different design of the rinse nozzle would be used in each case (see Chapter 7, Section 1 and Table 1).

Water quality is not relevant. Excellent cleaning was reported in deionized water, tap water, water produced by reverse osmosis (RO), water containing tramp insoluble soil, and water containing ~1% emulsion.

There is no practical limit to operating temperature. Issues which determine temperature are soil and substrate. Lower temperatures make emulsion formation and removal more difficult because the emulsion is more viscous. Higher temperatures cause cleaned steel surfaces to rust. Typical values of operating temperature are between 80°F and 125°F.

Holdup time under sonication (formation of emulsion) should be at least 15 seconds. Holdup time is a tradeoff between cleaning quality and productivity.

The number of steps of sonication is a tradeoff between cleaning quality and machine size or cost.

[27] Baker, J.Y. and Durkee, J.B., "Rethinking Cleaning Processes, Parts I, II, and III," *$A^2C^2$ Magazine*, 4–9 (October 1999), 4–6 (November 1999) and 39–40 (January 2000).

The No-Chemistry approach has yet to make a commercial impact.

### 6.1.8 Electropolishing: The Perfect Surface Finishing Method?

Electropolishing (EP) is normally used to produce a shiny finish on complex metal parts. That's probably why you know of it. Industries use EP for finishing of metal used in pharmaceutical, medical, semiconductor, and food processing applications.

EP may be the ultimate cleaning technique. EP can provide removal of imperfections such as stains or surface corrosion, heat discoloration, oxide films, localized stresses, weld marks, or scratches, as well as particles of all sizes, organic films, and biological debris. That outcome may not be cleaning, but it may be better!

#### 6.1.8.1 *Electrochemistry 101*

The piece to be electropolished/cleaned is immersed in a liquid acidic bath, after pre-cleaning because electropolish is a finish cleaning step. The piece is connected to the positive (anode) terminal. The negative (cathode) terminal is connected to a conductor.

A positive direct current from a rectifier (converts AC power to DC power) is introduced into the piece which is hung from a center electrode. The piece is surrounded by the cathodes (negatively charged, see Figure 6.8).[28]

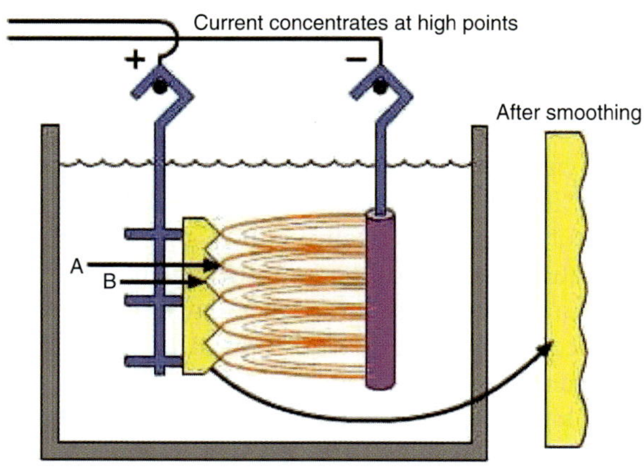


**Figure 6.8**

In electroplating, metal ions are deposited from the solution onto the work piece. With EP, also called "reverse plating," metal is removed by transforming it into ions soluble in an reducing acid bath.

#### 6.1.8.2 *Electrochemistry 201*

Faraday's Law of electrolysis generally requires that the amount of metal removed from a surface is directly proportional to the amount of electrical current flowing. This is calculated as current density multiplied by electrode surface area multiplied by exposure time – Equation (6.1):

$$\text{Metal removal} = \text{Current} \times \text{Time} \qquad (6.1)$$

Specifically, Faraday's Law of electrolysis requires one equivalent weight of a substance to be removed from the cathode during the passage of 96,487 coulombs of charge through an electrolytic cell. A coulomb is 1 ampere flowing, under any voltage, for 1 second.

Written simply, the EP process is the following:

$$\text{Acid} + \text{Metal} \rightarrow H_2 + \text{Metal salt} \qquad (6.2)$$

There are three unexpected significant and chemical consequences of this equation:

1. Acid in the bath is depleted, and must be replaced to maintain the same bath composition and EP quality.
2. A hazardous material, hydrogen gas, is produced[29] and must be safely contained or emitted. This is another reason why many EP operations are done in job shops.
3. A metal salt is produced, which absent other action is another waste product. This is the most significant negative concern.

#### 6.1.8.3 *Features and Benefits*

Typically, voltages are around 25 volts direct current (VDC). Voltage level affects quality of the surface finish, not the rate of its production (see Figure 6.9).

[28] Figure 6.8 courtesy of Delstar.

[29] Usually less than 1 cubic foot per hour per square foot of metal.

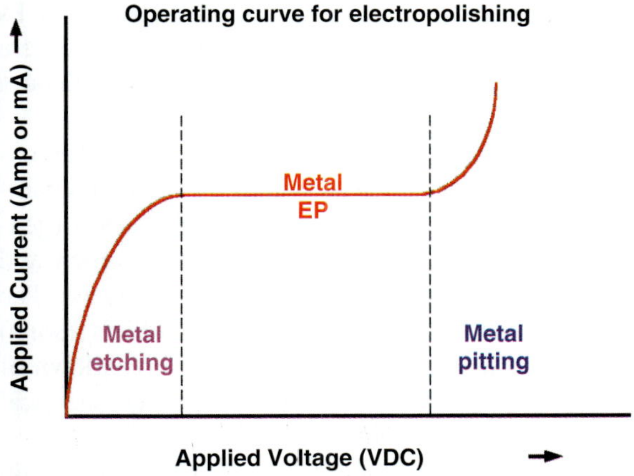


**Figure 6.9**

By removing the surface layer, EP also:

- Hygienically cleans the surface, making it resistant to bacterial growth due to removal of Hydrogen, and a smoother surface.
- Provides the most dense, durable, passive, and corrosion resistant film on stainless steel that it is possible to achieve. Users of cleanrooms demand non-contaminating and non-particulating surfaces. EP is the ultimate finish for cleanroom tables, chairs, waste containers, light fixtures, exposed electrical conduit and outlet boxes, manufacturing and processing equipment, and other metallic components used in cleanrooms.
- Removes surface stresses caused by mechanical working. Applications in dental science include EP for removal of surface stress which can later cause surface cracks.
- Removes micro-scratches to improve fatigue resistance.
- Smooths the surface, increasing reflectivity, and creates a "bright" appearance. To achieve that benefit is why most people use EP.

### 6.1.8.4 *Costs of EP*

This author doesn't know what EP costs. Operating costs are a closely held secret because so much EP work is done in competitive job shops.

The two major components are labor and management of waste products. Labor includes racking and unracking of parts, selection and creation of the appropriate cathodes, and learning time to develop the proper conditions of voltage, time, and bath composition.

The best estimates, obtained from several sources, are €0.10–0.50 per "part," with the "part" specifications being undefined. If you don't have a specific "part," use the value for EP of five cents per square inch. In this author's experience, that is cheaper than costs of critical cleaning, for a part not specifically defined.

### 6.1.8.5 *New EP Technology*

There are problems of waste disposal and process control. Both problems can be seen from the information in Equation (6.2). Both the metal salt and the Hydrogen gas are unwanted. And the acid must be replenished.

The EP bath contains a liquid mixture of several strong acids, soluble and insoluble salts, and perhaps some other chemicals. Strong acids are necessary to dissolve the metal salts liberated from the metal surface.

All of these chemicals are extremely hazardous. Many, such as acids, manifest several hazards. An EP bath is not as relatively benign as is an aqueous cleaning bath. Avoidance of these hazards is one reason why most EP work is completed in job shops.

Most users doing EP work, and most job shops, allow the metal salt to accumulate and precipitate. It is periodically removed as a sludge. Similarly, fresh acid is added to makeup that consumed. This is a brute force approach. It probably isn't adequate for the level of process control needed in critical cleaning.

Fortunately, some firms have developed better ideas. One[30] is to purge from the EP bath to a second solution tank in which the reverse of EP (electroplating) is conducted. Here, the metal salt is converted to acid and the metal is plated on another surface. The acid is returned to the EP bath. A US commercial firm[31] has a variant of this approach. They claim years of life without replenishing the EP bath.

[30] US Patent 5,882,500; among others.

[31] Global Stainless Technology.

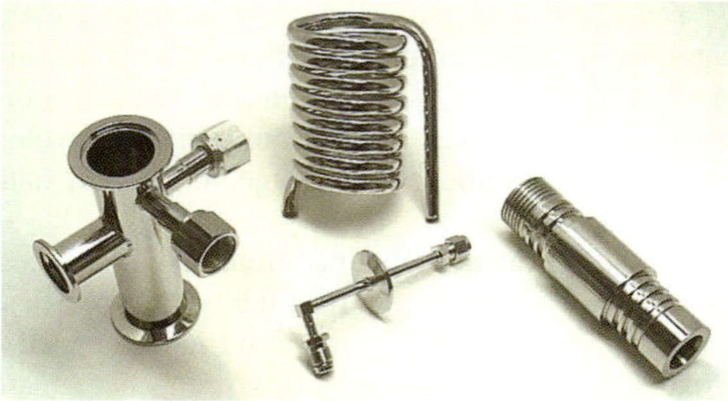

**Figure 6.10**

### 6.1.8.6 *Applications of EP*

Stainless steel and Copper are the most common metals to which EP is applied. Some work is now being done with Aluminum and Titanium. Industries where these metals are nearly always electropolished include food and beverage processing, medical and pharmaceuticals, nuclear, electronics, and any in which work in a cleanroom is required (see Figure 6.10).

The high conductivity of these metals makes the process more efficient and predictable. EP is being done with Silicon – an inefficient conductor – in the production of semiconductor wafers.

The tendency of EP to preferentially remove material at thin surface sections was recognized by many, and found to be an advantage in an uncommon application: sharpening of edges.[32] Unique examples are sharpening of wires[33] and needle electrodes.[34]

## 6.1.9 A Specific Method for Cleaning with Chemistry

Cleaning under vacuum with plasma does involve chemistry: oxidation. The oxidizing specie conventionally is atoms of Oxygen. The products of oxidation are those of biological oxidation, $CO_2$ and water.

Suitably sized parts are placed on shelves or racks into a vacuum chamber whose gaseous atmosphere may either be air, or the air may be replaced with pure Oxygen.

A radio frequency (RF) source is connected across a set of electrodes placed within the vacuum chamber.[35] The radiation energy excites the gases (generally Oxygen). Atomic fragments of Oxygen and ultraviolet (UV)/visible radiation are produced. The former reacts with hydrocarbon materials (soils) on the surface. The latter can enable breakdown of polymeric materials or initiate the desired chemical decomposition reactions.

Gaseous byproducts ($CO_2$, water, hydrocarbon intermediates) are removed as exhaust when the chamber is pressurized with air. The only heating is provided by the RF radiation. Since, the mass of gas is overpowered by the mass of the parts, little heating of the part occurs.[36]

Even metals which are easily oxidized such as Copper or Silver can be plasma cleaned of organic contaminants without discoloration of the metal by using non-oxidizing gases such as Nitrogen, Hydrogen, or Argon. The reacting specie is not Oxygen atoms but energetic free atoms of these other gases. Often a gas mixture is used.

The uniqueness of this process is its specificity. Only the soil is affected; there is little crazing or cracking of surfaces. The amounts of $CO_2$ and water produced, and the amounts of soil affected, are measured in grams. One doesn't need 5 gal of a water-detergent mixture or a quart of solvent. What one does need is capital.

Managers recognize plasma cleaning as a financial extreme. The financial requirement is to pay off significant investment used to purchase the vacuum, energy supply, and control equipment.[37] There is essentially no operating capital for consumables, though skilled and trained staff and electric power are essential.

Plasma cleaning with air, or other gases, is often a step in a surface treatment process. Since the work is

---

[32] US Patents 3,492,178, 4,406,759, 4,710,279, and 5,616,255.
[33] US Patents 4,375,396, 4,935,865, 3,697,403, and 2,434,286.
[34] US Patents 4,587,202, 4,777,096, 5,693,454, and 5,762,811.
[35] Hollahan, J.R. and Bell, A.T., *Techniques and Applications of Plasma Chemistry*. John Wiley & Sons, New York, NY, 1974.
[36] Paquin, D., "The Gas Plasma Alternative to Wet Cleaning," *Parts Cleaning Magazine*, March, 1994, pp. 45–49.
[37] That's why, more than any other type of cleaning process, plasma cleaning is conducted by contract cleaning services (job shops).

done under vacuum, deposition of metalized coatings is often a sequential operation.

Applications are:

- Few for aircraft wings and other large objects.
- More for optic components in guidance equipment.
- Many when the consequences of inadequate cleaning are critical.
- Well justified when complete soil removal is required, when part surface condition is fragile, or cleaning *in situ* is necessary.

#### 6.1.9.1 *Plasma Cleaning without Vacuum*

The advantages of plasma cleaning, while attractive, seldom outweigh the disadvantage of investment requirement. Research prompted by that situation has led to the development of a plasma-based cleaning process which does not require vacuum equipment, and can be used to continuously clean some structures.[38,39]

The plasma zone is controlled and contained between two sacrificial anodes. The part becomes the cathode. A DC voltage is applied across the gap between the cathode (part) and each anode. The zone is an electrically conductive foam produced by boiling an aqueous solution of sodium carbonate and water. Local surface temperatures are quite high because the electrically conductive foam is not thermally conductive.

Cleaning is done in two ways, by: (1) local surface melting and (2) surface disruption caused by collapsing bubbles and production of shock waves.

Successful applications[40] appear to be in cleaning stearate-based drawing fluids from wire and strip.

#### 6.1.9.2 *Summary About Cleaning without Chemistry*

The situations covered this chapter are outside the paradigm of traditional chemically enabled immersion and spray (low pressure) liquid-based cleaning.

Some are familiar (blast cleaning, pressure washing, and abrasive cleaning), some are developmental ("No-Chemistry" technology), some are uncommon (plasma cleaning or $CO_2$ snow cleaning), and some are not normally thought of as being cleaning technology (EP).

All can bring value and challenges.

## 6.2 REMOVAL OF PYROGENS IN BIOMEDICAL APPLICATIONS

Pyrogens are bacterial cell wall fragments. They are not bacteria. Typically, they are complex carbohydrates. Being chemically stable, pyrogens are not necessarily destroyed by conditions that kill bacteria.

Pyrogenic means to cause heat. If pyrogens are injected into humans, they can cause an increase in temperature (fever), and by doing so can influence research results when scientists use living subjects. Pyrogens may cause fever when injected, but are not a problem if ingested by humans.

### 6.2.1 Removal of Pyrogens from Water

There are two approaches generally followed to get pyrogens out of water. Neither is a chemical treatment.

Ultrafiltration (UF) is an excellent way of removing pyrogen contamination from water. Ultrafilters (positively charged nylon 66 membranes) are recommended for the final "polishing" of water already treated by deionization (DI) or reverse osmosis RO. Ultrafilters remove most organics over 1,000 weight-average molecular weight, such as pyrogens. Also, pyrogens preferentially sorb to alumina.

### 6.2.2 Removal of Pyrogens from Parts

Their chemical nature makes pyrogen removal problematic. Pyrogens are unusually thermally stable below the boiling point of water, and fairly insensitive to pH changes. High concentrations of acids or bases are necessary to destroy these complex carbohydrates within a reasonably short time.

---

[38] Ryabkov, D.V., "Process and Apparatus for Cleaning and/or Coating Metal Surfaces Using Electro-plasma Technology," US Patent 6,585,875, July 1, 2003.

[39] Gupta, P., Daigle, E.O., Tenhundfeld, G. and Calliham, B., "Next Generation Cleaning and Surface Modification Technology," *Wire & Cable Technology International*, November–December, 2003, p. 52.

[40] As this is written (2005), this proprietary technology can only be identified as developmental. For more details, see http://www.captechnologiesllc.com/.

Hence, the removal of pyrogens is generally done with mechanical force augmented with aqueous detergents. Removal can be done with only repeated detergent action for mechanically fragile parts (see Section 6.2.2.1).

Most successful processes for getting pyrogens off parts use cavitation produced by ultrasonic transducers. Obviously, the main concern is collateral damage to the part. This has been observed in cleaning of tiny, fragile stents.

Also, pyrogens are removed from medical equipment via the mechanical technique of hand scrubbing with an aqueous detergent. Obviously, this is technique is not suitable for continuous manufacture of medical parts. Pressurized sprays augment the scrubbing action.

All cleaning and rinsing would be done under immersion,[41] whether with chemical solvents or aqueous detergents. The basic sequence is probably clean and rinse, inspect; then sterilize and validate.

As with most cleaning situations, either aqueous or solvent cleaning can be employed. A manager should make the choice based on the nature of the parts, as they would any other cleaning situation.

#### 6.2.2.1 *Aqueous Detergents Used in Removal of Pyrogens*

Proteolytic[42] enzymatic detergents with a pH range between 6.0 and 8.0 should be used. These detergents have non-ionic surfactants.[43]

A common problem associated with detergent use is its presence. Detergents are not part of the manufacturing process and are only added to facilitate cleaning. The US Food and Drug Administration (FDA), and others, expect that *no* detergent levels remain after cleaning. That sets the requirement for the rinsing process.

DI water should be used to wet parts and for rinsing. Water hardness is a concern because of mineral deposits remaining after drying.

A commonly used process is to:

- Soak medical parts for a minimum of 5 minutes and a maximum of 10 minutes in enzymatic detergent.
- Rinse them thoroughly with warm DI water, making sure to irrigate the interface(s).

Without mechanical techniques playing a contributing role, the soak/wash/rinse procedure should be repeated several times per cleaning job.

#### 6.2.2.2 *Chemical Solvents Used in Removal of Pyrogens*

Isopropanol (IPA) is the solvent of choice because of its familiarity in medical applications. No other solvent has much standing in pyrogen removal.

Yet from a technical standpoint, IPA is not a favorable choice. The key to solvent selection here should be surface tension – the lower the better. The value of 22 dyne/cm (nM/mm) for IPA may not be low enough to clean crevices, cracks, and interstices. HFE-7100 has a surface tension below 15 dyne/cm and has been tried. Solvent boiling point is not a major factor because pyrogens are thermally stable.

### 6.2.3 Sterilization of Cleaned Parts

Cleaning is not sterilization. There is no level of cleaning which can be done which removes pyrogens and meets the FDA requirements – or those of any other agency requiring medical certification.

A unit based on UV light shouldn't be used for sterilization as other bacteria cells are not removed in it but are converted into pyrogens, making the situation worse.

Sterility is not synonymous with non-pyrogenicity. Each defect must be addressed separately.

A two-step approach is suggested based on temperature (for metal parts suitable to this exposure):

- **Step 1**: High-temperature pyrogen removal cycle: 270–275°F (132–135°C) with a minimum exposure time of 10 minutes.
- **Step 2**: A 1-minute purge with Nitrogen and at least 15 minutes of vacuum drying.

Notice the extent to which the temperature must be raised because pyrogens are so thermostable at ordinary temperatures.

---

[41] Note that spraying pressurized liquid under immersion requires a different type of nozzle than that used in most aqueous spray cleaning machines (see Chapter 7, Section 1 and Table 1).

[42] Proteolytic proteins are enzymes which digest proteins.

[43] See Chapter 2, Section 2.4.2.1.

Sterilization can be done afterwards in a single stage with specified chemicals. Ethylene oxide (EO), a carcinogen and mutagen, is a commercial choice. The sterilization cycle is:

- 100% EO at 131°F (55°C) for 60–180 minutes.

### 6.2.4 Analytical Issues

Pyrogens from bacterial cell walls (the most commonly encountered type of pyrogen) are referred to as bacterial endotoxin and are readily detected by gel clot and kinetic chromogenic Limulus Amebocyte Lysate (LAL) testing systems. This is a qualitative test for sub-gram quantities of bacterial endotoxin. LAL as supplied in individual reaction vials is to be reconstituted with the solution being tested. After incubation, and in the presence of endotoxin, gelation occurs; in the absence of endotoxin, gelation does not occur.

## 6.3 MANAGING CLEANROOMS AS IF THEY WERE AQUARIUMS[44]

Isn't an aquarium another kind of cleanroom? Is it possible that some of what managers might have learned about raising tropical fish could be lessons of value in management of cleanrooms?

Aren't an aquarium and a cleanroom both managed environments? Can a manager learn about managing one environment from managing another?

### 6.3.1 The Strategy of Over-Diligence

Both cleanrooms and aquariums function best when they are at equilibrium. That realization has cost the lives of too many fish, and made too much off-spec product.

Some managers feel that the best managed cleanroom is one which is always under change. That is, they are always striving to get the last particle, fiber, and micro-organism identified, located, and discarded. These managers are always changing the system in order to improve it.

That would be an excellent strategy if environments were linear and not multidimensional. But all that thrashing about liberates "sleeping" particles, fibers, and micro-organisms. This is dirt that was not causing quality problems and was located outside critical work areas. Environments are non-linear and multidimensional – one change produces another, and another.

### 6.3.2 The Equilibrium Approach

Lessons from managing aquariums can show why the opposite strategy may a better one. Here are some examples of why this author feels an equilibrium strategy is best for both aquariums and cleanrooms – even if the total level of housed dirt is more in the equilibrium case:

- **The over-manager**
  I had always wanted to keep Discus fish. The books all said the secret was pH: Keep it rock steady at 6.800, or so I was always sampling, testing for pH, adding base or acid, and sweeping up the dead Discus at €50 per fish. I was the above cleanroom manager – thrashing about. I killed the fish because I never let them get comfortable – at equilibrium.[45]
- **Watch the whole aquarium**
  I had raised knife fish to nearly 2 ft and Arowana fish to nearly 3ft long. They were the stars of my tanks, and I loved them! But I killed them as well – by introducing change. Their diet was small, live fish. Once, to save money, I substituted minnows for Zebra Danios. The minnows were raised in the wild, and were diseased. Within three days, all critters great and small were dead. I killed these fish because I made a change I didn't have to make. I introduced a new variable. I destroyed the equilibrium.
- **The wrong variables**
  Cichlids like cold, clear, acidic water. I had a tank of lovely blue and yellow Rams (Ramirez fish). Always chasing each other, they were happy and healthy (if you don't count split tails). My under-gravel filter was old and full of dirt. So I removed the fish to a hospital tank and cleaned their home.

[44] See also Chapter 1, Section 1.7.2.

[45] See Chapter 4, Section 4.12.1 about on-aim control. These two concepts are compatible, if the point of aim is a modest range and not a specific value.

On return to their fresh, cold, clear acid water, they all got sick. Why? A microbiologist friend explained that I had liberated a strain of bacteria that was under control by the organisms in the dirty filter. I was chasing the wrong variable. It wasn't "dirt." It was that strain of bacteria which was once under control.

- **Too many unknowns**
  Managers of cleanrooms deal with particle counts, air flows and distributions, and cumulative sum (CUSUM) plots of significant variables. They look around and feel secure. Who wouldn't, in the presence of all that data? Once, I managed my fish tanks that way. But I have seen my fish sicken (that strain of bacteria) and quality in cleanrooms decline (a cause that mysteriously went away) without perceptible changes in either environment. Some unknown variable caused both problems.

The lesson offered in Section 6.3 is simple: if you're satisfied with what you have, aim to maintain equilibrium. In other words, "If it ain't broke, don't fix it."

## 6.4 HUMAN FACTORS IN CLEANING OPERATIONS

This book was written to cover engineering and chemistry – chemical engineering. Another type of engineering and chemistry is often as least as significant in affecting the outcome desired by a manager. That is human engineering and human chemistry.

The components of success in most industrial, and other, situations are:

- An understanding of the true situation.
- Developing a suitable plan.
- Having the right technology (tools).
- Implementation of that plan, and most important.
- Support from the people who will implement that plan.[46]

People can control success by what they think, feel, know, do, and don't do. This section is about people doing cleaning work and how to manage them.

### 6.4.1 Lessons from the Chlorofluorocarbon Phaseout

The phaseout[47] of chlorofluorocarbons (CFCs) and other ozone-depleting compounds (ODCs) as cleaning agents caused a global revolution in the way products are manufactured and repaired. There was a paradigm shift in parts cleaning.

The loss to the global marketplace of these chemicals was severe, and was a "Full Employment Act" for cleaning consultants. These chemicals were cleaning processes – good solvents, free rinsers,[48] and rapid dryers. Vapor degreasing and cold cleaning were and had been the prime technologies: nothing else was needed.

Many conversions from ODCs to other cleaning agents failed. The reasons included:

- **Emotions**. Not all operating managers believed in the basic atmospheric science, and felt the Montreal Protocol was only "politically correct." At the time, many did not support the change.
- **Indecisive governmental actions/rumors**. Many believed until 1996 that the US Clean Air Act (CAA) would be changed/repealed.[49]
- **Fragmented industry**. The US cleaning industry is highly fragmented. The change became

[46] No amount of positive teamwork and support by workers can offset the negative effects of a leader (management or labor) who doesn't believe in a project and prefers to see it fail.

[47] CFC-113 and 1,1,1-trichloroethane were the chief cleaning chemicals affected (see Chapter 1, Sections 1.13.1 and 2.1.2, and Chapter 2, Section 2.3.5.2).

[48] This means that their surface tension was very low, 15 dyne/cm (nM/mm) and below.

[49] The 1988 Montreal Protocol lacked specifics. The 1990 CAA appeared both punitive and overly aggressive to many. In 1992 President Bush's speedup further raised tensions among users. The component of the CAA called "Labeling Law" bred uncertainty, and its rescission at the last minute caused confusion and anger among those who had sought to comply with it. The US EPA's Significant New Alternatives Program (SNAP) decisions included toxicity information and exposure limits, which had previously been under OSHA's purview. The result was confused and frustrated users.

In 1995, a US congressperson with a substantial reputation discussed with this author his view that the CFC phaseout was harming US industry and inquired about how it might be scaled back.

total: from solvent to aqueous technologies, from "bad" to "good" solvents.

- **Incredible margins[50] and expensive equipment**[51]. Many suppliers, not all now present, acted as if the phaseout was their retirement program.
- **Bad cleaning science**. The simplicity of using ODC solvents tranquilized many into that all cleaning agents performed as do ODC solvents. There was no unbiased source of education.[52]
- **Under-perform and over-promise**[53]. This became the effect mantra of many suppliers. Many users lost confidence.
- **Job losses**. Some were fired for making poor choices about replacement cleaning systems.[54]

At their root, all of these reasons derive from human concerns. The outcome of the CFC phaseout was angry and risk-averse users.

What this means to a manager is that mistakes have been made because of lack of the right information. As a manager, your decisions will be no better than the information on which they are based. Base your expectations on good information.

### 6.4.2 It's the Operator

The operator is the most important component in your cleaning system:

- Universally and independently, operators can cause your system to fail.
- Operators have more "hands-on" job knowledge than does any manager.
- Operators know the interface with the previous and next operations better than any manager does.
- Operators may resist change because of concerns about job security.

Change usually makes humans nervous. With a new cleaning system, the job of operators just got harder – or, at least, different. Without their permission, in some union plants, operators can refuse to do the work with the new system.[55]

### 6.4.3 The New System

When a new system is contemplated, operators must be included on the team which selects it. Their buy-in is crucial.

Your operator is your front-rank, in-house expert. Your operator has to solve the problems as they

[50] Banned products had been cheap – costing cents/lb. The replacements were formulations versus single-component cleaning agents which were priced at dollars/lb. Some of this pricing was reasonable because of the high cost of development and user support for new products. Unfortunately, competition led to the low formulation cost becoming known. The result was angry users.

[51] Multiple-stage aqueous systems will always be more expensive than single-stage solvent systems. And with aqueous technology, drying becomes important. In addition, process control became necessary. Floorspace requirements were much greater. Since many of the processes were new, prices had to include development and support costs which led to expensive equipment. The result was angry users.

[52] One example, from personal experience, was the engineer who recommended alkaline aqueous cleaning technology for cleaning mixed metals. There were two results: galvanic corrosion and a new employment situation.

Another example involved the US Air Force. A large airbase bought a big, new aqueous cleaning machine for €2.5 MM. Naturally, it was bought based on low bid. The supplier spoke about the aqueous waste: "Just put it in to the drain" because their low bid hadn't included facilities for waste treatment. The state and Federal EPAs naturally wouldn't provide a permit for that! The unit became known as the big €2.5 MM "boat anchor" or the "Blue Elephant." One key lesson was that waste treatment may cost more than cleaning.

Still another example is about the firm which wouldn't allow chlorinated solvents in their shop. They spent two years and significant funds trying various non-solvent cleaning technologies with no success. Later, their new management installed a new low-emission machine that cleaned with a chlorinated solvent. The key lesson is that "political correctness" doesn't solve cleaning problems.

[53] The general industry view was that phaseout would be a bonanza. Users were exposed to continuous claims of "we can do that." Unfortunately, multiple types of solution are possible with different consequences. So customers were set up for conflict with suppliers. Naturally, without trust, customers bought "on the cheap." The result was confused users.

[54] Many "conversion co-ordinators" ride to work in a different carpools. Management didn't understand the above situations or didn't listen. The result was fear of failure.

[55] Operators can maintain hidden stocks of old materials, which are known to work (!), and not accept the new system.

occur, 8,760 hours/year. Operators must know and control the factors affecting quality in new system.

Use of a replacement system will put their troubleshooting abilities to use. Remember, their past training won't work because the technology has changed.

Interaction with operators is always essential (if not "politically correct"). Managers must be in contact with the principal operators, or else operators will allow failure through ignorance or neglect. The operator must be empowered to give feedback or criticism to management (you). As a manager, you must be open and listening – or necessary truths will be lost.

### 6.4.4 The "Golden" Rules

This author has developed rules for making good conversions to replacement systems. Lessons about human behavior from the CFC phaseout are a substantial basis for these rules:

- **Rule 0: There are No drop-ins – Zero, Zilch, Nunca, Nada, Nichts, Non, Nein. …** About every new replacement system, something is different. And it will identify itself by teaching managers something they didn't want to know.
- **Rule 1: Build a Team[56].** The team must include at least one person who will do the work (be the "hands-on" person) and one who will be responsible for doing the job. There are two corollaries to Rule 1:
    1. If this team includes cohorts of another management group (such as your management), build a second operating team reporting to you.
    2. Input and buy-in from the "hands-on" person is crucial to success.
- **Rule 2: Set Some Time Horizon for Success.** Get your management to accept it. You can depend on your management to expect more than you can deliver, because they don't value or understand cleaning problems. Without acceptance of a time horizon for success, your replacement system will be a castle built on sand.
- **Rule 3: Don't Spend Time on Choosing Vendors.** Don't test all their offerings. Don't spend significant time optimizing the purchase cost. Spend your time on implementing your chosen system. That's on what you are being graded. Everyone who has tested everything is working everywhere else.
- **Rule 4: Choose ~ 2 Suppliers Who Can Integrate Systems.** Work with one of them. Prices of cleaning systems are normally competitive. Integration of equipment and chemicals is vital! *Choose the one supplier who most wants and can support your business.*
- **Rule 5: Don't Buy Anything without Testing It with Your Situation.** Believe that every situation is different than the one with which you must deal, until your situation has been made to work!
- **Rule 6: Understand What to Expect versus Your Previous System.** The new system will be different. Get some unbiased education to find out where, why, and how. Communicate this to management. Or update your resume.
- **Rule 7: Choose Aqueous Cleaning Technology If:**
    - The soil on your parts is compatible with water detergents (soluble, emulsifiable, etc.).
    - Your parts can stand contact with water.
    - Floor space is not an issue.
- **Rule 8: Choose Solvent-Based Technology If:**
    - Your parts have intricate sections and a fluid with a low surface tension is necessary to fully flush them.
    - You clean different parts with different soils.
    - Drying quality is important.
- **Rule 9: Be Careful with Parts Drying.** It is harder, will take more effort than washing, and will cost more than cleaning. It's a separate process. It's not cleaning. It's not well documented (see Chapter 1, Section 1.13).
- **Rule 10: Develop Some Written Standards.** These should include the quality specification, procedures, controls, training, and maintenance (see Chapters 4 and 5). If you don't, someone else will, or maybe, else no one will. Written means defined and accepted, not just committed to paper.

[56] See also Section 6.8.4.

### 6.4.5 What About Costs?

As a manager, your management properly expects you to control costs, have a budget, and keep them informed about deviations from it. The budget must be realistic or you will overrun it.

But cleaning costs are not well known (see Section 6.7). They are not published for reasons of security. It is certain that operating costs of a replacement system will be different than that the replaced system – so it is likely there is no useful baseline for budgeting.

Your best source for information is probably your supplier. They have experience accumulated from other customers. This can be a valid basis for choice of supplier. Consultants can help as well, but they are often bound by secrecy agreements.

As with operations, involve your management in development of your budget. Be candid about uncertainties. Be forthright about your plans. Let initial operation define metrics. Then develop plans to reduce known cost elements.

Budget failure may be more important than failure of quality or productivity because it may be valued more highly by your management.

That's a mistake. Cleaning should not be a major cost center in manufacturing or maintenance operations. The real cost of cleaning should be the cost of failure to complete subsequent operations when cleaning is done poorly.

### 6.4.6 Stuff Happens

No matter how fully automated and instrumented is the cleaning system you manage, human factors will at some time become more significant that technical factors.

This will almost always happen at the worst possible time – a crisis. That's when expectations by your management won't be fulfilled. That's when the capability and interaction of your operating team will be most strained. Will you have prepared them properly.

## 6.5 THE EXPONENTIAL COST OF DRAGOUT

As much as any other factor, rinsing dominates cleaning – especially critical cleaning.[57]

In rinsing operations, users "fog" clean (or almost so) water or solvent across all surfaces of a part. Rinsing is getting the soiled cleaning agent off parts. Rinsing is also keeping it off the parts while getting the soil out of the machine. And the No. 1 enemy of rinsing is dragout (see Chapter 1, Section 1.12.5 for the engineering basis underlying the unit operation of rinsing).

### 6.5.1 The Impact of Equilibrium Rinsing

It's simple: whatever soil materials are remaining on parts after cleaning must be removed by dilution rinsing.

The more material remaining, in whatever form, the more expensive in time, facilities, resources, and management involvement it will be to remove that material by dilution rinsing.

Consider this example:

- The parts being cleaned have 10 $in^2$ of surface.
- These parts are wet with a film of dirty water which is 1 mil (0.001 in) thick (so the volume of dirty water is only 0.164 ml and weighs only 0.164 g = 164 mg).
- The concentration of dirt in the water is 1% (so the mass of pure dirt is only 1.64 mg, and the mass of dirt plus water on the part is 164 mg).

So the soil concentration on the unrinsed part surface is 0.164 mg/si (1.64/10), or 25.4 $\mu g/cm^2$ in units more commonly used in cleaning.[58] (Figure 6.11).

Surely, it can't take much rinse volume to remove that tiny amount – 0.164 ml of dirty water!

[57] This is because critical cleaning, where inadequate cleaning is critical to the enterprise, nearly always means removal of minuscule amounts of soil. That's what poor rinsing leaves on parts.

Soil removal is much more straightforward than rinsing – mostly a matter of solvent or detergent selection. Cleanliness happens when the open and closed sections of the part have become free of the impurity defined as soil to the extent required in the definition of cleanliness. (see Table 1.1).

[58] Those parts might or might not be considered dirty. Soil levels on cleaned high-value parts should be around 1 $\mu g/cm^2$, or may be a lot less (see Table 5.3).

**Figure 6.11**

## 6.5.2 The Enemy of Equilibrium Rinsing: Dragout

That dirty water is dragout. Dragout is the fluid remaining on parts after cleaning.

Dragout, sometimes spelled "drag-out," is the mixture of soil, cleaning is the mixture of soil, cleaning agent, solvent or water, and particles left on the parts after cleaning is complete. In a very real sense, cleaning is *not* really complete at that point, because the parts certainly aren't clean when they are covered with dragout liquid.

Rinsing and dragout are joined in a "love–hate" relationship. The job of rinsing is separate the dragout from the parts.

If dragout is decreased, rinsing is made significantly easier.

There are three major factors which control dragout:

1. Surface tension is the chief contributor to dragout. It is the force holding liquid to parts. So, dragout is much less important with solvent cleaning (where surface tension is <30 dyne/cm) than it is with water (72 dyne/cm). One of the key values of CFC-113 was that its surface tension was quite low – around 17 dyne/cm.
2. The second major contributor to dragout is part character. A simple film around a single ball bearing is one extreme. A complex film contained within a multi-port injection valve is the other.
3. The third major contributor to dragout is part stacking. Suppose that ball bearing is the bottom one in a five-layer stack. Then, the rinse fluid from the bearings above, and much of that from adjacent bearings, must flushed before one can start rinsing the bottom bearing.

## 6.5.3 Removal of Dragout from Parts

Don't rinse dragout from parts!

Remove as much as you can before you start rinsing! Rinsing is the most expensive, time-consuming, and least efficient way to remove dragout.

Consider the method by which a wet dog dries itself: It shakes most of the water from its coat. The remainder is removed through gravity drainage and evaporation:

- That's how to most effectively remove dragout. Assist gravity to drain liquid attached to your parts.

The basic procedure for removing dragout from parts is to remove the fluid surrounding the parts by mechanical means. Do this before starting to rinse. In other words, minimize the difficulty of the rinsing job (Figure 6.12).

Here are some specific ways to do this. The choice is usually dependent on the structure and vulnerability of the parts. Remove liquid (water) by:

- Allowing the parts to drain at whatever pace they do (see Figure 1.5).
- Assisting that drainage with low-energy/low-frequency vibration (remember the example of the dog).
- Removing parts from their fixtures and repackaging them in a centrifugal dryer whose rotational velocity has been limited so it doesn't harm the parts (see Chapter 7, Section 12.6.1).
- Aiming air knives at the parts.[59]
- Wiping some water with a clean wiper.
- Use of vacuum tools (see Chapter 7, Section 7.12.6.3).

This is not drying. This is preparing to rinse by reducing the magnitude of the task. If the parts are fully dried, it will be necessary to remove the dried residue from their surfaces.

The aim of the above techniques is to remove *ca.* 60–85% of the water and leave the surface wet. The reason for this water-separation step is that rinsing is

[59] While being careful to avoid reinfecting the parts with water entrained in the spent air stream (see Chapter 7, Section 7.12.1).

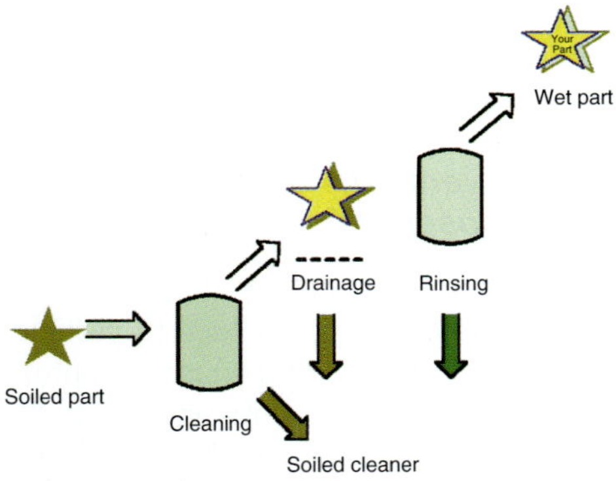


Figure 6.12

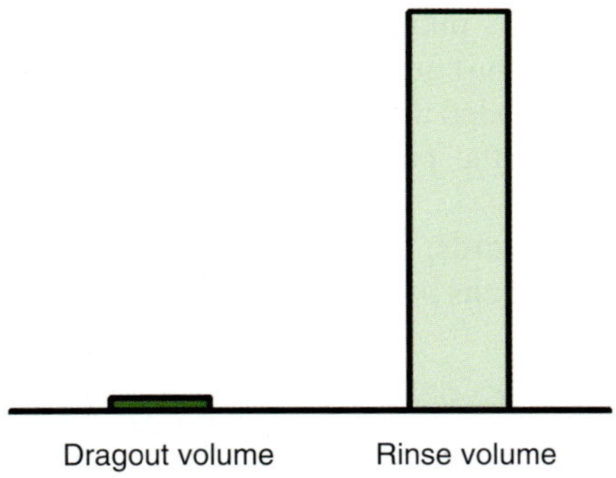


Figure 6.13

so inefficient – particularly at low levels of retained residue:

- For the example above, reduction of the soil level on the parts from 25.4 to 0.254 g/cm$^2$ (a 100 × dilution/reduction) requires use of at least 16.4 ml (100 × 0.164 ml) of water.

In practice, some soil-laden fluid is trapped in part crevices and not mixed with the pure rinse fluid. *Even more than 16.4 ml of clean rinse fluid must be used.*

In many cases, such as machined valves, porous castings, or heat transfer fins, the additional volume of rinse may be higher by an *additional* factor of 10–1000:

- If the parts are structures other than simple, clean ones such as ball bearings, *one might need 1640 ml of rinse fluid* (0.164 × 100 × 100) to dilute that film (0.164 ml) remaining from the wash bath to an acceptable level.

The volume of dragout fluid and the volume of rinse fluid typically necessary to dilute it are compared in Figure 6.13.

And that's the theme of Section 6.5 – rinsing is extremely costly of space and materials (water) because it is extremely inefficient.

## 6.6 PARTICLE REMOVAL

Whether trying to remove medical residue from glass, nuclear contamination from scrap metal, chips from screws after a machining operation, or chemical mechanical polishing/planarization (CMP) byproducts from semiconductor stock; one has to be concerned about technology trends[60] in managing removal of particles from surfaces.

This chapter will discuss the *operating limits* of particle removal by mechanical technology.[61] The technology is mechanical, because chemical technology generally doesn't work. These particles aren't soluble in solvents managers want to use.

### 6.6.1 Removal of Micron-sized and Sub-micron Particles

Particles can take almost any physical form as solid contamination (fibers, solid surface, flakes, biological matter, or scale) on an unclean surface. Granted, not all of these contaminants can be removed from surfaces by mechanically produced force; some require chemical action. But a great many can be, if the applied force is great enough.

In 2004, Sematech[62] published its International Technology Roadmap for Semiconductors, projecting minimum feature sizes[63] of 70 nm (0.07 μm[64])

[60] See Chapter 1, Section 1.6 about management of the trends in removal of particles, and the reasons for them.

[61] The obvious exception is hydrofluoric acid (HF), used to solubilize Silicon residues in the manufacture of semiconductors despite the hazards its use presents in the workplace.

[62] A research and technology-focused consortium who sees their mission as being the "world's catalyst for accelerating the commercialization of technology innovations into (semiconductor) manufacturing solutions."

[63] DRAM ½ size. Details of this comprehensive forecast can be found at http://www.itrs.net/Common/2004Update/2004Update.htm.

[64] One nanometer is 1 billionth of a meter. There are 25,400 nm in a mil, and 1000 nm in a micron. But there are 10 nm in the dimension used to describe atoms – angstroms.

as this book is published, and 55 nm in 2009. This author has heard others say that if the diameter of a particle exceeded about one-fourth of the minimum feature size, the particle could cause fatal device failure.

In other words, the particles to be removed in future operations will be very small particles.[65]

## 6.6.2 Processes for Removal of Particles: Yesterday

For many years, including the 1980s, cleaning involved two basic, and quite different, concepts around parts cleaning:

1. A tank of warm water and a detergent at an elevated or neutral pH, into which ultrasonic transducers[66] had been inserted.
2. A tall tank of boiling solvent, generally without ultrasonic transducers.

Both technologies performed, and still do, in a very satisfactory manner. They can, and are, being used for pre-treatment as a particle removal step.[67]

### 6.6.2.1 *Conventional Ultrasonic/ Megasonic Technologies*

Unfortunately, the technology developed by Bran and Puskas has limitations related to particle size. Approximately 1,000 nm (1 μm) is the approximate particle size where sonic (ultra- and mega-[68]) technology[69] becomes less effective than desired.

One reason is the boundary layer effect,[70] which keeps small particles hidden from high-velocity fluid fronts of force (sonically generated waves). That limit is about 600–900 nm, depending on wave frequency.[71]

Sonic technology is called "wet" technology because it is conducted in a liquid, usually water.[72]

This technology situation is summarized in Table 6.2.

---

[65] If it is said that men and boys are differentiated by the size of their toys, dimensional size of the contamination to be removed is a way of differentiating cleaning applications which are industrial, precision, and critical. Since these names are characterizations and not definitions, so be the classification of sizes of particles which are removed in each: industrial – ⩾10 μm (10,000 nm), precision – >0.5 μm (500 nm), and critical – <<0.5 μm. (see Table 5.3).

[66] Mario E. Bran and William L. Puskas are the leading patent holders for cleaning technology using some type of ultrasonic or megasonic transducer. Together, they hold 39 patents (22 for Bran and 17 for Puskas). Puskas's work is associated with the design of powered transducers; Bran's, with their implementation as mechanically driven cleaning systems.

[67] See Chapter 1, Section 1.6.1.

[68] The designation of frequency type here is artificial. But ultrasonic frequencies are typically those below about 250,000 cycles/s (250 kilohertz or 250 kHz). Megasonic frequencies are typically those somewhat above that level and less than 1,000,000 cycles/s (1,000 kilohertz or 1,000 kHz).

[69] Sonic-based technology is divided into at least two categories, based on the frequency of the pressure waves produced by the transducers, and their impact.

Sonic-generated pressure waves in a liquid produce cavitation at external surfaces of a part or surfaces wetted within it. Pressure waves whose frequency ranges around from 20 kHz to around 150–250 kHz are useful in producing cavitation forces on surfaces. Increasing frequency lessens the cavitation force produced by a single bubble, but substantially more of them are generated. The net effect is that more total energy is applied to surfaces but the effect of each bubble asymptotically declines. Beyond around 150–250 kHz, the effect of cavitation is thought to become negligible (see Chapter 7, Section 7.10.2.5 and Figure 7.3.9).

Still higher frequencies produce another effect which is somewhat less well understood. But it can be used to remove still smaller particles. The effect is called fluid streaming or acoustic streaming. Frequencies up to 1000 kHz (or higher) are employed. Here the pressure waves produce both local and regional flow fields which either detach particles or prevent them from reattaching to cleaned surfaces. Technical judgements are mixed about whether megasonic frequencies produce cavitation. But there are many successful applications where megasonic transducers aid in removing sub-micron particles from Silicon wafers, flat panel displays, hard disks, and other high-value surfaces. A good example can be found in this reference: Wu, Y., Franklin, C., Bran, M. and Fraser, B., "Acoustic Property Characterization of a Single Wafer Megasonic Cleaner," *Presentation & Proceedings, Electrochemical Society*, Honolulu, HI, October, 1999.

Major differences between ultrasonic and megasonic cleaning are: (1) waves produced by ultrasonic transducers move in all directions, waves produced by megasonic transducers travel are focused in one direction; (2) the boundary layer for megasonic pressure waves is considerably smaller than for ultrasonic pressure waves; and (3) pressure waves produced by megasonic transducers produce much less damage to surfaces. For a discussion about equipment associated with megasonic cleaning (see Chapter 1, Section 1.6.5 and Chapter 7, Section 7.10.10).

**Table 6.2 Limitations of Sonic Particle Removal Technology**

| Method | Limiting Particle Size (nm) | Wet/Dry | Applications | References |
|---|---|---|---|---|
| Ultrasonic-powered hydrodynamics (cavitation) | ≳1000–5000 (@40–100 kHz) | Wet | Nearly all precision and metal cleaning | US Patent 6,276,370<br>US Patent 5,911,232<br>US Patent 4,372,787 |
| Megasonic-powered hydrodynamics (acoustic streaming) | ≳100–1000 (@1500 kHz) | Wet | Semiconductor wafers | Northeastern University<br>US Patent 6,463,938<br>US Patent 6,048,405 |
| Mechanical Force: 200–500 kHz mega-sonics and brushes | ≳100 | Wet | Superconductive tape, Semiconductor wafers | Clarkson University<br>Northeastern University<br>US Patent 6,273,100 |
| High-velocity solution impact and filtration (no sonics) | ≳100 | Wet | Semiconductor wafers | US Patent 5,039,349<br>US Patent 6,235,145 |

[70] Baker, J. and Durkee, J., "C4: Hiding Particles in the Boundary Layer: Part 1," *$A^2C^2$ Magazine*, September, 2001. Both ultrasonic and megasonic technologies are limited by boundary layers. These are zones immediately adjacent to surfaces where all flow fields are in physical contact with the surface. Technically, this means that the fluid velocity at the surface is zero – for all external flow fields. Practically, this means there is a thin zone adjacent to all surfaces where no external high-velocity flow field can disturb the surface environment. This zone is called a boundary layer (see Figure 7.39).

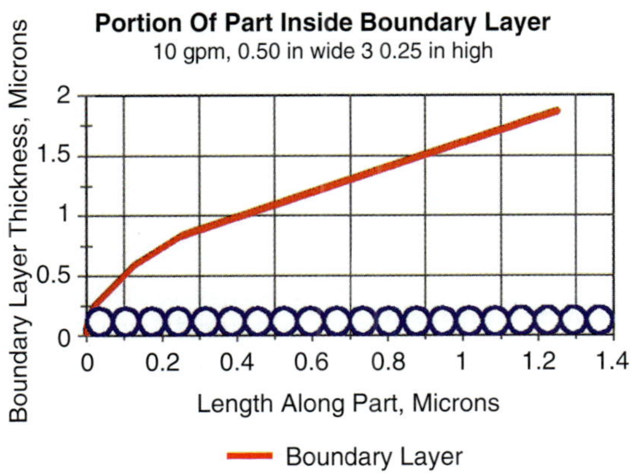


Particles smaller than this thickness can't be removed by flow fields of any kind – continuous or oscillating. Fresh cleaning solution can't penetrate this thin zone and reach the surface. So the surface remains uncleaned because cleaning methods based on fluid force can't penetrate surface boundary layers.

[71] Busnaina, A.A., "Nano and Microscale Particle Removal," Northeastern University Microcontamination Research Laboratory, 2002.

[72] High surface tension is the reason why water is the preferred fluid in which to conduct sonic-based operations. Larger cavitation bubbles contain more energy. Larger bubbles are more stable and easier to generate if the surface tension forces holding them together are also larger. Cavitation energy can be intense in water (surface tension of 72 dyne/cm [nM/mm]) and hardly so in fluorinated speciality solvents whose surface tension is below 15 dyne/cm.

Said another way, fluorinated speciality solvents (HFE-7200, HFC-43 10mee, and HCFC 225 ca/cb, etc.) may be excellent choices because their low surface tension allows penetration of narrow apertures, but they paradoxically limit the extent of cavitation for the same reason.

### 6.6.3 The Transition from Wet to Dry Methods

As this volume is written (2005), tactics and methods for particle removal are (and will be) in a state of flux. The transitions and reasons for them are discussed in Chapter 1, Sections 1.62–1.65 and Section 1.6.7.

Critical cleaning is gradually evolving[73] from wet to dry cleaning for some basic reasons. Dry cleaning is a different paradigm. Its use requires managers to think differently.

A significant problem with the new "dry" cleaning technologies is that no single technology is proven to rid surfaces of organic films, trace metallic elements, or particles *simultaneously*.

Critical cleaning is becoming more than putting the items on a rack, sequentially immersing the rack in three tanks of warm water-based chemicals with each tank containing sonic transducers, and blowing the parts dry with Nitrogen.

In a sense, managers are choosing to move from cleaning in a tank of water to cleaning in a conditioned-environment chamber. That transition must be co-ordinated with Imbesi's Law (see Chapter 5, Section 5.2.1.3): "Whenever something gets clean, something else gets dirty."

Said in other ways:

- For wet cleaning, the parts will be no cleaner than the last rinse fluid.
- For dry cleaning, there can't be more dirt in the cleanroom than there is on the parts.

### 6.6.4 Dry[74] Methods of Particle Removal

This a transient situation. Without doubt, during the useful life of this book, methods listed here will be both exploited and discarded, and new ones will be implemented.

#### 6.6.4.1 *Tradeoffs*

At least two major factors dominate particle removal. The two are interrelated via particle size, and are:

1. The need for sufficient mechanical force to dislodge the particle.
2. The possibility that the force may do unintentional damage to the substrate.

Removal of nano-sized particles is especially challenging. As particle size decreases, the average adhesion stress between a particle and a substrate increases by a fractional power law. The lift-off acceleration required for nano-sized particles can be as high as 1 kg in the rolling removal mode. Liquid speeds as high as 100 m/second are required to remove certain types of sub-micrometer particles. That level of force can be considered hazardous to the health of many substrates whose value is provided by features of miniature size.

#### 6.6.4.2 *Current Technology*

As of this writing, the technology listed below is commercially useful.

*Pulsed laser*: A pulse of light from a laser strikes a substrate. Some of the absorbed pulse energy heats what it strikes. That produces a rapid thermal expansion of the irradiated substrate. Elastic waves accelerate along the substrate. Naturally, the particles attached to the surface are accelerated as well. Particles are detached when the net force resulting from the acceleration exceeds the adhesion force binding the particle to the substrate.

This action is repeated via the pulsed laser contact. There are many variants of this approach. Surface damage is the concern as this technique is quite effective in removing small particles which require additional force for removal.

*Argon aerosol*: This technology is similar to $CO_2$ blast cleaning – covered in Section 6.1.3[75] – only, solid Argon (Ar) crystals are used.

A variant is to use Nitrogen crystals. The basic idea, versus $CO_2$ blast cleaning, is to avoid any solution effects produced by a liquid solvent[76] and to use a material which is quite inert.

Obviously, this approach requires cryogenic conditions versus $CO_2$ snow cleaning.

---

[73] Bardina, J., "Methods For Surface Particle Removal: A Comparative Study", *Particulate Science and Technology*, 1988, Vol. 6, p. 121.

[74] Dry in this sense means without immersion in liquid. These methods have nothing in common with removing stains with perchloroethylene.

[75] Solid $CO_2$ crystals are "shot" at surfaces and knock off particles by momentum transfer (see Footnote 16 in Section 6.1.5).

[76] Some cleaning effects have been explained as if solid $CO_2$, heated by friction, formed a metastable liquid which dissolved soils rather than removed them by impingement. Evidence here is unclear to this author.

**Table 6.3 Limitations of Particle Removal Technology**

| Method | Limiting Particle Size (nm) | Wet/Dry | Applications | References |
|---|---|---|---|---|
| Pulsed laser (thermo-elastic coupling ) | ≈500–1500 | Dry laser | Particles on flat surfaces | |
| Argon or $CO_2$ aerosol bombards surface | ≈<1000 | Dry | Oxides, metals, etc. | US Patent 6,036,581 |
| Laser-induced plasma | ≈500–1500 | Dry laser | Particles on non-flat surfaces | Vanderwood, R., *$A^2C^2$ Magazine* June, 2002 |
| Vaporization via laser heating | ≈500 | Dry laser | Microelectronic and opto-electronic manufacturing | Clarkson University, IBM, others |
| Energetic-focused cluster beam | ≈100 | Dry | Space propulsion, ultra-fine powder production, thick coatings, ink-jets and surface cleaning | US Patent 6,033,484 US Patent 5,796,111 |
| "Laser tweezer" | ≈10–100 | Dry laser | Living biologic materials (yeast cells, human cheek epithelial cells, wheat chancre cells, protein cells), large molecules | US Patent 6,416,190 |
| Microscopic gripping mechanism | ≈10–100 | Dry | Most any microscopic object | US Patent 6,398,280 |

The following three techniques raise significant concern about surface or substrate damage.

*Laser-induced plasma*: Here, a beam of light from a laser is focused to a point on the substrate. This causes rapid local temperature elevation. Consequently, a hot ionized gas (a plasma) is produced from the surface materials. The effect of the expanding plasma and resultant wavefront creates a localized flow that transports the detached particles away from the point on the surface.

*Vaporization via laser heating*: This is not similar to pulsed laser cleaning or laser-induced plasma. Particles are not removed via surface movement or local fluid movement.

Particles are removed by vaporization of the material of which the particle is composed. Obviously, biological debris are more easily vaporized than are metal oxides. This technology is similar to that of EP (see Section 6.1.8) in that in removal of particles, new surface is created.

*Energetic focused cluster beam*: High-energy clusters are beams of micro droplets (<1,000 nm in diameter). They are formed by pneumatically feeding a conductive fluid to the tip of a capillary emitter. A high-voltage electric field is applied to the capillary tip, charging the micro droplets (clusters). These relatively large clusters expend their energy[77] over an extended area of the surface, lifting off micron and sub-micron particles, organic film, and metallic contaminants. The fluid is a high-purity glycerol solvent doped with an electrolytic additive such as ammonium acetate.

The following two technologies apply to removal of single entities (particles).

*"Laser tweezer"*: The technology is based on the principle that dielectric particles (those that do not conduct electric current) experience forces that draw them to where the light is brightest. Optical tweezers are constructed using optical gradient forces from a single beam of light. Individual particles can be manipulated, and thus removed.

[77] This is an analogous technology to blast cleaning (see Section 6.1.3) except that both hydraulic and electrostatic forces are involved.

*Microscopic gripping mechanism*: This approach is purely mechanical. A nano-sized gripping mechanism and a complementary "handle" are created. The gripper can grasp any nano-sized mechanism, including a particle.

This technology situation is summarized in Table 6.3. They are listed in order of current commercial accomplishment.

#### 6.6.4.3 *Other Options*

Another approach which is being developed by some is to use chemicals about which there is less concern about cost, safety, disposal, and environmental affairs. Two examples are where oxidizing action is important to cleaning: use of $H_2O_2$ and ozone.[78] Obviously, if oxidation is not critical to debris removal, this approach is useless.

Supercritical $CO_2$ is a solvent about which there is considerable interest. However, its unique intermolecular forces limit its use to a minority of soils.[79]

#### 6.6.4.4 *Location, Location, Location*

Ultrasonic transducers are omni-directional: one doesn't have to know the location of any particle to use the technology. Megasonic transducers, however, are uni-directional: one does have to know in which direction from the radiating transducer the particles are located. So the location of one or many particles is not a dominant issue with wet cleaning.

Dry cleaning, on the other hand (except for Argon aerosol and perhaps cluster beam technologies), is very location-sensitive. One has to know specifically where the particle is so one can aim the laser or the tweezer at it (see Chapter 5, Section 5.2.1.3[80]).

Said another way, cleaning of particles becomes a two-step operation with many dry cleaning technologies:

1. First scan for an object somewhat larger than a molecule.
2. Remove it.

Obviously, integration of optical detection methods and software with some dry cleaning technologies is a necessary and forthcoming advance.[81,82]

#### 6.6.4.5 *New Technologies*

Two technologies were described at a recent technical conference.[83] Both are of the immersion type. Both are claimed to remove sub-micron particles. Use of neither requires knowledge of the specific location of debris. Both involve bubbles.

The science behind and explanations about either have yet to be published as of this writing (summer 2005).

6.6.4.5.1 *Gas-aided Megasonics* This technology involves use of high-frequency (>1,000 kHz) megasonics, added soluble gas, batch operation, and Silicon wafers arranged so that their carriers are not blocked by the directed pressure waves.[84] Significant (30–50%) particle removal efficiencies (PREs) are claimed for silica particles whose size is as low as 34 nm. Oxygen has been used as one added gas. PRE

---

[78] Song, J.-I., Novak, R., Kashkoush, I. and Boelen, P., "Using an Ozonated – DI-water Technology for Photoresist Removal," *MICRO Magazine*, January, 2001.

See Chapter 2, Section 2.5 where criteria pollutants in the US are described. Ozone is a criteria pollution. This is a situation described by the motto of real estate agents: location, location, location. The hazard of concern about criteria pollutants is smog in the troposphere. Dissolved in water at the 100-ppm level, ozone is not of concern as a criteria pollutant.

[79] Durkee, J.B. and Williams, L.L., "An Independent Evaluation of Cleaning with $CO_2$ – Where is the Value?," *Presented at the 9th International Symposium on Particles on Surfaces*, Philadelphia, June 17–18, 2004.

[80] The application of Heisenberg's Uncertainty Principle to cleaning operations – to get a particle off a surface, first one has to find both.

[81] Bunday, B, Godwin, M., Lipscomb, P. Patel, D. and Bishop, M., "Meeting Manufacturing Metrology Challenges at 90 mn and Beyond," *MICRO Magazine*, August, 2005, pp. 1–41.

[82] Rother, T., "Enhancing the Imaging Chain in X-ray Inspection," *SMT Magazine*, August, 2005, pp. 30–32.

[83] *Ninth International Symposium on Particles on Surfaces*, Philadelphia, June 17–18, 2004.

[84] Vereecke, G., Parton, E., Holsteins, F., Xu, K., Vos, R., Martens, P.W., Schmidt, M.O. and Bauer, T., "Investigating the Role of Gas Cavitation in Megasonic Nanoparticle Removal," *Micro Magazine*, April, 2004.

in megasonic operation without the added gas is as expected: zero.

There is another unexplained and unexpected effect as well – use of chemicals ($NH_4OH/H_2O_2/H_2O$) aid in particle removal. Apparently, despite known analytical difficulties of monitoring nanometer-sized particles, these results are repeatable. At least a second organization is doing research with Silicon wafers.[85]

This author's inference is that the added (20 ppm) soluble Oxygen gas[86] is vaporized because of the pressure fluctuations caused by the high-frequency megasonic transducers. Bubbles formed from the vaporized Oxygen gas collapse and their energy displaces particles. Apparently, vapor (water) bubbles are not produced by the pressure waves. This is opposite to behavior with ultrasonic transducers. Obviously, additional work here is necessary and is being done.

6.6.4.5.2 *Vacuum Cavitational Streaming* Even higher removal efficiencies are claimed for use of vacuum cavitational streaming (VCS) with particles sized from 10–150 nm.[87] A fluid added as a gas is involved as well. But megasonic transducers are not involved.

The substrates are placed in a vacuum chamber and a modest vacuum is established (~100–400 torr). The chamber is filled with either solvent or water (with or without chemistry). A small amount of noncondensible gas is injected to produce a small local pressure increase. The developer claims that the *noncondensible gas preferably collects at surface defects.* Gas-rich zones are nuclei for local boiling. The collapse of these bubbles creates a force at the defect which can remove it.

Excellent PRE with a variety of substrates is claimed by the developers. They also claim to have more questions than answers about the limits and mechanism behind this technology.

6.6.4.5.3 *Back to the Future* The reason for following these two developments is that they offer the potential to return to the past – where users can remove commercially significant (nano-sized) debris in a forgiving process.

Said another way, these two developments offer the potential to repeatedly clean nano-sized debris without knowing its location, or being concerned about substrate orientation within the process. Certainly, users greatly desire this outcome.

These two technologies, if either proves scientifically sound and commercially viable, could make removal of nano-sized debris from substrates considerably more simple and forgiving. On the other hand, perhaps their evolution may lead to a technical dead end. Without question, forgiving solutions to these cleaning challenges will be welcomed by users no matter what technologies are involved.

## 6.7 HOW MUCH CLEANLINESS CAN BE/SHOULD BE AFFORDED?

There is a tradeoff between cost and quality – in clothing, in food products, in jewelry, in automobiles, and in cleaning.

This tradeoff has different outcomes for the three types of cleaning operations identified as industrial (metal or gross) cleaning, precision, or critical cleaning.[88] The stereotypical outcome is that industrial cleaning is cost-driven,[89] critical cleaning is quality driven, and precision cleaning may be driven by either factor.

In practice, the tradeoff between cost and quality is a managerial prerogative. But those to whom managers report want the quality improvements which they have been supporting with budgets, AND significant reductions in cost. So in a sense, it's a false choice.

[85] Verhaverveke, S. and Gouk, R., "Single Wafer Megasonics Configurations: Parallel and Perpendicular to the Wafer Surface," *Presented at the 9th International Symposium on Particles on Surfaces*, Philadelphia, June 17–18, 2004.
[86] NOT material vaporized during the expansion cycle of the pressure waves.
[87] Gray, D. and Frederick, C., "Sub-Sub Micron Cleaning Using Vacuum Cavitational Streaming (VCS)," *Presented at the 9th International Symposium on Particles on Surfaces*, Philadelphia, June 17–18, 2004. See also US Patent 6,418,942 and 6,743,300.
[88] See Table 5.3.
[89] For example, cleaning costs per piece can't exceed €0.001 when a steel screw sells for €0.01–0.03 and produces a profit of ~€0.0005.

### 6.7.1 To Control Cost, It Must be Known

There are at least four reasons why cost control is difficult in *all* cleaning work.

1. The *absence of industry standard metrics* about expenses. Valid concern about keeping technology proprietary dampens nearly all dialog about details of the real expenses of completing an cleaning operation. A recently published book covering most facets of critical cleaning does not have the word cost in its index or table of contents.
2. The *variety of quality standards* used by managers. The price for improved quality is not linear with quality. The relationship between quality and degree of treatment becomes asymptotic as all contaminants are removed.[90]
3. The *variety of operations* conducted by managers. Solvent, aqueous, and $CO_2$ cleaning systems have major differences in required floorspace, compliance with environmental, safety and health protocols, capital investment, and operating labor.
4. At most sites, *information about distribution of utility costs is absent*. Electrical power, dry Nitrogen, water of various qualities, compressed air, waste treatment service, and staff operating/maintenance labor are used to support more operations than just the cleaning system. While there may be a summary cost sheet for each of these expenses across the operating site, a manager is unlikely to find a breakdown about the fraction of any of those cost elements which was consumed by the cleaning machine for which they are responsible.[91]

So the cost of cleaning operations will have to be managed not in the usual way enterprises are operated, that is, without direct and complete information.

### 6.7.2 Five Steps to Cost Management

The five-step approach below is recommended because it has the virtue of having been successfully been implemented by a variety of organizations who have managed cleaning (and other) processes:

1. *Seek the backing, partnership, and team support* of the person empowered as the arbiter of quality. The best argument is that neither of you is likely to have a job unless your quality output can be profitably sold. Remind them of the disk drive industry where a product can't be sold which fails to meet basic cleanliness quality standards, but where competitive pressure to reduce costs has driven firms to business default. The backing of this arbiter is determinant of success in cost reduction where a tradeoff in quality is made.[92]
2. *Decide how to charge* cost against the cleaning process. Choose based on piece throughput (€/piece), or elapsed on and off-time (€/hour or €/cycle) or some other basis.[93]
3. *Identify*[94] the single largest single cost element:
   - Then identify the second largest single cost element, then the third, and so on through the fifth

[90] This author uses a simple guideline for planning of critical cleaning operations, but cannot defend it with data:

- Quantum changes in critical cleaning quality will require up to a decade of change of process factors such as time, system volume, floorspace, analytic requirements, and thus – cost. That is, reduction of non-volatile residue (NVR) from ~5 to ⩽1 mg/SF will require up to a decade more work, including cost.

[91] Change in this situation can only come by fiat from an enterprise leader (manager). If this person values knowledge about control of cleaning costs, they must be willing to commit additional resources (which don't produce a direct return) to purchase and implement the metrology necessary to learn what that cost breakdown is. Included are meters for utilities (power, etc.), "sign-in" procedures so staff work with the cleaning system can be distinguished from other work they do, and segregation of costs for overall waste treatment. Also included are the services of an accountant to manage and validate the reported cost value.

[92] A change which affects cost always affects quality – otherwise it would have been previously implemented.

[93] This is possibly the most significant decision because it separates fixed from variable costs. Fixed costs are applied to all units of production. Variable costs are only those needed to clean an additional unit of production. Any valid system for cost management must allow differentiation and segregation of cost elements among the fixed and variable types.

An approach many have used is based on time. Two types of time are defined: hours ON when the unit is operating, and hours OFF when it is not. The manager responsible for the unit's operation is responsible for that differentiation in every circumstance. Fixed costs (power, operating labor, investment amortization, insurance, maintenance, etc.) are charged against hours OFF. Fixed costs and variable costs (ingredients, energy, waste disposal, etc.) are both charged against hours ON.

An outcome might be that it costs €300/hour to own and operate a cleaning unit and €100 to own it when it is not operating. Airline companies, faced with the same sort of economic outcome, have learned to maximize hours in the air every day. That's why Southwest Airlines is noted for trying to keep their planes flying every hour of the day.

[94] The second most difficult challenge is the decision about just which work is associated with the cleaning process and which work is not.

largest single cost element. Most cleaning processes have between *three and five significant cost elements*.[95] Decide how to charge all against the cleaning process. They will probably include items from this short list:

- Ingredient consumption (forcing the study of recycle of chemicals, including rinse water).
- Capital investment in cleaning machines.
- Disposal of hazardous materials (this can be 5–100 times the cost of materials purchased).
- Labor (both operating, which is a fixed cost,[96] and maintenance, which is probably a variable cost[97]), and possibly.
- Floorspace (this may include overall costs such as insurance, taxes, corporate Research and Development [R&D], etc.).[98] This is the ultimate fixed cost.

4. *Accumulate current total costs* based on those three to five elements. If you don't know what they all are you may not be able to recognize success, or non-success. Accumulate elements of cost and charge them per the decision made in No. 2.
5. *Review the cost accumulations* with the arbiter of quality. A single cost summary may provide little insight. This would be because there is no point of reference. Typically 3–6 months of experience must be summarized to provide a sound basis for examination and criticism. There may be effects of seasonality (business or weather). The arbiter of quality and yourself must jointly present this cost information to the enterprise leader.

After cost is understood and accepted, seek to reduce or replace an element without a negative impact on quality. The best way to do this is by using a designed experiment (see appendix to Chapter 4) organized by a team sensitive to the effects of change on human psyches (see Section 6.4).[99]

### 6.7.3 Controlling the Cost of Cleaning Operations

The following are proven methods for cost reduction with cleaning systems:

1. A manager can save real money by *processing only full batches of parts*.[100] The boundary condition is usually the production or inspection schedule, or possibly delays imposed to reduce dragout (see Section 6.5).
2. *Recycle of treated rinse water*. Where it has been tried, water recycle has generally been more successful than expected:
   - **Start small.** Get the arbiter of quality to agree to a target of ~5% recycle after treatment. That's a puny amount. Later, after success the target can be gradually increased as justified.
   - **Be analytical.** Do mathematical modeling to show the buildup of trace materials, known or unknown. Find some analytical characterization non-volatile residue (NVR), bacteria counts, conductivity, pH, etc. to correlate with line quality, and your modeling results. The latter can provide a norm for comparison. Strive to not-stop recycle at the first process "bump."[101]

---

[95] This means not all cost will initially be identified. Don't let a search for "perfection" allow the "good" to be discarded. Identification, accounting, and benchmarking of costs is an evolving quest. When the manager better understands their cleaning process they will be able include more than five elements of cost.

[96] Because operators usually aren't fired when the cleaning unit is shut down.

[97] Because maintenance workers are presumed to be assigned to another site operation when the cleaning unit is shut down.

[98] Where the element is the cleanroom environment, many argue it is the least controllable. If the cleanroom is used for just finish cleaning and packaging, the cleanroom should be viewed as part of the cleaning process for cost control. Use of the cleanroom for preparation, manufacture, inspection, and cleaning means cleaning is only charged with a portion of the cost.

[99] An example of why both are necessary is recycle of treated rinse water. This proposal is usually greeted by the arbiter of quality with the enthusiasm shown by guests on the television show "Fear Factor." Recycle of treated materials can be done if it is done scientifically and with the knowledge of everyone affected (see Section 6.7.4).

[100] Please be not fooled. This advice only saves variable cost. Fixed cost elements continue unabated (see Section 6.7.2).

[101] Managing recycle of ingredients is necessary to manage costs. It should be treated as one would manage risk. There will be no cost reduction (for items 3) unless some materials are recycled. And there is a limit to degree of recycle which can only be identified by exceeding it. Further, as investment risk is not constant (prices rise and fall without our permission), the risk of ingredient recycle is not constant – the limit may change with time. Managers are paid to manage risk, not avoid all risk.

3. *Don't "go cheap"* on purchase of the cleaning machine. Don't save 20% on its purchase price if that compromises control of or detection of quality. Consider buying the model with the next higher capacity and grow into it, as per item 1.[102]
4. Ingredients, *and their disposal*, can be a dominant cost. This is because of their classification by type of waste. In most countries, the classification "hazardous" because of flash point or presence at a concentration[103] of a toxic pollutant, is a trigger which changes the cost of disposal. Reclassification of waste type from one level of hazard to a lesser one can save significant money. Managers do that by process analysis leading to a different waste treatment scheme – by removing the toxic material.

In addition, *adopt Deming's strategy* of continuous improvement (see Chapter 4, Sections 4.9.1.1 and 4.9.2). This strategy basically challenges managers to identify improvements through trial and adopt those which provide benefit. Metrics are not gained, but cost improvements can be.

### 6.7.4 Infringement upon Freedom

Decisions about costs often are limited by factors outside the constraints of the cost versus quality tradeoff. Degrees of freedom may be constrained by FDA, Association for the Advancement of Medical Instrumentation (AAMI), Controlled Environment Testing Association (CETA), Good Manufacturing Practice Regulations (GMP)[104] or other requirements for the details of a specific process.

These requirements speak not to quality or cost, but to procedures, and both cost and quality are derived from procedures. In these situations, adherence to procedures and maintenance of quality versus standards take primacy. Cost is probably not a factor which can drive significant management action.[105]

### 6.7.5 Benchmarking the Cost of Cleaning with Metrics

The business landscape is replete with situations where firms failed by not controlling cost relative to both in-kind[106] and non-in-kind competition. Reliable metrics are vital, yet as pointed out in Section 6.7.1, not normally available. Sadly, this author has no storehouse to open for inspection.[107]

So, how are managers to manage costs when blind about competitive situations?

Answers to this crucial question are necessarily bland, may not appear nourishing, but some have been proven to provide value:

- *Do the best that can be done now.* Examine and criticize enterprise history. Build a body of information (a spreadsheet or database) about costs, the conditions in place when they were incurred, and what changes were made before and since. Use that experience to construct metrics of what has been experienced, and understand why that is so. But don't believe that any other enterprise is

[102] The opposite guidance can also be argued. A manager might purchase a cleaning machine known to be too small to meet the expected capacity – as per item 1. This opposite guidance is sensible only in cases where operation will be short term, it is certain expected capacity won't be growing, or where insufficient investment capital is available.

[103] Dilution won't (or shouldn't) fool the US EPA's Toxic Characteristic Leaching Procedure (TCLP), and change the classification of a waste. The US EPA has identified 40 toxic chemicals that can cause harm when products containing them are disposed of in landfills and the chemicals leach out. If the amount of a particular chemical released under test conditions exceeds regulatory limits, the waste is defined as hazardous. It must be handled according to regulations governing hazardous waste (see http://www.epa.gov/epaoswer/general/orientat/). Details are outside the scope of this book.

[104] Promulgated by the US FDA under the authority of the Federal Food, Drug, and Cosmetic Act (see http://www.gmp1st.com/gmp.htm).

[105] There is an aphorism in the discipline of project management which applies here. One can control any two of the three factors which define a project – cost, timing, and quality. But that expression may not have much impact on guidance from enterprise management.

[106] The US metal finishing industry is an excellent example, where off-shore competition with lower labor costs captured (and continues to do so) significant market share in commodity metal polishing, plating, and coating applications.

[107] Consultants have details of cost information from specific customers. But this author has signed many Non Disclosure Agreements (NDAs) stating that all information including cost, wouldn't be shared.

experiencing the same circumstances. Competition will be doing both better and worse. Assume the former.

- *Think and act boldly* – in partnership with the arbiter of quality. Call changes you both propose experiments (See Appendix 1). If they don't produce the amount or type of benefit for which you hope, reversing them will not consume organizational or managerial prestige.[108]
- *Network.* Attend meetings of trade associations and technical societies – if for no other reason than to share/glean/trade information about how other firms manage situations similar to those you mange.
- *Interface* with suppliers and especially customers. The reason is the same as above – to learn, because metrics are so scarce. Let suppliers know you are open to spend capital to make worthwhile gains in operating cost. Customers will speak if they have been offered a competitive price.
- *Use your sales staff* to arrange a plant tour of a foreign competitor's site or an information exchange arrangement, and offer reciprocity. This may be possible where there is not significant mutual participation in markets and where cleaning cost is less significant in the overall cost structure.

In summary, information which produces cost metrics will come from both internal and especially external sources.[109]

## 6.7.6 If It Doesn't Add Value, Don't Do It

Cleanliness should not be a cost element; cleanliness should be a value generator.

If cleaning operations don't add value, don't do them.

The amount of cleanliness which can be afforded is:

- *No more than the amount of value which its absence loses.*

The reason for adding a cleaning process step should only be that it adds more value than its cost. Determine its cost from the material in Section 6.7:

- Determine its value from not doing it and evaluating the consequences via an experiment.

Answers to the question, "How much cleanliness can be afforded?" can especially be found in the suggestions for developing sources of external information in Section 6.7.5:

- While it may be difficult to identify amount a competitor spends to complete a cleaning step, *it is legitimate and possible to purchase their product and analyze it* and its components for cleanliness, using the methods of Appendix 2.

### 6.7.6.1 *Setting Cleanliness Standards*

The first step in setting these standards should be to recall that the only goal of cleaning operations is to assure integrity of downstream operation.

The second step is to collect and review past operating data of cleanliness – and downstream operation:

- Use the statistical techniques from Appendix 1 (especially the *t*-test).[110]

The third step should be to set a trial standard. This should not be a decision made in haste, by a person responsible for only one aspect of a product, without team buy-in,[111] and for perpetuity.

For a new or recently started operation, numerical cleanliness standards for the three types of cleaning work might be: industrial (>25 mg residue/SF surface), precision (1–25 mg/SF), and critical (immeasurable to <1 mg/SF).[112]

Clients have usually found it is better to start too clean, and then loosen a standard in an experiment. That experiment must show that there is a corollary benefit (ingredient cost or production volume or maintenance savings) associated with the acceptance of poorer cleaning. Otherwise there is no reason for accepting the change.

---

[108] Recalling the advice from someone wiser than this author, it is always easier to ask for forgiveness than to ask for permission.
[109] It is not trivial to write that this information should be thoroughly vetted before acceptance.
[110] Establish that the values of proposed cleanliness standard associated with both populations of acceptable and unacceptable downstream operation are different given the level of statistical confidence you are willing to assume.
[111] See Section 6.4 about human factors associated with cleaning (and other) operations.
[112] See Table 5.3 for general guidance.

## 6.8 HOW TO PURCHASE CLEANING EQUIPMENT

When managers make a selection to solve a cleaning problem, they have at least three chances to make mistakes. They can choose the wrong process. They can choose the wrong chemical. They can choose the wrong equipment. The purpose of this section is to help managers make none of those mistakes.

### 6.8.1 Stuff Happens

Some examples from this author's experience as a professional cleaning consultant include:

- Clients ask to make a spray booth work like an immersion bath (or the reverse). It really can't be done. No client wants to hear that if they own a spray booth (or the reverse).
- A client bought an aqueous belt conveyor machine to clean parts which must be tumbled so all sides are exposed to cleaning action. That can't be done with a belt conveyor.
- Another client needed help in rebuilding a set of wash tanks to allow for soil removal. The client never considered what happened to the soil once it was liberated from their parts. The client's tanks were full of soil.
- Another client wanted help to dry a hydrocarbon-based cleaning agent from cleaned parts. The cleaning was done well. But the drying equipment was designed to evaporate water, not evaporate a high-boiling material. The parts appeared soaked with oil.
- Still another client was trying to rinse parts by immersing them in a water bath. The parts were held in a non-porous basket that had been used to store parts – not rinse them. It had about 25% open area. That inexpensive basket didn't allow adequate circulation of the rinse fluid through the parts. Rinsing was poor, but it improved when a basket with around 60% open area was used.[113]

Here's an example from a supplier:

- "I lost a sale of an inline to a rotating table batch washer. The customer's justification was price and quality: '€15,000 vs €75,000, and it cleans just as well.' However, after installation, the customer soon realized that cleaning 'just as well' applied to ONE part. After one or two parts, the small size solvent sump was spent, full of crap, and had to be cleaned out."

### 6.8.2 The Common Denominator

The thread connecting all of these situations is that the user didn't understand how cleaning was done. So they bought equipment ill-suited to their job.

An understanding of the principles affecting cleaning work is the secret to avoiding that unpleasant visit with your boss or your employment counselor.

### 6.8.3 Suppliers: Are They to Blame for Failure?

Suppliers can be contributors to a poorly completed application. Their representatives can and do sell you equipment which isn't well suited to meeting your cleaning needs. That has happened, does happen, and will happen.

But they aren't responsible for most application failures where the wrong equipment is used.

This is because suppliers, generally, have too much at stake to deceive managers by misrepresenting the capabilities of their equipment and the principles underlying its design.

Equipment suppliers will occasionally, or often, overstate the capabilities of their products. Those selling the products manufactured or repaired (and cleaned) by your enterprise occasionally do so as well. Its the nature of competitive business to exaggerate. This is because many products are very similar. Representatives of a few suppliers do this without mercy, concern, or restraint. Most do not.

Representatives may get paid by selling equipment badly matched to your needs. But suppliers know if they do that, they are not likely to sell you anything else as a replacement. They're also not likely to sell anything to those with whom you associate.

The business plan of suppliers is to prosper over the long term. Intentionally selling users equipment (or cleaning chemicals) ill-suited for their application is not likely to be a component of that strategy.

[113] See Chapter 7, Section 7.7.

### 6.8.4 Roles, Goals, and Who's Got the "D"?

A philosophy of management that this author has used successfully is to make sure everyone knows their *role*, their *goal*, and *who is empowered to make a decision* (and who is not). This approach is known as "Roles, goals, and who's got the '*D*'."[114]

In the purchase of cleaning equipment, as a manager:

- Your *role* as a consumer/purchaser is to understand what you need and how what you purchase will fulfill those needs. You are nearly certain to make the wrong choice of equipment if you can't do this. And today that can lead to finding another career opportunity by visiting http://www.monster.com.
- Your *goal* is make the cleaning difficulty disappear via selection and implementation of the appropriate cleaning process (and procedures). Since many firms see cleaning as a value-subtracted operation, and not one which adds value, you may not be perceived as adding value to your organization when you are often seen gazing into the cleaning machine.

In the purchase of cleaning equipment:

- The *role* of the supplier's representative is to inform you of how offerings they represent can meet those needs, in a competitive environment. The normal supplier's representative treads a narrow path. On one side there is failure because of misapplication. On the other side is failure to feed their family. They seldom stray from that path because their family may want a warm meal next month.[115]
- The *goal* of the supplier's representative is to get paid via selling and installing equipment which they can use as a reference when their family needs additional food. They can't do that by selling you a lemon.

It is the *manager, you, who has the decision.* Don't leave it in the possession of another enterprise manager, the financial controller, the supplier's representative, or anyone else.

As a manager, you are responsible for the success of the application. It is you who will be held responsible if it is not successful. You make the decision. This book is a resource to help you do that.

### 6.8.5 A Key Principle

Forget selling price.

There are two reasons for this. First, the market for cleaning equipment is highly competitive – on a national basis. There are few large suppliers. Most firms manufacturing and selling equipment are small. They don't have the financial muscle to manipulate prices. Very occasionally, one of the few large suppliers will try to "buy an application" via offering a factory-sponsored price which can't be matched by other firms. But don't wait for this event.

Second, the cost of your failure is usually much more significant than the value of your success. Poor cleaning performance causes product quality to be poor. This usually leads to lost business, production shutdowns, unsold inventory, or political disruption. Granted these items show only indirectly on an area cost sheet. But it is sure that their impact can be easily located on a business balance report.

In summary, pay the price to buy the right equipment for the job.

### 6.8.6 And Yet…

Don't expect prices for similarly designed immersion or spray parts washers to be the same from one supplier to another.

Suppliers differentiate their offerings by providing different features, quality, support, or warranty.[116] All require a cost to the supplier. You will

[114] The management technique is also described in more detail as broad-based approach in Chapter 1, Section 1.7.2.

[115] In nearly every situation in which this author has been involved, a supplier has not answered a direct question with a lie. If a manager asks the right question, a supplier will give the correct answer to it. But if a manager doesn't ask necessary and pertinent questions, they won't get answered.

[116] See Chapter 7 where components used in cleaning machines are identified as being found in good, better, and best quality systems. That's why this chapter is in this book!

pay different amounts for a machine if you choose to value these items. And likely, you should do so.

Since most supplier firms are small, without financial muscle, they may seek to differentiate their offering from that of the competition. They'll offer better quality at a higher price or lesser quality at lower price. Examples include:

- A pump with magnetic seals may last longer or leak less. But it will cost more for the supplier to purchase and install than a pump with graphite-packed seals (See Table 7.2).
- A blower with less volumetric air throughput, or a heater with a lower wattage, can substantially change the price of a parts dryer. Can your production quotas and quality specification be met at that price?
- Stainless steel skins (outside covers) make a great impression. But they add cost versus glossy painted metal. A 2B finish to stainless steel skins looks even better, but isn't cheaper.
- Polypropylene tanks may do the job just as well stainless steel – though they don't appear as fashionable (see Table 7.3).
- Tanks with cove corners are more easy to flush (they trap less undrained liquid), but they cost significantly more to manufacture.
- Computer-controlled ultrasonics cost more than do those started by a switch.
- Automated handling of materials can double or triple the price of a cleaning machine versus one requiring more human participation.

Prices for cleaning machines do vary because what is offered in return for them varies. Make sure you get what you need, and only that.

### 6.8.7 Anatomy of a Cleaning Machine

When you purchase cleaning equipment, boxed machinery will arrive on your loading dock.

What you have really purchased is at least four items: a design, the components used in the design, assembly of the components to complete the design, and support during your use. This true whether you purchase an automobile, video game, or cleaning machine.

You can't separate these four items. You can't choose a cleaning machine with a brand of pump you use at your site, *and* a design of parts racks you have found useful, *and* the longest warranty, *and* workmanship equivalent to that done at your site. If you did choose all those items, you couldn't afford the custom machine which would be the outcome.

A cleaning machine is a package – as are most purchases. If you choose a cleaning machine with a certain brand of pump, you will have to accept whatever design is used, whatever warranty is supplied, and whatever workmanship is demonstrated.

The persons who market (not sell) each cleaning machine make conscious decisions about these four items to maximize their yearly profit. They have (or should have) targeted a type of customer with certain needs and values. You make the best choice of cleaning machine when you recognize what are your needs and values, and select a supplier of machines who has organized his offering to meet those needs and values.

Choice among the four items depends on you. In general, this author ranks the four items in this order of importance: design, support, components, and workmanship. On a scale of 1–10, design and support are 10s, components are a 6, and workmanship is a 2.

### 6.8.8 The Fallacy of Cleaning Tests[117]

The standard sales tactic is "Let us test your parts and see if we can clean them ..."

This author has been on all three sides of this tactic: supplier, customer, and consultant. This has led to one of the author's rules about cleaning:

> *No supplier has ever returned dirty parts (at least deliberately) to a client after a cleaning test.*

Cleaning tests are usually a waste of time: yours and the supplier's. And they are a huge sales cost for suppliers – a sizable commitment of equipment and manpower. Their impact on sales price is certain.

[117] This chapter does not refer to the material in Chapter 5. Rather, it refers to the customary (almost obligatory) request by suppliers to users to provide parts for evaluation in their processes.

There are two problems with traditional ("Let us test your parts and see if we can clean them") cleaning tests:

1. They are *not definitive.* Often this is the fault of the user – managers send only one (literally) part, or a handful of parts. A supplier can't develop *and confirm* a process using that small number of parts. Alternately, if you send a keg, crate, case, or boxcar of parts, the supplier probably doesn't have the time and resources to test that amount. Suppliers aren't in the business to develop processes and test them for every customer at that amount of parts.

   Further, the effectiveness of these tests is self-limiting. The offer to test is made, in good faith, to nearly every customer. A free test is hard to decline. Nearly all users accept it. Suppliers then have to conduct tests for three-five-eight-ten potential customers for every sale, because they get only a fraction of the orders on which they quote. But the cost of all those tests is incorporated into the price you pay for your unit.
2. Something you may value more than cleanliness is *not tested.* That is throughput. The reasons are above. A legitimate measure of productivity requires a significant amount of parts/time/commitment – whose cost can't be justified by the supplier for every sales opportunity.

The outcome is usually that the sales-related cleaning tests produce only cleaned and returned parts.

But they are almost certain to not produce all the information about the application that a manager needs to support design, completion, and operation of a cleaning machine, to the level of quality and productivity they require.

### 6.8.9 Another Way

This author advises another strategy – don't accept the free cleaning tests.

*Buy* the test you want:

- Screen suppliers and choose one with whom you are fairly certain you want to work (see Section 6.9).
- Offer them €1,000 to do a test.
- Require them to use enough parts to establish that their proposed cleaning machine will meet your stated cleanliness standards *and* production requirements.
- Witness that test. Observe soil management.

The €1,000 (or so) that you spend will be returned to you as increased certainty about having chosen the right equipment (and supplier) to meet your needs. You'll get more than €1,000 of the attention of your chosen supplier – they'll know you are serious and likely to buy from them if they can demonstrate they can meet your two needs.

Accounting for overall project cost will show the €1,000 to be an item of inconsiderable financial consequence. But if the cleaning equipment fulfills your cleanliness and productivity requirements, it will be of practical consequence to your business.

### 6.8.10 Cleaning is Just One Process Operation

Equipment for metal finishing must perform as well as must equipment for manufacturing articles with metal.

It doesn't matter if the customer rejects the article because it has been made with poor size control, a low-quality alloy, or the surface is contaminated with "water spots." The outcome is identical in each case – loss of business.

Cleaning equipment has the same value as does any equipment necessary to retain the customer's business.

### 6.8.11 Summary: Equipment Purchases

Purchase of anything which costs thousands or tens of thousands of dollars and which can have direct impact on the health of your business is worth doing well. Cleaning equipment fulfills that criterion.

A manager can buy the right equipment for their application – that which meets quality and productivity needs. A manager can also not do so, and risk failure. The difference is in the approach to the situation.

A manager's approach should be that stated by writers more literate and technologists more knowledgeable than this one:

- *Figure out what your needs are and meet them.*

Said another way, don't let the normal issues of supplier competitiveness/price, proof testing, or desire to have or avoid uniqueness sidetrack you from this mission.

Purchase of equipment is simple: buy what you need.

## 6.9 HOW TO SELECT A SUPPLIER

If the selection of which equipment to buy should be made according to what you need (as in Section 6.8), the selection of from whom to buy it should be equally straightforward.

Purchase your next cleaning system from the firm who *demonstrates they most want your business* by:

- Their interest and ability to support your system *after it has been purchased and operated successfully.*

Assume your system won't work in the future, even if it were properly designed at the time of purchase and its condition hasn't deteriorated. Cleaning systems usually work as well as the problems they were designed to solve are understood. And problems change, as:

- A chemist tests a new soil.
- A production manager increases the throughput.
- A financial manager needs a cost reduction.
- A customer tightens their quality specifications.

These changes challenge the design basis for a previously made purchase decision.

The answers to those questions will identify just a few suppliers on whom you should concentrate your attentions:

- Which supplier firm best understands what mechanisms are necessary to solve those cleaning problems?
- Which firm's equipment designs are based on implementation of mechanisms, rather than on certain features?
- Which firm convinces you that they can best solve your future problems because they've solved others in the past?

You can choose among them – at a trade show.

[118]See Section 6.13.

## 6.10 HOW TO ATTEND A TRADE SHOW

Some argue that trade shows are anachronistic that they are a carryover from medieval village fairs, where the local artisans hawked their wares. It is true that the Web alone can provide you with access to some product information faster, cheaper, and better.

### 6.10.1 King/Queen for a Day

Trade shows are managed for the benefit of managers like you. Very few trade shows exist which don't cater to end-use customers.

When your feet get tired, when you get lost for the third time, when you are bored while standing in a security line, or when a salesperson acts too forward, when the flight home is delayed, try to remember that.

Suppliers pay exorbitant exhibit fees and commit major staff resources to interest managers like you in their offerings. A trade show may consume 20% or more of a firm's annual advertising budget. For a manager, a trade show may be a relief from the daily grind, an opportunity to learn how other firms do something of importance to them, or a challenge to sieve advertising claims and recover information they value. But for exhibitors, a good trade show will make their year.

For a few days, nearly all the resources of a supplier's organization are focused on *you.* As a potential customer at a trade show, you are king (or queen)! Even Rodney Dangerfield would get respect (or at least attention) at a trade show.

The price for the attention is bias. You are not likely to find unbiased sources of information at a trade show. That's why its first name is "trade." This situation is normal, satisfactory, and expected.

If you don't learn what you intended to learn, it's probably due to choices you made. The purpose of this sub-section is to help you make those choices.

### 6.10.2 Trade Shows versus Internet[118] As Sources of Information

A broadband Internet connection cannot provide all the information a manager needs to do their job. Neither can a trade show.

A manager should use trade shows to obtain information not available on the Web, explore in more depth technology to which they were introduced on the Web, and eliminate/sort/prioritize options.

At a good trade show, covering technologies, products, and services of interest, a manager will find at least these opportunities not well done on the Internet:

- *Comparative demonstrations.* The odor of a cleaning solvent, the visual effect of ultrasonic agitation on particles, the ease or difficulty of using a control panel, the connection between cleaning machines and parts conveyors, and the quality inherent in a cleaning machine as seen in its fit and finish are all aspects of an offering which a manager can examine at a trade show.

  What's significant is that a manager can compare multiple offerings by simply walking to another booth! You may want to rethink your decision if the lines are non-existent at your chosen supplier and long at a competitor's booth.
- *Personal discussions.* Especially with senior members of a supplier's staff. A manager may learn of forthcoming products, or that a firm is not really dedicated to earning your business.
- *The experiences of others.* Eating pizza and drinking beer while standing with four others around a table the size of a bicycle wheel is bound to stimulate conversation.
- *Sales.* "Show Specials" frequently offer savings of 5–10% or immediate delivery. Yet, a small cost saving shouldn't be a deciding factor in your purchase – but it's a nice bonus if you have already made a decision.

Most importantly, at a trade show a manager can meet and interview staff from a firm and determine if they wish to do business with them. This is the most powerful reason for attending a trade show.

### 6.10.3 Meeting your Organization's Needs at a Trade Show

The technical and organizational details of four situations[119] are described in Table 6.4.

Each is similar to those faced by most managers – complicated, with more than one issue.

Consider the following, which is derived from Table 6.4:

- Issues are resolved in situations by identifying needs of the enterprise, not by implementing any specific technology.
- A trade show *can*[120] bring together many suppliers, allowing options for resolution of issues to be evaluated in a most efficient manner.
- But competitive interests, and complexity of situations, can make attendance at a trade show be a total waste of resources – if sufficient preparation isn't accomplished before attendance.
- Multiple disciplines are often involved in resolution of cleaning situations.
- Information is exchanged at trade shows, but that can include proprietary information that some firms don't wish to be exchanged.

## 6.11 TEN PRINCIPLES[121] FOR SUCCESSFUL CLEANING WORK

This is not a short list because of the diversity of cleaning applications and situations.

### 6.11.1 Manage Involvement

Get those who do the work involved in the selection process. The unit operator or shop foreman understands the cleaning situation and its needs. Listen well to them. But as the manager responsible for success, make your own decision.

[119] All four represent situations faced by actual clients.

[120] Some useful trade shows for those managing cleaning operations are: American Electroplaters and Surface Finishers Society's (AESF) SUR-FIN, National Association of Corrosion Engineers (NACE), Powder Coating, National Plant Engineering & Maintenance Show and Conference, NEPCON, SEMICON, Cleanrooms East and West, International Manufacturing Technology Show (IMTS), and Clean-Tech.

It should be apparent that cleaning technology is a component of many trade shows which are organized to support industries where cleaning is but one of many unit operations.

[121] See also Table 1.4.

**Table 6.4 Solving Application Issues at Trade Shows: Four Hypothetical Examples**

| Application Details | Requirements | Before the Show/At the Show | Lessons Learned |
|---|---|---|---|
| • To clean a fatty acid release agent from delicate parts.<br>• The parts are made using powder metallurgy and a precious metal. The pores in the parts trap the solid acid.<br>• The enterprise insists on solvents not being used.<br>• Tests show ultrasonic cavitation can damage the fragile bond between the precious metal and the porous surface, while dislodging the acid. | • Remove the acid.<br>• Not damage the parts.<br>• Thoroughly flush the pores to remove acid and cleaning agent.<br>• Dry all the water without overheating the parts and leaving an oxide film on the precious metal.<br>• Obey the corporate directive. | • Seek those producing aqueous cleaning chemicals. Learn as much as you can about composition of their cleaners. Thoroughly interrogate them about why their offering has, can, and will repeatedly remove the fatty acid.<br>• Ask why competitive offerings can't. Arrange for a demonstration.<br>• Then, seek out those selling aqueous cleaning machines. Inquire about their experience in applying the cleaning agent you favor.<br>• Inquire about other firms who supply equipment to use vacuum in drying parts whose structure is similar to yours. | • One may need to stay more than 1 day at the show to investigate more than one technology – and multiple suppliers. |
| • To clean steel frames which are to be galvanized, then painted for cosmetic effect.<br>• The frames come from a hydraulic press with a wax-based material used as a lubricant. The crystalline wax melts at 180°F.<br>• Aqueous cleaning is preferred since a galvanizing process is downstream.<br>• There is a concern about stability of organic detergents at that high temperature. | • Heat the frames.<br>• Remove all the wax.<br>• Inspect for wax removal since residue will frustrate the galvanizing process.<br>• Not interfere with other downstream operations.<br>• Use aqueous technology. | <u>Prior to the trade show</u><br>• Evaluate methods for heating the frames. As a large metal mass, they will reduce cleaning bath temperature by absorbing heat from the bath. If bath temperature declines below 180°F, the wax will precipitate or not be removed. The point here is: before attendance at a trade show to study cleaning, there is another problem to solve.<br><u>At the trade show</u><br>• Investigate options for a choice of heating system.<br><u>At a trade show about cleaning</u><br>• Seek those producing aqueous cleaning chemicals useful for removal of high-temperature waxes. Learn as much as you can about composition of their cleaners.<br>• Thoroughly interrogate them about why their offering can and has successfully remove(d) the wax. Ask why compctitive offerings can't.<br>• Arrange for a demonstration.<br>• Then, seek out those selling aqueous cleaning machines. Inquire about their experience in applying the cleaning agent you favor and drying parts of structure similar to yours.<br><u>At a trade show – NOT about cleaning</u><br>• This trade show is dedicated to metallographic inspection.<br>• Investigate methods of inspection for organic films on metal surfaces. | Preparation for attendance is essential. Trade shows covering multiple disciplines are likely to provide optimal value. |

| | | | |
|---|---|---|---|
| • Screw machine parts are made using a specific cutting fluid which is only partially soluble in water.<br>• The oil is a proprietary blend developed by your firm to achieve certain unique ends.<br>• Your firm considers the blend composition to be a trade secret.<br>• Your site values recycle of chemicals to avoid exposure to future environmental regulations. | • Examine the oil composition and choose an approach to remove it – aqueous, aqueous emulsion, solvent, or something else.<br>• Identify and prove a process for removing the oil.<br>• Remove the oil from the parts.<br>• Remove the cuttings from the parts.<br>• Recover and recycle the water (if used).<br>• Possibly recover the proprietary oil for reuse. | <u>Prior to the trade show</u><br>• Understand which cutting fluid components are the carriers for the other components that reduce friction and enhance heat transfer. It is likely that if you can remove the carrier component(s) from your parts via aqueous detergents or solvents, you can remove all components of the cutting fluid.<br><u>Before you speak with any suppliers</u><br>• Choose a suitable solvent and a type of aqueous-based detergent. You are going to evaluate both solvent and aqueous cleaning.<br><u>At the trade show</u><br>• Search for and screen suppliers of the candidate solvent and detergents.<br>• Arrange for demonstrations, so you can make a choice of approach (solvent or aqueous).<br>• Don't allow suppliers to make that choice by revealing the soil recipe.<br>• After **you** make that choice, seek recommendations from the supplier about process equipment in which their products have been used successfully.<br><u>At a later trade show</u><br>• Search for and screen suppliers of suitable process equipment.<br>• Thoroughly interrogate them about why their offering can and has used with the chosen cleaning agent, AND removed cuttings, AND recovered cleaning materials and soils. Ask why competitive offerings can't. Arrange for a demonstration. Inquire about their experience in drying parts of structure similar to yours. | A manager:<br>• Without preparation, will gain little from attendance at *any* trade show.<br>• May have to attend more than one trade show because the choice of cleaning technology and process equipment may be made separately.<br>• May be tempted by excellent opportunities to reveal proprietary technology. |
| • A crew from your enterprise has used vapor degreasers for many years.<br>• The cleaning needs are undemanding.<br>• There will be a switch to aqueous technology.<br>• A process has been developed, tested, and found satisfactory for cleaning parts. | • Retrain workers in the use of aqueous cleaning technology as it is completely different than solvent cleaning technology. | <u>At the trade show</u><br>• Search for and screen suppliers based on their knowledge, ability to communicate, and willingness to provide a course of training.<br>• It is especially significant if they are familiar with both aqueous and solvent cleaning technologies.<br>• Many firms are uncomfortable with retraining workers. Be prepared to pay fairly for it.<br>• Take someone from your operating crew with you to the trade show. | • The value of attendance at a trade show is that it's unlikely that you could select a supplier, based on these criteria, in any manner more efficient. than this one. See Section 6.11.1 |

### 6.11.2 Don't Play the Aqueous/Solvent Game

Both aqueous and solvent cleaning processes can be made to clean almost any parts. A manager's choice will probably will be made between them based on factors other than expected part cleanliness.[122]

### 6.11.3 Spend Time Doing, Not Choosing

Winnow choices to one or two suppliers using published information and referrals. Then witness a cleaning trial by the supplier you most prefer. Don't waste time in the selection process (see Section 6.8). Spend your time making that selection work well.

### 6.11.4 To Manage Cleanliness, Manage Soil

Take the time to understand how the chosen cleaning machine manages the process of soil transportation. Cleaning is nothing but moving soil from where it isn't wanted (on parts) to where it should be (a dumpster, recycle tank, etc.). It is usually not difficult to get soil off parts. It is more difficult to move soil out of the cleaning bath, out of the rinse bath, not into the drying zone, out of the cleaning machine, and into some receiver.

### 6.11.5 Choose Quality Over Quantity

Quality is *usually* of more value to a manager's organization than is productivity. If the downstream user won't accept poorly cleaned goods, the goods are worthless to the manager of the cleaning system.

Enterprise management will insist on both. *But* they are likely to settle for some acceptable product rather than inventory much unacceptable product, and *then* insist on more of the acceptable product.

That's why cleaning machines are normally chosen and purchased with more emphasis on if and how good work can be produced. There is usually less emphasis on how much of it can be produced.[123]

### 6.11.6 The Cleaning Step is Just the First Step

Rinsing usually more difficult to do than is cleaning (see Section 6.5). This is because the concentrations of soil are low in the rinse system. Thus the chemical or physical driving forces for their control are significantly lowered versus cleaning. Equipment for rinsing can take several times more floor space than is needed for tankage where cleaning is done. Cycle time is also likely to be stretched if good rinsing is necessary.

### 6.11.7 Don't Forget Drying: the Next Step

If cleaning is removal of soil from parts, if rinsing is separation of dirty cleaning agent from more pure cleaning agent, then drying is separation of parts from pure cleaning agent.

If rinsing may impose more burden than cleaning (see Principle No. 6), drying is likely to impose still more burden than is rinsing. That burden is at least floor space, investment for equipment, cycle time, attention, and energy cost.

As with Principle No. 4, dry to no lower level of retained cleaning agent than is necessary. Ask why "dry-to-the-touch" isn't satisfactory. Consider non-evaporative methods of separating cleaning agent from parts (see Section 1.13.5)

### 6.11.8 Keep Your Cool

Do all cleaning, rinsing, and drying work at as low a temperature as possible. It is true that solubility of soils is increased, soils are more fluid, and the drying cycle is shortened when parts are kept at an elevated temperature.

But the price to achieve those benefits may not be worth the benefit. That bill includes heightened concern about part damage, increased utility (heating *and* cooling) costs, additional safety equipment and procedures, reduced life of surfactants, increased solvent loss and associated environmental concern, and a new control set point (temperature). Further, cycle time is stretched to allow both energy transfer in both directions.

---

[122] See Chapter 1, Section 1.5 about management of choices among cleaning processes.

[123] Yet, in some cases, the opposite has been clearly true.

### 6.11.9 Nothing Lasts Forever

The useful lifetime of cleaning equipment is 3–5 years. Don't use a longer time to amortize an investment. After that time, the unit may have "rusted out," been made obsolete by environmental regulations or higher quality standards, be incompatible with then-current business plans, or competitive versus new technology. When the unit is financially amortized, a decision about replacement is more easily understood.

### 6.11.10 Use Other Chapters of this Book

Define the quality of cleanliness and dryness needed (see Section 6.7.6). Do this by understanding what will be next done with the cleaned parts. Insist on finding some cleanliness test which quantitatively mimics that next step (see Chapter 5). Use the test in a statistically sound way (see Chapter 4).

## 6.12 TEN SOLUTIONS FOR SPECIFIC CLEANING PROBLEMS

This also is not a short list because of the diversity of cleaning applications and situations.

### 6.12.1 Keep the Velocity Up

A most effective way to get soil off parts using aqueous technology is to impact the parts with cleaning solution at a high velocity and volume. Low velocity probably won't get the job done.

### 6.12.2 Also the Heat

Waxy soils have to be heated and softened before removal is attempted by either solvent or aqueous technology. But remember Section 6.11.8.

### 6.12.3 Flush Thoroughly

Blind holes are best cleaned with a continuous flushing action by either aqueous or solvent technology:

- Don't use a high-velocity jet or a spray-based process.
- Use an immersion process, with either aqueous or solvent technology, and low-intensity local turbulation to achieve flow into and out of the blind holes.

The parts must be fixtured (supported) so that the flow is aimed into *all* blind holes. Alternately, one can use ultrasonic transducers to produce the turbulation, but flow circulation is still needed.

### 6.12.4 Don't Clean More than Once

It's one thing to remove particles from parts, but quite another to permanently remove the particles from the cleaning bath without having those particles be redeposited on to the parts (see Section 6.6).

### 6.12.5 Follow Good Cleaning with Good Rinsing

Rinsing is most efficiently done if the parts are thoroughly drained of liquid before the rinsing operation starts. This is called reduction of dragout.

Remove all the liquid possible by non-evaporative means, impact by high-velocity air jets, vibration-enhanced drainage, or centrifugal force (see Section 6.5.).

### 6.12.6 Save Some Clean Rinse Fluid

Your parts will be no cleaner than the quality of the last rinse solution with which they were contacted (see Section 1.12.6).

### 6.12.7 Avoid Evaporation Unless Necessary

Evaporative drying of water takes about five times more energy than does evaporative drying of solvents.

### 6.12.8 It's Cheaper Not to Pollute Than to Remediate

If solvent cleaning is done in the absence of air, there will be little or no air pollution. Enclosed vacuum or pressurized cleaning systems separate air from solvent cleaning agents.

### 6.12.9 Buy A System, Not Just "Juice" and A Tank

A manager's satisfaction, and maybe their job, depends on the integration of at least both a cleaning agent and cleaning equipment. This is particularly true for aqueous technology, which is much less forgiving than solvent cleaning.

A manufacturer of cleaning machines will stand behind their machine if an "appropriately chosen cleaning agent" is used. A supplier of cleaning agents will stand behind their products as long as they are used in a "properly designed machine."

### 6.12.10 Keep Clean Parts Clean

Too often a manager responsible for the cleaning system considers success to be parts which pass a cleaning test. If those parts aren't transferred in that condition to the next user, the work of cleaning is wasted. In critical cleaning, packaging materials are used which are pre-processed to be cleaner than the specifications for parts produced by the cleaning system.

## 6.13 INFORMATION MANAGEMENT WITH THE INTERNET

An author can't write about this topic. It's changing too fast.

This author has written about, or given a talk about, experiences with Internet-based information every year since 1996. None of the past materials will be of value during the useful life of this book.

Yet, this chapter is a snapshot of resources and information I find useful. The links listed here are, hopefully, are apt to have some level of permanence.[124]

### 6.13.1 R&D for Free

The US Department of Energy (DOE) Office of Scientific and Technical Information provides searchable resources in the physical sciences and areas of interest to DOE. It's the best site this author has found for R&D sponsored by the US Government – and there is a lot of that. It's all free at:

- http://www.osti.gov/collections.html

Similar portals include:

- http://www.firstgov.gov/
- http://www.science.gov/
- http://www.google.com/unclesam

### 6.13.2 Best of Breed

In selecting stocks, steaks, dentists, and members of the opposite sex, managers should give consideration to using the title of this section as a guide.

The following, Table 6.5, are this author's selections for certain areas of information. No warranty is expressed or implied about their future capability or existence. Sites with commercial representation have been avoided where possible.

### 6.13.3 Add to Cart?

In 1998, this author's prediction that one would soon be able to purchase cleaning equipment on-line was greeted with the skepticism it probably deserved.

Today managers can purchase the following directly from web sites: ultrasonic transducers, small vacuum

[124] Many footnotes in this book contain Universal Resource Locators (URL) addresses. That was done only if there was no better way of directing readers to find the noted information. URLs are not permanent postal addresses – but they may be the best available identification method.

**Table 6.5 "Best of Breed" Internet Sites**

| Information Category | Web Site Address |
|---|---|
| Physical properties | http://ull.chemistry.uakron.edu/erd/ |
| Toxicology data | http://www.intox.org/databank/index.htm |
| Manufacturers Safety Data Sheets | http://physchem.ox.ac.uk/MSDS/<br>http://www.jtbaker.com/asp/Catalog.asp<br>http://www.rtnsoftware.com/links.htm |
| Free library | http://www.lib.umich.edu/ http://www.ipl.org/ http://vlib.org/ |
| Broad-based search engine | http://www.refdesk.com/ |

dryers, systems for water purification, many types of analyzers about surface cleanliness, single-stage cleaning machines, and other needed equipment.

Why? It keeps sales cost down. Today, we all expect to make purchases from the Internet.

#### 6.13.3.1 *What's on eBay?*

Want a 1990s "famous brand" solvent cleaning system? Need cleaning service on your optical equipment? Can you use one of several ultrasonic cleaning baths or a "famous brand" ultrasonic cleaning console? For only €14,999 you can purchase an apparently new multistage aqueous cleaning system made by a former client. As this is written (summer 2005), they're all there on eBay (http://www.ebay.com).

### 6.13.4 Patents, Anyone?

Every few weeks interested managers should take a quick look at what's been recently patented by whom in the US. What's significant is that they can also see what the competition is applying for patent coverage – *before the patent is granted!* The site is:

- http://www.uspto.gov/patft/index.html

Patents from other countries can be found at the following sites (all free) for: the UK, European countries, Canada, Japan, and a global patent database (respectively):

- http://www.patent.gov.uk/
- http://ep.espacenet.com
- http://patents1.ic.gc.ca/intro-e.html
- http://www.ipdl.ncipi.go.jp/homepg_e.ipdl
- http://www.wipo.int/ipdl/en/

There are paid subscription or pay-per-use patent services. There should be no reason to use them.

### 6.13.5 Fire!

Managers can read, for free,[125] every National Fire Prevention Association (NFPA) standard at http://www.nfpa.org. This can be very useful for budget-limited managers (and consultants). NFPA standards are recognized worldwide.

### 6.13.6 What About Suppliers?[126]

If a manager is considering a supplier, and the supplier don't have an informative web site, consider another supplier.

A manager should be able to download the manuals for all equipment as PDF files, view graphs of operating or calibration data, and read the theory behind unit design.

Don't be surprised if the most informative web sites aren't owned by US companies.

### 6.13.7 Whose Blog?

This author knows of none devoted to critical, precision, or industrial cleaning – yet.

---

[125] See Footnote 264 in Chapter 3, Section 3.22.

[126] It is not possible for this author to list favorite web sites sponsored by suppliers as that might constitute a recommendation on behalf of that supplier.

This is an area where persons involved with cleaning at any level could profit. Other than attendance at the odd technical conference, and previously developed personal relationships, there is little networking done around solving of common problems.

## 6.14 HOW AND WHEN TO HIRE A CONSULTANT FOR SUPPORT

In the interest of full disclosure, please note that the author is a professional cleaning consultant employed in industrial, precision, and critical cleaning.

### 6.14.1 Why Hire a Consultant?

As a consultant, I have heard my service described as a "lubricant" – not that it's oily, but that it makes things flow. A good consultant *enables clients* to do what they find difficult, time-consuming, or expensive.

Its that simple. Don't hire a consultant to impress the VP. Hire a consultant to do what you were going to do anyway, if you had the knowledge, experience, or time.

Specifically, hire a consultant to:

- Advise and implement a purchase decision.
- Demonstrate and teach a technology you need to know.
- Solve a problem with which your staff isn't familiar.
- Evaluate current operation and make serious recommendations about profitable improvements grounded in broad-based industry experience.
- Make a contact that would be inappropriate for your firm to make directly.

**Table 6.6 Different Viewpoints on the Same Issues**

| Requirement | A Manager's Point of View | View from the Consultant's Desk |
|---|---|---|
| Consultant's background | • A manager wants to be able to know everything, or all they need to know, about this consultant. That includes at least: area of expertise, area of non-expertise, pertinent work experience, education, past clients, character, work relationships, potential conflicts of interest, and referrals.<br>• The most important of these is character. | • A consultant expects to discuss the significant details of their background, the work situation, and have the client come to a decision within a reasonable time.<br>• A consultant doesn't want to feel like an object to be studied, but not seriously considered for professional service. |
| The situation | • A manager wants to be able to discuss the significant aspects, factors, and values about their problem without being charged.<br>• A manager also wants to receive a preliminary analysis without being charged.<br>• A manager would expect this from a lawyer or an advertising agency. | • A consultant wants to not be, or feel that they are, being "ripped off."<br>• A consultant doesn't want potential clients to try to extract every scrap of information they can without recognizing the consultant's right to provide it as professional service.<br>• A certain amount of information transferral is mandatory and expected; beyond that, the meter starts running. |
| Consultant's responsibility | • A manager wants to be able to set limits on the service which they receive, and for which they are charged.<br>• Managers don't want someone spending their time and money on something they don't want or haven't authorized.<br>• Consulting service is expensive on the basis of time; managers want good value. | • A consultant does not want to be, or feel that they are being, "ripped off."<br>• A consultant wants to be more than a question–answering machine and to be able to suggest ideas or approaches.<br>• A consultant wants to be trusted, and not viewed as a potential "thief." |

### 6.14.2 About Those Consultant Fees

The fee you pay to a consultant should be only a tiny fraction of what you hope to earn from his or her contribution.

A quote attributed to Red Adair is "If you think it's expensive to hire a professional to do the job, wait till you hire an amateur."

If a manager doesn't expect to earn €5 to €10 to €25 for every €1 paid a consultant, they don't need a consultant; they need an additional lower-paid staff member.

### 6.14.3 Identity of a Consultant

- An expert is someone from out of town who's made all the mistakes possible in a very narrow field.
- A recent TV commercial showed a consultant offering *very* general advice for a short time to a group, and then turning in a bill for €1,000,000. Many think that's the usual relationship between consultants and clients.

Perhaps the work of consulting isn't described by these two experiences, but both need to be understood by anyone considering investing in (or offering) consulting services.

There are two viewpoints which, when rationalized, produce a professional relationship valued by both the client and the consultant, or demonstrate that such a relationship isn't needed by the client and shouldn't be accepted by the consultant (see Table 6.6).

### 6.14.4 A Common Perspective

When these two perspectives overlap, there is a situation in which a manager should consider hiring a consultant, and a consultant should consider accepting the assignment.

The outcome should be a business contract, and a valued relationship.

7

# Equipment used in cleaning

**Chapter contents**



Only in the last decade has specialized equipment, not found in other industrial applications, become commonly used in cleaning applications. The demarcation seems to be particles sized above roughly 0.5–2 μm in major dimension. Laser-based technology (see Chapter 6, Section 6.6.3.5), designed to remove material sized from that range down to molecular dimensions, is not found in other cleaning applications.

With that major and evolving exception, components found in cleaning equipment are found in nearly all other fluid processing operations. What distinguishes a cleaning machine from a filter press, a distillation column, a bottling machine, a paint sprayer, or a deep fryer is (1) that other components are present and (2) how the common components are arranged to do the job for which the machine was designed: filter, distill, bottle, spray, fry, or clean.

This section will describe those components,[1] identify types among them, and recommend when each should be used.

The value of Chapter 7 should be to allow identification to managers of superior cleaning machines, and those which are less so.

This is because the performance and useful life of a cleaning machine is a function of the components used in its construction as well as the design[2] upon which the construction is based. It is assumed that the manager is or will be doing a performance test.

No manufacturer of cleaning machines produces its own components. Every cleaning machine is assembled from purchased components – available to every other manufacturer of cleaning machines (see Chapter 6, Sections 6.8.6 and 6.8.7).

Spray nozzles, and the pumps which supply them, are the most crucial components of cleaning machines, especially aqueous cleaning machines.

## 7.1 SPRAY NOZZLES

Spray nozzles are the fingers of cleaning machines. Their output touches soil and part surfaces. Every cleaning machine has at least two types of spray nozzles, used for multiple purposes.

### 7.1.1 Aqueous Cleaning Machines

It is the choice and aiming of spray nozzles which allows cleaning machines to perform as their owners intend. The cleaning machine may be expensive. The proper cleaning agent may be chosen. But:

- If the wrong nozzles are chosen, the desired cleaning process won't be implemented. The money and effort spent will net little return.

[1] Components used for part conveyance aren't discussed here because of their specificity to applications.

[2] See Section 7.1 for information about how cleaning, rinsing, and drying processes should be designed. See Chapter 6, Section 6.8.7 for their relative importance.

Typically, spray nozzles are chosen based on the nature of the parts, and how they are oriented. *When the list of parts to be cleaned in any aqueous cleaning machine is changed, the selection and placement of spray nozzles should be reviewed, and likely modified.* Replacement of nozzles is inexpensive and quick.[3]

- If the right nozzles are wrongly aimed, something will be cleaned. But it may not be the parts as is desired.

Typically, spray nozzles are aimed to produce a certain cleaning effect – wetting, impacting, rinsing. *Even worse, since nozzles are often used in batteries or groups* (see Figure 7.1[4]), *the entire set may be mis-aimed.*

By analogy, a surgeon can't work with welder's gloves on his hands. A rifle with bent sights is an expensive walking stick.

When an aqueous cleaning machine is inspected before purchase or use, management focus must be placed on the type and position of the spray nozzles, or the time spent is wasted. In every case, the inspection by management should include actual operation of the nozzles with water,[5] and perhaps actual parts.

Almost always the spraying action is done in air at some distance from the parts. Obviously, the nozzles must be close enough to the parts for the desired spray action to occur. For example, the force available from the solid stream nozzle rapidly dissipates with distance so that the nozzle is useless for its intended purpose beyond *ca.* 1-ft separation.

Three of the many available types of spray nozzles are described and illustrated[6] in Table 7.1. These three types are commonly used in aqueous spray cleaning where the parts are not immersed in liquid.

Not every selected nozzle requires the same input pressure and supply rate of liquid. Nozzles used for rinsing seldom require large input pressures but do require larger flow rates. After all, rinsing is dilution of material on a surface, not relocation of it.

**Figure 7.1** Spray nozzles in action

### 7.1.2 Solvent Cleaning Machines

Nozzles useful with immersion solvent cleaning technology are seldom the same nozzles found in aqueous spray cleaning technology. After all, the solvent medium, through which the sprayed materials must move, has completely different density and viscosity than does air.

There are two areas where spray nozzles are used in solvent cleaning.

First, for many rinsing or flushing applications involving immersion of parts, only replacement (turnover) on the part surface of the soil-rich liquid with clean liquid is desired. Pressure impact is not intended by the designers of the cleaning process. But movement of a large volume of fluid is intended.

Three examples of nozzles useful for immersion rinsing with solvents are shown in Figures 7.2 and 7.3 (see Footnote 6).

These nozzles are useful for more than short-range rinsing work. The parts can be within one-half to six inches from the nozzle tip.

In some cases, the nozzle can be an open pipe. But a designer would use such a nozzle where a specific part feature (such as a blind hole) must be flushed. As with aqueous cleaning, if the nozzle is not aimed so the nozzle jet covers the blind hole, soluble materials in the solvent will not be flushed from the blind hole. This is why nozzles are often organized in arrays.

[3] Most nozzles, in stainless steel, cost 25–100 euro each. They are attached by pipe threads, clip-on, or "quick-disconnect" fittings.
[4] Figure 7.1 is courtesy of Ransohoff Corporation.
[5] Should management personnel get wet, that is normal occupational hazard. Performance of an aqueous cleaning machine can't be understood from a computer terminal.
[6] Images in Table 7.1 are courtesy of Spraying Systems Corporation.

**Table 7.1 Types and Functions of Spray Nozzles Useful for Spray Cleaning**

| Type | Function | Example |
|---|---|---|
| Flat fan spray  | Wet surfaces, so cleaning agent solution or mechanical force can be applied.<br>These nozzles are especially useful for creating overlapping patterns of fluid on surface using a multiple nozzle header.<br>Some nozzles of this type can also be used for higher pressure operation and inflict significant mechanical force (pressure washing). |  |
| Solid stream 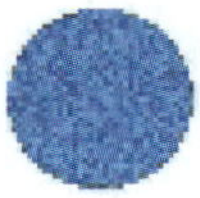 | Inflict focused mechanical force upon soil – vital to aqueous cleaning technology.<br>These nozzles have proprietary inner chambers which continuously "collect" fluid flow and provide integrity of the liquid stream after it has left the nozzle orifice. | 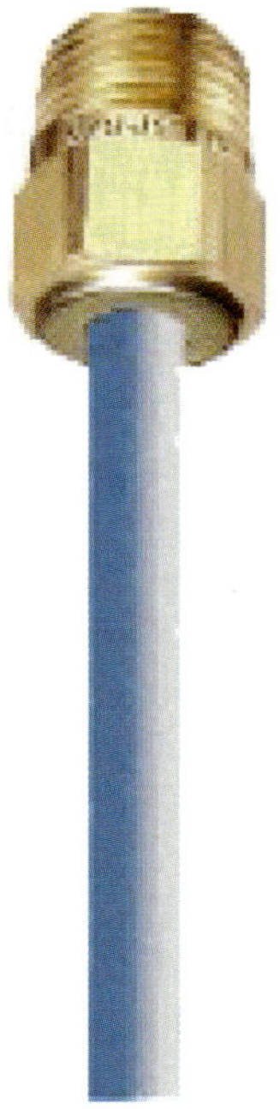 |
| Full cone 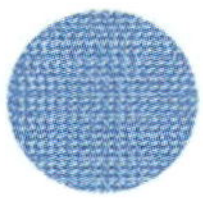 | Flush/rinse debris from partially cleaned surfaces. The spray pattern can be square, oval, or round.<br>Continuous rinsing can be established by moving the parts past a header containing multiple nozzles. |  |

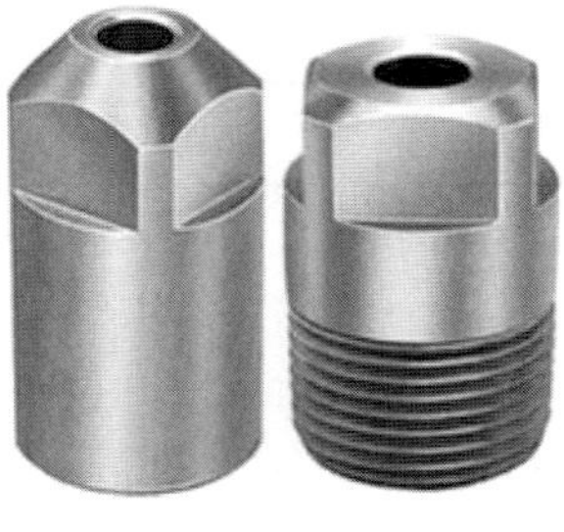

Figure 7.2

Figure 7.3

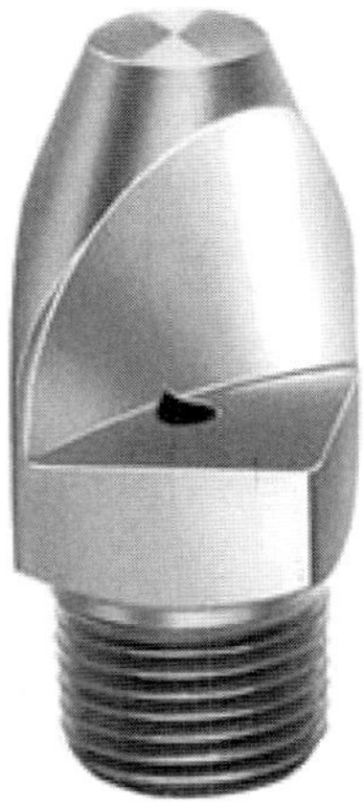

Figure 7.4

Figure 7.5

Second, solvent cleaning can involve more than flushing of immersed surfaces. A designer of cleaning processes may need to provide the capability to dislodge low molecular weight organic material swollen with solvent, high molecular weight material as surface skins, or particulate. Impact with high-velocity fluid can remove this debris.

Two nozzles typically used in these applications are shown in Figures 7.4 (narrow coverage angle) and 7.5 (broad coverage angle) (see Footnote 6). Note how the fluid exiting the narrow hole is "thrown" against the curved nozzle wall. It rebounds to become a focused wavefront.

These nozzles are useful only for short-range impact work.[7] The parts must be within one-half to less than two inches separated from the nozzle tip.

## 7.2 PUMPS

Pumps are the heart of a cleaning machine. They pressurize fluid between their intake (suction) and their discharge, and so they drive spray nozzles. Every cleaning machine has multiple pumps because machines normally have multiple sets of spray nozzles.[8]

The mechanical integrity of a cleaning machine will be no better than that of the pumps within it. If your supplier of cleaning machines uses low-quality pumps, you have a low-quality cleaning machine no matter its design or technical support. When shopping for a cleaning machine, the quality of the pumps within it must be a major concern.

[7] Another application in which these nozzles are found is in air spray conveyors.

[8] Wash fluid and rinse fluid are not commingled and are pumped separately.

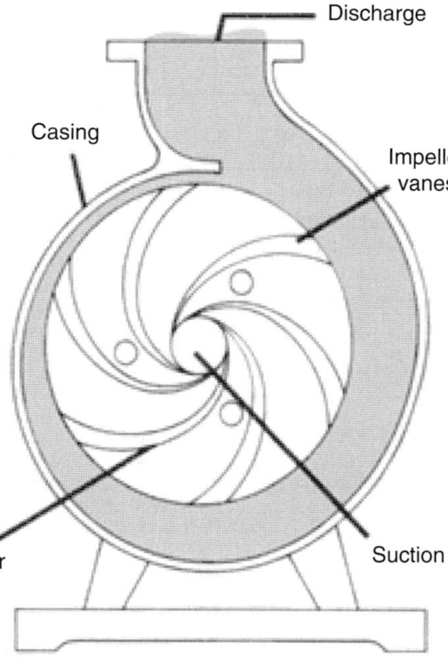


**Figure 7.6** Centrifugal pump

When the pump fails, and they do, the cleaning machine (and thus the overall process) doesn't succeed. One reason pumps in cleaning machines fail is that they seldom are fed pure fluids.[9]

There are more variables in the selection of fluid pumps than there are flavors of ice cream. Information in this section will allow you to evaluate a cleaning machine based on the quality of the pumps within it. It is not intended to provide design guidance for cleaning machines, but rather to enable recognition of the hallmarks of good, better, and the best ones.

Pumps used in cleaning applications are nearly always centrifugal pumps.[10] Here, a rotating disk applies centrifugal force to fluids and moves fluids from the intake and discharge. A cross-section view of a centrifugal pump is shown in Figure 7.6.[11]

Table 7.2 enumerates some of the specifications a manager should consider in evaluating the pumps which are supplied with a cleaning machine.

Pumps with these specifications[12] will cost at retail from 400 euro to more than 1500 euro. Since the global market for pumps is highly competitive with many suppliers, the supplier of your cleaning machine will have paid as little as 40% to 60% of those amounts.

Note the following differences:

- Specifications for solvent pumps are quite different than those for pumps used with aqueous cleaning technology. Pumps used with aqueous technology move more fluid at a higher pressure than do

[9]Pure rinse fluid would be the exception.

[10]Centrifugal pumps are extremely common, but only one of many types. Fluid is pressurized and moved by other pumps which use reciprocating pistons, rotating screws, vibrating diaphragms, rotating gears, rotating vanes, and even "massage" of a flexible tubing.

Centrifugal pumps move and increase the pressure of water and other fluids as they pass through a pump by the application of centrifugal force. The force is great because the velocity of disk rotation is that of the motor, which is usually either 3,550, 1,800, 1,200, or 900 revolutions/minute (rpm). Centrifugal pumps are direct drive as the rotating disk moves at the speed of the motor.

[11]Figure 7.6 is courtesy of Goulds Pumps, ITT industries. Pumps can be designed by users at, among other web sites, http://www.gouldspumps.com/pss.html

[12]A self-priming pump is a pump which will clear its passages of air if it becomes air bound and resume delivery of the fluid without outside attention.

NPSH is the level of pressure necessary to feed the pump at the rated flow rate in order for it to produce the stated flow (not starve) through the chosen diameter of inlet piping. In Table 7.2 an NPSH value of 6 ft of water column pressure (equivalent to 2.6 psi) means that the discharge of the feed tank must be at least 6 ft above the pump suction, with no flow-reducing fittings (valves or filters) between the feed tank and the pump. Please note this situation produces an exceedingly tall cleaning machine. Nearly all supply tanks in cleaning machines are not pressurized. Hence the pressure available to feed the pump is that created by the height of fluid in the tank above the pump. NPSH can be stated in absolute pressure units as well (psia).

Rated capacity, in horsepower (HP), is of the motor. Since motors are usually rated with integral values of HP, the next larger integral value is usually chosen. However, for the wash pump specified in a "good" cleaning machine, a 2-HP motor would almost certainly be used.

Figure 7.7 is courtesy of Ebarra Pumps.

**Table 7.2 Selection of Pumps for Cleaning Machines**

| Item | Good | | Better | | Best | |
|---|---|---|---|---|---|---|
| Material of construction (wetted parts) | 304 L stainless steel | | 316 L Stainless Steel | | | |
| Connections (Static Seals) | Screwed Pipe | | Flanged Pipe (Teflon-impregnated Gasket) | | Flanged Pipe ("O"-ring) | |
| Motor Enclosure for Aqueous Service | Open Drip-proof | | Encapsulated | | Totally Enclosed, Fan-Cooled | |
| Seals for Aqueous Service (Dynamic Seals) | Balanced Mechanical Seal | | | | Cartridge Mechanical Seal | |
| Motor Enclosure for Solvent Service | Explosion-Proof, Dust-Ignition-Proof | | | | | |
| Seals for Solvent Service (Dynamic Seals) | Balanced Mechanical Seal | | Cartridge Mechanical Seal | | Magnetic Connection (Sealless) | |
| Motor Coupling | Motor and Pump as Separate Units | | Motor and Pump as Integral Unit | | | |
| Arrangement | Horizontal | | | | Vertical | |
| Feed | Gravity | | Self-Priming | | | |
| Suction Connection Diameter | Matching Pump Suction Diameter | | | | Larger Than Pump Suction Diameter | |
| Manufacturer's Warranty | 1 yr | | 3 yrs | | 5 yrs | |
| **Capacity issues** | | | | | | |
| **Aqueous applications** | **Wash** | **Rinse** | **Wash** | **Rinse** | **Wash** | **Rinse** |
| Rated capacity HP | 2.1 | 1.6 | 2.9 | 1.8 | 3.9 | 2.6 |
| Minimum Net Positive Suction Head (NPSH), ft of water (psi) | 6 (2.6) | 6 (2.6) | 4 (1.7) | 4 (1.7) | 2 (0.9) | 2 (0.9) |
| Rated Head, ft of water (psi) | 100 (43) | 80 (34) | 130 (56) | 80 (34) | 160 (69) | 100 (43) |
| Rated Flow, gal/min (gpm) | 10 | 10 | 15 | 20 | 20 | 30 |
| **Solvent applications** | **Wash** | **Rinse** | **Wash** | **Rinse** | **Wash** | **Rinse** |
| Rated capacity HP | 2.6 | | | | | |
| Minimum Net Positive Suction Head, ft of water (psi) | 6 (2.6) | 6 (2.6) | 4 (1.7) | 4 (1.7) | 2 (0.9) | 2 (0.9) |
| Rated Head, ft of water (psi) | 80 (26) | 80 (26) | 80 (26) | 80 (26) | 80 (26) | 80 (26) |
| Rated Flow, gal/min (gpm) | 10 | 10 | 15 | 20 | 20 | 30 |

pumps used with solvent technology. This is because mechanical force plays a dominant role in soil removal with aqueous cleaning technology, and much less so with solvent cleaning technology (see Chapter 1, Section 1.2.2 and Figure 1.7).

- Specifications are not the same for the wash and rinse pumps in a cleaning machine. This is because the needs for washing and rinsing are different. Mechanical force (pressure) is more needed for washing and volume (flow rate) is more needed for rinsing.

In summary, use the perceived quality of included pumps as a significant factor to assay the quality of the cleaning machine considered for purchase. Certainly, the machine won't perform better than the specification of the components from which it is constructed.

**Figure 7.7** Pump and motor as integral unit

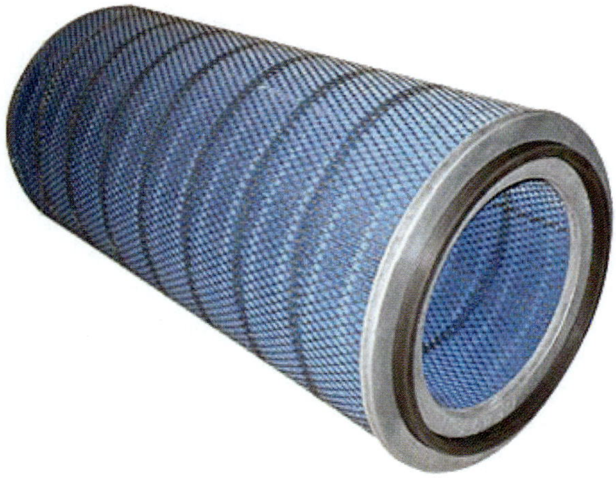

**Figure 7.8**

## 7.3 FILTERS

Filters are like condoms. They protect against infection, of pumps and nozzles with solid contamination.

### 7.3.1 Anatomy of a Filter

Most filters used in cleaning equipment are cartridge filters. The cartridge is a metal tube contained on or supported by a metal cylinder. Layers of fiber, often polymeric or cotton, are wound or wrapped around the surface of the inside tube. The winding does not completely block flow, as pores remain between adjacent fibers. The winding and its pores are the filtration media.[13] An end view of a cartridge element is shown in Figure 7.8.[14]

The outside tube, called a housing, contains the process fluid. A selection of filter housings is shown in Figure 7.9.[15]

For applications involving large flows rates or high solids loadings, the filter element may be a fiber bag.

Solid contamination may be large to fit through the filter pores and will be retained within the cartridge. When most of the cartridge pores are filled with contamination, flow stops. The cartridge must be replaced.

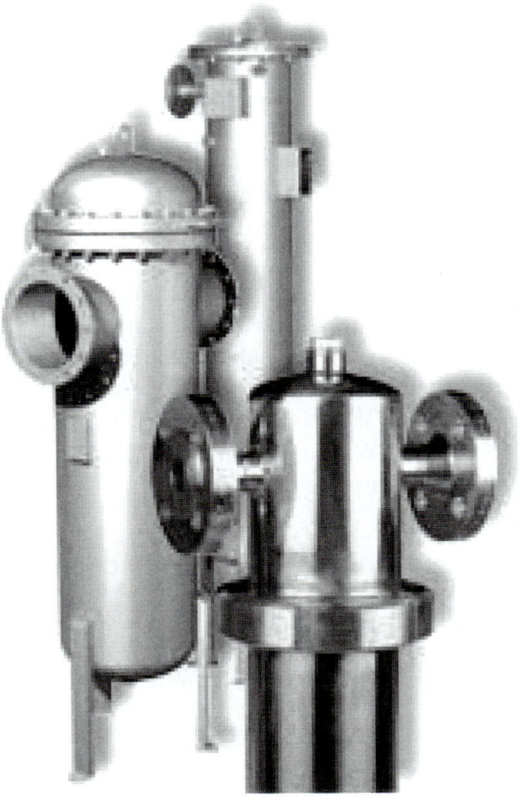

**Figure 7.9**

[13] Filtration is done in depth. Particles may penetrate one pore between adjacent surface fibers only to strike another fiber beneath the surface pore. That surface pore allows fluid flow but retains some particulate and allows some particulate to pass to a deeper layer of fiber. The process is repeated through successive fiber windings until the bulk flow reaches the inside tube.

[14] Figure 7.8 courtesy of Zhangjiagang Duty-Bonded Area Filter International Trade Co., Ltd.

[15] Figure 7.9 courtesy of Techno-Filt International.

Filter cartridges are rated based on the largest particle size supposedly not to fit through the filter pores. The size is usually given in microns.[16]

### 7.3.2 Filter Cartridges

Filter cartridges capable of blocking the smallest particles (usually around 0.1 μm) have minimal volumetric capacity to hold debris. Filter cartridges which block only larger particles have larger volumetric capacity to hold debris.

Consequently, filtration is often done in stages. The first stage removes the largest particles. Stages later in the sequence are rated to retain smaller particles. Thus the presumably greater volume of larger-sized debris is first removed. This protects the smaller capacity (and more expensive) cartridge which is rated to retain the smallest particle size.

In nearly all cases, the filter in a cleaning machine should have multiple stages. If the filter is designed to remove only large debris (say ~100 μm or ~4 mil), then use of a single filter element is justified. Otherwise, a two-stage filter (~10 and ~50 μm[17]) should be provided in a quality cleaning machine.[18]

As to retail cost, it's hard to spend more than 1,500 euro on a filtration system.

There is a second and more insidious cost. Fluid flow through a tortuous path, such as a network of fiber windings, is achieved only with the loss of fluid pressure. Frictional work is done by the fluid moving through the filter. With a clean filter, the loss may be only a few psi.[19] With a partially blocked filter, the loss may be one or more dozen psi.

### 7.3.3 Protection, Protection, Protection

Location of the filter in the cleaning machine is both crucial, and a tradeoff.

If the role of a filter is to provide protection, shouldn't it be located upstream of the devices being protected? That is, upstream of (before) both the pumps and spray nozzles – so they could be protected?

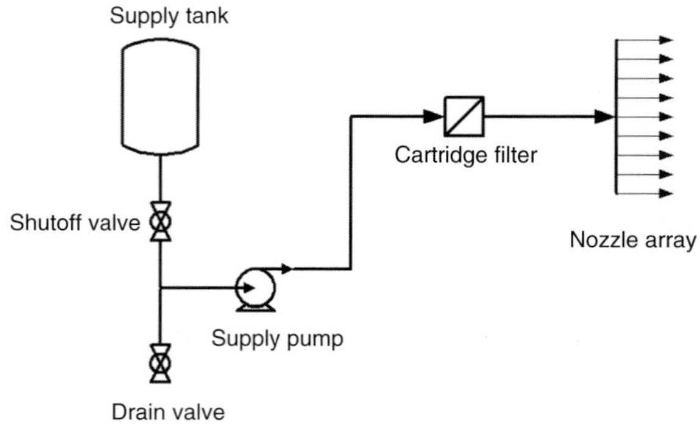


**Figure 7.10**

Yes, filters should be located upstream of pumps. But they normally aren't. The reason is found in the specifications for pumps in Table 7.2. Please compare the needed NPSH value – 2–6 ft of water column – with the pressure loss noted above to be expected during flow of fluid through a clean filter, a few psi. Since these two values are similar, (see Footnote 12) there is an excellent chance that the cartridge filter will starve (limit the flow to) the supply pump – even when the filter is clean. With a used filter, there is no chance of expected operation. The pump will be starved for fluid. It will not pump the required volume of fluid. The spray nozzles won't have the intended cleaning effect.

If you are considering the purchase of a cleaning machine, of either the aqueous or solvent persuasion, please note the relative locations of the feed tank, cartridge filter, and supply pump. They should be as in Figure 7.10, unless the supply tank is pressurized:[20]

- There should be unions between the various components so they can be efficiently disconnected.[21]

---

[16] 1 μm = $1 \times 10^{-6}$ m, 0.0394 mil, $3.94 \times 10^{-5}$ in, or 1,000 nm.

[17] The values are provided for illustration. Actual sizes depend upon actual contamination.

[18] More than occasionally, three-stage filtration systems are found in cleaning machines, especially where the major cleaning task is removal of small-sized particulate.

[19] Pounds per square inch is a unit of pressure (force/area). Expressed as height of a fluid column, pressure in feet of water equivalent is 2.3 × psi. Consequently, 2 psi are equivalent to 4.6 ft of water column.

[20] Pressurized supply tanks can be found in some vacuum vapor degreasers, but never found in open-top vapor degreasers or aqueous cleaning machines.

[21] A pipe union is a connective fitting (not an association). It allows piping to be disconnected (broken) so that components (pumps, tanks, filters, etc.) can be accessed without all of the piping by which they are joined having to be displaced.

**Table 7.3 Comparison of Cleaning Tanks**

| Feature of Tank | Good | Better | Best |
|---|---|---|---|
| Material of construction | Plastic (polypropylene) for aqueous cleaning | Steel for aqueous cleaning | Stainless steel for aqueous cleaning |
| | Stainless steel for solvent cleaning | | |
| Volume | Twice the volume of the parts basket, when filled | Three times the volume of the parts basket, when filled | |
| Geometry | Twice the area of the parts basket | Three times the area of the parts basket | |
| Needed floorspace | As provided | Tanks vertically stacked to conserve floorspace | |
| Pipe fittings | Screwed pipe | Flanged pipe (teflon-impregnated gasket) | Flanged pipe ("O"-ring gasket) |
| Heating coil | Outside of tank, so as to not restrict interior liquid circulation | Inside of tank, for improved heat transfer | |
| Cover | Probably none | Essential, to minimize emissions | |
| Bottom corners | Square | | Cove (rounded) |
| Finish (if metal) | As provided | Weldments ground smooth | Polished to a 2B finish |
| Bottom | Flat | | Graded to drain to one side |

- The drain valve should be located beneath the supply pump, so both can be drained. The shutoff valve allows the pump to be removed for maintenance without the supply tank being drained.
- The cartridge filter should be located downstream of the supply pump, to protect the spray nozzles.

The purpose of the cartridge filter is not to protect the centrifugal pump.[22] The purpose is to protect the spray nozzles from plugging with suspended material, and not having the intended cleaning effect.

## 7.4 TANKS

Tanks contain and/or allow use of cleaning solution or rinse fluids. Cleaning work can't be done without them. They can be the most expensive single component in a cleaning machine. But in general, their quality of manufacture has only a minor effect on cleaning quality. Occasionally, a manufacturer of cleaning machines will fabricate their own tanks.

### 7.4.1 Tanks in General

Table 7.3 shows some of the specifications one should consider in evaluating the cleaning tanks which are supplied with a cleaning machine.

These specifications apply to either aqueous or solvent cleaning machines.

Many cleaning machines sold at low prices have tanks made of plastic, often polypropylene. Obviously, this choice contributes to the lower prices, and can make good sense. But there are at least two items of concern: (1) that the plastic sidewalls not be used for mechanical support of the fluid mass and (2) the tank not be used to contain aqueous cleaning agents at a

[22] That purpose is abandoned because of inadequate NPSH to feed the pump at full flow. This is why centrifugal pumps, which don't have tight clearances as do piston pumps, are used in cleaning machines.

temperature above the design limit[23] for the plastic. Support must come from external metallic side braces.

### 7.4.2 Self-Cleaning Tanks (Bottom)

Tanks contain the cleaning process. They also contain the insoluble debris cleaned from parts.[24] Chemically this material can be normally soluble soil which is not soluble because a solubility limit was exceeded; normally insoluble material[25] dislodged from parts by mechanical action; or soil materials (tramp soils) present which were not expected by the system's designers.

Most such debris collects on the tank bottom because it is insoluble and more dense than the cleaning agent. But tank walls can be infected too. Visually, this debris often resembles metal-laden mud.

Debris accumulation occurs with both aqueous and solvent cleaning technologies.

Unless removed at the rate it enters, that debris will accumulate on the tank bottom and the cleaning tank will become a storage silo for soil materials. Please remember, cleaning is soil management.

There are two ways to remove bottom debris, manual and automatic:

1. **Manual** cleanout is simple. The cleaning machine is shut down. The cleaning agent is emptied, and the bottom debris removed, usually by hand labor and/or vacuum tools. Frequency of cleanout can be once every month to every 2 years – often depending upon whether needed pre-cleaning of parts is done (see Table 1.9). Almost always, cleaning quality improves immediately after a system cleanout – that's why it's done (see Chapter 4, Section 4.12.4 about on-aim control).
2. **Automatic** cleanout involves continuous, or occasionally periodic,[26] removal of debris. This

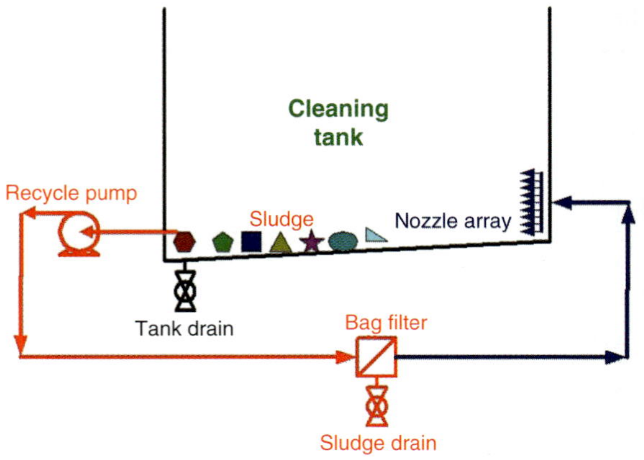


**Figure 7.11**

is done by facilities designed into the cleaning system. A flow diagram of the equipment components included in these facilities is shown in Figure 7.11.[27]

These facilities are used by flushing the bottom of the cleaning tank with filtered cleaning agent. Insolubles are entrained or forced by fluid momentum to travel down the slight grade of the tank bottom to a pickup point. There, a low-pressure/high-volume centrifugal pump collects the liquid debris (sludge) and forces it through a bag filter.[28] Solid-free liquor returns to the nozzles.

Automatic sludge removal facilities are provided only in more expensive cleaning machines. They are not needed by every user. Those cleaning machined parts with attached chips, drilled parts with attached burrs, or molded parts with attached scrim are good examples of those who do.

### 7.4.3 Self-Cleaning Tanks (Top)

Not all debris is heaver than the cleaning agent, and sinks to the tank bottom. Some is immiscible and

---

[23] This is apparently not well defined. Literature references cite values of temperature limit for polypropylene as high as 82–100°C (212°F), but this author's experience is to witness: (1) sidewalls of both polypropylene homopolymer and copolymer tanks being exceptionally flexible at 60°C (140°F) and (2) puncture of one thin-walled polypropylene cleaning tank at 82°C (160°F). Ask for a mechanical design review.
[24] Thus debris also usually contains a few parts unintentionally dislodged from baskets.
[25] Good examples are chips from machining operations, fines from grinding operations, or metal burrs liberated by cleaning chemistry.
[26] Once per day, for example, because accumulation of bottom debris fortunately occurs slowly.
[27] The nozzle manifold array is shown only in concept, as is the overall figure. Other components and equipment configurations are used by manufacturers of cleaning systems.
[28] All this piping is of large diameter, usually around 6 in. A bag filter is used, instead of a cartridge, because it will accommodate the large flow rate of sludge liquor. In some systems, the bag filter is located upstream to protect the recycle pump. The pump is not one found in Table 7.2. The flow rate is only that needed to entrain solid material.

Figure 7.12

Figure 7.13

lighter than the cleaning agent. That debris[29] floats to the top surface of the tank. Usually, it remains there.

This is only a negative outcome if clean parts being removed from the cleaning tank pass through this surface. Which, of course, they nearly always do!

Consequently, clean parts are made dirty when removed from the cleaning tank. This situation is quite common in industrial aqueous cleaning, where soils are organic materials which are immiscible with water.

A familiar example is waste motor oil and water. Never a pretty sight, one example is shown in Figure 7.12.[30]

Sold under many trade names and incorporating various similar concepts, the device which can alleviate this situation is conceptually identified as an oil skimmer.

All systems useful in industrial cleaning must acceptably provide two functions: (1) collection of the debris and (2) recovery of it from water without reflecting the cleaned parts.

## 7.5 COLLECTING THE DEBRIS

Collection is usually done with a proprietary fixture that either floats on the fluid surface or is attached to a tank wall and functions as an overflow weir.

However, floating debris may not all be located on the fluid surface. Some may be stratified just beneath the surface.[31] In this case, a floating device with a shallow pickup may not remove insolubles at the rate they enter the separation tank (see Figures 7.14–7.16).

One commercial pickup (collection) fixture is shown in Figure 7.13. There are many other proprietary models.[32]

### 7.5.1 Separating the Collected Oil from Water

Immiscible oil will normally separate from water because of difference in density, but the time to do so may delay the cleaning cycle.[33]

Conceptual behavior of immiscible oil droplets in a cleaning tank (see Footnote 99) is shown in Figure 7.14. Oil droplets rise because of differences in density, but also move horizontally and vertically with bulk fluid movement. Larger oil droplets (particles) rise sooner.

There are several commercial methods by which oil is separated from water, and which are used with cleaning equipment.

Normally, like pumps, nozzles, and tanks, these separation devices are purchased as add-on systems

[29] The debris is known as a rag, skim, scum, free-floating, or just dirt. Not by any means is this debris purely organic. Dust, fines, and other particulate are often trapped within the floating layer.

[30] Figure 7.13 is courtesy of Slickbar.

[31] This material is likely present as very fine droplets, or an oil-rich emulsion.

[32] See US Patent 5,498,348; US Patent 5,580,450; US Patent 5,679,265; US Patent 6,488,841; US Patent 6,287,260; and Figure 7.16.

[33] Separation, because of the difference in density between water and oil, may take minutes to hours. This is because the difference in density between oil and water is not great (~0.05–0.2 g/ml). Bubbles rise faster in beer (~1 g/ml density difference) and steel shot falls faster in water (~≫1 g/ml density difference).

Smaller oil droplets always rise more slowly than do larger oil droplets. The rate of rise is roughly proportional to the square of the droplet diameter.

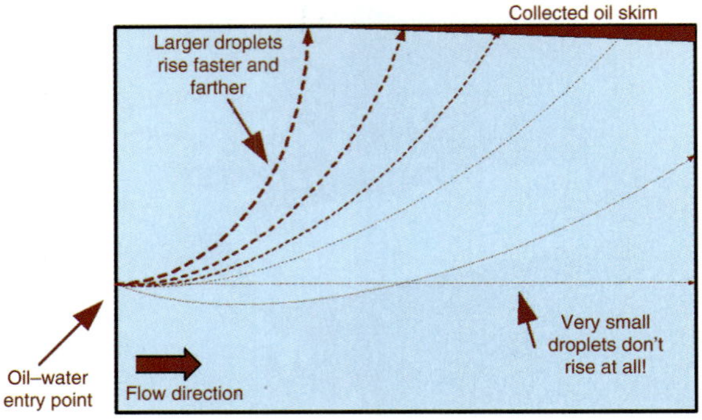


Figure 7.14

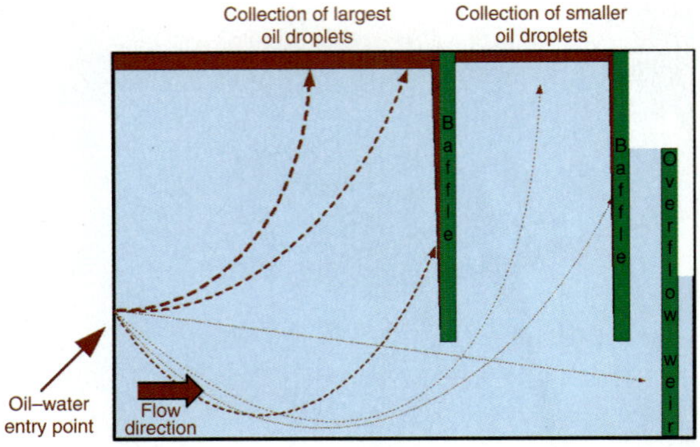


Figure 7.15

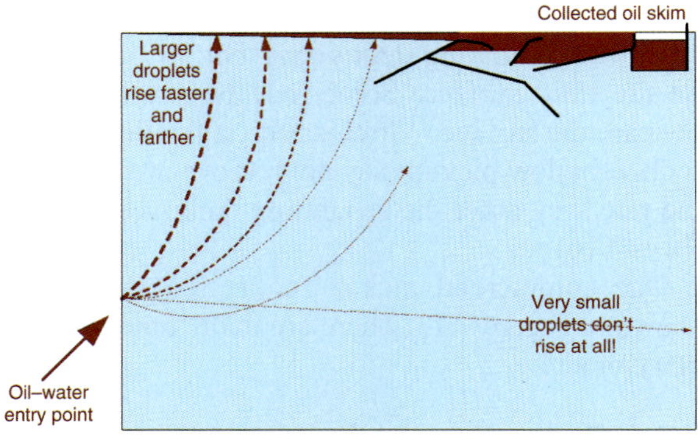


Figure 7.16

by the manufacturer of the cleaning machine, and not manufactured by them.

### 7.5.2 Enhanced Gravity Separation

Here various designs of baffles increase opportunities for oil droplets to rise through the fluid in which they are immiscible.[34]

Conceptual behavior of oil droplets in a baffled system is shown in Figure 7.15. Considerably more than two stages are possible. Please note that the surface oil must then be collected using a device similar to the one in Figure 7.13.

Passive gravity-based systems can require significant amounts of real estate (floorspace). This can be one of the drawbacks to aqueous cleaning technology.

The inherent limitation on gravity-based separators is the size of the oil droplet. Larger oil droplets,[35] which have less surface area per mass of oil, rise faster for two reasons: (1) there is less frictional force opposing the rise and (2) there is more inertial force (mass) causing it. Droplets larger than ~150 μm can be well managed in gravity-based systems.

A unique patented design of enhanced gravity-based separator has been witnessed by this author.[36] It adds a unique collection function and eliminates most requirements for floorspace (see Figure 7.16).

Droplets of oil/aqueous cleaner must rise to the surface to be collected between the gaps, called valves. Nearly water-free oil can be recovered. And a high efficiency of collection is claimed of droplets which do rise to the surface.

### 7.5.3 Centrifugal Separation

Mechanical forces, other than gravitational, can be used to separate oil soils from aqueous cleaning agents.

These and similar devices are commonly also used outside of parts cleaning – industrial waste water treatment, coolant cleanup, environmental control on oil production platforms, and treatment of marine bilge water (see Figure 7.17[37]).

Generally, they are static devices (no moving parts) called hydrocyclones.[38] Oily water is *tangentially*

---

[34] See US Patent 5,236,585.

[35] Unfortunately, the nature of the cleaning application determines the size of the oil droplets and the system designer can do little to promote larger oil droplets. Centrifugal pumps commonly degrade oil droplet size.

[36] See US Patent 6,287,460 and http://www.lovasc.nl/

[37] Figure 7.17 is courtesy of VortexVentures.

[38] Centrifuges are dynamic separation devices with a rotor as the moving part.

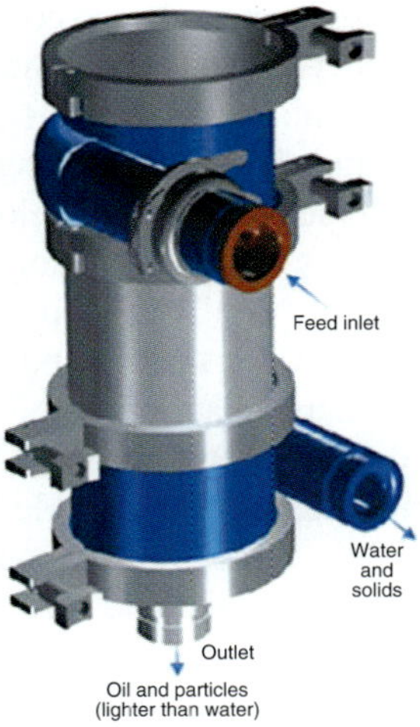


**Figure 7.17**

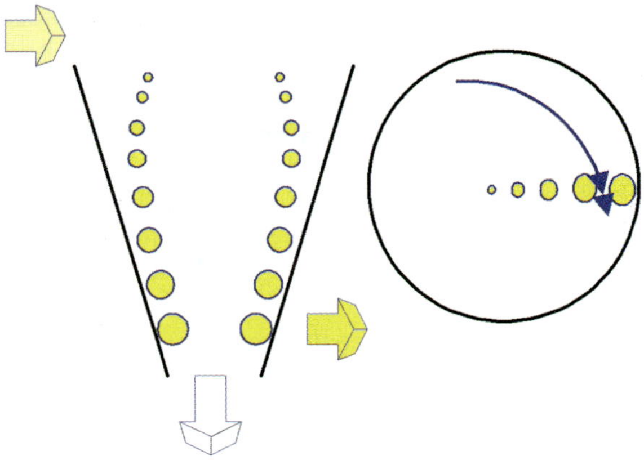

**Figure 7.18**

pumped into the closed circular chamber at 30–50 psig. The diameter of the chamber is larger at the top and smaller at the bottom, forming a downward-pointing cone.

Tangential entry causes the fluid stream to rotate (spin) – applying centrifugal force to the two-phase mixture:

1. Heavier elements are pulled to the outside, where they "ride" down the outside, walls of the circular chamber. Hence, the effluent at the outside of the chamber is richer in heavier particles.
2. Lighter elements (smaller oil droplets) are not as strongly pulled to the walls. Hence, they remain in the effluent from the center of the chamber.[39]

This is shown in Figures 7.17 and 7.18. Oil-free water (hopefully), the heavier fluid, exits from the side.

Because a greater level of force (centrifugal versus gravitational) can be applied, centrifugal separators remove smaller oil particles (droplets) than can be removed by enhanced gravity separators. That's the good news.

The bad news is that the oil-rich stream from a hydrocyclone contains a significant amount of water, which contains once-expensive cleaning chemicals.

### 7.5.4 Coalescers

These devices collect oil from water using static structures that take little floorspace. Collection is done based on the characteristics of materials known as oleophilic and hydrophobic.[40]

[39] A good test for the efficacy of using a hydrocyclone is to observe settling by gravity of oil–water mixtures in a transparent container. If there is little phase separation after a few minutes, a hydrocyclone is likely to be of little use. Then, separation based on intermolecular forces may add value (see Section 7.5.4). A good Internet-based reference is http://www.hydrocyclone.com

[40] Oleophilic does not describe someone who has eaten too much margarine and hydrophobic does not describe someone who is afraid of Hydrogen. An oleophilic material is one which is "oil-loving," "preferring" to associate with oil. An hydrophobic material is one which is "water-hating," "preferring" to not associate with water. These characteristics attest to the chemical structure of the molecules of which the material is comprised. Oleophilic materials "look like" oil. Hydrophobic materials don't "look like" water. Oleophobic materials don't "look like" oil and hydrophilic materials do "look like" water.

See the figures below, where the chemical structure of oil is similar to that of polypropylene, and not that of water. Is it any wonder why oil films adhere to polypropylene, and oil and water are immiscible?

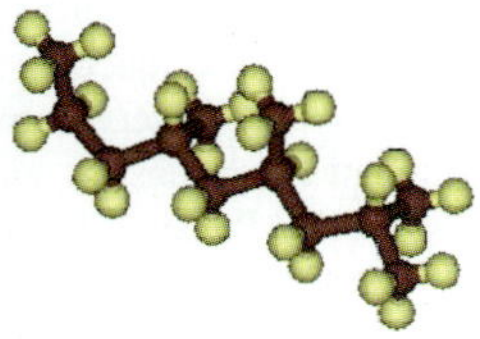

Representation of polypropylene

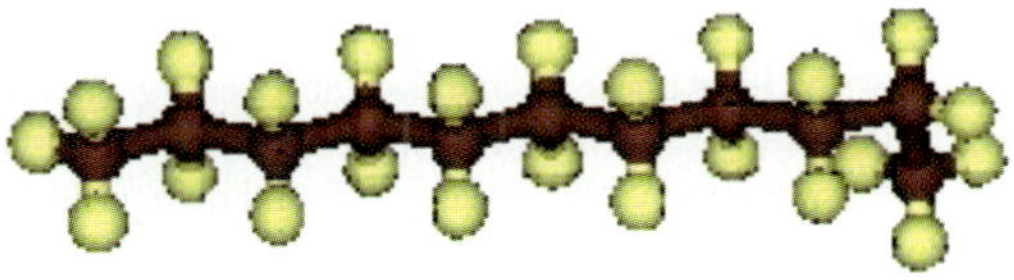

Representation of oil

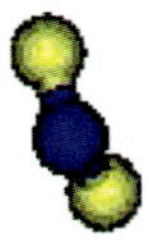

Representation of water

Oil is drawn from water by intermolecular forces to oleophilic materials such as polypropylene.

Usually surface area of the plastic limits the quantity of oil which can be removed. Consequently, common plastic structures used in coalescers are usually fibers or spheres. Some coalescing elements do resemble cartridge filters, bag filters, or packed beds. Other elements are wide belts which move through the oil–water interface.

A set of polymeric coalescer elements is shown in Figure 7.19.[41]

Mechanical action is needed to overcome the relatively weak intermolecular forces and displace the oil from the oleophilic structure. This force is commonly applied via a fluid jet or a mechanical scraper, after the oleophilic structure is removed from the oil–water interface.

**Figure 7.19**

A commonly used approach is to construct a bed of polypropylene pellets and pump oily water downward through it. Regeneration of the bed (i.e. removal of the oil) is accomplished by pumping some water upward[42] through the bed at a high velocity.

Other approaches involve pulling a "rope" or "disk" or "drum" or "belt" or "mop" of polypropylene through an oil-water surface and wiping the oil off via some mechanical action.

Coalescing devices aren't perfect. Oils which are chemically emulsified, or are soluble in water, will not be effectively removed by coalescing. The emulsified oils, or "emulsions," are comprised of oil, detergent, and water. The individual components of emulsions do not naturally separate from each other when allowed to settle, and consequently intact[43] emulsions usually must be disposed of as hazardous waste. Also, it can be difficult to remove trace amounts of oil using a coalescing element.

Some oil soils, synthetic motor oil, for example, which contain both oleophilic and hydrophobic structures, won't be well separated by coalescing devices.[44] Here, chemical structure of the coalescing element must be tailored to the chemical structure of the soil:

- The good news is that coalescers can make excellent separations in small spaces at low cost.
- The bad news is that this performance is application-specific. One changes the temperature or adds another soil component, and then may need another coalescer device!

### 7.5.5 Separating the Separators

Droplet size of immiscible oils in effluents from cleaning baths is not a factor controllable by the operator or designer of the aqueous cleaning system.

[41] Figure 7.19 is courtesy of AFL Industries, Inc.

[42] Flow directions are chosen because oil is less dense than water. Its natural tendency is to rise in water.

[43] The general approach to recovering soil and cleaner from an emulsion of cleaner/water/soil is to first "break" the emulsion. This can often be done via an increase of temperature in collected spent emulsion.

[44] This point begs the question: what happens to the aqueous cleaning agent? Obviously, it too must exhibit both oleophilic and hydrophic behavior – or it won't dissolve in water and attract oil. Some firms claim, and can demonstrate, application-specific dual cleaning agent/coalescer technology. They manage sequential separations. First oil is separated from water – the aqueous cleaning agent being chosen to partition with either phase. Generally, the cleaning agent is separated in a second coalescer from the phase in which it has partitioned. Both coalescers have different composition and are operated at different conditions.

An excellent example, which both enjoyed commercial success and found difficulties, is the technology in US Patent 5,849,100. Where successful, both the cleaning agent and the oil could be reused. The preferred cleaning agent solution was identified as "contains about 0.9 lb/gal of sodium meta-silicate pentahydrate, about 4.1 lb/gal sodium xylene sulfonate, about 0.94 lb/gal of a non-ionic surfactant, and the balance of the gallon is deionized water."

Managers of cleaning systems are pleased to remove all oil (non-water soluble soils) from parts and don't normally care what physical size the oil takes in the waste water.

So managers select separation systems based upon the nature of oil distributions they *find* within and around aqueous cleaning machines.

Oily water separation efficiency for *all three* separator types is highest with large oil droplets. Very small droplets are more difficult to separate.

This is a reason to prefer one type of oil–water separator over another:

- That is the ability of systems to recover smaller-sized droplets of oil from a "slurry" of oil particles in water. This is shown in Figure 7.20, where the performance of gravity-based, hydrocyclonic, and coalescing units are compared on a consistent basis.[45]

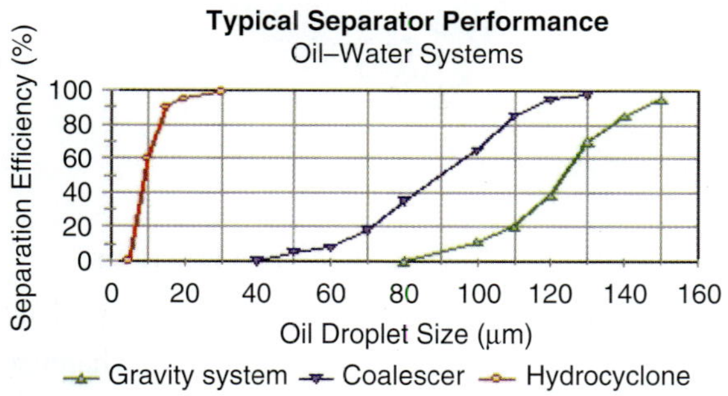


**Figure 7.20**

An important aspect of performance quality of aqueous cleaning machines is recognized by their designers[46] (see Table 7.4).

In summary, this author recommends that a hydrocyclone system (pump and tubular separator) be retrofitted to every aqueous cleaning system *in which recycle of water to the cleaning bath is crucial.* But secondary treatment, perhaps with an enhanced gravity system, may be necessary to minimize the volume of oily water to be disposed.

Alternatively, this author recommends that an enhanced gravity system be retrofitted to every aqueous cleaning system *in which it is crucial to recover the oil or reduce the volume to be disposed.*

Coalescer systems can and do provide good value though they are not forgiving. Finally, retention of the low-cost gravity skimmer system should be avoided as would be a holed umbrella.

## 7.6 LESSONS FROM THE BIRDS

What does a mother bird teach a young one about keeping their nest clean? In every cleaning situation there is an opportunity to learn and practice that lesson.

Imagine parts just removed from an aqueous cleaning bath or the cleaning sump of a vapor degreaser.

**Table 7.4 Comparison of Oil–water Separation Systems**

| System Feature | Poor | Good | Better | Best |
|---|---|---|---|---|
| Ability to recover oil droplets smaller than 20 microns at a high level of efficiency | Gravity | Coalescer | Enhanced gravity | Hydrocyclone |
| | Figure 7.14 | Figure 7.19 | Figure 7.16 | Figure 7.17 |
| | ≈50–500 Euro | | ≈5000 euro | |
| Material treated by separator | Fluid on the surface of a tank | Any fluid contacting device | Fluid on the Surface of a tank | Any fluid fed to device |
| Quality of water-rich stream | Relatively free of oil | | | Very free of oil |
| Quality of oil-rich stream | Relatively free of water | Can be quite free of water | Very free of water | Not free of water |

[45] Data courtesy of Ultraspin (http://www.ultraspin.com.au/Tutorial-4.htm) who manufacture hydrocyclonic oil–water separators.

[46] Most often, any of these three separation systems are not manufactured by the manufacturer of the cleaning machines. All three types of separators are purchased from OEM (original equipment manufacturers) firms. However, a few manufacturers of cleaning systems do manufacture proprietary gear.

Films of soil-laden liquid cleaning agent cling to all surfaces. And some soil-laden liquid may be trapped in crevices. Drainage of this liquid, called dragout, immediately starts when the parts are removed from the cleaning tank. Good practice is that (see Chapter 6, Section 6.5 and Chapter 1, Figure 1.5) time in the cleaning cycle should be allotted for a substantial portion of this liquid to separate itself by gravity from parts.

Into which tank should this drainage flow, the cleaning or the rinse tank?[47]

- The wrong answer is the rinse tank. When that happens, the rinse fluid, which is supposed to be clean, gets dirty. Rinsing will be then done with more dirty fluid. Soil in the dragout fluid can reinfect parts.
- The right answer is the cleaning tank. When that happens, dirty fluid is placed back into the tank from which it came. That's where it belongs.

A well-designed cleaning machine will incorporate the right answer – dragout will drain or be diverted back to the cleaning tank from which it came.[48]

Various schemes are commercially employed. The simplest one is to not move the parts basket from above the cleaning tank until the drainage period is complete. Another scheme is to use a movable pan to collect the drainage and cycle it to the cleaning tank.

This capability is another hallmark of an excellent cleaning machine, and is far too often not provided. Failure to provide this capability is not an issue of cost.

To return to the opening question, the mother bird told their young ones to not make messes in their nest.

## 7.7 PARTS BASKETS

A poor choice of parts basket can doom an otherwise well-designed cleaning machine to failure. Yet the parts basket is probably the least expensive component in the machine.

If spray nozzles are the fingers and pumps, the heart of cleaning machines, parts baskets are the hands of cleaning machines.

**Figure 7.21**

**Figure 7.22**

A parts basket must provide two functions:

1. Support the individual parts so that all surfaces can be exposed to the cleaning action, whatever that may be.
2. Allow *all* fluid cleaning materials to drain from the parts.

Normally, the first function is provided by plastic or metal fingers which arrange the parts so they face the direction of the cleaning action. The orientation of these fingers is similar to that of home or commercial dishwashes (see Figure 7.21).

This arrangement can be so effective that many cleaning machines practicing the semi-aqueous process are organized into a dishwasher facility.[49] The reason is that the cleaning cycle could be made so repeatable.

A typical parts basket used in metal cleaning work is shown in Figure 7.22.[50]

[47] Please remember, again, cleaning is soil management.

[48] Stiveson, S., "Alleviating Production Cleaning Constraints Through Efficient Design," *Metal Finishing Magazine*, September 2003.

[49] Albeit one with stainless steel interiors, multiple-stage filtration, and a plethora of spray nozzles. Parts cleaned are printed circuit boards. These dishwasher machines usually have no removable parts baskets. Rather, the parts are arranged in sliding trays with the same internal structure of a basket.

[50] Figure 7.21 is courtesy of Aqueous Tech; Figure 7.22 is courtesy of Bowden Industries.

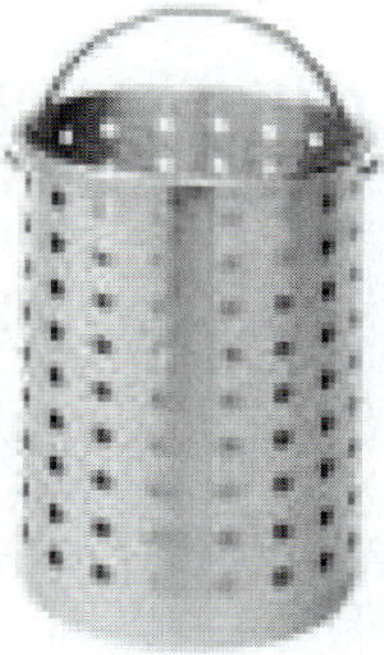

Figure 7.23

Figure 7.24

These effective arrangements come at a cost: operating labor, and auditing the effectiveness of that labor. Parts not properly arranged may as well not be placed in the basket.

The second function is normally provided by holes in the parts basket which allow effective drainage of cleaning and rinsing agents.

Percent open area is the key parameter. Consistent with structural support, holes should comprise at least 50%, and hopefully more, of the external area of the basket. Values of 70–80% can and should be achieved:

- Parts drain through the holes in parts baskets.
- Parts are contacted by forcing fluid through the holes in parts baskets.

Open external area allows both efficient drainage and effective contact. Parts baskets, such as those in Figures 7.23 and 7.24,[51] trap and obstruct fluid movement, and their use should be avoided.

Examples of useful parts baskets, with high levels of open area, are shown in Figures 7.25–7.27.[52]

Figure 7.25

Figure 7.26

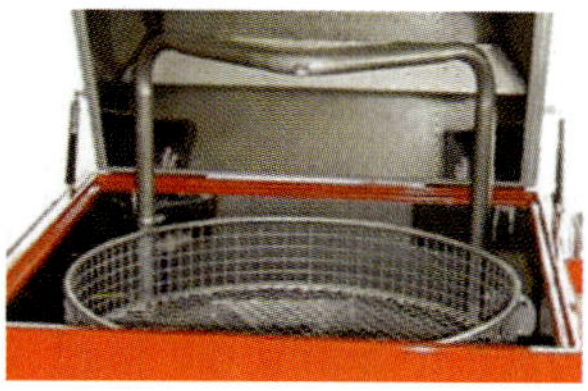

Figure 7.27

It's hard to spend 150 euro on a parts basket. Sometimes it's hard to spend 25 euro. This author wonders why cleaning machines are sometimes sold with poorly chosen parts baskets.

## 7.8 PARTS HOISTS

If spray nozzles are the fingers, pumps are the heart, and parts baskets are the hands, then parts hoists are the arms of cleaning machines.

These devices insert the load, parts baskets filled with parts, into the cleaning baths. In one sense, they are simply muscle – often used to lift large heavy

[51] This item is commonly, and ineffectively, used in plating baths, and the cleaning tanks which precede them.

[52] Figures 7.23–7.27 were collected from general advertisements on the Internet.

Figure 7.28

(and greasy) parts into vats of aqueous cleaning agents. Several of a similar type are shown in Figure 7.28.[53] These units are used as needed, based on the weight and balance of the part. The quality of industrial grade models is usually satisfactory for less than 500 euro. This item is not a differentiating item in choice of cleaning machines.

### 7.8.1 Programmable Hoists for Batch Solvent Cleaning Machines

Less commonly used with aqueous cleaning technology, their use is essential (and almost mandated in the US) with solvent cleaning technology.

Here the value is not muscle. Rather the value is speed control.

Parts inserted into the tall, narrow chamber that is a vapor degreaser act like a piston.[54] They displace vaporized solvent upward. Since the top of the degreaser is open to admit the load, this displaced vapor is usually emitted from the machine.

Workers are unnecessarily exposed to additional solvent fumes. Emission of volatile organic compounds (VOC) may increase, depending upon the solvent used. The US EPA's engineering standard[55] requires suppliers and managers to consider use of a programmable hoist for some[56] solvent degreasers to avoid this emission. The "speed limit" to be enforced[57] by the hoist is 11 ft/min (5.6 cm/s).

Figure 7.29

This author strongly recommends their use for nearly all batch solvent cleaning operations.[58] This feature does allow differentiation among cleaning machines with various quality levels. A model, with a small parts basket attached, is shown in Figure 7.29.[59]

One can't spend more than 2,500 euro on a microprocessor-controlled two-axis hoist, and frequently can spend 1,000 to 1,500 euro for a perfectly acceptable model.

## 7.9 HEATERS

Heaters are seldom a differentiating factor in recognizing one cleaning machine as superior to another.

But there are real differences among them (consider Table 7.5).

[53] Figure 7.28 is courtesy of Craneveyor Corporation.

[54] Good design principles suggest that the insertion area be no more than one-half of the exposed area of the solvent tank, and less if possible. The parts basket should never "just fit" into the open area that is the top of a solvent vapor degreaser. The basic idea is not to entrain upward or displace downward solvent vapor by moving the parts basket at a high rate of travel (see Tables 1.3, 4.13, and 4.14).

[55] This rate is a limit required for emission control by the US EPA's NESHAP (US CFR Vol. 65, No. 197, September 8, 2000, or Guidance Document EPA-453/R-94-081) for vapor cleaning equipment. A complete summary of applicable regulations is available at http://www.epa.gov/ttnatw01/degrea/halopg.html. Suppliers and managers have a menu of choices, including a programmable hoist.

[56] Technically, this standard or regulation only applies to chlorinated solvents including 1,1,1-Trichloroethane, chloroform, carbon tetrachloride, methylene chloride, perchloroethylene, and trichloroethylene.

[57] Human nature is to speed, to increase productivity, to shorten cleaning cycle time when performance lags the production schedule. This leads human operators to "drop" parts baskets into vapor degreasers, causing unwanted emissions. The purpose of the automated/programmable parts hoist is to replace that human tendency with predictability and control. Obviously, it is assumed that the operator, or the manager, won't reprogram the hoist to defeat the intent to restrict emissions and slow entry rate!

[58] Whether the gain is reduced pollution, cost savings when expensive solvents are used, avoidance of hazardous situations, or just improvement in the quality of the work environment, a powered hoist with programmable speed control should be strongly considered in every purchase of a solvent cleaning machine.

[59] Figure 7.29 is courtesy of Unique Equipment Corp.

**Table 7.5 Comparison of Heaters**

<table>
<tr><th>Feature of Tank</th><th>Good</th><th>Better</th><th>Best</th></tr>
<tr><td rowspan="2">Material of construction</td><td>Steel for aqueous cleaning (not Copper)</td><td>304 stainless steel for aqueous cleaning</td><td>316 L stainless steel or incaloy for aqueous cleaning</td></tr>
<tr><td colspan="3">Stainless steel for solvent cleaning</td></tr>
<tr><td>Connection to tank</td><td>“Over the side”</td><td>Screwplug</td><td>Flanged</td></tr>
<tr><td>Open space available for removal with single wrench</td><td colspan="3" rowspan="3">Required</td></tr>
<tr><td>Thermal overload protection</td></tr>
<tr><td>Electrically grounded</td></tr>
<tr><td>Electrical</td><td>Single phase</td><td colspan="2">Three phase (to reduce current value)</td></tr>
<tr><td rowspan="2">Watt density, W/in$^2$ (lower densities last longer)</td><td><60 for aqueous cleaning</td><td><40 (size permitting), for aqueous cleaning</td><td><25 (size permitting), for aqueous cleaning</td></tr>
<tr><td colspan="3"><25 (size permitting) for solvent cleaning</td></tr>
<tr><td>Voltage, AC</td><td colspan="3">As available; higher voltage reduces current load</td></tr>
<tr><td>Enclosure, NEMA (National Electrical Manufacturer’s Association) rating</td><td>NEMA 4 (water-tight, and not submersible), NEMA 4X (water-tight, corrosion resistant, and not submersible)</td><td>NEMA 6 (submersible)</td><td>NEMA 9 (explosion proof)</td></tr>
<tr><td>Certification</td><td colspan="3">UL (Underwriters Laboratory) listed and CSA (Canadian Standards Association) certified. Both accepted worldwide</td></tr>
</table>

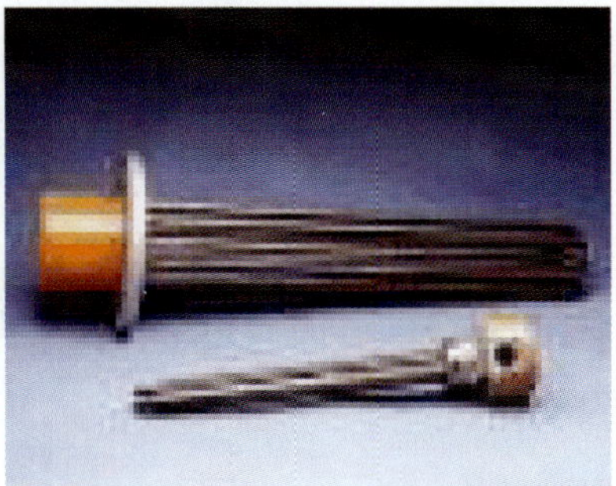

**Figure 7.30**

Heaters are not expensive. One can purchase several for 1000 euro, or less (see Figure 7.30).

Replacement models should be stock items at most supply houses. That point is worth investigation before purchase of a cleaning machine as heaters do have useful lives and do fail.[60]

## 7.10 SONIC (ULTRA OR MEGA) TRANSDUCERS

These equipment components are used in both aqueous and solvent cleaning applications. Chiefly used for removing solid particulate matter, they are agents of agitation which can dislodge soil components that can't be removed solely by chemical action. In common use for decades, they are becoming (or have become) commodity equipment products despite the best efforts of suppliers to provide differentiation.

[60] Failure of heaters often occurs when an excessive burden of soil is imposed on the cleaning machine. The mode of failure is usually burnout caused by deposition of soil elements on heater surfaces (fouling). Here, heat transfer rate to the cleaning solution is limited by the insulating soil elements while the heat supply rate hasn't been reduced. The result is that the surface or sheath temperature increases and approaches its design level. So the thermal cutout switch disconnects the heat supply so as to protect the overall cleaning machine.

### 7.10.1 Vibrating Diaphragms

Ultrasonic transducers produce waves of fluid pressure which bombard part surfaces (and all surfaces under immersion). The waves are produced by diaphragms which vibrate under immersion in fluids.[61] The device producing the vibration is a transducer.[62]

Frequency of vibration is high, from tens of thousands to hundreds of thousands of oscillations (cycles) *per second (cps or Hertz)*.[63] Consequently, the effect of each cycle of vibration is negligible, but their cumulative and continuous effect can be either positively or negatively dominant.

There are two methods by which transducer diaphragms are caused to vibrate.

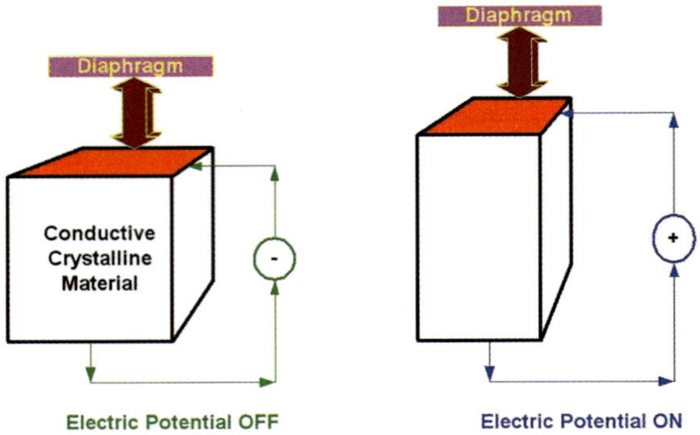


**Figure 7.31**

**Figure 7.32**

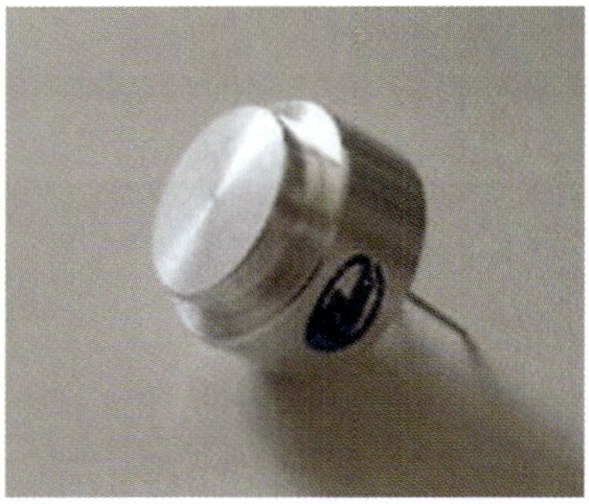

**Figure 7.33**

#### 7.10.1.1 *The Piezoelectric (Curie) Effect*

A piezoelectric material[64] has two unusual and interrelated characteristics. They are basically the reverse of one another:

- When a force is applied to a piezoelectric material, a tiny electric current is produced.[65]
- When an electric current is passed through piezoelectric materials they deform, changing in size (volume) by a few percent.

It is the latter characteristic which produces a vibrating diaphragm. A rigid connector (arm) causes the diaphragm to move slightly when the piezoelectric material changes shape upon application of an electric current. This is shown in Figure 7.31.

Repeated application of the electric current, followed by its relaxation, enables a diaphragm to move forward and backward in one dimension.

---

[61] Please note that ultrasonic transducers are not used in air. They must be immersed in a fluid (liquid). Consequently, spray-in-air cleaning does not involve sonic agitation.

[62] The technical definition is a device which converts one form of energy to another. In this case, electrical energy which is used to drive the diaphragm is converted to rapid motion (mechanical energy).

[63] A common frequency of vibration is 40,000 cycles/seconds or 40 kHz.

[64] This effect was discovered by Pierre Curie in 1883. It is also linear – the crystal expansion is proportional to the applied charge. The word piezo is Greek for "push." Piezoelectric solids typically resonate within narrowly defined frequency ranges. Materials which exhibit this effect are quartz, $SiO_2$ (used for precise frequency reference in radio transmitters) and ceramics. Barium titanate, lead zirconate, and lead titanate are ceramic materials which exhibit piezoelectricity, and are used in ultrasonic transducers (and microphones).

[65] This effect has become quite valuable in creation of industrial sensors. Automotive airbags can be activated by piezoelectric materials. The force of an impact on the piezoelectric material produces (transduces) an electrical current through the material which activates external devices, including inflation of the airbag.

Single transducer elements with piezoelectric materials are shown in Figures 7.32 and 7.33. The images represent similar products from two different manufacturers.[66]

Most piezoelectric materials are ceramics,[67] many of which contain silicon, lead,[68] aluminum, or titanium oxides.

#### 7.10.1.2 *The Magnetostrictive (Joule) Effect*

There is a magnetic analog to the piezoelectric effect.

A ferromagnetic material (magnetic Iron) will respond mechanically to magnetic fields. This effect is called magnetostriction.[69] Magnetostrictive materials transduce or convert magnetic energy to mechanical energy.

As with the piezoelectric effect, the reverse is also true.

When a magnetostrictive material is magnetized it changes dimension in one direction.[70] As in Figure 7.31 that dimensional change can be used to cause a diaphragm to move,[71] though driven by a different factor.

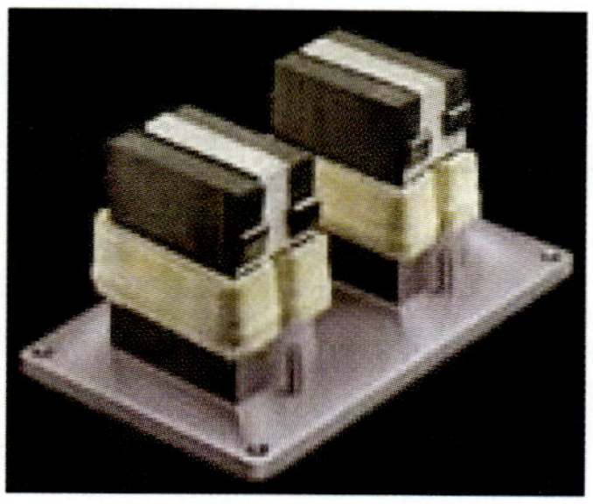

**Figure 7.34**

Most magnetostrictive materials are metal alloys of Nickel or contain significant quantities of Nickel[72] compounds.

Single transducer elements with magnetostrictive materials are shown in Figures 7.34 (two transducers).[73]

Magnetostrictive transducers are not used at frequencies above around 30 kHz. The main reason is that the difficulty and cost of controlling the motion[74] of the material associated with magnetostrictive transducer elements becomes too severe at frequencies above that level.[75]

---

Phonograph cartridges have long used this effect. As a stylus made of a piezoelectric material moves within a corrugated groove, an electric tiny current is produced. The current is amplified, and used to drive a speaker. Positioning of a probe for a scanning tunneling microscope along a surface is done with a piezoelectric ceramic wafer.

[66] Figure 7.32 is courtesy of Blackstone Ultrasonics and Figure 7.33 is courtesy of Branson Ultrasonics.

[67] These forms or pieces are made from spray-dried ceramic powder which are fired in an oven, and then machined after shrinkage to the desired dimensions. Then, Silver electrodes are screen printed on them. Standard frequency tolerance can be as low as ±5%. Most forms can be produced (rods, disks, plates, rings, etc.), which enables the ability to make transducer elements for unusual applications. These transducers are also known as Langevin-type transducers.

[68] PZT is an acronym for lead zirconium titanate – a common ceramic material exhibiting piezoelectric behavior. In some publications, PZT refers to any piezoelectric material, without regard to its specific chemical composition.

[69] This effect was discovered by James Joule in the 1840s. Joule identified the change in length of an Iron sample as its magnetization changed. There is also a reverse Joule effect where a material can be compressed (causing its length to change) and a magnetic field is created.

[70] More specifically, "... When a magnetic bias is applied to magnetostrictive material, the magnetostrictive material constricts (gets shorter). Basically the magnetic field makes all the molecules want to get closer together. In a generic 20 kHz magnetostrictive transducer, this change in dimension is about 0.0005 in. The commercial limit is about 30 kHz. About half of that movement is actually driving the diaphragm, the remainder is in free air". Personal communication from J. Paulhus, FMT Inc., January 2006.

[71] The rate of movement is surprisingly high – at 20 kHz the rate is 5 in/s in total movement. Personal communication by J. Paulhus, FMT Inc., January 2, 2006.

[72] Nickel maintains its magnetostrictive properties on a constant level longer than do ceramic oxides.

[73] Figure 7.34 is courtesy of Blue Wave Ultrasonics.

[74] Since the velocity of sound in the Nickel-based material is constant, the frequency is changed by decreasing or increasing the length of the Nickel laminations. For example, at 20 kHz, they are 5⅜-in thick; at 16 kHz, they are 6¼-in long; and at 25 kHz, they are 4⅞-in long. "At the higher frequencies, with shorter Nickel laminations, the amount of constriction of the Nickel reduces with diminishing length. This reaches the point where the dimensional constriction is no longer effective in driving a loaded diaphragm". Information and quotation courtesy of J. Paulhus, FMT, Inc., January 2006.

[75] A similar situation applies with ultrasonic transducers, but at a significantly higher frequency. One can't manage controlled oscillation of the same mass of piezoelectric transducer at a higher frequency (170 kHz) than at a much lower frequency (40 kHz).

#### 7.10.1.3 *Comparison of Piezoelectric and Magnetostrictive Transducers*

In its simplest form, the control system for a sonic transducer of either type applies a tiny current for a very short duration, and then stops that current flow for an equivalent short duration so that the material can return to its original shape or dimension, and the diaphragm can return to its original position:

- The current causes the piezoelectric material to deform and changes the position of the attached diaphragm.
- The current passes through a wire coil which generates a magnetic field that magnetizes the magnetostrictive material and changes the position of the attached diaphragm.

Both types of materials cause diaphragms to vibrate. The surface to which that diaphragm is attached is also caused to vibrate. This is the container (housing) wall into which the transducer is mounted. In use the container is immersed into liquid.

In other words, application of an electric current causes a housing wall immersed in liquid to vibrate so that pressure waves are spread within the liquid.

In practice, the housing is populated with multiple transducer elements as shown in Figures 7.32 to 7.35, and *the entire populated assembly* is referred to as the transducer.[76] Such an assembly is shown in Figure 7.35.[77] Each transducer element consumes about 50 W of power. The assembly in Figure 7.37 is rated for 600 W because it contains six rows each containing two transducer elements.

Useful sonic transducers are produced using both types of materials. However, there are substantial differences (see Table 7.6).

Suppliers may inform managers that the choice is between the higher purchase price and longer maintenance life of magnetostrictive transducers versus the opposite for piezoelectric transducers, or to achieve a lower level of operating noise with piezoelectric transducers.

**Figure 7.35**

That's a false choice. The choice should be totally based on the character of the parts:

- No one would consider use of magnetostrictive transducers for cleaning of disk drive components, where piezoelectric transducers are commonly used. The components would "dance" in the water bath and be destroyed with piezoelectric transducers.
- No one would consider use of piezoelectric transducers for removal of scale prior to painting of small engine blocks for lawn mowers. Nothing would be removed.

### 7.10.2 What Is the Frequency, Kenneth?[78]

The two prefixes normally attached to the word sonic are *ultra* and *mega*:

1. *Ultra* refers to frequencies above those identified by humans, above ~18 kHz. Ultrasonic transducers

---

[76] This is because the individual transducer elements are buried within the container (housing) and are never (hopefully) seen by users.

[77] Figure 7.36 is courtesy of Blackstone Ultrasonics. Individual transducers are also known as "horns."

[78] This attempt at humor refers to a personal experience told by former CBS News anchor Dan Rather, and the song by R.E.M. Rather was mugged by an unknown assailant who uttered the phrase "What is the frequency, Kenneth?" The assailant was later apprehended and found to be mentally disturbed, believing the media was "beaming" signals into his head.

**Table 7.6 Comparison of Piezoelectric and Magnetostrictive Transducers**

| Item | Piezoelectric | Magnetostrictive |
|---|---|---|
| Location relative to liquid tank | Within (as an immersible unit) or externally attached to sides or bottom | |
| Material | Ceramic | Metal |
| Normal attachment of diaphragm to housing wall | Epoxy bonding, has been replaced by brazing with Copper under vacuum | Brazing with Silver |
| Relative mass of transducer element | Low | High |
| Normal thickness of housing wall | Thin (typically 14 gauge [1.7 mm]) | Thick (1/16 [5 mm] or greater) |
| Normally used frequency range | 20, 30, 40 kHz, and up | 20, 25 and 30 kHz |
| | This is the frequency of alternating current supplied to the transducer element by a matched and provided power generating unit. The power generating unit is supplied locally with 60 Hz electricity. | |
| Potential for part damage via excessive sonic action | Much less, because frequencies are higher | Much more, because frequencies are lower |
| Applied energy level – at the same frequency | Lower | Higher |
| Applications by mass of parts | Used with smaller part masses, where energy absorption is not significant. If a larger part mass is added, it will absorb sonic energy and that available for surface cleaning will be reduced. Hence, it is important to match the transducer to the mass being cleaned. | Used with larger part masses, which absorb significant quantities of supplied sonic energy. |
| Application by type of parts | Small assemblies, machined fasteners, rolled sheet and strip, compact disks, etc. | Foundry castings, extruded pipe, injection molds, or parts which are highly resistant to damage. |
| Application by type of soil | Single particles, spores or other biological contamination | Mill scale, some coatings |
| Concern about useful life | General maintenance life is quite long, but failures to occur due to:<br>• Stress corrosion cracking,<br>• Cavitation damage on radiating surface,<br>• Fatigue of the epoxy bond,<br>• Cracking of the ceramic.<br>Users should expect a useful life of at least two years. | Some manufacturers warranty the transducer and housing for whatever period the purchaser desires to use it. |
| Purchase price for the same frequency and power level, €/W | ≈€3/W, and up | ≈€5–6/W, and up |
| Author's characterization of noise level | An electric razor | A machine gun (hearing protection should ***always*** be used by workers) |

are the type most commonly used, with frequencies above 20 kHz, and below ~250 kHz. A manager purchasing a ultrasonic system without specifying the frequency would probably receive one operating at 40 kHz.

2. *Mega* is not scientifically defined in this context. A commonly accepted limit is frequencies exceeding 250 kHz. Megasonic transducers are chiefly used for removal of low levels of fine particles from valuable surfaces.

#### 7.10.2.1 *Ultrasonic Operations*

The reason waves (fluctuations) of pressure are valued is that they produce cavitation bubbles.[79] Collapse of those bubbles releases high levels of energy which can interrupt local collections or networks of debris (soil). That's cleaning!

Larger bubbles, which will ultimately release more energy per bubble when collapsed, are formed when there is more time for them to do so, this means when the frequency is low. Said another way, a lower frequency generates wavefronts with a longer time interval between them, thereby allowing more time for bubble growth. Smaller bubbles are produced when the frequency produced by the transducer is higher. Calculated bubble size versus frequency is shown in Figure 7.36.

**Figure 7.36**

But there is another factor affecting energy release. That's the number of bubbles produced. More bubbles are produced at higher frequencies because there are more opportunities to do so, more cycles of compression and rarefaction.

Essentially, the energy released to do cleaning work on surfaces is the product of the volume of each bubble times the number of bubbles. In other words, for the same power input from the transducer to the liquid tank:

- A low frequency will produce fewer cavitation bubble implosions each with higher release of energy.
- A higher frequency will produce more cavitation bubble implosions each with lower release of energy.

Two different types of operation with the same power level are illustrated in Figure 7.37.[80] Which would you prefer?

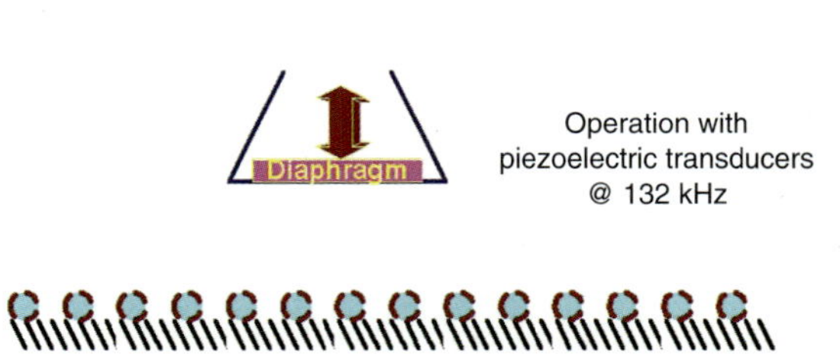


**Figure 7.37**

[79] Pressure waves are rarefactions (negative pressures) and compressions (positive pressures). They produce pockets, bubbles, cavities, or zones where fluid vapor exists. The vapor is evaporated liquid, not air. The bubbles are called cavitation bubbles.

Vapor bubbles can be stable, or unstable and collapse, depending upon their size and the nature of the pressure waves surrounding them. Bubble size is determined by a force balance between surface tension forces which are trying to collapse the vapor volume and buoyancy (differential pressure) forces which are trying to expand it.

In any case, bubble lifetime is measured in fractions of seconds. The waves, naturally, propagate at the velocity of sound.

Collapse of these bubbles, (implosions) releases a shock wave which radiates in a "jet" from the point of collapse.

[80] The crucial difference between these two modes of operation **is not** that one is activated with a piezoelectric transducer and the other is activated with a magnetostrictive transducer. Rather the crucial difference is in what each produces, a different size distribution of cavitation bubbles. The bubbles do the cleaning work!

#### 7.10.2.2 *Choosing the Right Ultrasonic Frequency*

What's significant is that the cleaning capabilities will be quite different in these two examples, and that the value of that difference will depend upon the nature of the cleaning work to be done:

> *The right ultrasonic frequency is that which best matches the cleaning capability to the needed cleaning performance.*

Please recall that ultrasonic cleaning technology involves generation of vapor bubbles and management of their collapse upon the soiled surface. Successful applications involve release of energy (producing mechanical force) at the point of bubble collapse sufficient in type and amount to dislodge the soil from the surface, without harming the surface.

Some have referred to this action as being "pecked to death by ducks." This is because other mechanical actions such as blast cleaning with solid media or impact from a pressurized fluid jet apply such different stress to soil elements and the surface on which they lie. To complete this analogy, blast and pressurized jet cleaning technologies might be thought of as being "eaten by a T-Rex dinosaur." Consider Table 7.7 in which this analogy is presented in a generalized visual form.

The point of this presentation is that the transducer frequency should be chosen to match the nature of the cleaning task. Each choice of frequency will be more useful when applied to a specific type of soil material, and will have different effects on the underlying surface. Said another way, use the right tool (frequency) for each job (cleaning situation).

And how is the right tool to be identified? Managers should organize and witness cleaning demonstrations using actual soiled parts with facilities provided by suppliers. These parts should be cleaned using several transducer configurations and the performance evaluated by the normally used cleaning test (see Chapter 5). Let the details of the application reveal the right choice of frequency.

#### 7.10.2.3 *Multiple Choice*

Some cleaning situations involve multiple soils. This can be where there is a distribution of sizes of soil materials, where there are layers of contamination that are sequentially removed, or where soil materials are degraded in the cleaning process. In all of these circumstances, what criteria would a manager use to select the frequency of ultrasonic-produced agitation to use?

The same criteria would be used as above: match the frequency and the capabilities its use provides to the characteristics of the soil materials. If that means multiple transducer frequencies are required, so be it. Commercial facilities exist to implement that choice.

Operation with multiple frequencies has become a featured commercial capability as suppliers seek competitive advantage via replacement of commodity products with specialties. While early efforts in the 1990s promised more than could be delivered, technologies available to managers today are achieving respected performance.

**Table 7.7 Visualization of Various Ultrasonic Transducer Applications**

| Magnetostrictive (kHz) | Piezoelectric (kHz) | | | |
|---|---|---|---|---|
| 20 | 40 | 58 | 104 | 170 |
|  |  |  |  |  |
| T-Rex eating man | Pelican eating fish | Hornbill eating beetle | Cockatoo eating nuts in shell | Finch eating seeds |

There are two approaches, and each can be implemented in several different ways. The assumed situation is that it is desired to establish three different frequencies of ultrasonic-produced surface agitation within a single tank (see Tables 7.8 and 7.9).

A major concern with the technology in Table 7.8 is the interference, both positive (constructive) and negative (destructive), between the pressure waves produced by the individual transducers.[81]

This is illustrated by the calculated outcomes in Figure 7.38. Please note that the intensity of combined

**Table 7.8 Multiple Transducers Immersed in a Tank**

| Arrangement Diagram | Number of Transducers | Is Each Frequency Continuously Present? | Total Power Level |
|---|---|---|---|
| 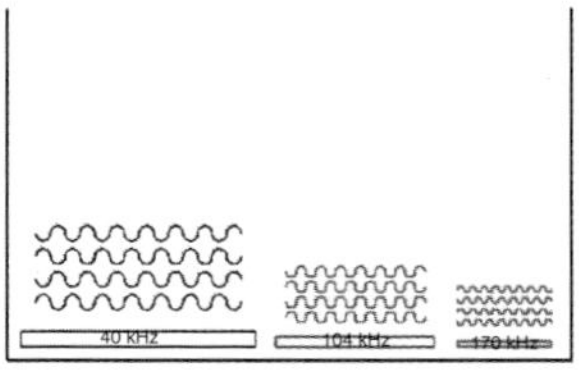 | 3 | Yes | 1.5 to 3X |

*Implementations*

1. Output of transducers directed to specific soil locations.
2. Different frequency transducers were spaced in equilateral triangles with each triangle containing the three frequencies and with no frequency next to itself. Parts placed within the space circumscribed by the triangle.
3. In US Patent 6,653,760, various harmonics of a chosen frequency are radiated into a cleaning tank. More than three are impractical.

**Table 7.9 Single Transducers Immersed in a Tank**

| Arrangement Diagram | Number of Transducers | Is Each Frequency Continuously Present? | Total Power Level |
|---|---|---|---|
| 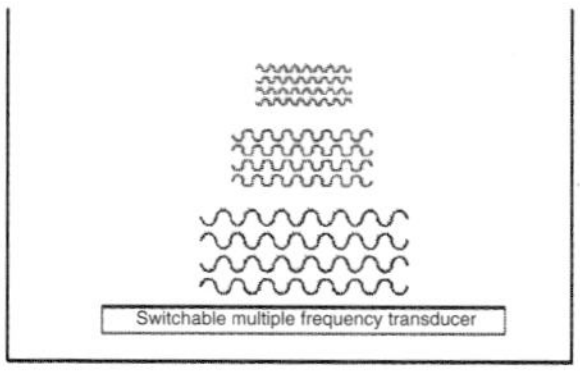 | 1 | No | 1X |

*Implementations*

1. Programmable electronic switches, or mechanical relays, enable power to be fed to single transducers in a programmed sequence.
2. Application of a selected frequency of pressure waves followed by implementation of another cleaning technique, which is then followed by application of a different selected frequency of pressure waves. An example is where force is sequentially applied by ultrasonic transducers and then by jets of rinse flow. Here particles are loosened, and then washed away (as would be done by a dentist). The rinse flow must be separated in time from the application of the ultrasonic power because ultrasonic-generated pressure waves can be redirected or their effects canceled by high-velocity bulk flows.

[81] This is known as production of a "beat" frequency.

pressure fluctuations can increase between 2 and 3 times of the normal pressure fluctuations.[82]

More intensity of pressure fluctuations is not necessarily beneficial, especially if the parts are fragile.

A second concern with this approach is power allocation. If each of several transducers is provided the normal amount of power, then the total level of power must be increased by the number of transducers present. More power means more cost, more concern about damage, and more heat buildup.

Table 7.9 differs from Table 7.8 in that only a single multi-functional or universal transducer surface is used to impose multiple pressure waves on liquids.

A major advantage of the technology in Table 7.9 (multiplexing) is that the same power level is applied at each frequency, from the same transducer surface.

A second advantage is reduction of concern about part damage because constructive and destructive interference of pressure waves is impossible. A "beat" frequency can't happen.

A third advantage may be that sequential[83] application of different waveforms destroys soil structure by first removing one size of material, then another, and so on.[84] Obviously, the order at which frequencies are applied can be chosen and programmed.

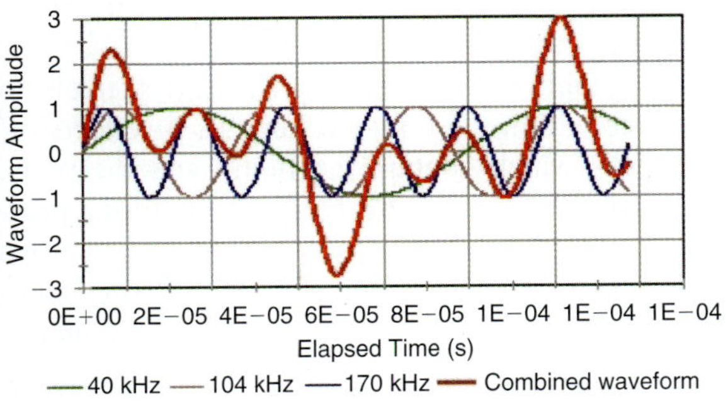


**Figure 7.38**

The chief drawback to the technology in Table 7.9 is that treatment cycle time may be greatly lengthened if only a single frequency is optimum for removing most soil components and the additional frequencies are only optimum for minor soil components. Here a second treatment tank might be more appropriate.

As mentioned in Section 7.10.2.2, a trial with actual soiled parts should decide the issue. Unfortunately, multiple suppliers are likely to be involved because single suppliers are likely to offer only one approach. One outcome is certain: each supplier will have performance data from its own laboratory, or from a customer's site, showing that their approach can be quite successful in cleaning parts. However, the outcome will apply to only the application tested.[85]

#### 7.10.2.4 *Are Multiple Frequencies of Value to Managers, or Just an Option?*[86]

It's always easier for a manager to decide what they *could* do. It's significantly more difficult to decide what they *should* do.

Systems capable of providing multiple frequencies are more expensive to buy than those providing a single frequency of pressure waves (size and amount of cavitation bubbles). The premium varies with the application, but an increase of 2–3 times above the cost of a single frequency isn't unusual:

- Can multiple frequencies enable superior cleaning results? Yes.
- Can multiple frequencies enable cleaning results which couldn't be obtained via any other method? Yes,[87] but probably not in general. Additional other

[82] To some extent, as described in US Patents 6,019,852 and 5,865,199, this interference can be overcome by adjusting the spacing between transducers to be at least a certain amount.

[83] The period of application of a single frequency may be only seconds, and the period when no frequency is applied is typically a small fraction of a second.

[84] One firm has claimed, via US Patent 6,313,565; US Patent 6,462,461; US Patent 6,538,360; and US Patent 6,822,372, that seven different frequencies can be separately applied to soiled parts using a single transducer.

[85] Don't expect comparative data. For competitive reasons, such studies haven't been and likely won't be done.

[86] The same question can be asked about "designer waveforms." Development of custom ceramic materials and special electrical circuits has allowed suppliers to further differentiate their offerings from commodities. Instead of pressure fluctuations being implemented in a sinusoidal fashion over time, the pressure waveform versus time can be a square wave, one with variable amplitude over some period, one with variable frequency over some period, one with a monotonic change in frequency over some period, one with modulation of both frequency and amplitude over same period, one with significant periods of dead time over some period, or whatever else can be imagined. To both "designer waveforms" and multiple frequencies should be directed the same level of scrutiny about value received.

[87] Certainly, every supplier will be able to present case histories where this has been so.

facilities, or additional cycle time or additional labor, can often enable improvements in cleaning performance, but at an additional cost as well.

So the answer to the opening question for managers resolves, as questions always do, to what level of cleaning performance is necessary to meet downstream requirements, and what is the value of doing so? This consultant's belief is that the cost increment to provide multiple frequencies will be justified in only a small minority of applications. But yours may be one!

#### 7.10.2.5 *In the Limit*

Removing specific particles smaller than about 1 μm can only be done via application of mechanical forces which can penetrate the boundary layer.[88]

The similarity of the relationship between calculated boundary layer thickness and ultrasonic frequency, and the same for calculated size of cavitation bubbles, is shown in Figure 7.39.[89]

Please note the major difference in physical size between the thin boundary layer and the much larger size of cavitation bubbles. Cavitation bubbles are 10–30 times larger than the aperture into which they must fit (the boundary layer[90]). But that ratio declines at higher frequencies.

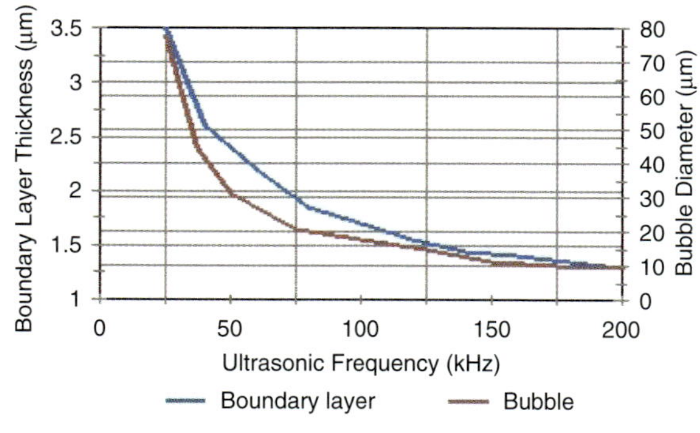


**Figure 7.39**

There is a point of diminishing return:

- Increase of frequency produces smaller sizes of cavitation bubbles, but each bubble releases only lesser amounts of energy when collapsed.
- Increase of frequency does allow some access to smaller particles hiding in boundary layers adjacent to part surfaces, but the outcome is not completely satisfactory.

Use of cavitation bubbles generated by high-frequency ultrasonic transducers to remove sub-micron particles might be analogous to trying to pocket pool balls with beach balls, mow grass with hand grenades, kill mosquitoes with hammers, or whatever. The point of these extravagant analogies is that one could remove some sub-micron-sized particles, but not all, and there would be serious concern about damage to the underlying surface.

#### 7.10.2.6 *A New Frequency Sweeps Clean*

Selection of a transducer which radiates pressure waves into fluid and onto part surfaces at a selected, constant, and fixed frequency may solve cleaning problems (as above), but also create concern about part integrity.

Any single wave frequency can and is likely to resonate within the liquid volume as it reflects off the walls which contain the liquid and the parts. Resonance is the term for combination of the pressure amplitudes which occur at the constant wave frequency.[91] Here, as in Figure 7.38, pressure values (amplitudes) can combine if the wave frequency doesn't change.

This isn't bad, if there isn't some threshold pressure which can harm the parts. But delicate parts will fracture when excited into resonance. This outcome was catastrophic for those removing particles from fragile parts such as those used in disk drives.

The solution developed was to force the transducer frequency to vary over a small range by changing the

---

[88] See Chapter 6, Section 6.6.2.1 for a discussion of fluid boundary layers. If force can't reach a particle, the particle can't be consistently and uniformly removed.

[89] Please note that this figure involves two vertical axes. The information about bubble size is the same as plotted in Figure 7.36.

[90] See Chapter 6, Section 6.6.2.1.

[91] Resonance occurs when a chamber will hold an integer number of pressure wavelengths. Since these waves propagate at high frequency, their wavelengths are very short. So a chamber of any size larger than one holding a few drops of liquid will effectively hold an integer number of pressure wavelengths. In other words, all such pressure waves will resonate and amplify themselves.

frequency of the alternating current supplied to the piezoelectric crystal. This prevented wave resonance and application of unwanted high-pressure forces to fragile parts.[92]

Deliberate variation of frequency around a central value is known as sweep. The amount is usually 1 or 2 or 3 kHz for a transducer designed to produce pressure fluctuations at 40 kHz.[93] This capability is now a standard feature of nearly all commercial ultrasonic transducer systems – whether to be used with fragile disk drive components or used with sturdy drive gears.[94]

### 7.10.3 Power to the Parts

It is a human characteristic to believe "more is better." This characteristic is reflected in the financial advice, "Bears make money, bulls make money, and hogs get slaughtered."[95] Another example of this characteristic is the choice by many users of ever-larger power ratings for sonic-powered transducer systems.

There are at least three factors to be considered by a manager when choosing the power level for the ultrasonic transducers in a cleaning system. The factors are parts, cycle time, and tank size.

#### 7.10.3.1 *Effect of Parts*

A *generalized* relationship between cleaning effectiveness and power for a properly designed system is illustrated in Figure 7.40.[96] Note that the relationship is "S-shaped":

- Modest application of ultrasonic power has only minor effects. This is because an adequate number of cavitation bubbles of sufficient size hasn't been produced.
- At some level of applied power, the ultrasonic cleaning system performs well, as designed.
- When a high level of cleaning performance has been achieved, there is little gain by applying additional ultrasonic power. In this situation, if removal of the small levels of remaining soil is necessary, a secondary cleaning process should be employed rather than try to force this process to perform beyond its capability.

In summary, without regard to the character of the parts, there is a suitable range of power levels. There is no point to paying for more power and no point to trying to economize by paying for less. For the fictitious situation described in Figure 7.40, that amount is 1000 W.

### 7.10.4 Effect of Cycle Time

Cycle time (contact time with ultrasonic agitation) should be viewed similarly. Cleaning quality will have the same general ("S-shaped") relationship[97] versus time as seen in Figure 7.40. Parts just "dipped" into the ultrasonic tank will not be cleaned. Parts cooked as some like their steak to be well done will not be cleaned to a premium level.

Doubling the cycle time will not double the cleaning quality. For a properly designed cleaning system, if the production rate is raised and the associated cycle time shortened, cleaning quality will suffer.

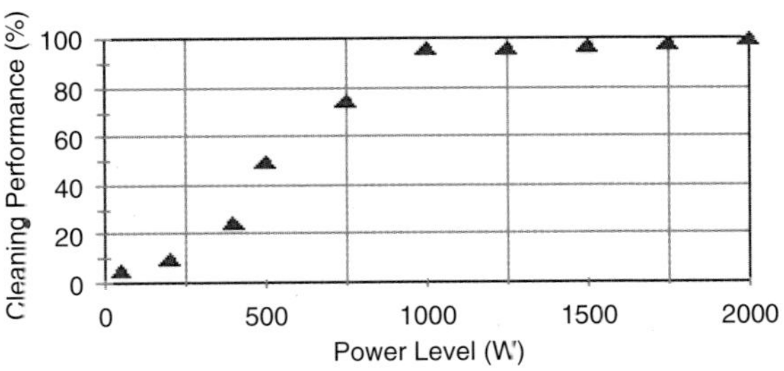


**Figure 7.40**

---

[92] As expected, this change in frequency also changes the expected bubble size. So the sweep capability enables a narrow distribution of cavitation bubble sizes.

[93] Extent of sweep is a function of frequency. At 132 kHz the sweep frequency might be 6 or 8 kHz and at 170 kHz it might be 8 or 10 kHz.

[94] Many commercial ultrasonic systems will also continuously vary the amount of sweep around the central transducer frequency node. This is known as "sweeping the sweep."

[95] James J. Cramer.

[96] This figure represents that operation has occurred for a constant period of time in a certain cleaning tank for each *fictitious* data point.

[97] For the same cleaning tank size, ultrasonic power loading, applied chemistry, etc.

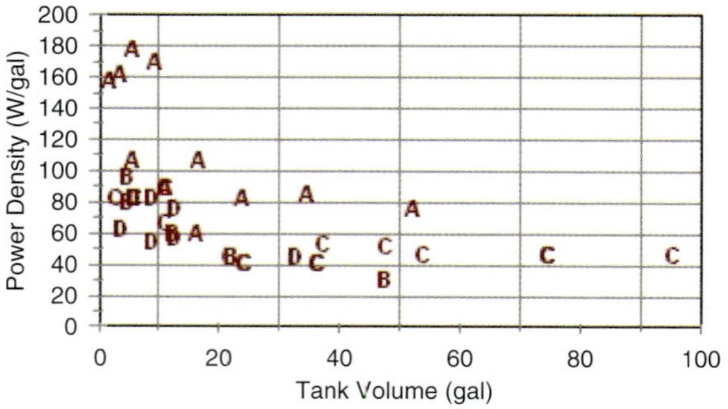


**Figure 7.41**

As benchmarks, a cycle time of 2 min contact would be quite short, but perhaps satisfactory. A cycle time of 10 min would be quite long, but perhaps necessary.

### 7.10.5 Effect of Tank Size

Size – (fluid volume) of the tank in which the cleaning work is being done – matters. *Less power is used in tanks with smaller volumes.*

Ultrasonic power level is normally specified as a density, power per volume. Specifications for standard systems produced by four major US suppliers have been collected. The suppliers are identified only as "A," "B," "C," and "D."

The power density provided in standard systems is graphed in Figure 7.41.[98] Please note that these values are standard, provided without any definition of the parts being cleaned or of the cycle time.

Please recall that in Figure 7.41 supplier "A" is not necessarily providing superior cleaning systems because their systems have a higher power density. But for a similar price, this author would prefer small cleaning tanks provided by Supplier A rather than these provided by Supplier B.

### 7.10.6 Effect of Part Size

Ultimately, all mechanical energy added to a cleaning or rinsing tank by ultrasonic transducers is converted to heat. The mechanical energy is consumed in doing frictional work – either against the mass load of parts, against the walls of the tank, within the water, or as heat and additional frictional forces produced by the collapse of cavitation bubbles.

Consequently, if the parts are a large, dense mass of metal, more ultrasonic power will be required to compensate for that absorbed by the metal. If parts are left over-long within an ultrasonic-powered cleaning tank, they and the fluid within the tank and the tank walls will become warm.

And, if the parts occupy a large amount of the volume within a tank, it is likely that internal surfaces may not be effectively cleaned.

Some suppliers recommend that the weight of parts in a ultrasonic cleaning tank be no more than about one-third to one-half of the weight of water in the tank. This author's experience favors the lower value.

Such a recommendation doesn't mean that more large systems be purchased; it may only mean that multiple loads be processed in a smaller- and lower-cost machine.

### 7.10.7 Test Test Test

A manager's objective, in every demonstration with a supplier's ultrasonic (or megasonic) facilities, should be to identify the power level and the cycle time which should be used to design a commercial system:

- Excess power has negligible value. A good manager should not pay for that.
- Excess cycle time is a waste of productivity. A manager should not stand for that.

All the generalized relationships and specific recommendations above matter not at all relative to actual performance data.

### 7.10.8 Replication Can Be Hard to Reproduce

Performance of sonic-powered cleaning system, *for a given set of parts*, is related to much more than the choice of frequency and sweep rate, tank size, and power level. Chemicals, and their concentration, used in the operation can affect performance.

---

[98] Some suppliers use the rules that ultrasonic power level should be around 100 W/gal for tanks less than 20 gal, and around 50 W/gal for larger tanks. This is consistent with the apparent asymptotic relationship displayed in Figure 7.40.

But there are other factors which can be significant, or not, which are not so obvious. Some observed by this author are:

- Tank configuration – depth[99] versus open area.[100]
- Tank configuration – presence of unusual shapes where waves aren't reflected back onto parts.
- Positioning (racking) of parts within the open volume of a tank.
- Location of transducers within a tank.
- Operating temperature.
- Residual gas (air[101]) content.
- Water quality.[102]
- Smoothness of the part surface.[103]
- Fluid circulation[104] within the tank.
- Waveform of the ultrasonic-produced pressure pulses.
- Anything present on the part surface which would prevent it from being wetted (and submerged).
- Accumulation of debris within the tank.

It isn't that ultrasonic cleaning in static tanks isn't reproducible. It very often can be and is so. *Ultrasonic cleaning is reliable very often.*

But specific results (claims by single vendors of superior performance in unique applications) can often be difficult to reproduce in ultrasonic systems provided by other vendors.

In other words, if a supplier can back up a claim with repeatable performance data with your parts, a manager should give great priority to that supplier in the selection process.[105] This recommendation is consistent with the recommended approach for vendor selection in Chapter 6, Section 6.8.

### 7.10.9 Sonic Cleaning without Cavitation

One can extend to high frequencies the trends displayed in Figure 7.39 and attempt to predict at what high level of frequency cavitation bubbles will be sized small enough to fit within a boundary layer, so they can be used to dislodge specific sub-micron particles from surfaces.

That's the problem of extrapolation beyond data, or of extrapolation from a regime in which one mechanism dominates to another regime controlled by a different mechanism.

While the empirical evidence conflicts about identification of the exact ultrasonic frequency at which cavitation bubbles aren't produced by pressure waves from ultrasonic transducers, there is little question that this is true at some upper frequency.[106,107]

The reason is that there is inadequate time between compression and refraction stages for sufficient local heat and mass transfer to occur so that a bubble can be produced.[108,109]

---

[99] Please remember that the top fluid surface of an ultrasonic cleaning tank reflects pressure waves back into the tank at least as well as does a metal wall.

[100] Most ultrasonic cleaning tanks are shaped so that their length to depth ratio is around 3/2, and their ultrasonic power to open area ratio is around 3½ watts/in$^2$. The information in Figure 1.41 and Table 1.5 reflects this basis.

[101] Cavitation bubbles are not composed of air; they are composed of vaporized fluid. The rarefaction stage of a pressure wave doesn't produce air vapor, it's already in that phase.

[102] Exceptionally pure water will have fewer imperfections (suspended solids, etc.) and thus fewer sites for nucleation of cavitation bubbles. Further, a significant number of particles can cause a sound wave to be scattered or reflected (dissipated).

[103] An exceptionally smooth surface will have fewer pockets of surface roughness which can be nuclei (sites) for growth of cavitation bubbles.

[104] Please remember that fluid circulation is simply another set of pressure waves, though of a much greater magnitude and much lower frequency.

[105] A corollary to this approach is that the operating conditions in test equipment used successfully in a demonstration test should be reproduced as closely as possible in use of that equipment after purchase. In other words, manage the purchased system as was the test done.

[106] Busnania, A.A. and Gale, G., "Ultrasonic and Megasonic Particle Removal," *Precision Cleaning Symposium*, # 247, 1995. This paper comments that cavitation has been observed not to exist at frequencies around 360 kHz and above (Figure 7.39). Current thinking is that the demarcation between ultrasonic (cavitation-based) and megasonic (based on fluid streaming) occurs around 250 kHz. This differentiation only matters if one is a sub-micron particle or a fragile surface.

[107] Esche, R., *Acustica*, 1952, Vol. 2, p. 208.

[108] Schwartzman, S., Mayer, A. and Kern, W., *RCA Review*, 1985, Vol. 46, p. 81. Pioneering data presented here showed that there was inadequate time for bubble formation at 850 kHz.

[109] Mayer, A. and Schwartzman, S., *Journal of Electronic Materials*, 1979, Vol. 8, p. 855.

What's produced is a local pressure fluctuation (called streaming), without a phase change.

In other words, one can't produce with ultrasonic transducers cavitation bubbles that are sized small enough to fit within a local fluid boundary layer.

### 7.10.10 Megasonic Operations[110]

Megasonic cleaning is that done with high-frequency pressure waves, where cavitation is not involved.

Application of megasonic force by discontinuous (and continuous) fluid movement (streaming) is very suitable for selected applications:

- Where prevention of failure of parts due to mechanical damage is critical.
- Where sub-micron sized[111] particles are found within the boundary layer adjacent to surfaces.[112]
- Where rinsing of delicate parts is required (versus cleaning).[113]
- Where the transducers can be aimed at the entire area to be treated.[114] This is more obvious when Figure 7.42[115] is examined.

Exposure time and level of applied megasonic power are the most significant variables affecting megasonic cleaning.

While it might be expected that additional exposure time (cycle time) aids particle removal by megasonic-enabled fluid action, that is not always so:

- As opposed to the asymptotic behavior illustrated in Figure 7.39, sub-micron sized particles are too small to settle or be easily collected. Thus cleaning performance may worsen with increased cycle time as cleaned surfaces are made dirty by redeposition of previously liberated particles.

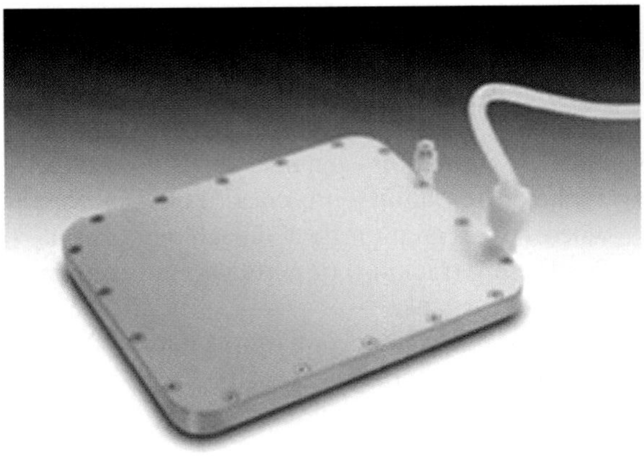

**Figure 7.42**

Megasonic technology is not "opposite" to ultrasonic technology employing megasonic transducers. It is different. Both involve pressure waves. But ultrasonic technology (no matter at what intensity of power or frequency) involves production and collapse of bubbles. Megasonic technology involves production of local turbulent eddies, and no bubbles.

### 7.10.11 Transducers Aren't in Boxes Anymore

When Norman Branson[116] constructed ultrasonic transducers more than two generations ago, they were rectangularly shaped as in Figures 7.35 and 7.43.[117]

These are the type most commonly used in metal cleaning operations, found in commercial aqueous cleaning machines. Some suppliers provide transducers only in this configuration.

These transducers radiate pressure waves from a flat surface as a moving curtain or a flat front. This is schematically shown in Figure 7.44 (derived from Figure 7.31).

[110] See Chapter 6, Section 6.6.2.1 for an expanded discussion of the fluid dynamic differences between ultrasonic and megasonic cleaning technology, especially as both relate to particle removal.

[111] Particles whose size would be identified by having a characteristic dimension measured in nanometers (nanoparticles) are not likely to be removed by megasonic action. This is because the mechanical force required is larger than can be provided by fluid streaming forces. Further, it is often necessary to know the specific location of these particles to accomplish their removal.

[112] Applications include cleaning of Silicon waters and substructures, laser optics, and super conductive tape.

[113] With rinsing, fluid displacement is more significant than application of fluid force.

[114] Ultrasonic transducers apply force (pressure waves) in an omnidirectional pattern. Megasonic pressure waves are applied in the direction faced by the megasonic transducer.

[115] Figure 7.42 is courtesy of ProSys.

[116] In 1946, Norman Branson helped to develop the "Audigage," an ultrasonic thickness-gaging instrument that utilized ultrasonic resonance techniques to measure workpiece thickness from one side. Later, a company he founded produced ultrasonic transducers for industrial and precision cleaning applications.

[117] In Figure 7.43, the radiating surface faces up. Image courtesy of Blackstone Ultrasonics.

Figure 7.43

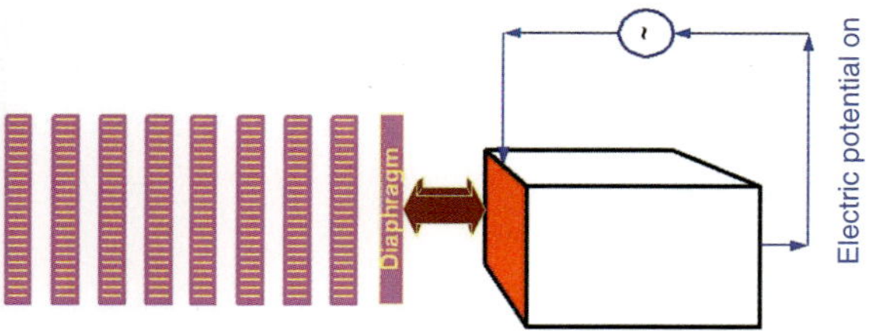


Figure 7.44

These pressure waves will strike other surfaces (container walls, part surfaces, or air–fluid interfaces), reflect,[118] and continue to strike other surfaces.

### 7.10.11.1 *Two Types of Radial Transducers*

However, pressure waves can be organized to radiate in radial directions, versus the horizontal direction of Figure 7.44.

Ultrasonic transducers, whether energized via magnetostrictive or piezoelectric crystals, can take other shapes. Some have been developed to provide improved performance or reliability. Others have been developed to enable completion of unusual applications.

Figure 7.45

Figure 7.46

There are two types. One directly produces radial pressure waves. The other indirectly produces radial pressure waves.

7.10.11.1.1 *Direct Radial Transducers* A radial transducer[119] of the direct type is shown schematically radiating outward in Figure 7.45.

Here the active material is formed as a cylinder. Both types of transducer materials can be used (see Table 7.6) as elements arrayed radially around the circumference of a cylinder. In this case, the "tank" is the fluid contained within the channel whose walls are the radial transducer. Internal diameter is around 3 in and lengths of each cylinder are around 1 ft, though obviously multiple units can be arranged in series.

One example of this development is the cylindrical-shaped transducer shown in Figure 7.46 (40 kHz, piezoelectric). Another is shown in Figure 7.47 (20 kHz, magnetostrictive). Both are made by the same manufacturer, whose name was mentioned in Section 7.10.11.

[118] The angle of reflection is twice the complementary angle.

[119] The new products are often called resonators (versus transducers). Because they do convert electrical energy to mechanical energy (repeated motion), even though they may produce continuous wavefronts and be said to resonate, they are functionally transducers.

Figure 7.47

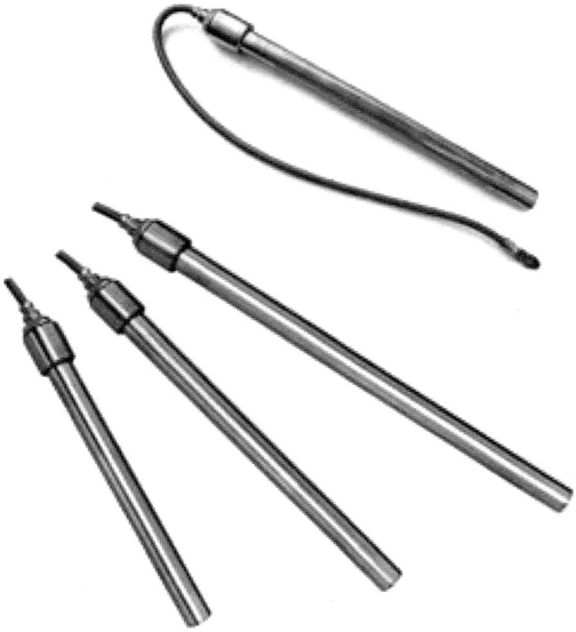

Figure 7.48

Applications[120] include removal of drawing soaps and lubricants from drawn wire, extruded metal forms, or cable; removal of metal oxide or scale following heat treat operations; conditioning of *continuous*-filament woven products; and cleaning of metal strip.

Note the word *continuous*:

- Flat transducers are nearly always used in batch processes, tanks without continuous work flow. These direct radial transducers are used in *continuous* working involving generation of cavitation bubbles. Line speeds for wire of up to 100 ft/min have been claimed for multiple transducer systems.

7.10.11.1.2 *Indirect Radial Transducers (Tube Resonators)* Indirect radiating transducers are referred to as a tube resonators. They are transducers, converting electrical energy into mechanical energy. However, the horizontal or linear mechanical motion is secondarily and indirectly converted into radial motion.

Tube resonators are assembled of three parts – a metal tube with small piezoelectric transducers mounted at each end. Distance of separation between the transducers can be from around 6–24 in.[121] Various tube resonators are shown in Figures 7.48[122] and 7.49.[123]

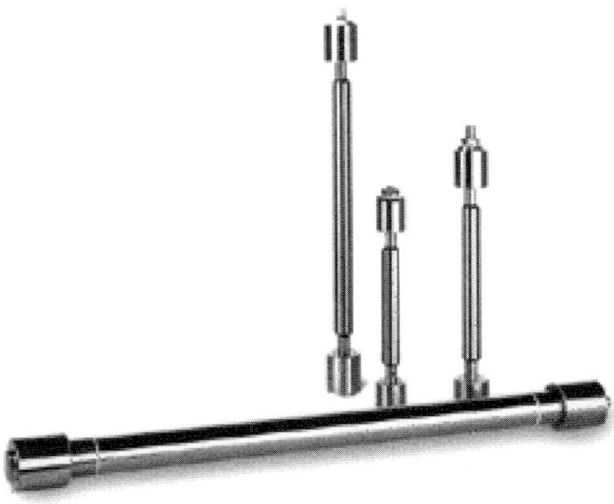

Figure 7.49

Each transducer moves horizontally. The horizontal movement is timed so that one transducer "fires" while the other is temporarily dormant.[124] So, the tube moves a tiny distance in one direction. Then the other transducer "fires," causing the tube

[120] This author has tested continuous operation with small parts conveyed in a moving stream of water. The part-laden fluid continuously flows through a channel which is the bore of a continuously radiating transducer similar to that in Figure 7.46. Power requirements can be huge – several thousand watts.

[121] The tube is not randomly chosen. It is a length which is an integral multiple of ½ of a predetermined wavelength for vibrations.

[122] Image courtesy of Martin Walter (Crest Ultrasonics).

[123] Image courtesy of Telsonic.

[124] Motion may be adapted to operate in phase or in phase opposition.

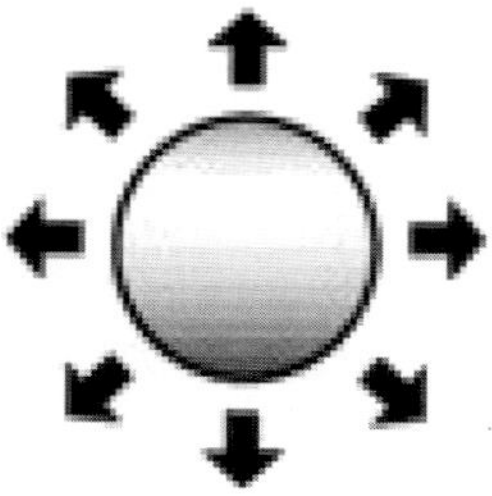

Figure 7.50

to move a tiny distance in the other direction:

- The direct net result is that the tube assumes a reciprocating motion along its axis.
- The indirect result is that this reciprocating motion forces (pulls or pushes or drags) fluid away from the surface of the tube – first in one direction, and then the reverse.[125]

Such indirect movement creates pressure waves whose focus is centered at the center of the tube's length. The tube can be hollow[126] or solid,[127] and there are advantages claimed for each.[128]

Please note that there is no free end of this transducer (resonator) system from which pressure waves can be radiated. All waves are radiated radially (see Figure 7.50).[129]

Applications are unique, derived from the shape of the resonator. Managers can apply cavitation energy in small diameter chambers (pipes) or through small diameter openings (ports in tanks). Applications have been to: clean filters while in-line, clean tanks over extended periods of time,[130] provide fluid agitation (emulsifying and dispersing) in pipelines, and aeration treatment of sewage sludge.

This author's experience has been that these outward-radiating radial transducers can provide a more uniform distribution of cavitation energy than do flat transducers when the applications involve small parts in small tanks.[131]

### 7.10.12 Ultrasonic Transducer Systems in Cleaning Machines

A manager probably won't be allowed to choose the characteristics and features (quality) of the ultrasonic transducers provided in a purchased cleaning machine. A business relationship between the manufacturer of the cleaning machine and of the transducers will define those limitations.

As in previous sections, the quality of a cleaning machine with ultrasonic transducers can be inferred from the quality of that component. Recommendations are given in Table 7.10.[132]

---

[125] This could not happen were there is no frictional forces between the tube surface and the fluid. In other words, these resonators would not provide radial pressure waves in a medium such as liquid $CO_2$ (or air), which has negligible viscosity.

[126] US Patent 4,537,511, *Apparatus for Generating and Radiating Ultrasonic Energy*, August 27, 1985. Assignee is Telsonic AG Für Elektronische Entwicklung Und Fabrikation. The hollow tube can be possibly supplied with a fluid. In this arrangement, irradiation occurs inwardly, which results in very high acoustic intensities, due to the focusing effect. Shapes can be round, square, or multi-sided. Use of both piezoelectric and magnetostrictive elements are claimed.

[127] US Patent 5,200,666, *Ultrasonic Transducer*, April 6, 1993. Assignee is Martin Walter Ultraschalltechnik G.m.b.H (Crest Ultrasonics).

[128] A solid resonator has the advantage of greater durability since it is not subjected so much to cavitational erosion as a hollow bodied resonator is. On the other hand, a hollow resonator provides for greater vibration amplitudes and is therefore somewhat more effective than a solid resonator (see Footnote 119).

[129] Figure 7.50 is courtesy of Martin Walter Ultraschalltechnik G.m.b.H (Crest Ultrasonics).

[130] The US Department of Energy's NICE (National Industrial Competitiveness through Energy, Environment, and Economics) program has supported development as a way of reducing production of waste cleaning chemicals. See http://www.eere.energy.gov/industry/chemicals/pdfs/dupontmerck.pdf

[131] A common problem with all ultrasonic transducers used in liquid cleaning applications is known by acoustic engineers as impedance mismatch. Chemical engineers, as this author is professionally registered, would describe this problem as where the vibrating transducer surface (made of metal) produces more kinetic energy of motion than the cleaning bath (liquid) can absorb.

This means that the product of density times velocity$^2$ is different between solid and liquid by around a factor of around 17–50 for radiating transducers made of Aluminum and stainless steel (respectively) immersed in water. Between cavitation bubbles and liquid water, the mismatch is even more extreme.

One strategy to incorporate more kinetic energy of motion into aqueous cleaning baths is to use more radiating surface (see Figure 7.35). A way to implement this strategy is to use tube resonators from which motion is applied to the liquid over nearly all the surface of the resonator tube.

[132] See Table 7.6 about selection of transducer material, Section 7.10.3 about selection of power level, Section 7.10.2 about frequency, and Table 7.8 about use of harmonics. All issues should be determined based on the details of individual applications.

**Table 7.10 Selection of Ultrasonic Transducers for Cleaning Machines**

| Item | Good | Better | Best |
|---|---|---|---|
| **Piezoelectric transducer** | | | |
| Material of construction | Metal | Metal oxide | Ceramic |
| **Transducer body** | | | |
| Material of construction | Stainless steel (s/s) | 16 gal, 316 s/s | 14 gal, 316 L s/s (to avoid stress corrosion cracking) |
| Radiating surface | Material of construction | | Hard chrome plated |
| Welding method | Inert gas-welded, with fillet-welded lap joints | | |
| **Power generator** | | | |
| Manufacturer's warranty[133] | 1 year | 2 years | 3 years |
| Sweep control | Fixed frequency and rate | Variable frequency range and fixed rate | Variable frequency range and rate |
| Power generator | Meets FCC Rules and Regulations Conforms to the EN 60-204-1 Standards | | |
| | Single frequency | Operable at multiple frequencies | Programmable to operate at multiple frequencies |
| | Manual power amplitude control | Variable power amplitude control | Programmable power amplitude control |

## 7.11 EQUIPMENT USED IN RINSING

The same equipment components used in construction of cleaning stages (see Sections 7.1–7.4 and 7.7) are also used in construction of rinsing stages (see Chapter 1, Section 1.12).

However, the components (pumps, tanks, nozzles, heaters, etc.) are selected and arranged to meet the different needs of either stage of work (see Table 7.11).

One can combine sugars, flours, eggs, dairy products, and salts to produce different culinary confections: cakes, pastas, breads, cookies, or baked items containing fillings. A different recipe is used with different techniques in each case.

One can also combine pumps, tanks, heaters, and nozzles to do either cleaning or rinsing work. Each combination involves a different design. The design for a good, better, or best rinsing system is Equation (1.1), the decision associated with Table 1.15, the allocation of cycle time to removal dragout described in Chapter 1, Section 1.12.5, and the "Central Rinsing Theorem" of Chapter 1, Section 1.12.6.

The tables referenced in Table 7.11 should be used as the "good, better, best" recommendation for the components to implement a design for either cleaning or rinsing.

A cleaning machine in which the same pumps, nozzles, and tanks are used for both cleaning and rinsing operations is most common. It may be cheaper for the manufacturer to construct and for the user to maintain. But it is likely not to be produce the best cleaning *and* rinsing performance.

### 7.11.1 Divers Do It Deeper

Experience as a certified scuba diver provides a perspective.

After an ocean dive, equipment is always thoroughly flushed to remove residual salt. Since the salt is removed via solutioning and dilution with

[133] Magnetostrictive transducers should be selected based on support, service, *and* length of warranty.

**Table 7.11 Comparison of Components Used for Better Performance in Rinsing vs Cleaning Operations**

| Component | Rinsing | Cleaning |
|---|---|---|
| General need | Dilution with fresh fluid of soil-rich fluid materials remaining on the part surfaces | Application of mechanical force (aqueous technology) or holdup for solution (solvent technology) |
| Pumps (see Table 7.2) | Higher volume, low pressure | Low volume, higher pressure |
| Nozzles (see Table 7.1) | Fan or flood patterns are commonly used to spread fluid over the entire surface | Flow is targeted in pattern with narrow aperture for maximum fluid impact (force) |
| Tanks (see Table 7.3) | Larger, because complete mixing of soil components within clean fluid is required and the presence of parts inhibits complete circulation of fluid | Smaller, because there is no need to homogenize the multiple components and phases present |
| Heaters (see Table 7.5) | Probably used[134] | Commonly used |
| Parts baskets (see Section 7.7) | The same basket is used in both operations | |

fresh water, this operation is analogous more to rinsing[135] than to parts cleaning.

In this author's experience of having logged more than 300 dives in several countries, equipment is always rinsed by agitated *immersion* in preference to spraying, assuming a tank of water can be made available.[136]

The reason is simple: immersion provides a longer and more thorough contact than does fluid spray:[137]

- Immersion rinsing contacts dive equipment with at least gallons of somewhat-salted water for a period which can (and should) easily span several minutes.[138] Some dilution and solution can be accomplished with that volume of water, and that time.
- Spray rinsing contacts surfaces of dive equipment with at most a few ounces of fresh water during momentary contact. That's not enough water, or time, to accomplish significant dilution or solution (see Chapter 1, Section 1.12.1).

Further, the consequences of failure are different:

- Inadequate spray rinsing can leave an unwetted salt crystal in an "O" ring gland.
- Inadequate immersion rinsing can leave a partially solubilized (and presumably smaller) salt crystal in the same gland.

As a diver, this author knows which failure he would be more willing to accept.

### 7.11.2 An Optimum Washing/Rinsing Process for Aqueous Technology

Recreational diving experience leads to another perspective.

[134] Rinse tanks may be heated to: (1) accelerate the drying process, (2) avoid foaming of rinse, brightener, or rust-preventive chemicals, and (3) allow for continued cleaning (see Table 1.9). Specifically, rinse tanks are usually heated: (1) in plating or other operations where the parts are not to be dried of water, (2) where the succeeding step is done at ambient temperature, or (3) when energy conservation is paramount.

[135] This includes metal components and elastomeric components, as well as the recesses into which they fit, of a regulator system or underwater camera, elastomeric fabric from which a buoyancy compensation jacket is made, and personal gear made of plastic. Consequences of poor rinsing range from fatal in the case of a seal between a pressure hose and an air cylinder, to expensive in the case of a camera housing, to unpleasant in the case of personal gear.

[136] When a tank of water isn't available, equipment is sprayed with water before drying, but good practice is to re-wet the equipment via immersion prior to next use or storage. A common wash tank on the dive boat, containing salt from the equipment of many divers, is preferred over a spray rinse with fresh water.

[137] A past client who manufactured devices for insertion into human tissue insisted upon a spray rinsing process because it would conserve floorspace and best use existing facilities. This client, subsequently, was subjected to lawsuits for selling contaminated goods, and the goods had to be recalled.

[138] The manual for one underwater housing owned by this author speaks to soaking the housing for several minutes before opening to remove the camera. Spraying is not mentioned.

For many applications, the best selection of equipment components for maximum cleaning/rinsing performance (though not necessarily cost or floorspace) is to use *spray components for aqueous cleaning operations and immersion components for rinsing operations*:

- High-pressure spray application of aqueous cleaning agents would maximize implementation of mechanical force which is a crucial component of aqueous cleaning technology (see Chapter 1, Section 1.4).
- Dilution of retained soil components by immersion would maximize surface quality from the combined operation.

This author has never seen a commercial aqueous[139] cleaning machine with that configuration, though numerous user-constructed facilities have been so organized in applications where the consequence of soil retention was critical.

## 7.12 EQUIPMENT USED IN DRYING

See Chapter 1, Section 1.13 for general information about the process of parts drying.

Managers devote scrutiny to the facilities in a cleaning machine associated with cleaning and rinsing of parts. Facilities associated with drying generally gather little interest. But the value provided by a cleaning machine is not fully realized unless all three unit operations are successfully completed.

This section will cover facilities for two general different methods of drying parts, evaporative and non-evaporative. Two different types of equipment will be described for each method.[140]

### 7.12.1 Air Knives

To paraphrase the title of a favorite novel by Raymond Chandler, this is "The Big Knockoff."[141]

**Figure 7.51**

This is blast cleaning (see Chapter 6, Section 6.1.3) using high-velocity air to remove liquid cleaning agent, instead of solid media, to remove soil materials.

The air knife produces a thin curtain of concentrated violence over a target range of about 3 in and the width of the air knife.

An encounter for a fractional portion of a second is sufficient to locate, dislodge, and drive essentially all water droplets or films from a surface.

This incident is shown dramatically in Figure 7.51.[142] Air from the circular chamber is forced through a thin aperture (the knife edge). Directed toward parts on a conveyor belt, it wrecks havoc on liquid retained on any parts it contacts.

Drying with air knives can be extraordinarily effective. This author has successfully developed parts drying systems based on air knives for a broad variety of part shapes. Some guidelines for evaluating the quality of cleaning machines which use air knives are:

- Only one[143] well-aimed air knife should be needed and used. If multiple air knives are provided, it should be clear that there is significant separation between them (at least 12 in) and that the outfall from one does not rebound upon contact with a surface and recontact the parts (see below).

[139] A solvent cleaning machine would not be so organized because both the cleaning and rinsing operations depend upon immersion for success.

[140] Durkee, J.B., "New Process Developments in Replacement Cleaning Systems," Presented at the *International CFC and Halons Conference*, Washington, DC, October 25, 1995.

[141] The reference is to the novel *The Big Sleep*, which may have been the best Bogart/Bacall movie.

[142] Figure 7.51 is courtesy of Air Blast Corporation. This image was chosen from many because it depicts the local violence produced by an air knife. Please note the deflection of the conveyor belt. Air knives do remove and relocate pieces of water; they can also relocate parts and other objects.

[143] This refers to number of times air from a knife contacts the parts. Obviously, on a wide belt, multiple knives, each *ca.* 12-in wide, will be needed to cover all the belt width.

- The aperture of the air knife should be accessible so that it can be cleaned when necessary, which will be certain.
- Pressure instrumentation is necessary on the air feed line. This will allow detection of blockage[144] in the knife's aperture, when the gage pressure rises beyond normal.
- There must be a filter on the air feed line to remove particles, else the air knife will be fouled, or clean and dry parts will be infected with particles.
- Inspect for shims. These are thin strips of plastic which can be inserted in the knife aperture. They close the already thin gap (perhaps 0.010 or 0.060 in), thus increasing the linear velocity of air but simultaneously restricting the volume of flow. Shims allow customization of the drying effect an air knife produces.

Drying quality can also be totally inadequate:

- Air knives always remove water where the air stream impacts. But if the air stream doesn't impact the underside of a part drying will be incomplete.
- If the air stream is mis-aimed for whatever reason, the parts won't be dried. This is a key point. Fixturing of the air knives so they are aimed as desired is essential (see Figure 7.52), and *must be checked on a continuing basis* by the manager using these facilities. This should be included on whatever *daily* check sheet for quality control is employed by the manager (see Chapter 4, Section 4.14.1 and Appendix I).
- Outfall from use of an air knife, a hurricane of air and water, is difficult, if not impossible, to quarantine.

  If it strikes containing surfaces it will reflect from them and possibly produce a second encounter with the temporarily dry parts. This is called reinfection. Avoidance requires that air knives be used in an open area without surfaces which can reflect the air stream and produce re-wetted parts (this is also shown in Figure 7.52).
- The hurricane can also disturb and entrain debris – particles, shop dirt, fibers, etc. Consequently, parts can also be re-wetted *and* soiled! A filter in the air supply line (see above) will not prevent this.

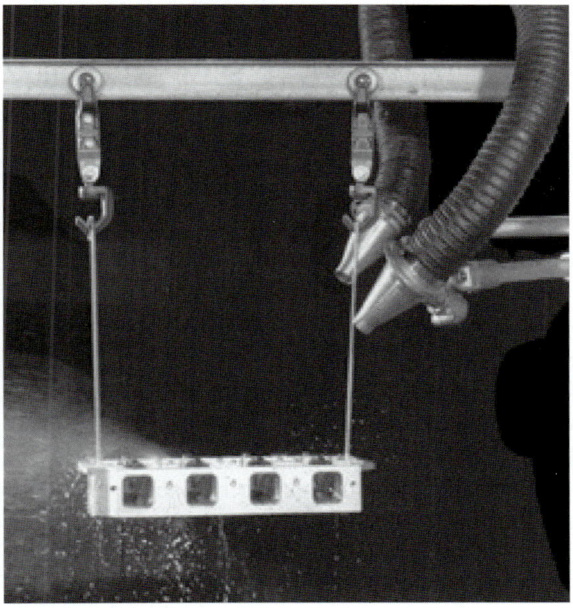

**Figure 7.52**

**Figure 7.53**

The material of construction is nearly always Aluminum alloy, although stainless steel can also be procured from many suppliers. Five hundred euro will easily buy several air knives (see Figure 7.53[145]).

The quality of dryness expected by managers who use air knives should be that described as "dry to the touch." This means that *all surfaces* of the parts feel dry. Quantitatively, this means that around 95 wt% of the moisture has been removed.

[144] A wise manager will record details about the air knife on the cleaning machine in which they have an interest. Subsequently, the air knife manufacturer should be contacted for a recommendation about air pressure and volume flow of air for optimum operation. The manager should then witness a test of this cleaning system operated at those values. Some suppliers, in an effort to reduce noise level of air knives and save cost, have reduced air supply by incorporating an undersized air compressor or a centrifugal blower. The result may not be the balance a manager desires from a cost-quality tradeoff.

[145] Figure 7.53 is courtesy of Spraying Systems, Inc.

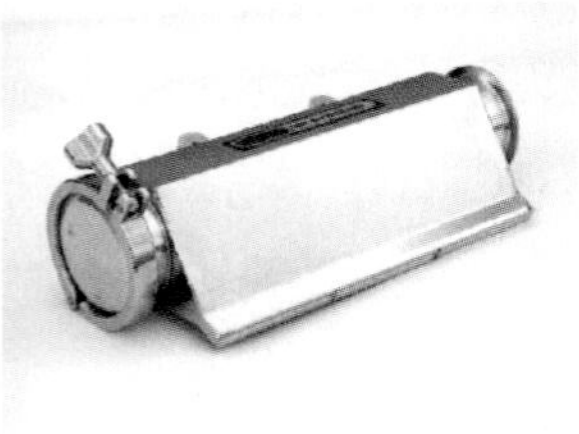

**Figure 7.54**

Air knives are not quiet. The rush of vibrating air produces sound. Levels around 90 dBA (decibels of sound amplitude) can be achieved, and are probably in violation of local, provincial, or federal regulations.[146] Even worse, if centrifugal air blowers are used, the "whine" produced in their operation can minimize concern about the sonic effect produced by air knives.

Air knives are seldom if ever used in drying of solvents. The result would be a mist or aerosol of solvent, almost certain to be a highly flammable mixture[147] (see Chapter 3, Section 3.9).

Air knives are used in far more drying applications than for drying of cleaned parts. Some applications include belt wipers, can driers on a bottling line, fruit and vegetable driers (see Figure 7.54), and for drying of plated parts.

A manager should insure that personal protective equipment always be used by those in direct exposure to the noise levels and air velocity typically produced by an air knife. This includes ear and eye protection.

There are two general methods of providing high-velocity air for an air knife: (1) centrifugal blowers and (2) air compressors.

### 7.12.2 Centrifugal Blowers

The fans (rotors) in centrifugal blowers rotate at the speed of a hard disk drive, sound like a jet airplane, and can be vulnerable to mechanical damage from a variety of sources.

They produce a high volume of air at a low level of pressure elevation, as a fan (impeller) with 30–50 vanes (pockets) rotates at speeds of several thousands of revolutions per minute.

The analogy to a hard disk drive is apt. Both are simple in concept: a disk rotates at a very high speed in a housing. Both are mechanical marvels and almost any flaw can produce failure. Both are made by many suppliers, and global price competition is keen. Both are in common use.

The two critical components of a centrifugal blower are the rotor and the bearings on which it is supported. Design and materials of construction are of concern:

- Vaned rotors (must be precision balanced to avoid vibration, which is typically fatal).
- Bearings (typically ceramic based to maintain operation at high temperatures due to frictional heating).
- Housings (typically made from cast Aluminum alloys).
- Shafts (typically made from precision-ground steel).
- Motor pulley (typically made from Nickel-plated steel).

Specification for a unit used in many cleaning machines would be:

- Up to 1000 SCFM[148] air flow at up to 80 in WC maximum pressure.[149,150]
- 7.5 HP requiring 5.5 kW and 19 amperes of three-phase 220 VAC power.
- Noise levels between 85 and 90 dBA.

The apparent tradeoff between an increase of volumetric flow rate in standard cubic feet per minute

[146] In the US occupational noise exposure is limited by the Occupational Safety and Health Administration (OSHA) as 90 dBA for 8 hours continuous exposure. See CFR 1910.95(b)(1). Please recall that the dBA scale is logarithmic, not arithmetic.

[147] This statement is written without regard to the published flash point or explosive limits of the solvent. An aerosol of solvent in air is not the condition produced in either a flash point or explosive limit test setup. The massive amount of liquid surface area exposed to air (oxygen) in an aerosol removes all limits of mass and heat transfer in limiting reaction rates. Combustion, once initiated, will cease only when a reactant (solvent or oxygen) is depleted.

[148] This volumetric flow is rated as standard cubic feet per minute, and is equivalent to ~28 $m^3$/min.

[149] This is referred to as 80-in WC (water column) pressure, and is equivalent to ~2030-mm WC, or 149 mm Hg pressure. The abbreviation WC refers to the pressure equivalent to a height of water column. This is the pressure measured inside the air knife (see Footnote 153).

[150] Both the maximum pressure and volume values cannot be achieved *simultaneously* with the motor specifications given.

(SCFM) and decrease of output pressure in inches WC is not a tradeoff at all. Increase of volumetric flow rate is actually an increase in pressure *applied to surfaces*. The pressure equivalent of volumetric air flow is given by Equation (7.1):[151,152,153]

$$\text{Pressure} = \frac{\text{Air density} \times \text{Air velocity}^2}{2 \times 32.174 \times 5.197} \qquad (7.1)$$

The performance curve shown in Figure 7.55 can be replotted using Equation (7.1), as Figure 7.56. It shows how the equivalent pressure applied to surfaces through the air knife increases with increasing volumetric flow rate.

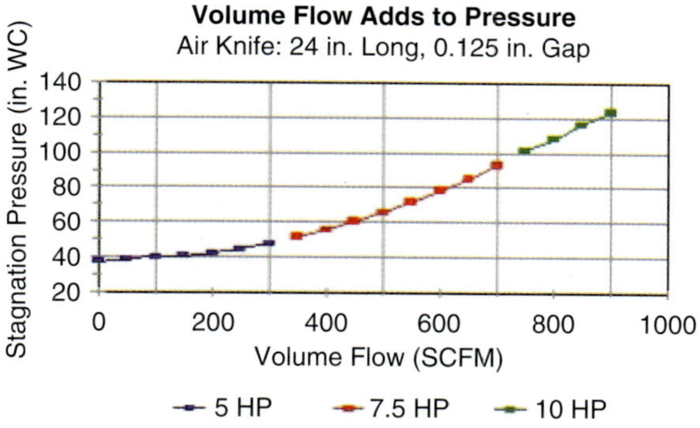


**Figure 7.56**

Please note the ~300% increase of pressure[154] applied to surfaces as velocity through the air knife is increased. Naturally, this is provided by ~200% increase in applied motor power (HP).

If a user wants *both* high velocity (pressure) and high volume flow rate, the price is a very high requirement for motor power and noise reduction facilities.

Quality is less recognized in centrifugal blowers by specifications of pressure and volumetric flow rate, and more recognized by design, workmanship, and materials of construction. These are manifested in the length of the manufacturer's warranty and the company's reputation.

A practical gage of quality is in the sound level produced at the desired level of output, and the vibration felt on the overall assembly. Centrifugal blowers are extremely noisy with a high-pitched whine. Some

Volumetric flow is dependent, for the same supplied motor power and rotational speed, upon the desired level of output pressure. Naturally, less output pressure (called backpressure) allows more volumetric flow rate, and the reverse – with an attendant increase/decrease in motor power. This is shown, for a generalized unit, in Figure 7.55, and represents the characteristic performance curve for a centrifugal blower driven at increasing rotational speed (requiring additional motor power). Three levels of power are shown.

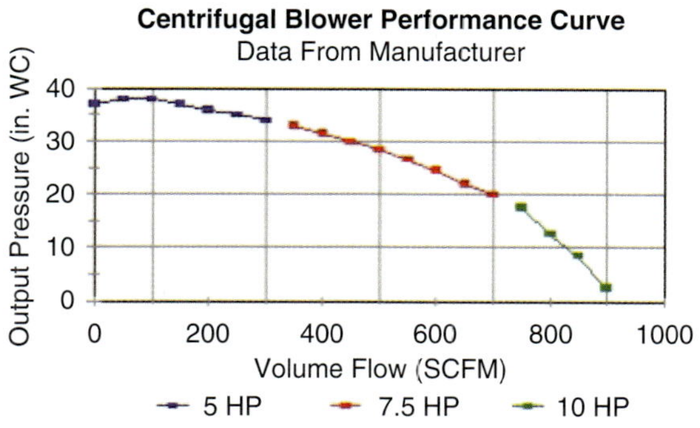


**Figure 7.55**

[151] Please note the exponent on the velocity term in Equation (7.1). The two constants in Equation (7.1) are used to convert units to a consistent set. They are 32.174 $lb_{mass}$ ft/$lb_{force}$ − $sec^2$ which is used to convert from mass to force, and 5.197 which is used to convert pressure from $lb_{force}$/SF to WC. Air density is given in $lb_m$/CF, and air velocity in ft/s.

[152] When a volumetric flow of air (in cubic feet per minute) is forced through the narrow gap of an air knife, a velocity is produced (ft/s). The relationship between volumetric flow rate and velocity is given by Equation (7.2). The value 60 converts time from minutes to seconds and the value 144 converts from inches to feet. Length and width are in inches.

$$\text{Velocity} = \frac{\text{Volumetric flow rate} \times 144}{\text{Area of air knife, length} \times \text{width} \times 60} \qquad (7.2)$$

[153] The total pressure applied to surfaces from an air knife is the sum of the discharge pressure from the centrifugal blower plus the pressure contribution of velocity calculated from Equation (7.1). The sum is also referred to as the stagnation pressure which is the equivalent pressure applied to surfaces as if the air were not moving. When high-velocity air stream exits the air knife, essentially all of this stagnation pressure is converted to velocity (via Equation (7.1)). In other words, the stagnation head of the blower is converted into the velocity head of the jet.

[154] Please consider the practical effect of applying pressures of this level to a water droplet on a surface. From Figure 7.56, at 500 SCFM, the applied pressure is equivalent to about 60-in WC. In effect, the drop wetting the surface is struck by another from a height of 5 ft above the surface. This explains the effect shown in Figures 7.51 and 7.52.

Centrifugal blower

find it unacceptable to work around them,[155] even with hearing protection.

A third differentiating factor centrifugal blowers found in various cleaning machines is local availability of a replacement machine. It should be anticipated, though not hoped for, that a centrifugal blower will fail at least once during the lifetime of a cleaning machine.[156]

In no case should a manager purchase a cleaning machine using air knives for drying that are driven by a centrifugal blower[157] *without witnessing a "hands on" (ear protection on) demonstration[158] of unit performance*. Participation by a staff member expected to operate the machine is essential as well, to understand the effect on them of the noise level (with hearing protection).

### 7.12.3 Air Compressors

The situation is different with air knives powered by air compressors.

The most significant difference is that *the purchaser of the cleaning machine is expected to supply their own air compressor*,[159] whereas the centrifugal blower is integral to the cleaning machine. The reason for this difference is the need to avoid pressure loss in tubing connecting the device powering the air knife to the air knife.[160]

The second difference is within the air knife. Since the high velocity used to dislodge water droplets is produced at the knife tip, a different design is used.

[155] One differentiation, other than economics as in Section 7.12.6, is that air compressors can support air knives located within a cleaning machine, and be located remote from the cleaning machine. Thus workers are not exposed to high levels of noise at the cleaning machine. The compressor and the cleaning machine are connected through a header pipe. However, because of frictional energy losses due to high velocities, it is quite inefficient to locate centrifugal blowers in remote locations. Workers located around the cleaning machine must be exposed to the high levels of noise produced at the centrifugal blower. This author is one who finds that noise level objectionable (with hearing protection).

[156] This author assumes no useful cleaning machine will have a maintenance life more than 5 years.

[157] The image of the centrifugal blower is courtesy of Paxton Corporation.

[158] This demonstration must include all of the various types of parts expected to be processed with the proposed drying system. Further, this demonstration must employ the actual methods expected to be used of: part support (racking), organization of part orientation, and collection of wet discharge air.

The obvious aim of this test is to learn if all parts can be dried to the degree required and if any reorganization (of support method or direction relative to the air supply) is necessary. A subliminal, and no less important, aim is to learn if some parts will be reinfected with water after being dried with the proposed facility.

[159] Air compressors are seldom purchased to support single cleaning machines. Rather, noisy large-scale machines, rated at hundreds of HP, are purchased to support the needs of factories. They are located remote to where they are used.

Compressed air is fed in steel pipes to local machines where it is expanded to do work. However, size of this pipe used for flow distribution can be crucial to the success of a drying application. Normally diameters of header pipe are 3–6-in NPT (National Pipe Thread), and air pipes feeding local machines are sized at least three-quarters to 1 in NPT. This author has witnessed the discomfort of clients who received unexpectedly poor drying quality when an air knife was fed with a pipe sized three-eighths NPT diameter. Please recall that pressure loss is proportional to pipe diameter raised to the *fifth power*, whereas pressure loss is only linear with pipe length.

[160] Pressure loss is nearly a function of velocity to the second power. The velocity used to dislodge water droplets is produced by rotation of the high-speed centrifugal blower. That high velocity, and relatively low pressure, exists through all connections between the centrifugal blower and the tip of the air knife.

The opposite is true with air knives driven by air compressors. Here, the velocity is produced by expansion of air at the tip of the air knife. Velocity is modest in the tubing (piping) which connects the remotely located air compressor and the air knife.

In this case, the air knife is basically an expansion nozzle. The gap or aperture is considerably thinner when the air is supplied by an air compressor.[161]

A third difference is in the temperature of the air striking the parts. This should have no effect upon the drying rate,[162] but the parts will be at a different temperatures.

### 7.12.4 The Transvector

This is another type of air-based drying tool, which can be powered by air compressors. The transvector, also known as an air amplifier, uses compressed air to suck (pull, not push) air from a zone.

Basically a venturi nozzle, a flow diagram of a transvector is shown in Figure 7.57.[163]

Compressed air expands across the nozzle and entrains atmospheric air, increasing the total volume of flow by a factor of 50–500%. Basically, operation is a tradeoff of pressure for volume.

The value of a transvector is that of a "broom" to clean up "mess."

A transvector can immediately remove the debris (water droplets, particles, moist air) from the zone downstream of where an air knife has been used for parts drying. This protects parts from reinfection. If properly organized, with the aid of some "dead volume," a transvector can remove supersaturated humid air produced by the action of a centrifugal blower.

Heated aqueous cleaning baths are a second application. Here natural evaporation "pollutes" the work environment with humidity and raises ambient temperature.[164] This emission can be collected via a transvector and directed to a mist eliminator device.

No transvector costs more than 100 euro. An aqueous cleaning machine whose designer is thoughtful enough to include this feature has probably provided a quality cleaning machine.

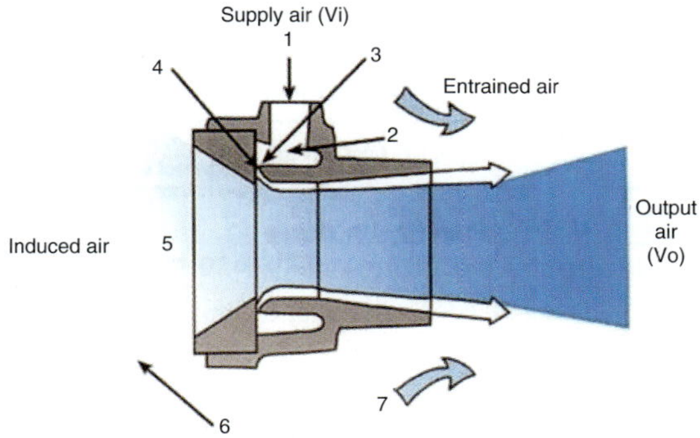


**Figure 7.57**

### 7.12.5 It's Always the Economics

Useful in both drying and rinsing operations, the capital investment in air knives is a bargain. One can't spend 500 euro on a several of modest size.

But that's not the true cost picture. It's not the knife which dominates the cost of air knives. It's the air.

Capital and energy requirements for each can be severe. Usually, the choice between high-velocity blowers and high-pressure (relatively) compressors is based on economics.

Suppliers of both blowers and compressor systems claim their offering produces a superior economic position. This author's recent experience has produced a comparative economic analysis of using both methods to drive a modest installation of air knives in a cleaning machine. The results are shown in Table 7.12,[165] for comparable battery limits costs of power,[166] maintenance, etc. All prices are retail, in euro.

Electric power is the major cost element – the major reason to select a blower versus a compressor.

---

[161] Typical operating parameters are 60–100 psig pressure and 40–80 SCFM.

[162] Evaporation isn't involved. Drying is by impingement. Parts are cooled when struck with cold air produced by an air knife powered by an air compressor.

[163] Image courtesy of Tech Sales, Inc.

[164] Use of transvectors also proves the adage that there is "no free lunch." Production of humidity around an aqeous cleaning tank means water has been evaporated at the expense of 1,000 BTU of energy consumed per lb evaporated. If this humidity is continually removed in a transvector, to improve working conditions, it will be continually replaced in order to maintain equilibrium between the liquid in the cleaning bath and the vapor in the working environment. Thus, the price of improving the working environment is a continual consumption of energy which provides no cleaning benefit.

[165] Often a site will have an air compressor with excess capacity, as air is used to drive other machines and activate instruments. In that case, use of that capability is the more secure choice as minimum capital investment is required.

[166] Power requirements are based on the HP rating of the device.

**Table 7.12 Comparison of Economics for Operation of Air Knives**

| Compressor | | Blower | |
|---|---|---|---|
| 55 SCFM @ 80 psig | | 150 SCFM @ 68″ WC | |
| 2 12-in Air Knives @ €120/each | €240 | 4 6-in Air Knives @ €150/each | €600 |
| 15 HP 80 psig Air Compressor @ €6439 | €6,439 | 5 HP Blower @ €5500 | €5,500 |
| Air Dryer and 50 CF Receiver @ €1700 | €1,700 | Inlet Filter Silencer @ €235 | €235 |
| Water Filter @ €245 | €245 | Low Loss Hose (25 ft) @ €480 | €480 |
| Air Filter @ €231 | €231 | | |
| Pipe Fittings @ €500 | €500 | | |
| **Investment cost** | €9,355 | | €6,815 |
| Maintenance Contract @ €900/yr | €900 | Maintenance Contract @ €450/yr | €450 |
| Hrs operation/yr | 8,000 | Hrs operation/yr | 8,000 |
| Years for Capital Payout | 5 | Years for Capital Payout | 5 |
| Power cost, €/KWH | €0.07 | Power cost, €/KWH | €0.07 |
| **Annual cost** | | | |
| Capital Repayment | €1,871 | Capital Repayment | €1,363 |
| Maintenance | €900 | Maintenance | €450 |
| Power | **€6,266** | Power | **€2,089** |
| **Total** | €9.037 | | €3,902 |

#### 7.12.5.1 *Choosing Between Centrifugal Blowers and Air Compressors*

This is a choice which can be made by a manager seeking to purchase a new cleaning machine. Pertinent items are shown in Table 7.13.

This author has no recommendation as the choice is basically a tradeoff between cost versus quality of environment.[167]

### 7.12.6 Other Equipment for Drying Without Evaporation

Several other types of process equipment have some currency in drying of water from parts.

#### 7.12.6.1 *Centrifugal Dryers*

Centrifugal dryers are an excellent choice, if a manager's parts are small enough to fit into a dryer. This technology has been in use for many decades. But its use is relatively rare in the cleaning industry.[168]

Parts are loaded into a cylindrical mesh or plastic basket (see Figure 7.58[169]). There may be discrete sections for holding individual parts if the parts would be damaged by contact. The basket will be open if the parts can be mixed.

Largest basket size has a diameter and height of ~30 in; smallest is 6 in by 6 in.

The parts are spun at 900 rpm for a cycle of 30 sec to 10 min, depending on the part configuration or degree of dryness needed.

[167] Two other viewpoints are found in: (1) Wilson, J., "Air Knife/Blower versus Compressed Air System," *Drying Times*, Vol. 1, No. 2 and (2) VanderPyl, D.J. and McGlothlan, K., "Precision Drying Completes Precision Cleaning," *Precision Cleaning Magazine*, March 1995.

[168] One firm has commercialized an aqueous cleaning machine within the facilities of a centrifugal dryer. Centrifugal force is used for both cleaning and drying.

[169] Image courtesy of Nopal.

**Table 7.13 Comparison Between Air Knives Driven by Centrifugal Blowers and Air Compressors**

| Item | Centrifugal Blower Supplied | Air Compressor Supplied |
|---|---|---|
| Drying Effect | Equivalent | |
| Equipment supplied with cleaning machine | Integral to cleaning machine | Customer purchased |
| Required capital investment | Lower | Higher, mitigated by air compressor capacity already available on-site |
| Operating cost, with cost of power exacerbating the difference | Lower | Higher |
| Function of air knife | Flow direction | Fluid expansion |
| Noise level | Much higher (~>85 to 90 dBA) | Much lower (~70 to 80 dBA) |
| Cleanliness of feed air | Needed to protect blower | Needed to avoid reinfection of parts |
| Concern about maintenance life | High | Low |
| Air Temperature | Hot, due to frictional heating in the high-speed blower | Cold, due to expansion at the air knife tip |

**Figure 7.58**

For drying of solvent *or* aqueous cleaning agents, *there is no need for air flow or heat supply.*[170]

Liquid is recovered for reuse at the bottom of the dryer:

- Advantages: no VOC emissions for use with solvent cleaning agents, little floorspace required (~5 ft × 5 ft), very little investment (€5,000 or less) and operating cost (see below).
- Disadvantages: not all parts will fit into available basket sizes; cylindrical dryer baskets are round while cleaning baskets are traditionally square, so labor is needed to move parts from the cleaning to the drying basket; processing is done in batch mode.

This purchase is, of course, an add-on to a purchased cleaning machine. Length of manufacturer's warranty is the only basis for recognition of quality.[171]

### 7.12.6.2 *Unrealized Fear*

Many managers fear use of centrifugal dryers. Part damage is their concern.[172]

This author strongly believes this is a unfortunate attitude that deprives managers of the inherent benefits produced by centrifugal dryers – excellent energy savings and rapid speed of drying.

The fear of part damage is exaggerated, in this author's experience. After parts are correctly fixtured, the centrifugal force which pulls the water films from parts also pulls parts into adjacent fixture

[170] The author has witnessed demonstrations where the heater on a centrifugal dryer is turned off and has seen no change in drying quality or drying time versus when the heater was used. This should be expected – because water is not removed by evaporation, but by centrifugal force.

[171] One US manufacturer does offer a centrifugal dryer with a lifetime warranty, though the purchase price is roughly 3 times that of a unit with a standard 1- or 2-year warranty.

[172] Durkee, J., II, "Parts Drying Made Easy," *Products Finishing Magazine*, February 1995, Vol. 59, No. 5, p. 63.

**Figure 7.59**

elements – *protecting them from movement which could cause damage*. Several parts baskets are shown in Figure 7.59.[173]

#### 7.12.6.3 *Removal of Water Films by Vacuum Entrainment*[174]

This technology is suited only for very regular part sections, flat surfaces, or wires. Air is pulled by a vacuum device[175] across a narrow opening, which creates a high velocity. The opening (nozzle) is moved across the work (or the reverse), and liquid is entrained in the moving air stream. The work is usually dry to the touch with one pass of the nozzle. A demister recovers the liquid for reuse.

Design parameters vary with the custom application. There is no commercial "drop in" equipment. Yet, local construction should not be expensive. This method makes good sense for strip or wire, but not for most other shapes.

### 7.12.7 Equipment for Drying via Evaporation

The second and the most common method of drying parts is by evaporation of the liquid upon and within them. All solvent cleaning operations produce dry parts by evaporation of the solvent after rinsing in the freeboard area of a vapor degreaser.

Traditionally for continuous aqueous cleaning operations, this was done in an oven integral to the cleaning machine. The oven was heated by forced hot air. Still in common use, hot forced air drying is expensive of energy and time. Managers are urged to consider non-evaporative drying technology as described in Chapter 1, Section 1.13.5.

Facilities for drying of parts are often "bolted on" to an otherwise excellent aqueous cleaning machine because some suppliers believe users value soil-free parts versus clean and dry parts. Because of this situation, it is quite common for a site to construct its own drying equipment.[176]

#### 7.12.7.1 *Forced Hot Air Systems*

Use of these systems requires a compromise among three major factors:[177]

1. Drying time (cycle time).
2. Drying quality (specification).
3. Costs of operation,[178] chiefly energy costs. This factor requires evaluation of two subfactors:

---

[173] Image courtesy of Nobles Manufacturing.

[174] This is not vacuum drying (see Section 7.12.8).

[175] The device is usually a venturi nozzle powered by compressed air. Operation is based upon the same concept as the transvector (see Section 7.12.4), but with different internal geometry.

[176] The strength of industrial offerings for drying systems is poor. The web sites of major global suppliers of cleaning systems have few or no listings for products as stand-alone drying systems. Further, there are limited or no descriptions given about drying capability of their integrated cleaning systems.

Limited vision by industry suppliers will be enhanced when the soaring of energy costs ignites demand by managers for additional choices. To a limited extent this has happened as a few firms supply infrared (IR) ovens for parts drying, and a few users employ abrasive materials such as cob grit in mass finishing operations to both smooth part surfaces, as well as absorb moisture from them.

Microwave drying (which produces internal frictional heat) of dielectric (non-conductive) wood and plastic "parts" has already been pilot tested. See Hansson, L. and Antti, A., Design and Performance of an Industrial Microwave Drier for On-Line Drying of Wood Components, *8th International IUFRO Wood Drying Conference*, 2003.

[177] Managers experienced with project management will recall the dictum that there are three factors associated with any project (cost, timing, and quality), and that one can simultaneously have control over just two.

[178] Optimization of energy factors is beyond the scope of this book. But a useful recent paper which covers energy optimization (including the two subfactors) is Bousquet, A. and Ladoux, N., Flexible versus Designated Technologies and Inter-Fuel Substitution, *Working Paper Series of the Institut d'Economie Industrielle (IDEI)*, May 13, 2004.

**Table 7.14 Guidelines for Forced Hot Air Drying Systems**

| Parameter | Good | Better | Best |
|---|---|---|---|
| Application of hot forced air streams | Impacting parts from a direction perpendicular to their length | | Parallel to, or along with, the length of parts |
| Location of heating source | Beneath parts | | Around parts |
| Heater rating[182] | 3–5X | | 6X |
| Electrical power supply | 220 VAC | | 440 VAC |
| Power control | Variable | | |
| Fan capability, ft/min face velocity | 500 | 800 | >1,000 |
| Maximum air temperature | 300° F | | ~230° F (see Figure 1.11) |
| External insulation[183] | Adequate to meet site and local regulations | | |
| | NOT capable of drying parts on re-startup | | Capable of drying parts on re-startup |
| Internal construction | Steel | | Stainless steel |
| Flexible barriers/curtains to retain heat while allowing parts entry and exit | Single layer on exit | Single layer on entry and exit | Double layers, on entry and exit[184] |
| Instrumentation: temperature control, timer | Yes | | |
| Inlet and outlet humidity sensor | No | | Yes |
| Viewport | No | | Yes |

(1) the choice of fuel – natural gas or electrical energy and (2) startup strategy – continuous or periodic operation.

Only recently, as energy costs have soared and a dislocation has emerged in the marketplace between the energy prices of electricity and natural gas,[179] has the third factor become a dominant one.[180] Traditionally, the compromise was between productivity (cycle time) and quality (dryness specification).

Whether a drying system is integral to the cleaning machine or is a stand-alone forced hot air drying system (oven), there are some features which a manager should recognize as differentiating good from better from best[181] choices (see Table 7.14).

It should be apparent that most of the items in Table 7.14 are associated with controlling the cost of energy consumption versus controlling the quality of parts drying.

7.12.7.1.1 *Don't Let the Wheels Come Off* When natural gas cost €0.25 per million BTU, the exhaust from a parts dryer could be discarded to a stack because it wasn't worthwhile to recover the energy

---

[179] This book was written in late 2005. While the dislocation between the price of energy (as supplied by natural gas or electricity) is related to supply/demand, environmental, political, and other issues, this author believes that this dislocation will continue on a local or regional if not global basis.

[180] And yet, a manager may find this is no choice at all. First, only one utility may be available at a site. Natural gas is not universally available. Second, site distribution of 440 VAC electricity may not be allowed because of safety concerns or employee training, even though device current loading at 220 VAC might be excessive). Third, if waste steam is available from allied operations, it should always be used for heating of drying ovens as its energy cost is usually free. Finally, the traditional guidance that natural gas heating was more expensive to use but provided more rapid heatup may not be true, because of dislocation of energy prices among natural gas and electricity.

[181] Readers must recognize that unlike other components of cleaning machines described in this book, the "best" drying systems may not be commercially available.

[182] This is relative to the energy requirement needed to evaporate the estimated amount of water clinging to the parts in the desired cycle time. For example, the energy requirement needed to dry 10 parts each with 10-in$^2$ surface wet with about 7 mil of water film in 5 min is about 1 kWh. An excess allows for rapid heatup time.

[183] Ability of the system insulation to retain the heat provided for drying, after 1 h in standby condition. This stipulation is recognized only by this author.

[184] While this author knows of no commercial parts dryer with this feature, its cost is negligible.

value of that stream. Today, no employed manager should permit that inaction – yet some do.

This change is dramatic in scope, but subtle as it manifests itself over short time periods.[185] One view of it is to compare electrical power costs as a portion of the total cost for operating an aqueous cleaning machine. This is shown in Figures 7.60 and 7.61.[186]

*The cost of energy necessary to dry parts by evaporation of water will become or has become the single largest component of operating cost.[187] Manage your parts dryer today as if that were true – because it is or soon will be so.*

Recycling of energy is imperative if parts are to be economically dried by forced hot air. This can be done efficiently through the use of a heat wheel.[188]

A heat wheel is a device which recovers heat from one fluid stream, stores it for some period, and then transfers it to another stream. In other words, one stream becomes colder while another becomes hotter.[189] A diagram of the functional use of a simple heat wheel is shown in Figure 7.62.[190]

Hot forced air, containing some moisture (at 250° F in Figure 7.62), is passed through the heat wheel. The hot air heats the metal wheel element, which in turn heats cool incoming air.

A photograph of a heat wheel, used as a "cassette," is shown in Figure 7.63.

Note that not all of the heat content of the air from the dryer can be transferred to incoming cool air.

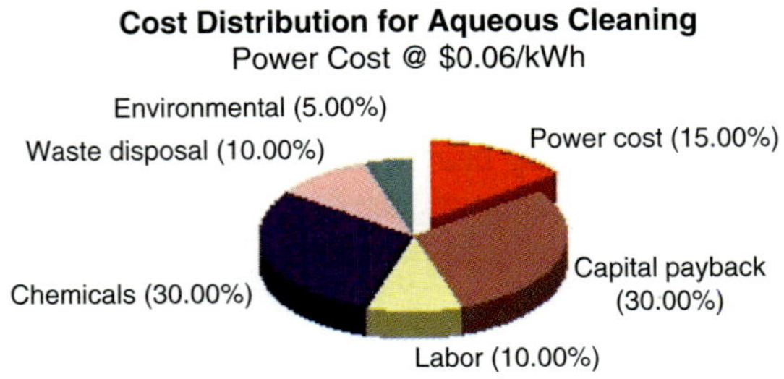


**Figure 7.60**

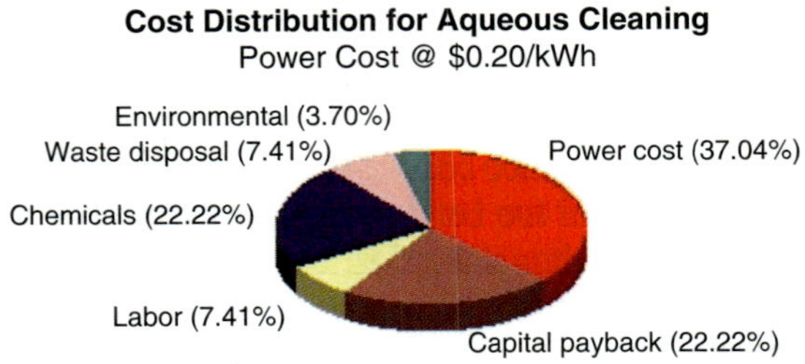


**Figure 7.61**

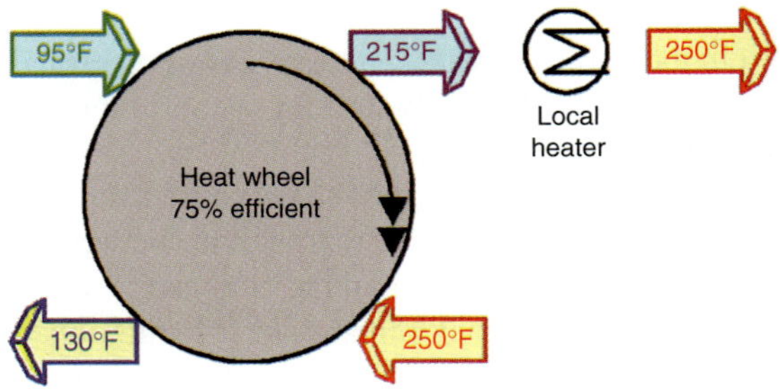


**Figure 7.62**

[185] In most plants, operators work on jobs other than just the cleaning system. Electrical, steam, and compressed air are supplied to more than one machine. The waste water treatment plant accepts waste water from more than one area. And the parts washer may be a central washer, cleaning parts of various types from various plant operations. So it is difficult for a manager to know the true costs of operating their cleaning system.

[186] The information in Figure 7.60 was published by the author in May 2000 in *Metal Finishing Magazine*. It was based on work with clients in the 1990s with an average cost for electrical power of around €0.06/kWh. Ignoring the effect of the increase in energy prices on cleaning chemicals and other components of cost, the increase in the invoice for electric power will be noticed by every organization's financial manager. The price of €0.20/kWh is currently being experienced by some users.

[187] It is interesting to note how the cost of parts drying is often ignored in evaluation of one cleaning alternative versus another. An excellent example of this blindness can be found in the US EPA's monograph on Aqueous Parts Cleaning for Fleet Maintenance, November 1999. It can be found at http://www.epa.gov/region09/cross_pr/p2/autofleet/fleetclean.pdf

[188] Laundry cleaning shops have used heat wheels for generations. There is a method for testing air-to-air heat exchangers – ANSI/ASHRAE 84-199.

[189] A fine point: heat wheels are not direct heat exchangers. Sections of a heat wheel become cooler and then are later heated (or the reverse). In a traditional heat exchanger, all sections are at some (but different) equilibrium temperatures. Said another way, a heat wheel involves unsteady state heat transfer.

[190] The heat wheel described in Figure 7.62 would be made of Aluminum alloys, about 5 ft in diameter and 10-in thick, rotate at around 20 rpm, and cost less than 10 000 euro. Payback time, based on years of industrial experience, is touted as being between 1 and 2 years.

**Figure 7.63**

That's not thermodynamically possible. About three-quarters of the sensible heat energy is passed to the incoming air. The remaining heat is provided by a local heater:

- The outcome of this operation is that the moisture removed from the parts is removed from the drying system.[191] Power cost to heat air for drying can thus be reduced by approximately three-quarters (see Figure 7.61). Only around one-quarter of the hot air fed to the parts dryer must be heated.

While not in common use in US cleaning operations, heat wheels do have some currency in other operations. Globally, especially in India, there is no shortage of suppliers. Availability of experienced maintenance and support is the chief factor which should drive a selection decision in the US. Significant design features of heat wheels are: (1) measured efficiency of energy recovery, which should be at least 75%, (2) quality of static seals between the rotating wheel and the cassette liner, and (3) pressure drop, which should be less than 20 in WC.

Any material that attracts and holds water vapor is a desiccant.[192] Developers have impregnated a desiccant into the rotor of a heat wheel. In this way, there is no need for a purge of water from the system – that happens as the desiccant heat wheel normally functions. While useful in non-cleaning applications, it is not clear whether this technology is required in parts cleaning work.

7.12.7.1.2 *The Future Is Not So Hot* In the useful lifespan of this book, this author believes major changes in parts drying technology *must* be developed and implemented, if aqueous cleaning technology is to play the role it currently does.

While it goes without saying, although it was written in Chapter 1, Section 1.13.8, a manager should always choose to dry part to the minimum level necessary – if not to reduce cycle time, then to control energy costs.

Those energy costs, when recognized by users, will drive the complacent to abandon the evaporation step required to complete aqueous cleaning work, or implement alternate drying technology. This may include equipment options listed in Table 7.15.

These equipment options won't be seriously considered until managers recognize the total cost and distribution of cost associated with their cleaning systems, and the local price of energy demands action.

#### 7.12.7.2 *Vacuum Dryers*

Vacuum evaporation is only a polishing technique used to get moisture content down to the range of 5–100 ppm. The best procedure is to dry the parts via some other method to the "dry to the touch" level.

Yet, vacuum drying is fast and effective, especially if parts have blind holes or some structure which blocks forced flow of hot air.

Vacuum levels are 1 Torr and above. Temperatures are room temperature to ~250°F. Cycle times can be 1 hr or greater. The needed equipment is expensive, large, heavy, and not often used in metal finishing work.

### 7.12.8 Drying with Solvents in Vapor Degreasers

Many boiling solvents can act as good drying agents, usually for water. This is because the intermolecular forces within some solvents cause them to display a

[191] Air exhaust from a parts dryer is not very wet. The calculated relative humidity content of air represented by operation in Figures 1.8–1.10 is around 1% or 2%. Consequently, that air may be reused for some additional drying work. But at some point, the water must be removed from the drying system or the rate of drying will decline to zero. This means there must be a purge stream to remove the moisture accumulated from wet parts.

[192] Desiccant heat wheels are used for broadly different applications outside of parts drying, such as production of moisture-sensitive foodstuffs, pharmaceuticals, Lithium batteries, and drying of thermoplastic resins. The desiccant heat wheel is regenerated with an amount of fresh dry hot air equal to about 20% of the normal air moist flow.

**Table 7.15 Options for Energy Management of Parts Drying Associated With Aqueous Cleaning Technology**

| Approach | Current Situation | Possible Situation |
|---|---|---|
| Better information control | Systems currently available for selection based on Table 7.14 will probably provide only: (1) an on–off switch for energy, (2) an on-off switch for the fan, and likely (3) measurement of hot air temperature which can be adjusted by regulating the energy supply. | • Sensors for differential humidity and dryer profile<br>• On-line energy consumption<br>• Optical detection of dryness<br>• Control program for optimization of energy use while meeting dryness specifications |
| Energy recovery | One-pass forced hot air systems with 100% fresh feed | • Recycle of humid hot air, using above optimization technology<br>• Purge of hot, moist air through a desiccant or standard heat wheel |
| Evaporative drying without heating | Not commonly used | • Gross removal of liquid water<br>• Finish drying with high volumes of unheated air (humidification)<br>• Dehumidification of unheated air using conventional heating ventilating and air conditioning (HVAC) technology |
| Evaporative drying | Industry standard approach | Implementation of the non-evaporative drying technology described in Sections 7.12.1–7.12.6 |

remarkable antipathy for water, while others do display the opposite.

The equipment by which different solvents are used to dry water is a (possibly modified) vapor degreaser.[193]

Both displacement and alcohol driers can allow removal of water from nests of intertwined parts. That's a difficult drying job!

#### 7.12.8.1 *Displacement Drying*

Displacement rinsing[194] and displacement drying[195] are the same processes, with different purposes and different fluids used in similar equipment:

- In displacement rinsing, a light material (often a hydrocarbon) is displaced with a heavier fluid (one of higher density). The heavier fluid, relative to the fluid being displaced, can be water, a traditional chlorinated solvent, *n*-propyl bromide, or a recently developed "designer" solvent (HFE-7200, HFC-43 10mee, OS-10, etc.).
- In displacement drying, the same happens. Here the lighter fluid is water, and the list of heavier fluids is also usually one of the same solvents.[196]

Displacement drying is solvent cleaning[197] – where the soil is water, and the solvent is a non-solvent for water.

In displacement drying of water, there is no evaporation of water. Hence, there are no remaining mineral deposits ("water spots").

[193] Durkee, J.B., *On Solvent Cleaning*, to be published in 2007 by Elsevier, ISBN 185617 4328.

[194] See Chapter 1, Section 1.12.3.2, about displacement rinsing.

[195] Stagliano, S., "Displacement Drying," *Precision Cleaning Magazine*, April 1991, pp. 29–31.

[196] See US Patent 4,618,447, US Patent 5,256,329, US Patent 6,365,565, or US Patent 6,956,015. Surfactants and stabilizers are often added to the drying fluid to enhance rejection of water and promote solvent life.

[197] Wet parts are inserted into the "cleaning" sump where water is displaced. The water, as the lighter phase, then rises to the top of the sump where it is decanted, usually with a weir. Collected water is decanted from expensive solvent in a second stage of separation, and the water is discarded and the solvent recycled.

#### 7.12.8.2 *Alcohol Driers*

Drying with boiling alcohol is also solvent cleaning – where the soil is water, and the solvent is a solvent for water. Isopropanol is the commonly used solvent.[198,199]

A concern not present with displacement drying equipment is recycle of soluble water. As with dragout,[200] water-laden alcohol is a soil. The undiluted water will probably remain on the parts after completion of drying, as a remaining "soil."

A second concern, which may not present with displacement drying equipment, is flammability. While some "designer" solvents are not flammable because there is no measurable flash point (e.g. HFC-43 10mee, HFE-7100), *all alcohol drying solvents are flammable*. This is why drying with alcohols is less commonly practiced.

#### 7.12.8.3 *Good Equipment*[201]

A manager should select displacement or alcohol driers from the same supplier they would choose for vapor degreasers.

A vapor degreaser is considered to be a commodity product. An alcohol drier is rarely used and would be considered by many suppliers as a specialty product. That distinction won't decrease its purchase price.

A sound approach during purchase would be to ask a chosen supplier for a demonstration test of a favored flammable-rated vapor degreaser to be used to "clean" parts using isopropanol as a solvent.

Not present with normal vapor degreasing equipment is containment or packaging to retain the integrity of the water drying process.[202] Granted, a manager may locate a conventional vapor degreaser in an open area, but if the same machine is used to dry water from parts the machine should not be located in an open area where atmospheric humidity can contaminate the dried part surfaces.

### 7.12.9 Comparison of Specific Equipment for Parts Drying

Table 7.16[203] gives specific recommendations for drying some common parts. Caution should be used in blindly following them as some of your local conditions haven't been incorporated. Examples are the soil being cleaned or the next processing step.

## 7.13 WATER, WATER EVERYWHERE

The quality of dried parts may sometimes depend less upon the process chosen, its design, or the equipment components from which the process implementing it was assembled, and more upon the quality of the water which was used for the rinsing work done previously.

---

Parts are then moved to the "rinse" sump, where they are contacted with pristine (water-free) solvent. Dry parts are produced as usual in a vapor degreaser by allowing/causing the low-boiling solvent to evaporate in a hot vapor zone above the "rinse" sump.

Please don't assume that only two stages of contact are involved. There may be, and often are, multiple "cleaning" stages.

[198] Wet parts are inserted into the "cleaning" sump where water is solubilized as it would be with another appropriately designed cleaning process. Rinsing with water-free alcohol and normal drying produces water-free (and dry) parts.

Please note, versus Section 7.12.1, that there are no weirs for elimination of supernatant water in alcohol driers. The water separator is external, and the feed probably is chilled.

[199] Other alcohols may bring value in drying operations as well. Tertiary butyl alcohol (TBA) and ethanol (EtOH) are useful because they form binary azeotropes with water (2% and 5% water, respectively). Please note that isopropanol also forms a similar azeotrope with water, so removal of water from parts may either be by solution if large amounts of water are present, or by azeotropic distillation if tiny amounts are present (as there should be). Methyl acetate (MeOAc), which is exempt as a VOC in the US, also forms an azeotrope with water (5% water). Unfortunately, all of these mixtures are likely to be flammable.

[200] See Chapter 1, Section 1.12 and following about elimination of dragout.

[201] Apologies to Alton Brown of *Good Eats*.

[202] Similarly, as pre-cleaning is used to remove high levels of soils from parts before entry into a solvent cleaning machine, so should all supernatant liquid water be removed by blowoff before either displacement or alcohol drying is commenced. These drying techniques are not intended for removal of water which can be removed in any simpler or cheaper way.

[203] This table is a companion to Table 1.17, which offers more general and broad-based recommendations.

**Table 7.16 Recommendations About Drying Equipment**

| Description of Parts | Cleaning Agent (Chosen Based on Soil Type) | Recommended Drying Equipment for "Dry to the Touch" Performance |
|---|---|---|
| Batches of: bundles of tubes 15-in long and ½-in ID; airplane wing struts; truck axles and wheels | Aqueous or semi-aqueous | Hot forced air |
| | Solvent | Displacement |
| Batches of: tubes 1.5-in long and ½-in ID; 2-in gate valve bodies; brass castings; plated cosmetic shells; circuit boards; eyeglass lens; small frames | Aqueous or semi-aqueous | Centrifugal (angle tube slightly in dryer, mount parts, shells, boards, and lens in fixtures) |
| | Solvent | Centrifugal (as above) |
| Batches of: mixed screws, nuts, rivets; electronic connectors; plastic parts | Aqueous or semi-aqueous | Centrifugal |
| | Solvent | Centrifugal |
| Continuously moving small engine assemblies | Aqueous or semi-aqueous | Hot forced air |
| | Solvent | High-velocity air blowoff |
| Batches of coiled tubing, any size | Aqueous or semi-aqueous | Hot forced air, or displacement |
| | Solvent | Displacement |
| Batches of flat stock, sheet; rolled sheet | Aqueous or semi-aqueous | Hot forced air |
| | Solvent | Entrainment by vacuum |
| Continuously moving wires | Aqueous or semi-aqueous | Hot forced air |
| | Solvent | Entrainment by vacuum |

After all, not everything in water can be evaporated, certainly not the components which aren't water and aren't volatile.

And what can't be evaporated is left behind – as retained imperfections, defects, or flaws.

This is an absolutely crucial situation in the manufacture of semiconductor and optic components, and of no interest whatever to managers involved with extrusion of Aluminum bars into arrows for archery.

Some guidelines for water purification systems are given in Table 7.17.[204] *Please remember these values are for general guidance – application to some specific applications (such as critical cleaning) may be hazardous to your professional health.*

Please remember that the cost of producing water of the quality indicated in Table 7.17 is not included in these guidelines.

And there can be several types of rinse water. It is only the water which last contacts the parts which is of greatest concern in minimizing water spots.[207] That's why the last step before drying in many cleaning processes is to displace surface rinse water on parts with a small volume of water of higher purity – the water whose quality is described in Table 7.17.

## 7.14 VAPOR DEGREASING EQUIPMENT

### 7.14.1 Batch Open-Top Equipment

As with other cleaning equipment, the quality of a vapor degreaser is also dependent upon both the quality of the components (nozzles, pumps, tanks, filters, etc.) from which it was assembled, as well as the design upon which it is based.

Guidelines for selection of the design of batch open-top vapor degreasers should basically be to minimize solvent emissions[208] (see Table 7.18).

[204] Quitmeyer, J.A., "Aqueous Cleaning: When to Rinse and Dry," *Precision Cleaning Magazine*, June 1995, pp. 14–17. This "ancient" but excellent article is reprinted in http://www.p2pays.org/ref/02/01825.htm

[205] Water hardness is taken to mean metal salts, which will remain on parts as spots.

[206] Less than 50 may be necessary to avoid water-spotting.

[207] Please recall the Central Rinsing Theorem from Chapter 1, Section 1.12.6.

[208] Requirements are well defined by the US EPA in their NESHAP for halogenated solvent machines. See http://www.epa.gov/ttn/atw/degrea/haloguid.pdf. Information in Table 7.14 should apply to use of any solvent – whether for reasons for environmental, control personnel exposure, or cost control.

**Table 7.17 Guidelines for Rinse Water Quality**

| Quality Parameter | US EPA Secondary Drinking Water Standard | World Health Organization Guideline for Drinking Water | Rinse Water Quality | | | Type 1 Water (Water for Laboratory Use ISO 3696:1987) | | | |
|---|---|---|---|---|---|---|---|---|---|
| | | | Good | Better | Best | ISO 3696 | ASTM D1193 | NCCLS | US Pharmacopoeia |
| pH | Between 6.5 and 8.5 | | Between 5.5 and 8.5 | Between 6.0 and 8.0 | Between 6.5 and 7.5 | No specification. Quality defined by the process of production (see Chapter 4, Section 10) | | | |
| Total hardness (ppm) | | | <100 | Between 60 and 100 | <40[205] | | | | <0.1 (total heavy metals) |
| Ionic Chloride (ppm) | | <250 | <80 | Between 80 and 20 | <20 | | 1 | | |
| Sulfate (ppm) | 250 | <500 | <100 | Between 40 and 100 | <40 | | | | |
| Total dissolved soils (ppm) | 500 | | <400 | Between 100 and 400 | <100 | ~<1 | ~<10 (as TOC) | <0.1 | ~<50 (as TOC) |
| Conductivity Siemens/cm | | | <750 | Between 800 and 200 | <200[206] | 0.1 | | ~<20 | <1.3 |

**Table 7.18 Guidelines for Batch Open-Top Vapor Degreasing Systems**

| Parameter | Good | Better | Best |
|---|---|---|---|
| Compliant with U.S. EPA NESHAP idling emission guidelines | 0.045 lb/hr – $ft^2$ solvent interface area (<0.22 Kg/h−$m^2$) | | |
| Compliant with U.S. EPA NESHAP overall emission guidelines | <30 lb/ft2 solvent interface area–month (<150 kg/$m^2$) | | |
| Acceptability by U.S. EPA for permit | Maximum Achievable Control Technology (MACT) | | |
| Automated parts handling hoist | YES (see Chapter 1, Section 13.9.1) | | |
| Freeboard refrigeration device | YES | | |
| Lip exhaust routed to Carbon adsorption canister | NO | YES | |
| Maximum air blanket temperature | <30% of normal solvent boiling point[210] | | <25°F[211] |
| Superheated vapor (SV) drying system | NO | Temperature in SV zone to be 10°F above normal solvent boiling point | Temperature in SV zone to be 20°F above normal solvent boiling point |
| Secondary freeboard refrigeration device | | NO | YES |
| Freeboard ratio[209] | 1.0 | 1.5 | 2.0 |

## 7.14.2 Vacuum Vapor Degreasers

These machines are also known as "airless," or "air-tight," or "machines which don't have a solvent–air interface." They were developed during the 1990s to allow the use of halogenated cleaning solvents – consistent with nation, regional, and local regulations.

Since the 1990s, vacuum vapor degreasers have been modified to clean and dry with OS-2, HFC-43 10mee, HFE-7100, and similar solvents.[212] Here the value they bring, against which to justify their increased investment (at least doubled) versus open-top machines, is cost control.[213–215]

**Figure 7.64**

[209] Freeboard ratio is the distance from the solvent interface to the top of the machine divided by the smaller internal dimension (width, height, or depth) of the machine.
[210] For solvents which don't form azeotropes with water. The value should be 50% for water–solvent azeotropes.
[211] For solvents which don't form azeotropes with water. The value should be 40% for water–solvent azeotropes.
[212] Durkee, J.B., *On Solvent Cleaning*, published in 2007 by Elsevier, ISBN 185617 4328.
[213] Gray, D. and Durkee, J.B., "Enclosed Cleaning Systems, Chapter 2.11, p. 305, of *Handbook for Critical Cleaning*, Kanegsberg, B. and Kanegsberg, E., CRC Press, 2001.
[214] *High Vacuum Vapor Degreasers*, TURI (Toxic Use Reduction Institute) Energy Efficiency Case Study, 2004.
[215] Rasmussen, J., "Finding a Balance: Texas Instruments Makes Cleaning Better for the Environment and the Bottom Line," *Precision Cleaning Magazine*, May 2000, pp. 12–18.

Applications are generally covered by patents.[216–219] One application is patented with water.[220] Other applications involve only parts drying.[221] Some applications involve multiple processing chambers[222,223] (see Figure 7.65[224]).

Guidelines for selection of the design of batch open-top vapor degreasers should basically be to minimize solvent emissions (see Table 7.19). Selection based on quantity of components used (pumps, tanks, nozzles) should be of secondary concern.

**Table 7.19 Guidelines for Vacuum Vapor Degreasing Systems**

| Parameter | Good | Better | Best |
|---|---|---|---|
| Material of construction | 304 stainless steel | | |
| Compliance capability | US EPA NESHAP Emission Guidelines[225] | Los Angeles, California, South Coast Air Quality Management District (SCAQMD)[226] | Able to produce local exposure within all regulations,[227] and receive exemption from permit to operate wherever desired |

[216] Tanaka, M. and Ichikawa, T., US Patent 5,193,560, *Cleaning System Using a Solvent*, March 16, 1993. Assigned to Tiyoda.
[217] Tanaka, M. and Ichikawa, T., US Patent 5,051,135, *Cleaning Method Using a Solvent While Preventing Discharge of Solvent Vapors to the Environment*, September 24, 1991. Assigned to Tiyoda.
[218] Grant, D.C.H., US Patent 5,106,404, *Emission Control for Fluid Compositions Having Volatile Constituents, and Method Thereof*, April 21, 1992. Assignee is Baxter International.
[219] Turicco,T., US Patent 5,449,010, *Pressure Controlled Cleaning System*. September 12, 1995.
[220] Nafzifer, C.P., US Patent 5,301,701, *Single Chamber Cleaning, Rinsing, and Drying Apparatus, and Method Therefor*, April 12, 1994. Assignee is Hyperflo.
[221] Though, the possibility of conducting cleaning work is not restricted. See Miranda, H.R. and Dye, M., US Patent 6,959,503, *Method and Apparatus for Removing Liquid from Substrate Surfaces Using Suction*, November 3, 2005.
[222] Gray, D., US Patent 6,743,300, *Multistep Single Chamber Parts Proceeding Method*, June 1, 2004.
[223] Gray, D., US Patent 6,783,602, *Multistep Single Chamber Parts Processing Method*, August 31, 2004.
[224] Figure 7.64 is courtesy of Serec-Tiyoda.
[225] Monthly emission limit $<85.5 \times$ (chamber volume in CF)$^{0.6}$. US EPA. Guidance Document, Part 2, Section 1.2, 1995.
[226] Rule 1122, available as http://www.aqmd.gov/rules/reg/reg11/r1122.pdf
[227] Two examples around exposure limits involve *n*-propyl bromide and chlorinated solvents. The American Congress of Governmental Hygienists (ACGIH) threshold limit value for the former was established in December 2004 as 10 ppm. This exposure almost certainly requires a vacuum vapor degreaser versus an open-top machine. In the EC, the VOC Directive (1999/13/EC) requires exposure limits of 2 mg/m$^3$ to allow solvents with a R45 risk phrase rating. These exposure limits can be achieved by a combination emission control by the cleaning machines supplemented by abatement treatment from Carbon absorption systems.

The EC VOC Directive (Annex IIA par. 4), for example, allows exemption for emissions of VOC chemicals when less than 2200 lb (1 metric ton) of chlorinated cleaning solvent are used per year. Here, use is calculated (not measured) as purchase volume less volume returned for recycle. This exemption, per Annex 2A par. 4, also means exemption from all threshold values for losses.

The point here should be clear – the best vacuum vapor degreasers are those which allow freedom from compliance with environmental regulations.

Appendix 1

# Statistical procedures for management of cleaning (or other) operations

**Chapter contents**



This book is about management of cleaning, or other, operations. It's not a substitute for a treatise on statistics.[1–3] But the material in this appendix is provided for readers who value additional details about the material in Chapter 4.

The core value of this appendix is to demonstrate the use of simple spreadsheets as statistical analysis tools for managers. Example spreadsheets are provided here for construction of "R", X-bar, and CUSUM control charts. Common data are used for all three types. Design of experiments is explained and demonstrated through the use of example spreadsheets.

Significant management responsibility can be discharged with the technology implemented in the example spreadsheets of this appendix.

## A1.1 BACKGROUND

A fundamental value of the science of statistics is that its use fosters more efficient use of information. Today, use of statistics should generate increased interest because of our accelerating demand for information and analysis of that information.

Statistics allows extrapolation of scale and time. The science enables a user to "go from the one to the many" – from a small body of information to a broad experience.

Lost in the often-confusing nomenclature and jargon is the core idea:

- A single (or a few) observation(s) may not correspond to the reality of all experience, but one can infer a significant amount about that reality by making educated choices about the method of making those observation(s).

[1] Useful books on this topic, in alphabetical order by author, are:

A. Abelson, R.P., *Statistics as Principled Argument*, Lawrence Erlbaum Associates 1995, ISBN: 0-80-580528-1.

B. Hinton, P.R., *Statistics Explained. A Guide for Social Science Students*, Routledge 1995, ISBN: 0-41-510285-5.

C. Montgomery, D.C., *Introduction to Statistical Quality Control* (4th ed.), Wiley 2001, ISBN: 0-47-31648-2.

Said more simply:

- Statistics can help managers make the right observations which will lead to the right decisions.

The methods behind that "education" and those decisions are called statistical methods, or statistical science or statistics. We see those methods used in endeavors outside of cleaning, such as:

- The first or hundredth vote counted in an election may or may not have voted with the majority. News analysts will survey a few voters throughout the election period and forecast an outcome.
- Many read trade journals to learn about salary values and variations for various professions and specialized work within those professions. Most believe they are underpaid (or not) by comparing their lot with those surveyed who are represented by those values and variations.
- After a first encounter, and a few subsequent ones, we project that a person will be a suitable friend or life partner for us. Our information isn't (generally) numerically processed, but we infer future lifetime happiness from observations we make about characteristics we value.

One can observe (measure cleanliness) in a few representative situations (cleaning operations); understand circumstances (cleaning performance); take action (make adjustments or shut down) to affect them; have a quantifiable explanation (means; deviations, *t*-statistics, etc.) to justify that action; and have a quantifiable probability that the explanation derived from a few situations to applies all operation (confidence limits or values). That's statistics applied to management of cleaning work.

Managers don't need statistics if results don't matter or if everything is tested. The bulk of experience usually falls between those limits.

## A1.2 NOMENCLATURE USED IN THIS APPENDIX

$x$ ≡ An individual result of cleanliness measurement.

$x_i$ ≡ Since there are multiple (hopefully) measurements, $x_i$ is the $i$th measurement in a series of cleanliness measurements.

$n$ ≡ The number (the last value of $i$) of cleanliness measurements made of selected parts.

$N$ ≡ The number of parts cleaned (items in the population). Except where the parts have a high value, $n \ll N$.

$\Sigma$ ≡ The sum of values (usually measured values).

$\bar{x}$ ≡ The mean of value of part cleanliness for all measurements tested (*sampled*), where:[4]

$$\bar{x} = \frac{\sum(x_i)}{n} \tag{A1.1}$$

$\mu$ ≡ The mean[5] value of part cleanliness for all parts cleaned (the *population*).

D. Ross, S., *Introduction to Probability and Statistics for Engineers and Scientists*, Elsevier 2004, ISBN: 0-12-598057-4.
E. Urden, T.C., *Statistics in Plain English*, Lawrence Erlbaum Associates 2001, ISBN: 0-80-583442-7.
A useful book on design of experiments is:
Antony, J., *Design of Experiments for Engineers and Scientists*, Elsevier 2003, ISBN: 0-7506-4709-4.

[2] Useful internet sites are:
http://www.richland.cc.il.us/james/lecture/m170/.
http://davidmlane.com/hyperstat/.
http://www.meandeviation.com/tutorials/stats/.

[3] Statistics should be taught by statisticians. Though this author doesn't qualify, the principles behind statistical methodology are correctly used in the material in Chapter 4 and this appendix. They should be the basis for action (or not) by all readers of this volume.

The material in this volume has been "simplified" – not "dumbed down," but *simplified*. The material in this volume would not be found in a statistics handbook because it is not a general and complete description of fundamental methods such as the *t*-test. The material in this book has been prepared to be a "instructional cookbook." Basic concepts are explained. But the implementation of these methods is specific to problems faced by managers doing serious cleaning work.

This appendix is provided to support use of the material in Chapter 4. It is recommended that readers not interested in learning more about the science of statistics use the spreadsheet functions found there.

[4] This calculation is easily done via a spreadsheet. In Excel the @ function is =MEAN(…) and in Quattro-Pro the @ function is @MEAN(…).

[5] An average is a measure of central tendency. An arithmetic mean (each value weighted by the number of times it occurs) is a type of measure of a central tendency. Mode (the most common value) and median (the middle or halfway value) are some others.

Standard Deviation
≡ The most commonly used measure of statistical dispersion of measurements around the mean of those measurements:[6]

- Large values of standard deviation signify that the individual values are widely displaced from the mean of those values.
- Small values of standard deviation signify that the individual values are closely packed around the mean of those values.

$s$ ≡ Standard deviation for the cleanliness measurements of those parts tested (*sampled*), where:[7]

$$s = \sqrt{\frac{\sum (x_i - \bar{x})^2}{(n-1)}} \qquad \text{(A1.2)}$$

$\sigma$ ≡ Standard deviation for the cleanliness of all parts in the *population* (most of which haven't been tested for cleanliness), where:[8]

$$\sigma = \sqrt{\frac{\sum (x_i - \mu)^2}{N}} \qquad \text{(A1.3)}$$

SE ≡ Standard error (deviation) for the difference between the mean of the measured cleanliness tests and the mean of all parts.

$t$ ≡ Student's "$t$"

$$t = \frac{(\bar{x} - \mu)}{\sqrt{\frac{(s^2 - \sigma^2)}{n}}} \qquad \text{(A1.4)}$$

## A1.3 NORMALIZING DATA

A convention used by many is to normalize data. This means each measurement is divided by a chosen constant. The effect is that all measurements are distributed around 1.0. The constant is chosen once, and not changed for any reason. The constant can be any number. It does not have to be a goal value, and is often the average value.

The virtue of this convention is ease of understanding and communication. Most statistical formulae are based on use of normalized data.

## A1.4 PRELIMINARY ANALYSIS OF HOMOGENEOUS[9] DATA: AVERAGES (MEANS) AND STANDARD DEVIATIONS

The mean and the standard deviation of a data set go hand in hand and are usually reported together.[10] In a certain sense, the standard deviation is the "natural" measure of statistical dispersion if the center of the data is measured by the mean (see Table A1.1).

Consider in Table A1.1 the measurements of NVR, taken over a period of 8 hours of production

---

Readers of this book, not statisticians, should consider average and mean to be the same. The spreadsheet functions in Quattro-Pro and Excel are @AVG or =AVG, @MEAN or =MEAN, @MODE or =MODE, and @MEDIAN or =MEDIAN.

[6] For example, the sets {0, 5, 9, 14} and {5, 6, 8, 9} each have a mean of 7, but the second set has a much smaller standard deviation.

[7] The spreadsheet formula is @STDS(...) in Quattro-Pro and =PURESTDS(...) in Excel.

[8] The spreadsheet formula is @STD(...) in Quattro-Pro and =PURESTD(...) in Excel.

[9] The word homogeneous is used to mean that cleanliness test data is treated as a group with all members having the same character: precision, accuracy, and repeatability. A homogeneous data set may have outliers (flyers).

Heterogeneous data would be that which was produced from multiple sources, or data in which the conditions which produced it were not similar or constant.

[10] There is no point in knowing the mean (average) of a set of values if one doesn't also know the standard deviation. It is the standard deviation which informs about the quality of consistency of the data.

Consider these measurements of depth of water into which you are about to step: 50, 100, 150, 200, and 250 cm. The average is 150 cm, but the standard deviation is around one half of the average. Consider another set of measurements of water depth: 130, 140, 150, 160, and 170 cm. The average is the same, but the standard deviation is less than ten percent of the average. If you can't swim, which standard deviation value suggests you may be about to step into water well above your height?

during which the overall production rate was 20 parts/hour (160 parts produced).

**Table A1.1 NVR Cleanliness Measurements**

| Sample ID | NVR Value (mg/SF) |
|---|---|
| 1 | 6.3 |
| 2 | 11.0 |
| 3 | 4.2 |
| 4 | 9.2 |
| 5 | 15.0 |
| 6 | 4.7 |

The mean ($\bar{x}$) of these cleanliness measurements is given by Equation (A1.1) as:

$$\bar{x} = \frac{(6.3 + 11.0 + 4.2 + 9.2 + 15.0 + 4.7)}{6} = 8.4$$

The standard deviation of these cleanliness measurements is given by Equation (A1.2) as:

$$s = \sqrt{\frac{\begin{bmatrix}(6.3 - 8.4)^2 + (11.0 - 8.4)^2 \\ +(4.2 - 8.4)^2 + (9.2 - 8.4)^2 \\ +(15.0 - 8.4)^2 + (4.7 - 8.4)^2\end{bmatrix}}{(6 - 1)}} = 4.1$$

## A1.5 THE HAZARD OF USING ONE DATA POINT

This simple example illustrates the danger of working with one data point.

Suppose a user is cleaning small stamped parts. There are about 25 parts in one pound (~18.144 g/part). A dirty part has about 250 mg/SF of soil (17.36 mg of soil if each part has about 10 SI area). These values are absolutely typical – they are the data from a client this author worked with in June 2002. The three steps below constitute a cleaning test:

1. Choose one part, and weigh it. Suppose it weighs 18.16163 g.
2. Run it through the cleaning process. Weigh it again. Suppose it weighs 18.14443 g. So the cleaning machine removed 17.20 mg of soil.[11]
3. Run that same cleaned part through an extraction process using isopropanol which is *believed* to remove all soil. Suppose now it weighs 18.14420 g. An additional 0.23 mg of soil was removed.

If it is *believed* that the isopropanol extraction process removes all soil, then the weight of this soil-free part is 18.14420 g.

Calculate the percent clean after the cleaning machine as 17.20/(17.20 + 0.23) = 98.68%.

**Table A1.2 IPA Extraction Main Cleaning Test**

| Number of Samples Tested | *Cumulative* PERCENT Clean via IPA Extraction (%) | Goal Percent Clean (%) | Meeting Standard? |
|---|---|---|---|
| 1 | 98.68 | 98.5 | Yes |
| 5 | 98.53 | 98.5 | Yes |
| 10 | 98.45 | 98.5 | No |
| 25 | 98.27 | 98.5 | No |
| 50 | 98.26 | 98.5 | No |
| 100 | 98.23 | 98.5 | No |

[11] Please note that this is not cleaning data! All that is known is how much soil has been removed. It's not known how much more soil is on the parts. It's not known if the parts are clean. Unremoved soil could weigh another 17, 0.17, or 170 mg. It's just not known yet.

Suppose the required standard of performance is to remove at least 98.50% of the soil. Is the cleaning machine meeting that standard? With just this one sample, the answer is "Yes, why do you ask?"

Suppose, the user uses 99 additional samples and performs the same test. Then they average (take the mean of) the percent clean data. Suppose the information in Table A1.2 summarized the *cumulative* average of all individual measurements, and compares that to the goal value.

With between 5 and 10 (or more) samples, the user can see that their process is not meeting the standard of performance of 98.5% soil removal. The parts are not clean! What are the chances of that being true? We'll answer that below with the aid of the *t*-test.

In summary, one measurement is very unlikely to represent an entire population (see footnote 14).

## A1.6 THE *t*-TEST

### A1.6.1 General Background

Student's distribution[12] arises when (as in nearly all practical work) the population standard deviation is unknown and has to be estimated from the data. This is because managers seldom test every part in a population.[13]

Gosset discovered that when individual observations follow a normal distribution, confidence intervals for population means could be constructed in a manner similar to that for large samples.

Said another way, Gosset published materials which showed that the relationship between a sample and the population from which it was taken was not related to the absolute size of either the sample or the population.

This means that managers don't have to process more samples to learn about larger populations. Managers only have to establish that *both the sample and population are normally distributed.*[14]

The number of samples needed to learn about the population is not related to the size of the population, but is a function of:

- The error in measurement associated with the samples,
- How well the user wanted to know if the sample was representative of the population.

[12] The derivation of the *t*-distribution was first published in 1908 by William Sealey Gosset (an Irish brewery worker) in a paper written under the pseudonym Student. The Guinness brewery in Dublin, Ireland wouldn't allow him to publish his work under his name, so he used the pseudonym "Student." The *t*-test and the associated theory became well-known through the work of R.A. Fisher, who called the distribution "Student's distribution."

[13] The only reasons for testing the cleanliness of every cleaned part are: (1) ignorance of statistical principles or (2) a requirement that cleanliness be established at 100% certainty. If some lesser level of certainty can be accepted by the enterprise (such as the commonly used 95% value), then only a small portion of the population of parts must be tested for cleanliness. See Chapter 4, Section 4.3.3 for a discussion of how many samples need be tested as a function of certainty (and expected errors of measurement).

[14] A normal distribution is one which is unbiased. This means that only those variables which "normally" affect the outcome do affect the outcome. There are no unexpected or spurious variables which affect the outcome. In other words, there are no surprises. If outcomes are different among samples produced at the same values of controlling variables that is so because of random variations in the controlling variables.

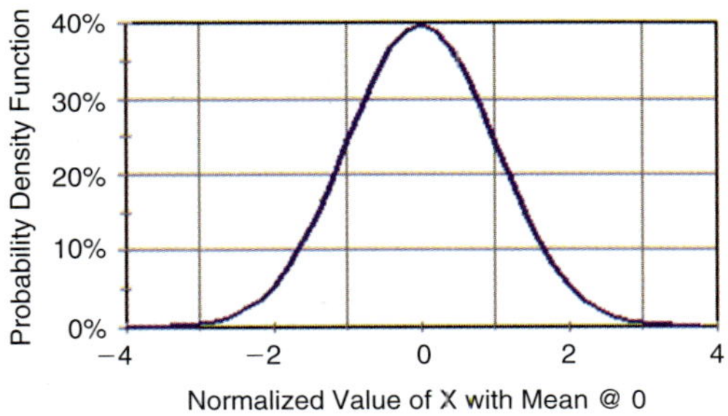


An example of outcomes in cleaning work which are not normally distributed is where an operator occasionally "goes for a smoke" and terminates a cleaning cycle based on their convenience and not on procedure. Here, some parts are cleaned in a shortened cycle, while others are cleaned using the standard time cycle. When cleaning quality is a function of cycle time, measured cleanliness may be dependent upon the personal needs of the operator as well upon the variables which normally affect cleaning performance.

The equation for the standard normal distribution for a population is below. X can be any measured quantity, including measured cleanliness value. The distribution has a mean at X = 0, has the familiar "bell" shape, is symmetrical about 0, is continuous, and never reaches the x-axis. The distribution is plotted at right.

$$\textit{Probability} = \frac{e\,[-(x)^2/2]}{\sqrt{2 \times \pi}}$$

Testing for cleanliness of a fixed percent of cleaned parts is a waste of resources. Don't do it. Use Gosset's "*t*-test" instead.

## A1.6.2 Specific Background

The *t*-test is useful for deciding if some sort of treatment was effective compared to the control group.

That's what cleaning professionals do. We want to know if some parts are cleaner than a standard part or other parts. Familiarity with the *t*-test is critical for all who work with data-especially those who do parts cleaning (Figure A1.1).

There are some internet sites[15] from which you can learn about and use on-line calculators (applets) for the *t*-test.

The primary assumptions for the *t*-test are:

- Population data from which the sample data are drawn are normally distributed.
- The variances[16] of the populations to be compared are equal.

Actually, empirical studies[17] of the "*t*-test" have demonstrated that these assumptions can be violated to an amazing degree without substantial effect on the results.

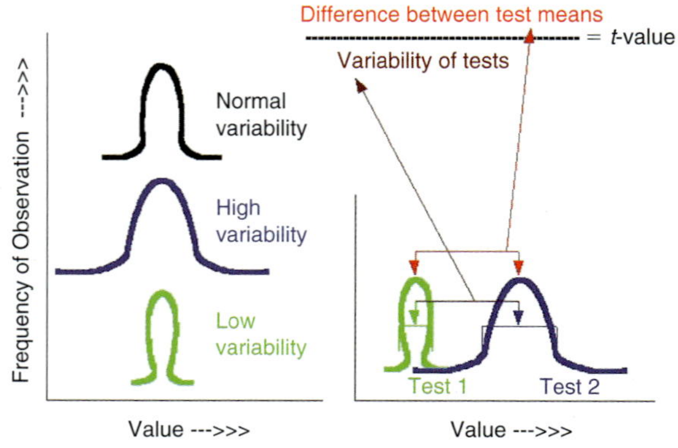


"*t*" is the difference between the means of two groups of data divided by the combined variability of both groups.

A higher ratio occurs when the means are more easily identified – a greater difference, or a lower probability of misidentifying one mean for the other.

**Figure A1.1** The "*t*-test"

In other words, unless you know that your data are unstable (have some area of operation which can't be [or always are] achieved), or are of two different types (continuous analog or distributed digital), use the *t*-test.

Use of spreadsheet functions (@TTEST or =TTEST) to complete the "t-test" is shown in Chapter 4, Section 6.

---

Values on the horizontal axis are of a number of standard deviations below (minus values) or above (positive values) the mean. Values on the vertical axis are of the calculated distribution, which is the probability that the measurement is X standard deviation units different from the mean. The area under the calculated curve is the cumulative probability that the measurement X is valued between any two values of standard deviation from the mean. About 95.4% of the area under the normal distribution curve is between −2 and +2 standard deviations from the mean.

In other words, the probability is about 40% that any single measurement will be the mean (average) value, and 95.4% of all normally distributed data lie within 2 standard deviations from the mean.

[15] http://www.physics.csbsju.edu/stats/t-test.html and http://fonsg3.let.uva.nl/Service/Statistics/2Sample_Student_t_Test.html have on-line calculators.

Other useful internet sites are:

http://bmj.bmjjournals.com/collections/statsbk/7.shtml
http://mathworld.wolfram.com/Pairedt-Test.html
http://www2.mc.uky.edu/Athletic_training/publications%20folder/JacobsATT.pdf
http://techniques.geog.ox.ac.uk/mod_2/week_5/lecture-5.htm
http://www.ruf.rice.edu/~bioslabs/tools/stats/ttest.html
http://www.richland.cc.il.us/james/lecture/m170/ch08-mu.html
http://davidmlane.com/hyperstat/
http://en.wikipedia.org/wiki/Student%27s_t-distribution
http://www.mne.psu.edu/undergrad/CaseStudy/Statistics/t.htm
http://laxmi.nuc.ucla.edu:8234/P234_97/234_7_o.html
http://techniques.geog.ox.ac.uk/mod_2/week_5/lecture-5.htm

[16] Variance is the sum of the squared deviations from the mean.

[17] Hays, W.L., *Statistics*, Holt, Rinehart and Winston, New York, 1963, pp. 319–323.

### A1.6.3 Use of "*t*-Test Using a Spreadsheet"

Here is an example where the *t*-test can be used to provide valuable information which could hardly be obtained in any other way.

Consider, in Section A1.5 that the IPA extraction process may not remove all soil. In that case, the user must complete another independent test to learn if the first extraction does remove all soil. This is called validation.[18] Here the user is proving, or disproving, that their cleaning test (the IPA extraction) is a valid test for part cleanliness.

The validation test might be another extraction, using the hexane–isopropanol azeotrope to be certain of removing both polar and non-polar soils.

After completing the validation test the user has the results in Table A1.3, for the single and multiple samples:

- Table A1.2 is repeated from Section A1.5 – with the normal cleaning test results. This is data set 1 (only the measurements in the center column).
- Table A1.3 shows the results of the validation cleaning test IPA isopropanol extraction. This is data set 2 (only the measurements in the center column).

The *t*-test permits analysis of cleanliness validation, cleanliness testing, and cleaning operations – see Table A1.2. The analysis is shown in Q&A format in Table A1.4.

**Table A1.2 IPA Extraction Main Cleaning Test**

| Number of Samples | Cumulative percent Clean Via IPA Extraction (%) | Goal Percent Clean (%) |
|---|---|---|
| 1 | 98.68 | 98.5 |
| 5 | 98.53 | 98.5 |
| 10 | 98.45 | 98.5 |
| 25 | 98.27 | 98.5 |
| 50 | 98.26 | 98.5 |
| 100 | 98.23 | 98.5 |

**Table A1.3 Hexane – IPA Extraction Validation Cleaning Test**

| Number of Samples | Cumulative percent Clean Via Hexane-IPA Extraction (%) | Goal Percent Clean (%) |
|---|---|---|
| 1 | 100.166 | 100 |
| 5 | 99.973 | 100 |
| 10 | 100.000 | 100 |
| 25 | 99.996 | 100 |
| 50 | 99.960 | 100 |
| 100 | 99.990 | 100 |

### A1.6.4 Use of "*t*-Test" with Equations

Some, by preference or situation, don't use spreadsheets. Long before Dan Bricklin invented the first spreadsheet (*Visicalc*) in 1980, managers (and statisticians) were using fundamental statistical equations to compare data sets.

The following explanation describes how two data sets can be compared. The data sets can be one of measured cleanliness data and a "Golden Lot," or they can be two measured sets of cleanliness data. The explanation describes the technical basis of the spreadsheet function @TTEST (or =TTEST).

There are three necessary steps in this comparison. The purpose of the comparison is to learn if the two data sets come from the same population of data, or do not.

1. **The *t*-statistic must be computed from both data sets.** The *t*-statistic value is given as a function of the total number of data points and the value of each using Equations A1.4 to A1.6 below. The data are the values in Tables 4.3 and 4.4.

   The defining equation for the *t*-statistic is Equation (A1.4A).

$$t = \frac{(\bar{x} - \mu)}{\sqrt{\dfrac{(s^2 - \sigma^2)}{n}}} \tag{A1.4A}$$

Equation (A1.4A) can be restated for the situation involving two general sets of data, instead of one set of data and a population.[19] This becomes

[18] See Chapter 5, Section 12.

[19] Equation (A1.4A) is the general form of the definition of the *t*-statistic.

**Table A1.4 Hexane – IPA Extraction Validation Cleaning Test**

| Question | Answer |
|---|---|
| **Did the Hexane–IPA validation extraction process remove all the soil as expected?**<br><br>The answer is learned via comparing the cumulative average of measured cleanliness with the goal (100%) | From Table A1.3, measured cleanliness values are indistinguish able frome100% clean – especially after ten measurements.<br><br>However, this is not an situation where the "t-test" should be used to compare the mean of the cleanliness measurements with the goal values. This is because the "t-test" includes differences between means and the variability within each – and the goal value has no variability.<br>If a "t-test" is performed using the measured data in the center column and the goal value in the right-hand column of Table A1.3, the result is 53.8%. This result of (1-@TTEST), being remote from ~95%, clearly reflects that the measured and goal values of cleanliness aren't from the same population. That shouldn't be expected because the measured values have vari ability, and the goal value has none. |
| | A "t-test" should not be used to compare a measured data set to a constant-value limit. That's why, in Chapter 4 Section 6.1, past data of successful operation are recommended as a benchmark for cleanliness, rather than a fixed number. **Use of a fixed value as a benchmark renders the "t-test" useless, and it's too powerful a tool to discard!** |
| **Did the normally-used IPA extraction process remove all the soil as expected?**<br><br>The answer is learned by comparing the center column in Table A1.2 with the center column in Table A1.3 (which is known from the above question to be 100% removal). | For the raw data which produced Table A1.2, @TTEST is 13.4% or (1−@TTEST) is 87.6%.<br>This means that there is an 87.6% chance that the cleaning process IPA extraction (center column of Table A1.2) results *are different* than the validation (right-hand column of Table A1.2) test results.<br>**Thus the answer is: there is only an 87.6% chance that the IPA extraction is a useful cleaning test.** That's not good enough. Another routine cleaning test is needed. |
| **How many cleaning tests are needed to show that the cleaned parts were not cleaned well enough with the IPA extraction test?**<br><br>The answer is learned by comparing the center column in Table Appendix 1.2 to the center column in Table Appendix 1.3 – at various numbers of cleaning tests increasing from one. | For the raw data which produced Table A1.2 and A1.3, (1−@TTEST) is 99.50% at 10 samples.<br>This means that the 10 results which produced an average in the center column of Table A1.2 of 98.45% clean are different from the 10 results which produced an average in the center column of Table A1.3 of 100.00% clean.<br>**So, if 10 parts are tested, there is only a ½% chance that the cleaning work with IPA is done to the 100.00% level of soil removal.** |

Equations (A1.4B) and (A1.4C) which are the same fundamental equation as Equation (A1.4A).

$$t = \frac{\left[\dfrac{\text{Difference of means}}{\text{Standard error of difference} = \text{SE}}\right]}{\sqrt{\text{Weighted number of data points}} = \text{NM}} \tag{A1.4B}$$

$$t = \frac{\left[\dfrac{\text{Difference of means}}{\text{SE}}\right]}{\sqrt{\text{NM}}} \tag{A1.4C}$$

Equation (A1.5)[20] allows computation of the standard error (SE) associated with the difference between means of the data sets.

[20] The reason for the deductions of 1 and 2 in Equation (A1.5) is that the SE is not an "independent" statistic. It is the difference between two means. Each mean is not an "independent" statistic either. Each mean is dependent upon the data from which it was

**Calculations of Standard Deviations Based on Data in Tables 4.3 and 4.4**

| | |
|---|---|
| $S_1$ = the standard Deviation of data from Table 4.3 | $n_1 = 6$ and $\bar{x}_1 = 8.4$ |
| $s_1 = \sqrt{(6.3 - 8.4)^2 + (11.0 - 8.4)^2 + (4.2 - 8.4)^2 + (9.2 - 8.4)^2 + (15.0 - 8.4)^2 + (4.7 - 8.4)^2/(6 - 1)} = 4.16$ | |
| $S_2$ = the standard deviation of data from Table 4.4 | $n_2 = 6$ and $\bar{x}_2 = 8.13$ |
| $s_2 = \sqrt{(8.8 - 8.13)^2 + (8.2 - 8.13)^2 + (8.4 - 8.13)^2 + (7.7 - 8.13)^2 + (7.7 - 8.13)^2 + (8.0 - 8.13)^2/(6 - 1)} = 0.43$ | |

$$SE = \sqrt{\frac{(n_1 - 1) \times s_1^2 + (n_2 - 1) \times s_2^2}{(n_1 + n_2 - 2)}} \quad (A1.5)$$

Equation (A1.6) allows computation of the weighted number of total data points (NM).

$$\sqrt{NM} = \sqrt{\frac{1}{n_1} + \frac{1}{n_2}} \quad (A1.6)$$

For the data, from Chapter 4, Section 4.5, the "*t*" analysis starts with the calculation of the standard deviations to be used in Equation A1.5 for data sets 1 and 2. Please see the Table above.

The value of SE, for *both* data sets, is computed from Equation (A1.5).

$$SE = \sqrt{\frac{(6 - 1) \times 4.16^2 + (6 - 1) \times 0.43^2}{(6 + 6 - 2)}} = 2.96$$

The value of weighted number of *total* samples (NM), for *both* data sets, is computed from Equation (A1.6).

$$\sqrt{NM} = \sqrt{\frac{1}{6} + \frac{1}{6}} = 0.577$$

**Table A1.5 Decision Table for t-Test**

| t-Statistic from Data > Reference t-Statistic | t-Statistic from Data < Reference t-Statistic |
|---|---|
| Data sets DON'T represent the same population | Data sets DO represent the same population |
| 0.156 > **0.064** | |

The value of the *t*-statistic[21] for the two data sets is computed from Equation (A1.4C).

$$t = \frac{\left[\frac{8.4 - 8.13}{2.96}\right]}{0.577} = 0.156$$

2. **The reference *t*-statistic** must be computed at the same number of samples (12), and the desired level of confidence (95%). This is the target value to be able to show that both data sets represent the same population.

   The reference *t*-distribution is often presented as a table, where number of samples and levels of confidence are the boundaries, and *t*-value are the table entries. Such an organization is shown in Table A1.5.[22]

   From Table A1.5, to achieve 95% certainty that the two data sets represent the same population

---

calculated. A value of 1 is subtracted from the number of data points from which each mean is based. A value of 2 is subtracted from the total number of data points when the difference between two means is tabulated. Most commonly, managers seeking to compare two means will use the procedure of Table A1.5 if they don't directly use the spreadsheet functions.

[21] Values of the *t*-statistic are always expressed as positive numbers. It does not matter which mean is subtracted from the other. Only the difference matters.

[22] **ATTENTION: This is not the standard presentation found in statistics books. Since this volume is about cleaning, the science of statistics is used but not covered here.** Table A1.5 simply uses the number of total test values and the required probability that the two means represent the same population.

The "correction" for "degrees of freedom" has already been made in the construction of the table. Please note that since there are two means being compared, there is no value of the *t*-statistic for only two total test values.

**Table A1.5 t-Test Values**

| No. of Total Data Points | Percent Chance that Two Means Represent the Same Population | | | | | | | | | | | |
|---|---|---|---|---|---|---|---|---|---|---|---|---|
| | 85% | 87.5% | 90% | 91% | 92% | 93% | 94% | 95% | 96% | 97% | 98% | 99% |
| 1 | | | | | | | | | | | | |
| 2 | | | | | | | | | | | | |
| 3 | 0.215 | 0.178 | 0.142 | 0.128 | 0.114 | 0.099 | 0.085 | 0.071 | 0.057 | 0.042 | 0.028 | 0.014 |
| 4 | 0.206 | 0.171 | 0.137 | 0.123 | 0.109 | 0.095 | 0.082 | 0.068 | 0.054 | 0.041 | 0.027 | 0.014 |
| 5 | 0.202 | 0.168 | 0.134 | 0.120 | 0.107 | 0.094 | 0.080 | 0.067 | 0.053 | 0.040 | 0.027 | 0.013 |
| 6 | 0.199 | 0.166 | 0.132 | 0.119 | 0.106 | 0.092 | 0.079 | 0.066 | 0.053 | 0.040 | 0.026 | 0.013 |
| 7 | 0.197 | 0.164 | 0.131 | 0.118 | 0.105 | 0.092 | 0.078 | 0.065 | 0.052 | 0.039 | 0.026 | 0.013 |
| 8 | 0.196 | 0.163 | 0.130 | 0.117 | 0.104 | 0.091 | 0.078 | 0.065 | 0.052 | 0.039 | 0.026 | 0.013 |
| 9 | 0.195 | 0.162 | 0.130 | 0.117 | 0.104 | 0.091 | 0.078 | 0.065 | 0.052 | 0.039 | 0.026 | 0.013 |
| 10 | 0.195 | 0.162 | 0.129 | 0.116 | 0.103 | 0.090 | 0.077 | 0.064 | 0.052 | 0.039 | 0.026 | 0.013 |
| 11 | 0.194 | 0.161 | 0.129 | 0.116 | 0.103 | 0.090 | 0.077 | 0.064 | 0.051 | 0.039 | 0.026 | 0.013 |
| **12** | 0.194 | **0.161** | 0.129 | 0.116 | 0.103 | 0.090 | 0.077 | **0.064** | 0.051 | 0.038 | 0.026 | 0.013 |
| 13 | 0.193 | 0.161 | 0.128 | 0.115 | 0.103 | 0.090 | 0.077 | 0.064 | 0.051 | 0.038 | 0.026 | 0.013 |
| 14 | 0.193 | 0.160 | 0.128 | 0.115 | 0.102 | 0.090 | 0.077 | 0.064 | 0.051 | 0.038 | 0.026 | 0.013 |
| 15 | 0.193 | 0.160 | 0.128 | 0.115 | 0.102 | 0.089 | 0.077 | 0.064 | 0.051 | 0.038 | 0.026 | 0.013 |
| 16 | 0.192 | 0.160 | 0.128 | 0.115 | 0.102 | 0.089 | 0.077 | 0.064 | 0.051 | 0.038 | 0.025 | 0.013 |
| 17 | 0.192 | 0.160 | 0.128 | 0.115 | 0.102 | 0.089 | 0.076 | 0.064 | 0.051 | 0.038 | 0.025 | 0.013 |
| 18 | 0.192 | 0.160 | 0.128 | 0.115 | 0.102 | 0.089 | 0.076 | 0.064 | 0.051 | 0.038 | 0.025 | 0.013 |
| 19 | 0.192 | 0.160 | 0.127 | 0.115 | 0.102 | 0.089 | 0.076 | 0.064 | 0.051 | 0.038 | 0.025 | 0.013 |
| 20 | 0.192 | 0.159 | 0.127 | 0.115 | 0.102 | 0.089 | 0.076 | 0.064 | 0.051 | 0.038 | 0.025 | 0.013 |
| 21 | 0.192 | 0.159 | 0.127 | 0.114 | 0.102 | 0.089 | 0.076 | 0.063 | 0.051 | 0.038 | 0.025 | 0.013 |
| 22 | 0.191 | 0.159 | 0.127 | 0.114 | 0.102 | 0.089 | 0.076 | 0.063 | 0.051 | 0.038 | 0.025 | 0.013 |
| 23 | 0.191 | 0.159 | 0.127 | 0.114 | 0.102 | 0.089 | 0.076 | 0.063 | 0.051 | 0.038 | 0.025 | 0.013 |

with a total of 12 data points, the *t*-statistic *must be less than* ***0.064***.[23] See the number in **red**.

3. **The *t*-statistic computed from the data and the reference *t*-statistic are compared, and the following rule is used to make a judgment.** Said without the table, the *t*-statistic from the data must be less than the reference *t*-statistic for the two data sets to have come from the same population.

   The *t*-statistic value calculated in step 1 from the two data sets is 0.156 which is greater than the reference *t*-statistic computed in step 2 of **0.064**. Thus it *cannot* be stated with 95% certainty that the two data sets represent the same population.

   The negligible payback associated with increased testing is clearly shown in Table A1.5. Beyond a number of 10, the *t*-statistic does not significantly change. This is equivalent to a total of 12 cleanliness tests for both means.[24] In other words, more data isn't going to change the conclusion that the data in Tables 4.3 and 4.4 don't come from the same populations.

---

[23] Please note: for two means to represent the same population, they must be closely valued. Thus the *t*-statistic must be a small number – for any level of certainty and number of measurements. If two means are not closely valued, it will take many measurements or an acceptance of low certainty (confidence) to show they represent the same population – probably because they don't.

[24] See Equation (A1.5). The difference between 12 and 10 is the number 2 in Equation (A1.5). This is the number of means being compared.

**Table A1.6 Part Weight Measurements**

| ID for Set A | Value (g) | ID for Set B | Value (g) |
|---|---|---|---|
| 1 | 20.0294 | 1 | 20.4349 |
| 2 | 20.2208 | 2 | 19.5271 |
| 3 | 19.5884 | 3 | 19.5403 |
| 4 | 19.5703 | 4 | 19.7645 |
| 5 | 20.1712 | | |
| 6 | 18.9993 | | |
| 7 | 21.1315 | | |
| 8 | 18.9875 | | |

**Table A1.7 Confidence Limits**

| Sample Size | Possibility About Confidence Limits |
|---|---|
| 1 | Be serious |
| 2 | Waste of time |
| 4 | Little or no idea |
| 16 | Good enough for industrial work |
| 64 | Extremely high level of confidence |
| 256 | Essentially perfect certainty |
| 1024 | Waste of resources – too many samples taken |

If the value of 0.156 is entered in Table A1.5, for 12 total samples, one can see that the certainty that the two data sets represent the same population is slightly greater than 87.5% ($t$-statistic = **0.161**).[25] See the number in **purple**.

The same result is obtained, easier, using the spreadsheet functions. That's the result given in Chapter 1, Section 4.6, where the spreadsheet formula @TTEST is used (87.91%).

This author believes that managers make better decisions when they use statistical methods. Chapter 4 shows how to that. Appendix 1 shows the basis behind it. Spreadsheet functions allow that and dilute the opportunity to be confused by nomenclature.

### A1.6.5 Additional Use of the *t*-Test

Data other than cleaning test data can be examined with the $t$-test.

Manufacturers of small stamped parts will often weigh parts as a way of comparing their overall thickness – assuming that the metal density is a constant.

Using the methodology of Table A1.5:

- Count the number of measurements in each data set ($n_1 = 8$ and $n_2 = 4$). The data sets can be interchanged.
- Calculate the means of each set using Equation (A1.1) ($\bar{x}_1 = 19.8373$ g and $\bar{x}_2 = 19.8167$ g).
- Calculate the two standard deviations of each data set using Equation (A1.2) ($s_1 = 0.7099$ g and $s_2 = 0.4263$ g).
- From Equation (A1.5), SE = 0.5175 g.
- From Equation (A1.6), $\sqrt{\text{NM}} = 0.6124$:

$$t = \frac{\left[\left(\dfrac{19.8373 - 19.8167}{0.6382}\right)\right]}{0.6124} = 0.065$$

Using Table A1.5, with number of data points as 12 (= 8 + 4),[26] for 95% confidence, the $t$-statistic is 0.064. Consequently, the probability that these two data sets represent the same population is about 95%.

---

[25] Actual certainty that the two data sets represent the same population can be calculated via the spreadsheet formula =TDIST(0.161,10) or @TDIST(0.161,10). Converted to a percent, the result is 87.91%.

[26] Please note that when using Equations A1.4–A1.6, the number of data points can be different in each set ($n_1$ versus $n_2$). But when using the spreadsheet function @TTEST, the number of data points must be equal.

Please also note that the averages, standard deviations, and standard error values all have the native units of the original data. They are not dimensionless. The statistic "t" can be computed with the spreadsheet function @COUNT, @AVG, and @STD. The following formula, based on these functions, can be used to calculate "t" with data sets contain different numbers of measurements: "t" = ((@AVG(DataSet#1)-@AVG(DataSet#2))/((@SQRT((((@COUNT(DataSet#1)-1)*(@STD(DataSet#1)^2))-((@COUNT(DataSet#2)-1)*(@STD(DataSet#2)^2)))/((@COUNT(DataSet#1)1@COUNT(DataSet#2)-2)))*((@SQRT((1/@COUNT(DataSet#1)+(1/(@COUNT(DataSet#2))))))

## A1.7 CONFIDENCE LIMITS

The concept of confidence[27] is employed in the use of the *t*-test.[28] There, the concept represented the user's desire to know whether a set of measurements was representative of a population of measurements.

A numerical value of confidence level is the probability that the sample (with a stipulated level of measurement error) is representative of the population of measurements. A confidence level is not a hope or a guess. Rather, it is based on the assumption that the measured data are random – not produced with a bias. That is, unexpected or spurious variables don't affect the outcome.[29]

Confidence limits are derived from the same assumption. They are used, not to compare sets of measurements as with the *t*-test, but to learn the range over which data without bias will be distributed at a stated level of certainty. The level of certainty is the level of confidence.

Data not included within a confidence limit is considered at 100%, less the percent level of certainty, to not represent the population of data:

- A quantitative example statement might be that one can be 95% certain that data are within 6.8% on either side of the data mean. Said another way, the 95% confidence limit is between 0.932 and 1.068 for data whose the mean is 1.000. The confidence limit (interval) is the likely range of the true value.[30]
- An equivalent statement is that a data point whose value was 0.915 or 1.072 should only be expected to be part of the measurements no more than 5% (100 less 95%) of the time.[31]

The units of confidence limits are the native units of the data.

Confidence limits (intervals) become more narrow when more measurements are made, as shown with assumed information in Table A1.7.

The equation used for calculation of confidence limits is Equation (A1.4), restated as Equation (A1.7A).

$$\bar{X} - \left\{ t \times \frac{s}{\sqrt{n}} \right\} \leq \bar{X} \leq \bar{X} + \left\{ t \times \frac{s}{\sqrt{n}} \right\} \quad \text{(A1.7A)}$$

In other words, the range of confidence limits on either side of the mean is given by Equation (A1.7B).

$$\text{Range} = \pm \left| t \times \frac{s}{\sqrt{n}} \right| \quad \text{(A1.7B)}$$

The *t*-statistic, used for calculating confidence limits in Equation (A1.7B), is computed using its definition as Equation (A1.4). Since the definition of the *t*-statistic included $n$ (the number of observations), and the formula for the range of confidence limits also includes $n$, it is convenient to combine them.

Results are given in Table A1.8.[32] The sample standard deviations is multiplied by the value in Table A1.8 to compute the range.

[27] See Chapter 4, Section 4.2.9.

[28] Snedecor, G.W. and Cochran, W. G., *Statistical Methods* (8th ed.), Iowa State University Press 1989.

[29] Sources of bias might be non-random sampling (measuring the next six produced clean parts, selecting parts for testing when a certain operator is on duty, testing only the high-value parts, or sampling just before or after a chemical change out). Another source of bias might be taking only a few samples, (see Chapter 4, Section 4.3.3).

[30] If the confidence interval overlaps a data point, the data point is said to be statistically significant within the stated percent confidence.

[31] Said another way, one could exclude (discard) the 0.915 or 1.072 values from the data set and be 95% certain that they would not distort the meaning of the measurements. The risk of that action is that 5% of the time those values contribute to the meaning of the measurements, and to discard them would distort that meaning. One can never discard measurements outside any limits and be 100% certain that they were not part of the expected population.

[32] **The basis for use of Table A1.8 is not the same as for Table A1.5.**

The left-hand column in Table A1.8 is the total number of test samples. When the percent confidence is chosen, the value read is the number of standard deviations on either side of the mean into which the chosen percentage of test values is expected to occur. The parameter in Table A1.8 includes the number of test values in both the numerator (implicitly) and the denominator (directly). It is given in Equation (A1.7B).

$$\pm \frac{t}{\sqrt{n}} \quad \text{(A1.7B)}$$

**Table A1.8 Confidence Limits as Number of Standard Deviations Around Mean**

| No. of Total Data Points | 85% | 87.5% | 90% | 91% | 92% | 93% | 94% | 95% | 96% | 97% | 98% | 99% |
|---|---|---|---|---|---|---|---|---|---|---|---|---|
| 1 | | | | | | | | | | | | |
| 2 | 2.945 | 3.555 | 4.464 | 4.968 | 5.597 | 6.405 | 7.480 | 8.985 | 11.239 | 14.994 | 22.501 | 45.012 |
| 3 | 1.317 | 1.476 | 1.686 | 1.792 | 1.917 | 2.066 | 2.250 | 2.484 | 2.799 | 3.258 | 4.021 | 5.730 |
| 4 | 0.962 | 1.057 | 1.177 | 1.235 | 1.303 | 1.381 | 1.475 | 1.591 | 1.741 | 1.948 | 2.270 | 2.920 |
| 5 | 0.795 | 0.866 | 0.953 | 0.996 | 1.043 | 1.098 | 1.163 | 1.242 | 1.341 | 1.475 | 1.676 | 2.059 |
| 6 | 0.694 | 0.752 | 0.823 | 0.856 | 0.894 | 0.938 | 0.989 | **1.049** | 1.125 | **1.226** | 1.374 | 1.646 |
| 7 | 0.624 | 0.674 | 0.734 | 0.763 | 0.795 | 0.832 | 0.874 | 0.925 | 0.987 | 1.069 | 1.188 | 1.401 |
| 8 | 0.572 | 0.616 | 0.670 | 0.695 | 0.723 | 0.755 | 0.792 | 0.836 | 0.890 | 0.960 | 1.060 | 1.237 |
| 9 | 0.531 | 0.571 | 0.620 | 0.643 | 0.668 | 0.697 | 0.730 | 0.769 | 0.816 | 0.878 | 0.965 | 1.118 |
| 10 | 0.498 | 0.535 | 0.580 | 0.601 | 0.624 | 0.650 | 0.680 | 0.715 | 0.758 | 0.814 | 0.892 | 1.028 |
| 11 | 0.470 | 0.505 | 0.546 | 0.566 | 0.587 | 0.612 | 0.639 | 0.672 | 0.711 | 0.762 | 0.833 | 0.956 |
| 12 | 0.447 | 0.479 | 0.518 | 0.537 | 0.557 | 0.579 | 0.605 | 0.635 | 0.672 | 0.719 | 0.785 | 0.897 |
| 13 | 0.427 | 0.457 | 0.494 | 0.511 | 0.530 | 0.552 | 0.576 | 0.604 | 0.639 | 0.682 | 0.744 | 0.847 |
| 14 | 0.409 | 0.438 | 0.473 | 0.490 | 0.507 | 0.528 | 0.551 | 0.577 | 0.610 | 0.651 | 0.708 | 0.805 |
| 15 | 0.393 | 0.421 | 0.455 | 0.470 | 0.487 | 0.506 | 0.528 | 0.554 | 0.585 | 0.624 | 0.678 | 0.769 |
| 16 | 0.379 | 0.406 | 0.438 | 0.453 | 0.469 | 0.488 | 0.509 | 0.533 | 0.562 | 0.599 | 0.651 | 0.737 |
| 17 | 0.367 | 0.393 | 0.423 | 0.438 | 0.453 | 0.471 | 0.491 | 0.514 | 0.542 | 0.578 | 0.627 | 0.708 |
| 18 | 0.355 | 0.380 | 0.410 | 0.424 | 0.439 | 0.456 | 0.475 | 0.497 | 0.524 | 0.558 | 0.605 | 0.683 |
| 19 | 0.345 | 0.369 | 0.398 | 0.411 | 0.426 | 0.442 | 0.460 | 0.482 | 0.508 | 0.541 | 0.586 | 0.660 |
| 20 | 0.335 | 0.359 | 0.387 | 0.399 | 0.414 | 0.429 | 0.447 | 0.468 | 0.493 | 0.525 | 0.568 | 0.640 |
| 21 | 0.327 | 0.349 | 0.376 | 0.389 | 0.402 | 0.418 | 0.435 | 0.455 | 0.479 | 0.510 | 0.552 | 0.621 |
| 22 | 0.319 | 0.341 | 0.367 | 0.379 | 0.392 | 0.407 | 0.424 | 0.443 | 0.467 | 0.496 | 0.537 | 0.604 |
| 23 | 0.311 | 0.333 | 0.358 | 0.370 | 0.383 | 0.397 | 0.413 | 0.432 | 0.455 | 0.484 | 0.523 | 0.588 |
| 24 | 0.104 | 0.111 | 0.119 | 0.122 | 0.126 | 0.131 | 0.136 | 0.142 | 0.148 | 0.157 | 0.168 | 0.187 |
| 25 | 0.103 | 0.110 | 0.118 | 0.122 | 0.126 | 0.130 | 0.135 | 0.141 | 0.148 | 0.157 | 0.168 | 0.186 |

*The parameter in Table A1.8 is that of Equation (A1.7B), not Equation (A1.4).*

Here are some examples:

- For the data set in Table 4.3, the mean is 8.4, the standard deviation is 4.16, and there are six measurements. The number of standard deviations for the 95% confidence interval is **1.049** from Table A1.8 – as the total number of tests is 6. So, 95% of all measurements from the population are expected to fall between 4.03 and 12.7 as the range is computed as from [8.4 − (**1.049** × 4.16)] to [8.4 + (**1.049** × 4.16)].
- For the data set in Table 4.4, the mean is 8.13, the standard deviation is 0.43, and there are 6 measurements. The number of standard deviations for the 97% confidence interval is **1.226** from Table A1.8 – as the total number of tests is 6. So, 97% of all measurements from the population are expected to fall between 3.29 and 13.50 and as the range is computed as from [8.4 − (**1.226** × 4.16)] to [8.4 + (**1.226** × 4.16)].

## A1.8 FREQUENCY OF TESTING

The basic idea is that managers want to spend only enough resources to test parts for cleanliness so they can tell if one group of parts is more or less clean than some other group of parts. The other group of parts is a standard of performance or a reference group.

Table 4.2 is based on the definition of the *t*-statistic, Equation (A1.4) restated as Equation (A1.8). For simplicity, all measurements are assumed to be normalized.[33]

The details are as follows:

$$t = \frac{\left[\dfrac{\text{Difference of means}}{\text{SE}}\right]}{\sqrt{n}} \quad \text{(A1.8)}$$

| | |
|---|---|
| Difference of means | = The expected error in Table 4.2. For a given level of confidence, you choose the difference within normalized data that you want to be able to recognize. |
| SE | = The normalized standard error of measurement, or how well the measurements of cleanliness correctly represent the actual cleanliness condition of the parts. |
| $n$ | = The number of measurements needed. |
| % Confidence | = Percent of the time the true value will lie within the estimate given. Said another way, the chance of being right. |
| $t$ | = Computed from the spreadsheet function @TINV (or =TINV). The arguments in this function are 1 − % confidence, and $n$. |

Here is an example. Suppose you want the amount of residue found on parts to be no more than 17.56 mg/SF. This is the standard of performance. Further, you want to be 97.5% sure that the level of NVR does not exceed 18.07 mg/SF. Analysis of past cleanliness testing has shown that the standard deviation for NVR measurements is 0.87 mg/SF.[34] How many samples must be tested to achieve this level of surety?

**Application 1**

Start by normalizing the data. Divide all NVR values by 17.56 so that the standard of performance is 1.00. Then:

| | |
|---|---|
| Difference of means | = (18.07 − 17.56)/17.56 or 0.03. |
| SE | = 0.87/17.56 or 0.05 |
| % Confidence | = 97.5% or 0.975 |
| $t$ | = @TINV(1−0.975,$n$), and |
| $t$ | = $[(0.03)/(0.05)]/\sqrt{n}$ |

The number of samples to be tested ($n$) is determined by solving the two expressions above for the *t*-statistic. It is convenient to do this in a table, and that's Table 4.2.

For an SE of 0.05, 97.5% confidence, and an expected error (difference of means) of 0.03, 17 samples must be tested[35] – see the numbers in brown.

**Application 2**

Suppose the reproducibility of residue measurements isn't as well done as above. Suppose the sample standard deviation was 1.26 mg/SF of residue. Now the measurement error (SE) isn't 0.05, it's (1.26/0.87) × 0.05 or 0.072.

This means more samples must be tested for cleanliness to recognize the same difference of means with the required level of confidence.

Table 4.2 is linear with respect to SE. It can be used with values other than 0.05 for expected measurement error.

Multiply the desired difference of means (0.03) by the ratio of the old and new SE values. The calculation is 0.03 × (0.87/1.26) = 0.0207, or 0.021.

Enter Table 4.2 at 97.5% confidence, read down until 0.021, and read left to **31** tests – see the numbers in dark purple.

The poorer reproducibility, roughly 50% worse, causes the required number of cleanliness tests to be nearly doubled.

[33] Whose mean is 1.00.

[34] This corresponds to excellent performance.

[35] Enter Table 4.2 at 97.5% confidence and read down until 0.03 and read left to **17**.

## A1.9 PROBABILITY DISTRIBUTIONS OF DISCRETE DATA

When using discrete data of cleaning performance (binary GO or NO GO outcomes), it is necessary to segregate performance. Because the data do not have a continuous range of outcomes, an analysis cannot be done continuously as operation proceeds.

The approach here produces an estimate of the *mean failure rate* over or during the units of cleaning work which have been segregated for analysis.

When another accumulation of cleaning work is analyzed, their mean failure rates can be compared via the methods of Table A1.4 above or Chaper 4, Section 4.6 (*t*-test).

The choice of size of the data set is somewhat arbitrary. Depending upon operating rate, one shift is probably too small. One month is too large. Consider one or a multiple of a day's operation.

### A1.9.1 Example of Failure Rate

The sampling interval of an operation was chosen for convenience as 100 minutes so that the operation included 15 intervals ($15 \times 100/60 = 25$ hours).

Operating rate was 550 units per hour, so that the data set included 13 750 units ($15 \times 550$).

Data of cleaning performance during about one day's operation were organized, and are displayed as Table A1.9.

The right hand column is the number of NO GO results from a cleaning test which were obtained in that interval. The data are plotted as well (Figure A1.2).

The plot shows an asymmetrical distribution of NO GO results – parts which were not satisfactorily cleaned.

This is not a situation where the number of NO GO results is roughly the same during every sampling interval! More the opposite.

This is a core concept.

- *When managing situations where the rate of failure is low,*[36] *relative to the number of items processed, the distribution of failures will be completely non-uniform over the sampling period.*

**Table A1.9 Probability Distributions of Discrete Data**

| Interval Number | Number of NO GO Cleaning Tests |
|---|---|
| 1 | 1 |
| 2 | 5 |
| 3 | 9 |
| 4 | 11 |
| 5 | 10 |
| 6 | 7 |
| 7 | 5 |
| 8 | 3 |
| 9 | 2 |
| 10 | 1 |
| 11 | 1 |
| 12 | 0 |
| 13 | 0 |
| 14 | 0 |
| 15 | 0 |

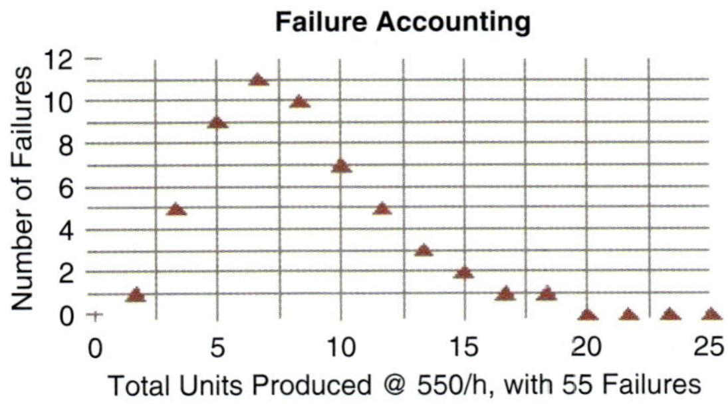


**Figure A1.2** Failure incidence at low-failure rate

Don't expect results during one sampling interval of a set of tests to be approximately equivalent to the preceding or succeeding one. The folly of testing during just one time interval should also be obvious.

There is a maximum number of NO GO results – which occurs here early in the operating period. *More failures are measured in sampling intervals after the maximum (29) than before it (15)*. That is a general result for this type of distribution of failures.

[36] To learn whether the failure rate in a situation is low, generate the above plot using current data. If the distribution of failures is relatively uniform, the failure rate is not low and the approach here is not necessary.

The Poisson Distribution[37] adequately represents this type of performance.[38] The Poisson Distribution assumes that failures are not the result of bias but are the result of random chance as in Chapter 4, Section 4.7.

The Poisson Distribution is the probability of an exact but very small number of events happening, given a average (mean) rate at which they might happen. The equation for the Poisson Distribution is given as Equation (A1.9). Conventionally, the two parameters are λ and $n$, where:

$$\frac{e^{-\lambda} \times \lambda^{n}}{n!} \tag{A1.9}$$

λ = The distribution mean.[39] That is the value sought from this analysis. The units of λ are number of NO GO results per sampling interval.

$n$ = The number of events to be considered. $n = 1$ is the first NO GO result in the sampling interval, $n = 2$ is the second NO GO result in the sampling interval, and so on.

The spreadsheet function which represents Equation (A1.9) is @POISSON(λ,$n$,1).[40]

Equation (A1.9) yields the probability that there are *exactly n* NO GO results during a sampling interval, when the overall failure rate is λ failures per sampling interval. $n$ is simply a counting parameter, while λ is a parameter to be adjusted to match the measured data.

The data above and the expected value [@POISSON(7,$n$,1)] are plotted in Figure A. 13 from $n = 1$ to $n = 25$. The units of λ in this example are an integer number of NO GO results/1000 units processed.

Please note that the predicted distribution of NO GO results does not match the measured values.

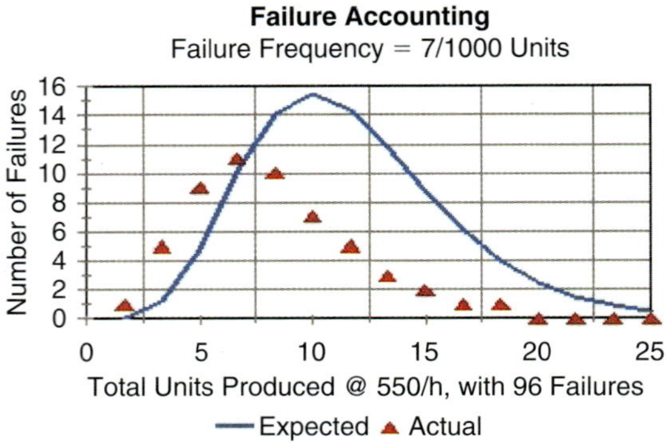


**Figure A1.3** Poisson distribution for seven failures

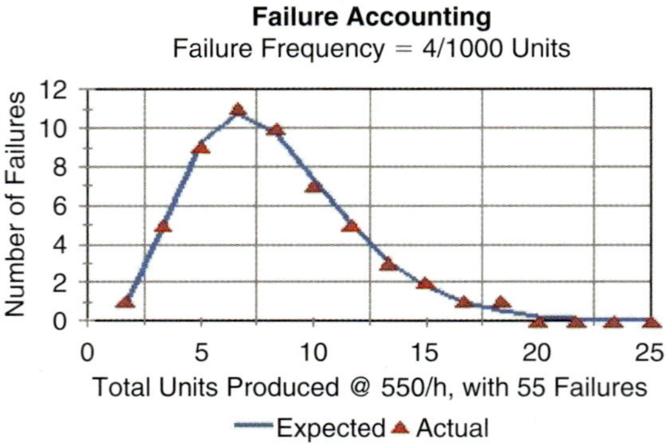


**Figure A1.4** Poisson distribution for four failures

This means that the correct value of λ (failure rate) is not 7 the assumed value of (Figure A1.3).

A spreadsheet allows easy choice of a better value of λ. Another plot shows that the value of failure rate which best describes the measured results is 4 NO GO results per 1000 units cleaned (Figure A1.4).

*This result, λ, should be recorded in a log. It represents the failure rate for one recent day (25 hours) of production.*

---

[37] Siméon-Denis Poisson (June 21, 1781 to April 25, 1840), was a French mathematician, geometer and physicist. The Poisson distribution is sometimes called the "law of small numbers." It is the probability distribution of the number of occurrences of an event that happens rarely but has very many opportunities to happen. That's how managers want failures of cleaning performance to occur.

[38] For time-distributed events, the Erlang distribution is the probability distribution of the amount of time until the nth event – the next failure of cleaning performance. The Poisson distribution is similar but different – the probability distribution of the number of events that would occur within a preset time. It will be used here because the Erlang Distribution is not provided as a spreadsheet function. Results would be similar if either distribution was used (see Figure 4.4).

Both distributions find substantial applications in modern technology: management of telephone switching systems, automobile traffic control systems, sizing of computer servers, management of rare diseases, and error analysis.

[39] The choice of interval sizes must produce a value of $\lambda > 1$.

[40] The value 1 is used to choose an instantaneous distribution. A value of 0 would produce the cumulative distribution. In Excel, the spreadsheet function would be =POISSON.

Successive values of λ for successive days of production should be compared to learn if the cleaning process is performing consistently.

This can be done with a separate plot over time, or using the @TTEST function described in Chapter 4, Section 4.6 or in Table A1.5.

As described in Chapter 4, Section 4.3.4, sampling should be done over all production – not just that before or after a solvent change-out, or some other event.

Evidence that sampling was not done over all production or that a special cause influenced operation would be that the Poisson Distribution did not describe the incidence of cleaning failures during the intervals chosen:

- *That's the reason for making the plot (Figure A1.4) of failure incidence versus Poisson Distribution – to determine if operation is unexpectedly being influenced by special causes.*[41]

## A1.10 CONTROL CHARTS FOR CLEANING PROCESSES

The are more than a half-dozen common techniques for creating control charts to manage product quality.[42] The few techniques selected from experience for this book are chosen because of the somewhat unique process control characteristics of cleaning operations:

- **Nothing happens quickly (generally).** The holdup time of a cleaning or rinsing bath, whether solvent or aqueous technology is practiced, is measured in multiples of minutes. The cycle time for cleaning is typically between five and 25 minutes for either technology.

  Detection and response within a fraction of a second, as would be required in a fighter aircraft or video game, is not needed. Control algorithms can detect change and respond to it within one hour.

  Relatively few measurements are necessary to obtain a useful average (mean) value. More than ten is never necessary. Often, four to six measurements are quite satisfactory. Occasionally, two or three will suffice.
- **Nothing happens in large measure (generally).** Elevation of boiling point due to soil intrusion is (or should be) only a few degrees temperature. Flows are either on or off. Compositional changes of soil or stabilizer as seen by direct or implied measurements don't exceed a few absolute percent in any increment of change. Part transport is on or off.
- **Required action is usually simple.** This comprises: start up, shut down, purge to distillation, increase purge to distillation or decanter by 5–20%, raise or lower heat or coolant input by 5–20%, etc. A finely judged, non-proportional response is never necessary.
- **Perfection is usually not needed.** One hundred percent cleaning quality is seldom required. While the yield loss of poor cleaning must be avoided, the consequences of that outcome are only financial and not life or environmental threatening.

In summary, control of a solvent (or aqueous) cleaning process is not "rocket science" and does not strain the capability of process control technology. Well-proven common technology will do just fine!

The proper tool for construction of "R," X-bar, and Cusum control charts is a spreadsheet. Familiarity

---

[41] Calculation of λ (failure rate) as 4 can be done via algebra – 55 total failures from Table A1.9 divided by 13,750 total units produced = 4 discrete failures per thousand units.

[42] Some types of control charts not covered in this volume are:

- Run charts which are line graphs that show data points plotted in the order in which they occur. Figure 4.6 is a run chart with hourly data.
- Median charts in which the center element of a data set is plotted. No arithmetic is required.
- An Exponentially Weighted Moving Average (EWMA) chart is used when it is desirable to detect out-of-control situations very quickly. The formulas involved are somewhat complex.
- A "P-chart" is one in which the measured probability of non-conformance is plotted. The chart is used to detect when external events change the rate of failure. The groups being compared can be of unequal size.
- An "np chart" is a P-chart when the groups are of equal size.
- A "C-chart" is a P-chart where the rate of probability of non-conformance is measured per unit of work.
- A "U chart" is a P-chart where rate of non-conformance is measured per inspection – when it is not possible to have an inspection unit of a fixed size.

with a spreadsheet is nearly universal as most homes have at least one computer. Examples of control charts constructed with a spreadsheet are provided in this book.

### A1.10.1 Elements of Control Charts

Control charts are plots of something on a vertical axis versus time[43] on a horizontal axis (Figure A1.5).

There are four elements which must be displayed against the vertical axis. They are:

1. **The actual value, at each time value, of the measured characteristic about the cleaning process or its performance.** This might be bath temperature, percent soil, or something else noted in Table 4.6 (inputs for "product by process" operation as in Chapter 4, Section 4.10) or Table 4.7 (outputs). Values are plotted as single points, usually connected in sequence by lines.
2. **A horizontal line representing the average value of the characteristic over the operation time covered by the control chart.** Typically, this defines the center of the vertical axis.
3. **A horizontal line representing the upper control limit** (UCL).
4. **A horizontal line representing the lower control limit** (LCL).

   The more certain the measurement the less is the span (distance or width or gap) between control limits.

The width of the span between the upper and lower control limits is inversely proportion to the number of measurements included in the average value of the characteristic. That is, the limits are more close to the average value when more measurements are included in that average.

The span between the upper and lower control limits is directly relates to the level of confidence required for determination that the average characteristic exceeds the control limit. The span is broader when the level is confidence in higher.

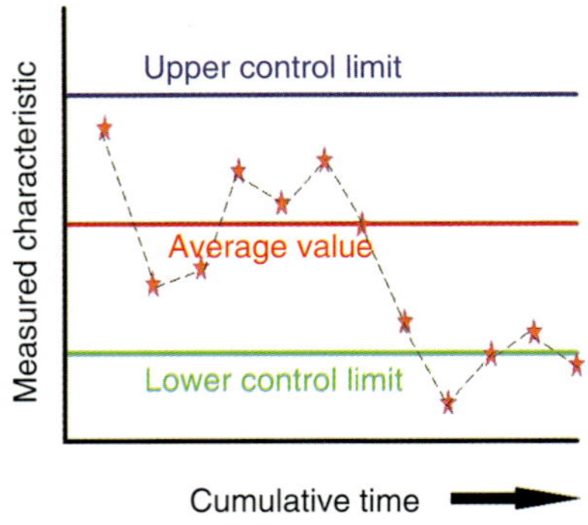


**Figure A1.5** Sample control chart

In general, the equation defining the control limits is:

$$\text{Either control limit} = \text{Average} \pm \text{Constant} \times \text{Standard deviation} \quad (A1.10)$$

In terms of symbols, where $L$ is a generalized constant:

$$\text{UCL} = \bar{X} + L \times \sigma \quad (A1.11)$$

$$\text{LCL} = \bar{X} - L \times \sigma \quad (A1.12)$$

If the process is in control, nearly all measured values of the chosen characteristic will lie between the two control limits. Points outside the control limits should be interpreted as that the process is out of control.

This construction, with all four elements, is shown in Figure A1.3. The data are fictitious. Please note how the latest four average measurements suggest that some cause (special or common) is at work.

### A1.10.2 Sample Data for Control Charts

A data set for use in constructing all three control charts is in Table A1.10.[44] Values are oil concentration in the solvent – measured in volume percent. The data reflect hourly measurements taken over

[43] Time means cumulative operation – hours on-line, for example. Time is often expressed, and plotted, implicitly. For example one can plot on the horizontal axis: consecutive lot or part number, cleaning cycle number, or number of parts cleaned. These latter items allow construction of a control chart when there are different time intervals associated with each unit of cleaning work.

[44] It was generated by the same model spreadsheet which generated the other figures which illustrate the principles expressed in Chapter 4, Sections 4.4 to 4.6. The data are shown in this packed table in order to conserve space. Hours are read horizontally. Days are read vertically.

**Table A1.10 Data Set for Construction of Control Charts (in Percentage)**

| Hour/ Day | 1 | 2 | 3 | 4 | 5 | 6 | 7 | 8 | 9 | 10 | 11 | 12 | 13 | 14 | 15 | 16 | 17 | 18 | 19 | 20 | 21 | 22 | 23 | 24 |
|---|---|---|---|---|---|---|---|---|---|---|---|---|---|---|---|---|---|---|---|---|---|---|---|---|
| 1 | 0.136 | 0.201 | 0.262 | 0.321 | 0.379 | 0.433 | 0.487 | 0.537 | 0.586 | 0.633 | 0.679 | 0.723 | 0.765 | 0.806 | 0.844 | 0.881 | 0.917 | 0.951 | 0.985 | 1.016 | 1.047 | 1.076 | 1.105 | 1.132 |
| 2 | 1.158 | 1.183 | 1.207 | 1.231 | 1.254 | 1.275 | 1.295 | 1.315 | 1.334 | 1.352 | 2.996 | 2.949 | 2.905 | 2.861 | 2.820 | 2.781 | 2.743 | 2.706 | 2.671 | 2.637 | 2.604 | 2.572 | 2.542 | 2.513 |
| 3 | 2.485 | 2.458 | 2.431 | 2.406 | 2.383 | 2.360 | 2.338 | 2.317 | 2.297 | 2.277 | 2.258 | 2.239 | 2.222 | 2.204 | 2.188 | 2.173 | 2.157 | 2.143 | 2.129 | 2.116 | 2.103 | 2.606 | 2.574 | 2.542 |
| 4 | 2.513 | 2.484 | 2.457 | 2.431 | 2.405 | 2.381 | 2.357 | 2.334 | 2.312 | 2.291 | 2.272 | 2.253 | 2.234 | 2.217 | 2.200 | 2.183 | 2.168 | 2.153 | 2.139 | 2.125 | 2.112 | 2.099 | 2.086 | 2.074 |
| 5 | 2.063 | 2.052 | 2.042 | **2.031** | 2.022 | 2.013 | 2.004 | 1.996 | 2.471 | 2.443 | 2.416 | 2.390 | 2.366 | 2.343 | 2.320 | 2.298 | 2.278 | 2.258 | 2.239 | 2.220 | 2.203 | 2.186 | 2.169 | 2.154 |
| 6 | 2.139 | 2.125 | 2.111 | 2.099 | 2.086 | 2.074 | 2.062 | 2.050 | 2.040 | 2.030 | 2.020 | 2.010 | 2.001 | 1.993 | 1.984 | 1.976 | 1.969 | 1.961 | 1.954 | 2.251 | 2.232 | 2.213 | 2.195 | 2.179 |
| 7 | 2.162 | 2.147 | 2.132 | 2.117 | 2.104 | 2.090 | 2.078 | 2.065 | 2.053 | 2.042 | 2.032 | 2.021 | 2.012 | 2.003 | 1.994 | 1.985 | 1.977 | 1.969 | 1.962 | 1.955 | 1.948 | 1.941 | 1.935 | 1.929 |
| 8 | 1.923 | 1.917 | 1.912 | 1.907 | 1.903 | 1.899 | 2.209 | 2.192 | 2.174 | 2.157 | 2.141 | 2.125 | 2.110 | 2.096 | 2.083 | 2.071 | 2.059 | 2.047 | 2.036 | 2.025 | 2.015 | 2.005 | 1.996 | 1.986 |
| 9 | 1.978 | 1.970 | 1.962 | 1.955 | 1.947 | 1.941 | 1.934 | 1.928 | 1.922 | 1.917 | 1.911 | 1.906 | 1.900 | 1.896 | 1.892 | 1.887 | 1.883 | 2.145 | 2.130 | 2.115 | 2.100 | 2.086 | 2.073 | 2.060 |
| 10 | 2.049 | 2.037 | 2.027 | 2.017 | 2.007 | 1.997 | 1.988 | 1.979 | 1.970 | 1.962 | 1.954 | 1.947 | 1.940 | 1.933 | 1.926 | 1.920 | 1.914 | 1.909 | 1.904 | 1.899 | 1.895 | 1.891 | 1.886 | 1.882 |
| 11 | 1.878 | 1.875 | 1.871 | 1.868 | 2.413 | 2.384 | 2.356 | 2.329 | 2.303 | 2.279 | 2.256 | 2.234 | 2.212 | 2.192 | 2.173 | 2.154 | 2.137 | 2.121 | 2.106 | 2.091 | 2.077 | 2.064 | 2.051 | 2.038 |
| 12 | 2.027 | 2.016 | 2.005 | 1.995 | 1.985 | 1.976 | 1.967 | 1.958 | 1.950 | 1.942 | 1.935 | 1.928 | 1.922 | 1.915 | 1.909 | 2.191 | 2.171 | 2.153 | 2.135 | 2.118 | 2.102 | 2.087 | 2.073 | 2.060 |
| 13 | 2.047 | 2.034 | 2.022 | 2.011 | 2.001 | 1.990 | 1.981 | 1.972 | 1.963 | 1.955 | 1.946 | 1.939 | 1.931 | 1.924 | 1.918 | 1.912 | 1.905 | 1.900 | 1.895 | 1.889 | 1.885 | 1.881 | 1.876 | 1.873 |
| 14 | 1.869 | 1.865 | 3.595 | 3.505 | 3.418 | 3.336 | 3.257 | 3.183 | 3.112 | 3.045 | 2.981 | 2.920 | 2.863 | 2.808 | 2.756 | 2.706 | 2.659 | 2.614 | 2.571 | 2.531 | 2.493 | 2.457 | 2.423 | 2.390 |
| 15 | 2.359 | 2.329 | 2.301 | 2.274 | 2.248 | 2.224 | 2.201 | 2.180 | 2.158 | 2.139 | 2.120 | 2.104 | 2.087 | 2.095 | 2.080 | 2.064 | 2.050 | 2.037 | 2.023 | 2.011 | 2.000 | 1.988 | 1.978 | 1.968 |
| 16 | 1.959 | 1.950 | 1.942 | 1.935 | 1.927 | 1.920 | 1.914 | 1.907 | 1.901 | 1.896 | 1.890 | 1.885 | 1.880 | 1.875 | 1.871 | 1.867 | 1.862 | 1.858 | 1.855 | 1.852 | 1.848 | 1.846 | 1.843 | 1.841 |
| 17 | 1.813 | 1.812 | 1.812 | 1.811 | 1.810 | 1.809 | 1.808 | 1.807 | 1.806 | 1.805 | 1.804 | 1.804 | 1.804 | 1.804 | 1.803 | 1.803 | 1.802 | 1.802 | 1.801 | 1.801 | 1.800 | 1.800 | 1.800 | 1.801 |
| 18 | 1.801 | 1.801 | 1.801 | 1.800 | 1.800 | 1.801 | 1.800 | 1.800 | 1.799 | 1.799 | 1.799 | 2.510 | 2.468 | 2.429 | 2.392 | 2.356 | 2.324 | 2.293 | 2.263 | 2.235 | 2.210 | 2.186 | 2.162 | 2.141 |
| 19 | 2.121 | 2.102 | 2.083 | 2.066 | 2.049 | 2.034 | 2.019 | 2.006 | 1.993 | 1.981 | 1.970 | 1.960 | 1.950 | 1.941 | 1.932 | 1.924 | 1.917 | 1.911 | 1.903 | 1.896 | 1.890 | 1.884 | 4.055 | 3.920 |
| 20 | 3.792 | 3.673 | 3.559 | 3.452 | 3.352 | 3.258 | 3.169 | 3.085 | 3.007 | 2.932 | 2.863 | 2.797 | 2.736 | 2.678 | 2.624 | 2.573 | 2.525 | 2.480 | 2.438 | 2.398 | 2.360 | 2.325 | 2.291 | 2.260 |
| 21 | 2.231 | 2.205 | 2.180 | 2.155 | 2.132 | 2.111 | 2.090 | 2.072 | 2.054 | 4.199 | 4.050 | 3.910 | 3.779 | 3.656 | 3.540 | 3.431 | 3.328 | 3.231 | 3.140 | 3.054 | | | | |

a period of about 3 weeks – five hundred values in all.

The purpose of this appendix is to construct three different control charts, and analyze the situation which produced these data. The control charts are "R" (*special causes*), X-bar (*common causes*), and CUSUM (process control).

There are some preliminary issues around this sample data set which need to be resolved before it is used:

1. Operation described in this table involves a startup from when the cleaning bath contained pure solvent. Initial concentrations of oil (soil) are quite low. Since the purpose of this analysis is to manage the cleaning bath to produce acceptable cleaning quality after startup, this initial data will be discarded. The first 99 data points (hours) will not be used. In other words, operation during the first ~four days will be neglected because oil concentration accumulated up to that point poses no threat to the desired level of cleaning quality. Both the decision to eliminate startup operation and the definition of startup length are arbitrary.

> *Analysis in this appendix*:
>
> The hundredth data point will be the first used (2.031% oil).

2. Data sets always have flaws – fliers, outliers, "snowbirds," etc. In order to maintain the quality of every data set, some action needs to be taken to diminish the influence of a single data point whose authenticity may be open to question. BUT just which ones are flawed?

   To deal with the flawed data, do not use personal bias to filter the data. It is conventional procedure to average individual measurements, and then use the average values in analysis with control charts. But what averaging protocol should be used?

   The choice is a tradeoff between response time and certainty.

- If more hourly data are included in an average, the effect of a single spurious value is diminished.
- But fewer hourly data will better represent a transition (change, expected or unexpected).

The chosen number of measurements in the hourly average is usually between two and ten. Common choices are three, four, five, and six values.

Don't discard fliers. Trust the statistical techniques you use.

> *Analysis in this appendix*:
>
> Given the occasional difficulty of collecting a well-mixed sample of oil in solvent and the occasional operator error in measurement of refractive index,[45] compounded by the occasional presence of a tramp soil component, this author prefers the larger choice for oil concentration. Five consecutive measurements were averaged and that homogenized value is the one used with control charts in this appendix.
>
> When there is less concern about errors, fewer values should be averaged. For example, two or possibly three values might be averaged for bath temperature. Three or four values of acidity or stabilizer content might be used for an average. Four or five values might be used for an surface tension value. And for measurements of cleanliness, three to five values might be used to generate an average value.

3. The construction of "R" and X-bar control charts here is being done after nearly 3 weeks of fictitious operation. They will be artifacts. That's OK. They won't be used for process control. A CUSUM control chart should be prepared on-line and used for that purpose.

[45] It is assumed that the content of oil, or other soil, in solvent is estimated via a direct measurement of refractive index and comparison to a calibration curve.

> *Analysis in this appendix*:
>
> The Construction of "R" control chart will be used to identify *special causes*[46] of variation and the X-bar control chart will be used to identify the existence of *common causes* of variation.

4. Data are not filtered for precision. The reported number of significant digits will be accepted.

## A1.11 CONSTRUCTING "R" CONTROL CHARTS

The "R" (range) control chart should be first constructed, and used. This is because special causes (massive oil intrusion) are usually more common, more easily identified, and can wreak more havoc than can common causes.

It is easier to process these calculations in a spreadsheet with the values in tabular format. The first thirty hourly values (after removal of the 99 startup values) from Table A1.10 are rearranged in Table A1.11.

The five values used to calculate each range, average, and standard deviation are arranged from left to right. Any experienced spreadsheet user can organize their measurements in this manner.

Population standard deviation of the range of the 397 (500 less 99 less 4) averaged oil concentrations in Column J is 0.159% Population average (mean) value of the range in Column J is 0.355%. Blanks are not included in these calculations.

The first action to take with most any set of data is to make a graph of it! This is shown in Figure A1.6 – range of oil concentration (Column J) in solvent vs time (Column B) or 5-value group number (Column A).

Any untrained observer can recognize the existence of some *special causes* – throughout the entire period of operation. Something has happened when range increases from the average value to ~three to five times the average!

No limit lines or statistical parameters are necessary to see that![47]

### A1.11.1 Specification of Control Limits

The two control limit lines are horizontal – one lower (LCL) and one upper (UCL). Each is symmetrical to the average (mean) range (0.355%). The two lines are plotted at a number of standard of standard deviation units (each of 0.159%) below and above the average range.

The general formula for either the upper or lower control limit is Equation (A1.10) – for range (R-bar), average ($\overline{X}$), or CUSUM (process control) charts. The question is: what constant should be used in Equation (A1.10)?

$$\text{Either control limit} = \text{Average} \pm \text{Constant} \times \text{Standard deviation} \tag{A1.10}$$

This author recommends use of the *t*-test to identify the values above and below the control limit lines that are not from the same population as the average (mean) value.

This approach allows inclusion of a chosen level of confidence in the selection of control limit lines, as well as a consequence of the choice of number of values in the hourly average.

To implement this choice, use Table A1.8 above. For 95% confidence, and 5 values in each average, the number of standard deviations between two means from different populations (different means) is found in Table A1.9 as 1.242:

- The UCL is a horizontal line based on the average 0.355% *plus* 1.242 × 0.159% or 0.553%.
- The LCL is a horizontal line based on the average 0.355% *less* 1.242 × 0.159% or 0.157%.

[46] The terminology of chance and the concept of different types of causes was developed by Dr. Walter Shewhart. Special causes have also been identified within the literature by and around Shewhart as assignable causes. A process where assignable causes (*special causes*) can be demonstrated is considered to be a process out of control See Chapter 4, Section 9.
Similarly, common causes have also been identified as chance causes.

[47] Writing both directly and whimsically, the observer should immediately seek to identify and eliminate those special causes rather than complete construction of this "R" control chart. When they return, the "R" control chart can be completed via addition of the limit lines.

**Table A1.11 Calculation of Averages**

| A # | B Hours | C Value (%) | D 1 (%) | E 2 (%) | F 3 (%) | G 4 (%) | H 5 (%) | I 6 (%) | J Range (%) | K Average (%) | L Std (%) |
|---|---|---|---|---|---|---|---|---|---|---|---|
| 1 | 100 | 2.031 | | | | | | | | | |
| 2 | 101 | 2.022 | | | | | | | | | |
| 3 | 102 | 2.013 | | | | | | | | | |
| 4 | 103 | 2.004 | | | | | | | | | |
| 5 | 104 | 1.996 | 1.996 | 2.004 | 2.013 | 2.022 | 2.031 | | 0.036 | 2.013 | 0.014 |
| 6 | 105 | 2.471 | 2.471 | 1.996 | 2.004 | 2.013 | 2.022 | | 0.475 | 2.101 | 0.207 |
| 7 | 106 | 2.443 | 2.443 | 2.471 | 1.996 | 2.004 | 2.013 | | 0.475 | 2.185 | 0.248 |
| 8 | 107 | 2.416 | 2.416 | 2.443 | 2.471 | 1.996 | 2.004 | | 0.475 | 2.266 | 0.244 |
| 9 | 108 | 2.390 | 2.390 | 2.416 | 2.443 | 2.471 | 1.996 | | 0.475 | 2.343 | 0.197 |
| 10 | 109 | 2.366 | 2.366 | 2.390 | 2.416 | 2.443 | 2.471 | | 0.105 | 2.417 | 0.041 |
| 11 | 110 | 2.343 | 2.343 | 2.366 | 2.390 | 2.416 | 2.443 | | 0.100 | 2.392 | 0.040 |
| 12 | 111 | 2.320 | 2.320 | 2.343 | 2.366 | 2.390 | 2.416 | | 0.097 | 2.367 | 0.038 |
| 13 | 112 | 2.298 | 2.298 | 2.320 | 2.343 | 2.366 | 2.390 | | 0.092 | 2.343 | 0.036 |
| 14 | 113 | 2.278 | 2.278 | 2.298 | 2.320 | 2.343 | 2.366 | | 0.088 | 2.321 | 0.035 |
| 15 | 114 | 2.258 | 2.258 | 2.278 | 2.298 | 2.320 | 2.343 | | 0.085 | 2.299 | 0.033 |
| 16 | 115 | 2.239 | 2.239 | 2.258 | 2.278 | 2.298 | 2.320 | | 0.081 | 2.279 | 0.032 |
| 17 | 116 | 2.220 | 2.220 | 2.239 | 2.258 | 2.278 | 2.298 | | 0.078 | 2.259 | 0.031 |
| 18 | 117 | 2.203 | 2.203 | 2.220 | 2.239 | 2.258 | 2.278 | | 0.075 | 2.239 | 0.030 |
| 19 | 118 | 2.186 | 2.186 | 2.203 | 2.220 | 2.239 | 2.258 | | 0.072 | 2.221 | 0.029 |
| 20 | 119 | 2.169 | 2.169 | 2.186 | 2.203 | 2.220 | 2.239 | | 0.069 | 2.203 | 0.027 |
| 21 | 120 | 2.154 | 2.154 | 2.169 | 2.186 | 2.203 | 2.220 | | 0.067 | 2.186 | 0.026 |
| 22 | 121 | 2.139 | 2.139 | 2.154 | 2.169 | 2.186 | 2.203 | | 0.064 | 2.170 | 0.025 |
| 23 | 122 | 2.125 | 2.125 | 2.139 | 2.154 | 2.169 | 2.186 | | 0.060 | 2.155 | 0.024 |
| 24 | 123 | 2.111 | 2.111 | 2.125 | 2.139 | 2.154 | 2.169 | | 0.058 | 2.140 | 0.023 |
| 25 | 124 | 2.099 | 2.099 | 2.111 | 2.125 | 2.139 | 2.154 | | 0.055 | 2.126 | 0.022 |
| 26 | 125 | 2.086 | 2.086 | 2.099 | 2.111 | 2.125 | 2.139 | | 0.053 | 2.112 | 0.021 |
| 27 | 126 | 2.074 | 2.074 | 2.086 | 2.099 | 2.111 | 2.125 | | 0.051 | 2.099 | 0.020 |
| 28 | 127 | 2.062 | 2.062 | 2.074 | 2.086 | 2.099 | 2.111 | | 0.049 | 2.086 | 0.019 |
| 29 | 128 | 2.050 | 2.050 | 2.062 | 2.074 | 2.086 | 2.099 | | 0.048 | 2.074 | 0.019 |
| 30 | 129 | 2.040 | 2.040 | 2.050 | 2.062 | 2.074 | 2.086 | | 0.046 | 2.063 | 0.018 |

Other approaches can be and are used to choose the LCL and UCL.[48]

The control limit lines, average range value, and hourly average range value are plotted in Figure A1.7, a conventional "R" range control chart.

This "R" control chart focuses attention on the many instances where the range of oil concentration has greatly changed from the average range – because of the assumed temporary 25% increase of oil loading which occurs for one hour each day. The on-aim

[48] Well-respected and traditional volumes about statistics, such as those mentioned in the first footnote, suggest use of a parameter that is independent of the required level of confidence as the multiplier of standard deviations used to locate the limit lines. A lucid derivation is found in Montgomery, pp. 210–211.

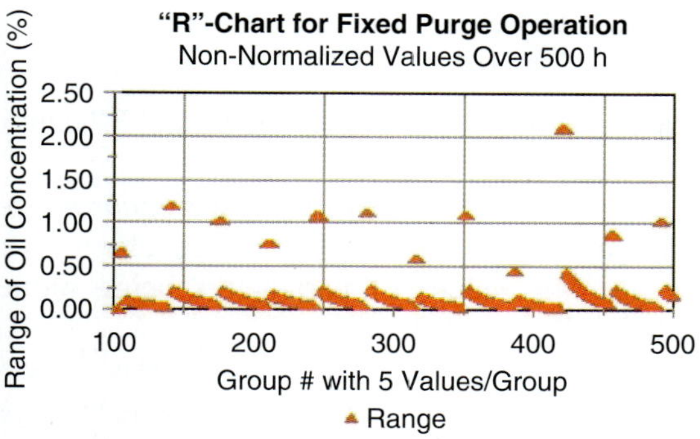


**Figure A1.6** Range values of oil concentration

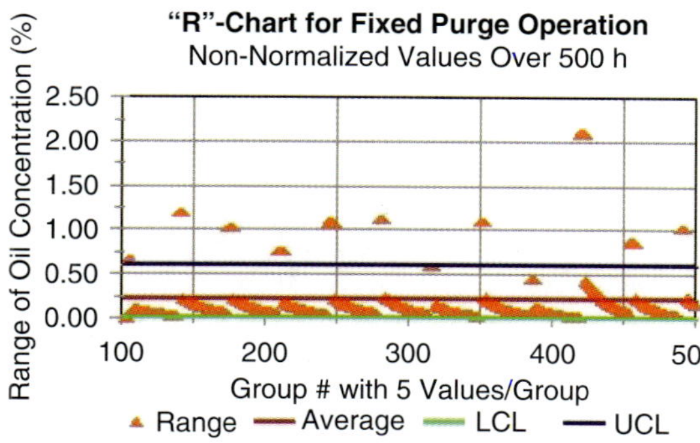


**Figure A1.7** "R" control chart of oil concentration

control of oil concentration via purge to distillation (or waste) is not capable of coping with the uncontrolled *application of stamping oil in upstream operation.*[49]

It is clear that the operating (simulated) unit which produced the data in Table A1.10 is in need of attention – especially upstream of the solvent cleaning unit.

This example shows the value of the "R" control chart – it galvanizes action. It should also suggest the timing of when a special cause(s) is present, and may allow deduction about their specific identity or general type.

## A1.12 CONSTRUCTING X-BAR CONTROL CHARTS

Please recall the X-bar is intended to illuminate *common causes* of process variation.

The X-bar is also constructed using the information in Table A1.11. Just have your spreadsheet graph the data in the column marked time (column B) or 5-value group number (column A) on a horizontal axis and the data in the column marked "average" (column K) on the vertical axis. The result is Figure A1.8.

---

The parameter suggested by Montogomery to be chosen as a constant which is a quotient. $L$ is calculated from the required chance that a range measurement won't exceed either control limit line, divided by the square root of the number of values in the hourly average. Montgomery's suggested equation is Equation (A1.13):

$$UCL_{range} \text{ or } LCL_{range} = \bar{X}_{range} \pm \mathrm{L}/\sqrt{n} \times \sigma_{range} \tag{A1.13}$$

The value of $L$ is chosen from the table at right. The traditional choice is for $L = 3$. That is where 99.73% of measurements are expected to lie within the control limit lines. This is the well-known as the 3-sigma limit. The also well-known 6-sigma limit requires 99.999998% of all measurements to be within control limits.

This author is not qualified to instruct a course in statistics. Nevertheless, there are four reasons why the above recommendation is made: (1) both approaches produce similar outcomes [the value 1.242 above would become 1.342], (2) it is worthwhile to be able to choose a level of confidence other than 95%, (3) it is simpler to use a consistent methodology to identify the significance of small differences, and (4) the focus of the "R" control chart should be more about the displacement of range values above or below the average range and less about the proximity of extreme ranges of any limit line.

Constant (**L**) used to calculate LCL & UCL

| **Change of Measurement** | | |
|---|---|---|
| **Inside Control Limit** | **Outside Control Limit** | **L** |
| 68.27% | 31.73% | 1 |
| 99.45% | 4.55% | 2 |
| 99.73% | 0.27% | 3 |
| 99.9937% | 0.0063% | 4 |
| 99.999943% | 0.000057% | 5 |
| 99.99998% | 0.000002% | 6 |

[49] It is the assumption which this author used to produce the data within this "R" control chart.

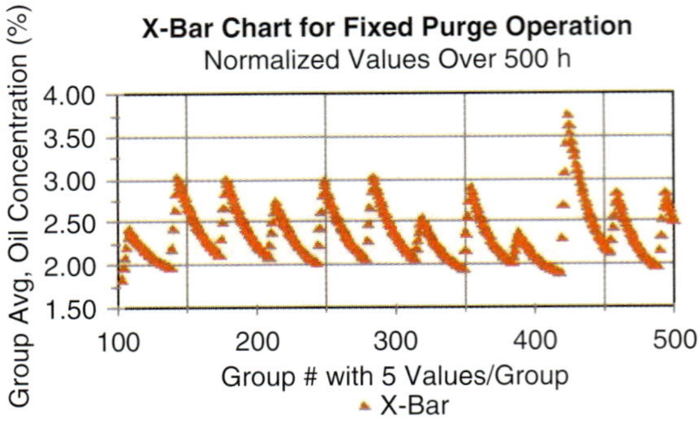


**Figure A1.8** Average values of oil concentration

Each point is an average of 5 measurements of oil concentration, plotted at the group number (time) of the last-measured member of that group. There is no homogenization of these measurements over a time period more than 5 hours.

The overall average of the 5-hour average values (column K) is 2.167% and the standard deviation is 0.355%.[50]

A significant change in any data point means a significant change has occurred over the 5-hour period associated with that data point. What standard is to be used to define significant? That's the purpose of the control limit lines.

The same approach is used as was used with the "R" control chart. The *t*-test is used to identify the values above the control limit lines that are not from the same population as the average (mean) value.

The value of $\{t\sqrt{n}\}$ as determined from Table A1.8 has not changed from 1.242 because the number of measurements being considered hasn't changed from 5 or the % confidence changed from 95%.

Consequently, the LCL is 1.726%. That's 1.242 × 0.355% less than the average of 2.167% (Figure A1.7).

Consequently, the UCL is 2.608%. That's 1.242 × 0.355% more than the average of 2.167%.

## A1.12.1 Difference of X-Bar Versus "R" Control Charts

*Actual averaged data* are plotted in an "R" control limit chart.

To make shifts easier to understand, *normalized values* are plotted within X-bar control charts – versus *absolute* values of parameters being plotted within "R" control charts.

The center line in an X-bar control chart is always 1, as the value used to normalize all time-averaged values is the overall average.

In this case that is 2.167% oil.

In addition to the overall average line, the individual 5-hour average values, the LCL, and the UCL values are all divided by 2.167% to produce normalized values.

Thus, the LCL becomes 0.797 (= 1 − [0.355% * 1.242/2.167%]) and the UCL becomes 1.203. (= 1 + [0.355% * 1.242/2.167%]).

Again this is easily done with a spreadsheet, as shown in Table A1.11. The result of normalized 5-hour average values and control limit lines is shown in Table A1.12.[51]

These control limit lines[52] are plotted, along with the overall average value and the individual averages from the above figure. The result is Figure A1.9.

There are major and minor exceedances of the UCL line, as well as major and minor exceedances of the LCL line.

This example also shows the value of the X-bar control chart. X-bar control charts, too, should galvanize action to identify and remove the *common causes* which have produced the deviations outside of control limits.

## A1.12.2 Use of X-Bar and "R" Control Charts

The X-bar control chart shown in Figure A1.9, with or without limits (Figure A1.8), is *useless*.

[50] Please note this is the same standard deviation as that calculated for use with the "R" control chart. The reason is that the same data were used.

[51] Please note that all values in the % oil column in Table A1.12 are identical to all values in the average column (K) of Table A1.11.

[52] Please note that as with "R" control limit charts, another approach to computing the UCL and LCL is described in well-respected and traditional volumes about statistics.

As above, the "*t*"-statistic divided by @SQRT(n) from Table A1.8 is used to multiply the overall data standard deviation to compute the LCL and UCLs. The same four reasons apply as noted above. The values of multiplier would change from 1.242 to 1.342 approach described in Montgomery, pp. 210–211, be used.

**Table A1.12**

| # | Hours | Value (%) | x̄ | Average | LCL | UCL |
|---|---|---|---|---|---|---|
| 5 | 104 | 2.013 | 0.929 | 1 | 0.797 | 1.203 |
| 6 | 105 | 2.101 | 0.970 | 1 | 0.797 | 1.203 |
| 7 | 106 | 2.185 | 1.008 | 1 | 0.797 | 1.203 |
| 8 | 107 | 2.266 | 1.046 | 1 | 0.797 | 1.203 |
| 9 | 108 | 2.343 | 1.081 | 1 | 0.797 | 1.203 |
| 10 | 109 | 2.417 | 1.116 | 1 | 0.797 | 1.203 |
| 11 | 110 | 2.392 | 1.104 | 1 | 0.797 | 1.203 |
| 12 | 111 | 2.367 | 1.092 | 1 | 0.797 | 1.203 |
| 13 | 112 | 2.343 | 1.081 | 1 | 0.797 | 1.203 |
| 14 | 113 | 2.321 | 1.071 | 1 | 0.797 | 1.203 |
| 15 | 114 | 2.299 | 1.061 | 1 | 0.797 | 1.203 |
| 16 | 115 | 2.279 | 1.052 | 1 | 0.797 | 1.203 |
| 17 | 116 | 2.259 | 1.042 | 1 | 0.797 | 1.203 |
| 18 | 117 | 2.239 | 1.033 | 1 | 0.797 | 1.203 |
| 19 | 118 | 2.221 | 1.025 | 1 | 0.797 | 1.203 |
| 20 | 119 | 2.203 | 1.017 | 1 | 0.797 | 1.203 |
| 21 | 120 | 2.186 | 1.009 | 1 | 0.797 | 1.203 |
| 22 | 121 | 2.170 | 1.001 | 1 | 0.797 | 1.203 |
| 23 | 122 | 2.155 | 0.994 | 1 | 0.797 | 1.203 |
| 24 | 123 | 2.140 | 0.987 | 1 | 0.797 | 1.203 |
| 25 | 124 | 2.126 | 0.981 | 1 | 0.797 | 1.203 |
| 26 | 125 | 2.112 | 0.975 | 1 | 0.797 | 1.203 |
| 27 | 126 | 2.099 | 0.969 | 1 | 0.797 | 1.203 |
| 28 | 127 | 2.086 | 0.963 | 1 | 0.797 | 1.203 |
| 29 | 128 | 2.074 | 0.957 | 1 | 0.797 | 1.203 |
| 30 | 129 | 2.063 | 0.952 | 1 | 0.797 | 1.203 |

This is because it neither *adds new information* nor impels a sense of urgency beyond that provided by the "R" control chart.

The reason is that they are both drawn from the same data set – a data set which is larded with *special causes* of process variation. That ranges are usually smaller in absolute value than are average measured values shouldn't conceal that the same periodic behavior is shown in each control chart.

This leads to an important lesson in process management. It is that there is no point in drawing or studying a X-bar control chart until the *special causes* which are highlighted in the "R" control chart are eliminated from having an effect upon the operation. In other words,

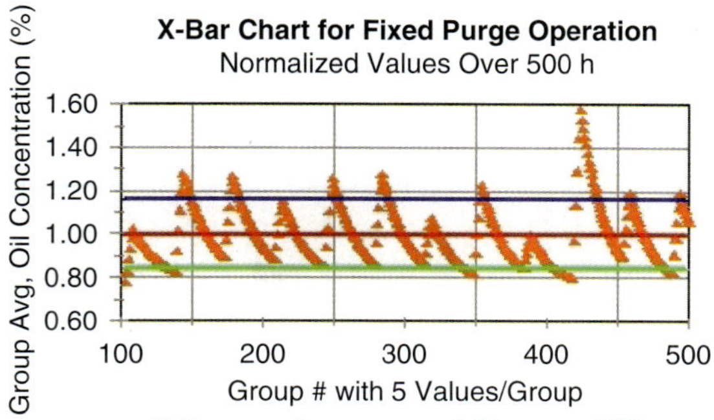


**Figure A1.9** X-bar chart with periodic variation

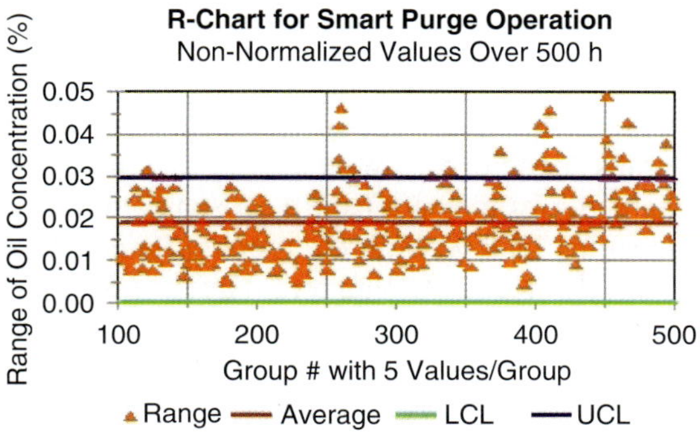


**Figure A1.10** "R" chart for smart purge operation

- One can't seek *common causes* until the special causes have been eradicated.

Once the periodic infusion of tramp oil from upstream operation is brought under control, the effect of common sources of process variation can be sought. The "R" and X-bar charts below are produced after upstream mis-operation is corrected.

The "R" control chart (Figure A1.10) shows no *special causes* are likely to be acting:

- The control limits for range are quite narrow. Thus the standard deviation for range is quite small. In other words, range values are nearly constant.
- There is some exceedance of these narrow control limits, but it doesn't suggest a substantial *special cause* is present.
- There is no correlation with time – the same causes appear to act at every instant.

The X-bar control chart (Figure A1.10) and the R control chart (Figure A.11) – neither based on Table

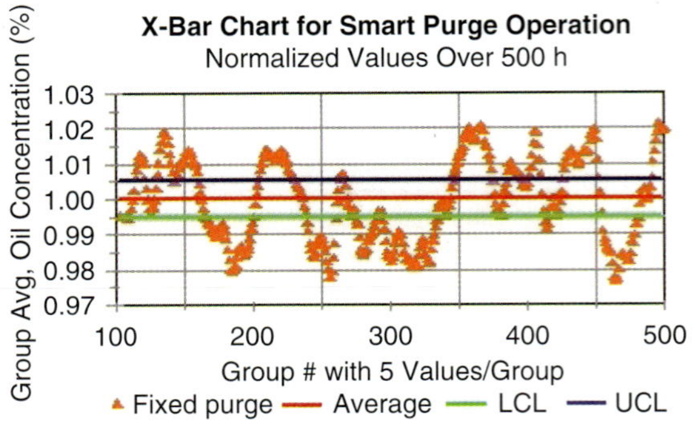


**Figure A1.11** X-bar chart for smart purge operation

A1.11 – plainly show *common causes* are certain to be acting:

- Control limits are routinely exceeded – yet the limits are quite narrow, indicating that these *common causes* may not have a substantial effect on cleaning quality. *BUT the probability is at least 95% that these exceedances represent operation with a different average performance than that within the control limit lines*. That may be ignored as inconsequential, but it has to be recognized because it's true.
- There is a pattern of behavior with time – outcomes appear to change sequentially. The oscillation suggests that the *common causes* seem to be self-correcting but their presence isn't being inhibited.

In fact, both control charts are produced by a system model with no periodic behavior and a small level of random noise in the measured parameter (oil concentration). This is not Table A1.10.

## A1.13 CUSUM CONTROL CHARTS

X-bar (Shewhart) and "R" control charts are valuable and proven tools for identification of special and common causes. In general they are not tools for on-line control of cleaning machines.[53] This is because cleaning processes don't (hopefully) rapidly change their state.

The CUSUM[54] control chart is especially effective where it is desired to detect small shifts or trends in composition of a cleaning bath. Cleaning baths generally don't decay from useful to impaired to just a few hours.

The two types of control charts are compared in Table A1.13.

X-bar (Shewhart) control charts are better used (in place of CUSUM control charts) for on-line control when larger changes are expected to occur over shorter periods of time.

## A1.14 CONSTRUCTING CUSUM CONTROL CHARTS

Actually, the equation noted in Table A1.13 is an oversimplification.

There is a feature which is unmentioned, unique, and valuable. There is also an unrevealed and necessary computational procedure.

The actual equation is:[55]

$$C_i = \sum_{j-1}^{i} \{\mu_j - \overline{X} - [k \times \sigma \times S] + C_{i,j-1}\} \tag{A1.14}$$

where $C$ = CUSUM
$i$ = One of two sums, either plus or minus
$\mu$ = Population average
$j$ = Each hourly average
$k$ = Constant between 0 and 1
$h$ = Constant between 3 and 10
$\sigma$ = Measurement standard deviation
$S$ = Number of standard deviations recognized before response
$C_{i,j-1}$ = The previous value of the sum

Use of the CUSUM technology to manage conditions in and around aqueous or solvent cleaning baths is highly recommended to readers of this book.

---

[53] One might have to use an "R" or X-bar control chart for on-line process control. In that case, what values of overall average and standard deviation should be chosen? The issue is that several hundred values are not likely to be available to be homogenized – as with the example above. In this situation, the best possible choice should be made – use all the available data to compute a running average and a standard deviation. Please note, this will cause the control limits to vary somewhat over time.

[54] The name CUSUM is an abbreviation for Cumulative Sum.

[55] Ross, S., *Introduction to Probability and Statistics for Engineers and Scientists*, Elsevier (2004), ISBN: 0-12-598057-4.

**Table A1.13 Comparison of "R," X-bar, and CUSUM Control Charts**

| "R" and X-bar Control Charts | CUSUM Control Charts |
|---|---|
| **Information content** | |
| Uses information about the process contained in the most recent hourly average value. | Uses information from essentially all past behavior. |
| **Intended uses** | |
| A "R" or X-bar control chart is intended to provoke *long-term action* to seek and eliminate special and common causes of variation. | A CUSUM control chart is intended to provide *short-term action* to adjust the controllable process parameters to produce a continuing stream of well-cleaned parts. |
| Both "R" and X-bar control charts are intended to instigate action, or provide balm to the enterprise's nervous system when action is not needed. | A CUSUM control chart is fundamentally different that either a "R" or X-bar control chart because it is intended to only be used in real time to monitor small shifts. |
| **Sequence of use** | |
| "R" control charts should be first prepared and used to eliminate *special causes* of process variation. Then X-bar control charts should be prepared to identify *common causes* of process variation. | CUSUM control charts should be prepared after "R" and X-bar control charts are used for their intended purpose. |
| **Equation being plotted** | |
| $(x_i - x_j)$, or $\bar{x}$ where $\bar{x}$ = hourly average value | $\sum\{\mu - \bar{X}\}$ |

Equation (A1.14) is not difficult to implement with any spreadsheet program. Simple directions and specific formulae to be inserted in spreadsheet cells are provided below. Once the spreadsheet is prepared, it can be reused by just erasing old data and replacing it with new.

## A1.14.1 Step 1

**Collect the data**[56] to be managed. Enter it into the spreadsheet. This shown as columns B and C in Table A1.14. The initial row is #8:

- Column A contains an identification – date/time, run/group number, counter, or any non-repeating designation which can be converted to a number and plotted on the horizontal axis in a graph as in step 7.
- Column B contains operating data about the cleaning process. This is the data[57] to be analyzed by a CUSUM procedure.

In real-world operations, it is likely that someone will key in two values every time a measurement is made. First, the identification number is entered[58] into an empty cell. Then, to the right, the measured[59] value is entered into another empty cell.

The remainder of the spreadsheet is updated automatically and replots the CUSUM graph as is shown in step 7.

[56] This is a different data set than that from which the "R" and X-bar control charts were generated – though it was computed from the same model of a solvent cleaning tank where olive oil is removed from Aluminum stamped parts using trichloroethylene.

In this data set, there are no *special causes*. But there is a randomization of all values of oil concentration. This should be typical of an operating unit where *common causes* may be acting.

[57] Measured oil concentration in the cleaning solvent is the value. The first value at group #100 is 2.210% oil. As above, initial values representing startup operation have been omitted. The highest group number is 500.

[58] To be consistent with the nomenclature in this appendix, the initial identification number should be entered into cell A. The Initial measured value should be entered into cell B.

[59] Or values from a calibration curve if the variable to be monitored is inferred via an indirect measurement (such as refractive index) and a calibration curve.

**Table A1.14 Spreadsheet for CUSUM Calculations**

| | A | B | C | D | E | F | G | H | I | J | K |
|---|---|---|---|---|---|---|---|---|---|---|---|
| 1 | | | | | | | | | | | |
| 2 | | | | | | | | | | | |
| 3 | | | | $\mu$ | $\sigma$ | *S* | *k* | *h* | *n* | | |
| 4 | | | | 2.275% | 0.030% | **1** | **0.5** | **8.61** | 2 | | |
| 5 | | | | | | | | | | | |
| 6 | | | | | | | | | | | |
| 7 | *Point* | *Data* | | | *Average* | *Range* | *Std* | $C^+$ | $C^-$ | LCL | UCL |
| 8 | 100 | 2.210% | | | | | | 0.000% | 0.052% | −0.257% | 0.257% |
| 9 | 101 | 2.206% | 2.206% | 2.210% | 2.198% | 0.006% | 0.004% | 0.000% | 0.099% | −0.257% | 0.257% |
| 10 | 102 | 2.220% | 2.220% | 2.206% | 2.208% | 0.009% | 0.001% | 0.000% | 0.138% | −0.257% | 0.257% |
| 11 | 103 | 2.222% | 2.222% | 2.220% | 2.219% | 0.006% | 0.000% | 0.000% | 0.180% | −0.257% | 0.257% |
| 12 | 104 | 2.213% | 2.213% | 2.222% | 2.221% | 0.003% | 0.005% | 0.000% | 0.223% | −0.257% | 0.257% |
| 13 | 105 | 2.221% | 2.221% | 2.213% | 2.219% | 0.000% | 0.007% | 0.000% | 0.266% | −0.257% | 0.257% |
| 14 | 106 | 2.213% | 2.213% | 2.221% | 2.220% | 0.000% | 0.003% | 0.000% | 0.310% | −0.257% | 0.257% |
| 15 | 107 | 2.219% | 2.219% | 2.213% | 2.225% | 0.007% | 0.002% | 0.000% | 0.347% | −0.257% | 0.257% |
| 16 | 108 | 2.228% | 2.228% | 2.219% | 2.228% | 0.003% | 0.009% | 0.000% | 0.380% | −0.257% | 0.257% |
| 17 | 109 | 2.227% | 2.227% | 2.228% | 2.225% | 0.001% | 0.001% | 0.000% | 0.411% | −0.257% | 0.257% |
| 18 | 110 | 2.231% | 2.231% | 2.227% | 2.229% | 0.007% | 0.001% | 0.000% | 0.439% | −0.257% | 0.257% |
| 19 | 111 | 2.231% | 2.231% | 2.231% | 2.239% | 0.007% | 0.006% | 0.000% | 0.470% | −0.257% | 0.257% |
| 20 | 112 | 2.227% | 2.227% | 2.231% | 2.243% | 0.002% | 0.002% | 0.000% | 0.506% | −0.257% | 0.257% |
| 21 | 113 | 2.222% | 2.222% | 2.227% | 2.238% | 0.005% | 0.010% | 0.000% | 0.540% | −0.257% | 0.257% |
| 22 | 114 | 2.230% | 2.230% | 2.222% | 2.233% | 0.002% | 0.004% | 0.000% | 0.571% | −0.257% | 0.257% |
| 23 | 115 | 2.229% | 2.229% | 2.230% | 2.230% | 0.002% | 0.002% | 0.000% | 0.602% | −0.257% | 0.257% |
| 24 | 116 | 2.228% | 2.228% | 2.229% | 2.227% | 0.002% | 0.004% | 0.000% | 0.632% | −0.257% | 0.257% |
| 25 | 117 | 2.232% | 2.232% | 2.228% | 2.227% | 0.003% | 0.002% | 0.000% | 0.661% | −0.257% | 0.257% |
| 26 | 118 | 2.231% | 2.231% | 2.232% | 2.227% | 0.003% | 0.000% | 0.000% | 0.685% | −0.257% | 0.257% |
| 27 | 119 | 2.241% | 2.241% | 2.231% | 2.226% | 0.001% | 0.002% | 0.000% | 0.708% | −0.257% | 0.257% |
| 28 | 120 | 2.234% | 2.234% | 2.241% | 2.227% | 0.000% | 0.006% | 0.000% | 0.736% | −0.257% | 0.257% |
| 29 | 121 | 2.228% | 2.228% | 2.234% | 2.221% | 0.007% | 0.006% | 0.000% | 0.762% | −0.257% | 0.257% |
| 30 | 122 | 2.239% | 2.239% | 2.228% | 2.222% | 0.008% | 0.009% | 0.000% | 0.781% | −0.257% | 0.257% |
| 31 | 123 | 2.242% | 2.242% | 2.239% | 2.230% | 0.004% | 0.000% | 0.000% | 0.795% | −0.257% | 0.257% |
| 32 | 124 | 2.252% | 2.252% | 2.242% | 2.232% | 0.001% | 0.002% | 0.000% | 0.810% | −0.257% | 0.257% |
| 33 | 125 | 2.238% | 2.238% | 2.252% | 2.229% | 0.003% | 0.004% | 0.000% | 0.831% | −0.257% | 0.257% |
| 34 | 126 | 2.239% | 2.239% | 2.238% | 2.233% | 0.008% | 0.002% | 0.000% | 0.850% | −0.257% | 0.257% |
| 35 | 127 | 2.244% | 2.244% | 2.239% | 2.237% | 0.002% | 0.001% | 0.000% | 0.868% | −0.257% | 0.257% |
| 36 | 128 | 2.241% | 2.241% | 2.244% | 2.240% | 0.006% | 0.001% | 0.000% | 0.886% | −0.257% | 0.257% |
| 37 | 129 | 2.243% | 2.243% | 2.241% | 2.244% | 0.000% | 0.004% | 0.000% | 0.905% | −0.257% | 0.257% |

## A1.14.2 Step 2

**Make a decision about the degree of response** you want from the CUSUM procedure.

One decision has already been made. But there are two additional choices to be made. It is likely that you may want to adjust them over time. Spreadsheets make that easy!

1. The decision already made is based on an personal observation of behavior in Tables A1.11 and A1.12. The observation is that measurement of oil concentration is fairly repeatable. The decision is that only **2** values, versus 5 in previous data sets, will be used to calculate average values. Cells C8 and D8 in Table A1.14 are blank because there is no other value at that time from which to calculate an average.
2. The first choice concerns the out-of-control average (mean) to be quickly detected by the CUSUM procedure. The choice is about how much change from the average value is it desired to detect before the CUSUM procedure recognizes that change.

One unique feature of CUSUM control charts is that they are silent about small changes in an average value which are smaller than a level of change in which you have an interest to detect. Essentially some lower levels of variation are dampened so that somewhat higher levels of variation can be more quickly identified.

The choice is specified in units of the standard deviation of the measured quantity. The parameters are:

$S$ = The integer number of standard deviations displaced from the average that it is desired to detect. $S$ is commonly chosen as 1.[60]

$k$ = A multiplier applied to $S$. $k$ is often chosen as being halfway between the overall average and the out-of-control average. This would make $k = 0.5$.[61]
When $k = 0$, every deviation from the overall average affects the CUSUM outcome. When $k = 1$, only those deviations from the overall average which are greater than $S$ standard deviations affect the CUSUM outcome.

This is a choice typically found in process control situations – a choice between too noisy and too dead. Halfway between noisy and dead is a useful starting choice. A new value of k can be chosen at any time, and the spreadsheet will recalculate.

### A1.14.3 Step 3

**Calculate two basic statistics** whose values change whenever new information is keyed into the spreadsheet. The statistics are the overall average (population average = $\mu$) and the overall standard deviation (population standard deviation = $\sigma$). In both cases, these statistics are based on all data available.

$\mu$ = The population average of all measurements entered to date is calculated via the spreadsheet function @PUREAVG(range). One can include blank cells in the range over which the average is to be taken.[62]

$\sigma$ = The population standard deviation of all measurements entered to date is calculated via the spreadsheet function @PURESTD(range). One can include blank cells in the range over which the average is to be taken.[63]

The product $\{S \times k \times \sigma\}$ is subtracted from the CUSUM value before it is compared to limit lines. In this way, the CUSUM calculation can be responsive to the quality of the underlying data and the needs of the user. This is not normally done with X-bar (Shewhart) control charts.

### A1.14.4 Step 4

**Make a decision about the level of change** of the CUSUM statistic that you consider to be out of control:

- This second choice is also specified in units of the standard deviation of the measured quantity. Implementation is via the parameter $h$.
  $h$ = The number of standard deviations from the normal average that are the maximum and minimum values the CUSUM statistic can attain without action being recommended.

This author recommends use for $h$ of two times the value of Student's "$t$."

The spreadsheet function used to calculate "$t$" is @TINV(1−[%confidence/100], number of points

[60] $S$ should be entered into cell F4.
[61] $k$ should be entered into cell G4.
[62] $\mu$ should be entered into cell D4.The statistic is based on current and previously-collected data.
[63] $\sigma$ should be entered into cell E4.The statistic is also based on current and previously-collected data.

in a single average). For 95% confidence and two measurements per average, "$t$" is 4.303. Thus, $h = 2 \times$ "$t$" $= 4.303 \times 2 = 8.61$.[64]

The parameters $k$ and $S$ and $h$ serve two different functions in a CUSUM control chart:

1. $k$ and $S$ are intended to allow variations to be recognized only if they are above a threshold. Otherwise, the CUSUM control chart is silent.
2. $h$ is intended to allow determination of whether the recognized variation is of a level which should prompt action, or not.

The product $\{S \times k \times \sigma\} = 1 \times 0.5 \times 0.030\% = 0.015\%$. This is the amount of change in the 6-hour average oil concentration which will not be included in the CUSUM statistic.

### A1.14.5 Step 5: Write the Key Equation (A1.14)

**There are two CUSUM terms. They are the positive and negative sums.** Both are plotted. Conventional nomenclature is $C^+$ (positive) and $C^-$ (negative).

Equation (A1.14) for each sum is:

$$C^+_{i+1} = \text{MAX}[0, ((\bar{X} - \mu - \{S \times k \times \sigma\}) + C^+_i)] \quad \text{(A1.14A)}$$

$$C^-_{i+1} = \text{MAX}[0, ((\mu - \bar{X} - \{S \times k \times \sigma\}) + C^-_i)] \quad \text{(A1.14B)}$$

### A1.14.6 Step 6: Construct the Spreadsheet

Table A1.14 shows the basic organization and specific location of required information:[65]

- Make the following data entries in row 4:
  - Overall population average ($\mu$), 2.275%, in cell D4.
  - Overall population standard deviation ($\sigma$), 0.030%, in cell E4.
  - The integer number of standard deviations displaced from the average that it is desired to detect ($S$), **1**, in cell F4.
  - The multiplier applied to $S(k)$, **0.5**, in cell G4.
  - The number of standard deviations from the normal average that are the maximum and minimum values the CUSUM statistic can attain without action being recommended ($h$), **8.61**, in cell H4.
  - The number of measured values included in each average ($n$), **2**, in cell I4.

---

[64] The value of $h$ should be entered in cell H4.

A general value for $h$ of 5 is recommended in most textbooks about statistics, including reference 1D.

The methodology of footnote 48 may also be used. This is based on reference 1C and Woodall, W.H., and Adams, B.M., "The Statistical Design of CUSUM Charts," *Quality Engineering*, Vol. 4, No. 5 (1993), p. 564, Table 2. One chooses a value of the percent of measurements which are expected to lie within the control limit lines. Then one reads a value of $h$ from the table at right. The value above of 5 is an approximation of 4.77 for a 3-sigma limit.

| Constant (**h**) used to calculate LCL and UCL | | |
|---|---|---|
| **Chance of Measurement** | | |
| **Inside Control Limit** | **Outside Control Limit** | **h** |
| 68.27% | 31.73% | 1.56 |
| 99.45% | 4.55% | 2.07 |
| 99.73% | 0.27% | 4.77 |
| 99.9937% | 0.0063% | 26 |
| 99.999943% | 0.000057% | 2390 |
| 99.99998% | 0.000002% | 68004 |

Please review the author's reasoning for selection of control limits for "R" and X-bar control charts in footnote 48. The reasons are similar for the same in CUSUM control charts. It is worthwhile to be able to: (1) narrow or widen the gap between control limits as the CUSUM becomes more or less stable when more or fewer measured values are included in the average, (2) use a consistent methodology to identify the significance of small differences, (3) choose a level of confidence other than 95%, and (4) focus on the reasons for choice of the control limits rather than repeatedly choose a general value.

[65] Obviously, any other arrangement that is convenient may be used. BUT the formulas must be adjusted – especially those in cells H8 and I8.

- Make the following formula entries in row 8:
  - +@MAX((E8−$H$4−($E$4 * $F$4 * $G$4)+H7), 0) in cell H8
  - +@MAX(($H$4−E8−($E$4 * $F$4 * $G$4))+I7, 0) in cell I8
  - +(−($H$4 * $I$4)) in cell J4
  - +(+($H$4 * $I$4)) in cell K4
- Make the following formula entries in row 9:
  - +B8 in cell C9
  - +B9 in cell D9
  - +@PUREAVG(C9..D9) in cell E9
  - +@ABS(C9−D9) in cell F9
  - +@PURESTD(C9..D9) in cell G9
- Copy cell range H8..K8 to H9..K9.

Then copy cell range C9..K9 down through columns C through K to include all data available or expected – say C9..K9 to C500..K500.

### A1.14.7 Step 7: Plot the CUSUM Graph

The ranges are:

- X-range is A8..A500
- Y-range for $C^+$ is H9..H500
- Y-range for $C^-$ is I9..I500
- Y-range for the LCL is J9..J500
- Y-range for the UCL is K9..K500

## A1.15 USING "R," X-BAR, CUSUM CONTROL CHARTS TOGETHER

Each control chart has a different purpose. And there is an order in which these purposes should be fulfilled:

- First "R" to identify and eliminate *special causes*.
- Then X-bar to identify and eliminate *common causes*.
- Finally CUSUM to control the process through managing the effect of the most elusive *common causes*.

All three control charts which act intermittently are shown in Table A1.15.

Please:

- Recall the assumptions behind the model which produced the data in Table A1.14. They were no special causes, and a modest amount of random variation. This situation happens routinely. That's why the CUSUM technique is so useful – it allows routine operation to be examined, analyzed, and hopefully improved.

CUSUM plots exaggerate the impact of minor *common causes* so focus can be directed to these causes and they can be eliminated.

But CUSUM plots are all but useless until all *special causes* have been eliminated. Such plots are rich with line movement and can resemble paintings by Salvador Dali.

- Recall Chapter 4, Section 4.12.1 (on-aim control). The aim when controlling any degreaser is to produce consistently clean parts. If cleaning quality is acceptable, and then for a short time gets better than expected (such as when additional soil is removed), that's not necessarily good! Why? There are two reasons:

  1. Unless the improvement is permanent, cleaning quality will soon be worse than previous. Thus downstream operation won't be as expected, either.
  2. Better represents a change, and some downstream operations may be vulnerable to any change.

- Note that it is possible, around group ~300, for both $C^+$ and $C^-$ sums to both be non-zero (or outside expectation). This happens when operation is either above or below expectation, and there is rapid change in the opposite direction.
- Non-zero values for both $C^+$ and $C^-$ sums are an essential feature of CUSUM technology. They allow rapid response to conditions producing rapidly shifting measurements. The reason is that one of the sums always[66] starts from zero – which may be more closely located to a control limit than is be the other sum close to zero. This happens in the CUSUM control chart above around group 275.

[66] Or nearly so. Change between non-zero sums can happen for some processes whose condition is oscillating between two extremes. Fortunately, cleaning systems aren't those processes.

**Table A1.15 Use and Comparison of Control Charts**

<table>
<tr><th>Control Chart</th><th>Comments</th></tr>
<tr><td>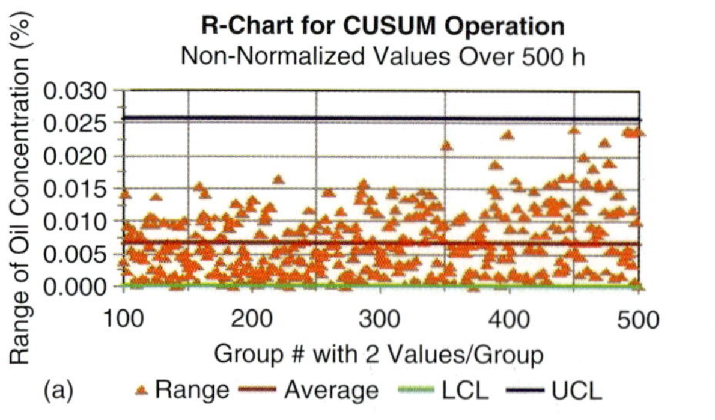
</td><td>There are no special causes acting on the cleaning process if the "R" control chart is the one at left.<br>The control limits are very narrow, which means that variation in range is almost negligible.<br>There is no correlation with time – the same causes appeal to act at every instant.</td></tr>
<tr><td>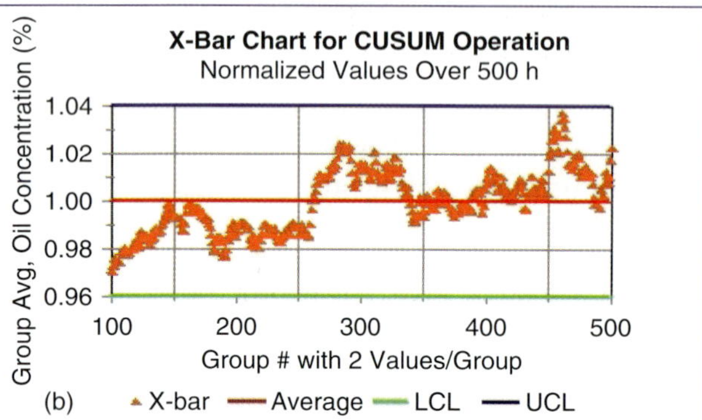
</td><td>Control limits are not exceeded. There is little oscillation.<br>But operation is not stable:<br>• Oil concentration is generally increasing as overall input rate slightly exceeds purge rate over the 400-hour period.<br>• There are at least three local instances where the purge rate exceeds input rate (groups ~110, ~300, and ~460).<br>This suggests that common causes aren't being eliminated, but their presence may be somewhat self-correcting.</td></tr>
<tr><td>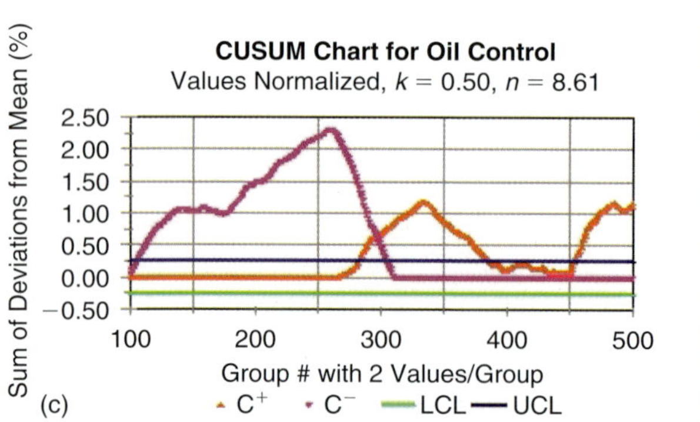
</td><td>Control limits are nearly always exceeded!<br>The shifting effect of common causes is easily recognized:<br>• Prior to group 110, oil accumulation (and therefore cleaning quality) is below the level (better cleaning quality) one would expect more than 99% of the time.<br>• Around group 300, and again around group ~450, that status is reversed.<br>Only groups ~390 to ~450 represent operation one would expect 99% of the time. Common causes are clearly afoot!</td></tr>
</table>

Please note that, when the measurement is rapidly increasing, the $C^+$ sum exceeds the UCL ~20 groups before the $C^-$ sum reaches zero.

- That's why two sums are used. A simple cumulative sum of deviations from average would over-represent extreme behavior and not recognize correct of same.
- There are a variety of CUSUM techniques. Mainly they are designed to recognize process change still more rapidly. Some approaches are to set the initial values of $C^+$ and $C^-$ at other than zero or to more heavily weight more recent operation. Reference 1C is excellent here.

## A1.16 HISTOGRAMS

Suppose you had Non-Volatile Residue (NVR) cleanliness data with the maximum value being 25 mg/SF, and the minimum value being 15.

Calculation of the average NVR as 19.88 and comparison of that to the goal value of 20.0 should ensure your satisfaction. The standard deviation of around 2 is within expectations. So what's not to like?

You might plot that hourly data as a run chart (without control limits). It would be in the form of Figure A1.12.

If you did that, your level of learning from that graph would be tiny.

A histogram might greatly enhance that level of learning. It is a simple graphical display of tabulated frequencies. A histogram can give some insight into the operating processes which produce the data in the table. A histogram can illuminate behavior hich might not otherwise be seen.

That is, a histogram is the graphical version of a table which shows what proportion (frequency) of cases fall into each of several or many specified categories.

A histogram is produced from a "tally sheet" where entries are made on the sheet each time a value is between chosen limits:

- You might count the number of values between 15 and 16, and enter that total on a "tally sheet.[67]
- Then you might do the same count for values between 16 and 17, and enter that number in the "tally sheet."
- You might continue this with the final entry into the table on the "tally sheet" being the number of NVR values between 24 and 25.

That would produce the table ("tally sheet") in Table A1.16.[68]

If you graphed this table, with NVR ranges on the horizontal axis and frequency on the vertical axis, you would have the histogram (Figure A1.13).

This plot clearly shows that there is no operation at the average NVR value of 20.

Yet, there are two types of operation – above NVR goal which is off-quality, and below NVR goal which is within the quality goal of 20.

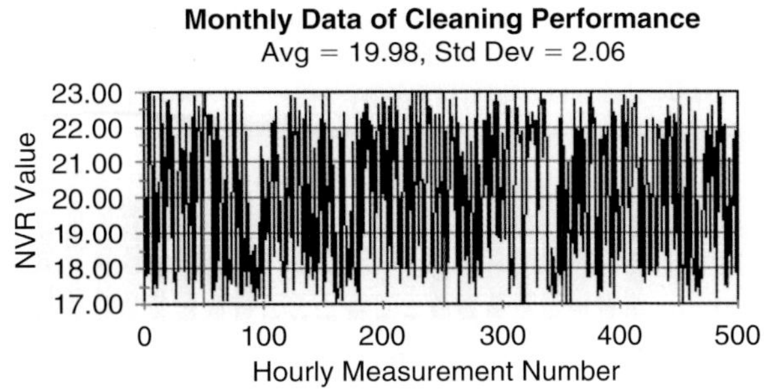


**Figure A1.12** Hourly NVR data

**Table A1.16 "Tally Sheet" for NVR Data**

| Range lower limit | Number Within Range |
|---|---|
| 17.0 | 35 |
| 17.3 | 25 |
| 17.5 | 30 |
| 17.8 | 29 |
| 18.1 | 31 |
| 18.4 | 29 |
| 18.6 | 30 |
| 18.9 | 17 |
| 19.2 | 0 |
| 19.5 | 0 |
| 19.7 | 1 |
| 20.0 | 0 |
| 20.3 | 0 |
| 20.5 | 0 |
| 20.8 | 28 |
| 21.1 | 36 |
| 21.4 | 42 |
| 21.6 | 40 |
| 21.9 | 26 |
| 22.2 | 32 |
| 22.5 | 33 |
| 22.7 | 36 |

In other words, the cleaning process which this data represents has a split personality. It can meet cleanliness requirements. But it doesn't always do so. In fact, it NEVER operates at the alm value! That's not apparent from the linear time plot.

[67] The classification of NVR value between 15 and 16, or any other range, is often called a bin.

[68] To enhance the appearance of this histogram, the "tally sheet" was actually prepared with 0.25 NVR units being as the separation between bins – versus the value of 1 in the general description above.

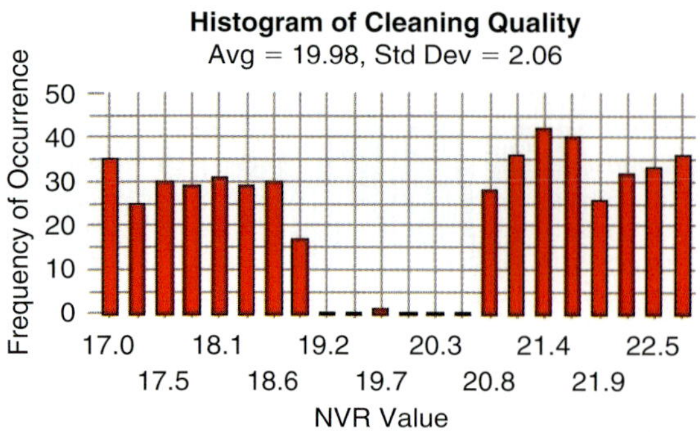


**Figure A1.13** Histogram of NVR data

The reason is that the off-quality (high NVR) operation must be a *special cause – because it's not always in effect*.

Histograms are easy to construct – given the availability of a spreadsheet. The macro function called HISTOGRAM is useful for either Quattro-Pro or Excel spreadsheet programs. The macro command is given by:

{HISTOGRAM range of all measured values, range of values for "tally sheet"}

## A1.17 CHECK SHEETS

A check sheet is an enumeration of the problems noted, the number of times they occur, and when they occur. Yet again, a check sheet is an operating log – expressed in tabular format.

Operating personnel design and use check sheets to identify the type of problems (defects) which must be eliminated. If operation were uniformly acceptable, there should be little need for a check sheet.

Supervisors and operators would normally complete a log of events for every shift worked. A check sheet could be prepared by anyone, supervisor or operator, from that information.

To avoid a sea of blank spaces, the frequency of entry is usually weekly, or monthly. The two items entered are the "incident" and the frequency of its occurrence over that period.

The word "incident" is used in a general sense. Both quality and operating issues would be "incidents." Operating issues would include: equipment failures, personnel actions, significant maintenance performed, safety or environmental incidents, training or procedural changes, or general observations.

The nature of the "incidents" of concern must be specified prior to use of the check sheet. New types of incidents should be added after they occur. A check sheet is a "living" document – not an income tax form!

The purpose of a check sheet is to enable preliminary analysis for trends about quality, area accounting, or safety/environmental problems. The check sheet should be the basis for performing more detailed analysis. Only occasionally is the check sheet a source of convincing evidence relating cause and effect. But it strongly suggests where to seek that evidence, and perhaps with what level of deligence.

The above check sheet in Table A1.17 shows some typical outcomes:

- Operators are often retrained after an inspection.
- "Government" jobs (personal tasks) do exist.
- A change of solvent quality (color or odor) usually provokes action – including examination of the stabilizer condition and the distillation system.
- Cleaning quality can be related to solvent quality.
- Unexpected incidents do occur – when a cause is yet to be identified.

## A1.18 PARETO CHARTS

The Pareto chart produced from the information in the check sheet is shown in Figure A1.14.[70]

A Pareto[69] chart is the sibling to the check sheet. A Pareto chart is used to graphically summarize and display the relative importance of the differences

[69] This analysis tool is named for Vilfredo Pareto (July 15, 1848 to August 19, 1923). He made several important contributions to economics, sociology and moral philosophy, especially in the study of income distribution and in the analysis of individuals' choices.

The 80/20 rule is often named for him because of his observation that 80% of the property in Italy was owned by 20% of the population. Said another way, the 80/20 assumption is that most of the results in any situation are determined by a small number of causes.

[70] To avoid confusion, the labels on the horizontal axis are printed in small text to avoid overlap. They can be identified from the sibling check sheet. The right-hand column in the check sheet is the value plotted.

**Table A1.17 Weekly Check Sheet**

| | Week Number | | | | | | | | | | |
|---|---|---|---|---|---|---|---|---|---|---|---|
| **Incident** | **1** | **2** | **3** | **4** | **5** | **6** | **7** | **8** | **9** | **10** | **Total** |
| NVR > 20 | 1 | 2 | 2 | 5 | 2 | 1 | 2 | | 6 | 2 | 23 |
| NVR > 25 | | 1 | 1 | 2 | | 1 | 1 | | 1 | 2 | 9 |
| NVR > 30 | | | | 1 | | | | | | 2 | 3 |
| Unexpected film | | | | 1 | | | | 1 | | | 2 |
| Fail salt spray test | | | | 1 | | | | | | | 1 |
| "Salty" residue | | | | | | | | | 1 | | 1 |
| Heavier part | | 1 | | | 1 | | | | | | 2 |
| Lot # X000 | | | 1 | | | | 1 | | 1 | | 3 |
| New refractometer | | | | 1 | | | | | | | 1 |
| Still cleanout | | | | | | | 1 | | | | 1 |
| Aroma?? | | | | | 1 | | | | | | 1 |
| Solvent change | 1 | | | 1 | | | 1 | | | | 3 |
| Hoist jammed | | 1 | | | | | | | | | 1 |
| OSHA inspector | | | | 1 | | | | | | | 1 |
| Operators retrained | | | | | 1 | | | | | | 1 |
| Stabilizer checked | 1 | 1 | 1 | 1 | 1 | 1 | 1 | 1 | 1 | 1 | 10 |
| Stabilizer replaced | | | 1 | | | 1 | | | | | 2 |
| Coolant intrusion | | | 1 | | | | | | | | 1 |
| "Govt." job | | | | | | 1 | | | | | 1 |
| No cover | 1 | | 1 | 1 | | | | | | | 3 |
| "Pink" solvent | | | | | | 1 | | | | | 1 |
| New basket | | 1 | | | | | | | | | 1 |

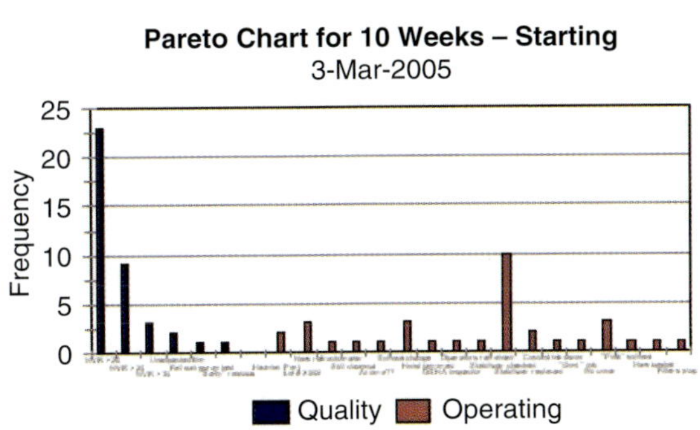


**Figure A1.14** Pareto chart from check sheet

between groups of data – often those provided within a check sheet.

Significant questions can be answered by a Pareto chart. They include:

- What are the largest[71] issues involved with the cleaning machine (or any other part of any system)?
- Where should efforts be focused to achieve the greatest[72] improvements?
- What 20% of sources are causing 80% of the problems (80/20 Rule)?

[71] Please note here that the word *largest* refers to those issues which occur most frequently, and not to those issues which have the most impact upon the enterprise.

[72] As above, the word greatest refers to frequency of occurrence and not to moment.

Please note that the variable of time is not included in a Pareto chart. Consequently, cause and an associated effect are unlikely to be related in a Pareto chart. But that's not why they are used. A check sheet is one tool better-used for that purpose. That's why there are multiple analysis tools.

## A1.19 CAUSE-AND-EFFECT DIAGRAMS

Cause-and-effect relationships govern everything that happens and as such are the path to effective problem solving.

The cause-and-effect diagram (C&E) is the brainchild of Kaoru Ishikawa,[73] who pioneered quality management processes in the Kawasaki shipyards. It was created so that all possible causes of a result could be listed in such a way as to allow a user to visually (graphically) show these possible causes. From this diagram, the user can sometimes define the most likely causes of a result.

The C&E diagram is also known as a:

- *Fishbone diagram* because it was drawn to resemble the skeleton of a fish, with the main causal categories drawn as "bones" attached to the spine of the fish.
- *Tree diagram*, resembling a tree turned on its side.

The C&E Figure A1.15 illustrates the general approach. The horizontal line is the main logical analysis (stream of causes) about some outcome or effect (favorable or unfavorable). The six angled lines are selected major categories of effort. The outcome being studied is likely being produced by a collection and/or interaction of causes within those categories. The short horizontal lines, which intersect the six angled lines, represent each possible sub-cause.

For cleaning work, these six categories of cause should be considered:

1. Cleaning *machine* (or process).
2. *Materials* (cleaning agent) used.
3. *Methods* and procedures used.
4. *Measurements* made.
5. *Personnel* involved.

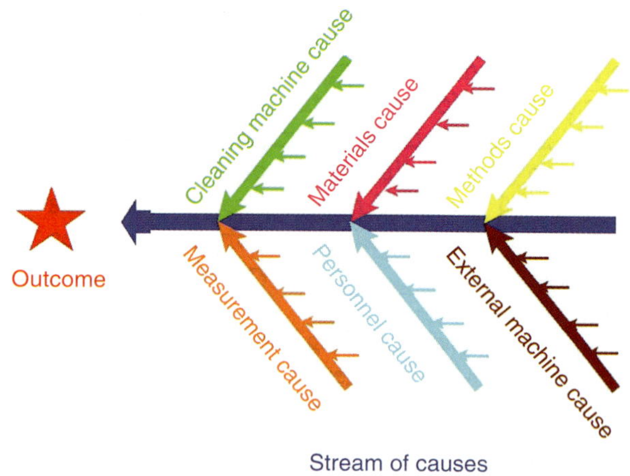


**Figure A1.15** Cause-and-effect diagram

6. The *external* systems, upstream and downstream, with which the cleaning system interacts.

That there is an event or outcome (favorable or unfavorable) being studied suggests that the causes lie within those six general categories. If users believe additional categories should be added because of local circumstances, than that should be done. But removal of any one of the six categories is risky!

Construction of a C&E diagram can be characterized as similar to a comedy writer's composition of a joke.[74] But in every case,

1. A team MUST be involved. No single person can conceive/recognize/accept/ignore all possible causes. Experience, perspective, and prejudice are limiting factors.
2. The problem, effect, or outcome must be defined – both specifically and generally – to the satisfaction of *every* member of the team.
3. The team will contribute possible causes to each category. Each possible cause must be ranked by being likely to impact the situation.
4. Even when the conundrum has been unraveled and its causes identified, action (corrective or otherwise) must be taken. Otherwise, the C&E effort is wasted.

---

[73] A pioneer of quality engineering. The career of Kaoru Ishikawa in some ways parallels the economic history of contemporary Japan. One of Ishikawa's early achievements contributed to the success of quality circles. He was a member of the committee for the Deming Prize.

[74] Long enough to cover the subject, and short enough to make it interesting.

Actually, this author recommends an alternative to C&E diagrams. This author prefers tables of causes rather than graphical constructions. There are three reasons:

1. A table of categories and causes is more easily reused in other situations, whereas drawings are considered situational.
2. C&E drawings can become more complex than is necessary to produce the desired understanding.
3. C&E drawings can be valued as the end product of some teamwork, rather than the means by which a problem is solved or a gain is retained.

A table with categories of causes is listed in Table A1.18. Ten specific possible causes are listed within each category of cause. The sixty individual causes don't contribute to each effect being analyzed. They are just possibilities to be considered by the team.

C&E tables for three different types of defects are shown in Tables A1.19–A1.21. They are:

1. High NVR.
2. Failure in safety/health/environmental administration (SHEA).
3. Poor productivity.

Some of the causes can affect all three defects. Others can affect just one, or two.

The three C&E tables shown as examples can serve a starting points for your team's analysis.

Please note how Table A1.18 enumerates causes your team believes are general, and then individual causes deemed not pertinent by your team are discarded leaving a list of hypotheses to be evaluated in the operating area. This author has found that: (1) it is useful to prepare Table A1.18 during the calm prior to the existence of a specific problem, and (2) a team can more easily focus on a solution by discarding specific potential causes rather then stretching their imagination to suggest what no one else has suggested.

## A1.20 DEFECT CONCENTRATION DIAGRAMS

If you don't already own one, this can be your justification to purchase a digital camera.

A check sheet can become a catalog of general system failures. Similarly there is a need to catalog instances of specific failures – where cleaning quality was not acceptable. A collection of color images on a CD[75] should be that catalog.

Authors of books about SPC call that catalog a Defect Concentration Diagram. When defect data, (cleaning failures) are portrayed on a Defect Concentration Diagram over a sufficient amount of production, one can learn about the causes of these defects. After all, failure repeated should become failure uncovered and prevented.

Here, unlike the C&E diagram, an image supplies real value – considerably more than a table or check sheet. The value is often simplicity and accuracy.

The following steps are suggested to develop a digital Defect Concentration Diagram:

1. Use the camera to record the appearance of parts which fail the required cleaning test.[76] The recording should be from all sides – not just a scenic view.
2. Store the images on a CD. Use a file name which describes the cleaning defect and when/where it was produced.

[75] A manager should organize the following software tools with a personal computer and a digital camera: (1) means of transferring images from the latter to the former, (2) means to copy digital images to a CD or other storage media, (3) means to annotate digital images, and (4) means to display or show individual digital images and those which have been annotated. These means can be supplied with the Windows XP operating system and Microsoft Office, Corel Office Suite, Adobe Photoshop Elements or other software.

[76] *One must be careful to record the true situation*. Please remember that completion of the cleaning test may affect the appearance of the part. For example, a validation test should remove all soil.

Therefore, the photographs may be of parts sequentially produced.

BUT they may not have failed the cleaning test. For example, consider the situation where only parts located at certain positions in a basket are poorly cleaned. Here, since the next-removed part may not be from the critical location in the basket, judgement and patience are required.

**Table A1.18 Categories of Causes**

| Cleaning Machines | Materials | Methods | Measurements | Personnel | External Machines |
|---|---|---|---|---|---|
| Individual Causes Within Categories | | | | | |
| Use aqueous (or solvent) machine | Wrong detergent (or solvent) | Cycle too short | Thermocouples fouled | No operating records | Soil intrusion |
| Machine too small | Detergent (or Solvent) degrades | Cycle too long | Hoist timing off – dwell | Rapid turnover | Parts hot |
| Excessive emissions | Multiple solvents needed | Solvent not boiled | Refractometer uncalibrated | Poor training | New downstream use |
| Sump(s) dirty | Wrong stabilizer | Solvent boiled dry | Insensitive cleaning test | Unmotivated | Polymeric soil |
| Distillation dirty | Solvent odor | Short drainage period | Densitometer uncalibrated | Inaccurate records | Environment dirty |
| Hoist misaligned | Hazardous solvent | Not fully immersed | Wrong cleanliness test | No specific training | New specification |
| Multiple stages needed | High surface tension | Pre-cleaning needed | Degradation not detected | Illogical analysis | Parts cold |
| No sonics provided | Affinity for water | Nozzles un- or mis-aimed | Wrong emission measurements | No regulatory records | Unexpected parts |
| Baskets ill-designed | Flammable solvent | Basket overfilled | Hoist timing off – lifting | Motivated to harm | Unexpected soil |
| Poor maintenance | Poor distillation decanting equipment | Wrong protective equipment | Level or filter fouled | Staff too busy | Inorganic soil |

**Table A1.19 C&E Table for High NVR Defect**

| Cleaning Machines | Materials | Methods | Measurements | Personnel | External Machines |
|---|---|---|---|---|---|
| Use aqueous (or solvent) machine | Wrong detergent (or solvent) | Cycle too short | Thermocouples fouled | No operating records | Soil intrusion |
| Machine too small | Detergent (or Solvent) degrades | Cycle too long | Hoist timing off – dwell | Rapid turnover | Parts hot |
| Excessive emissions | Multiple solvents needed | Solvent not boiled | Refractometer uncalibrated | Poor training | New downstream use |
| Sump(s) dirty | Wrong stabilizer | Solvent boiled dry | Insensitive cleaning test | Unmotivated | Polymeric soil |
| Distillation dirty | | Short drainage period | Densitometer uncalibrated | Inaccurate records | Environment dirty |
| Hoist misaligned | | Not fully immersed | Wrong cleanliness test | No specific training | New specification |
| Multiple stages needed | High surface tension | Pre-cleaning needed | Degradation not detected | Illogical analysis | |
| No sonics provided | Affinity for water | Nozzles un- or mis-aimed | | No regulatory records | Unexpected parts |
| Baskets ill-designed | | Basket overfilled | Hoist timing off – lifting | Motivated to harm | Unexpected soil |
| Poor maintenance | Poor distillation decanting equipment | | Level or filter fouled | Staff too busy | Inorganic soil |

**Table A1.20 C&E Table for SHEA Defect**

| Cleaning Machines | Materials | Methods | Measurements | Personnel | External Machines |
|---|---|---|---|---|---|
| | Wrong detergent (or solvent) | | | No operating records | |
| Machine too small | | | | Rapid turnover | |
| | | | | Poor training | |
| | | Solvent boiled dry | | Unmotivated | |
| | Solvent odor | | | Inaccurate records | |
| Hoist misaligned | Hazardous solvent | | | No specific training | |
| | | | | Illogical analysis | |
| | | | Wrong emission measurement | No regulatory records | |
| | Flammable solvent | | | Motivated to harm | |
| | | Wrong protective equipment | | Staff too busy | |

**Table A1.21 C&E Table for Poor Productivity Defect**

| Cleaning Machines | Materials | Methods | Measurements | Personnel | External Machines |
|---|---|---|---|---|---|
| Use aqueous (or solvent) machine | Wrong detergent (or solvent) | Cycle too short | Thermocouples fouled | No operating records | Soil intrusion |
| Machine too small | Detergent (or Solvent) degrades | Cycle too long | Hoist timing off – dwell wrong | Rapid turnover | Parts hot |
| | Multiple solvents needed | Solvent not boiled | | Poor training | New downstream use |
| Sump(s) dirty | Wrong stabilizer | Solvent boiled dry | Insensitive cleaning test | Unmotivated | |
| Distillation dirty | | Short drainage period | Densitometer uncalibrated | Inaccurate records | |
| Hoist misaligned | | Not fully immersed | Wrong cleanliness test | No specific training | New specification |
| | High surface tension | Pre-cleaning needed | Degradation not detected | Illogical analysis | Parts cold |
| No sonics provided | Affinity for water | Nozzles un- or mis-aimed | | No regulatory records | Unexpected parts |
| Baskets ill-designed | | Basket overfilled | Hoist timing off – lifting | Motivated to harm | Unexpected soil |
| Poor maintenance | Poor distillation decanting equipment | | Level or filter fouled | Staff too busy | |

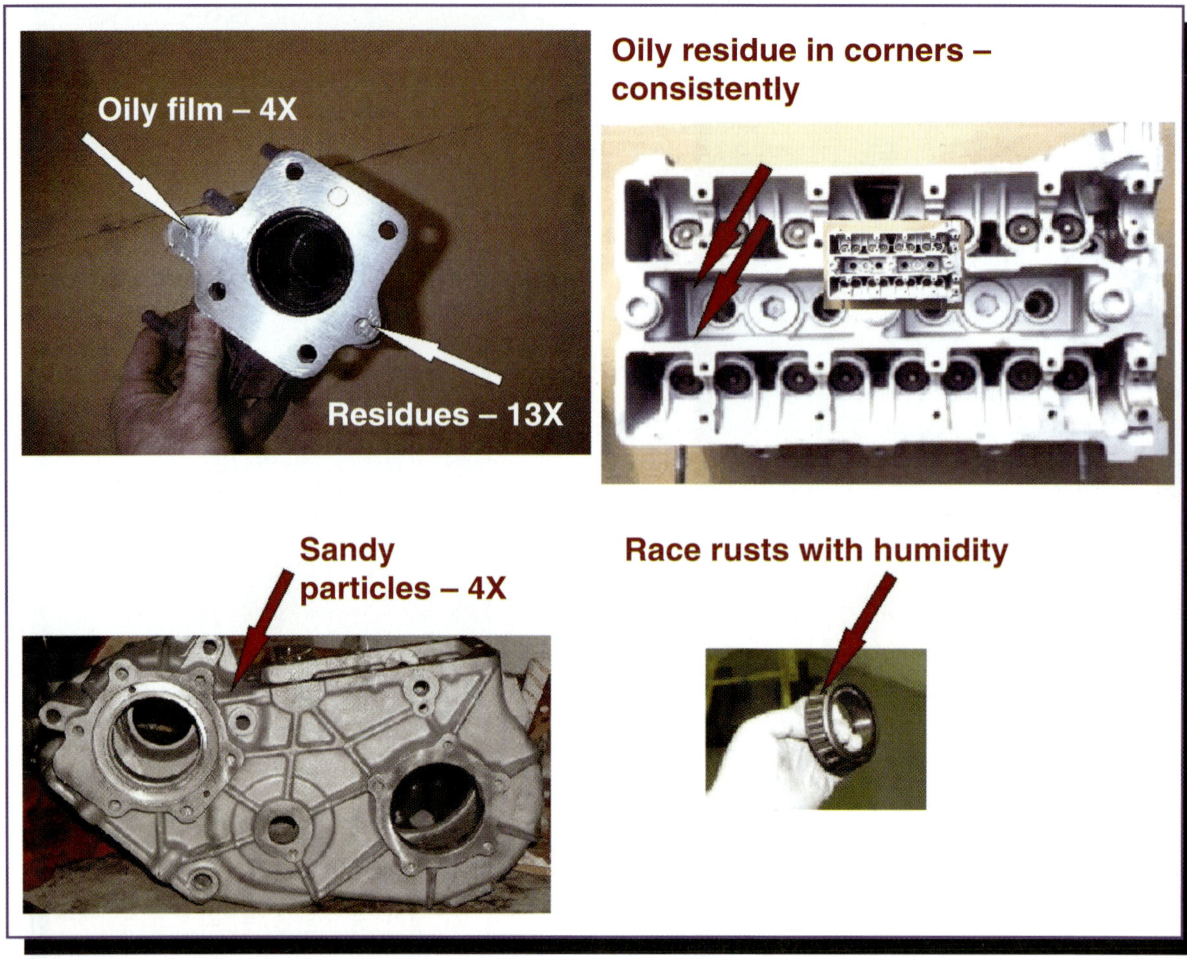


**Figure A1.16 Digital defect concentration diagrams**

3. Designate a set of master images – one for each different view of each different part. Each master image is to be the log where the census of defects is recorded.
4. Annotate each master image with a mark indicating where a defect was noted.

Four examples are displayed in Figure A1.16. There is no standard approach here (× means the number of times the defect has been noticed). Use what allows your organization to clearly understand the situation.

## A1.21 SCATTER DIAGRAMS

Scatter diagrams (plots) are graphs of performance data. Usually, one operating variable is plotted against some measurement of cleaning performance.

The basic idea is to investigate if there is a cause and effect relationship between the operating variable and cleaning performance – by visually inspection of the graphs:

- In a sense, this is a great idea. This author does it all the time. A graph almost always reveals more than inspection of the data table from which it was constructed.
- In another sense, it's a terrible idea. Cause and effect is only established at a level of confidence or statistical certainty. Inspection of a graph often leads to impulsive action based on a presumed cause and effect relationship which is only 80 or 85 or 90% certain based on statistical analysis versus random behavior. That's not normally enough certainty to justify action. The result can be confusion

or mistakes because what is observed to be true is not statistically true.

The best use of scatter plots is as a tool within a toolkit. Other components of that toolkit are the techniques covered above, design of experiments (DOE) covered below, AND your mind. *The latter is the most important.*

### A1.21.1 Use of Multiple Scatter Plots

Here are some examples where scatter plots can used to both confuse and inform. The data are from modeled experiments where four solvents (methyl acetate [MeOAc], trichloroethylene [TCE], Soy Gold 1000 [SoyGold], and HFE 7200 [HFE 7200]) were used to clean a variety of common soils using a vapor degreaser.

Additional variables were use of ultrasonics, contact time, and multiple stages of contact. Cleanliness performance could be taken as NVR, Optically-Stimulated Emission Electron (OSEE), and wt.% clean (see Appendix 2).

There are several points to be made about use of scatter plots.

The scatter plot (Figure A1.17) shows cleanliness quality produced with TCE in removing linseed oil. The intent was to learn if:

- Additional stages of cleaning contact produced improved cleaning quality (lower NVR) and/or
- Ultrasonic transducers aid in improving cleanliness.

From Figure A1.17, it appears as if cleanliness is improved if multiple contact stages are provided, and that use of ultrasonics may be of value.

Yet, an experienced person would discard use of TCE as a candidate solvent for cleaning of linseed oil – because cleanliness quality was poor, even with *four* contact stages. Note that the NVR values are all 18 mg/SF or above. That's not very clean!

> The *point is that an apparent relationship in a scatter plot may not have any practical meaning.*

The scatter plot (Figure A1.18) was prepared with similar intent, using a different solvent.

MeOAc appears to be a better choice for a solvent to clean linseed oil. NVR values around 10 can be achieved with four stages of contact.

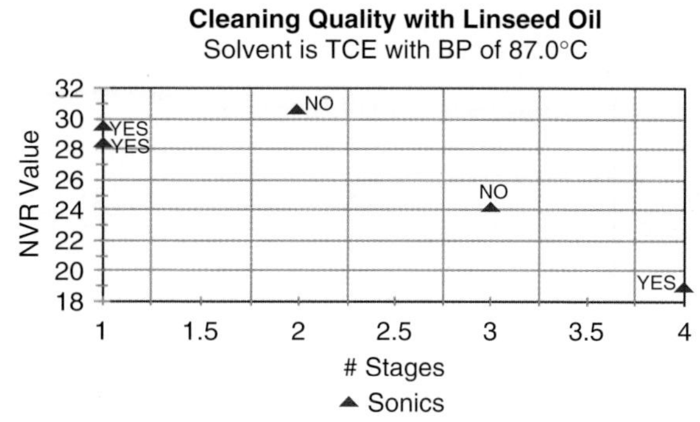


**Figure A1.17** Solvent cleaning with TCE

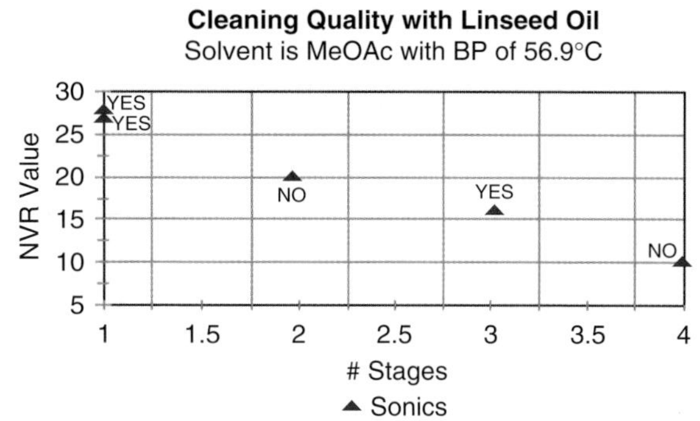


**Figure A1.18** Solvent cleaning with MeOAc

The visually apparent correlation within this scatter plot of cleanliness with additional contact stages appears to define a successful application.

Yet an experienced person would question if this application were useful – because NVR levels are quite high (above 25) with ultrasonics after a single stage of solvent cleaning. No one wants to purchase and maintain four cleaning and rinsing stages with a more useful choice of solvent.

The same point is repeated: an apparent relationship in a scatter plot may not mean anything. *Your experience and your judgment are at least as important as are scatter plots.*

The scatter plot (Figure A1.19) was also prepared for the same purpose – with still another solvent. HFE 7200 is that more useful choice to clean linseed oil.

But Figure A1.19 doesn't illustrate that conclusion – *until you note that the NVR values are all around zero.*

The point here is that multiple views of data using scatter plots should be the normal procedure – rather

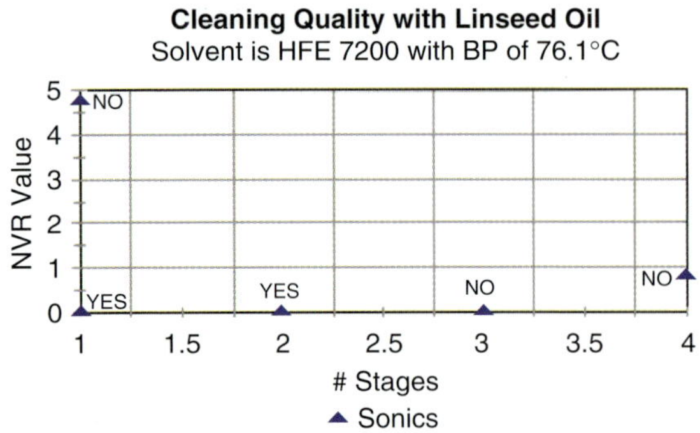


**Figure A1.19** Solvent cleaning with HFE 7200

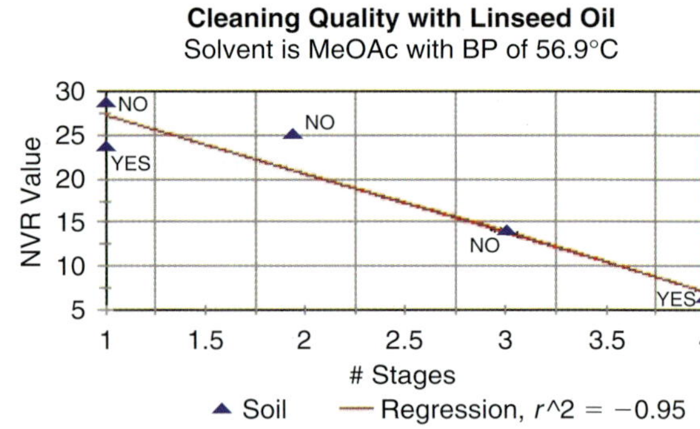


**Figure A1.20** High correlation coefficient

than purchasing equipment or cleaning agents or services after observing the first scatter plot.

### A1.21.2 Correlation within Scatter Plots

Figure A1.20, with MeOAc used to clean linseed oil is repeated at right (Figure A1.20).

It appears there may be a useful mathematical correlation between use of additional contact stages and improved cleanliness – even if there are better process choices.

A linear relationship between an observation and a cause is common. In fact, many are suspicious of more complex non-linear relationships.

The equation for a linear relationship between an input variable (operating variable) and an output variable (cleaning performance) is:

$$\text{Output variable} = \text{Constant 1} + (\text{Constant 2} \times \text{Input variable}) \tag{A1.15}$$

The standard approach for determining the two constants is by the method of *least squares*.[77] With this technique, *each* of the two constants are chosen so that this sum of squared deviations is minimized.[78]

Fortunately, spreadsheets make this work easy. Both Corel's Quattro-Pro and Microsoft's Excel have "wizards" which automate the process. Only the input and output variables need to be identified as ranges with the same number of values.

In addition, there are spreadsheet functions from which the constants can be calculated as values in a single cell. Three of these functions are:

1. Constant 1 = @SLOPE[79](Range of Output Data,Range of Input Data).
2. Constant 2 = @INTERCEPT(Range of Output Data,Range of Input Data).

A linear equation fit to the cleanliness data in Figure A1.18 is also plotted as a line in the scatter as Figure A1.20. The calculated correlation coefficient[80] for that regression line is 0.95.

---

[77] Here one calculates the deviations between the values calculated by Equation (A1.15) and the true values of the output variable. These deviations are squared so that positive deviations don't algebraically cancel negative deviations. Then these squared deviations are summed over all values of the input variable.

[78] It is a small point, but the constants are separately determined so that their individual values are chosen to each produce a minimum in the sum of squared deviations. It is not true that both constants are chosen together to produce a minimum. Said another way, the constants are chosen to each individually produce a minimum (the minimums may be different) and not a zero value of the sum of the squared deviations.

Equations can be derived by which each constant can be calculated as a function of the data values. The derivation and the equations are found in any of the references given on the first page of this appendix.

[79] In Microsoft Excel, the equivalent spreadsheet functions are =SLOPE, =INTERCEPT, and =CORREL. See Section A1.23

[80] A correlation coefficient is a measure of how much of the variability in the input and output is in common to both. A value of unity means that all of the variability in the input is found the output. In simple language: when the correlation coefficient is +1, the input increases and the output always increases; when the correlation coefficient is −1, the input increases and the output always decreases, and when the correlation coefficient is 0, the input increases and the output does as it wishes.

The value of a correlation coefficient[81] should be treated as similar to a percent confidence.[82] Reliable and useful correlations have correlation coefficients around 0.95.

3. Correlation Coefficient = @CORREL(Range of Output Data,Range of Input Data).

In this case, the correlation coefficient evaluates the response between cleanliness (produced when an additional stage of vapor degreasing with MeOAc is added) and addition of stages of cleaning.

Yet, when that value of $r^2$ is 0.95, it can be meaningless. The information in the scatter plot above may be incomplete. It may miss the point.

Such is the case in the scatter plot shown in Figure A1.21.

Here the solvent is HFE 7200, which has already been shown (Figure A1.19) to produce excellent cleanliness removing linseed oil in a single stage of vapor degreasing.

The correlation coefficient in Figure A1.21 is low. This means that the change in cleanliness when an additional stage of vapor degreasing is added with HFE 7200 does not correlate well with that change in number of stages.

The reason should be obvious. The NVR, with ultrasonic transducers, is already about zero after one stage of cleaning. All the soil has already been removed. No more soil can be removed via additional stages of cleaning. *There is no variation to correlate.*

Another example is illustrated by Figures A1.22 and A1.23. The data plotted were produced when cleaning another soil with the same solvents. Correlation coefficients suggest that additional stages of vapor degreasing are highly certain to remove additional soils.

Yet, closer inspection reveals that more than ten stages of vapor degreasing might be required to produce NVR levels below five!

Plainly, TCE and SoyGold 1000 are not the right choices to clean cottonseed oil.

> *A high correlation coefficient is not a guarantee that the relationship it describes is worth anything in practical terms.*

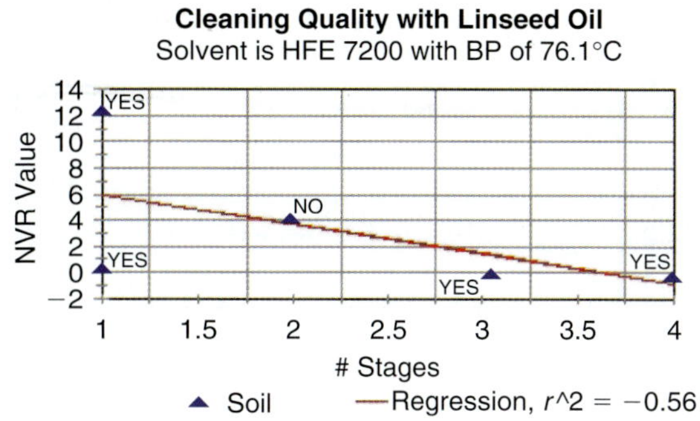


**Figure A1.21** Low correlation coefficient

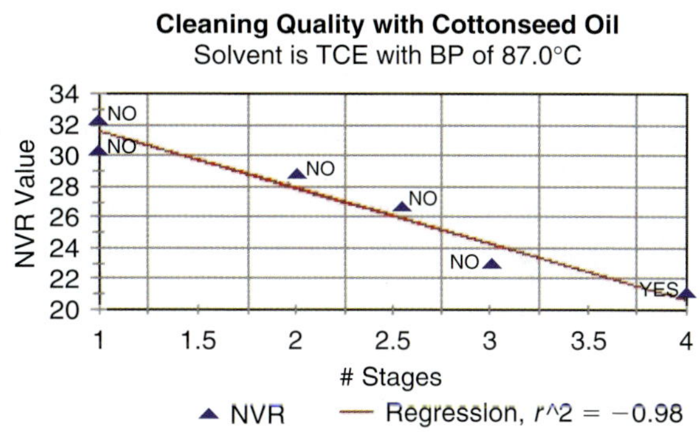


**Figure A1.22** TCE – not the right solvent choice

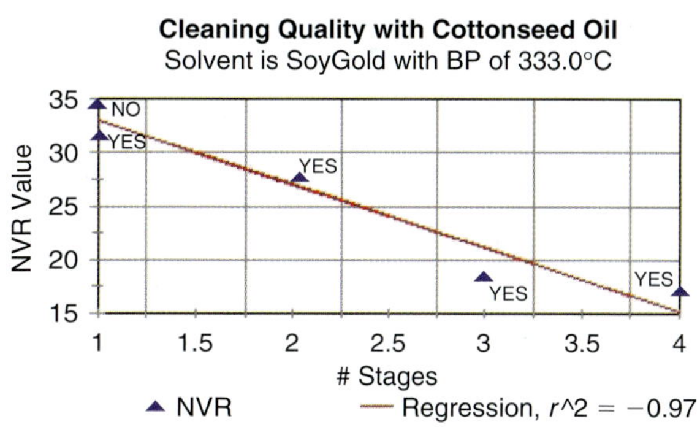


**Figure A1.23** SoyGold – not the right solvent choice

The scatter plot as Figure A1.24 shows that methyl acetate is a much better choice for cleaning solvent for cottonseed oil because all the NVRs values are low.

But as with cleaning linseed oil with HFE-7200 in Figure A1.21, the correlation coefficient with

[81] Nomenclature for the correlation coefficient is $r^2$. The correlation coefficient measures the degree to which two things vary together ($0 < r^2 < 1$) or oppositely ($-1 < r^2 < 0$).

[82] A correlation coefficient is NOT a measure of statistical probability.

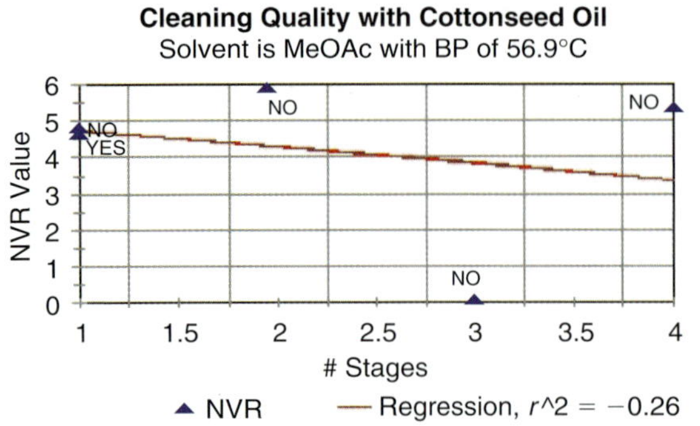


**Figure A1.24** Methyl acetate – a better solvent choice

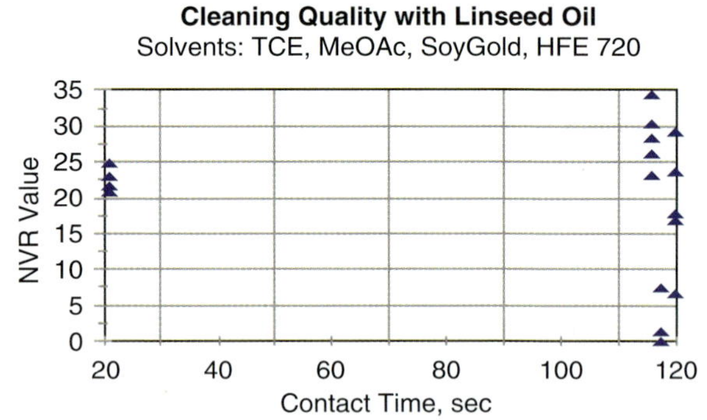


**Figure A1.25** Scatter plot with the wrong data

increasing number of stages is worthless because the soil has already been removed in the first stage of contact.

The values of correlation coefficients associated with data in these figures have no meaning. By analogy, it's like the driver of an automobile making good time but not moving in the right direction. Don't value relationships derived from data only by the values of correlation coefficients associated with those relationships.

The point here is that values of correlation coefficient are preferred when they are around 0.95, but (1) those values don't guarantee any understanding about the meaning of the data on which they are based, and (2) lower values occasionally can be even more preferred.

### A1.21.3 Scatter Plots with the Wrong Data

Scatter plots are just displays of data. It is not true that within each scatter plot is a meaningful conclusion. Said another way, "Sometimes a cigar is just a cigar."[83]

Figure A1.25 is an excellent example of a scatter plot which ought to suggest a useful conclusion – but doesn't because there isn't one within the plotted data. Cleanliness data are plotted for four solvents versus the cleaning cycle time for vapor degreasing (Figure A1.25).

The problem is that one of the solvents really isn't a solvent for vapor degreasing as are the other three. SoyGold 1000 boils at over 330°C. The others boil at between 55 and 90°C. Cycle times when SoyGold 1000 is used as a vapor degreasing solvent are often very short because most parts cannot survive long exposure at that high temperature.

One-quarter of the data within this scatter plot doesn't belong there. That may be relatively easy to spot – because these are examples from model data. With real data, the suitability of all data is much more difficult to spot.

### A1.21.4 Summary About Scatter Plots

They can be efficient tools for learning about relationships. This author probably overuses them. Spreadsheets make their creation easy. Undoubtedly, they enable visual communication.

While concern about their use may have been artificially exaggerated in this chapter, the hazards of their use should now be more obvious.

## A1.22 DESIGN OF EXPERIMENTS

Most managers don't design experiments. More than several generations after useful and proven statistical techniques were developed, managers don't use them. They use the OFAT (One Factor[84] at a

[83] Sigmund Freud.

[84] Outside of statistical literature, the words variable or parameter can be substituted for the word factor.

Time) approach to manage cleaning (and other) exploratory[85] process tests.[86]

Design of experiments (DOE) refers to the body of statistical technology used by some (but not many)[87] managers to better understand:

- How a certain operation "really works."
- What levels of which control variables produce an optimum process result.

The OFAT approach is based on the usually untrue idea that there is no interaction among variables – that peanut butter and jelly don't make a better sandwich than either separately does, that a string quartet doesn't produce a better sound than do the four instruments when separately played, that the yellow light will somehow be produced if green light and red light are separately shown on a screen, or that children can be produced by each parent acting separately.

The basic idea in DOE is to devise a small set of experiments, in which all pertinent variables (factors) are varied systematically. Said another way, the idea is to arrange an experiment as a set of partitioned sub-experiments. Each differs from the other in having one or several factors or treatments applied to them. Statistical analysis[88] is used to identify how the combinations of different variables in the sub-experiments permit differences in the measured outcome.

### A1.22.1 Plan First, Test Later

The first word in DOE is the key – design. The tests or experiments must be planned before the first one is done. Said strongly, *don't use statistical processes with ad hoc or "found" data.*

The reason for this negative guidance is that the four principles listed in Section A1.22.3 may not have been incorporated in the choice of which experiments were completed. The two data analysis techniques listed in Section A1.22.4 won't produce the desired results if the four principles in Section A1.22.2 are not incorporated into the data collection plan – and *managers won't know that.*

### A1.22.2 Expectations for Designed Experiments

The best than can be produced by a designed experiment is certainty (or uncertainty) about a relationship.[89]

Managers should expect DOE to:

- Identify if there is a relationship between a variable(s) or factor(s) and some outcome, as well as the probability that this relationship is not misidentified.[90] *The relationship is assumed when both variables have the same variability in their means.*
- Identify if there are interactions between two or more variables and an outcome. Here the outcome may or may not be produced by either variable, but *is produced* by their interaction. As above, the outcome may be a tasty sandwich or yellow light, but neither jelly and green light or peanut butter and red light will produce it.

Correlating equations, correlation coefficients, scatter plots, etc. can all be produced once the certainty of a relationship is assured. Additional data may be required – but it's justified when the certainty of a relationship has been established. Why waste resources developing data for a correlating equation between a variable and an unrelated outcome?

### A1.22.3 Principles Used in DOE

There are four general principles (concepts) used to select which data should be collected. That is the design for the experiments. Listed in alphabetical

---

[85] In this topic, experiments are not tests for cleanliness. Experiments, in this topic, are those efforts where operating parameters/variables/factors are deliberately varied so as to learn what outcomes are produced by their change.

[86] Anderson, M., "Design of Experiments," *The Industrial Physicist Magazine*, September 1997, p. 24.

[87] Various reasons are attributed. This author casts no aspersions.

[88] DOE is attributed to Sir Ronald Aylmer Fisher (1890–1962) who was a researcher and teacher in the field of agriculture. Fisher's first significant publication was *Statistical Methods for Research Workers* (1925).

[89] Additional details about the certainty of relationships can be found on *The Jerry Springer Show.*

[90] This misidentification is known as a Type I error – rejecting a valid relationship that does exist. A Type 2 error is the opposite – believing that a relationship exists when it does not.

order, and summarized in non-statistical language, they are:

1. **Blocking.** A manager using this principle divides their experiments into subgroups (blocks) using a system associated with the variables being tested. For example, in testing the effect of ultrasonic transducers on removal of particles, one might test at two levels of ultrasonic input power and leave the frequency sweep control[91] both on and off. Here, Frequency sweep on and off are "treatments" and each level of ultrasonic power is a "block."

   *Choose a few variables and test at two (or so) levels of each.*

2. **Orthogonality.** Rather than right angles, this principle relates to selection of variables as being independent – at least as is known prior to the experimentation. A crucial point, in a true orthogonal experiment, the manager can separate the influence of each independent variable upon the dependent variable. In the example above, if transducer frequency had been substituted for ultrasonic power level as an independent variable, the effects of frequency sweep and frequency level on particle removal might not have been separable. This is because both variables would have been manifestations of a single one: frequency.

   *The variables must be totally independent of one another.*

3. **Randomization.** Time is often an implicit (implied or unrecognized) controlling variable. If a test is to be conducted once per shift, and the sample is always collected during the first shift hour (which is common), the result may be abnormal. This is because shift operators often (and incorrectly) make adjustments at the start of their responsibility so as to "correct mistakes" made by their predecessors. A superior approach would be to choose the hour for sampling each shift by a random draw.[92]

   *Identify specific times, samples, locations, etc. by random choice.*

4. **Replication.** This principle is an attempt to minimize the impact of "fluke," "flier," or "snowbird" data without arbitrarily discarding a data point. One deliberately replicates (repeats) an experiment to gain some estimate of the variability between experiments, and identify possible and unexpected interactions among variables.

   *Repeat some tests – it will strengthen your data set.*

### A1.22.4 Techniques Used in DOE

Simplicity should be the "order of the day."

- **Factorial experiments.** In a factorial experiment each sample/measurement/test is characterized by more than one factor (not the OFAT approach). In the example above, a measurement of particle contamination is associated with a level of ultrasonic power and whether or not the frequency sweep is turned on.

  *Identify key variables which influence an outcome by testing each at two or three levels – zero[93] and some other value(s).*

- **Analysis of experimental results.** A manager does this by analyzing the deviations of individual measurements from group average values. The technique is called analysis of variance (ANOVA).[94]

  *Spreadsheet functions do the job without confusion.*

---

[91] This is a feature, formerly optional but now nearly always standard, in which the main transducer frequency is deliberately varied over a small range. The purpose is simple – to avoid a constant frequency wave in the tank. Such a wave might resonate within the parts, be amplified, and case part damage. See Figure 7.38.

[92] The spreadsheet function @RAND in Quattro-Pro [or =RAND in Excel] will generate random numbers between zero and one. These can be scaled to fit any needs. Alternately, the function @RANDBETWEEN(minimum value, maximum value) or [=RANDBETWEEN] can be used. If the times, sample numbers, or locations, etc. can be assigned numbers, a random choice can be made.

[93] Testing at the zero level is usually worth the effort – as a baseline from which change can be recognized. Medical researchers would call it a placebo (lacking therapeutic effect but reassuring just the same).

[94] A collection of statistical procedures (models) by which the observed variance within the data is partitioned into components. Each component is due to different factors which are estimated and/or tested.

**Table A1.22 Significant Output From ANOVA Spreadsheet Functions**

| Output | Identification | Source | Utility |
|---|---|---|---|
| F-Test | Ratio of the mean variance *between* samples to the mean variance *within* samples | The spreadsheet function ANOVA calculates this result from the data provided by the user – values of variables (factors), the number of them, and outcomes | If there *is a statistically significant relationship* between the measured outcomes and one or two variables (factors): *F-Test > F-Critical* |
| P-Critical | Actual level of confidence (statistical significance) associated with the calculated F-Test value | | This is the level of confidence (significance) that there is a relationship between the measured outcomes and one or two variables (factors) |
| F-Critical | F-Test value for chosen level of confidence (statistical significance) and numbers of data values | The spreadsheet function ANOVA calculates this result – though it can be found in textbooks | If there *is no statistically significant relationship* between the measured outcomes and one or two variables (factors), *F-Test < F-Critical* |

### A1.22.5 Spreadsheet Functions[95] Which Conduct ANOVA

In general, the purpose of ANOVA is to test for significant differences between many means. If there were only two means, the *t*-test would be all that's needed.

ANOVA allows the effects of multiple variables to be examined – more than two – *and interactions between at least two variables to be identified.* It is the latter value which many statisticians feel is the keystone value of ANOVA.

ANOVA is traditionally classified as "one-way," "two-way," "three-way" (three independent variables), etc. While this does not refer to how chili is prepared in Cincinnati,[96] it does refer to the number of independent variables being considered in the experiment.

ANOVA[97] functions in both Microsoft's Excel and Corel's Quattro-Pro spreadsheets allow "automatic"[98] evaluation of ANOVA for one-way and two-way situations. Three-way and more complicated situations are not evaluated by spreadsheet functions, and so will not be considered in this volume.[99]

There are only three outputs of interest to users of spreadsheet ANOVA functions (Table A1.22).[100]

There are three ANOVA spreadsheet functions in each of both brands of spreadsheet. They can be

---

[95] The approach used in this appendix is written in Footnote 3. Non-statisticians find the language and methodology of DOE complicated, and may use that reason to avoid the use of this valuable management tool. So both will be avoided here. As with *t*-tests, spreadsheet functions and their implementation will be described in this appendix.

[96] Chili in Cincinnati, OH, US, is referred to as: one-way (chilli only), two-way (chilli over spaghetti), three-way (two-way with grated cheddar cheese), four-way (three-way with chopped onions), and five-way (four-way with kidney beans).

[97] Analysis of variance (ANOVA) tools help determine if two or more samples come from the same population by calculating the F-statistic, the ratio of the mean variance between samples to the mean variance within samples. The ANOVA tools are commonly used in cases involving more than two samples (where *t*-tests cannot be used).

[98] Here, "automatic" means that nearly all of the computational details are hidden from the spreadsheet user. Only the results, and some intermediate results, are provided.

[99] For those who wish to learn more, but don't wish to become immersed in one of the reference books in Footnote 3, there is an alternative. Professor Richard Lowry of Vassar College, has established a web site titled "Concepts and Applications of Inferential Statistics." It is well-recommended and can be found at http://faculty.vassar.edu/lowry/webtext.html.

[100] Though others, such as intermediate results, are provided.

**Table A1.23 Spreadsheet ANOVA Functions (and their Macro Commands)**

| Function | Name | Type of Situation | Purpose |
|---|---|---|---|
| ANOVA1 | One-way | One set of outcomes<br>One independent variable<br>One data point per variable | Determine if change in the outcome is related to change in the variable |
| {ANOVA1 InBlock, OutBlock, <Grouped>, <Labels(0\|1)>, <Alpha>} | | | |
| ANOVA2 | Two-way, with replication | One set of outcomes<br>Two independent variables<br>Multiple data points per variable pair | • Determine if change in the outcome is related to change in the variables<br>• Determine if the change in the outcome is related to the combined or interactive effect of change in the two variables |
| {ANOVA2 InBlock, OutBlock, SampleRows, <Alpha>} | | | |
| ANOVA3 | Two-way | One set of outcomes<br>Two independent variables<br>One data point per variable pair (no replication) | Determine if change in the outcome is related to change in the variables |
| {ANOVA3 InBlock, OutBlock, <Labels(0\|1)>, <Alpha>} | | | |

executed within a single cell as a macro,[101] or can be implements within an automated process called a "Wizard" (Table A1.23).

### A1.22.6 Use of ANOVA Spreadsheet Functions

Use of ultrasonic transducers to remove particles, using water or aqueous chemistry, is a common situation. Experimentation is seldom done. Some examples are provided below where those experiments are designed and the significant relationships identified via ANOVA.

Experiments should be designed to answer questions – using data developed in the experiment (conducted *after* the questions have been posed), and statistical evaluation of that data. Selection of the experimental design, and the data, depends upon the question.

#### A1.22.6.1 *Question 1:[102] Does the Level of Applied Power Matter in Removal of Particulate?*

**Methodology to Develop an Answer:**

1. Prepare an appropriately-sized (10 gallons) ultrasonic tank with a low concentration of a suitable aqueous detergent. The tank should have an immersed ultrasonic transducer oscillating at 40 kHz with a "full on" power level of 25 to 100 watts/liquid gallon. The transducer should have one power level setting considerably different than the "full on" value, and frequency adjustment (sweep) to avoid standing waves of single frequency. Use ambient conditions.
2. Prepare a second tank of the same size with pure water – as a rinse tank. Since particles are more likely to be removed in the rinsing operation,

[101] Syntax for these macros is simple, with examples to follow, and copied from the Quattro-Pro software help file. This author finds the Wizards more efficient than operation using the macros as written in Table A1.23.

InBlock: Input cells containing two or more sets of numeric data arranged in columns or rows.
OutBlock: Upper left cell of the output cells. The Wizard will not overwrite existing data.
Grouped: "C" to group results by column or "R" to group results by row; the default is "C."
Labels: 1 if labels are located in the first column or row of the input cells; 0 if the input cells do not contain labels; the default is 0.
Alpha: The significance level at which to evaluate values for the F-statistic; the default is 0.05 (95% confidence).

[102] Each question in this appendix is artificial. The data from which the questions are answered is also artificial, though developed by a spreadsheet model based on experience with client systems.

conditions in that tank will be those adjusted in experiments. Conditions in the cleaning tank will be constant. Use equivalent facilities in both tanks.

3. Immerse soiled parts (e.g. stents for medical implant) in the sonicated[103] wash tank for from two to ten minutes. Allow parts to drain for one minute. Immerse cleaned parts in the sonicated rinse tank for the same time. Allow parts to drain for two minutes. Flush parts with Nitrogen and dry for twenty minutes under vacuum. Measure uncleaned particle content via counting them within a specified area under magnification. Conduct experiments where:

   The ultrasonic transducers are not powered in the rinse tank, and the frequency sweep feature is not powered. This will answer a question about whether ultrasonic transducers in the rinse tank do provide any value, and another question about if the frequency sweep feature (available at extra cost) is justified.

   Both questions can be answered in the same or separate experimental designs. The former will be done first to illustrate the ANOVA1 spreadsheet function.

   The basic data is shown intable A1.25 in the order taken. The same data are collected for analysis by the ANOVA1 spreadsheet function in Table A1.25. The spreadsheet in which the analysis is done is Table A1.26.

Please note that:

- The block of experimental data is located in cells B3..C7.[104] of Table A1.26. There are four repeats (replications) per factor (variable).
- The upper left corner of the analysis results is cell E3.
- The default organization by columnar data was accepted as the two levels of the factor are listed in columns and not rows,
- Column labels (cells B3 and C3) were used.
- The default significant level of 0.05 (95% confidence) was accepted. While this result was produced using the spreadsheet ANOVA1 Wizard, the following macro command could have been used to produce the same effect:

ANOVA1{B3..C7,E3,C,1,0.05}

**Table A1.24 Basic Date – Effect of Ultrasonic Action**

| Power | | | Particle Count per SI |
|---|---|---|---|
| Rated | % | Applied | |
| 500 | 100 | 500 | **50** |
| 500 | 0 | 0 | **88** |
| 500 | 100 | 500 | **50** |
| 500 | 0 | 0 | **77** |
| 500 | 100 | 500 | **45** |
| 500 | 0 | 0 | **66** |
| 500 | 100 | 500 | **41** |
| 500 | 0 | 0 | **55** |

**Table A1.25 Data for ANOVA1 Analysis (Particle Count/SI)**

| 500 W | 0 W |
|---|---|
| 50 | 88 |
| 50 | 77 |
| 45 | 66 |
| 41 | 55 |

The spreadsheet is shown in Table A1.26.

The desired results are in cells I16, J16, and K16. The other cells provide preliminary results, but are not the reason for doing the analysis. Since the value of F (11.33) is greater than F-Critical (5.99), ultrasonic power level does have an effect upon particle count. And the probability of power not having an effect upon particle count is only 1.51% (cell J16).

*So the answer to Question 1 is that the level of applied power does matter in removal of particulate, and this outcome is 98.5% certain.*

**Please note the answer to Question 2 below.**

This work could have been done with the *t*-test above. But the following question could not have been answered.

[103] A tank with operating ultrasonic or megasonic transducers.

[104] *This is a crucial point* – please note the labels are included in this block.

**Table A1.26 Spreadsheet Showing ANOVA1 Analysis Procedure and Results for Question 1**

| | A | B | C | D | E | F | G | H | I | J | K |
|---|---|---|---|---|---|---|---|---|---|---|---|
| 1 | | | | | | | | | | | |
| 2 | | | | | | | | | | | |
| 3 | | *500 W* | *0 W* | | ANOVA One-way | | | | | | |
| 4 | | 50 | 88 | | | | | | | | |
| 5 | | 50 | 77 | | *Summary* | | | | | | |
| 6 | | 45 | 66 | | | | | | | | |
| 7 | | 41 | 55 | | *Groups* | *Count* | *Sum* | *Average* | *Variance* | | |
| 8 | | | | | 500 W | 4 | 186 | 46.5 | 19 | | |
| 9 | | | | | 0 W | 4 | 286 | 71.5 | 201.67 | | |
| 10 | | | | | | | | | | | |
| 11 | | | | | ANOVA | | | | | | |
| 12 | | | | | | | | | | | |
| 13 | | | | | *Source of variation* | | | | | | |
| 14 | | | | | | | | | | | |
| 15 | | | | | | *SS* | *df* | *MS* | ***F*** | ***P-value*** | ***F-Critical*** |
| 16 | | | | | *Between groups* | 1250 | 1 | 1250 | **11.33** | **1.51%** | **5.99** |
| 17 | | | | | *Within groups* | 662 | 6 | 110.33 | | | |
| 18 | | | | | | | | | | | |
| 19 | | | | | *Total* | 1912 | 7 | | | | |

### A1.22.6.2 *Question 2: Does the Level of Applied Power and/or Sweep of Applied Frequency Matter in Removal of Particulate?*

This question has two parts. The reason for answering both in the same experiment design is to reduce the number of experiments, and answer a third question (unasked) about whether the ultrasonic power level and sweep function interact to affect particle removal.

The same methodology is used.

Data collected simultaneously with that above, but not enumerated, is added in cells C8 to C11. This is from operation with frequency sweep on, with, and without power.

The following macro command could have been used, instead of a Wizard:

ANOVA2{C3..D11,F3,4,0.05}

The desired results are in Table A1.27, as spreadsheet cells J32 through L36. The other cells provide preliminary results.

Since:

- $F_{Sample}$ (7.92) is *greater* than F-Critical$_{Sample}$ (4.74), so frequency sweep does have an effect upon particle count – at more than the 95% level of confidence.
- $F_{Column}$ (8.14) is *greater* than F-Critical$_{Column}$ (4.74), so ultrasonic power level does have an effect upon particle count – at more than the 95% level of confidence.
- $F_{Interaction}$ (0.62) is *less* than F-Critical$_{Interaction}$ (4.74), so there is not an interaction between ultrasonic power level and frequency sweep on particle count – at the level of confidence of 95%.

### A1.22.7 The Need to Be a Manager

But there is a problem. When managers use any statistical analysis tools, they must remember they are managers first – responsible for observing and thinking and not simply "pouring" numbers into a spreadsheet.

**Table A1.27 Spreadsheet Showing ANOVA2 Analysis Procedure and Results WITH FLAWED DATA FOR QUESTIONS 1 AND 2**

| | A | B | C | D | E | F | G | H | I | J | K | L |
|---|---|---|---|---|---|---|---|---|---|---|---|---|
| 1 | | | | | | | | | | | | |
| 2 | | | | | | | | | | | | |
| 3 | | | *500 W* | *0 W* | | ANOVA: two-way with replication | | | | | | |
| 4 | | *Sweep OFF* | 50 | 88 | | | | | | | | |
| 5 | | | 50 | 77 | | *Summary* | | | | | | |
| 6 | | | 45 | 66 | | | | | | | | |
| 7 | | | 41 | 55 | | | *500 W* | *0 W* | *Total* | | | |
| 8 | | *Sweep ON* | 87 | **132** | | | | | | | | |
| 9 | | | 63 | **100** | | | | | | | | |
| 10 | | | 67 | **68** | | | | | | | | |
| 11 | | | **72** | **55** | | Count | 4 | 4 | 8 | | | |
| 12 | | | | | | Sum | 186 | 286 | 472 | | | |
| 13 | | | | | | Average | 46.5 | 71.5 | 118 | | | |
| 14 | | | | | | Variance | 19 | 201.67 | 220.67 | | | |
| 15 | | | | | | | | | | | | |
| 16 | | | | | | *Sweep ON* | | | | | | |
| 17 | | | | | | | | | | | | |
| 18 | | | | | | Count | 4 | 4 | 8 | | | |
| 19 | | | | | | Sum | 284.94 | 341.63 | 626.57 | | | |
| 20 | | | | | | Average | 72.29 | 88.79 | 161.07 | | | |
| 21 | | | | | | Variance | 104.90 | 1178.16 | 1283.06 | | | |
| 22 | | | | | | | | | | | | |
| 23 | | | | | | *Total* | | | | | | |
| 24 | | | | | | | | | | | | |
| 25 | | | | | | Count | 8 | 8 | | | | |
| 26 | | | | | | Sum | 470.94 | 627.63 | | | | |
| 27 | | | | | | Average | 118.79 | 160.29 | | | | |
| 28 | | | | | | Variance | 123.90 | 1379.83 | | | | |
| 29 | | | | | | | | | | | | |
| 30 | | | | | | | | | | | | |
| 31 | | | | | | | | | | | | |
| 32 | | | | | | *Source of variation* | *SS* | *df* | *MS* | ***F*** | ***P-value*** | ***F-Critical*** |
| 33 | | | | | | | | | | | | |
| 34 | | | | | | Sample | 1493.23 | 1 | 1493.23 | **7.921** | **1.562%** | **4.747** |
| 35 | | | | | | Column | 1534.40 | 1 | 1534.40 | **8.140** | **1.454%** | **4.747** |
| 36 | | | | | | Interaction | 117.26 | 1 | 117.26 | **0.622** | **44.559%** | **4.747** |
| 37 | | | | | | Within | 2262.08 | 12 | 188.51 | | | |
| 38 | | | | | | | | | | | | |
| 39 | | | | | | Total | 5406.97 | 15 | | | | |

Please note in cells D8 to D11, the two highest particle levels (the 132 and the 100) are found when the ultrasonic power is OFF, and frequency sweep is ON (vs the 88 and the 77). And, two of the values reported (the 68 and 55) for the same experimental conditions are less than those reported when the power is ON and the sweep is OFF (the 87 and the 72). That's not reproducibility!

This doesn't make physical sense. When the ultrasonic power is off, the frequency sweep cannot have any effect! If there is an effect, the power cannot truly be off!

The F-Test analysis above is worthless because it was based on apparently flawed data.

Since **all** data were (properly) taken at the same time, they **all** must be discarded. This a core principle of DOE:

- **Managers must treat *all* data alike. One flawed – all flawed!**

More than that, the answer to *Question 1 must also be discarded* as it was obtained with the same data set:

- **Managers must treat *all* data alike. If one conclusion is flawed from some data – all conclusions from any of that data must be viewed as flawed!**

An inspection of facilities revealed the condition which produced the flawed data.[105]

Revised data, with the flaw repaired (Table A1.28).

As above, the desired results are shown, *with correct data,* in Table A1.28 (cells J32 through J36, K32 through K36, and L32 through L36). The other cells provide preliminary results. Since:

- $F_{Sample}$ (4.33) is *less* than F-Critical$_{Sample}$ (4.74), so frequency sweep does *not have* an effect upon particle count – at the 95% level of confidence.[106]
- $F_{Column}$ (1.90) is *less* than F-Critical$_{Column}$ (4.74), so ultrasonic power level does *not* have an effect upon particle count – at the 95% level of confidence.
- $F_{Interaction}$ (7.63) is *greater* than F-Critical$_{Interaction}$ (4.74), so there is an interaction between ultrasonic power level and frequency sweep on particle count – at higher than the level of confidence of 95% (98.3%).

What an interesting outcome! It isn't either of two process features which provides effective particle removal. It's the interaction of two process features (frequency sweep and ultrasonic power level) which enable effective particle removal! With the ultrasonic power turned on, and the frequency sweep turned off, particle removal isn't effective.

*The answer to Question 1 is that the level of applied power does NOT matter in removal of particulate – at the level of statistical confidence which managers normally require (95%).*

*The answer to Question 2 is that sweep of frequency does NOT matter in removal of particles.*

*However, the answer to an unasked question is that there is an interactive relationship between level of applied power and application of sweep to frequency that affects removal of particles. That interaction is highly certain (100 − 1.716 = 98.3%) to be valid – based on the data presented.*

How else could this critical and unexpected interactive relationship have been determined – except by a properly designed experiment and ANOVA2? It could not have.

This is the proper role of DOE – to learn about relationships, expected and unexpected.

### A1.22.7.1 *Question 3: Does the Frequency of Applied Power Matter in Removal of Particulate?*

**Methodology to Develop an Answer:**

1. Prepare the same facilities as used in Question 1.
2. Two additional piezoelectric transducer systems are required for the rinse tank. This makes the

---

[105] Please recall these data are artificial – created by a model, for the purpose of this analysis.

[106] There is a significant effect of frequency sweep on particle count, but only at about the 94% (100% − 5.9%) level of significance. Many mangers would accept this "close enough for industrial work."

**Table A1.28 Spreadsheet Showing ANOVA2 Analysis Procedure and Results With CORRECT Data for Questions 1 and 2**

| | A | B | C | D | E | F | G | H | I | J | K | L |
|---|---|---|---|---|---|---|---|---|---|---|---|---|
| 1 | | | | | | | | | | | | |
| 2 | | | | | | | | | | | | |
| 3 | | | *500 W* | *0 W* | | ANOVA: two-way with replication | | | | | | |
| 4 | | *Sweep OFF* | 8 | 1229 | | | | | | | | |
| 5 | | | 109 | 842 | | *Summary* | | | | | | |
| 6 | | | 172 | 803 | | | | | | | | |
| 7 | | | 19 | 965 | | | *500 W* | *0 W* | *Total* | | | |
| 8 | | *Sweep ON* | 85 | 13 | | | | | | | | |
| 9 | | | 28 | 49 | | | | | | | | |
| 10 | | | 293 | 8 | | | | | | | | |
| 11 | | | 173 | 8 | | Count | 4 | 4 | 8 | | | |
| 12 | | | | | | Sum | 460 | 2500 | 2960 | | | |
| 13 | | | 111 | 490 | | Average | 78 | 960 | 1038 | | | |
| 14 | | | | | | Variance | 5887 | 37 070 | 42 957 | | | |
| 15 | | | | | | | | | | | | |
| 16 | | | | | | *Sweep ON* | | | | | | |
| 17 | | | | | | | | | | | | |
| 18 | | | | | | Count | 4 | 4 | 8 | | | |
| 19 | | | | | | Sum | 796 | 114 | 909 | | | |
| 20 | | | | | | Average | 145 | 20 | 164 | | | |
| 21 | | | | | | Variance | 13 315 | 395 | 13 710 | | | |
| 22 | | | | | | | | | | | | |
| 23 | | | | | | *Total* | | | | | | |
| 24 | | | | | | | | | | | | |
| 25 | | | | | | Count | 8 | 8 | | | | |
| 26 | | | | | | Sum | 1255 | 2614 | | | | |
| 27 | | | | | | Average | 222 | 979 | | | | |
| 28 | | | | | | Variance | 19 202 | 37 465 | | | | |
| 29 | | | | | | | | | | | | |
| 30 | | | | | | | | | | | | |
| 31 | | | | | | | | | | | | |
| 32 | | | | | | *Source of variation* | *SS* | *df* | *MS* | *F* | *P-value* | *F-Critical* |
| 33 | | | | | | | | | | | | |
| 34 | | | | | | Sample | 262 777 | 1 | 262 777 | **4.335** | **5.940%** | **4.747** |
| 35 | | | | | | Column | 115 330 | 1 | 115 330 | **1.903** | **19.295%** | **4.747** |
| 36 | | | | | | Interaction | 463 011 | 1 | 463 011 | **7.638** | **1.716%** | **4.747** |
| 37 | | | | | | Within | 727 389 | 12 | 60 616 | | | |
| 38 | | | | | | | | | | | | |
| 39 | | | | | | *Total* | *1 568 507* | *15* | | | | |

total of 500 W transducers used in this cleaning operation to be four. The two new ones will operate at both higher and lower frequencies in the rinse tank. The two new frequencies are 25 and 68 kHz.[107]

3. The same process as practiced in step 3 above is practiced. It has to be practiced a total of three times, as each of the three frequencies (25, 40, and 68 kHz) must be employed in the rinse tank.
4. Two levels of power will be used (500 W and none). Conduct limited experiments – only six experiments are necessary.
5. The spreadsheet with the developed data and the ANOVA3[108] analysis is shown (Table A1.29).

The following macro command could have been used, instead of a Wizard:

ANOVA3{B7..E9,G8,1,0.05}

The desired results are shown in Table A.29 as spreadsheet cells K23 through M24. The other cells provide preliminary results.

Since, from the ANOVA3 spreadsheet above:

- $F_{Sample}$ (1.11) is *much less* than F-Critical$_{Sample}$ (18.51), so frequency level does ***Not*** have an effect upon particle count.
- $F_{Column}$ (0.14) is *much less* than F-Critical$_{Column}$ (19.00), so ultrasonic power level does ***Not*** have an effect upon particle count.
- There is no value of $F_{Interaction}$ because there are no replications. Interactions cannot be statistically determined without additional data.

*So the answer to Question 3 is that the frequency level of these ultrasonic systems does NOT matter in removal of particles – based on the data presented.*

*Further, as in the ANOVA2 spreadsheets, the level of applied power does NOT matter in removal of particulate.*

### A1.21.7.2 *Question 4: If More Testing Were Done (If More Data Were Taken), Would There Be an Interaction of Frequency Level with Power Level?*[109]

**Methodology to Develop an Answer:**

1. Prepare the same facilities as used in Question 3, and operate them in the same manner.
2. Since the value of frequency sweep (when used in combination with ultrasonic power level) has been shown above (in the ANOVA2 spreadsheet), it will be used in the ON state for all future operation.
3. Complete an extensive, and probably extravagant, test program where eight levels of ultrasonic power between 0 and 500 W are applied in the rinse tank, and three levels of frequency are used. In addition, there are six repeats (replications) for each of the 24 (8 power levels × 3 frequency levels) experimental conditions. That's a total of 144 (24 conditions × 6 replicates).

   The 144 conditions were plotted and shown as Figure A1.26. From this scatter plot (with lines connecting the data points), very little can be determined with confidence. That's why the ANOVA spreadsheet functions are so valuable – they can reduce some complexity to simplicity.

   The spreadsheet with the developed data and the ANOVA2 analysis is shown as Table A1.30.[110]

---

[107] Transducer frequency can play a significant role in removing particles. The key to ultrasonic cleaning is the physical implosion (cavitation) of vapor bubbles in a cleaning solution. The higher the frequency, the smaller the cavitation bubble size (see Figure 7.36.) It is a thesis of this experiment that those smaller cavitation bubbles will be more effective in removing smaller particles than with larger bubbles produced with the transducers operating at a higher frequency.

[108] With the ANOVA3 analysis plan, there are two variables producing a single outcome. There are no repeats (replications) of experiments which would add robustness to the data set and strength to the veracity of the conclusions.

[109] The data in the ANOVA3 spreadsheet show one reason for asking this question. It is that there appears to be a substantial level of unexplained variance (error) in these results. Please note that because there is no replication within this experimental design, the largest source of variance in error (876 units for error, versus 486 for frequency, and 129 for sweep).

[110] Some blank rows have been omitted to meet space requirements.

**Table A1.29 Spreadsheet Showing ANOVA3 Analysis Procedure and Results of Question 3**

| | A | B | C | D | E | F | G | H | I | J | K | L | M |
|---|---|---|---|---|---|---|---|---|---|---|---|---|---|
| 1 | | | | | | | | | | | | | |
| 2 | | | | | | | | | | | | | |
| 3 | | | | | | | | | | | | | |
| 4 | | | | | | | | | | | | | |
| 5 | | *Table completed manually* | | | | | | | | | | | |
| 6 | | | 25 kHz | 40 kHz | 68 kHz | | | | | | | | |
| 7 | | *Power* | *Frequency* | | | | | | | | | | |
| 8 | | 500 | 81 | 194 | 896 | | ANOVA: two-way without replication | | | | | | |
| 9 | | 0 | 506 | 505 | 131 | | | | | | | | |
| 10 | | | | | | | *Summary* | *Count* | *Sum* | *Average* | *Variance* | | |
| 11 | | | | | | | | | | | | | |
| 12 | | | | | | | *500 W* | *3* | *215.00* | *390.46* | *194 860* | | |
| 13 | | | | | | | 0 W | 3 | 161.00 | 380.66 | 46 917 | | |
| 14 | | | | | | | | | | | | | |
| 15 | | | | | | | 25 kHz frequency | 2 | 138.00 | 293.76 | 90 268 | | |
| 16 | | | | | | | 40 kHz frequency | 2 | 122.00 | 349.64 | 48 417 | | |
| 17 | | | | | | | 68 kHz frequency | 2 | 116.00 | 513.28 | 292 955 | | |
| 18 | | | | | | | | | | | | | |
| 19 | | | | | | | ANOVA | | | | | | |
| 20 | | | | | | | | | | | | | |
| 21 | | | | | | | *Source of variation* | *SS* | *df* | *MS* | ***F*** | ***P-value*** | ***F-Critical*** |
| 22 | | | | | | | | | | | | | |
| 23 | | | | | | | Rows | 486.000 | 1.00 | 486.00 | **1.110** | **40.26%** | **18.513** |
| 24 | | | | | | | Columns | 129.333 | 2 | 64.67 | **0.148** | **87.14%** | **19.000** |
| 25 | | | | | | | Error | 876.000 | 2 | 438.00 | | | |
| | | | | | | | *Total* | *1491.333* | *5* | | | | |

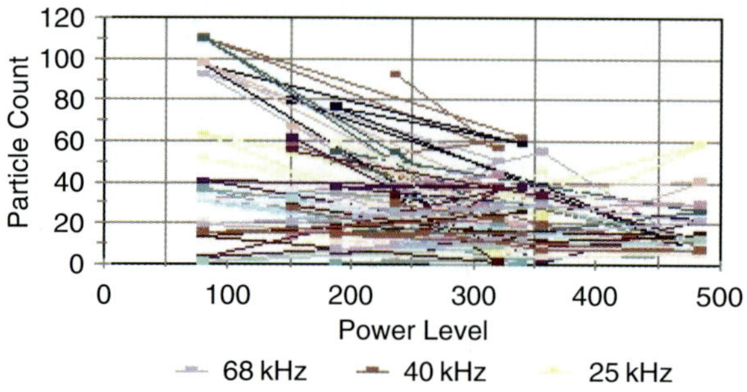


**Figure A1.26** Two-way experiment with power and frequency

## A1.22.8 The Boss ANOVA

The following macro command could have been used instead of a Wizard:

ANOVA2{B8..J26,L8,6,0.05}

The desired results are in cells P44 through R48 of Table A1.30. The other cells provide preliminary results.

Please note the number of replications (six in this example) and the number of levels (eight for power level and three for frequency level) is essentially unlimited[111] when using the ANOVA2 (or its associated Wizard) spreadsheet function.

Please also note how the experimental data are ordered from left to right in block B8..J26. This is the order in which the experiments were conducted.

The order, in terms of power level, was chosen at random, as were the exact levels of power selected. Managers need to be aware of the tradeoff between random selection of experimental states and selection numerical order or some other scheme:

- Random selection insures that time or spacial effects are eliminated in an experimental design. A statistician would prefer this selection.
- But a manager has to be prepared in the event the experiment is not completed and the experiment can't be restarted or redone – for reasons outside the control of the manager. This contingency[112] argues for competition of the first, last, and center conditions before the remaining intermediate conditions are completed. This selection preserves some information in the event the experiment can't be completed as designed.

Since:

- $F_{Sample}$ (395) is *much greater* than F-Critical$_{Sample}$ (3.07). Ultrasonic power level does have an effect upon particle count. The effect is significant at nearly 100% (100%–~0%) confidence.
- $F_{Column}$ (64) is *greater than* F-Critical$_{Column}$ (2.08). Ultrasonic frequency level does have an effect upon particle count. The effect is significant at nearly 100% (100%–~0%) confidence.
- $F_{Interaction}$ (4.72) is *greater* than F-Critical$_{Interaction}$ (1.77). There is an interaction between ultrasonic power level and ultrasonic frequency level on particle count. The effect is significant at nearly 100% (100%–~0%) confidence.

*So the answer to Question 4 is that the frequency level of these ultrasonic systems DOES matter in removal of particles – when sufficient data are presented.*

*Further, as in the ANOVA2 spreadsheets, there is an interactive relationship between level of frequency and level of applied power that affects removal of particles. That interaction is also highly certain to be valid.*

A crucial lesson from these examples is that the design of the experiment and the data produced by it do matter:

- Mistaken or incorrect conclusions can be avoided.
- Unexpected interactions between variables can be uncovered.

## A1.22.9 General Limitations of ANOVA

ANOVA is *not limited*[113] to linear relationships between some variables (factors) and an outcome.

A comparison of where $F_{Sample}$ is numerically greater than $F_{Critical}$ *does not speak to why or how* much makes the independent variables affect a dependent variable.

[111] In tests, the author has not been able to identify a practical limit for the number of replications or levels.

[112] This author can almost guarantee this contingency will occur in a production environment.

[113] Variance is a statistical measure of deviation – not personal, but numerical. It is calculated by summing each deviation squared of a measurement from the mean of measurements. Each deviation is squared so that positive and negative deviations don't innocently and algebraically cancel. When the sum is divided by the number of tests (less 1), the result is the statistical term – variance.

**Table A1.30 Spreadsheet Showing ANOVA2 Analysis Procedure and Results for Question 4**

| | A | B | C | D | E | F | G | H | I | J | K | L | M | N | O | P | Q | R | S | T | U |
|---|---|---|---|---|---|---|---|---|---|---|---|---|---|---|---|---|---|---|---|---|---|
| 1 | | | | | | | | | | | | | | | | | | | | | |
| 8 | | 68 kHz | 70.4 W | 425.0 W | 462.4 W | 196.5 W | 346.9 W | 459.6 W | 491.1 W | 83.3 W | | ANOVA: two-way with replication | | | | | | | | | |
| 9 | | 68 kHz | 136 | 53 | 62 | 109 | 95 | 45 | 41 | 158 | | | | | | | | | | | |
| 10 | | 68 kHz | 139 | 65 | 69 | 116 | 61 | 47 | 54 | 161 | | *Summary* | | | | | | | | | |
| 11 | | 68 kHz | 161 | 69 | 71 | 103 | 73 | 47 | 43 | 111 | | | | | | | | | | | |
| 12 | | 68 kHz | 122 | 55 | 64 | 108 | 96 | 57 | 58 | 117 | | | *70 W* | *425 W* | *462 W* | *196 W* | *347 W* | *460 W* | *491 W* | *83 W* | *Total* |
| 13 | | 68 kHz | 165 | 61 | 44 | 89 | 66 | 58 | 44 | 100 | | | | | | | | | | | |
| 14 | | 68 kHz | 137 | 59 | 49 | 109 | 67 | 46 | 57 | 137 | | *68 kHz* | | | | | | | | | |
| 15 | | 40 kHz | 180 | 125 | 69 | 181 | 128 | 85 | 87 | 213 | | | | | | | | | | | |
| 16 | | 40 kHz | 186 | 79 | 79 | 135 | 160 | 93 | 77 | 210 | | Count | 6 | 6 | 6 | 6 | 6 | 6 | 6 | 6 | 48 |
| 17 | | 40 kHz | 181 | 83 | 93 | 163 | 114 | 105 | 73 | 264 | | Sum | 403 | 495 | 937 | 522 | 601 | 676 | 735 | 313 | 4682 |
| 18 | | 40 kHz | 184 | 104 | 100 | 206 | 126 | 71 | 91 | 267 | | Average | 143 | 60 | 60 | 106 | 76 | 50 | 50 | 131 | 675 |
| 19 | | 40 kHz | 177 | 84 | 70 | 172 | 120 | 89 | 96 | 197 | | Variance | 270 | 35 | 116 | 84 | 236 | 31 | 58 | 641 | 1472 |
| 20 | | 40 kHz | 228 | 88 | 75 | 211 | 161 | 81 | 73 | 215 | | | | | | | | | | | |
| 21 | | 25 kHz | 424 | 154 | 121 | 354 | 178 | 157 | 146 | 363 | | *40 kHz* | | | | | | | | | |
| 22 | | 25 kHz | 393 | 199 | 130 | 257 | 229 | 122 | 148 | 315 | | | | | | | | | | | |
| 23 | | 25 kHz | 440 | 146 | 182 | 269 | 196 | 177 | 117 | 422 | | Count | 6 | 6 | 6 | 6 | 6 | 6 | 6 | 6 | 48 |
| 24 | | 25 kHz | 327 | 217 | 152 | 266 | 226 | 124 | 146 | 352 | | Sum | 642 | 835 | 1457 | 950 | 856 | 1101 | 1 242 | 547 | 7632 |
| 25 | | 25 kHz | 430 | 198 | 158 | 273 | 203 | 184 | 152 | 389 | | Average | 189 | 94 | 81 | 178 | 135 | 87 | 83 | 227 | 1074 |
| 26 | | 25 kHz | 445 | 207 | 174 | 333 | 233 | 122 | 125 | 416 | | Variance | 366 | 308 | 162 | 796 | 426 | 135 | 96 | 900 | 3189 |

(Continued)

**Table A1.30 Spreadsheet Showing ANOVA2 Analysis Procedure and Results for Question 4 (Continued)**

| | A | B | C | D | E | F | G | H | I | J | K | L | M | N | O | P | Q | R | S | T | U |
|---|---|---|---|---|---|---|---|---|---|---|---|---|---|---|---|---|---|---|---|---|---|
| 27 | | | | | | | | | | | | | | | | | | | | | |
| 28 | | | | | | | | | | | | *25 kHz* | | | | | | | | | |
| 29 | | | | | | | | | | | | | | | | | | | | | |
| 30 | | | | | | | | | | | | Count | 6 | 6 | 6 | 6 | 6 | 6 | 6 | 6 | 48 |
| 31 | | | | | | | | | | | | Sum | 1104 | 1354 | 2448 | 1566 | 1609 | 1910 | 2017 | 831 | 12,838 |
| 32 | | | | | | | | | | | | Average | 410 | 187 | 153 | 292 | 211 | 148 | 139 | 376 | 1915 |
| 33 | | | | | | | | | | | | Variance | 1984 | 854 | 570 | 1679 | 485 | 812 | 204 | 1657 | 8246 |
| 34 | | | | | | | | | | | | | | | | | | | | | |
| 35 | | | | | | | | | | | | *Total* | | | | | | | | | |
| 36 | | | | | | | | | | | | | | | | | | | | | |
| 37 | | | | | | | | | | | | Count | 18 | 18 | 18 | 18 | 18 | 18 | 18 | 18 | |
| 38 | | | | | | | | | | | | Sum | 2149 | 2684 | 4842 | 3038 | 3066 | 3687 | 3994 | 1691 | |
| 39 | | | | | | | | | | | | Average | 742 | 341 | 293 | 576 | 422 | 285 | 271 | 734 | |
| 40 | | | | | | | | | | | | Variance | 2619 | 1197 | 848 | 2560 | 1148 | 978 | 358 | 3199 | |
| 43 | | | | | | | | | | | | | | | | | | | | | |
| 44 | | | | | | | | | | | | *Source of variation* | *SS* | *df* | *MS* | ***F*** | ***P-value*** | ***F-Critical*** | | | |
| 45 | | | | | | | | | | | | | | | | | | | | | |
| 46 | | | | | | | | | | | | Sample | 710 493 | 2 | 355 246 | **395** | **0.00%** | **3.072** | | | |
| 47 | | | | | | | | | | | | Column | 401 617 | 7 | 57 374 | **64** | **0.00%** | **2.087** | | | |
| 48 | | | | | | | | | | | | Interaction | 66 174 | 14 | 4727 | **5** | **0.00%** | **1.775** | | | |
| 49 | | | | | | | | | | | | Within | 108 033 | 120 | 900 | | | | | | |

When, $F_{Sample}$ is numerically greater than $F_{Critical}$, there is a relationship between the independent conditions and the dependent outcome which may be simple or complex. See the Section A1.23.

This de-concatenation of variance into components requires:

- The data to be truly random, and not biased.
- Outcomes must be distributed normally.
- Individual samples must be independent (orthogonal) of one another.

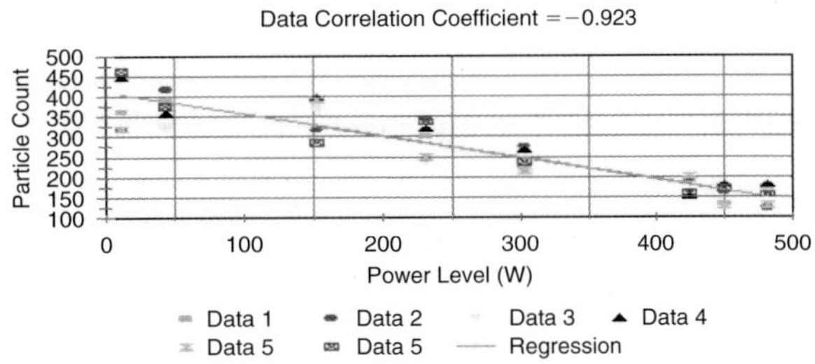


**Figure A1.27** Particle removal @ 25 kHz

### A1.22.10 Limitations of ANOVA Spreadsheet Functions

Please note these limitations and assumptions behind ANOVA spreadsheet functions:

- They can be used for one or two independent variables (one-way or two-way), and a single outcome.
- They can be used with multiple levels of each of two variables.
- They can be used with multiple (many) replications with the multiple levels of each of two variables.
- They cannot be used with more than two independent variables. The same methodology is implemented with more than two independent and non-linear variables in various commercial computer programs for statistical analysis. A list is given in by Anderson in the article referenced as Footnote 86.

## A1.23 REGRESSION ANALYSIS

The use of spreadsheet functions @SLOPE (or =SLOPE) and @INTERCEPT (or =INTERCEPT) was introduced in equation A1.15 with scatter plots:[114]

- @SLOPE is sometimes referred to as the *gain* of a relationship. It is the ratio of linear change in the outcome divided by change in the independent variable. Units are the units of the outcome divided by the units of the independent variable.[115]
- @INTERCEPT is sometimes referred to as the offset of a relationship. It is the difference from zero that an outcome has when the independent variable is zero-valued. Units are the units of the outcome.

These spreadsheet functions are the logical tools used after a valued DOE.

Identification of statistically significant relationships naturally leads to a desire to use them. Often that is done with a simple linear equation whose constants are identified via @INTERCEPT and @SLOPE.[116]

The data produced to answer Question 4 are plotted in Figures A1.27, 28 and 29. At each of three transducer frequencies (each graph) there are eight power settings and six replications of particle measurement. Thus forty eight data points were included in each calculation for the linear relationship between transducer power level and particle count.

It is important to recall that the large value of $F_{Sample}$ (relative to $F_{Critical}$) does not speak to the details of the relationship between an independent variable and an outcome. For example, values produced by @SLOPE can be large or small without regard to the extent to which $F_{Sample}$ exceeds $F_{Critical}$.

The basic premise in analysis of variance is that variance is a linear combination of individual variances each of which can be attributed to specific causes or random error. Said another way, variance can be partitioned – divided up.

[114] Please note that the information upon which the scatter diagrams (plots) were produced was information produced in a statistically-sound designed experiment.

[115] The values of @SLOPE are often negative. For example, increased transducer power better removes particles – fewer of them are present on the part surfaces.

[116] Identification of @SLOPE and @INTERCEPT is often referred to as "curve fitting" (see Equation [A1.15]).

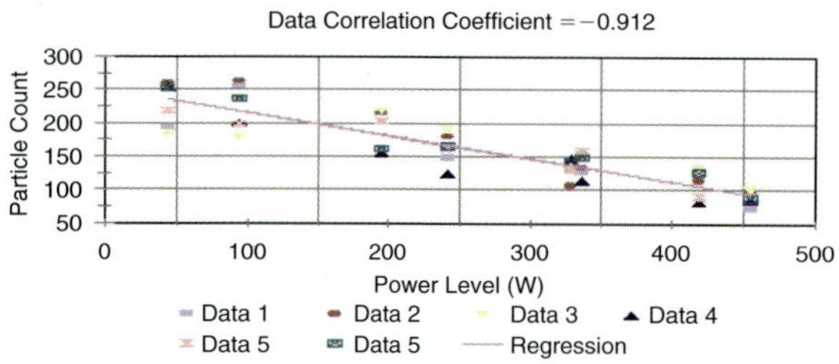


**Figure A1.28** Particle removal @ 40 kHz

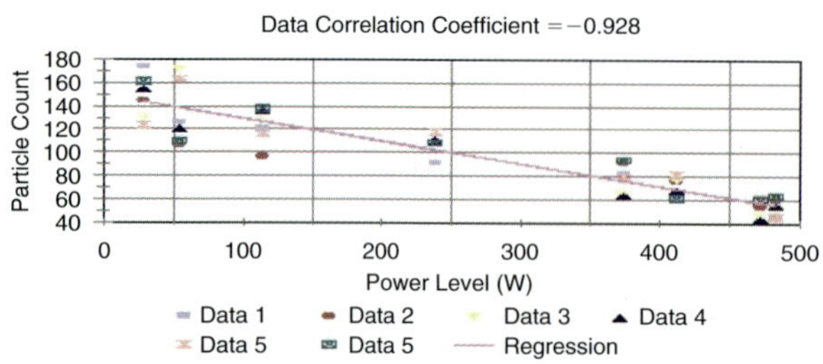


**Figure A1.29** Particle removal @ 68 kHz

### A1.23.1 Correlation Coefficients

A simple measure of the ability of these two constants to interpolate between[117] the data is the correlation coefficient – computed via the spreadsheet function @CORREL. It speaks to how well the variance in the measurements (data) is explained by the equation.

Normally statisticians reject equations for which the correlation coefficient is not at least 0.95, or more. Usually, that's sound advice.

*In cleaning work, perhaps that's too conservative.* Counting of particles on surfaces[118] is a good example where the statistician's normal quality standard should be combined with a little forgiveness. In the opinion of this author, the quality of the correlation between the data and equations demonstrated in the three graphs above is excellent – despite the fact that the correlation coefficients don't meet the 0.95 standard.

---

[117] Extension beyond the bounds of the measurements must always done with caution. There is no statistical basis to support use of any equation beyond the data upon which it is based. For example, @CORREL has no meaning outside the range of the data upon which it based.

But that doesn't mean managers don't extrapolate – and successfully too. Many predict their next raise from their past history of salary treatment.

[118] Many cleaning tests do not have a high level of repeatability. This is especially true outside of critical cleaning. A major reason for this situation is that cleaning processes are hydraulic. Fluid contact on some surfaces is masked or blocked by other surfaces. So the same level of cleaning contact is not applied to each and every surface each time. For industrial (metal or gross) and precision cleaning, this discontinuity is not a significant problem. Where it is, that cleaning is referred to as critical cleaning.

Appendix 2

# Description of analytical procedures for cleanliness testing

**Chapter contents**



The following procedures/technologies/instrumentations are commonly used for measuring the cleanliness of part surfaces.[1] Segregated into two types, these procedures can be:

- Used directly for cleanliness testing.
- Used to characterize a surface, but less likely to be used as cleaning tests.[2]

This segregation is by no means absolute. Some do use atomic force microscope (AFM) technology to qualify atomic integrity of semiconductor surfaces, or gas chromatograph mass spectrometers (GCMS) to determine when cleaning solvents are suitably soil-laden and ready for distillation.

Within the two types, they are listed[3] in the order of increasing capability to resolve increasingly smaller amounts of contamination. Naturally, cost and complexity also increase.

## A2.1 ANALYTICAL PROCEDURES USEFUL FOR CLEANING TESTS

### A2.1.1 Nonvolatile Residue Analysis (NVR)

There is no general analysis procedure for NVR.[4] This is because NVR may be a solid or a slow-evaporating liquid, and of most any chemical type. NVR is soil which won't disappear "on its own." Its vapour pressure is exceedingly low.

Three analytical approaches are common:

1. Weigh parts[5] considered to be clean,[6] and compare their weight with soiled parts. An analytical balance which can reliably and repeatedly recognize

[1] Useful references are:
Anderson, J.L. and Benkovich, M.G., "Measurement of Organic Residues on Surfaces to a Low Fraction of a Monolayer," *Precision Cleaning Magazine*, May 1996.
Skoog, L., *Principles of Instrumental Analysis* (4th ed.), Saunders College Publishing, Fort Worth, 1992. The Evans Analytical Group web page is http://www.cea.com/index.htm

[2] Complexity is the chief reason why this second group of measurement techniques is not likely to be useful for direct cleanliness measurements. Here complexity means capital investment needed, time involved, training necessary, and expertise needed for interpretation of results.

[3] Techniques described in Chapter 5 are not listed in this appendix.

[4] The difficulty of a reliable analysis for NVR is illustrated by the fact that in 1980 the ASTM produced Standard: F332-75 – Method of Test for Amount of NVR in Trichlorotrifluouroethane (Meseran Procedure). The standard was withdrawn in 1982.

[5] See Section 5.4.7.

[6] Cleaning tests, and validation of cleaning tests, are covered in Section 5.12.

0.1 mg is minimum essential. Useful information can't be obtained unless the part[7] weighs less than about 100 g with less than about 100 SI surface areas.
2. Extract the part[3] in a boiling solvent, and analyze the solvent.[8] Use a method below (likely GCMS or FTIR [Fourier Transform Infrared Spectroscopy]) to estimate the mass amount of soil removed in the extraction process. Presumably, at least that amount of soil was on the parts.[9]
3. A surface analysis technique is selected from those below.

### A2.1.2 Optically Stimulated Electron Emission (OSEF)

UV light is employed. When high-energy UV light hits a surface, electrons are emitted. This energy flow can be detected (not seen) as a "current." That reflected "current" can be measured. A clean surface will give the highest return current, so any drop in current will represent contamination.

OSEE can detect low levels of contamination (both ionic and nonionic). But OSEE suffers the same shortcomings as UV photoelectric emission (see Chapter 5, Section 4.2.):

- OSEE can detect contamination, but not identify different species.
- OSEE doesn't produce *absolute* quantitative information, but output can be calibrated.
- OSEE can't identify cleaning agent as different from soil.
- OSEE also cannot detect particulate.

OSEE is best used to detect organic films.

In the latter application, OSEE can be an efficient and rapidly responding approach. The radiation emitter is hand-held. Response is instantaneous on a calibrated meter. OSEE is an excellent and low-priced inspection tool for detecting hydrocarbon (organic) based films on a continuing flow of material.[10]

### A2.1.3 Total Organic Carbon/Total Oxygen Demand Analysis (TOC/TOD)

These thermal techniques can yield quantitative information about soil level, but not information about soil type. They can be used very effectively to determine the presence of organic contamination.

It is assumed in both methods[11] that the soil material is organic and can be reacted with Oxygen when heated to a high temperature. This is called pyrolysis. The amount of $CO_2$ producted during pyrolysis (total organic carbon, TOC) or the oxygen consumed (total oxygen demand, TOD) consumed during pyrolysis is measured.

The sample is nearly always a liquid–water, an acid or a base, and never an organic solvent. Consequently, these two methods may be useful in aqueous cleaning technology to monitor the progress of a rinsing process. But they are generally not useful with solvent cleaning technology because the solvent is nearly always an organic material.

### A2.1.4 Particle Counting/Evaluation

In a sense, no surface is free of particles. For all but the most meticulously prepared surfaces, particles exist in various sizes from 1000 μm (~40 mil) boulders to micron-sized debris to sub-micron-sized contamination to objects whose dimensions are multiples of the sizes of atoms. That's what makes measurement of particle contamination so difficult: one needs a smaller and smaller scale to complete the task.[12]

Methods are both indirect and direct, with the former being more common:

- *The indirect method* involves examination of the cleaning fluid last in contact with the surface. Alternately, instead of the last-used working solvent, one may examine a liquid sample produced by contacting the surface in question under a level of agitation thought adequate to release nearly all

[7] Multiple parts may be used if the results are not attributed to any single part.
[8] A blank should always be used. This solvent is not used for extraction. The impurity level in the blank should be subtracted from the level in the solvent extract.
[9] Obviously the problems with this approach are: (1) selection of an appropriate solvent(s), and (2) certainty that the extraction process removed all soluble soil.
Under no circumstances should the extraction solvent be used for cleaning. The extraction process must be differentiated from the cleaning process!
[10] Chawla, M., "Measuring Surface Cleanliness," *Precision Cleaning magazine*, June 1997.
[11] Not performed in the same instrument.
[12] See Section 6.6.

particles from the surface into the fluid. In practice, sequential samples (often called "extracts") are taken and examined. Sampling stops when the measured particulate level is either negligible, or consistent and negligible.[13] The outcome is the total recovery of particles over all samples.

Particle counting (PC) methods[14] are based on light absorption and/or light scattering using samples collected as above. A photodetector diode detects the degree of light scattering (smaller particles scatter less light).

Obviously, all indirect methods can be insecure because it is seldom certain that all particles of all sizes have been liberated from the surface.

- *The direct method* involves surface examination at the scale of debris about which there is concern. One might use the 15X loupe to directly inspect for particles produced in a grinding operation or one might use an AFM to evaluate contamination as molecular-sized imperfections. Output from such examinations is "manual:"
  - A trained person counts the number of objects above a criterion size in a specified area from a statistically selected number of areas. This is commonly done with medical goods used as implants for humans, where a personal examination is required by a regulation, or where collateral contamination is also to be recognized.
  - An optical scanner and image-recognition software are used to characterize (inspect) a moving cleaned surface (strip, belting, fibers, etc.). This technology is common and rapidly being enhanced for reasons involving both security and productivity.

## A2.2 ANALYTICAL PROCEDURES NOT USEFUL FOR CLEANING TESTS, BUT USEFUL FOR SURFACE CHARACTERIZATION

There are at least three types of associated analytical methods: thermogravimetric analysis (TGA), differential scanning calorimetry (DSC), and thermomechanical analysis (TMA).

All three are based on a simple core idea, when a material is heated something may happen to it, and following the progress of those events will assist in characterizing it. All three methods are less useful for measurement of soil level than they are useful for soil characterization. The latter can be extremely useful in the formulation of a solvent cleaning process, but generally not useful in operation of one:

- **TGA**. In TGA, the weight of the sample is monitored continuously during heating. Initially volatile fractions are released. Then the weight of the sample will become constant. More significantly, whenever a reaction occurs, the weight will either increase or decrease. This can be seen by plotting the sample weight versus temperature. The temperature at which any weight change happens and the size of the change can be used to characterize the nature of the transformation.
- **DSC**. Here a sample is heated while placed in a calorimeter, a device that measures the release or consumption of heat by the sample. As long as the sample stays in one phase and there are no reactions taking place in it, the amount of heat change per degree will be a constant amount that is equal to the weight of the sample times its specific heat. Whenever a phase change or reaction occurs, the amount of heat will either increase or decrease, depending on whether the change is endothermic or exothermic. Almost all phase changes that occur during heating are endothermic, but reactions may be either.
- **TMA**. In TMA, the amount of expansion resulting from an increase in temperature of the sample is monitored in one axis at a time. As long as the sample stays in one phase and there are no reactions taking place in it, the amount of expansion per degree will be a constant amount. Whenever a phase change or reaction occurs, the amount (not rate) of expansion will either increase (generally) or decrease.

---

[13] Flush fluid quantity for sampling should be no less than 100 ml (~¼ pint) per 0.1 $m^2$ (1 $ft^2$) of surface area

[14] Two excellent references are:
SAE-ARP-598, "SAE Aerospace Recommended Practice for the Determination of Particulate Contamination in Liquids by the Particle Count Method" and ASTM F24-04 Standard Method for Measuring and Counting Particulate Contamination on Surfaces.

### A2.2.1 Gas Chromatography/Mass Spectrometry (GC/MS)

"Older than dirt," and very effective at identifying many kinds of dirt! These are two techniques (GC and MS) which have been linked because they are complimentary.

Organic compounds are separated via GC and are then identified, based on molecular weight, by MS. There are three processing steps, all of which must be properly designed and conducted to produce valuable information.

The contaminated surface is contacted with a selected solvent (or solvents). Soil materials are extracted by solution into the solvent(s). Then the solvent/soil mixture is flash-vaporized producing a gas.

The gaseous solvent/soil mixture is swept up by an inert carrier gas (often Helium) and forced through a temperature-controlled open tubular column of selected absorbent. Soil components preferentially absorb on the absorbent based on affinity. Separation occurs as the components partition themselves between the stationary phase on the inner wall of the column and the mobile phase (the carrier gas). In a standard GC (or gc), the column is flushed with carrier gas at a different temperature and the individual components are eluted those less strongly absorbed exit first. A detector that measures thermal conductivity responds differently to each component in proportion to the amount present. Obviously, a calibration is necessary.

Components elute from the GC column into a MS. Here, energetic electrons bombard the component molecules, ionizing some of them. The ions are then accelerated by an electric field and enter a mass analyzer, where they are separated according to their mass-to-charge ratios. By plotting the abundance of ions as a function of mass-to-charge ratio, a mass spectrum is generated. The mass spectrum can be a unique "fingerprint," allowing identification of unknown compounds through use of a database about unique chemical compounds.

Integration of GC and MS technologies has worked well for decades. GCMS analysis can fail, without suggesting jeopardy, if the extraction doesn't remove all of the soil components. Obviously, the extraction operation destroys all information about the location of soil components.

### A2.2.2 Scanning Electron Microscopy (SEM)

This is another approach, which detects the effect or the fact of electron flow. SEM utilizes a beam of electrons that is passed over a very small area of a surface. This beam scatters (deflects) when it strikes the surface, outlining the topography of the surface. The "back-scattering" carried by the return beam of electrons is measured by the microscope. The result is a finely detailed (at least 100,000X), 3-D image of the surface being scanned.

SEM, however, doesn't *detect* contamination, but produces an image of the surface on which there may be contamination.

A skilled operator and an SEM can be well suited for identifying particulate and potentially nonuniform or thick films of contaminant. The operator can directly count particles of various sizes, which can be very time-consuming. Frequently a small section is examined with an SEM and the result used to characterize the complete surface.

An SEM cannot efficiently "look at" a very large area, and is not low-priced.

### A2.2.3 Electron Spectroscopy for Chemical Analysis (ESCA)

In this familiar technique, surfaces are irradiated with monochromatic X-rays which cause the ejection of photoelectrons from the surface. The electron binding energies are measured by a high-resolution electron spectrometer. Due to the penetrating nature of X-rays, the substrate surface is commonly analyzed by grazing the X-ray beam across the surface.

The measured energy levels are used to identify the elements present. In many cases, this data also provides information about the valence state(s) or chemical bonding environment(s) of the elements thus detected. The depth of the analysis is typically the outer 3 nm of the sample. ESCA is considered nondestructive and has the potential to be very useful in identifying organic compounds.

### A2.2.4 Fourier Transform Infrared Spectroscopy

FTIR aids in the study of chemical structure. Individual chemical bonds, as well as groups of bonds, vibrate at characteristic frequencies. When exposed to infrared (IR) radiation, molecules selectively absorb radiation at frequencies that match those of their allowed vibrational modes.

FTIR analytical instruments allow measurement of the absorption of IR radiation by a surface as a function of IR wave frequency. The output is a spectrum that can be used to identify functional groups and consequently structure. FTIR spectra can be obtained in air (versus vacuum).

FTIR provides specific information about chemical bonding and molecular structure, making it useful for identifying organic materials and certain inorganic materials.

### A2.2.5 Grazing Angle FTIR

Enhanced resolution is the general differentiating factor versus standard FTIR analysis. This allows more effective resolution of the amount of a contaminating chemical versus only identification of it. Resolution improves for two reasons:

1. The grazing (versus near-normal) angle increases the length of the light path through a thin layer of contamination.
2. The *p*-component of the incident radiation undergoes a substantial phase shift which allows it to be separated from the *s*-component of the polarized radiation. A portable device for evaluation of contamination has been developed by the US Navy.[15]

### A2.2.6 Secondary Ion Mass Spectroscopy (SIMS)

Secondary ion mass spectrometry (SIMS) is yet another analytical procedure where the character of surface contamination is derived from the "reaction" of that contamination to being bombarded by something, which is usually either subatomic-scale particles or radiation of a specific character.

Here, an energized primary ion beam is directed at the surface, resulting in the ejection of surface atoms as secondary ions. This process is known as "sputtering." Ions in the primary beam can be $Ar^+$, $Cs^+$, $N_2^+$, or $O_2^+$.

SIMS can detect both positive and negative ions, establishing the nature of the charge of the contamination, if any. The typical sampling depth ranges from 2 to 6 Å. SIMS cannot identify bonding character as can ESCA.

### A2.2.7 Total Reflection X-Ray Fluorescence (TXRF)

Energy is added to metal ions on a surface when struck by high-energy X-rays. That energy produces a reaction.

With TXRF X-rays impinge the sample surface and excite the electrons on atoms in the top few monolayers of the sample, causing them to emit photons (fluoresce). The photons emitted by the surface atoms have energies that are characteristic of the particular element being struck by the X-ray.

### A2.2.8 Auger Electron Spectroscopy (AES)

This sophisticated and sensitive technique identifies elemental compositions of surfaces by measuring the energies of Augar electrons.[16,17]

Augar electron spectroscopy (AES) measures the energy of the Auger electron, which is unique to each particular atom. AES is a destructive analysis but is useful in looking for concentrated areas of contamination. It can also be used quantitatively. The typical sampling depth for AES is 20–50 Å. Auger analysis is surface-specific.

---

[15] See Hoffard, T., Technical Report, TR-2217-SHR, "Grazing-angle Fourier Transform Infrared Spectroscopy for Surface Cleanliness Verification," March 2003.

[16] In the Auger ("oh-jay" or "ah-jay") emission process, an external excitation beam removes the first electron from a core level of an atom. A vacancy is produced in the electron shell. A second electron falls from a higher level into the vacancy with release of energy. The resulting energy is carried off with the third (Auger) electron which is ejected from a higher energy level.

[17] See: Carlson, T.A. *Photoelectron and Auger Spectroscopy*, Plenum Press, New York, 1975, or Czanderna, A.W. ed., *Methods of Surface Analysis*, Elsevier, New York, 1975.

The common use of AES is to detect, and perhaps quantify, mineral/metal contamination on metal surfaces, raw semiconductor materials, and finished electronic devices. AES can distinguish between Si, $SiO_2$, SiO, and $Si_3N_4$ in a 10 nm layer on a Silicon wafer.

### A2.2.9 Atomic Force Microscopy (AFM), Scanning Tunneling Microscopy (STM), Magnetic Force Microscopy (MFM)

These are variations on a method of imaging surfaces with atomic or near-atomic resolution, collectively called scanning probe microscopy (SPM). They represent the most advanced, practical technology for surface analysis. Contamination as individual atoms can be recognized.

In AFM, STM, and MFM, a small tip (at the end of a cantilever) is scanned across the surface of a sample in order to construct a 3-D image of the surface. Fine control of the scan is accomplished using piezoelectrically-induced motions.

If the tip and the surface are both conducting, the structure of the surface can be detected by tunneling of electrons from the tip to the surface (STM).

Any type of surface can be probed by the molecular forces exerted by the surface against the tip (AFM). The tip can be constantly in contact with the surface, it can gently tap the surface while oscillating at high frequency, or it can be scanned just minutely above the surface.

By coating the tip with a magnetic material, the magnetic fields immediately above a surface can be imaged (MFM).

Image processing software allows easy extraction of useful surface parameters.

### A2.2.10 Phase Imaging

Phase imaging is a surface-mapping technique that extends the power of AFM techniques. With AFM, the cantilever is excited into resonance oscillation with a piezoelectric driver. With phase imaging, the phase lag of the cantilever oscillation, relative to the signal sent to the cantilever's pizeoelectric driver, is simultaneously monitored.

The phase lag is extremely sensitive to variations in material properties such as adhesion and viscoelasticity. The amplitude of the phase lag signal varies with changes in the surface topography. Surfaces are illustrated as light and dark areas. This technique can locate areas of surface contamination, which will stand out as topographically distinct.

# Index

*Note*: Italicized page numbers denote entries are taken from figures and tables while those from footnotes are in bold.